P9-DUY-971

Next Generation Wireless Applications

Next Generation Wireless Applications

Paul Golding
Magic E Company, UK

John Wiley & Sons, Ltd

Published by John Wiley & Sons etc.

Email (for orders and customer service enquiries): cs-books@wiley.co.uk
Visit our Home Page on www.wileyeurope.com or www.wiley.com

Other Wiley Editorial Offices

John Wiley & Sons Inc., 111 River Street, Hoboken, NJ 07030, USA

Jossey-Bass, 989 Market Street, San Francisco, CA 94103-1741, USA

Wiley-VCH Verlag GmbH, Boschstr. 12, D-69469 Weinheim, Germany

John Wiley & Sons Australia Ltd, 33 Park Road, Milton, Queensland 4064, Australia

John Wiley & Sons (Asia) Pte Ltd, 2 Clementi Loop #02-01, Jin Xing Distripark, Singapore 129809

John Wiley & Sons Canada Ltd, 22 Worcester Road, Etobicoke, Ontario, Canada M9W 1L1

British Library Cataloguing in Publication Data

A catalogue record for this book is available from the British Library

ISBN 0-470-86986-0

Typeset in 10/12pt Times by TechBooks, New Delhi, India
Printed and bound in Great Britain by MPG Ltd, Bodmin, Cornwall
This book is printed on acid-free paper responsibly manufactured from sustainable forestry in which at least two trees are planted for each one used for paper production.

To Mum and Dad.
To my wife and kids.

Contents

Preface

I have been involved with wireless technology for a long time. I have worked in many stimulating areas such as DSP chip architectures for GSM, fuzzy-logic interference cancellation, mobile virtual reality, device usability, Internet protocols, software engineering and many others. Certain events led to me to write a training course for the technical support team at Three UK. I gave it the rather intrepid title of "The Complete Anatomy of a Wireless Application", a 3-day course attempting to cover the entire constellation of current wireless technologies, from 3G CDMA principles to the mysterious inner workings of a J2EE application server: stubs, skeletons and all of that "under the hood" magic that makes service delivery platforms (SDP) work. This book started with an idea to write an accompanying text to the course, although more substantial in details. That's basically what I have written, using the knowledge I have acquired from a variegated involvement with so many interesting fields, trying to pass it on in a usable form to the readership.

The point of both the course and now the book is to give wireless practitioners a means by which to anchor themselves in the increasingly complex world of next-generation technologies, an attempt to contextualise each of the paradigms applicable to the rapidly expanding wireless cosmos. There can be no denying that these are exciting days for mobile applications development. The availability of new wireless technologies coincides with many other exhilarating areas of technology that are now within the practitioner's toolkit, such as accurate location-finding, interesting Java variants, smarter devices, useful protocols and many other mega-components. Possibly the most enticing aspect of these technologies is their ability to integrate into the massively connected world of The Internet, allowing new levels of service integration. This is thanks to the vast array of "internet aware" software tools, methods and services that continued to develop even after the dot.com collapse.

The rapid rise of so many any-time any-place possibilities is proving to be a challenge in terms of how to chart the dense cloud of ideas that can be utilised in providing new and exciting services. Perhaps a programmer working on an application for a Java-capable handset is unable to grasp how best to incorporate location-finding techniques, or how to think about the security implications of their design. It is this "think about" step that I am trying to aid with this book. It is not, as is often alleged, that we need to undertake a "paradigm shift" that the dot.com days supposedly demanded. Instead, we simply need to identify and then connect existing paradigms in order to make interesting things happen. For example, it has been possible for a long time to deploy interesting network services in the telephony part of the mobile network, such as selectively diverting calls depending on the time of day and the caller's number. This is the old paradigm of intelligent networks. Doing this programmatically is not very new either, although suggesting an industry standard is relatively new, such as the Parlay specifications, which probably constitute another paradigm. However, opening these possibilities to an application running anywhere on the net is perhaps the newest paradigm, one suggested by the nascent web-services revolution. Combining all three of these paradigms leads to a bristling array of possibilities.

Revealing such potentials is really the intent of this book, which is why the contents are not organised into a more conventional menu of subjects, but largely grouped into paradigms that collectively could fit within a single mental model when thinking about mobile applications, hence why I feel justified in talking about J2EE and CDMA in the same book. Exactly which mental model you adopt or formulate is up to you and beyond the scope of the book: that is for you to invent by inserting whichever of the paradigms herein that you see fit. From those models, many patterns will emerge. Watch this space!

I hope that you succeed in your wireless endeavours and that this book provides you with at least one gem of an interesting idea or with the origins of a useful trail of thought, perhaps a new mental model that helps you to "think mobile" in your own way. Please send me your considered feedback, experiences, questions, or even ideas by visiting my website.

Paul Golding, 2004.
Website: www.paulgolding.info

Acknowledgements

I would like to thank all those who have supported and assisted me in writing this book, especially my wife and kids for their encouragement, patience and many kind words at just the right moments. I thank all my brothers (Vince, Adrin, Chris) my in-laws, and various friends for their supportive words and for repeatedly asking, "Is your book finished yet?" This is a question I grew to hate, but which proved an essential motivator, if only so I could reply in the affirmative! I would like to thank the team at Wiley who backed the proposal to write this book and who paid for tea at Polly Tea Rooms: Birgit and Sally. I would like to thank Barry King and David Gumbrell for their intellectual stimulation (and many Pizzas) and Ahmed Gofur, Hassan Morrison and Khalid Taha for their generous support. I would like to thank all those companies who provided me with permission to refer to their materials, particularly the guys at Digit Wireless who also helped to boost my enthusiasm towards the pursuit of "out of the box" ideas. I would also like to thank all those who attended my course "The Complete Anatomy of a Wireless Application" and gave me their critical feedback, particularly Arif (of Three UK) who contacted me to write the course in the first place. Going back many years, I would also like to thank my University tutor – Dr Whitehouse – to whom I used to say that I have very little talent, but I don't mind working hard. He replied that working hard is the talent. Perhaps he was right, but plenty of people work hard, so if at all my hard work has yielded any fruit, then I thank God for that.

The author would like to acknowledge the use of the IXI Mobile handheld communication device in Figure 2.6 and in all subsequent figures thereafter.

Paul Golding

Abbreviations and Acronyms

2G	Second Generation
3G	Third Generation
3GPP	3G Partnership Project
AD	Analog–Digital
ADPCM	Adaptive Differential Pulse-Coded Modulation
A-GPS	Assisted GPS
AI	Application Interface
ALU	Arithmetic Logic Unit
AMR	Adaptive Multirate
AMS	Application Mgmt System
API	Application Programming Interface
ARPU	Average Revenue Per User
ASCII	American Standard Code for Information Interchange
B2B	Business to Business
B2C	Business to Consumer
B2E	Business to Employee
BEEP	Blocks Extensible Exchange Protocol
BER	Bit Error Rate
BSC	Base Station Controller
BSS	Base Station Sub-system
BT	Bluetooth™
BTS	Base Transceiver Station
CA	Certificate Authority
CAB	Cabinet
CB	Citizens' Band
CC/PP	Composite Capability/Preference Profits
CD	Client Domain
CDC	Connected Device Configuration

CDMA	Code Division Multiple Access
CEF	Common Executable Format
cHTML	compact HTML
CLDC	Connection Limited Device Configuration
CLR	Common Language Run-time
CMS	Content Management System
codec	Coder–Decoder
COM	Communication Port
CORBA	Common Object Request Broker Architecture
CPI	Capability and Preference Information
CPU	Central Processing Unit
CRM	Customer Relationship Management
CS	Client–Server
CSS	Cascading Style Sheet
CVM	C Virtual Machine
DBMS	Database Management System
DCT	Discrete Cosine Transform
DD	Download Descriptor
DDP	Packet Data Protocol
D-GPS	Differential GPS
DHCP	Dynamic Host Control Protocol
DIA	Device Independence Activity
DMA	Direct Memory Access
DNS	Domain Mechanism?
DNS	Domain Name System
DOM	Document Object Model
DSP	Digital Signal Processing; Digital Signal Processor
DTD	Document Type Definition
E911	Enhanced 911
ECC	Error Correction Correcting
EFI	External Functionality Interface
EIS	Enterprise Info Services
EJB	Enterprise Java Bean
E-OTD	Enhanced OTD
FCC	Federal Communications Commission
FDMA	Frequency Division Multiple Access
FML	Film XML
FOAF	Friend Of A Friend
FTP	File Transfer Protocol
GGSN	Gateway GPRS Service Node
GIS	Global Information System
GMLC	Gateway Mobile Location Centre
GNU	GNU Not Unix
GPRS	General Packet Radio Service
GPS	Global Positioning System
GSM	Global System for Mobile Communications

GUI	Graphical User Interface
GWLC	Gateway Location Center
H2C	Human to Content
H2H	Human to Human
H2M	Human to Machine
HDML	Handheld Device Markup Language
HLR	Home Location Register
HTML	HyperText Markup Language
HTTP	HyperText Transfer Protocol
HTTPS	HTTP Secured
ICMP	Internet Control Message Protocol
IDE	Integrated Development Environment
IETF	Internet Engaging Task Force
IF	Intermediate Frequency
IIOP	Internet Inter-ORB Protocol
IM	Instant Messaging
IMAP4	Internet Mail Application Protocol 4
IMSI	International Mobile Subscriber Identity
IN	Intelligent Networking
IP	Internet Protocol
IrDa	Infrared (a firm)
ISDN	Integrated Services Digital Network
ISP	Internet Service Provider
IVR	Interactive Voice Response
J2EE	Java 2 Enterprise Edition
J2ME	Java 2 Micro Edition
J2SE	Java 2 Standard Edition
JAAS	Java Authentication and Authorization Service
JAD	Java Application Descriptor
JAR	Java Archive
JAXP	Java API for XML Processing
JDBC	Java DataBase Connector
JDE	Java Developer Environment
JIT	Just In Iime
JMS	Java Messaging Service
JNDI	Java Naming and Device Interface
JNI	Java Native Interface
JPEG	Joint Picture Experts Group
JRE	Java Run-time Evironment
JSP	Java Server Page
JVM	Java Virtual Machine
JXTA	Juxtapose
KVM	Kilo Virtual Machine
LAN	Local Area Network
LBS	Location-Based Service
LCD	Liquid Crystal Display

LDAP	Lightweight Directory Access Protocol
LIF	Location Interoperability Forum
LMU	Location Measurement Unit
LOS	Line Of Sight
LTN	Long Thin Networks
M2M	Machine to Machine
MDB	Message-Driven Bean
MD5	Message Digest (algorithm) 5
MID	Mobile Information Device
MIDP	Mobile Internet Device Profile
MIME	Multipurpose Internet Mail Extension
MIPS	Millions of Instruction Per Second
MLP	Mobile Location Protocol
MM	Multimedia Message
MMC	Multimedia Message Centre
MMI	Man–Machine Interface
MMS	Multimedia Messaging Service
MMSC	Multimedia Messaging Service Centre
MMX	Multimedia Extensions
MOM	Message-Oriented Middleware
MP3	MPEG-1 Layer 3
MS	Mobile Station
MSC	Mobile Switching Centre
MVC	Model View Controller
NACK	Negative ACK
NAT	Network Address Translation
NTP	Network Time Protocol
O&M	Operation and Mgmt
OBEX	Object Exchange
ODBC	Open DataBase Connector
OMA	Open Mobile Alliance
OMC	Operations & Maintenance Centre
OOP	Object-Oriented Programming
ORB	Object Request Broker
OS	Operating System
OSA	Open Services Architecture
OSI	Open System Interconnection
OTA	Over The Air
OTD	Observed Time Difference
P2P	Person-to-Person, Peer-to-Peer
PABX	Private Automatic Branch Exchange
PAM	Pluggable Authentication Module; Presence and Availability Management
PAP	Push Access Protocol
PCMCIA	Personal Computer Memory Card International Association
PCU	Packet Controller Unit
PD	Personal Domain

PDA	Personal Digital Assistant
PDP	Packet Data Protocol
PDU	Packet Data Unit
PILC	Performance Implications of Link Characteristics
PIM	Personal Information Management
PKC	Public Key Cryptography
PKI	Public Key Infrastructure
PMG	Personal Mobile Gateway
PNG	Portable Network Graphics
POI	Point of Interest
POIX	Point of Interest Exchange Language
POP3	Post Office Protocol 3
POTS	Plain Old Telephone System
PPG	Push Proxy Gateway
PPP	Point-to-Point Protocol
PSAP	Public Safety Answering Point
PSTN	Public Switched Telephony Network
PT	Predictive Text
PTT	Push To Talk
QoS	Quality of Service
RADIUS	Remote Authentication Dial In User Service
RDF	Resource Description Framework
RF	Radio Frequency
RFC	Request For Comments
RISC	Reduced Instruction-Set Computing
RMI	Remote Method Invocation
RNC	Radio Network Controller
RPC	Remote Procedure Call
RSA	Rivest, Shamir and Adelman
RSS	RDF Site Syndication
RTOS	Real Time Operating System
RTP	Real-time Transport Protocol
RTSP	Real Time Streaming Protocol
RTTTL	Ringing Tones Text Transfer Language
Rx	Receiver
SACK	Selective ACK
SAR	Segmentation and Reassembly
SD	Social Domain
SDK	Software Developer's Kit
SDMA	Spatial Division Multiple Access
SDP	Service Delivery Platform
SGSN	Serving GPRS Service Nodes
SI	Service Indication
SIM	Subscriber Identity Module
SIMD	Single Instruction Multiple Data
SIP	Session Initiation Protocol

SISD	Single Instruction Single Data
SM	Spatial Messaging
SMIL	Synchronized Multimedia Integration Language
SMLC	Serving Mobile Location Centre
SMPP	Short Message Point-to-Point
SMS	Short Message Service
SMSC	Short Message Service Centre
SMTP	Simple Mail Transfer Protocol
SOAP	Simple Object Access Protocol
SP-MIDI	Scalable Polyphony MIDI
SQL	Structured Query Language
SSE	Streaming SIMD Extensions
SSL	Secure Socket Layer
SVG	Scalable Vector Graphics
Sync ML	Synchronization Markup Language
TCP	Transmission Control Protocol
TDD	Time Division Duplexing
TDMA	Time Division Multiple Access
TID	Transaction ID
TLS	Transport Layer Security
UAProf	User Agent Profile
UDP	User Datagram Protocol
UI	User Interface
UML	Unified Modelling Language
UMTS	Universal Mobile Telecommunication System
URI	Uniform Resource Identifier
URL	Uniform Resource Locator
USB	Universal Serial Bus
UTF	Unicode Text Format
UTM	Universal Transverse Mercator
UTRAN	Universal Terrestrial Radio Access Network
VB	Visual Basic
VC	Visual C++
VGA	Videographics Array
VLR	Visitor Location Register
VoIP	Voice over IP
VPN	Virtual Private Network
VR	Virtual Reality
W3C	World Wide Web Consortium
WAN	Wide Area Network
WAP	Wireless Application Protocol
WBXML	WAP Binary XML
WCDMA	Wideband CDMA
WCSS	WAP CSS
WDP	Wireless Datagram Protocol
W-HTTP	Wirelss-profiled HTTP

WJMS	Wireless Java Messaging Service
WLS	Web Logic Server
WML	WAP Markup Language; Wireless Markup Language
WSDL	Web Services Description Language
WSP	Wireless Session Protocol
WTAI	WAP Telephony Application Interface
W-TCP	Wireless-profiled TCP
WTLS	Wireless TLS
WTP	Wireless Transport Protocol
WURFL	Wireless Universal Resource File
WYSIWYG	What You See Is What You Get
XHTML	eXtensible HyperText Markup Lang.
XHTML-MB	Mobile Profile
XML	eXtensible Markup Language
XSL	XML Stylesheet Language
XSLT	XML Stylesheet Language Transformation
XUL	eXtensible User Interface Language

1

Prelude: 'How Did I Ever Cope without My Commie?'

1.1 PRELUDE TO THE PRELUDE – OPENING THE MIND

In a moment we will take a fictional tour of what the forthcoming *mobile services* might entail for the end user. As is often the case, new ideas are met with some degree of resistance, especially in the mobile world where there often seems to be more scepticism than enthusiasm for the emerging technology, particularly following the 'WAP[1] experience'. However, perhaps the following anecdote is worthy of some reflection as an aid to approaching this extensive and exciting topic.

While delivering a course called 'The Complete Anatomy of a Wireless Application', the inspiration for this book, I met some taut resistance to some of the applications scenarios that I presented as the potential future of mobile services. The struggle was with an employee of a well-known vendor of 3G (Third Generation) infrastructure equipment.[2] I was delivering the course on one of their R&D (research and development) sites in the UK.

I habitually quiz my course attendees concerning their reactions to existing or possible wireless services. The trouble with the reluctant attendee was his perceived lack of need for many, or most, of the services that I suggested. His common reply was 'I don't need that'. This was a chilling reaction if it rang true for the majority of potential customers for next generation services, a group of people we still know so little about.

During idle chit-chat over coffee the same attendee casually mentioned how useful he found being able to surf the Web for the opinions and reviews of others concerning various

[1] WAP stands for Wireless Application Protocol, a set of specifications to enable access to Internet resources from small mobile devices. Its initial underachievement in the market caused a lot of scepticism toward the forthcoming golden age of mobile data. Read this book and form your own view about the future of mobile.

[2] I don't want to say who, but I have noted the strange phenomenon where technology providers are frequently staffed by individuals who neither use nor evangelize their own wares.

Next Generation Wireless Applications P. Golding
 ISBN: 0-470-86986-0 (HB)

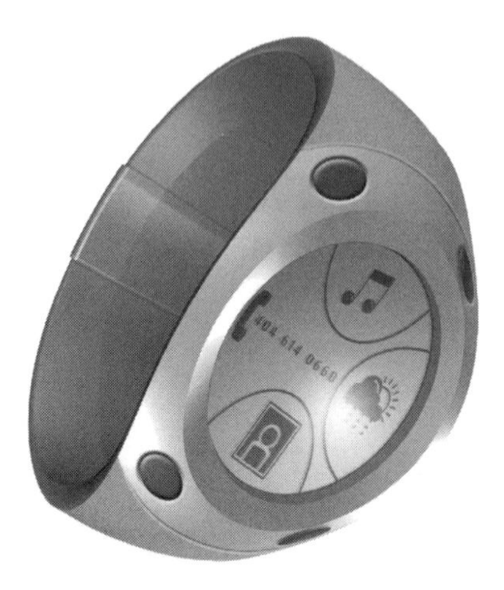

Figure 1.1 Smart watch. Reproduced by permission of IXI Mobile.

products, such as a domestic washing machine or other household item. The opinions of fellow shoppers had proved valuable in assisting his purchasing decisions on several occasions. I had already observed this interesting use of the Web, which seems to be an increasingly popular trend.

My comment to the attendee was something like[3] as follows: 'What if, say five years ago or before you had used the Internet, someone was to tell you that before you buy a washing machine you would log on to a computer, connect to a network (imagine we didn't know it was called the Web) and see what other people thought of the same model. What would you have said?'

He replied, musingly, 'I don't need that.'

1.2 THE FUTURE'S BRIGHT, THE FUTURE'S UBIQUITY

Let's take a trip into the very near future. It's a fine day for lunching outside the sultry office. Although warm and slightly humid outside the air stirs with a wonderfully cool breeze that could easily carry away the morning blues. Imagine you are standing in a bustling shopping plaza at one of the popular meeting points where you are soon going to meet three of your closest friends for lunch. You have in mind a cool crisp salad, perhaps a *salade niçoise,*[4] but as yet you don't know in which restaurant you are going to eat, just when or thereabouts.

The meeting time was suggested after a flurry of interactive picture messages that popped up on everyone's communicators when earlier you had sent a 'let's eat' invitation. One of your friends received the invitation via their wrist-based 'sleek device' (see Figure 1.1) which connects using Bluetooth™ to their personal mobile gateway (PMG), usually carried snugly in their pocket, looking like a brushed-aluminium cigarette case. This is why the word *communicator* has replaced the word phone, it no longer being the best word to describe some of the new devices, often barely resembling phones at all. When exchanging

[3] I don't recall the exact words.

[4] A French dish: tuna salad with olives and boiled egg.

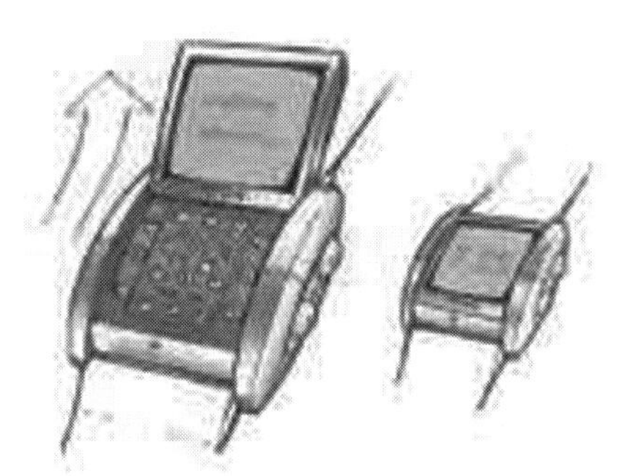

Figure 1.2 Smart watch with Fastap™ interface. Reproduced by permission of DigitWireless.

information, sending a message or placing a call, you can never be sure what device is on the other end of 'the line'; it might be someone's watch, a pendant, a pen or a PDA (personal digital assistant), or something else. Some people, inspired by the language on 'the street', have taken to calling these devices 'commies', short for communicators.

While standing still for a moment the flurrying farrago of shoppers swarming around you, your own wristpad commie alerts that you have an incoming call. It gently squeezes your wrist and vibrates, the outer ring glows and an agreeable glide of tones emits forth. You notice hordes of other cohabitants of the meeting place, also seemingly engaged with their wrists, bracelets, pendants, eyewear and other forms of commies – all shapes, sizes and colours, by now a perfectly familiar sight. Being preoccupied with another world is no longer a novel sight; across all ages there is no shyness. Your wristpad is also a 'sleek device' and is a fabric cuff that wraps neatly around your wrist and has a Fastap™ keyboard (Figure 1.2) and a curved semi-flexible pop-up display. Like your friend's connected watch, it also breathes via a Bluetooth™ lifeline to your PMG. Unlike the old-fashioned world of binary call handling in the Global System for Mobile Communications (GSM) – accept or busy – you have all kinds of options to process the inbound call. This is thanks to Internet Protocol (IP)-Telephony and the new world of SIP (Session Initiation Protocol).

With SIP your device doesn't just alert you with the call and the caller ID you also get a tagline introducing the call, like the subject field in an email. From the tagline you know that it's your personal fitness training instructor calling 'to discuss changes to tonight's programme' – words that tell you that this is a call you would like to take, words that your trainer probably didn't type, but were most likely translated from voice to text before he or she started the call.

There is no awkward conversation with your cuff as you have a latest design Bluetooth™ earpiece snuggled quite discreetly over and in your ear, much smaller than the earlier models. This one has been custom-sculptured like a prosthetic for your ear, just like the earpieces that stage musicians, such as Bono, have used for years, but cheaply done using a laser-modelling scanner at a walk-in booth in one of the mobile phone shops. It is light and has a long battery life thanks to the latest generation Zinc–Air cell, which is convenient as you use your earpiece often. You make calls with it in the office via your tablet laptop that has Bluetooth™ built in and can handle audio from the headset to a SIP client running on Windows™ – total wireless freedom.

You sometimes wonder why you need to come to work as you are most of the time talking, texting, messaging, chatting and emailing with people who are somewhere else.

The detachment can sometimes be quite unbalancing, as time and space seems to have taken on a different feel than before, merging 'there' with 'here', 'private' with 'public' and other such 'fuzzifications'.

Just prior to arrival, in the taxi ride to the plaza, you had received a call from a colleague who textually announced the call with 'wanting to discuss the latest sales figures', which you didn't have time enough to process as the taxi was close to pulling into the taxi rank. You were tempted to hit the 'call back in one hour' response that would have sent the same words back to the caller's SIP client (probably on their desktop PC), but you diverted (forwarded) the call to a subordinate who you had previously primed with the task of looking over the figures. As it happened your subordinate was in a meeting and had handled the call by converting it to an instant messaging (IM) session (still using SIP) for a brief and discreet chat with the caller, deferring the voice call to later. You then switched your *presence state* to 'out at lunch' to let callers know ahead of time that now might not be a good time to talk shop.

After forwarding the call from your colleague you remembered that you had not yet reconciled the latest sales figures for Europe. The easiest way to inform your colleague was to bring up their number on your wrist commie, simply by flicking your wrist back and then hitting 'send voice message'. Speaking into your Bluetooth™ headset you recorded a voice message that would go straight into their voicemail box without the need to chat.

Before setting out to lunch a hot button labelled 'let's eat' initiated the upcoming gathering. You pressed it, entered a start time and an end time and then selected invitees, which you did easily from your 'cool friends' folder (the default 'VIP' folder that you have renamed to be your 'cool friends' folder).

When browsing the folder you also noticed that one of your friends had sent you an email, yet unread, and another had left you a voice message, while another had posted to their weblog a new entry that you ought to read. On the way through you marked these tasks with the 'do next' command so that you didn't have to interrupt your current task flow. Later on, you may choose to resume these 'do next' tasks or receive a gentle reminder to look at them in your 'to do' queue.

Your 'cool friends' folder is not confined to people you have met in person, far from it. There are people you have met online, via email, chat rooms and by tracking their daily weblog entries. There are people whose reputation you value, perhaps as particularly good soulmates on Amazon™ with a similar taste in books and films or those with a similar taste in domestic products like washing machines – the most mundane information requirements are often the most valuable. These virtual soulmates will later be handy when you drop in the bookstall while waiting for your friends to arrive.

Next you clicked on the 'where?' option and you could elect to choose a place from your favourite places list, you could write free text, you could choose the 'nearby places' option or you could go for 'place finder'. You went for 'nearby places' and found Orbital Plaza, drilled down, went for 'meeting places' and chose 'dancing fountain', which is one of those synchronized water fountains that kids can run through. It's a popular meeting place and exists in the menu because other users of the 'let's meet' service have used it before: social influence constantly shapes the user experience.

It didn't matter if your friends had never been to the fountain before, they'll be able to use the digital map to home in on this regular rendezvous point. On their commies they received a 'let's share food' invitation which popped up with your photo, a happy composure combined with a 'let's eat' speech bubble and a brief, matching audio snippet, which is the way you had configured your invites to be sent out (using MMS[5] technology). Others get to choose their own invitation style too – perhaps a picture of a steaming pile of appetizing spaghetti. The message beneath the photo said, 'Want to eat at 1 p.m. at Orbital Plaza?' which you didn't have to type, 1 p.m. being your suggested start time. Your friends could select 'available', 'not available' or 'change time' in response to the invite. If they changed the time, then it could only fall within the range you had suggested.

Once all the votes are in, the acceptances and the convergence time are notified to you for your confirmation. Twenty past one is the agreed union time, which is why you arrived early as you already decided you would start your lunch break from 1 p.m. anyway, office fatigue having arrived early that day.

You have arrived early at the rendezvous point, having found where to go using the map and directions on your phone, not having been to Orbital Plaza before. Sitting on the seating next to a children's play area you first decide to read a few emails. In the play area mums are pointing their camera phones, taking short video clips of little kids whooshing down slides. Moments later a clip pops up in Daddy's email box at work, perhaps on the other side of the world. Dads everywhere are proud and happy.

You already know that an important client has sent you mail. Earlier a notification pinged up on your phone accompanied by a soothing glide of musical notes with a softly announced 'you have mail from ... ' spoken in a natural voice, biometrically tuned to be assuring and comfortable to your psycho-acoustic appetite. The notification occurs only because your phone automatically sensed that you are away from your desk where you would otherwise be checking your mail. You read the email and reply. Back at the office or home the reply is sitting in your 'sent items' folder and the email is marked as read: everything fully synchronized without the need to cradle.

To kill time you quickly review your wish list on Amazon. Via the world of Web Services and XML (eXtensible Markup Language) descriptors of one form or another, your list neatly appears in your task manager, appropriately formatted for the device display – no messing around navigating through from the homepage of Amazon, clicking through oversized pages and wasting valuable page credits on your mobile account, each one costing a few pence. You are spared the nefarious assault from technobabble like 'Web Services' and 'XML'. Your wish list just appears where you expect it to, perhaps to your pleasant surprise, as you didn't even have to configure it to be where it is. There it is – 'Amazon Wish List' – a glowing entry in your lists folder.

Armed with wish list and keen for your next read you eagerly scuttle over to the nearby bookshop. You visually scan the bookshelf for the title you're after. It doesn't appear to be available but you notice something that seems similar. However, taking risks on buying books is something of the past. You consult your 'cool' friends again, mainly those members of your online social network whose opinion you value, mostly people you have never met before but whose reputations figure highly in your estimation from past cyber-encounters. They are most likely fellow reviewers on Amazon and elsewhere. Your phone is equipped with an optical barcode reader[6] and you scan the book's code and wait for a response.

Three opinions are available, one from 'Max32' who is a valued virtual soulmate on growing cottage gardens (informal gardens stocked with colourful flowering plants). She – or he – gives a five-star approval and, even though you can see that Amazon is reporting a

[5] MMS stands for multimedia messaging service. It is explained later in the book.

[6] The camera used to take pictures doubles as an optical scanner.

saving of 20% on the shop price, the book from them is on a 3–5-day lead. Your urge to read tells you that the extra amount is worth it: instant information, instant decision and instant gratification. Thankfully, anxieties about buying goods are ephemeral these days, leaving plenty of time for you to worry about life's other uncertainties in the *postmobile* age.

Before buying the book you remember that you have a coupon from the store previously sent to your phone. Showing the coupon to the nearby coupon kiosk[7] causes a paper version to be printed, suitable for laser scanning at the cash desk.

Time marches on and you notice that the meeting time is approaching. You would like to pop into the clothing shop next door, but you are not sure if there's enough time. You pull up your 'buddy finder' map on the screen and can see that your nearest friend Joe is showing on the map five minutes away from the rendezvous point. There's just enough time to have a quick perusal of the latest fashion lines, but just in case you get distracted you hit 'push to talk' on your buddy's icon and, just like with a walkie-talkie, your voice blurts out on their device (or earpiece).

'Hi Joe – I see you're nearly here. I'm just popping into EJ's. See you at the fountain or in EJ's.'

Joe replied, 'OK – see you in a minute.'

Neither of you had to dial and wait for a connection. 'Push to talk' happens instantly. For Joe the sound came out on their loud speaker built into their PMG. You heard the response in your headset.

You feel satisfied that you have this degree of control that let's you confidently decide that time can be used productively while waiting for the friendly come-together.

You return to the dancing fountain next to the children's play area and the lunchtime gathering convenes. Next, your party talks about where to eat. There's a new bistro just opened along one of the polished walkways in the plaza. There is a consensus to try it out, but first you all check for opinions or 'epiphanons'. Finding the bistro in the food listings is easy as it's immediately accessible via the 'nearby places' menu, your phone having sensed where you are. Looking at the entry for the bistro, one of your friends sees that one of her acquaintances has already tried it out. Unfortunately, it gets a low rating due to a scant vegetarian menu, and this worries your friend, but, thankfully, her friend has posted an alternative venue that has already been checked out with a fantastic array of vegetarian options, so you all decide to head there instead.

In the eatery the food is great, and you all agree that it was worth a visit. It's time to pay the bill and get back to work. In the past you would have spent the next five minutes working out how to split the bill. This time, thanks to m-payments, it is not a problem. The bill has a barcode printed on it and you all take it in turns to scan the code into your phones. You each enter your pin code to clear the payment (see Figure 1.3).

Before your group can leave the restaurant the seating host uses a wireless scanning device to get a payment confirmation message on its backlit topside display. On clearance, courtesy messages pop up on everyone's commie announcing a coupon offer on future meals. The coupon is only valid if within the next day *all of you* post a review of the restaurant. Communal offers like these are common. These reviews will be pinned in space[8] near the restaurant. Other friends may soon be reading them.

[7] The kiosk has an optical scanner (not a laser one) which can recognize a coupon on a commie display.

[8] See Section 13.8 for a discussion of *spatial messaging*.

Figure 1.3. Phone with integrated barcode scanner.

For a while you mingle outside the restaurant with the usual reluctance to part with each other and head back to stuffy corporate-blue cubes in the office. Two of you agree to meet later in the week for a book-reading event at a nearby bookshop. You 'beam' the appointment, along with its whereabouts and details, to your co-bookworm. Swapping electronic items in person has become a common social habit thanks to Bluetooth™, but it doesn't have to be done face to face. Exchanging nuggets of information can take place just about anywhere and anytime, but often doesn't have the same exciting buzz generated in those irresistible spur-of-the-moment exchanges face to face.

You say your goodbyes and everyone melts back into the teeming milieu of shoppers. On your way back to the taxi rank your wrist commie alerts you with a 'throbbing brain' graphic. A tiny, animated cartoon brain pulsates below a glowing light bulb. This tells you that someone with a potentially similar interest profile is nearby. It could be someone potentially useful to 'mesh with' – the term used to indicate techno-assisted socializing purely for the sake of social networking. As with many other pools of personal information you have assigned part of your contacts folder to be publicly accessible. Things have come a long way since the private and jealously guarded pride-and-joy address book. You have many public information zones, including much about your professional and social profile.

Tapping the graphic reveals more information about the alert from your 'neighbourhood watch' agent. Many times before, you had made interesting new acquaintances in your nomadic neighbourhood, a circle of interest that floats around like an imaginary X-ray beam illuminating the area all around you, virtually penetrating into the 'public persona' of willing socialites who happen to pass through your cloud with their mobile devices.

On inspection, you see that the nearby cohort is actually someone selling an item on your 'things to buy' wish list. It seems an interesting opportunity, so you call them without having to know their number. The product is not the exact specification you are looking for, so this time you don't take it any further. You carry on with your business.

While pondering on extending your stay in the shopping plaza an alert causes your smart watch to slightly contract, thereby squeezing your wrist just enough for you to sense that this is an urgent message. A colleague is requesting to know your whereabouts via the track-a-phone service that you signed up to as part of your employment conditions, primarily as part of a health-and-safety edict, but all manner of other types of uses have arisen, just as long as you give consent. Pressing the 'reveal' on the alert will enable your colleague to know where you are, which you gladly do, not suspecting any malign intent. Finding you're at the plaza they request that you bring them back the latest copy of a particular hobbyist magazine, which you gladly do.

Stepping into the air-conditioned taxi that displays your name on its roof-mounted electronic signboard, having just hailed it using your commie, you sit back with the sun on your face and relax for the 20-minute journey back to work. You download the latest track from a new album you noticed advertised inside the cab. You jump straight to the download page by tapping the ad's 'dial-ad' number into your device.

You reflect on the events of that lunchtime with a sense of satisfaction – you seem to get so much done in a lunch break. You start wondering how you ever got by without the wireless plumbing of so many mobile services.

Feeling perplexed and amazed you ask yourself 'how did I used to cope without my commie?' As hard as you try you simply can't recall the time when you didn't need it.

2

Introduction

2.1 THE FUTURE IS NOW, NOT TOMORROW

The sampler of wireless services outlined in the prelude was not a tour of fluffy 'futuristic' concepts. If you think that's what it was, then you are out of touch with what really is possible *now* with existing technologies or within the immediate future. This book will put you in touch. I will describe in varying degrees of detail all the techniques underlying the tour. In techo-jargon, this boils down to exploring the meaning behind acronyms like 3G, GPRS, WAP, MMS, J2EE, J2ME, XML, SMIL, MLP, A-GPS and many others. We are an industrial people obsessed with acronyms – something the ancients didn't even know about,[1] let alone use.

Just to reassure you that we are not going to tread the path of hyperbole, let's ponder on what we didn't 'do' on the imaginary tour. In our sightseeing tour we did not wing past walls of shops morphing into personalized ads,[2] nor was there 'smart dust[3]' painted onto the soles of our shoes, computing our every move and even helping to calculate our calorie burn rate in real time. This would be useful I guess, perhaps ensuring we walked far and fast enough on the way to our fat-drenched burger meal. We did not walk on pressure-sensitive *Smart Floors*[4] that tracked our every move. There were no high-tech, rimless glasses to view the plaza, with all kinds of visual overlays augmenting the scenery, courtesy of a wearable computer, even though the separation of man–machine interfaces from the personal mobile gateway (PMG) is clearly a step in that direction.

[1] In many cultures and languages, acronyms were not used in the past and are even absent as a lexical concept in some languages today.
[2] Watch any of numerous sci-fi films for that – most recently *Minority Report* by Spielberg.
[3] B. Warneke, M. Last, B. Leibowitz and K.S.J. Pister, 'Smart dust: Communicating with a cubic-millimeter computer'. *Computer Magazine*, Jan. (2001) 44–51 (IEEE, Piscataway, NJ).
[4] Hightower, J. and Borriello, G. 'Location systems for ubiquitous computing'. *Computer*, **34** (2001) (IEEE Computer Society Press, Las Alamitos, CA).

Next Generation Wireless Applications P. Golding
 ISBN: 0-470-86986-0 (HB)

Figure 2.1 Wearable display technology. Reproduced by permission of MIT Media Labs.

These things do exist in the labs, such as the MIThril wearable computer,[5] as shown in Figure 2.1.

It's not that I don't 'believe' in these edge-of-future things, notwithstanding any qualms about some of the possible social consequences that we cannot yet fully grasp. In fact, I do accept that mobile and augmented reality technologies with tremendous impact are going to arrive soon; I started research into mobile virtual reality concepts back in 1994, so you can make of my 'futurologist' opinions what you will. Some of these aforementioned 'exotic' technologies are being tested at this moment, even in pre-production stages, but mostly they are not yet economically possible on a large scale. However, one thing that has startled many engineers is the rate of conversion to mass production from previously costly one-off experiments.

It is not that I'm not interested in writing about 'futuristic stuff', I certainly am and wish that I *was* writing about the more crazy reaches of technology, as some of what's coming would be exciting to have in our midst right now and is worth examining in the public discourse. This book is more concerned with bridging the relatively short-term gap between the technologies being available *today* and interesting services being available

[5] http://www.media.mit.edu/wearables/mithril/

tomorrow, or with integrating the new technologies in a smart way in order to realize exciting services? For now, we want to be 'smart integrators' rather than 'smart inventors', or that's the intended mode of thinking for users of this book in any case.

The relatively limited set of service ideas alluded to in the prelude do not matter in terms of their realism *per se* (down to earth though they are in the relatively near term), but what matters is making mobile services, other than 'plain' voice, more pervasive in our lives. We could say that this is the real goal of 3G (Third Generation), but we shall look later at such terms and both the intended and accidental definitions. Throughout this book I'm interested in how to be a 'smart integrator' and how to make you, the reader, think like one and become one. This book is not just about technology, it's about *thinking* about the technology and provoking useful ideas.

2.2 THE NEXT 'NEXT GENERATION'

Back in 1997/8 at the behest of a particularly switched-on marketing director and friend Dick Snyder, who was then working for Lucent Technologies, I designed and implemented what I still believe to be the first[6] working wireless portal. I called it Zingo (not to be confused with Zingo in the UK, which is a mobile taxi-hailing service that I have nothing to do with[7]). At the time, on the trade show caravans Zingo caused some consternation in the mobile community simply because many industry executives couldn't get their heads around it; it was not voice, so what was it? Was it a service or a product? Who gets to decide what the customer sees, who owns the customer? Do people really want to do all this on the move? It was clear that the brave new world of the net was not going to converge easily with the relatively staid and conservative world of the mobile operator, nor were these worlds going to clash; more like slowly blend via osmosis (the weaker solution being on the operator side). At the time even going in the opposite direction got a similar reaction. When Dick and I contacted AOL to talk about wireless instant messaging (IM) ideas, bafflement is what greeted us. That was then. Today, some six or seven years later, wireless IM is only just gathering pace, albeit relatively slowly, but there are fundamental limitations that we still have to overcome, particularly in how to type quickly, which has led to my current interest in novel input designs like Fastap™ (discussed later in the book). Moreover, on the cusp of 3G *availability* (not success or anywhere near its zenith) I just received a questionnaire from an operator asking its development partners for ideas about what they want from 3G. Therefore, things are still moving slowly.

We should not think for a moment that 3G was a direct response to the popularity of the Internet, especially the Web. This is a misconception about the origins of 3G due to its new clothes in the post-Internet (or post-Web) age, sometimes making it seem a deftly[8] timed technology. This it might still be, given the events of recent years and the 3GPP's[9] apparently agile response to absorb so many of the newer technologies, like the

[6] Such claims are always subject to definitions, most often deliberately contrived in order to make the claim. In this case I believe that nothing like Zingo had been demonstrated, so in that sense we were the first.

[7] Strange how they came to choose the same name – when I made up the name I was convinced that no one else would think of it, although I subsequently found a bubblegum-vending machine with the same name.

[8] Though, most commentators argue that it is either too early or too late.

[9] 3GPP stands for 3G Partnership Project, the association of interested parties who produce the specifications for 3G mobile networks like UMTS (Universal Mobile Telecommunications System).

rapidly emerging plethora of XML[10] vocabularies to describe data circulating in the Internet cosmos. The 3G effort was already a work in progress[11] that simply espoused the view that just as the wired datacomms world had progressed from 64-kbps pipes (e.g., ISDN) to 2 Mbps pipes (e.g., E1/T1), then surely the mobile world have to follow sooner or later, as an inevitability, if only to provide mobility for the 'predictable' forthcoming broadband services. The widespread availability of broadband services was also an assumption made early on, mostly encouraged by the advent of multimedia in the early to mid-1990s.[12]

Except for the ideas of a few visionaries, there were few, if any, imaginative concepts (never mind a consensus) for what this high-speed wireless world would fulfil for its end users when it became available. Even today there remains much uncertainty. In fact, there were very few visions beyond considering the mobile as just another viewing device for the Web and email. With Zingo we took a more progressive stab at guessing the future, replete with location-finding examples, people-centric messaging,[13] remote home automation (including message-driven fridge panels) and many other features. We simulated the 2-Mpbs link using a wireless LAN (local area network) technology, an early Lucent interpretation of the 802.11b standard. For a PDA (personal digital assistant) we used a Toshiba Libretto micro-laptop, a product idea (tiny laptops) that has come and gone several times since, further indicating the uncertainty or fickleness of the portable computing market.

I have mentioned the Zingo story because in my opinion this particular history matters, if only because it is a little bit of my history and inevitably shapes my outlook and subsequently how I think about topics examined in this book. I want to convey that mobile operators used not to 'get it', much to the exasperation of shareholders, software vendors and many potentially symbiotic 'players' looking for a piece of the semi-mythical *X* billion mobile market.[14] This is presumably a huge market potential implied by selling services to such an incredibly mammoth subscriber base that has nearly reached its zenith in size, or so we assume.[15] Many traditionalists knew not (know not) how to move the industry from voice to data, other than at the very crudest levels of grasping what the standards bodies were working on to enable information to fly reliably and quickly across the air. Knowledge of how to use this potential has been notably absent. The coincidence of voice market saturation with negligible mobile data revenues has sorely exposed this knowledge gap. The average revenue per user (ARPU), the favoured metric that has come to signify the vigour of the mobile industry, rightly or wrongly, no longer looks good. The ARPU is in urgent need of new services to prop it up (until we find a new metric).

To be fair to all involved the number of stars in the star map of possible mobile technologies has become overwhelming. A recent slide I saw from one of the operators sums up the problem nicely and I have produced my own version to signify the challenge, as shown in Figure 2.2.

Figure 2.2 shows the explosion in technology activities that we now have to embrace or deal with in moving from the 2G (Second Generation) voice-only world to the

[10] XML stands for eXtensible Markup Language and we cover it later and throughout the book.
[11] Many huge standards activities of this kind precede their realization by as much as 10 years.
[12] Multimedia-authoring and multimedia-capable PCs were beginning to take off by the mid-1990s.
[13] With people-centric messaging the entire messaging space was organized around people, not inboxes or fax trays, etc. – an unremarkable, yet effective idea that is still absent in many products.
[14] *X* = Unknown quantity, but hopefully large. Algebra is a wonderful thing, but business analysts and business plan authors should insert real numbers.
[15] Based on one device per person many markets have saturated, but multiple devices per person is a likely future development, as we shall find out.

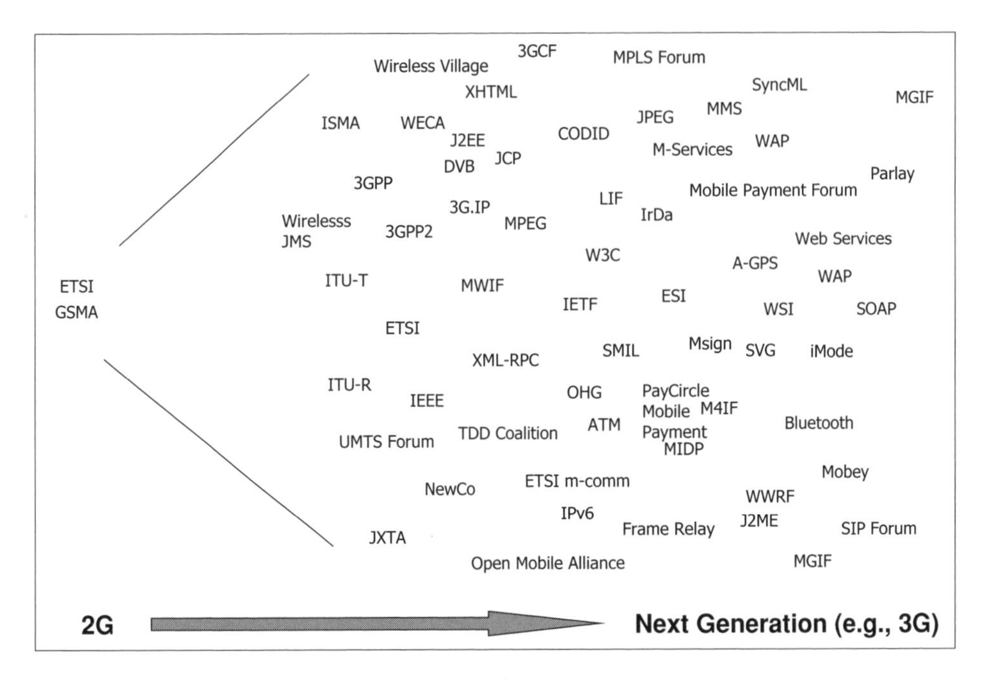

Figure 2.2 Explosion in technical activities relevant to mobile services.

Internet-connected 3G universe. The figure highlights a couple of interesting points. First, there is massive convergence of mobile technology with other activities, particularly the Internet, which itself is converging with all other kinds of activities and constantly spawning new spin-off protocols and concepts. What this means is that any notion of 'mobile services' has to be inclusive of all that is happening in these related fields. Hence, when we talk about *next generation mobile services* we are really talking about a huge gamut of technologies, standards and industry bodies, such as those shown in Figure 2.2. Next generation mobile services and applications encompass much more than just 3G[16] mobile technologies, such as the Universal Mobile Telecommunication System, or UMTS, the 2-Mbps cellular standard. All of the technologies in the figure are next generation – 3G – technologies, and this is a point often overlooked.

The second point gleaned from Figure 2.2 is that, while some of us have been busy trying to connect the dots to construct interesting and potentially profitable mobile services, many more dots have appeared on the map and continue to emerge at a pace that eludes keeping track. Most operators have understood smartly that involving partners is the best way forward. Smart operators have concluded that partnerships are the *only* way forward. However, partnerships are much bigger than partnership programs that dish out a few specifications and the occasional incentive to partner. The nature of collaborating has changed or needs to change. The old division-of-labour axiom of 'we're good at what we do [i.e., mobile customer care] and you're good at what you do [i.e., developing apps]' is not sufficient here to connect the dots. It is a step in the right direction and a big step for the historically customer-possessive operators to take, but to believe that this will lead naturally to the

[16] We shall discuss the topic of 3G later in the book.

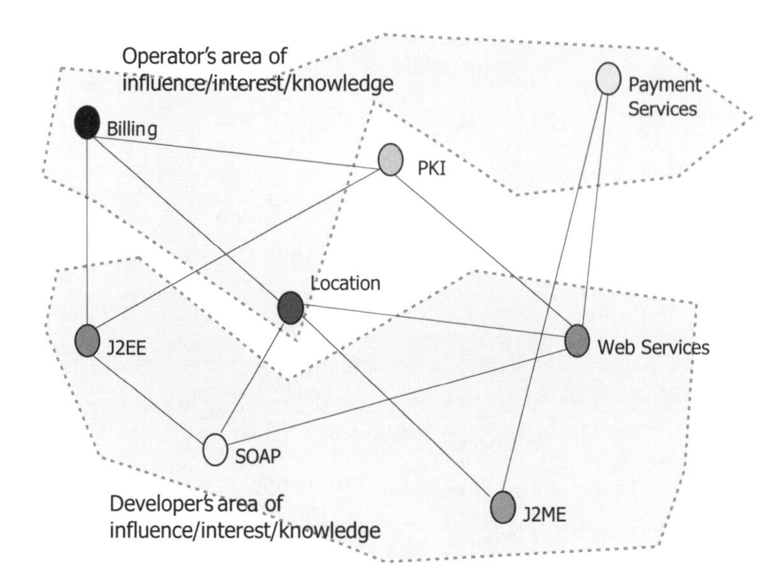

Figure 2.3 Connecting the dots in a mobile service.

emergence of exciting services is perhaps wishful thinking. It is not sufficient to have an 'over the wall' approach where each partner grasps little of what the other is doing. Both sides of the ever-shifting partnership boundary need to understand the other side's technological and knowledge base, and new ways of partnering have to be found. This seems increasingly important, if not vital.

Figure 2.3 shows a possible map of technologies and their connections, such as a particular mobile service might require. From it we can gain some insight into the challenges of developing next generation mobile services. The area of capability for a given partner and the operator might be the dashed-line boundaries shown. However, looking at the interconnectedness of the dots across the boundaries, each traversal implies a requisite technological, human and commercial understanding between the partners. Furthermore, there will often be areas, in this example PKI (Public Key Infrastructure[17]), which another party has to provide and has potentially different interface implications for both parties. Although the diagram is contrived it illustrates that for collaboration of this order to succeed a grasp of 'the big picture' (i.e., the entire dot map) is a useful perspective to acquire by all parties. Otherwise, we are really hoping for the best, that everything will come together in the end. Remarkably, this is still a common approach, but a costly one and unlikely to have significant impact on our desire to increase ARPU soon. The most likely victim will be the end user who ends up with a poor experience and an increased resistance to engaging with new services.

This book is not about WAP, 3G, J2ME and all the other acronyms in Figure 2.2. There are plenty of books about these specific technologies. I hope that this book is a guide to understanding the essence of these technologies, not as individual dots, but as neighbouring

[17] PKI is the management of encryption keys by trusted parties, such as those required for a mobile device to authenticate a service provider before the user is happy to pass sensitive details like credit card information. Public key cryptography is described later in the book.

dots on the map of mobile services. This is so that we gain sufficient contextual awareness to enable us to better connect the required dots together. Some of these dots are in the 'Internet camp', perhaps in its newest areas like Web Services. Some of these dots are in the 'software technologies camp', like Java 2 Micro Edition (J2ME) and Java 2 Enterprise Edition (J2EE). Some of these dots are in the 'RF [Radio Frequency] network camp', like location-based services (LBSs). It would probably be fine just to write about each of these in what I refer to as 'cookbook' fashion, but cookbooks only tell us how to prepare dishes, not how to make a meal. They are a list of instructions for making something, but do not tell what each ingredient does to the dish, what the alternatives might be and then how to bring all the dishes together in a harmonious fashion.

In this book I aim to show how these dots might be connected. I am not aiming to do this in a proscriptive manner, but in a thought-provoking manner, leading you to the edge of an exciting idea that formulates in your own bath time or coffee break. Because of this intent the structure of the book is not as obvious as one might expect at first. For example, within the context of LBSs is where I introduce *Java Messaging*, a key component of J2EE, not within the context of a general discussion of J2EE. I hope that you will see why when you get there. Similarly, parts of multimedia messaging service (MMS) are discussed when we introduce WAP, but a good deal of it is also discussed when we look at LBSs, because we attempt to construct some interesting application ideas using LBSs and MMS, so they are better dealt with together.

2.3 IT'S ALL ABOUT 'GETTING IT', NOT 'BRANDING IT'

One key thought I want to plant at this early stage is that we should stop thinking about discrete generations of the mobile evolution, be it 3G, 4G or whatever. We should start thinking about 'next', as in what shall *we do* next? It is a continuous process. The culture of discrete generational outlook in technology is outdated and not very useful. Thinking in terms of eras is unproductive. Its worse outcome is a kind of mass hypnosis, a suspended 'wait and see' mindset where everyone is waiting for everyone else (the proverbial 'they') to do something. We also need to get away from technological 'branding' that may lead to the wrong sort of mindset altogether. If we start to think of things like 3G as brands instead of technological possibilities, then we may falsely convince ourselves that one brand is superior to another in a void of constructive thinking about the services we are trying to concoct. Debates like '3G versus WiFi' are hollow in this vacuum and prevent us all from getting on with making a success. Calling something '3G' is actually an abuse of language, because a plethora of ideas and technologies cannot be so easily summarized with a two-letter acronym. Aristotle warned us of the dangers of not defining out terms when he said, 'How many a dispute could have been deflated into a single paragraph had the disputants dared to define their terms' – useful advice indeed!

You, the reader, could well be one of the people who will show us all what to do next and unleash the killer app.[18] We know that the electrifying stuff has yet to emerge, still the dormant seed in many an entrepreneur's and techie's minds, but I can confidently cry that

[18] The 'killer app' is the notion that there is a single application that does so well as to justify the existence of the technological and commercial apparatus that made it possible. Some people say that email or the Web is the killer app of the Internet. We examine this concept later in the book.

never before has there been such a colourful array of interesting technologies from which to engineer a masterpiece. Somewhere in the dots, the dots that exist and are emerging *today*, there lay not one, but *many* exhilarating connections that will underlay services that will become the norm in many of our lives. This book is my contribution toward making that happen sooner rather than later.

In this book I'm not really that concerned with the somewhat fallacious, often specious question, 'will 3G work?', whatever that really means. That is the concern of pundits, flop fanatics, so-called analysts and various commentators who can blog about it ad nauseam on their weblogs if they like, me included whenever I fancy the diversion.[19] For the record I am already thoroughly convinced of the future success of 3G (next generation) and I offer the following reframing thought from industry sage Toni Ahonen, taken from his book *m-Profits*:[20]

> *Many of the issues in this book* [m-Profits] *relating to new services and their usages may seem silly, strange or against all logic to those of us over 40 years of age. Remember that 3G is all about the future, and those who will be living it to the fullest are about age 20 today.*

In other words, just because you or I can't grasp the mobile future and see it in our heads (irrespective of our age[21]) that doesn't mean it doesn't exist or that it is not already in the heads of others. However, I don't agree with Ahonen's assertion that these disconnects are due to age alone.[22] This seems to be a popular idea, noting that some developer forums have gone too far with their 'youth-orientated' (childish) ideas[23] and language. I think Ahonen is cautioning us in general not to restrict our scope in imagining new services.

2.4 BECOMING SMART INTEGRATORS

There are actually no historical precedents for much of what we are attempting to build, but this should not deter us from forging ahead. I believe that the ingredients for potential success are now mostly here. With the right amount of background knowledge, like this book offers or leads you to, and with a good deal of whatever gets your creative juices flowing, we will begin to notice very interesting patterns in the dots. Incredible ideas will spring out, and I hope to illustrate this throughout the book. Furthermore, we will be able to build these ideas, knowing that the products will be deliverable to a vast end user community thanks to the large user base and the willingness of so many vendors and operators to use the same open standards.[24] I urge you to follow your ideas and your instincts with enthusiasm. The time is now right to build interesting mobile services. I hope that by reading this book you will understand that this really is the case. Most of what we need to build attention-grabbing

[19] My weblog is at URL (uniform resource locator) http://www.paulgolding.info

[20] T. Ahonen, *m-Profits*. John Wiley & Sons, Chichester, UK (2002).

[21] I think Tony's point is that many of the decision makers in this industry are possibly over 40 and they are not qualified by their experience to comment on the validity of ideas that appeal to future generations in a world that is moving so rapidly.

[22] In any case I don't like the obsession with age that mass consumption has engendered.

[23] One forum would not allow me to sign up without choosing a cartoon character to represent me in the online chat rooms.

[24] Open standards are those whose specifications and recommendations are openly (publicly) available.

mobile services is on hand now or will be within a short time frame. Those of us trying to build services several years ago were in a pause mode, waiting for one technology or another to appear or mature; nothing was quite usable enough. It seems that the waiting is now mostly over.

However, due to a mixture of reasons even the simplest of mobile service ideas can take an exceedingly long time to appear on the market. This delay is a luxury we cannot afford, given the dwindling 2G mobile revenues and the large holes burning in the pockets of all those who have invested in the 3G future. One of the reasons for delayed availability is the poor technical approach, often caused by a lack of knowledge somewhere in the technical management chain or by lack of knowledge as a collective. This lack of knowledge will frequently lead to overly complex projects that take too long to complete and cause excessive, even punitive, support implications.

Most of what we need exists already and all we need to do is piece it all together. This is where we need to be smart; hence, why I use the term 'smart integrators'. We should be seeking to minimize the amount of technology that we have to implement ourselves on the project in hand. This is an obvious point, but seldom tackled as aggressively as it could and should be, which is why even now many projects are dogged by overly complex undertakings to develop in-house technology. There are two reasons for this. The first is that, due to poor understanding of what's going on, senior management can be duped into authorizing work by techies who are seeking to gratify fantasies and fulfil their techno-voyeuristic desires. The second is that, due to the ever-widening palette of relevant technologies that we have available, it becomes gradually more difficult to figure out how best to select from the colours and brushes on offer. This book is seeking to address both these issues by charting the landscape and offering ways to connect the dots from one technology to another. We want to become 'smart integrators'. To do this we have to master not one, but several technologies and then understand how they fit together. It is fair to say that to be effective we should know not one, but several house-building skills, like plumbing, bricklaying and carpentry, and know something about architecture too. Many commentators have spoken on the need to be somehow both specialist and generalist at the same time.

Everything described in the prelude to this book (Chapter 1) is available now, or will be in the very near future, and most likely economically realizable with the gadgetry, features and capabilities now emerging in the early phase of 3G and all its sister technologies. Expectedly, there are much better things to come as the technology matures, and I will mention some of the emerging technologies as we proceed through the book. In the prelude it is assumed that the use of mobile services will be very much an integral part of life, just as mobile telephony and now text messaging have become. Underlying this assumption is the powerful ability to gain access to content, all kinds of content, with a higher degree of control than ever before, thanks to the Internet of course and all the services that are riding on the back of its golden protocol – HyperText Transfer Protocol, or HTTP. Not surprisingly, we focus heavily on this paradigm throughout the book, and the reason why will become obvious, if it isn't already to the more astute and informed reader.

This book is about wireless applications of the sort we tasted in the prelude, explaining how to construct them now and in the near future. It is about how we use existing and emerging technologies to realize these services, assuming that we want to build and deploy them now or in the next few years. To do that we need to know the next few layers of technical detail sitting behind the display on the handset all the way down to the content sitting in a database somewhere. As a rule, interesting information, or content, sits in a database. However, as we shall soon find out, it may no longer be necessary to know where that database resides, its structure, or even how to query it directly. In one sense we need a more constructive definition of what is content if we are going to take on board the entire spectrum of possible ways we can gather and process information in a mobile world empowered by the Internet.

2.5 THE HOLISTIC APPROACH TO WIRELESS

Sometimes we have to define a thing by saying what it is not. This book is not a detailed developer's tome, replete with pasted code fragments (and their errata). Indeed, fragments are exactly what I am trying to stay away from here. For the hardened code-munching 'techies' and 'hacks', there are many excellent wireless developer 'cookbooks' already in the market, covering all the areas that this book will take in; reinvention of the wheel is not my intention here. I will refer to other books you can use to complement your understanding and take certain topics to the next level. You will find references both in the book itself and in my online reference listing that you can follow on the Web as you read this book (the nearest I can get to an interactive book experience). Indeed, it is my hope that reading this book will enable you to pick up many other books and be better able to utilize their content having attained a good grasp of the overall picture. Where appropriate I shall go into some technical depth in those areas that I feel represent the essence of a particular approach and which often are glossed over. For example, with HTTP the protocol and its uses are widely discussed in many books. However, its support for downloading files one chunk at a time is seldom explained within the mobile context, especially versus other ways of achieving apparently the same thing (e.g., using an underlying protocol or an alternative method altogether). Such considerations are important for new services that involve downloading games.

The overall picture is often something we lack, yet it is a vital vision to have before us, emblazoned somewhere on our visual cortex. The origins of this book stem from a training course I wrote for mobile operator Hutchison 3G (branded 'Three' in the UK), the trail-blazing pioneers of 3G services in several markets. The grander depiction here is not the visionary kind that I look to the likes of Howard Rheingold for in his thought-provoking and very readable *Smart Mobs*.[25] The picture I am trying to paint for you is more Earthly, reviewing the *current technologies* that we can use for mobile services that might *really work*.

Let's take J2EE as an example. This technology is de facto the standard for building or deploying wireless applications. In this book I won't be attempting to explain the ins and outs and merits of the J2EE specification as a technology in its own right. Our starting point will be an examination of what we need from a mobile *service delivery platform* and then to explore which aspects of J2EE enable it. This approach is a common thread throughout the book. Given such a contextual approach, you can appreciate that I do not have any desire to deconstruct existing technologies, nor catalogue a list of historical developments, unless relevant to our discussion. This may leave some readers feeling adrift in a sea of complex technologies. After all, J2EE is a very complex set of technologies indeed, with

[25] Rheingold, H., *Smart Mobs: The Next Social Revolution*. Perseus Books, Cambridge, MA (2002).

an interesting history,[26] as are many of the constituents of next generation mobile services. However, I believe that understanding the potential uses of these technologies in a mobile context is the most important achievement that this book sets out to deliver. After this, it should be a lot easier to pick up a book on, say, J2EE, WAP, SMIL or MMS and backfill on important detail about the technology itself.

Let's first try to see how we can use these technologies – and shape them – to meet our service needs. As we shall see, the emergence of something called Wireless Java Messaging Service (WJMS) is one example of that approach – moulding J2EE to fit the particular needs of mobile service scenarios. Too often I hear statements like 'we need to get into this whole Java Messaging thing as that's the way things are moving in this industry.' That kind of rationale may have some legitimacy, but we need a more informed approach toward why that's the way things are moving (if indeed they are) and we should guard against pursuing technologies for their own sake, as it will waste too much time and could end up being a diversion on the road to increased ARPU.

Although I am a keen student of history, I am not too interested in giving many history lessons in this book, even though I find the history of various technologies quite fascinating. The history tour bus approach is one that I notice often in conventional 'how wireless works' courses, starting with the spark generator, taking in Marconi and then analogue, stopping off to look at how modulation works, the migration from analogue to digital and so on. OK, maybe I exaggerated about the spark generator, but I don't believe that the history lesson approach will serve us well here. As J.L. Synge said, commenting on the nature of scientific understanding after Einstein's Theory of Relativity had arrived:

> *To understand a subject, one must tear it apart and reconstruct it in a form intellectually satisfying to oneself, and that . . .* [27] *is likely to be different from the original form. . . .*
> *. . . In this age of specialisation, history is best left to the historians.*

Let's consider the HTTP as an example. While the story of its invention and its design rationale may be interesting, it is not that useful to know when considering its application to mobile device connectivity. An understanding of what types of problem it solves or could solve may be a more useful attainment, before embarking on a tour of its internal mechanics, so that we can better understand its usefulness in mobile applications. Better still, understanding how the HTTP paradigm could benefit mobile applications is a useful perspective. Given its prevalence and wide support in many application design technologies, there would then be no need to look into its historical origins and evolution. In other words, if HTTP solves a problem we face within the mobile context, then let's use it without deconstructing it first. If we understand its essence not only can we do that, we can also suggest improvements, should they be necessary.

2.6 DISTINGUISHING PUNDITRY FROM FUTUROLOGY

I have been involved with wireless for some time now, starting out in the wonderful world of GSM, looking at how bits and bytes whiz through the air while getting my hands dirty in the

[26] The history not in the J2EE development itself, but in the various software disciplines and techniques that fed into its development.
[27] My deletions.

Figure 2.4 Wristomo™ wristwatch. Reproduced by permission of NTT DoCoMo.

mathematical and hardware domains that made it possible for digital wireless to happen – an area called digital signal processing (DSP). The first cellular 'handset' I used in the labs of Motorola was the size of a filing cabinet, rapidly shrinking to the size of a briefcase and before long the size of a family-sized bar of soap. Now they are small enough to fit snugly in the pocket or even on the wrist, like DoCoMo's Wristomo™, as shown in Figure 2.4. We all know the story by now, as we are all users of these nifty devices from one manufacturer or another.

For most of the 1990s it used to take an exasperating amount of conversing with friends and associates to convince them about the merits of owning a mobile phone. Today, their reactions would seem foolish, for example, such retorts as 'what do I need one of those for?' and the equally amusing 'but there's a payphone on every corner'. With hindsight these stories seem amusing, perhaps like much of scientific progress and its perceived sociological impact. Only recently I read a weblog using similar arguments against 3G (a useful 'definition' of 3G comes later). The argument is a common one that roughly goes 'why do we need 3G when we have WiFi[28] hot spots?' That's a good question. It seems very similar to the public payphone spat against 2G (or even 1G), so one side of me feels a sagging feeling because we've been here before and maybe we're repeating old mistakes. The corollary though is the opposite response. Yes, we have been here before and look how amazing things turned out; nearly everyone owns a mobile phone. This should make us want to rush to the future by implementing ideas now that will flourish in an enveloping world of devices hooked into information networks and grabbing our attention with a plethora of useful services.

The 3G versus WiFi issue, real or imagined, important or not, is only one of many that colour the various views of the future for next generation wireless services and applications, tending to suggest that getting to the future is going to be a delicate balancing act. These issues will not only impinge on the currently interested parties. Even currently uninterested parties, like the unsuspecting public, are going to feel the influence of future mobile services; that is, those who haven't yet realized how important mobile technology is going to become, who haven't noticed the steady confluence of all kinds of interesting technologies and ideas that are gathering momentum on the pre-echoing ripples of 'next generation' wireless, be it 3G or WiFi. Perhaps like Chaos Theory predicts, one of these tiny ripples or perhaps several of them have the potential to become a tsunami that wipes clean the current analysts' predictions, eliminates current scepticisms and literally dumps the future of communications on all our shores, reconfiguring them for ever. From my own musings and experience in

[28] WiFi is the marketing name given to the IEEE standard 802.11b for short-range wireless LAN technologies.

this field I am confident this will be the case a lot sooner than we think. Reading the collected thoughts and observations in Howard Rheingold's *Smart Mobs* it is hard not to come to this conclusion. Later, when we look at the potential of combining some interesting technologies, like picture messaging, LBSs and semantic Web[29] concepts, some spectacular possibilities seem to emerge, all within the very near future.

Of course, predicting the future is not easy. As Robert Lucky, meritorious and long-time telecommunications sage, writes:

> *The history of telecommunications services in recent decades does not fill us with optimism that we know what society wants, or even that we know how to go about finding an answer to this question.*

Lucky says this having reminded us of the failure of the AT&T Picture Phone and its many successive emulators. A timely reminder that coincides with the launch of videophone services on some of the emerging 3G networks and, surprisingly,[30] even some 2.5G networks. To examine the issue of technological futurology is more of a philosophical quest than a technical one, and mostly I am concerned in this book with charting the technological landscape – the dots and how they might be connected. However, that is difficult to do without at least some leaning toward a future view. Toward the end of the book, after examining many technologies, I will discuss some examples of services that may be usefully provocative when contemplating the future direction of wireless, although my main aim is to show how the combination of existing technologies can lead to potentially powerful results. In that vein, what I aim to do in this book is cover all the materials in a way that highlights their potential connectedness. From experience I have noticed that often the most successful projects and people are all about combining existing and emerging technologies in an effective manner: 'smart integration'.

Ignoring the perils of futurology, I do recommend embracing boldness in trying to predict the future. With a basic understanding of the essence of several key technologies it is easy to posit some interesting service concepts, even though the details may remain sketchy for now. To an extent I heartily welcome such boldness, if only to evangelize an exciting future for mobile services, a future about which I am confident.

Rheingold goes on to examine the wider sociological implications of the future mobile technological landscape. He takes in such stimulating considerations as the future of democracy, the nature of privacy and even the philosophical view of self and others, in general submitting to the idea that next and future generation wireless technologies will become strong *social-shaping* forces, whether we like it or not (probably both for most of us).

Generally, I sense that there is a good deal of credence in what Rheingold says in his book and that, at the very least, I would agree that the wide-scale adoption of many new forms of communication is an unquestionable hypothesis still worth investment. In the absence of hard evidence, other than historical trends with the adoption of significant new forms of communication technology, I am generally an advocate of the mindset that Lucky describes as the 'Field of Dreams' approach. This is a philosophy that venerates the positive potential

[29] We discuss the semantic Web later in the book, but it is the notion of Web-hosted information being accessible to machine-based (i.e., software) intelligence.

[30] In practice, it would not seem that GPRS (General Packet Radio Service) has the bandwidth to support usable video and that its introduction would therefore be another reason for users to hold the promise of mobile data in contempt. This would only harm the future of 3G, not smooth the way to it.

of humanity's endeavours, as captured by the phrase 'if you build it, they will come[31]'. The most dramatic example of this approach has surely been the World Wide Web, which had no prior business case or analyst's model that would ever have predicted or dictated its success, but once it was built the people flocked to its vast landscape of colourful pages. In Chapter 3 I examine this analogy in some depth to see how useful it is in deciding on what exactly it is we should be building.

2.7 TECHNOLOGY OR SOCIOLOGY?

Another assumption for the future is that, along with new technologies, mobile operators are going to have to embrace new ways of thinking and new ways of doing business. Perhaps for the operators, the epithet 'if you build it, they will come' is probably talking about the innovative developers and their entrepreneurial associates whom we need to come before the end users. Get them to come and the users will surely follow. This requires a change of mindset and a new way of doing business. There is no doubt in my mind that, with the new technologies and features of next generation networks, there are myriad ways to build interesting services. The challenge is getting the operators to allow the services to run on their networks. An expression I have heard from operator quarters to justify a conservative approach is an articulation of not wanting to unleash anything that 'kills the network'. The symbolism conjures up ideas that applications are monsters in need of a leash and that developers are wanton assassins, not potential sources of ARPU-inflating ideas.

Operators own significant assets needed in the future of mobile information networks, they 'own' the spectrum and thus far they own all the potential customers. We cannot underestimate this legacy, no matter how much some of us may prefer it to be otherwise, and it isn't about to change anyway, so we all need to work in concert toward a common goal. Entrepreneurs with ideas for mobile ventures need the operators, but probably not as much as the operators need them. Instead of monsters to be unleashed, services should be viewed as cash cows in the waiting, in need of some fertile grazing land. Developers are the herdsmen who know how to get the cows to graze and let the milk flow, given half the chance.

At this time the only constant the operators can confidently assume is that people are still going to want to talk to each other and, more than likely, send messages, even though texting was a surprise to most – a pleasant one of course as it is practically a 'money for old rope' deal. I still recall vividly my trip to a major operator's offices back in 1997 when my company presented an idea to create an application programming interface (API) and toolset for third parties to send text messages easily from PCs to phones using the Internet. Their reaction was one of complete ennui leading to a dismissal of what they considered as an insignificant part of their business. Moving forward a few years, text messaging heralds as the great forerunner of mobile data services and accounts for a healthy portion of most operators' annual revenues; it is *the* original mobile data application and still gives operators hope that unexpected successes can happen in this business. Moreover, and this is perhaps what the US operators failed to grasp, text messaging is now a cultural phenomenon. This

[31] A phrase taken from the film *Field Of Dreams*, which is a semi-mythical tale about building a baseball park in the middle of a cornfield, inspired by the notion that something magical would happen when completed (http://www.fieldofdreamsmoviesite.com)

is something that operators have understood, and they are consequently looking at service proposals in a new light.

It isn't possible to mention mobile data without thinking of IT infrastructure and software. It is tempting to view the nascent shift in operator mindset to be away from the staid world of switched telecoms and the predictable, measurable billing of time consumed on the network toward the super-sexy world of net-connected IT and swishy nuggets of digestible content: 'business at the speed of thought' to use Bill Gates' adage. However, returning to Rheingold and his shuddering survey of the newly connected unconnected world, this swing of mindset is probably 'off message' (pun entirely intended). The ubiquitously connected world that brings people, communities and information together without time or space constraints should be analysed and understood at the social level rather than any technological one, thus predicating an altogether different and new world view for most of us. Deeper insights into consumer behaviour may be required along with a willingness to react quickly to the unexpected social consequences of new technologies. This will become more apparent when later in the book we look into the potential of LBSs.

Perhaps Rheingold got it right by naming his opening chapter in *Smart Mobs* 'Introduction: How to recognize the future when it lands on you' wherein he writes:

> *The 'killer apps' of tomorrow's mobile infocom industry won't be hardware devices of software programs but* social practices.[32] *The most far-reaching changes will come, as they often do, from the kinds of relationships, enterprises, communities, and markets that the infrastructure makes possible.*

These issues of uncertainty pose challenges to those trying to make headway in the new world of wireless services, a world in which we are no longer concerned with how we talk to each other, but with how people, information, things, places and ideas will interconnect in a global mesh of wireless networks. In societies where culture has become a process of constant change and no longer based on predictable traditions, the challenge of providing meaningful predefined services may prove too great. The better strategy might be to deliver service building blocks in ways that end users can make of them what they will.

2.8 AVOIDING THE 'RISC' OF NARROWBAND THINKING

The launch of 3G comes at a time when so many other technologies are developing at breathtaking pace. I don't mean the kind that make impressive demos in corporate labs or wherever the cutting edge is these days (probably not the mythical garage, or garden shed[33]). Of course, I am thinking of the progression of Internet-related technologies, with Web Services, XML-savvy software and the ongoing triumph of Moore's Law,[34] ensuring an affordable reach of the Internet into the most unexpected and tiniest of devices and places. With widespread wireless technologies, interconnectedness is about to reach new levels of ubiquity, and we may be forced to question old assumptions. For example, why do we need servers at all? That's a question that the provocateurs of the peer-to-peer computing

[32] The emphasis here is mine.
[33] In the UK the shed is historically a more romantic and traditional home for the inventor than a garage.
[34] Moore's Law suggests that silicon chips continue to get more powerful at a predictable rate of cost erosion. More on this later in the book.

world are trying to figure out, not to be confused with person-to-person and other P2P abbreviations (more on this later).

It seemed to me that an attempt to take the lid off of all that is happening in this giddying world would make an interesting and useful book, my personal mission being to maintain a hold on the big picture and the connectedness of all these new technologies. What are the major dots and how we might connect them? There are often so many ways to tackle a problem, but over-reliance on a particular technology leads us down a blind alley. I remember the financial controller in a major telecoms infrastructure company who insisted that he didn't want to deploy any new IT systems that involved risk. This was his defensive posture toward the IT rascals who were insisting that the future was to embrace risk, only they spelt it 'RISC' and it meant reduced instruction-set computing. This is another example of technology 'branding' I referred to in relation to promoting something just because it's 'WiFi' or '3G', irrespective of what it can do for us.

There are two lessons I gleaned from this story. One is to avoid any widely held dogma that the use of particular technology is unquestionably 'the way to go'; the brightness of various technological stars may often evoke a quasi-religious sentiment to follow their trail simply because that's what others are telling us to do. The solution is knowledge – know first and then act. We need to know what we're talking about and ensure it is understood by those who we wish to impose it on. A lot of time can be wasted by jumping on bandwagons only for the wheels to come off later and then for some bright spark to question its wisdom (known as 'genius' after the event, but 'renegade' prior). Let's consider contemporary examples – the XML and Web Services concept is currently promising to transform the way we implement networked software solutions. However, it shouldn't overly influence how we think – we should still be thinking about the problems we are trying to solve, not thinking in XML or any particular technology like it's a fashion. I have seen Web Services used as a 'cool way' to implement trivial interfaces where there is simply no need, but someone wanted to try it probably ('techno-voyeurism'). As we shall see, techno-voyeurism can be very costly when building large service delivery platforms for mobile services, where most of the effort is in 'doing XML' (or any other acronym you fancy) and not 'doing useful services'.

The RISC blunder is not the kind of faux pas I expect any of you to make, but surprisingly (unsurprisingly) I found that when asking the simple question 'what is 3G?' to those working 'in 3G', various peculiar answers arose. This led me to include this question as one of the opening slides in a training course[35] on wireless applications, to challenge what we thought we already knew, but apparently didn't. On a related note I am ceaselessly astounded and often worried that the makers of 3G kit seem to be the least likely to actually use the technology themselves. I have not yet fully grasped why. It is probably not important, although I sense an interesting psychological phenomenon that may be useful to analyse one day when I get time.

2.9 A LAYERED APPROACH

In order to introduce how we can proceed further in understanding next generation mobile applications we need to find a useful starting point. To enable the delivery of mobile services

[35] See http://www.wirelessguru.info for more details on training courses.

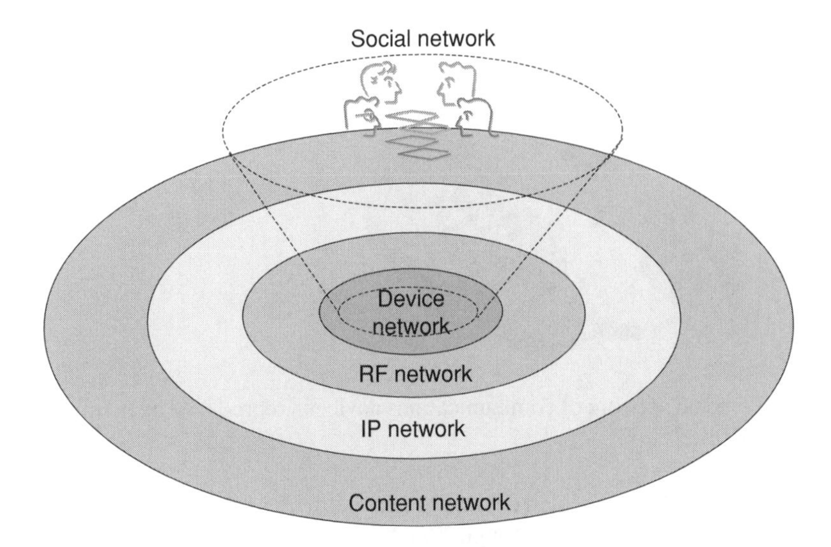

Figure 2.5 Mobile applications environment.

we need a network of some kind and software applications that interact usefully across the network. Part of that network will obviously be wireless, but what might this network look like end to end? Actually, a useful model by which to construct and understand ideas is to consider that mobile applications reside and operate within the context of several interconnected networks, as shown in Figure 2.5.

The entire applications or service environment is what I refer to throughout this book as the *mobile network*. This is in contradistinction to the actual RF network (sometimes referred to as the cellular network[36]).

There are four parts to the physical network, and these parts form the structure for the remainder of the book. All our technological discussions will be relevant to activities that take place within these networks, so let's introduce the networks next.

2.9.1 Device Network

This provides the primary *user interface* to the mobile services, although some services will of course have interface continuity into the wired world, delivered via the 'desktop'. We are used to having only one device, the mobile phone, but in the future we may well end up having a cluster of devices that can talk to each other using Bluetooth™ and talk collectively to the RF network, as shown in Figure 2.6. Out of the collection of devices in the device network, perhaps only one of them talks to the RF network and will form a PMG for the other devices, a concept demonstrated by companies such as IXI Mobile[37] who are pursuing the idea with alacrity. It is useful to note that not all devices may be required at once and need carrying together all the time. Personal circumstances and the required mobile tasks at any given time will determine the device or devices needed for the occasion.

[36] Cellular network is too narrow a definition, as the RF component can sometimes be noncellular. The notion of what is cellular will become clear later in the book when we look at the RF network in some depth.
[37] http://www.ixi.com

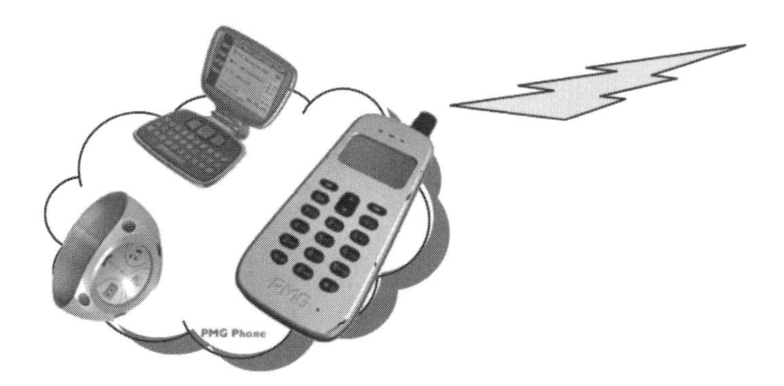

Figure 2.6 Personal network of communications devices. Reproduced by permission of IXI Mobile.

Even fashion may influence which devices to take out with us. Understanding those needs ahead of time may be a challenge though,[38] and we will think about this later.

A key understanding is that the device network may use RF technology for inter-device communication, but such pockets of RF coverage are completely different to the wide-scale RF networks that bring us ubiquity, as discussed next.

2.9.2 Radio Frequency (Wireless) Network

This is the critical wireless link between our device network and the earthbound hardware infrastructure, eventually connecting back to some content, person or machine somewhere on the planet. The job of the RF network is to enable the device network to interact with other networks irrespective of time and place. The only way to achieve freedom in these two dimensions is to bathe the landscape in RF signals which the mobile device can always pick up and reciprocate. In a wide area network we achieve coverage by dividing the landscape into cells, each served by a cellular base station. These stations mesh like a honeycomb and gather the RF signals, converting them into landline digital signals that ping to and from the Internet Protocol network which we introduce next. The network has to manage mobility by keeping track of which cell each mobile device occupies at any given moment.

There have been opinions uttered that local area wireless technologies, such as WiFi (802.11b[39]) will suffice, apparently leaving wide-area 3G systems dead in the water. Such arguments are difficult to interpret if we are only concerned with today's usage patterns and computing habits. No one would sensibly argue that if total coverage were possible, within whatever constraints we wish to place, then this level of ubiquity would not be desirable. Whether it would be necessary or not is often debated, but I think this debate will prove to be a waste of time. We have to heed Ahonen's wise advice not to limit our view of tomorrow's services based on today's prejudices, whether due to age or some other possible stricture. Today's heterodoxy is tomorrow's orthodoxy and I'm sticking with that – this is perhaps the only piece of historical wisdom that I am tempted to rely on throughout the book. We should however be aware that the RF network and the device network might blur

[38] In other words, even though we forgo taking a particular device out with us, we may still need some of its functions (if we still end up thinking of devices in these terms, which is doubtful).
[39] http://grouper.ieee.org/groups/802/11/

at the edges, with certain RF technologies being capable of realizing the required degree of connectedness both within our device cluster and beyond it. In other words, the 3G versus WiFi debate may prove to be the wrong focus. Understanding how to achieve effective use of *both* technologies, plus others, is perhaps a more productive approach to take.

One issue I want to address now is the common misunderstanding that might go 'because we don't really need lots of bandwidth on the move, 2.5G systems like GPRS are sufficient.' This is not the case. Something like GPRS has been grafted on the back of a voice-only network (GSM, or Global System for Mobile Communications) and cannot scale to widespread data services provisioning, so something better is needed regardless of whether the width of the pipe should be 2 Mbps (3G data rates[40]) or not. Whether or not a network provider can make a sustainable business case for sticking with 2.5G technology for the near future is another matter, but the types of services and technologies we discuss throughout this book are only viable on a mass adoption level within the context of newer RF network technologies, like 3G. I think this will become obvious throughout the book.

What is important is the impact that the nature of the RF network has on the possible service portfolio and how this in turn influences application design. I reiterate that I am interested in identifying the dots and suggesting how to connect them, not the realpolitik of which dots we should pin our flag on – so if WiFi works let's use it, ditto for any other RF connectivity solution, particularly if either technology (or others) have unique characteristics that make certain services more appealing.

Some commentators have tried to argue that there is no such thing as value-added mobile services, that all we need is a wireless connection to the Internet; this being the 'operator-as-data-pipe-provider' view. In my opinion, this is a rather naive view, even though I am attracted to its simplicity and I understand the temptations of others who would like to eliminate somehow the awkward slowness of unimaginative operators by expunging them altogether from the wireless future. However, this is unlikely and denies the operator a source of powerful assets, not least of which is the customer base. Other examples include the cellular network's inbuilt ability to detect the geographical location of a user. Location information can support all kinds of 'location-aware' services. This may prove vital in connecting the physical world with the virtual one in a mobile setting, as discussed at length by Rheingold in *Smart Mobs* and by Ahonen in *m-Profits*. We will look at this in more depth in Chapter 13 on location services, but for now we note that these types of services relate specifically to mobility and so contravene the simplified view that a wireless Internet connection is the sole function of an RF network. This doesn't negate the fact that mobile devices can detect location without the cellular network, but the existing cellular customer base carries with it a lot of weight. LBSs introduced to existing cellular customers on existing accounts will gain a lot more traction than new services using new devices.

The very context of being mobile leads to all kinds of service ideas that will not necessarily arise out of the Internet, if only because the business models for doing so may be hard to achieve without operator involvement and incentives. In addressing the myth of 'operator as pipe' we will consider various contraventions to this idea throughout the book, not forgetting that a good deal of services will include telephony functions and that these are still the preserve of the cellular network. So far, the Internet has not provided a viable

[40] 2 Mbps is the theoretical maximum for a first generation 3G base station, but this may be shared by more than one device – so the practical rates for users will be less than this and will vary. This is discussed later in the book when we look at the RF network.

model for commercially successful telephony. When we look at the RF network we will discuss how accessing the telephony features and other assets of the network, especially the ability to charge for items, makes cellular-bound applications potentially very powerful and exciting. This is especially so with interfaces like Parlay X which enable Web-based software techniques to access powerful cellular telephony assets. This is a momentous achievement which we have yet to see reflected in any service offerings and is possibly where many interesting developments are going to occur in what we are calling '3G'.

2.9.3 Internet Protocol (IP) Network

Not surprisingly, the means by which we connect into the wired world of information and information exchange is via the Internet. The rapidity at which the Internet has developed and its ability to support all kinds of interesting scalable end-to-end services makes it the obvious choice for providing the basic plumbing. Most of the software world has already made this decision anyway – it will become obvious why in later discussions. Indeed, we may argue that the Internet is the 'killer application'[41] for 3G (with hindsight, of course). Users' appetites for Internet services are such that extending their reach to the mobile setting is bound to be a popular move, notwithstanding the major pitfalls of portable device usability (which we shall discuss at length throughout the book). Perhaps we shouldn't think in terms of users wanting to see what they can see on the Internet, but wanting to do what they can do or wanting to do what they can't do!

The Internet is far from static. Its service agnosticism and apparent apolitical nature enables many different modes of communication and knowledge sharing to take place, constrained only by our ability to conceive of new applications. However, there are clearly limitations to extending existing services into a mobile setting, most notably the relatively constrained user interface possibilities of current and near future[42] mobile devices. This is another reason why simply providing a wireless pipe to the Internet is perhaps not a useful approach. We will talk about the mismatch between wired and wireless Internet later in the book when we look at markup languages and ways of presenting information on mobile devices.

Many flourishing Internet service concepts will work well in the mobile ecosystem and even provide particularly powerful, new opportunities specific to the mobile context. One such example is the ability to tap into *reputation networks* where we can gather trusted opinions about everyday products and services. This is a step toward imagining our mobile device as a personal pocket adviser, engendering new modes of emotional attachment to our devices – attachment which I think will become popular and indispensable, especially when combined with other modes of communication, like instant messaging, videophony and spatial messaging.

The golden protocol of the Internet is HTTP, and we will examine this in some depth, especially giving special emphasis to its fulfilment and performance in a mobile context. The limitations of earlier mobile data solutions, before 3G, meant that HTTP wasn't always the ideal protocol to support mobile services. This limitation is partly what brought about the

[41] Can't really call the Internet an application in the conventional sense, although it can be viewed as 'an application' for UMTS, as indeed the UMTS Marketing Study apparently does.
[42] 'Near future' is the next two years. I think beyond that we can expect to see many new interface innovations making their way into the marketplace.

Wireless Access Protocol, or WAP. Those of us who cut our teeth on WAP will recall the lack of market interest at the time, leading some pessimistic commentators to decry that 'WAP is crap'.[43] Well, 'crap' sticks; so those same commentators and many who heard their proclamations might not have heard that 'WAP is back'. Indeed, without WAP there is no picture messaging (MMS) as this technology has adopted WAP protocols as a bearer mechanism for delivering messages. In fact, WAP is a broad set of specifications that now embraces HTTP; so we shall be examining the uses of WAP when we look at IP-based protocols later in the book. We also look again at WAP when we discuss how to make applications secure, and we will scrutinize the myths and realities of wireless security.

2.9.4 The Content Network

Since the days of the rising popularity of the Internet, or more accurately, the World Wide Web, we have been hearing the mantra 'content is king'. It is an interesting observation, but one prone to misdirecting our attention from what is content. This sentiment arose from the idea that Internet users want to consume information as passive recipients, just like with television. This was in the dot.com days of aggressively pursuing 'eyeballs' on websites in the hope of future sales conversion opportunities. Some commentators have subsequently derided the notion of promoting, engineering and selling the Internet as a mass consumption phenomenon, realizing its essentially ephemeral nature, courtesy of the powerful click-and-leave hyperlinks (anchor tags[44]). David Weinberger, Internet sage and co-author of *Cluetrain Manifesto*[45] has been one of those to lay on the derision thickly. Like many other commentators he sees the Internet as a significant cultural phenomenon with people tuning in for much more than just 'a shopping experience', a topic he expands on in his latest book *Small Pieces Loosely Joined*[46] wherein he claims:

> *The Web – more an idea than a technology – is challenging the bedrock concepts of our culture: space, time, matter, knowledge, morality, etc.*

Weinberger's observations and ideas are often controversial, but he touches on similar topics to Rheingold in *Smart Mobs*, that the conjunction of the Net with new mobile technologies at the very least creates an effervescent pot of opportunities, which if asked about their significance these authors would point toward the revolutionary end of the social change spectrum. I suspect that mobile operators hope that this is true, even though they have learnt not to use words like revolution any more (since WAP) and probably don't yet know how such a revolution is going to take place, having already been surprised by the popularity of text messaging (except in the USA[47]). However, another school of thought says that maybe these operators just don't 'get it' and will attempt to foster a corporatist entertainment outlook of mobile Internet that would better fit with the 'AOL/Time-Warner' world view and probably eventually fail to engage most users.

At the heart of such differences of opinion is often an important battle of control. One view sees the Internet, and by extension the mobile Internet, as simply a 'dumb' network

[43] Forgive the apparent coarseness in the slang 'WAP is crap', but this was a widely used expression in the industry.
[44] We will look at anchor tags – '<a>' – when we discuss mark-up languages.
[45] http://www.cluetrain.com/
[46] http://www.smallpieces.com/
[47] Text messaging has not been so popular in the USA for a variety of reasons.

that provides the fabric for social exchange from which the users should be able to cut their own cloth. The other view sees the Internet as something to control in order to extend the reach and protect the interests of big, incumbent content providers. In the mobile world this latter view favours the operators themselves who are adamant that they own the customers (mobile users on their networks) and are therefore the rightful dictators of the entire mobile experience, monopolizing content decisions on behalf of the users.

The reason I mention these ideas at this early stage in this book is that I take a very broad view of content. The 'content is king' idea conjures a notion of syndicated content from traditional content providers, like news, entertainment, licensed ring tones and so forth. Such a view is too narrow in my opinion. Content should include any information that a mobile user wishes to handle (process) using their device. In this model even emails, contacts, Amazon™ wish lists and online opinions are content, as are the photos and video snippets taken by multimedia phones. Of course, this is not to say that the 'content is king' proponents are wrong, but the slight difference of emphasis is important. One leads to the notion of content flowing from a controlled centre, the other fosters user-defined content produced and consumed at 'the edges'. We probably need to accommodate both modalities.

For the most part the focus on content in this book is on the processes for delivering it into the mobile network and successfully presenting it to the users in a usable form. Of primary interest is determining success factors for deciding that this process has been effective. Thereafter, we shall spend quite a bit of time looking at how we build the content network, soon realizing that perhaps we require a unified view of the content network in order that content providers of all types can easily manage their content flows into the mobile network in a predictable fashion. The notion that there is a degree of commonality between services leads to the idea of *service delivery platforms* (SDPs) that offer common housekeeping functions to facilitate content delivery, such as charging mechanisms. Such platforms are beginning to take shape in the market and the relatively new term (i.e., SDP) has entered the lexicon, albeit not that well defined. In general, SDP products are those that provide 'out of the box' the common elements of a content network within the mobile context. We will see that the architecture of service delivery solutions naturally extends from the familiar client–server (CS) architecture and that many SDP service delivery mechanisms fall in line with what is happening in the Internet world with developments like *Web Services*.

2.9.5 The Social Network

Having discussed the four elements of our physical network we should be aware of a fifth network that I refer to in this book as the social network. Too often in books concerning communications networks of any form the approach is incredibly linear, mistakenly assuming that we can abstract our view of the network to a single user experience then simply extrapolate linearly to multiple users without special attention to the social dynamics of the user community itself. This is a mistaken approach.

Mobile networks have some essential differences compared with the relatively static Internet setting. The main difference is that with a mobile network the users can carry their interfaces with them everywhere they go, whereas in ordinary Internet activities they cannot (at least not in a convenient and therefore likely modality). Therefore, as our prelude to the book clearly illustrated, mobile users can interact physically with each other as well as electronically *at the same time*. Rheingold noticed this in his observations of mobile users convening in physical meeting places while simultaneously engaged with their mobiles, either with each other or with others outside of the physically present Social network. In my view, Rheingold did not examine this issue in enough depth and failed to analyse how this unique dynamic could provide advantages in enabling distinctively useful services and applications within the mobile context. We ponder on this later when many of the topics in this book culminate in looking at services based on the concept of *spatial messaging*[48] combined with messaging *semantics*[49], which enables the message pool itself to become aware of its own contents and association with senders and receivers at the psychological level.

Mobility implies movement of people. This migratory aspect, like all migration patterns, has a social dynamic. Therefore, we need to think of the entire mobile network as a mobile ecosystem to reflect its indispensable socio-technological nature.

Throughout this book I am attempting a holistic treatment of mobility; so, to ignore the social dynamics of the users would be folly. This is a point that has become more and more apparent to me in recent years, particularly with the emergence of 3G coinciding with an evolving view of what the Internet is able to offer to society and businesses. It is no longer a technological phenomenon, but a social one – the dynamics of its usage have overtaken the significance of the underlying technological developments, which were once the focus of those who frequented this world called the Internet. The reason for this relates back to the earlier discussion alluded to when introducing the content network (namely, because of the egalitarianism of its users who are the main content providers, not the corporate machines). Consequently, users are taking the Net in whichever direction takes their fancy and we are witnessing huge organic growth in ideas unfettered by any proscriptive higher powers.

There are some important issues to raise here about the increasingly organic nature of widespread electronic communications in both the wired and wireless worlds:

- Users themselves can be and often are content providers within the Internet cloud. Witness the huge, largely unexpected emergence of weblogs ('blogging'), the personal commentaries and diaries of anyone who wishes to maintain one, without confinement to corporately endorsed or consensus ideas. It is perhaps the emergence, at last, of the narrowcasting idea first discussed by Negroponte in *Being Digital*,[50] although it is more akin to publishing than broadcasting.

- Given the widespread availability of technologies that facilitate social exchange, users will form their own modes of communication and find new ways of socializing and forming social groups. This is what Rheingold refers to as 'swarming' and is so evident in the surprise success of text messaging, which itself has some surprising dynamics, as mentioned in *Perpetual Contact – Mobile Communication, Private Talk, Public Performance*[51] which has some academic studies into the usage patterns of 'texters'.

[48] Spatial messaging is the ability to send messages that are tagged with positional information, so that we can target people at particular locations.

[49] Usually, messages are treated as black box entities by the infrastructure, which remains blind to the contents. By adding semantics to describe what's in the boxes the infrastructure can handle the messages in new and interesting ways.

[50] Negroponte, N., *Being Digital*, Vintage Books, New York (1996). *Being Digital* is really easy to follow even for someone who is not all that computer-literate.

[51] Katz, J. and Aakhus, M. (eds), *Perpetual Contact – Mobile Communication, Private Talk, Public Performance*. Cambridge University Press, Cambridge, UK (2002).

- Despite an initial assumption that mobile telephones were solely 'speech delivery devices', witness the trends in personalization and emotional attachment to phones, suggesting an altogether different dynamic than was expected. It is tempting to draw parallels with the World Wide Web, Berners-Lee's initial vision being of a browser as an impersonal information delivery device, but one that has become a very personal information exchange device. The Web now includes a wide variety of both popular and niche social concerns and opinions, well beyond the academic context initially intended or envisaged. Indeed, it is safe to say that the Web has become a culturally relevant entity with apparent social-shaping powers well beyond its original technology-shaping boundaries.

From these issues we can see that the social aspects of mobile technology adoption are important and have the potential to shape not only society but also how the technology itself evolves through a socio-technological feedback loop. This has not been present insomuch that

At the more mundane level the importance of considering the Social network is that it represents another plane of information exchange that does not sit neatly in our concentric rings model of the physical network, as portrayed in Figure 2.5. Users are perfectly able to 'beam' content to one another in coalesced personal networks, using either Infrared or, increasingly, using Bluetooth™. There may well be other means of swapping information, such as physically swapping memory cards, games cards, SIM (subscriber identity module) cards and so on. In peer-to-peer computing networks (P2P), neighbouring devices may choose to swap information of their own accord. All of these exchanges are beyond the conventional network boundaries we are used to and involve new forms of social and information interchange in continually evolving network topologies.

2.9.6 Application Topologies

In order to build services we need to understand how applications might reside in the mobile ecosystem. There are two main planes of operation: CS and peer to peer (P2P), and there is no reason why there can't be hybrids of the two – something we will look at later. Figure 2.7 shows the two planes of collaboration.

Notice that we have deliberately shown the user interface as a separate application component. The reasons for this will become clearer later, but it will help us to maintain a degree of focus on the user interface as the tactile part of the application with which the user interacts. It also reminds us that the user interface and the actual bulk of the user-centred application could be running on separate devices within an interpersonal device network. Isolating the user interface in our model will force us to mull over the usability aspects of services and, thereby, give usability its due right on behalf of our target end users.

If we focus on the CS plane for a moment we notice that we are expecting the application to be in two parts. The server component of the application typically resides within the content network and we will appreciate that its essential role is to serve up content to the client on demand. However, we need to be aware that in mobile applications we have the possibility to push content to the client without demand. This latter scenario is what we call *asynchronous communication*, there being no deliberately sequenced and coordinated request and response for content – it can arrive at any time without concern for client readiness. True asynchronous communication is peculiar to the mobile ecosystem, and we

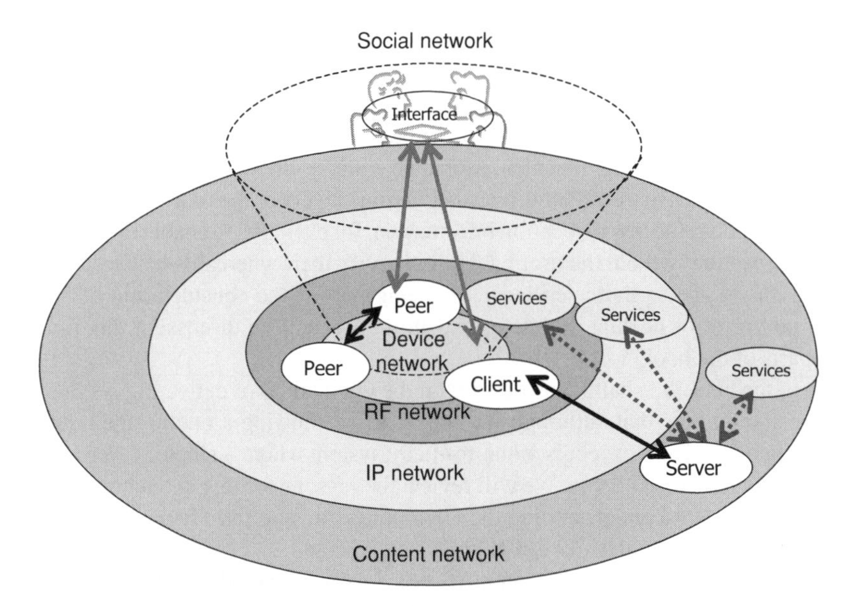

Figure 2.7 Application topologies in the mobile ecosystem.

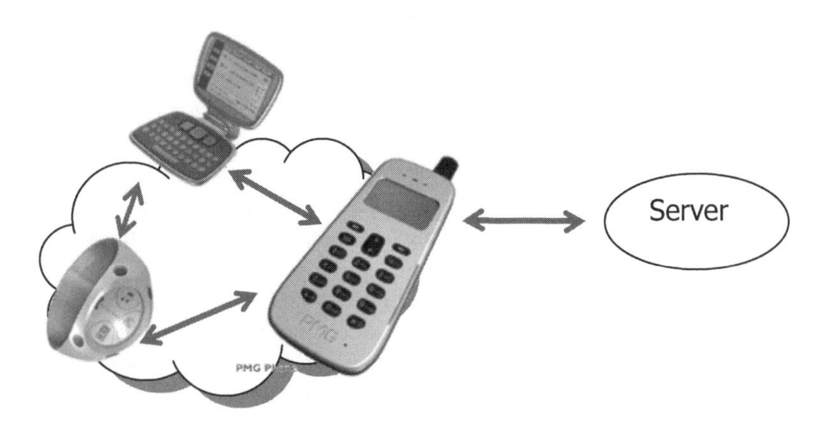

Figure 2.8 Application communication pathways in the device tier.

will examine enabling technologies like WAP Push and its powerful features and service-lending possibilities.

The client component of the application clearly resides somewhere in our device network, and it is entirely possible that there could be several client applications either on each device or spread throughout the cluster. Some applications may run on all devices acting in unison to realize a particular function or service. Generally, the client will be responding to user commands from the user interface and converting some of these into content requests for relaying back to the server. In a device cluster one of the devices could be acting as a proxy or gateway to relay requests to the server on behalf of other devices, perhaps solely because the gateway device has the necessary longer range RF communications capability to reach the server whereas the other devices don't, as is shown in Figure 2.8. Perhaps even the gateway function will jump from one device to another depending on its RF capabilities. A WiFi-enabled device may act as the gateway in a WiFi area and a 3G device act as a gateway in a 3G area.

The server part of the application tends to be the workhorse that has to perform many actions to manage a meaningful content exchange with the client. We can see from figure 2.7 that the server may have to call on services in various layers of the mobile ecosystem. Such services may include access to location information, for example, to enable tuning of content to end user location without the user having to specify their whereabouts. Clearly, we shall need a means to access these services, this being part of the consideration of building a services platform for hosting mobile applications. We will be discussing this in detail in several parts of the book.

Later in the book we shall focus heavily on the CS method of delivering services to end users. We should note that, although we appear to be showing a one-to-one relationship, what we really need and expect is a one-to-many system where a single server application can cope with numerous clients. We will review the reasons for this in the next section as a necessary precursor to understanding the CS architecture and introducing and elaborating on the enabling technologies like J2EE.

It is not necessary to ask for content from a server, especially in the cloud of device clusters where users can interchange information without intermediaries being required. This is the P2P mode of application exchange, and its simplest incarnation is 'beaming' of contacts, ring tones and pictures between phones, and recently games. However, we need not confine our understanding of P2P to being contingent on physical proximity. It is perfectly feasible and reasonable that two devices could communicate directly with each other using any of the layers above the content network, without any server involvement. We can think of this as a shifting of the conventional CS roles whereby one of the participating end user devices acquires a server role while the other becomes a client – and there is no reason why this can't be bidirectional. Such a configuration challenges many of the conventional understandings we may have about computing. Imagine opening a web browser to view a page on your friend's device, such as their contact list.

Figure 2.9 shows us two possible modes of P2P communication. The continuous lines indicate close proximity inter-device communication in the device network – these will commonly be Bluetooth™ connections but could also be infrared. These are both perfectly usable connection channels for P2P collaboration. The dotted arrows indicate a P2P session[52] taking place over a wider area network, possibly via the RF or the IP network layers. Note that text messaging in existing 2G mobile phones is an example of a P2P application involving the RF network layer (although strictly speaking there is a server involved called the Short Message Service Centre, but it is effectively transparent[53] as far as content exchange is concerned). In the figure we are showing the P2P session taking place via personal mobile gateways that are enabling the wide area connection.

In the later sections of this book, we shall examine in detail the underlying mechanisms needed to enable both CS and P2P applications. Let us first introduce some of the physical elements or building blocks in the network layers throughout the mobile ecosystem.

[52] We can think of a session as being an entire communications exchange, but we will formally define this term later as it has certain meanings that are assumed when discussing issues like transactions taking place via the Web (and HTTP).

[53] Inasmuch as it is really a message queue and does not provide any transformative function.

2.9.7 Physical Network Elements

For the rest of this book I will be aiming to interpret and explain the physical elements that we need to realize the networked mobile environment shown in Figure 2.5. Figure 2.10 offers a simplified view of some of the more important building blocks that we might expect to encounter.

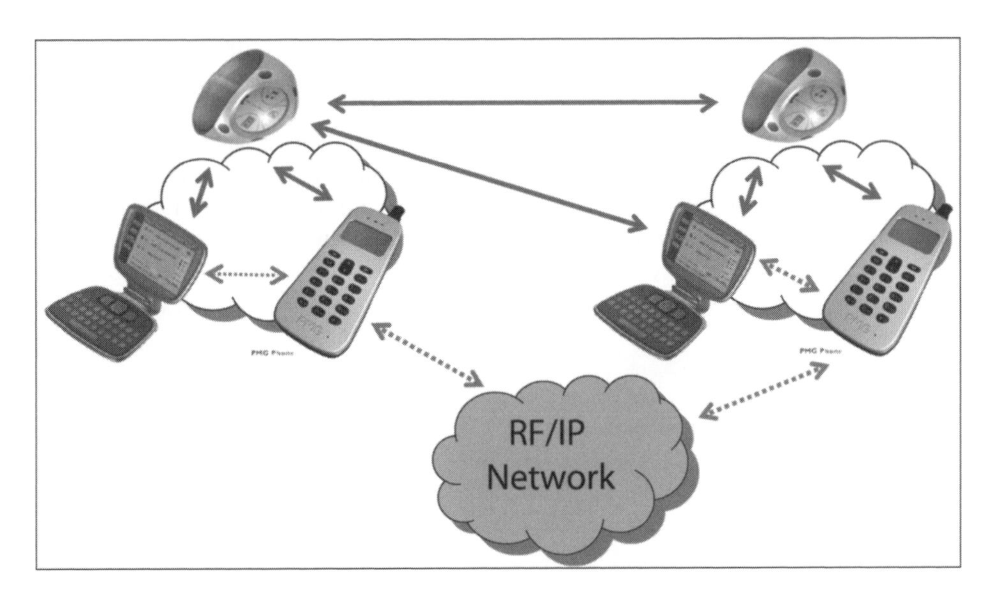

Figure 2.9 Peer-to-peer (P2P) communication between devices.

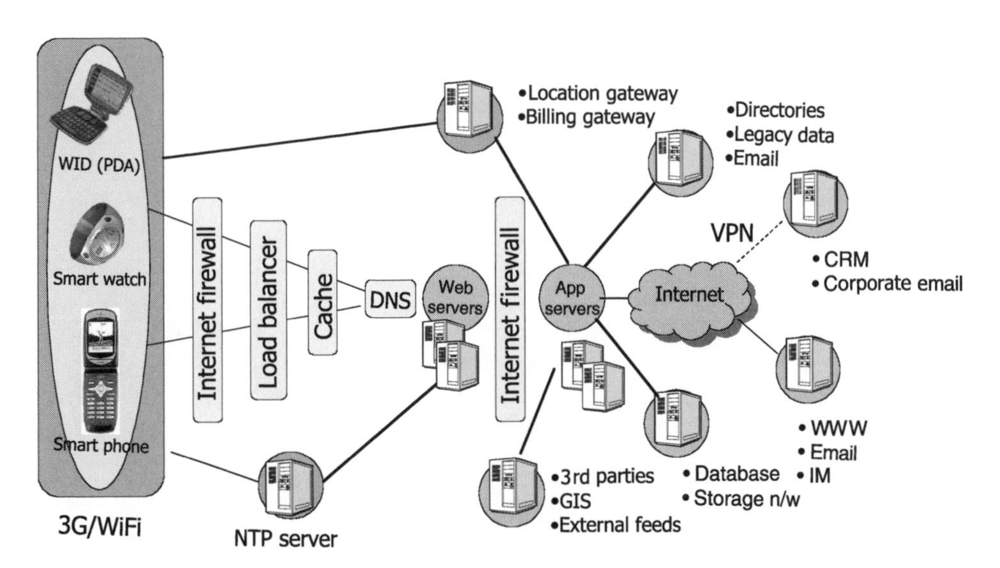

Figure 2.10 Simplified view of some important network elements.

The rest of this book will explain how these elements work within the mobile ecosystem, focusing on various levels of detail depending on how relevant the particulars are to our wider concern with mobile applications and services. In some areas we will need to drill down several layers of detail as the technologies involved are very specific to the mobile framework, such as how J2ME works on the devices, and the dynamics and systems issues of loading J2ME programs, called MIDlets, over the air (OTA), which means across an Internet Protocol (IP) connection supported by the RF network.

In other areas, like the IP network, we will focus mainly on the protocols and security aspects that are essential to understand in any mobile Internet application, but we will not look into too much detail about the infrastructural workings of an IP network, like the use of routers and so on. The only time we will delve into such discussions is when considering the 'edge effects' associated with the implementation and workings of the boundary between the IP network and the RF network, as the functional partition can begin to blur at this point; so, some clarification is necessary, like IP address assignment to devices in a mobile network.

3

Mobile Service Delivery

The following dictionary definition of 'interest' may prove valuable in this following discussion:

interest
Function: *transitive verb*
1 : to induce or persuade to participate or engage
2 : to engage the attention or arouse the interest of

3.1 THINKING MOBILE

Before delving into the mechanics of the constituent technologies in our mobile network we first need to examine exactly what mobile services are, their peculiarities and challenges. In this chapter we are going to start looking at what we need to build a platform to deliver our mobile services, and the rest of the book will expand on the enabling technologies needed to build mobile services.

We need to think carefully about various definitions of mobile services and not limit ourselves to any particular slant that will narrow our ability to design and build the most *useful* services possible. I deliberately use the word 'useful' here, as it is sufficiently generic and open-ended not to imply who the main benefactor might be or how they might benefit. As the previous chapter made clear, next generation mobile services can take all shapes and forms and involve all manner of users and all manner of partners in the delivery process, which means a multitude of possible beneficiaries. Our measure of usefulness will evolve throughout the following discussion.

Next Generation Wireless Applications P. Golding
 ISBN: 0-470-86986-0 (HB)

3.2 LEARNING FROM OUR MISTAKES

An early warning sign that we had not properly considered the nature of mobile services was the WAP fiasco.[1] This was characterized by a forlorn drive toward adopting a *technology* that could deliver information to mobile phones rather than a *solution* and without first pausing to consider what the end users might actually want to do with such a capability.

There have been many attempts to explain the early failure of WAP. It becomes tiresome to go over the possible reasons, especially when it becomes a pet chalice for the 'told you so' pundits to announce the premature death of 'mobile data'. I have no particular axe to grind, although I confess that I was doubtful of the apparent model evolving for WAP in its early stages, which was the 'Mobile Internet' analogue. Despite my own scepticism I was very keen for WAP to succeed and for it to become the first clear step toward the greatly anticipated mobile data revolution. In some sense this has materialized in the guise of the Multimedia Messaging Service (MMS), which uses WAP to transport picture messages – but more on this later. Many of us were optimistic about WAP and were even prepared to invest various amounts of time and money to make it succeed in the marketplace. I led several major projects that used WAP during its infancy and I learnt some useful lessons.

Let us return to defining mobile services. Presumably, we already have some kind of inkling of the definition of a mobile service, as we have been skimming the surface of the topic during the opening chapters of the book, including possible examples of services, as described in Chapter 1. I expect that most of us have our own idea of what a mobile service is without the need for a formal definition. However, I believe that by expressing the essential nature of mobile services via a definition we can engender some useful thought provocation. My intention in this chapter is to evoke a mindset about mobility that will serve as a reference point to keep coming back to as we consider which technologies to use when building services or, rather, how to approach the vast array of technologies in our map of dots discussed in Chapter 2.

It is worth contemplating the difference between 'mobile' and 'wireless' as they seem interchangeable words or concepts, but such substitution masks a subtle and important difference. Wireless does not necessarily infer or need to include mobility. We need to understand what is meant by the word 'mobile'. Traditionally, if we use such a word within the historical context of mobile telephony,[2] then an acceptable meaning is probably something like:

Mobile telephony definition

The ability to initiate, receive and sustain voice calls while freely roaming anywhere we are likely to go in conducting our usual, day-to-day business and social lives.[3]

Obviously, to achieve this goal implies the use of Radio Frequency (RF) – or wireless – communication technology, in order to remove any restrictions on our ability to *move freely*.

[1] WAP (Wireless Access Protocol), coined as the 'Mobile Internet' and the harbinger of mobile services failed to gain a significant sustained user base in most markets when first launched.
[2] The use of the term 'telephony' here refers strictly to voice communications.
[3] The exact wording is my own definition for the sake of argument, but it is a reasonable summary of what mobile telephony is.

I would like to stick to the above definition of mobility throughout the book, the emphasis being in the words 'while roaming *anywhere* we are likely to go'. In the network communications parlance we can think of this as 'wide area networking', but our definition not only includes networking anywhere we like *in situ* but also remaining networked while actually moving. I like to think of this as 'anywhere area networking', because it may be realized via local and wide area technologies in combination, so the use of these geographical terms can be misleading. In mobile communications parlance this constant connectivity regardless of location and speed[4] is called *roaming* and is one of the most basic and important features of a mobile communications network. In computing circles the 'anywhere area networking' is usually known as *pervasive computing* (some might say *invasive*). Generally, what we are talking about is *ubiquity*. There is often a debate about the need for ubiquity. It is not a need we can quantify now, just as we could not quantify the need for ubiquitous voice telephony before its possibility. However, I think by the time you reach the end of this book the need for ubiquity will become obvious.

The definition of mobility almost seems so obvious that we are in danger of being pedantic by examining such apparent minutiae. However, it depends on what our perspective is, and this is why I want to dwell on this point for just a moment, as I believe that being able to adopt and understand different outlooks leads more congenially to a richer set of service possibilities in the minds of the service creators. Additionally, it helps us to avoid blind spots in our understanding, which may undermine our ability to take ideas and building blocks from a wider pot of contributing technologies. There are some of us who are involved in technologies that are not conducive to ubiquity, so we end up justifying why we don't need ubiquity instead of maintaining an open mind.

While examining the topic of mobile service delivery using data networking the problem with our definition is that we start to run into difficulties when we think outside of the confines of voice. If we reflect on what we mean by mobile services, then an all-encompassing definition is not immediately obvious. I suggest you try it now. Try thinking about exactly what you think a 'mobile service' is. For example, consider the ability to play a game between two devices using short-range RF (e.g., Bluetooth™). Is this a mobile service?

Sticking to our definition of mobility for voice, the problem for services is thinking what exactly it is that we might want to do anytime, anywhere, other than talk to someone. Surely answering this question will influence what we mean by mobile service.

One approach is to start thinking of particular tasks, such as ubiquitous email access or nomadic access to the company local area network (LAN), and then work outward from there toward a general definition. However, what are the unspoken assumptions that we carry with us when thinking about mobile services? We may find that it depends on whom we ask. Operators tend to have a different view from enterprises about how things like email access should work. Fans of WAP have yet another view. Microsoft has its own view, as does IBM. Given the various perspectives, we should forgive potential customers for not understanding what mobile data really is. In terms of our discussion in the previous chapter about connecting dots on the mobile landscape each of these perspectives is arguing from the position of their own dot. The customer is often left to connect the dots. However, the possible connections are seldom obvious!

Another possible approach is to declare that a mobile service is *anything that a mobile phone facilitates* – end of story. This could well be a valid response, as it would seem that the

[4] The range of speed is usually from standing still to moving in a fast car (i.e., 0–120 km/h).

emerging market for mobile services is going to grow out of an existing user base of mobile phone owners by the accretion of new functions to the phones, which will eventually become communicators ('commies'). This evolution is already under way. However, the problem with this definition is that it is ruefully functional and does not make us think enough about utility or benefits; the value we are adding to the end user. Nor does it engender much original thought, as it gravitates toward existing phone usage patterns as the assumed starting point for newer services. It often seems that our thinking has already become stuck in a rut due to our phone-centric past, so newer perceptions might be useful to develop. Next generation services are going to alter our perception of mobile devices altogether. I am confident about this. So, I prefer to start from positions that veer off from the conventional as often as possible, even though we may come back to conventional means of achieving things if need be.

To arrive at a useful definition for this book, the word 'service' will be our focus, as its root meaning is 'to serve'. This is what we are trying to do: to serve end users, meeting their needs in useful and meaningful ways. The desired financial benefits for the service creators and suppliers should be a natural consequence of providing useful services.

Possible mobile services definition

The ability to interact successfully, confidently and easily with interesting and readily available content, people or devices while freely moving anywhere we are likely to go in conducting our usual, day-to-day business and social lives.

This definition is by no means definitive or exhaustive of all possibilities. Its purpose is to provide a focal point for determining the characteristics of mobile services, enabling us to *think critically* about their design and delivery. I have carefully worded the definition to induce certain ideas that would be useful to include in our thought processes when creating and delivering mobile services. We need to have an appropriate framework for thinking about mobile services before we look deeper into some of the implementation technologies in order to discern their relevance and usefulness as mobile service-enabling technologies.

3.3 ANATOMY OF THE MOBILE SERVICES DEFINITION

Let us walk through our definition of mobile services and critically examine its implications in order to convince ourselves that we have a useful definition, as we will want to return to it throughout the book. The defining keywords are:

- interact (interaction, interactivity);
- successfully;
- confidently;
- easily;

- interesting and readily available content;
- people;
- devices;
- day-to-day business *and* social lives

When looking at each of these keywords in turn we must not forget that they all relate to the qualifying part of the definition 'while freely moving anywhere we are likely to go in conducting our usual, day-to-day business and social lives'. Our definition should support the principle of roaming.

3.3.1 Interact

We engage with mobile services through human–device interaction. Mobile services are all about interaction. Interaction automatically implies that effort is required to act on something, that something being content, whether fetching it, sending it, reading it, editing it, composing it, listening to it, filing it, even playing with it (e.g., games) or any other mode of interaction that is possible.

Whenever humans expend effort to do something there is a perceived value in expending that effort. Mobility itself affects value perceptions. Interaction is likely to be more time-critical or time-sensitive than in static (immobile) situations, such as within the comfort of an office or a home, whether these time pressures are real or imaginary. In other words, the cost of time is more apparent. The user is forced to focus their efforts when using a small device because more concentration is required on the task in hand. This escalates the rate of expending effort, generally leading to lower tolerance of obstacles to progress. Perceptions of the financial cost of interaction also differ for mobile services, for several reasons. One of the more obvious reasons is that mobile device users today already accept paying for services, be it the cost of talking, sending a text message, downloading a ring tone, receiving an alert or some other tangible cost. This precedent is different from the Internet where there is a superficial[5] entry cost and thereafter everything is essentially free.[6]

3.3.2 Successfully

Successfully means we have achieved an objective through our interaction and are sufficiently satisfied with the result; hence, our interaction has been successful. We can think of our interaction as divided into tasks and that completion of the desired tasks is the goal, such as 'send a picture message to Bob'. As well as merely achieving the outcome of a task (Bob gets the message) the task itself involves a process to achieve it. This implies that there will be usability issues to consider. The degree of usability will inevitably impinge on the perceived success. For example, if it takes 10 arduous process steps and a lot of

[5] It is a real entry cost, of course – particularly if one includes the cost of a PC to gain access (assuming its cost is not amortized across other activities). However, in the mind of the user it is easy to consider the access costs as insignificant compared with getting everything for free thereafter.

[6] Of course, there are plenty of Internet services that now require payment (e.g., subscriptions), but the perception is still that a lot can be done for free on the Internet. This perception has no analogue in mobile markets.

time (money) to find out what was on television in the evening, then despite finding the information (successful outcome), the task may be deemed only moderately successful, possibly even unsuccessful.

3.3.3 Confidently

Many tasks carried out on a mobile device will involve a degree of interaction with a remote entity. These distant entities might be mailboxes, people, files, bank accounts, websites and so on. As with all electronic interaction and transactions that involve remoteness the distance we perceive between our part of the interaction and what is on 'the other side' can be problematic. It has nothing to do with physical distance, but something to do with dislocation. The physical limitations of current mobile devices exacerbate this feeling. In an extreme case, with only a few lines of text displaying a bank transaction and no friendly logo or face, the sense of detachment can become uncomfortable to the degree that confidence in the process is deeply affected.

If we walk into a bank to transfer funds we are very confident in that task as we can watch an attendant do their job, even though the task may be menial and no more sophisticated or skilful than what we could do via a mobile device. Standing in the bank itself also adds a significant degree of well-being (presumably). Although we may not feel comfortable in a bank, generally we feel confident about the services. We probably assume, quite rationally, that we are among self-assured people in a well-honed system. The whole affair feels familiar and we feel confident about it. In addition, there is potentially a high degree of intimacy at the cashier's desk and this adds a required feeling of security. If we now attempt all this via a mobile device, we are unable to feel the same emotions as in the bank. This could easily unnerve many of us, to the extent that we would rather not do it.

The detachment when using a mobile device is not only due to the low visibility of reassuring comfort cues. The environment is also different. It is different from interacting with the Web, for example, via a desktop PC. If I shop with a grocery market online, then I can visually see many things, such as the products, the branding and supplementary information. All this can be within my field of view at once and may well contribute to a feeling of familiarity and comfort. This will inevitably affect confidence. On a mobile device these additional cues are missing.

This is only one example of how the mobile setting can alter our confidence and the degree of trust we have in a mobile service. Additionally, we can expect that anything that involves a high degree of emotional energy in itself, such as handling money, will heighten our awareness of how confident we are while undertaking a particular task. There are other areas affecting confidence, such as how much we feel we are in control when using mobile technology. There is no logging on or logging off paradigm for most mobile devices. They are always on. Indeed, this is the new marketing mantra. However, this may prove disconcerting. For example, how do we know that our device is not doing things without us? This can become a problem if, for example, we feel that our device might be chatting to the network and costing us money (bandwidth) without us knowing about it. It is interesting to note that on most current General Packet Radio Service (GPRS) phones there is no meter to indicate network activity, thus adding to the sense of being completely adrift of what's going on with the device. The only indicator we get is the rather meaningless signal strength indicator.

Is a mobile device safe to use? I am sure that this question exists in many minds and not just the RF safety question.[7] The emotional and psychological implication of these issues needs proper examination. We are dealing with a huge spectrum of potential end users who have all kinds of perceptions about technology. User sensitivities need researching and taking into account when designing mobile services. Designing to instil confidence may lead to a completely new approach to mobile service and device design. This seems to me to be a critical requirement.

In considering the case of WAP, security scares were abundant early on in its development. Had its architects and proponents sat down and used our definition of a mobile service, taking into account *user confidence*, then this issue might have avoided. Of course, I understand the advantages of saying this with expert hindsight, but the point remains, as I do not recall seeing any design criteria for WAP (or its predecessor) that talked about user confidence. We should not forget that perception is important when dealing with highly subjective notions like confidence. As it happens, it was (and is) possible to configure a WAP system to be secure enough for financial transactions, but for our assurance the banks should have advertised this fact, not operators, nor security experts, nor the suppliers of the underlying security apparatus. After all, why should we trust them?[8] This may not have been possible, but some other plan to improve user confidence would have been useful instead of the muddled arguments of cryptography field engineers who apparently did not know what was outside of their dot on the map and hence made some wild claims about security. Security, cryptography and confidence are all different topics. Granted, they are related, but if the relationship isn't understood, then potentially disastrous ideas can surface. In any case the relationship might be (probably is) irrelevant to most end users.

3.3.4 Easily

Easily implies that the effort to carry out any task within a mobile service is perceived to be worth it and commensurate with reasonable user expectations. We should think in terms of temporal, psychological and financial assessments of ease, always from the end user perspective and never from the service designer's perspective. Simple, cost-effective end user testing (trials or even lab tests) can confirm the degree of user friendliness we have achieved. This is discussed later in the book when we look at user interface design.

Temporal and psychological considerations relate strongly to usability, not just in the classical terms of interface effectiveness but also in terms of the usability of the entire service. For example, in our prelude to this book we looked at how it might be possible to jump immediately to supplementary information and services relating to an advert we noticed on a billboard. Even though the actual user interface for the destination website might be well designed and easy to interact with, the wider and possibly greater consideration for usability is how easy it is to jump to the site in the first place. There could be several solutions to this problem, but perhaps the worst of them is what I have seen in some railway stations, which is to post a web address (URL, or uniform resource locator) on a wall poster. This is an extremely impractical proposition. How do we expect a busy and hurried traveller to

[7] There have been reports on and off that the RF radiation from mobiles can be a health hazard. Another of life's uncertainties for humankind.

[8] Confidence in an operator is not the same issue as confidence in a bank. We possibly perceive operators as sticking up aerials in fields, selling us phones and sending us bills, not handling our bank account and other sensitive information.

jot down a URL and manage to enter it into their device while on the move? It imposes a burden on the user that may well exceed their threshold of patience.[9]

In the end, we need to be highly self-critical in determining whether a service is worth using or not. Even though it may be groundbreaking, like WAP-based email was in terms of being able to offer the full range of email-related tasks, does the *service* adequately reward the user's psychological needs? The ease of use is a key influencing factor in answering this question. Ultimately, the user will decide how easy a service is to use and that in turn will affect our financial return from the service.

3.3.5 Interesting and Readily Available Content

'Content is king' was the mantra for the dot.com boom, but what is content? There often seems to be a tendency to view content as something that belongs to someone else and is available via syndication from a content provider, like the perennial stock prices or news bulletins. I believe this view is too restrictive and is symptomatic of viewing users like broadcast consumers, ignoring their potential role as freelance producers, even though on some networks the users already produce copious amounts of content in the form of text messages, presumably mostly original in nature (though not always[10]). I prefer to think of content as any information that the user will either generate or consume, regardless of format. Therefore, content could include *anything* read, watched or heard from the device interface regardless of its source and ownership, plus *anything* spoken, composed or otherwise entered into the device by the user, by whatever means. Content is the currency of exchange in the interaction process.

With a wide definition of content we can turn our attention to its peculiarities within the mobile habitat. Our first concern is for the content to be interesting, which means it has to engage the attention or arouse the curiosity of the user. It has to have emotional, social or financial capital. However, keeping in mind our desire for interaction to occur *easily*, especially within the mobile setting, then perhaps we should heed the cautionary advice from a leading Web usability guru Jakob Nielsen,[11] who advises us:

> *Ignoring users' immediate needs is certain death on the web.*

What Nielsen meant by this was that unless a user gets what they want immediately they are unlikely to stick around, which in the passage of Web surfing means to click off to another site. If that is true for the Web, then it is truer for mobile services. In the more demanding mobile setting it could mean to switch off and never come back again, which is what happened to many WAP users who signed up for the show, tried to get in and swiftly left the building, never to come back.[12] Nielsen demonstrated this fairly well in his controversial[13] WAP usability study.

[9] We discuss a solution to this problem later in the book.
[10] It is not widely known that a good deal of content is forwarded messages from one user to another (or several others), especially by young people passing around banal jokes (like office jokes were once passed via email).
[11] Jakob Nielsen's website is http://www.useit.com and is well worth a visit in order to understand usability issues in general.
[12] Many operators reported interesting sign-up numbers for WAP services, but later admitted that many of the subscribers never came back again after the initial experience (disappointment).
[13] Controversial with the WAP propagandists who felt it was unfair of Nielsen to use real world testing instead of contrived lab tests, the latter presumably more fixable than the former.

This reinforces that the service being interesting is only one dimension of the challenge (and our definition of a mobile service). The other, as we have stated in the definition, is that the content should be readily available. This is perhaps the greater challenge within the mobile setting in its formative stages, as we can appreciate that much of the existing content, whatever its source, did not arise with mobile access as a consideration. The specific problem being the more confined possibilities available for presenting the content on a mobile device, at least as they exist today. We still need to accommodate smaller information chunks – lower resolution, shallower colour depth, lower fidelity and so on. This makes maintaining the interest of the user more challenging. These restrictions are not necessarily to do with mobility *per se*, as we can be mobile while accessing content using a laptop fitted with a wireless modem card, but they are a consequence of portability. We generally need portability if we want to use mobile services *anywhere* we like, which relates to the final part of our definition that we will look at shortly.

However, this is the brave new world of unprecedented progress and changes in technology and our apparently insatiable demand for more of it, so providing content to mobile users should not concern us in terms of there being a motivation for making it available. The fact that in the UK alone billions of text messages are exchanged every month should tell us that mobile interaction is now a permanent fixture of our daily lives. People are ready for mobile interaction.

3.3.6 People

We must not forget people in our definition. The need for humans to interact and socialize will never diminish. We still very much envisage that human interaction will remain a popular use of mobile technology and services. Our definition here extends to include any mode of interaction, no longer limited just to voice calls. We could think of instant messaging, email, voice messaging, bulletin boards, picture messaging and even old ideas reincarnated, like ‘push to talk’ (walkie-talkie type of communication). Later in the book, when we look at location-finding capabilities within the RF network, we examine how human interaction can now have a geographical or spatial dimension that we have never seen before. The possibility for spatial messaging alone is giddying.

3.3.7 Devices

Here we are talking about the ability to interact with physical systems, such as the video recorder, oven, porch light, or any device, not necessarily restricted to the domestic environment, although home automation is an interesting possibility.

It is possible to consider devices as a source of content, but I think the need arises to discern between content abstracted from the physical world and the physical world itself. In essence, the mobile service in the latter case becomes a form of remote control, and commentators have talked about the concept of *telepresence*, the ability to be somewhere else in terms of sensual involvement (by sight, sound, touch, etc.) At first glance this seems a major diversion from the other types of service that involve more conventional or obvious arrangements for information exchange and processing. However, the increasing ability to web-enable devices of all shapes and sizes is enabling machines to play a much more active role in the networking fabric.

Domestic device control is not very difficult technically, but there are very few devices on the market offering this capability. Historically, there has not been much interest, as most devices lack a connection to a network in the first place. However, this is going to change. The key will be the Internet, of course. More and more houses and premises will be permanently hooked into the Net, and beyond that it's just a matter of interfacing devices to the network access point, something made all the more promising and likely by inbuilding wireless connectivity solutions of one form or another.

There is already a growing interest for wireless connectivity to devices, especially in the field of telematics and telemetry. It is increasingly common for vending machines to be hooked up to a central operations and maintenance centre using wide area wireless, even something as simple as a two-way text-messaging interface. For example, a drinks machine can notify its owners that it is running low on drinks or could alert diagnostics problems to the manufacturer or owner.

With future mobile technologies, users of services will be able to go further than just receiving alerts about the operational status of a physical system. It will be possible to control things remotely. What is useful to appreciate at this point in the discussion is that the trend is increasingly toward using the Internet, often the Web, as the backbone for control nervous systems of this type. This is a convenient model as it means that this particular class of mobile service may involve a similar software approach to more conventional information systems.

3.3.8 Day-to-day Business and Social Lives

Our business and social lives take place within and across many different physical spaces, seldom confined to one place for very long. We are constantly on the move to a lesser or greater extent, be that travelling to work, walking to school, working out in the gym, lunching in the shopping plaza, meeting for coffee, shopping for DIY items, attending a football game, sitting in a traffic jam, taking a train, catching a taxi and so on. Our lives are definitely mobile, so it makes sense to mobilize a good deal of our daily interactions. You can substitute my example list with your very own daily or weekly itinerary and think about what role a mobile device could play. But don't just think about your itinerary you should bear in mind that all your friends, kin and colleagues have their own schedules too, which are unlikely to be the same. We are all mobile!

Now, relating this back to accessing content, then there is a need to access content while in the different places we might occupy, depending on its relevance to the setting, circumstance and time. We could say that this is the nub of developing our understanding of a mobile service: to what extent is there a need, or interest, to access desired content at a given time and in a given place?

We can characterize people as being in one of three places at any given time, and this model fits well with some aspects of studies into patterns of socialization, such as Ray Oldenburg's idea of 'third places[14]'. In terms of our location at any time of the day we can be:

[14] Oldenburg, R., *The Great Good Place*. Marlowe & Company; 3rd edn (1999).

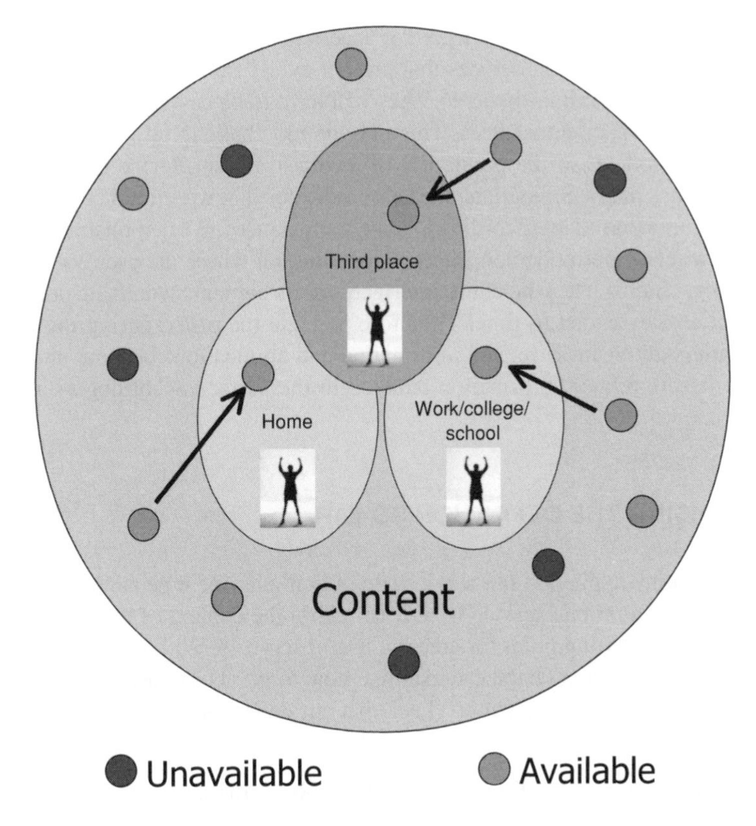

Figure 3.1 Usefulness of content depending on circumstance.

- *at home* – where we generally live, sleep, mix with family, etc.;
- *at work/school/college* – where we generally spend most of the middle part of the day, working, studying, etc.;
- *at a third place* – anywhere we like or need to be that is not one of the first two (sociologists and psychologists have mentioned that third places are important to us, but in recent times these places are being eroded simply because we don't have time to occupy them[15]).

Figure 3.1 shows us how we can think of the need for content as being situational. The need to interact with content at a time and place will depend on situational circumstances and their influence on our emotional need to interact with particular content types.

[15] Some psychologists refer to a concept of time compression where we are left with less time due to increasing pressures to be at work for longer hours, or at the home to compensate, or travelling between the two. This compression is squeezing out our existence in other places, and the existence of these places is also declining for various reasons (e.g., modern planning and building practices).

The potential demand for a high degree of access to content across all our social spaces implies the use of lightweight devices that are not overly cumbersome in their transport, such as a laptop would ordinarily be. Devices will have to be easy to carry and easy to use under a wide range of circumstances. This may be too challenging for any single device design to accommodate, so the possibility of having different device types for different occasions becomes likely. Situational need also indicates that we should be considering the design of interfaces that adapt according to our circumstances. We demonstrated this feature in the Zingo wireless portal demonstrator. Depending on where the user was (e.g., office, home, meeting, cinema, etc.) the entire interface and its contents would adapt accordingly. This concept also extended to time. Therefore, outside the office during the day or outside office hours all business-related information and applications become obscured from view, while leisure-related information bubbled to the surface according to location and circumstance.

3.4 APPRAISING THE DEFINITION SO FAR

The level of scrutiny applied to our services definition may seem pedantic. The intent was to understand what a mobile service is, at least within the context of this book, so that we have a constructive starting point for creating useful services. With this definition in mind we will proceed in the next chapter to examine how to go about building a useful mobile service, but first it might be prudent to check that our definition makes sense.

3.4.1 Benchmarking Our Definition – WAP Revisited!

Anyone designing a mobile service with our definition in mind will have to consider the success factors just described for measuring acquiescence to the basic user needs that we have assumed underpin the definition. With this in mind we can see why browsing Web pages on a phone (e.g., WAP) was always going to be problematic, especially if I quote from one of the White Papers[16] released by Unwired Planet (the inventors of what was to become WAP[17]):

> *This document provides a high level overview of the Unwired Planet product suite... that provides an Internet service model*[18] *over wireless networks.*

We see that the problem here is the assumption of a particular service model without really defining what it is. In other words, this is at the other end of the spectrum from a tight formal definition, which is to assume a service definition without stating what it is, never mind assessing its applicability to the mobile context. The intention was, as clearly stated, to adopt an Internet service model, not a *mobile* Internet service model. There seems to be an assumption that the latter is being facilitated by 'wireless-enabling' the former. However, in the absence of a definition of a mobile Internet service model, how will we know when it has been achieved? What are the success factors?

[16] These White Papers are no longer available online.

[17] Unwired Planet developed technologies that were to become the framework for WAP, though WAP itself was, of course, a collaborative effort of many companies.

[18] Emphasis mine.

The underlying assumption was probably that the success factors of the Internet would remain relevant and intact 'over wireless networks', and so there is little need for pedantic definitions and defining success criteria for the mobile context. I hope that the dissection of our mobile services definition has clarified that this is not necessarily the case, almost certainly *not* the case. Interaction in the Internet world is not the same as interaction in the mobile one, and Nielsen's study[19] brutally demonstrated this. Any mobile service has to embrace a distinct service model that takes into account the mobile context or else face possible annihilation in the marketplace.

We should clarify that, strictly speaking and contrary to the above excerpt, WAP was not trying to extend an 'Internet service model' to 'wireless', whatever each term really means in the above excerpt. The technological model in actuality was the World Wide Web, which is a very particular component of the wider Internet framework. The Web has its own peculiar factors for success, peculiar enough that when we consider their aptness in our mobile services definition we immediately run into problems. Let's reflect on these factors now as I believe they are so much taken for granted that we seldom think to identify and state them, which is possibly why we didn't rethink them in the mobile context. In my experience in the field many mobile service proponents continue to make this oversight. Perhaps, in some cases we still do not understand all the success factors of the Web. It is still a hotly debated topic in some circles,[20] and I certainly don't pretend to have all the answers. Weinberger[21] has made an interesting attempt to articulate the attributes of the Web that make it successful or unique.

Even at its worse, the Web is essentially *easy to use*, but in ways that do not transfer easily to mobile devices. Some key differences are shown in table 3.1.

We can see from Table 3.1 that by making WAP a close analogue of the Web we hardly manage to retain any of the Web's success factors. The Web model appears to break down in the mobile context and needs rethinking. Therefore, the WAP 'service model' was possibly ill-defined (if defined at all). We can see how far removed it is from fulfilling our simple definition of a mobile service, though I understand this contention could be argued from several alternative perspectives, especially taking into account historical constraints, speed to market and so forth. I am not trying to show the cleverness of our definition, just the value of having one and being able to use it. Looking at WAP under the microscope using our mobile services definition, we can see why it did not succeed as a mobile service and why 'mobile browsing' might still be problematic if we continue attempts to make it like the Web-browsing experience. The WAP service model seems incomplete as a mobile service, and a failure to identify the success factors early on meant it was difficult to achieve something useful with WAP in its early form.

To be fair to the WAP supporters and all of us[22] who attempted to deliver WAP-related products or services, myself included, there were other negatives. In my own experience, people were simply not ready to think about doing something on their mobile phone that was so far adrift of the established norms (namely, making a phone call in order to talk to someone

[19] Nielsen Norman Group Report, 'WAP usability report' (December 2000), available from http://www.nngroup.com/reports/wap/

[20] Typically, those who have made a career out of commenting on the impact of the Web on society and the evolution of IT, often referred to as the 'digerati', which has the connotations of belonging to the educated or 'knowing' class of people, as in the literati of society at large.

[21] http://www.smallpieces.com/

[22] Myself included: I cannot claim to have made such an insightful analysis at the time.

Table 3.1 Critically assessing WAP as a mobile service.

Web	WAP
Easy to skim-read the pages at a rate that allows large quantities of information to be processed, especially with a mixture of text and images. Limit of information browsing and consumption is human, not technical.	Skim-reading not possible due to small display size, slow display refresh rates (making scrolling slow). Absorbing information is an arduous task; technology interferes with human ability to consume information.
Plenty of computer power to process and display pages quickly such that dominant delay is the fetch time.	Many WAP browsers can be allocated a low computational resource that hinders display-rendering and compounds overall delay from demand to display.
Easy to navigate via URL entry, or 'point-and-click' linking, or via search engine, or all three.	URL entry extremely difficult – the 'grand' idea of namespaces not that great for mobiles. Early browsers didn't even include 'auto-complete' features to save on all the http://www thumb-tapping bit. What a drag, enough to kill the whole thing dead in the water all by itself.
Easier to 'deal with' poor performance: • by being in the relative comfort of sitting in a place without transit; • by clicking on other links to go elsewhere; • partial loading of pages gives sense of progress; • explicit back button offers some quick relief due to powerful cache.	Difficult to cope with poor performance: • being in transit lowers patience, to do with different sense of time; • not possible to click on links – have to scroll to them and not likely to be many – so 'jump-off satisfaction' less; • no partial page loading due to the way binary compression works; • there often is no explicit back button.
Relatively easy to locate information by: • search engines being available that can spew out productively digestible chunks of results with links to displayable Web pages; • the ability to 'jump off' and hop around easily from useful starting points.	Very difficult to find information: • lack of search engines, especially ones that can return results viewable in WAP browsers; • difficulty in skimming results from search engines;
The user interface capabilities of browsers is consistent or within a tolerable range of variation. New browsers can be installed to facilitate rapid improvement of user experience. Moreover, the initial adoption of the Web was a simple case of installing a browser, which itself was available free.	User interface capabilities of phone browsers are not standardized and widely diverse, thus causing problems with both production and display of content. Not possible to upgrade browsers in the field, and the initial adoption route was via a new phone.

on the other end). To go from talking to surfing was a big jump within the mobile context. The Web, despite its uniqueness, is a set of metaphors that are not so alien to PC users.

I am reluctant to use analogies, as the apparent likeness of two things can sometimes be misleading, but perhaps the transition from reading newsprint in papers to reading Web pages is more akin to the jump from voice to WAP. A shift in mindset and habits is required. The problem with mobile devices is that their usage is part of a mindset that largely still perceives them as a handier version of the 'plain old telephone system' (POTS). If this is true – which I suspect it is for the greater number of users (although picture messaging may soon change this perception) – this only reinforces the importance of realizing that mobile services are something very different from voice and demand an alternative and carefully considered approach: a different model.

3.5 WHICH SERVICES SHALL WE BUILD?

Having looked at defining mobile services we need to decide which services to build. It is fair to say that it is not clear which mobile services to offer. As we discussed in the opening chapter, next generation mobile services will break new ground. There are very few precedents to indicate how best to proceed. The uncertainty seems compounded by a commensurate lack of understanding of how to build new services. The service ideas described in the prelude to the book (Chapter 1) may already be in the minds of many creative people – I don't claim any originality for most of them, although I have been working on these sorts of ideas since the mid-1990s, so everything seems like an 'old' idea to me. The frustration for many would-be service providers arises from not knowing how to build such services. To date, about the only access point into the Radio Frequency (RF) network has been a data pipe. In terms of service access points then, it has been a gateway to submit text messages. With such restricted resources it is hard to see how complex and innovative services can be realized. However, this is changing, as you will find out as you proceed through the book.

Rather than analyse the business case for different ideas and approaches, I want to focus here on possible types of service, just to indicate the current trends in thinking about what services are likely to emerge. At the very least we can take the proposed categories of service offered by groups like the UMTS Forum,[23] as they have been instrumental in driving the mobile operator requirements for next generation wireless standards; so, we should at least look at the collective view of the mobile operators[24] to date. After all, these requirements have influenced the design of the new technologies and associated products, especially for 3G.

3.5.1 Killer Cocktails – Killer Tsunami

I want to stress that my own view on next generation services falls in line with what some commentators have called 'the killer cocktail' rather than the 'killer app'. The latter is the

[23] Standing for 'Universal Mobile Telecommunications System', UMTS represents an evolution in terms of services and data speeds from today's second-generation (2G) mobile networks. As a key member of the 'global family' of third-generation (3G) mobile technologies identified by the ITU, UMTS is the natural evolutionary choice for operators of the Global System for Mobile Communications (GSM) networks, currently representing a customer base of more than 850 million end users in 195 countries and representing over 70% of today's digital wireless market (source: GSM Association). See http://www.umts-forum.org

[24] The UMTS Forum is essentially an operator collective responsible for shaping requirements for 3G standards that arise from the 3G Partnership Project (3GPP).

notion that there is one service or application that is going to be a momentous service, which drives the widespread adoption of mobile services. The killer app of the Net is often thought to be email. The spreadsheet was the killer app for the desktop PC[25] in its formative years.

I only take exception to the use of the term 'cocktail' as it conveys a sense of only a few ingredients. I think that the eventual outcome of operators opening up their networks, especially with new service access points like Parlay X,[26] combined with the dizzying array of device technologies, will ultimately cause an overwhelming profusion of services, more akin to a tidal wave (tsunami) than a cocktail.

3.5.2 Ahonen's Five Ms

Tony Ahonen in his daringly titled book *m-Profits – Making Money from 3G Services*[27] has amply discussed the concept of killer cocktails and expounded on the possible recipe too. It is a thought-provoking book that is well worth reading and a follow-up to a previous book edited by him called *Services for UMTS*.[28] I will briefly mention here his five Ms of 3G services, which Ahonen introduces thus:

> *...five elements of 3G services that allow for value to the user, and profit to the operator.*

Ahonen then goes on to comment:

> *The 5 Ms can be used to create value to any mobile services on any mobile network technologies.*

So we don't have to confine our view solely to 3G in the current discussion, as indeed this book is about *any* next generation mobile service possibility – 3G or otherwise. We should understand that Ahonen's approach here is to *identify attributes* that could apply to any mobile service, which should help direct our thinking toward establishing success factors for our service propositions. In other words, each M is like an axis for adding value to the service idea being scrutinized. The more factors in our offering we can find to move along any of these axes, preferably all five, the more value we are potentially adding to our proposition.

Ahonen's five Ms:

1. *Movement* – escaping the fixed place.
2. *Moment* – expanding the concept of time.
3. *Me* – extending myself and my community.
4. *Money* – expending financial resources.
5. *Machines* – empowering gadgets.

[25] That is, the IBM PC.
[26] Parlay X is a programmatic means to access telecoms resources via the Web from our service providers' applications. It is discussed in the section on the RF network.
[27] Ahonen, M., *m-Profits – Making Money from 3G Services*. John Wiley & Sons, Chichester, UK (2002).
[28] Ahonen, M. and Barret, J. (eds), *Services for UMTS*. John Wiley & Sons, Chichester, UK (2002).

I will briefly summarize the five Ms, as I do not want to dwell for too long on the definitions as they are amply described in Ahonen's books. I would like to extract the salient points so that we have them to hand as our thoughts progress from our serviced definition toward service realization.

Ahonen is interested in how to arrive at meaningful service concepts that make sense commercially; so, I refer the reader to his book for more in-depth analysis within the commercial framework. I want to explore his view in order to extract useful trends or attributes that will enable us to understand better the implications for building our services and a service delivery platform(s).

Movement *Movement* is the ability for a service to continue operating in a meaningful way as the user roams, be that in their usual place of occupancy or wider travels, including going abroad. We already have this in our definition. However, Ahonen's key point is that if there is any degree of locality involved in the service itself, then the service should adapt to the area the user finds themselves in, such as local weather forecasts being appropriate when accessing the mobile weather service. We already discussed this point in the general terms of adaptive interfaces, which is still a relatively novel concept in mobile devices. There are some indications that this is within the thought processes of some device vendors, as some phones have situational settings to control the ring tone and alerting according to context. However, this is clearly a device-centric feature, whereas we need ultimately application or service-centric contextual adaptation, which is something altogether more challenging. For example, only displaying certain messages in my inbox according to my circumstances is something that potentially the device could do (via an email client) or a remote service would have to do, like a hosted WAP presentation of my inbox. In the next chapter we examine application topologies for mobile services, like client–server, standalone and peer-to-peer. We will see that the topology will have a lot to do with how we handle such issues as movement.

Moment *Moment* is the concept of being able to manage time in a manner adapted to the situation we are in. We saw this clearly in the prelude (Chapter 1) where the call-handling aspects of the mobile device changed according to the time of day, such as 'lunchtime' settings. Ahonen generalizes on the elongation of time as being afforded by management that involves both *postponing* tasks and providing the means to *catch up* later. The moment concept also extends to multitasking. Again from the prelude, a good example was the ability to shop physically (task 1) while simultaneously checking competitive pricing (task 2), consulting fellow shopper opinions (task 3) and at the same time (in background mode) tracking the whereabouts of our fellow diners (task 4). In fact, this example goes beyond postponement and catching up, as we can add 'bringing forward' as a possibility. The consultation of fellow shoppers and price checking are tasks that previously (i.e., without these mobile services) the shopper would have done later, such as when they get back to a desktop PC or a product catalogue. With the ability to consult prices there and then, this activity is brought forward in time. From another perspective, if we had planned to visit the shop in advance and would have previously had to check prices before shopping (the likely option), then we have in a sense postponed the checking until the actual moment of interest. The fact that we can do both is great because it means we can accommodate a wider range of human habits, which should be the general design aim of technology, not to shoehorn users into new habits. The latter approach has repeatedly failed historically.

Me *Me* is the concept of personalization, and this is a concept that has been well understood on the Web for some time and one that Web users are familiar with. When browsing Amazon we expect to see *our* history, *our* interests, *our* recommendations, not somebody else's. This aspect has a degree of overlap with our definition's requirement that content should be *interesting*, the implication being that it should be personally highly relevant. A high degree of relevancy suggests content be moulded to suit users' personal interests and needs. This is no different to the Internet personalization process except that interests and needs in the mobile context will change with time and place (the 'movement' and the 'moment'). This overlap is an important aspect of the personalization considerations for a mobile service.

Ahonen indicates that 'me' is the most vital of the five Ms: a critical ingredient for a successful mobile service. The more we can customize content to the individual's tastes the better. This implication is of paramount concern in this book: that we clearly need the ability to build a platform for services that is rapidly and easily configurable on a user-by-user basis, extending that idea readily to communities of users with common interests, tastes or goals. This means we need a flexible platform, which implies technologies that can support a high degree of customization, including the concept that services could almost be constructed on the fly, ephemerally emerging to meet our user needs on demand. On-demand service creation is a new element to service provision that is particular to the mobile context and yet again reinforces the uniqueness of the mobile services and the subsequent need to have distinct service models.

Money This is simply the support for financial transactions wherever the need arises. It is a reasonable assumption that, given much of our daily lives in an industrial society involves consumption, we frequently need to engage in financial transactions of one form or another. The more we begin to rely on mobile services the more sense it makes to link financial activities with our mobile tasks. Clearly, a framework is needed to make this work.

Machines The final M is the concept of enabling machines to communicate over a wireless wide area network (WAN). This is sometimes called M2M (machine to machine) and is often considered to have grown out of the idea of telemetry or remote measurement without the need for field service engineers.

The oft-given example is a vending machine, such as a drinks machine that can 'phone home' and report that it is running low on sugary brand *x*. A particular take on the vending theme is the brightly coloured photo booth machines that we frequently see in supermarkets, the ones used for passport photos. A divergence from passport photos is incorporating a fun element, such as posing alongside superstars. Not surprisingly, ideas have emerged – such as emailing the images – which require networking.

Networking of booths or kiosks presents all kinds of challenges: the usual ones to do with installing network access points, wiring them up and so forth. The solution is, of course, wireless. By adding a wireless modem to the booths, the need for fixed network points evaporates.

It is sometimes said that M2M usage of wireless will one day outstrip human-to-human interaction, although much of this would be embedded and hidden from us. However, the technologies are emerging to enable machines to network relatively easily and to interact with useful service points elsewhere in the hierarchy of networks. This is a fantastic opportunity and is how I interpret the significance of the 'machine' factor, which is to take any given process that a mobile service addresses and see where machines might be involved. This will usually require a deep level of inspection into the process, as we are usually oblivious to the involvement of machines in our daily lives. For example, a photocopier is a machine, a fax is a machine, an automatic door is a machine, a fridge is a machine, and a car central-locking system is a machine. Perhaps you get the point! Machines are everywhere, and we use them all the time. Imagine we can involve them in our mobile services. How might the end users benefit? That is a question worth poring over extensively during our mobile service creation processes.

3.5.3 What Do the Five Ms Mean to Us?

I am reluctant to mention any philosophy, theory or framework in this book without demonstrating its value explicitly. In his books Ahonen has plenty of examples of applying his 5M philosophy to real services, but I would like to mention one of my own here, as a means to gain independent confidence in the 5M approach.

I tend to view Ahonen's five Ms as a framework for brainstorming a particular mobile service idea, so it seems beneficial to take an idea and see all the different ways that each M attribute can be applied.

An example that I would like to consider, as it is something I would like to see more widely available, is the use of a mobile service to assist with car parking. When I visit a major city for a meeting I often will travel by car, which is my preference, but often I find that I do not know where to park. This is where we can invoke the first M of 'movement'. Let's imagine that I want to be able to ask my mobile device 'where should I park?', for it to know that I am in a certain part of London and answer accordingly.

As for the second M – 'moment' – then this becomes important when I am looking for somewhere to park for the next few hours. Suppose it is approaching the time when many car parks close for the evening, as they do in parts of London, for example. Therefore, I want to be told only about car parks that will be open for the next few hours. This aspect of 'moment' has to be designed into the system, and it definitely adds value. Had we not considered the 'moment' attribute we may have overlooked the fact that a parking requirement is clearly time-sensitive (and there are other time issues to consider too, like peak hours parking, etc.).

Having identified a suitable parking place based on 'movement' and 'moment', the service guides me to my destination using the location-finding capabilities of the mobile device,[29] combined with a directional navigation service based on a geographical information system (GIS[30]). Sensing that I am in a car the service automatically defaults to giving me audible directions, as a safety and convenience consideration. I listen to these via my Bluetooth headset. I can also get real video images of where I should be on my journey, just in case I need visual cues along the way. This is multimedia and its powerful capabilities as an enabling technology for the 'movement' 5M attribute of this particular service. It is not so 'far out' to expect a heads-up display capability built in to a mobile device

[29] Alternatively, the location-finding capabilities of the RF network, or the device and the network combined. We will discuss these possibilities in Chapter 12 about the RF network.
[30] GIS (Global Information System) – the kind of system that generates maps for sites like www.multimap.com, etc. We will look at these systems as part of the discussion on location finding in Chapter 13.

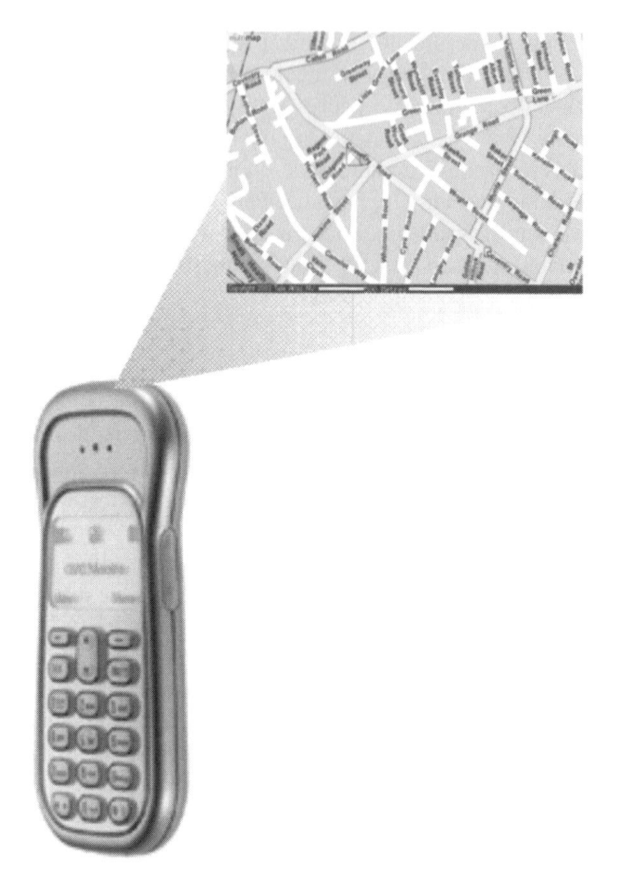

Figure 3.2 Projecting a map from a mobile device.

mounted on the dashboard, as shown in Figure 3.2. Companies such as Symbol Technologies are already demonstrating projection devices[31] small enough to fit into mobile devices.

On arrival I want to be able to book my parking for an allotted time and know that I can pay for it without fumbling around for coins and without having to leave my car to visit the nearest pay station, something that is often awkward or can feel insecure late at night. Therefore, I should be able to pay for the parking using my mobile device. The advantage of an integrated design approach is that the service already knows that I need to pay, as a natural consequence of traversing the task flow I have already gone through to find the car park. The application interface prompts me to 'pay by mobile' if I chose to visit the car park suggested by the results from my enquiry. The 'pay by mobile' option further prompts a range of payment modes, such as registered credit cards, PayPal™, mobile account or some other payment scheme that plugs into the mobile service framework. It is better than physically using a credit card at the payment station. We can think of the card swipe coming to me rather than me going to it (i.e., the pay station), and it already knows

[31] http://radio.weblogs.com/0114561/categories/hotProducts/2003/05/07.html

the price and product. This level of convenience is highly desirable, provided it is easy to use.[32]

This example has certain implications, as do many of the examples in Ahonen's own fruitful range of sample services. The principle implication is the need for integration with information services and customer-handling mechanisms that lie outside of the network operator's control. The car-parking idea only works if we have reliable information about parking spots and their opening times. This is the least level of information integration. However, this is only one aspect. For the payment method to work, then there has to be further levels of integration, not just at the informational level but also at the system level.

For this example and others like it we have two choices. Either we ask or rely on the RF network operator to provide an integration pathway into the car-parking vendor's system (and probably more than one vendor for the service to reach critical mass) or we allow the car-parking merchants to integrate into the operator network. There are other levels to position the service access points, such as the Internet Protocol (IP) network. The car park vendor's IT system could provide a web-based interface that is available to any registered business partner, irrespective of their intended end user service or application. In other words, the car park vendor need not be aware of the final mode of service delivery – that is entirely up to the service provider.

There is a further consideration. Ultimately, the success of next generation mobile services depends on being able to offer a huge spectrum (tsunami) of services that include as many of the five Ms as possible. It is not realistic that an operator is going to undertake significant integration efforts with hundreds, possibly thousands, of service partners. The partners need to be able to come to the operator with a service concept and simply plug in to the network and access the mobile install base, almost as if the operator were not directly involved. In fact, the less the operator is involved perhaps the better, except perhaps to aggregate some of these externally supplied services into service packages, and then only to make it easier for the average consumer not to have to select from a flat horizon of services that may be too wide and therefore too perplexing.

The implication for our platform is that there needs to be lot of behind-the-scenes connectivity to third parties, such as the car park owners in this instance. This all sounds like an integration nightmare that is not likely to happen overnight. If we leave it up to the mobile operator to implement, then perhaps it may never happen, certainly not in a hurry. However, it is probably better to make it easy for the car park owner to interface with the platform. This tells us that we need to build open interfaces, using technologies that are easy to implement and widely available. This is a major consideration for building the platform. We can take this a few steps further by offering the necessary incentives and tools to make the connectivity even easier, perhaps encouraging the task to be undertaken by an entrepreneurial agent who is neither the car park owner, nor the operator, but knows how to bring them together, perhaps overnight, literally!

Throughout the book, when examining enabling technologies we shall have this goal constantly in mind. The enablement is all about allowing any business to offer their wares using mobile services using any of the five Ms to shape their offering. How do we do this? That is the subject of this book.

[32] Ease of use is the key. I have not gone into how any equipment at the car park knows I have paid for a space, but optical recognition of my number plate is one possible scheme, although there are other possibilities, such as using Bluetooth™, to connect the mobile device with the entry barrier control system.

3.6 CATEGORIES OF SERVICE

Our mobile services definition enables us to understand some of the fundamental parameters for thinking about what to build. It is a benchmark against which to determine that our service concept is sensible. It helps us avoid service concepts that are not ideally suited to the mobile context, as WAP was in its early form.

Ahonen's five Ms have hopefully taken us a step further in refining our thinking about how to shape mobile services to realize our definition, while unlocking their full value potential. The next question to ask is what services might we actually end up building?

Of course, largely, we still do not know what services to build. The number of permutations of ideas and interpretations of how to use mobility is going to be huge, hopefully. However, it is possibly useful to categorize services, if only because this is precisely what the UMTS Forum have done as part of the process of defining service concepts for 3G in order to drive the experts designing the technology in the right direction, or so we hope; time will tell.

It is therefore pertinent to consider what the UMTS Forum has said about service categories. Again, these categories are flexible and certainly not cast in stone, and, most likely, the exciting services are going to be unexpected ideas that defy these categories. However, the UMTS Forum has envisaged certain patterns of service design and usage that might help us here in determining trends and boundaries for the underlying technology and its evolution.

3.6.1 UMTS Definition of Mobile Services

The UMTS Forum has published a range of reports on various aspects of 3G, including Report No. 9, which is titled 'The UMTS third generation market – Structuring the service revenue opportunities'. In that report the near-term demand for 3G services is categorized into six service areas that we mention below:[33]

1. *Mobile Internet Access* – A 3G service that offers mobile access to full, fixed ISP services with near-wire line transmission quality and functionality. Includes full Web access to the Internet as well as file transfer, email and streaming video/audio. (Consumer)
2. *Mobile Intranet/Extranet Access* – A business 3G service that provides secure mobile access to corporate LANs and virtual private networks (VPNs). (Business)
3. *Customized Infotainment* – A consumer 3G service that provides device-independent access to personalized content anywhere, anytime via structured access mechanisms based on mobile portals. (Consumer)
4. *Multimedia Messaging Service* – A consumer 3G service that offers real time multimedia messaging with always-on capabilities, allowing the provision of instant messaging. Targeted at closed user groups that can be service provider or user-defined. (Consumer)

[33] 'The UMTS third generation market – Structuring the service revenue opportunities': Report No. 9. UMTS Forum and Telecompetition Inc. Research (June 2000). Available from UMTS Forum at http://www.umts-forum.org

5. *Location-based services* – Business and consumer 3G services that enable users to find other people, vehicles or machines. It also enables others to find users, as well as enabling users to identify their own location via terminal or vehicle identification. (Consumer and business)
6. *Rich Voice* (Voice, Video, and Multimedia Communications) – A 3G service that is real time and two-way. It provides advanced voice capabilities – such as Voice over IP (VoIP), voice-activated net access and Web-initiated voice calls – while still offering such traditional mobile voice features as operator services, directory assistance and roaming. As the service matures it will include mobile videophone and multimedia communications. (Consumer and business)

These categories can be a bit vague as service definitions, but they are an attempt to segment the market. A useful exercise would be to take the five Ms and apply them to each of these segments to see if this reveals any patterns, particularly with respect to which M attributes figure more strongly in each category, assuming there are any patterns to be discovered. What we are eventually looking for is suggestions about what technologies we will need to build the service delivery platform and what shapes the platform might take. This would certainly help operators and service providers to understand what software services they need to start putting into place.

Before undertaking this exercise ourselves it is worth emphasizing again that the five Ms are *not* ways of categorizing services, even though there appears to be some overlap with the UMTS categories, such as the location services category, which would appear to represent the 5M attribute of 'movement'. When Ahonen first introduced the five Ms in his earlier work *Services for UMTS*,[34] he cautioned us to heed that:

> *The 5 M's are a useful way to explore attributes of UMTS services, but the 5 M's are not a tool for categorising services.*

This caution opens the chapter 'Types of UMTS services: Categorising the future'[35] to which I refer the reader for a more in-depth exploration of the UMTS categories, especially from a commercial perspective; our exploration is hopefully leading us to some conclusions about the underpinning technologies needed to realize useful services, whether they fit these categories or not.

3.7 APPLYING FIVE Ms TO THE MOBILE INTRANET/EXTRANET CATEGORY

Let us now take one of the UMTS Forum categories, *Mobile Intranet/Extranet*, to see what insights we can develop when we examine a potential mobile intranet service using the 5M analysis. Let us deliberately pick the category that appears to be the most narrowly defined and easiest to understand (namely, mobile intranet access). This is the ability to access the

[34] Ahonen, M. and Barret, J. (eds), *Services for UMTS*. John Wiley & Sons, Chichester, UK (2002).

[35] Ahonen, M. and Barret, J. (eds), 'Types of UMTS services: Categorising the future'. *Services for UMTS* (p. 187). John Wiley & Sons, Chichester, UK (2002).

company network from outside the network, in our case from a mobile device. The most obvious services in this category are wireless email and wireless Web, which could include internal websites, such as the online employee directory or some product support pages.

Wireless email is cited often as the most pressing need for an enterprise to 'make the transition to wireless', and this has some credence, though perhaps not as much importance as first anticipated. Another crucial area is the ability to access crucial business support schemes, such as a customer relationship management (CRM) system; but let's focus on email for now as it is more general and can therefore be more easily appreciated and understood.

3.7.1 The Breadth of Wireless Email

Wireless email access can include a wide range of modalities: from using a laptop with a wireless modem card (even within the enterprise, such as a WiFi-connected laptop) to listening to email via an ordinary mobile phone using text-to-speech technology. There are many solutions in-between, such as personal digital assistant (PDA) access or smart phones with local or browser-based clients. Some solutions use regular email clients with POP3 or IMAP4 access and some use more feature-rich and elaborate mechanisms, such as a constantly over-the-air synchronized solution, like the Blackberry from RIM.[36] Other solutions can involve the use of browsers, perhaps WAP, and can even tie directly into my Outlook™ information database rather than an email server, such as the solution from Seven.[37] Users can take their pick depending on preferences; we shouldn't think one size fits all, even for any given user who may find a variety of solutions useful depending on the circumstance. The ability to switch between different modalities is an idea that follows from our previous discussions about understanding the impact of mobility on our service models; but, let's now examine these issues in more depth using the 5M approach.

3.7.2 Wireless Email – Movement

In the earlier discussion of Ahonen's five Ms 'movement' carried with it the idea of escaping the fixed place, which is certainly a core concept in providing wireless access to email; increasingly, office workers are in another place rather than the office. There is little doubt about the benefits of wireless email in terms of providing increases in employee productivity while away from the office, especially now that email has become a major communications tool and users have typically come to expect a high degree of responsiveness from email recipients, meaning quicker turnaround times.

When parading early attempts at wireless email using text messaging[38] I coined the term *time-critical messaging* to refer to the ability to deal with information changes in an organization, such as an inbox, in a timely manner. Intellectually, we can all grasp there is at least an anecdotal relationship between improving cycle times and competitiveness. We have been hearing this mantra from business gurus for years, so let's assume that it's generally true. Figure 3.3 is an attempt to show the relationship in graphical form.

[36] http://www.rim.net
[37] http://www.seven.com/
[38] In a brave attempt to offer some mobility to email users, my company built a product to access Exchange using 2-way text messaging. Later on, we built one using WAP.

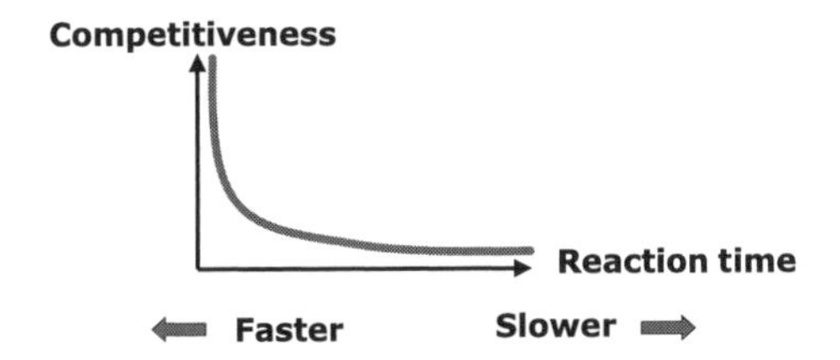

Figure 3.3 Graph to symbolize relationship between competitiveness and reaction time to information change.

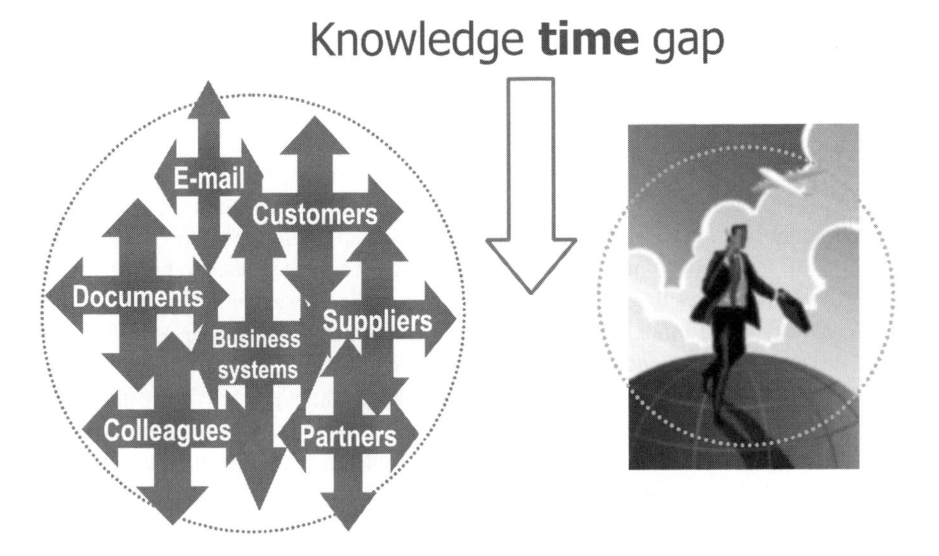

Figure 3.4 Distance between information change and information access equates to a knowledge gap.

The ability to react fast enough relates to the 5M of 'movement'. The distance away from an email access terminal or a modem telephony point represents the distance that needs to be travelled toward where critical data are made accessible, which translates to lost time. At the Windows Show in London (1998), in a presentation I gave, probably the first time ever that an enterprise wireless solution[39] was presented at the show, I referred to this distance as the *knowledge time gap*, as shown in Figure 3.4. Of course, this also relates to the 5M attribute of 'moment', as we will see, but let's first explore some of the more subtle aspects of movement.

Using mobile technology to facilitate *constant access* to email is the most powerful aspect of movement, and it is possibly important enough in this case that the 5M of 'movement' is the most imperative out of the five.

It is interesting to consider whether there are other ways in which the 'movement' attribute applies to this service. There are some interesting possibilities here. For example, the service should ideally be able to detect that I am away from the office and not logged on

[39] Referring here to the text-messaging interface to Exchange, as released by Xsonic (http://www.xsonic.com).

via my laptop, wherever I happen to be, like the hotel or conference room or at a customer's site. In this case, my movement away from my email client should automatically trigger email alerts to my mobile phone, so that even in these seemingly black spot areas, such as a customer site in the middle of a meeting, I am still in touch with the changing flow of information in my inbox; it's business as usual.

As WAP-Push[40] technology pushes the notification messages, I receive a clickable link to open the mail item or to respond to it. Furthermore, knowing that I am absent from my office, the notification system may also choose to interrogate my contacts database[41] on my behalf to see if the email sender has an available entry. If so the push message will automatically include their phone number, so that I have other ways to follow up there and then. One click in the coffee break, courtesy of the telephony interface capabilities of WAP,[42] and I'm talking to another client, who needs urgent attention, or putting my colleague in direct contact with the client using third-party calling techniques, courtesy of Parlay X3.[43]

3.7.3 Wireless Email – Moment

For a travelling businessperson, the 'moment' attribute obviously relates to movement. Conventionally, without wireless, email access would be sporadic depending on time, mostly because at different times I may not be in front of a PC connected back to my email server. Figure 3.5 clearly shows the problem. The out-of-touch areas are times away from the PC altogether, or where a modem can't be hooked up to access email. During the day then, these times are most likely due to meetings either on of off-site.

Ubiquitous wireless access goes a long way to addressing this problem. Even sitting in a meeting, with a wirelessly connected laptop, I have constant access to my email. However, constant access is not the only consideration. Again, just as with 'movement', if we reflect on the 'moment' attribute we may find that there any other ways we can make it work for us, keeping in mind our definition of a mobile service.

It would seem that a different mode of behaviour for email is possible during meetings or at certain times, and we explored this idea in the wireless journey described in the book's prelude (Chapter 1), such as the ability to handle important (VIP) contacts in an adaptive fashion. This would be especially useful for meetings. During these moments I may only want interruptions from emails arriving from important contacts, such as key clients, colleagues, associates, friends and family.

It is perfectly feasible for my email client to operate in this 'VIP mode' whenever I am in meetings, so this is a time-based filtering process. It is also possible for the VIP list to change according to circumstance. Later on, when next retrieving messages outside the meeting time, then all messages from all senders become visible again, after the filtering switches off automatically. Of course, as hinted at in the book prelude this kind of technique – called *presence management* – is also useful for controlling calls and other modes of communication, not just email.

[40] WAP Push gets discussed later in the book when we look at IP protocols for wireless.

[41] Interrogation of my contacts can be done in one of several ways, depending on how my personal information manager is set up.

[42] Using what's known as WTAI, or WAP telephony application interface, it is possible to click on a phone number in a message and for that number to be dialled. Later on, we will see that there are even more powerful possibilities, like making third-party calls via the Parlay X Web Services gateway.

[43] We discuss Parlay X in Chapter 12 on the RF network.

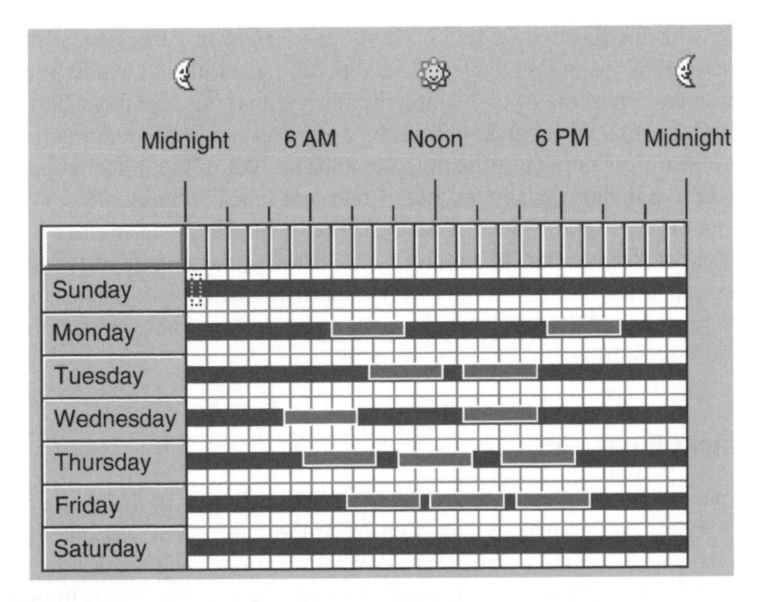

Figure 3.5 Email 'contactability' without wireless.

This temporal filtering mechanism can be applied in conjunction with movement, as was also seen in the book prelude, when in our imaginary (though not so unimaginable) journey things happened differently at lunchtime at the plaza as opposed to lunchtime in the office complex, or lunchtime in the taxi. It would be useful if lunchtime out-of-office communications could be kept to a bare and essential minimum, especially as control of private space becomes more and more critical as the private and public spaces in our lives begin to merge due to increasingly pervasive (invasive) technologies.

Thinking about time, we may wonder how our mobile service knows about meeting schedules, lunchtimes and so on. The obvious starting point is to interrogate the user's calendar, something easy to do in principle. Nonetheless, as we are all too aware, timetables seldom run according to plan, and lunchtime often ends up something we grab whenever we can get it, meetings often run late and, generally, flexibility is the name of the game. How then can we update mobile services with our activities?

Fortunately, there are mechanisms for doing this, and some we may already be familiar with if we use instant messaging (IM) services, like MSN Messenger™. Such tools have already introduced us to the concepts of *presence* and *availability* management. This is where we can inform the service about our availability, such as 'online', 'busy' or 'out to lunch'.

Presence management becomes more powerful with mobile technology as we usually carry our devices with us most of the time and so can regularly update our presence via a suitable interface (application) on the mobile device. This could well be our IM client or some other method, hopefully with a means to make the presence information known across all mobile services that need it, which is the likely scenario with next generation solutions.

Of course, with the location-finding capabilities inherent in geographically dispersed wireless access points (of any type) we have additional capabilities. It would be easy to set our presence to be 'out of the office' simply by sensing that this was physically the case: that we had left the office campus as defined by a zone on a map. By relating the moment with our movement it is not that difficult to combine an 'out of the office' condition with the hours of 12–2 p.m. and declare an 'out of office–at lunch' condition automatically or others like 'out of office–at meeting' or 'in the office–at meeting' and so on.

Once the 'out of office–at lunch' condition is detected by our mobile service, the email-alerting process is primed only to notify very important email events using our '9-5 VIP list' (or work-related list). Selecting which events cause notifications is a matter of personalization or the 'me' attribute.

3.7.4 Wireless Email – Me

The 'me' aspect is what Ahonen stresses as being probably the most important attribute of a mobile service. From what we have seen so far the 'movement' attribute is certainly very vital in the wireless email mobile service. We have also seen how essential 'moment' is and how it strongly relates to 'movement', as will often be the case within a mobile context. In both cases and where the overlap proves even more interesting, the concept of personalization also figures quite strongly, as we will examine.

The concept of filtering emails according to what we are doing is a 'me' aspect of the service that is driven by movement and moment. While the service is capable of automatically adjusting email filtering geo-synchronously, we are clearly only going to be interested in this being done for our own personal geo-temporal activities, not the actions of others, notwithstanding the possibility that our interaction with recipients could potentially be influenced by their own movements and circumstances. Therefore, for 'me' it is *my* calendar and *my* presence and *my* VIP list that matter, not yours, not anyone else's. This is personalization or customization of the service according to the 'me' element.

Again, trying to think beyond the obvious aspects of personalization: are there any others that we should consider? Tailoring the service to device type is conceivably another facet of customization according to 'me'. Adapting content delivery to the device characteristics is an oft-mentioned design consideration for mobile services. However, we are concerned here not only with content (or presentation of content) adaptation per device but also according to user preference.

Perhaps we have to drive to work or to meetings and during those times we would like to be able to use text-to-voice technology for listening to our messages. This would involve telling the service about our preferences for 'drive time' on the calendar (which might be automatically detectable by location-finding means). While driving to work the use of wireless devices is limited due to safety concerns. Let's imagine we only have an ordinary mobile phone with a Bluetooth headset, so text-to-speech conversion takes place in the network and the audio plays over a voice call connection. During a journey we could configure the service to call us every 10 minutes to read out the mail, assuming there have been new messages in the interim. We can use a similar service when driving back home. We might also use this time to review messages that did not cause alerts during the recent meeting and remained unread, as they were not from VIP senders.

This level of service is commensurate with our criteria for a mobile service allowing us to interact successfully and easily with our content while moving anywhere we like. Buying new gadgets or software has not been necessary. Our existing mobile phone is fine in this case and the service has made that essential allowance for 'me', for my way of doing things. If we had a PDA that could support Bluetooth audio profile, then we could have used a text-to-voice client instead or many other possibilities.[44]

3.7.5 Wireless Email – Money

At first glance the ability to spend money from the mobile terminal seems not to figure in a corporate wireless email application, but perhaps we can think of a few possibilities. Forwarding attached documents to a nearby fax machine or networked printer might be handy at a conference facility or while sitting in a meeting room.

While accessing email from a wirelessly networked laptop or even from a PDA or WAP-capable phone a crucial attachment may arrive in the inbox. It may be imperative to print out the attachment there and then – such as a document requiring a signature might demand – like a nondisclosure agreement from the legal department. There are several options for solving this immediate printing problem. One of them is to forward the email to an address of a nearby email-enabled printer, assuming that such an email-addressable printer exists and can be located nearby. There are several ways to do this, but let's relegate this discussion to elsewhere.

For WiFi users it would be relatively easy to download the document that requires printing and then upload[45] it to a local Web server, probably via the WiFi service provider homepage (access page), which gets displayed as part of authentication onto the local network. This might be part of a value-added services package available from the WiFi provider.

Another solution is to forward the attachment to a fax number, as most places will have a fax machine, even those where there are no printer services available. A quick glance at the conference literature gives the fax number. Email forwarding might still underpin this service, especially if we assume that the user does not have access to a networked fax server on their own office network, so remote faxing is not possible except using a gateway of some kind. This is one of the benefits of providing a hosted wireless email solution (namely, that the service provider can provide additional service features, like remote printing, without imposing special equipment needs on the users). In doing so, it is a chance for the service provider to earn more revenue.[46]

Whether via fax or a networked printer on the Web, what we need is a mechanism for charging for these printing services, so that users can easily take advantage of this service without cumbersome payment arrangements. If we have a micro-payment system in place, then we can charge per use of the service. The payment via the service makes life easier. For example, if we select the Web-printing option, then all we would have to do is collect the document from the office services bureau in the conference centre or hotel – or any location where the WiFi hotspot provider offers a printing service. On job submission the Web page issues a document number that we cite at the reception desk, where, most likely, we have to provide some identification to prove legitimate ownership of the print. We simply collect

[44] There are many ways that text to voice might work in a mobile email service. We have not mentioned many of them. However, the ability to handle all of them is the key to service uptake.

[45] Using HTTP (HyperText Transfer Protocol) post to a Web server that can print the document to a locally resident networked printer.

[46] Remote printing need not be restricted to email attachments, of course. Any document accessible to the user could be printed, such as submitting documents from 'My documents' on a Windows™ desktop.

the document without the concern of physical payment. The advantage of a mobile payment (*m-payment*) system like this is that the host of the printing service does not have to handle any money, so the whole idea lends itself to easy adoption.

There is no reason why such a printing service (i.e., via a Web page) could not be utilized in a 3G WAN. The additional challenge is one of service discovery. With WiFi the Web page used to access the hot spot is the means to advertise and access the printing service. For a 3G user there are no local access points that can bring neighbouring print services to the user's attention. A different means is required to advertise the service. One possibility is to use location-finding mechanisms, which brings us back to the 'movement' attribute.

We will discuss other uses of m-payment in our wireless email service in Section 3.7.6, where we look at interacting with automated services (machines).

3.7.6 Wireless Email – Machines

Sensing that you have arrived at their hotel and that you are using the conference facilities an email arrives in your inbox from the hotel IT system (a machine). They are offering you guidelines on how to order hotel services during your stay, using your wirelessly connected laptop. After all, they are hosting your wireless connection during your stay, so it makes sense to offer hotel services via that connection. In the welcome email message (or Web page) you are given an IM buddy address that you can use to contact the hotel discretely at any time, especially during meetings. The IM session is for the large part automated[47] by the machine, giving you submenus to navigate, like reply 0 for catering, reply 1 for front desk and so on. At the right moment, if required, a live operator will take over from the machine and interact with you, person to person.

During an arduous boardroom meeting you feel thirsty, but would rather drink herbal tea than the by-now rancid coffee stewing in the vacuum flasks. You also need an aspirin and mineral water to go with it; such are the heady vagaries of off-site meetings with a particularly awkward client. Using the IM session you place your order easily via a combination of automated menus. Had you not been able to find what you wanted, your IM session would have automatically connected to a real live assistant who would have engaged with you to offer their assistance. The automated system directs you to the appropriate person, as required, such as the catering assistant in this instance, although here you did not need their help. Following your order for refreshments you also interacted with the concierge and booked yourself a limo back to the airport, noticing that the meeting was running quite a bit ahead of schedule.

As you are not a resident in the hotel you do not have an account with them for your stay, so a method is required to pay for the refreshments. In this instance, the IM session automatically created a link to a Web page that confirms the order and then allows payment. In this case the items are added to your WiFi service bill or your host's bill – or you could have used a credit card or some other payment gateway. The WiFi service provider, as a hosted service, has provided the IM courtesy service and made the necessary arrangements to support payment.

[47] This can be done using a product called Auto Messenger from Dream Stream in the UK. See http://www.automessenger.net

3.8 THE USEFULNESS OF DEFINITIONS

What has been the usefulness of trying to define carefully a workable definition for mobile services and then a method for crediting its potential success to key attributes (namely, Ahonen's five Ms)? In addition, why choose Ahonen's five Ms and not some other lens through which to look at our service concepts?

To answer these questions we return to Section 3.1 titled 'Thinking mobile'. This has really been the focus of this chapter. We have not been too interested in establishing formal definitions and formal methods for approaching service creation. We could look at this, but that would be a topic for an entire book in its own right – and possibly someone needs to write such a book. Our aim was to enable us to realize that the mobile context is unique. It presents us with unique challenges. The rest of the book will examine these challenges from the technological perspective.

The good news about unique challenges is that they also present us with unique opportunities. We have seen how the mobile context enables us to move around in an interesting matrix of design considerations and possible outcomes. For example, with any mobile service we have the potential to contextualize the service according to 'movement' and 'moment', which are probably considerations we have not had the chance to consider before in nonmobile applications and services or in previous generation mobile solutions.

Next generation mobile services offer many exciting possibilities; the future's bright, the future's ubiquity.

4

The Mobile Services Technological Landscape

In this chapter we are going to examine at the highest level what components we need to build a mobile service end to end. The rest of the book elaborates on these components, their natures and technical details.

4.1 AVAILABLE APPLICATION PARADIGMS FOR MOBILE SERVICES

From the previous chapters we have understood that a mobile service is delivered to a user via some kind of device or multiple devices, wirelessly connected. We spent some time looking into the different kinds of services and examining their attributes in terms of identifying the core features that take into account the challenges of mobility. These challenges were implied in our definition of a mobile service, such as the ability to interact *easily*, *confidently* and so on. We further elaborated on a means to critically assess any service proposition, not only against our definition but also in terms of Ahonen's five Ms – movement, moment, money, machine, and me. This seemed to be a reasonable and potentially useful starting point for designing a mobile service while still at the brainstorming and whiteboard-scribbling stage.

The question now arises: how do we *build* successful mobile services? I am a firm believer in always starting at a very basic level of articulating a problem before diving in with all the technical detail (baggage) that we may have in our armoury. As Figure 4.1 shows, we have our roaming mobile user who wants a useful service and we have our service provider (in the general sense[1]) who wants to provide a useful service? What, as the question mark

[1] The use of the term *service provider* indicates any agent providing a mobile service. It is not restricted to the more traditional definition within the mobile telecoms markets.

Next Generation Wireless Applications P. Golding
 ISBN: 0-470-86986-0 (HB)

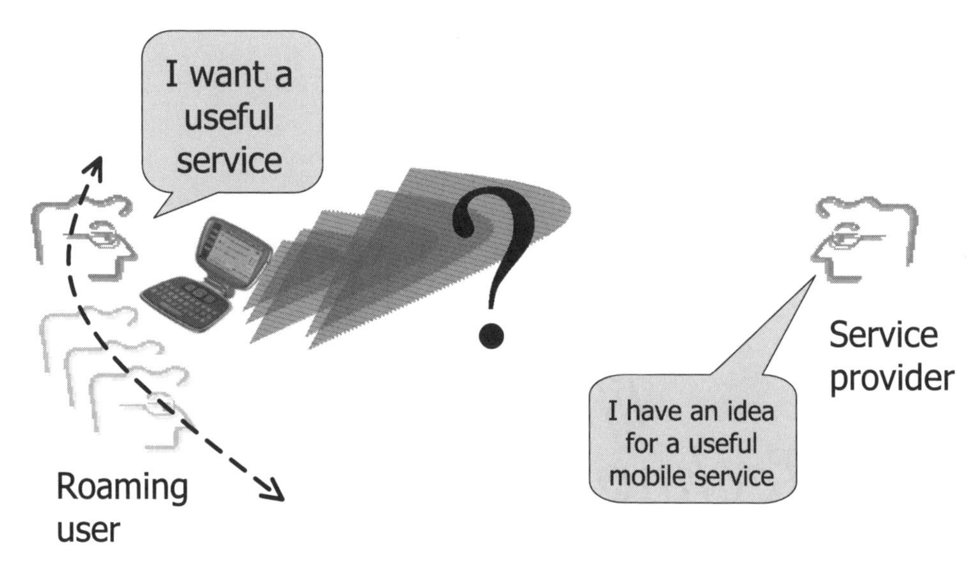

Figure 4.1 How do we connect the two ambitions?

is indicating, do we build in the middle to connect the ambitions of these two parties? The first line of enquiry is to consider the essential nature of mobile services, so that we have an idea of what it is we might be trying to build. What are its parameters, what ingredients are available, what are the limitations and so on?

Fortunately, we don't have an infinite variety of technologies with which to fill the blank. We already know a lot about what technologies and techniques can be used to enable mobile services, and much of this book is about describing the possibilities. Let's first look at the possible architecture of any solution: Figure 4.2 reminds us of the basic architecture of our mobile services network. Whatever service we build, we essentially have to traverse some or all of the layers of networks moving out from the device network to the content network. However, the ways of traversing the layers in this networked model are evolving all the time. For example, the dominant means of implementing services has been the classical client–server (CS) approach adopted from the world of networked computing.[2] The Web has taken this a step further by allowing for powerful interfacing solutions on the client, but in a way that enables many different data sources to be accessed from the server back end. This approach had been adopted by the Wireless Application Protocol (WAP[3]). However, we now have emerging technologies to enable powerful software applications to run directly on the mobile devices (e.g., Java made available for small devices, called J2ME[4]), and we have new ideas concerning software architectures, like peer to peer (P2P).

Often, and especially since the advent of WAP, mobile applications or services are assumed to have only one topology: namely, the CS architecture implied (imposed) by WAP or any Web-based solution. Many books will therefore discuss this type of approach

[2] The CS architecture arose out of the multi-user computing industry and has been adopted for many distributed computing solutions on the Internet; but, it was not the Internet that led to the CS approach – it had already emerged as an alternative to the mainframe or minicomputer approach.
[3] http://www.wapforum.org/
[4] J2ME stands for Java 2 Micro Edition, and we discuss it later in the book.

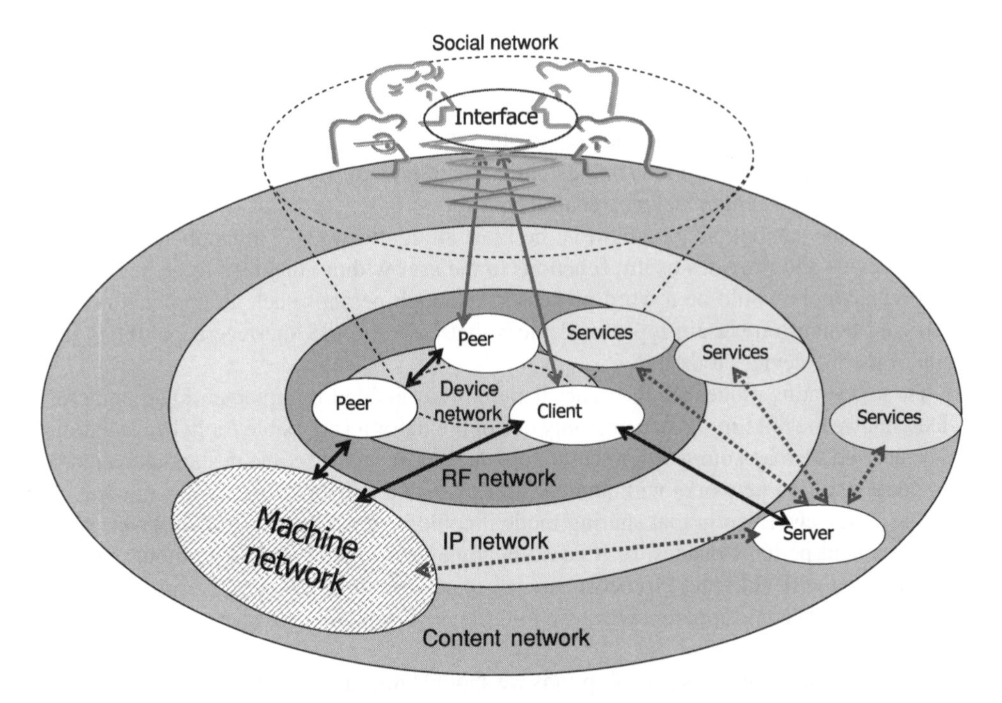

Figure 4.2 The assumed network architecture for mobile services.

at length as if it is the only viable topology for a useful mobile service. This may often be the case. However, as we have just alluded, this view is possibly too restrictive if we are truly interested in all the possibilities that next generation mobile services have to offer, taking into account all the various technologies available to support mobility and pervasive computing.

Therefore, what we aim to do in this chapter is to formulate the various possibilities for building services based on the network model shown in Figure 4.2. Each layer has its own attributes that need to be understood; so, we will describe the essence of each layer from here until the end of the book. As indicated in the figure and as we will see when we come to examine mobile devices in detail throughout the book, there are three main paradigms for mobile software:

Mobile application paradigms

1. Client Server (CS)
2. Peer to peer (P2P)
3. Standalone

CS is the classical architecture for many services, and there are good reasons for it. By the end of the book we will have examined all its facets in detail, especially *within the mobile context*. Notably, it usually involves a continuously available relationship between the device and a remote server in order for the service to be realized satisfactorily. The two ends are inextricably linked; so, naturally, what links them is important too and we will need to examine it.

Peer to peer is an emerging contender for enabling mobile services. As we will see, there is more than one way to network devices to other devices, not just in terms of the RF (Radio Frequency) connectivity options but also in terms of how to structure the communications in topologies that seem to lack any central control.

Finally, the standalone paradigm is the most straightforward. The application resides on the device and provides useful functions to the user without the need to be networked. A good example would be a word processor, although perhaps such an application is of restricted benefit on certain types of devices that have limited interface capabilities (the bulk of mobile devices today).

These operating modes are the major paradigms available for implementing mobile services, but we are not limited to using only one of them, for it is possible for hybrid solutions to be utilized. Perhaps our word processor operates most of its lifetime in standalone mode, but occasionally it networks with another user to enable collaborative document creation, editing or review. Even in that sharing mode it could be done via the content layer where our document perhaps resides, or it could be done peer-to-peer with or without the need to penetrate the IP (Internet Protocol) network to achieve the necessary peering infrastructure. This will become apparent when we look at networking modes, such as infrastructural versus *ad hoc*.

Because of the wide expanse of options we should know about and understand the many different approaches to service implementation. This book addresses how to understand the wide range of possibilities. It is my contention that most mobile services will ultimately involve a combination of programming techniques involving all the layers to a varying degree and all the above paradigms (multimode), also to a greater or lesser extent, probably in response to the dynamic influence of 'movement' and 'moment' (see Chapter 3 for discussion on the 5M attributes). Next generation mobile technologies and services are still in their infancy; so, we have yet to see this more liberal approach emerging. New tools, devices and operator-hosted service delivery platforms will make the transition to multimode applications easier and more than likely to happen.

Even a standalone application that runs only on the device (e.g., a single-player game), and, subsequently, never needs to interact with a back end server or another peer, probably will be deployed using an over-the-air (OTA) download mechanism (as discussed later in the book). In other words, the application itself would have originated from the content network and has to be sucked down to the device before it can be utilized; so, knowledge of more than one paradigm is still necessary. However, the production of purely standalone applications seems more and more unlikely. If multimode design is made easy, then most developers and service creators will find an excuse to incorporate powerful mobile-networking options wherever they can. For example, even an innocuous single-player game could become a shared network experience with a little bit of imagination. The incorporation of networking ideas will come from thinking of the five Ms and when the power of mobile networking becomes more apparent. New possibilities for user satisfaction and for service revenue will become more and more obvious as the next generation evolves.

In addition to the networking paths between user and content (or other users, or machines), the network model in Figure 4.2 shows us how applications running in the network can also draw on services from any layer – one or all of them. Of course, these services are themselves software applications, and we very much need to examine their natures so that we know what is available to empower our mobile service creation and how that empowerment takes place: how we plug our mobile application into these service access points. As an example, we can draw on telecoms services from the RF network layer, such as the ability to initiate a third-party call.[5] We could call on services in the IP network layer, such as a Public Key Infrastructure (PKI) solution or some kind of payment gateway. We could even call on services offered by other devices, such as shared diary applications or shared contact books.[6] All these services are supplemental to our own service. We are not talking about having to develop a shared calendaring solution or third-party calling server as part of our end user service offering. We simply want to be able to access such facilities as if they were subcomponents in the application or service. All of this is possible with next generation mobile services and is made even more powerful by having more than one application paradigm at our disposal along with the ability to contextualize our service according to the user's circumstances.

4.2 TAKING A BROAD VIEW OF THE POSSIBILITIES

We are not only concerned with analysing what is currently available to developers, such as WAP and Wireless Java (J2ME), and discussing how to use these ingredients. Rather, it will be more helpful to approach the problem going the other way, taking a heuristic approach toward uncovering the possibilities for mobile applications and services design and then understanding how various technologies can be used for implementation. Merely explaining the mechanics of constituent technologies used to build mobile services is what I earlier referred to as the 'cookbook approach'; this will not however enable us to appreciate the massive scope that exists for developing exciting next generation mobile services. The heuristic approach involves exploring possible problems and their solutions – which are presented by our own analysis of the mobile challenge – in order to uncover considerations for how to build services for the mobile context.

Our motivation is to supply fuel for innovation and to impart ideas and knowledge that may enable some readers to turn powerful service ideas into useful and lucrative services. A basic review of the mobile challenge will enable us to better understand how to utilize the bewildering array of technologies that are constantly being developed by the technology providers. In this current endeavour I do not want to be proscriptive about assuming a particular role or capability on behalf of you, the reader. This book is deliberately written for a wide audience who will have varying ambitions and intentions. For all I know, you may be a CEO of a major handset manufacturer and can implement wide-reaching ideas tomorrow that might transform how we can deliver exciting services to the users. Perhaps you are a leading protagonist of a particular technology and with the help of this book will uncover how to define a new offshoot for enabling electrifying mobile devices. Perhaps you work for an RF infrastructure supplier and can influence standards bodies to incorporate a simple change that will enable new breeds of services to flourish. Perhaps you are an inspired software developer looking to do something exciting and interesting for mobile users. In other words, I don't want to assume that your sphere of influence is limited to using the existing building blocks, like WAP or J2ME. Perhaps, with the right idea, you can

[5] A third-party call is placing a call between two (other) parties.

[6] See Chapter 13 for a discussion on sharing diaries within a location-aware application context.

develop new building blocks for us all to use in the near future. This is why the heuristic approach works better for this part of the book.

Our interest in next generation services implies that we are interested in building services that are at the forefront of the technically possible. Next generation mobile is definitely a brave new world of possibilities some of them radically different to what we have seen before in any field of technology, especially ones that can be placed in the hands of so many people. Possibilities that, most likely, will redefine our current view of mobile life and will further ensconce ubiquity and pervasive computing into our lives. Therefore, I am keen to explore first the essence of mobility and to lay bare many of the possible scenarios we can envisage for delivering interesting and exciting services with the current and emerging next generation technologies. This does not mean that, automatically, we are trying to invent ground-breaking technologies. It may well mean that we end up using current generation WAP or J2ME to deliver our services (there are other application paradigms too, as we will see). I don't have a problem with that; hence, one of the aims of this book is to survey and explain the rudiments of these technologies in any case. The power within next generation services probably lies in the judicial use of several technologies in clever combination; so, even a technology that yesterday may have not have seemed particularly powerful might become so if used as an ingredient within a powerful combination of technologies suited to the mobile context.

However, after dwelling on some of the architectural possibilities for delivering different types of mobile services, we may well decide that we need a new protocol or a new scripting language, or even a new software paradigm, to make interesting mobile services a reality. That would be a highly desirable outcome for me: if I could motivate fundamental technology provision into the mobile landscape, not for the sake of it, but for enabling useful services to come about. Therefore, just as we ended the last chapter and the tagline of my weblog[7] says – 'The future's bright, the future's ubiquity'.

4.3 PRIMARY CHALLENGES

In identifying what to deliver we have already examined some of the challenges. In the first instance these were concerned with making sure that we had defined a viable mobile service, so that we end up building something that would be useful and valuable, perhaps even transformative of our daily routines and habits.

We looked at two ways of reflecting on the usefulness of our service. First, did our service concept fall in line with our definition of a mobile service? We dabbled with definitions in the hope that we convinced ourselves that we had understood the essential nature of our cause. We scrutinized our definition to verify that it was, at the very least, likely to engender useful guidelines for framing our service ideas. We emphasized that there is nothing sacred about our definition: it merely acts as a focal point, or starting point, for our activities. As a second mode of reflection, we asked if we could clearly identify key attributes that would hopefully guide us toward successful services in the hands of the users. Our guidance came from the strong gravitational pull of Ahonen's five Ms? If we couldn't clearly identify meaningful 5M attributes, then perhaps we had not given enough thought to our service concept. Again, there is nothing sacred about the five Ms either. They are tools for thinking,

[7] http://radio.weblogs.com/0114561/

nothing more, but probably a good starting point to get us scribbling on a whiteboard while figuring out what our service is going to be all about.

There could be other ways for trying to ensure that our service is going to be useful, and some of those will emerge in the current discussion and throughout the book. What we need now is to construct a view of the technological landscape in which our mobile services can operate. This will lead us to an understanding of the ways in which we can turn our service concepts into a reality. How do we move from our service concepts and ideas toward an articulation of a real architecture and technical approach for realization?

So, what is the first step in the service creation process? Step 1 must surely be to define the service concept. This can be done and should be done without limitation. I consider this a brainstorming step; anything goes. Let's not limit ourselves from the start. We may feel that perhaps our idea is too ambitious, too crazy and 'off the wall'; so be it. We shouldn't adopt a restrictive mindset from the onset. There are now plenty of ways to get things done in the wireless world and plenty of potential partners if the idea is beyond our ability to realize by ourselves. Even if we could go it alone, partnering is probably still a good idea; so, we ought to feel comfortable with collaboration, but this indicates again that we need to know whom to talk to, what kinds of technologies are out there and how will our idea, our partnerships and the technological canvas fit together.

So the primary challenge is defining the service and brainstorming its potential, most likely using a framework like Ahonen's five Ms. But, what are some of the considerations that will help us along the way?

4.3.1 Who Are the Actors?

In pursuing our quest in designing mobile services we need to understand first who the players (actors) might be in delivery of mobile services. This is essential to our understanding as it will inevitably impinge on service delivery. I should stress here that by actors[8] I mean this metaphorically, not literally. An actor is any participating entity in our service that we can describe using a noun or a compound noun. Examples would be:

- office worker (an end user of the system);
- field worker (another potential end user);
- photocopier (a participating machine);
- sliding door (another participating machine);
- field service report (some content);
- daily task schedule (more content).

If there are several actors involved in delivering services, then we need to know what their natures are, their needs, objectives and limitations and then how do we cater for them. One could construe that this is possibly the first and greatest challenge. It may lead to the development of a set of questions that prove useful in the subsequent analysis. For example,

[8] The use of the term 'actor' comes from the Unified Modelling Language (UML) concept of determining software system requirements. We are not too interested here in the use of such formal methods, as our discussion remains at the superficial level of service creation, not the detailed mechanics of producing a requirements document for the intended service(s).

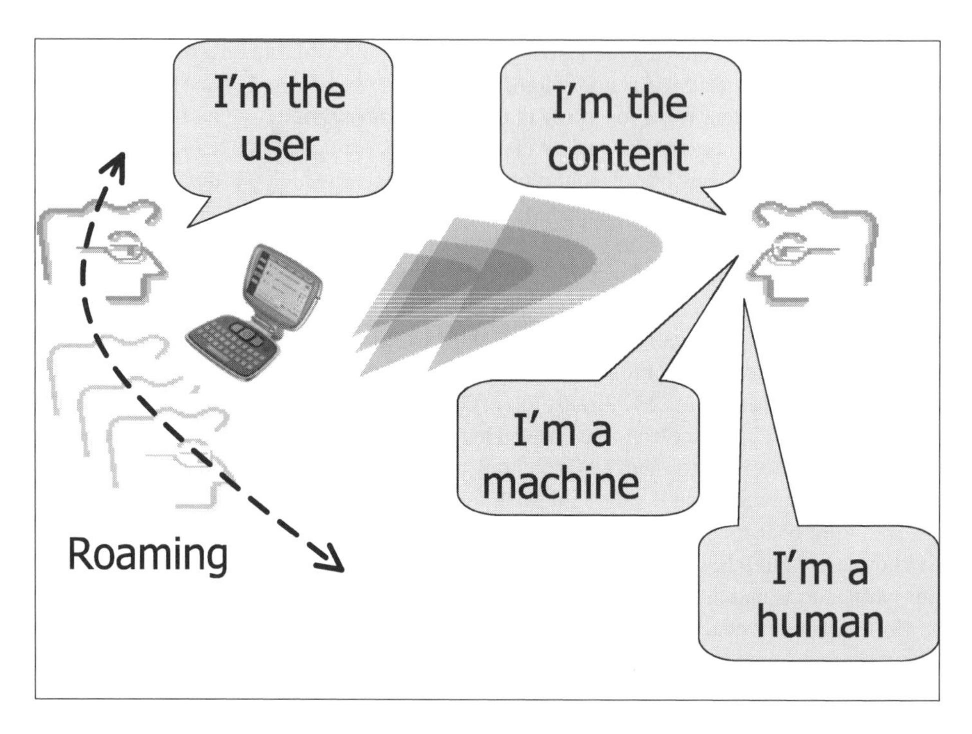

Figure 4.3 Essence of mobile interaction.

whom should we talk to about their needs from the service? Whom should we work with to enable the service? Who is going to pay for it? How is the average user going to discover it? Whom are they going to buy it from?

As we will find out, depending on what we want to do the answers to these questions will inevitably vary. This is because there is more than one way to deliver a service. Let us examine what the essence of a mobile service is in terms of how our actors would interact with each other to collaborate in the overall service delivery and usage.

4.3.2 How Do the Actors Interact in the Mobile Context?

In our previous discussions on mobile services and their definitions we focused on the highest level of description. Let us remind ourselves of the mobile services definition.

> **Mobile services definition**
>
> The ability to interact successfully, confidently and easily with interesting and readily available content, people or devices while freely moving anywhere we are likely to go in conducting our usual day-to-day business and social lives.

The essence of a mobile service is in the words '... ability to interact... while freely roaming...'. First, let's try to show pictorially the essence of this ability to interact within a mobile context. As Figure 4.3 shows, at the highest level of requirement in our mobile service we have roaming mobile users who wish to interact. They want to interact with content, machines or people; we have discussed examples of these in the previous chapter, but these are the actors we just mentioned. The diagram only illustrates the types of actors, but there could well be a multitude of each or all of them involved in the service concept.

Figure 4.3 is a useful abstraction of the mobile context seen from the user perspective, whereas the network diagram in Figure 4.2 is the context as seen from the service perspective, because only the elements of the service will know that they sit and traverse these network layers, which otherwise remain hidden from the users.

The possibilities for interaction are clearly limited to modes that can only involve the principal types of actors in a mobile service: content, machines or humans. It is useful to examine these modes and then later see how they might get mapped onto our layered mobile network model. In other words, if we start with the user-centric (or actor-centric) view and then move to the service-centric or network-centric view we will probably unearth the basic topologies and opportunities for building mobile services.

4.4 MODES OF MOBILE INTERACTION

It is possible to discern several distinct modes of interaction in a mobile service. There may be others, but I suggest there are four principal ones:

Possible modes of mobile interaction

- human to human (H2H);
- human to content (H2C);
- human to machine (H2M);
- machine to machine (M2M).

We can view these modes as being typed according to actors: human, content or machine. In all cases, at least one and sometimes both of the actors are roaming or able to roam. It is also possible that there are a multitude of actors involved, such as several humans interacting with the same machine in the case of shoppers using a mobile coupon fulfilment kiosk.[9] According to our definition, the roaming actors are able to roam *wherever they like*. An example of a roaming machine is a telemetry device attached to a goods delivery vehicle. Other examples will be considered later on.

M2M is an important type to include in our list as it reminds us that networking, and therefore mobile services, is not restricted to humans alone. Anything that can 'talk' via a networking protocol that will work over the RF network can become an active participant (actor) in a mobile service. In many cases the ability for a machine to roam and remain networked is a powerful enabler (e.g., like machines that move goods). They may want to report their activities to other machines that move goods or to static machines that want to

[9] A mobile coupon kiosk is a coupon-printing machine in a supermarket that converts e-coupons held in a mobile device into laser scan-compatible coupons (i.e., printed paper coupons)

track the movement of goods. Even goods themselves could be enabled to talk in a mobile network by virtue of RF-enabled packaging.[10]

In the case of H2H, this is clearly a social mode of interaction, but it does not have to be real time, like making a voice call. The interaction can be *asynchronous*. This means that both parties need not be involved in the communications process at the same time in order to interact successfully. Leaving a voicemail message or sending an email is a prime example of asynchronous interaction: we don't need the recipient to be available. Conducting an instant messaging (IM) session is an example of real time or *synchronous* communication where all parties (two or more) must be engaged in the process at the same time. In all cases of H2H interaction we are essentially interested in two-way *dialogue* or a degree of social interaction. The fact that we have synchronous and asynchronous modes of H2H communication is important, and it implies we need different services within our network to support real time and non real time communications. Any synchronous communication requires a dedicated pathway with a minimum level of quality (Quality of Service, or QoS). The infrastructure required to do this can be found in the RF network as well as in the IP network, but they have different characteristics. Accordingly, they also have vastly different implications for how we build our service application, and this will become abundantly clear when we discuss these parts of the network in the book.

As we will see later in the book, when we look at location-based services (LBSs), the ability to cope with asynchronous events is a major challenge and a critical need for LBSs. Consequently, we require software processes that can work on their own timeline, not just in response to a there-and-then need from a user (e.g., like a web page). The concept of software *messaging* emerges from any such consideration, which is the ability of software processes to talk to each other in a manner that is also asynchronous. Otherwise, the design prospect of keeping all our software processes in lockstep (synchronized) across the entire mobile network of networks is a daunting and probably impossible task, especially given the scale of the user base and its multifarious usage patterns. A similar challenge has existed for some time in financial systems, and this led some technologists[11] to develop an entire software intercommunications process based on messaging end to end, so that messages can be generated and consumed by the mobile devices, the machines and the content-hosting platforms.

H2C interaction involves a human wanting to gain access to any kind of content, the nature of which can be quite diverse, but essentially anything that is of use to the user via their device. This could include anything from train times to pictures used to decorate the device display (wallpaper). Usually, the fetching of content in H2C interaction does not require another human to be actively involved in a dialogue (though clearly humans may well have been involved in content production). In this sense H2C is a more passive mode of interaction than H2H and is more akin to a broadcast or publishing paradigm, but where the information is demanded, not actually broadcast. Information that has been made electronically available is demanded and consumed by a user. The process of making the content available has no direct connection with the user; usually, other people have made the content available for consumption.

It is often the case that providing mobile services is a business activity where ventures are trying to make money by offering the services or contributing to them. From the Internet

[10] This can be done by inserting an RF device into the packing carton.
[11] For example, Softwired and its mobile Java Messaging Solution. See http://www.softwired-inc.com, or see Chapter 11 of the book *Professional JMS Programming*, Wrox Press,

world, where Internet business models have been constantly proposed and refined for many years, the dominant model has been H2C for some time (though this is changing with the advent of Web services, which we will discuss later in the book). Within the predominantly H2C market a method of categorizing the interaction has arisen along the lines of identifying who the user is in relation to the business they are engaging with via the mobile service:

Classical categories of internet business applications

- business to consumer (B2C);
- business to business (B2B);
- business to employee (B2E).

In mobile service terms, B2C would be something like accessing an online bookstore to check a book price, place an order, check for a review and the types of activities that involve selling services or products to consumers. B2B would be something like a service to allow independent, onsite computer-service engineers from one business to find sources and prices for replacement parts as well as availability and delivery times from various parts suppliers. B2E is typically any service offered by a company to its employees: be that fairly horizontal (i.e., broadly of use to most employees), like email or time sheet submission, or vertical (i.e., of specialized interest), like field service reporting and job dispatch.

Arguably, these categories apply mainly to the H2C mode of interaction and are usually more useful in terms of identifying a business model, rather than offering us much to think about in terms of service architecture and attributes. When building our mobile services and surveying the different technical approaches, most of what we need to consider about interaction in our service is to understand which of the four modes apply – H2H, H2C, H2M or M2M – we don't really need to consider the relationship of the user to the business using or offering the service.

The problem with the classical Internet view of content consumption is that it is confined to a narrow notion of content. In our analysis of content in Chapter 3 we said that content is information that can be produced and exchanged across the mobile network: so, this definitely includes content produced by the users themselves or, we could say, produced by any of the actors in our network. For example, the drinks stock level alerts generated by a drinks-vending machine would be classified as content. Furthermore, such alerts might end up routed to other machines or to humans, so the interaction modes blur slightly if we no longer confine ourselves to thinking about content in a narrow sense. Similarly, a human may directly produce content, such as their own weblog. From the perspective of the reader of the log: is this now H2H or H2C? Actually, it doesn't really matter, because the main point to notice is that it is not some other mode of interaction not covered by one of these two. What we have really been trying to do is identify the unique modes of interaction as classes of interaction. Therefore, after we identify how each mode maps to the network and how they might then be realized using the available next generation technologies, we will have the necessary service platforms to cope with all eventualities. The classification of a particular service thereafter is probably just a matter of semantics. In the case of the weblog example then, any service delivery platform that allows successful H2C interaction should be able to cope with the weblog, so we can define it as H2C if we like.

Keep in mind that the heuristic approach that has produced these categories is simply an aid to thinking and for better interpreting the mobile technologies that we will describe throughout the remainder of the book. These modes of interaction as themes will become clearer as we proceed, and to an extent I am leaving it up to you to determine how you develop these themes in your own service creation activities. You may find that other variations work better. That's entirely up to you.

4.5 MAPPING THE INTERACTION TO THE NETWORK MODEL

In any of the identified modes of interaction, as our diagram in Figure 4.3 indicated, there are at least two end points, sometimes more (as in group chat applications, videoconferencing and the like). As the diagram also shows, we can imagine that the end points are participating sources of activity and information; they each have something to 'say' to each other and, so, in that sense they can be viewed as actors in a *sequence of events* that takes place to complete meaningful tasks within the mobile context. The use of the term 'actors' will be valuable in the following discussions, because an anthropomorphic[12] generalization of entities as 'people' that can talk to each other is a useful analytical technique. *Actor* is also a term that crops up in formal software analysis techniques, like the increasingly popular UML[13] notation; so, we should probably get used to it.

Importantly, the sequence of events and participating actors is not confined to the two end points, which themselves may change throughout the life cycle of the service. As Figure 4.4 shows, quite elaborate interactions can arise during the life cycle of a single session of tasks carried out by the mobile user. It is worth dwelling on this example sequence chart,[14] as it uncovers a lot of the issues that we need to address in building our mobile services.

The way to read the sequence chart is to follow it from top to bottom, which is tracking the flow of events in time. The basic intervals of the dialogue have been numbered to make it easier to follow. This sequence is a fictitious example of a user querying a location services application on their device to find out where they are. They subsequently request a map and then notice a restaurant nearby. Clicking on the number displayed for the restaurant causes a phone call to be initiated to the restaurateur and a table is reserved. Sounds exciting and already has my mouth watering! Later in the book, when we look at the RF network and then LBSs, we will see how to implement such a system.

By the time you have finished reading this book you will understand how every aspect of this sequence actually works. We can follow the number labels to see what's going on:

1. The user makes a request to the device application via its user interface. The application could be an embedded application installed on the device (we will look at this later in the book when we look at devices) or it could be a browser. Let's assume that it is a Java application (implemented using J2ME). Within the Java environment on the device we assume that there is a small program (an application programming interface, or API[15]) that is able to make location queries from the RF network, the

[12] Ascribing human characteristics to nonhuman things (except for our really human actors).
[13] UML is a means of thinking about how to analyse a proposed system for the purposes of writing down its required behaviour for subsequent implementation using software technologies. See http://www.uml.org/
[14] This is not a sequence chart in the formal sense in which they get used in UML notation or any other notation, but is illustrative of the concept.
[15] There is an activity within the Java Community Process to define such an API, called 'JSR-179 Location Services'. See http://www.jcp.org/en/jsr/detail?id=179

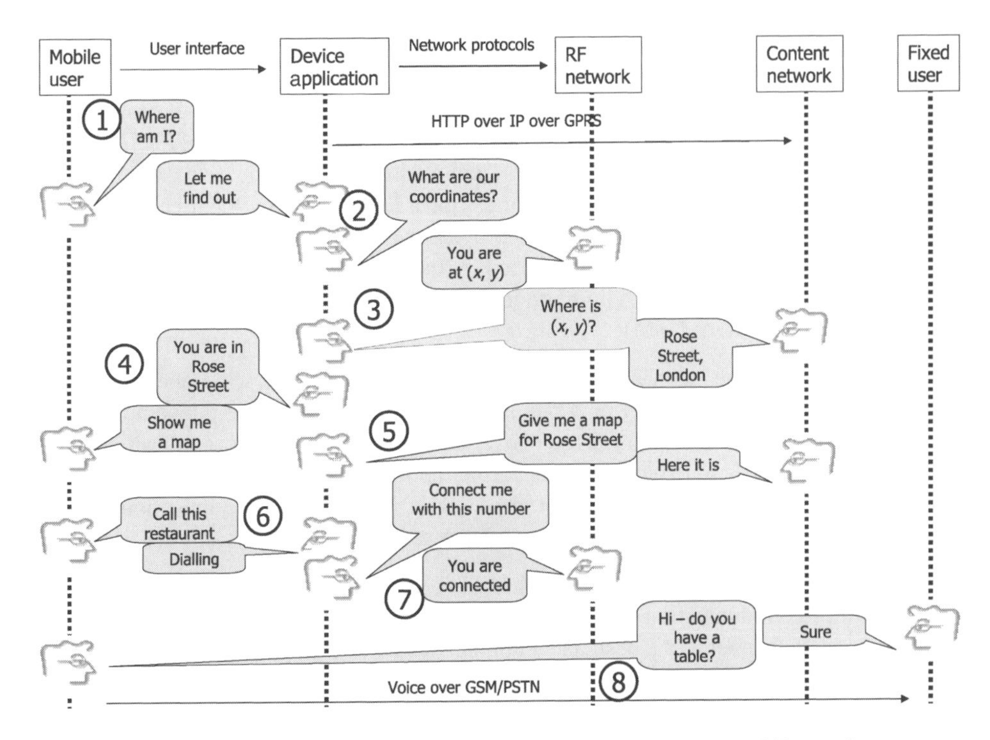

Figure 4.4 Sequence of events and participating actors in a sample mobile service.

network itself being more than capable of tracking device location, as we will find out later in the book.

2. In response to the request for location from the higher layers of the application the location services API on the device talks to the RF network, probably the Mobile Location Gateway (see Chapter 13 on mobile location services). The network responds with a coordinate pair[16] to say where the mobile is.

3. Having got the coordinates the device application then talks to a back end server, so we are now operating in CS mode. The server is a location information service delivery platform and can turn coordinates into geo-coded[17] information, like street names, which it duly does and returns the information to the application.

4. The user interface layer of the application displays the information to the user, who subsequently requests a map.

5. The application on the device goes back to the server and requests a map. As in the previous CS interaction, this connection relies on traversing several layers in the network. First, the application talks directly to the server using HyperText Transfer Protocol (HTTP), which we introduce in Chapter 5 and subsequently examine in

[16] As we will find out, a coordinate pair by itself is unlikely or not that useful. We most likely also need and would expect a radius of uncertainty that we can rotate around the coordinate pair to define a zone that the device is probably located within. This is discussed in Chapter 13 when we look at location-based services in detail.
[17] 'Geo-coding' is turning any coordinate information into an alternative geographical reference that is more suitable for subsequent processing, such as postal codes.

some detail in later chapters. An alternative would be to use the Wireless Session Protocol (WSP) from the WAP family, and this would have introduced yet another actor: the WAP gateway (proxy) to convert optimized[18] WSP messages to standard HTTP ones which our location information server can understand, HTTP being the lingua franca of most content servers deployed on IP networks (i.e., web servers). The HTTP messages are passed courtesy of a lower level protocol called Internet Protocol (IP), which in turn is able to provide packet-based communication courtesy of the mobile data channels established by the RF network; in this case we assume GPRS (General Packet Radio Service, an adjunct to the Global System for Mobile Communication, or GSM).

6. After fetching the map via HTTP from the location information server the user notices a nearby restaurant. Fancying a bite to eat, the user clicks on the phone number for the restaurant, which is hovering above the map, and the dialler application in the mobile device is initiated and starts to dial the requested number. This step assumes that numbers can be dialled programmatically from within an application. This is something we will understand in more depth when we look at devices in Chapter 10.

7. The dialler calls the low-level call-handling program on the device, which in turn initiates communication over a signalling channel on the air interface, requesting a voice call channel to the designated number. The dialler application gets a confirmation that the signalling was successful and a 'ringing'-state event is passed back up to the dialler, which interprets it graphically on the user interface and via an audio indication.

8. Finally, the restaurateur answers the destination phone and a voice call circuit is established; the user is now in a voice call and asks for a table to be booked.

As we have seen, for what appears to be a simple task from the user's perspective we have used a lot of powerful resources in our mobile services network, at all layers in the network. There are interactions between layers, so this implies that we need defined interfaces. Furthermore, these interfaces need to be programmatically accessible so that our software can take advantage of them, the concept that we first showed in Figure 4.2 where we indicated that applications can call on services from each layer in the network.

The availability of interfaces is a key area of understanding that we need to develop as service creators. In the above sequence we can conceive of various enhancements to the mobile service. For example, we could have had the restaurant send a text message to the user to remind them later on that their table reservation time is approaching. This helps the customer to remember their booking and it helps the restaurant to ensure that people turn up on time or at all, both being good for business. Let's say we wanted to implement such a feature: how would we do it? Figure 4.5 gives us some clues.

The most crude and least automated way is for the restaurant to send a text message from a mobile phone at the appropriate time (manual method). Another way is to enable the restaurant to schedule a reminder at the time of taking the booking. This has the added advantage of being an activity that can be incorporated into the booking process and, therefore, not likely to be forgotten. There are several ways we can tackle this problem. If

[18] We will look at the natures of the optimizations that WSP offers when we discuss IP-based protocols for mobile services in Chapter 7; but, WSP is essentially a compressed and more compact version of HTTP.

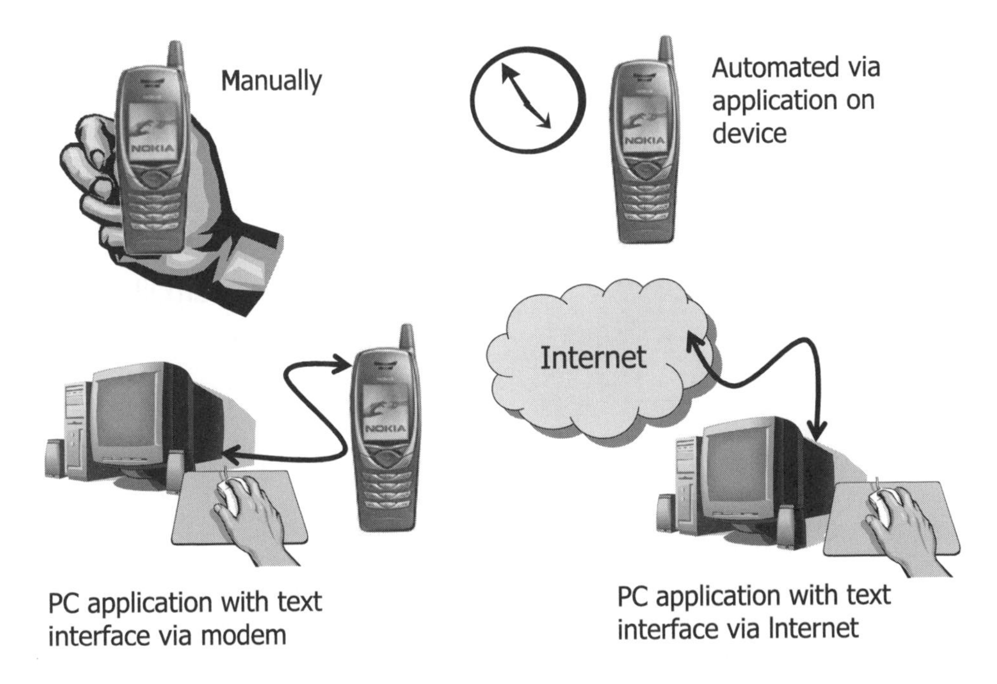

Figure 4.5 Various solutions to sending a text reminder. Phone image reproduced by permission of Nokia.

we have a PC-based booking system, then we can develop software to send a text reminder by scanning the diary and sending out reminders at the appropriate time. We could use a mobile device, like a phone, attached to the PC to send the message. Alternatively, we could submit a message via a CS link across the Internet to a messaging gateway application, which either the RF network operator provides or a third party who knows how to link the gateway up to the text-messaging infrastructure in the operator network.

This illustrates that frequently there are many ways to implement a mobile solution. Usually, the different methods involve different actors, so the piece of the network that we have to interface with is different in each instance. When designing mobile services it is crucial to understand which resources are available to us and how can we take advantage of them. Because there are many ways to build an application, it is important that we try to survey as many as possible. This book takes a wide view of the horizon rather than a narrow specialist perspective. There are already many books that specialize in WAP, J2ME and so on, but very few that tell us about the bigger picture.

4.6 INTERACTION VIA AN INTERLOCUTOR

In all the aforementioned modes of interaction within a mobile service we can envisage that the actors at either end are engaged in a conversation. As such, it requires that, just like an ordinary conversation, a common language is required and a means to engage in the dialogue. Our actors are physically separate and may well not understand how to communicate to each other natively. For example, in the case of H2C the content is probably stored in a

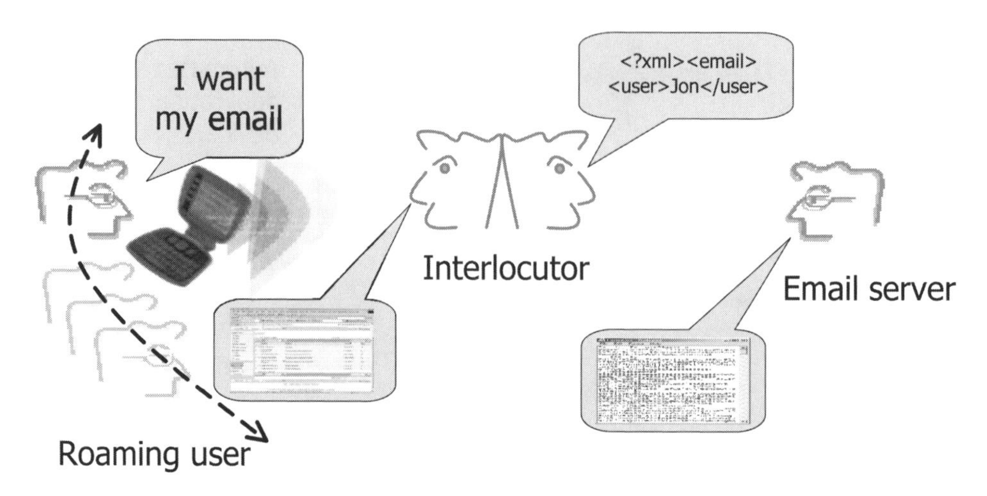

Figure 4.6 Mobile service as interlocutor.

certain format in a database. By virtue of the way that the data are hosted, it could be that the only way to access that content is using something called SQL, which stands for structured query language.[19] Typically, a mobile device will not know or understand SQL. Even if our device could issue SQL commands, it will not know where to look for the database server in the first place.

What we need – the essence of any service – is an interlocutor for our interaction, to carry the conversation between the two end points. Still thinking anthropomorphically, we can think of the interlocutor as a highly intelligent being: someone who is able to interpret what is going on at both ends and carry important messages and commands between them, as shown in Figure 4.6.

In Figure 4.6 we are showing a request from a mobile user to view their email inbox. We can quite easily imagine the mobile user demanding, 'I want my email.' Perhaps, a clever enough automated voice interpreter could understand such a request given verbally. However, we are interested in more than just interpretation. We are also interested in execution; we require the conversion of our interpreted request into tasks that the interlocutor then carries out. This is another aspect of the interlocutor role: namely, to convert requests (or state changes[20]) to meaningful tasks to undertake, followed by carrying them out to our satisfaction.

In the email example, the interlocutor receives the original request and somehow understands that it has to go and find the user's email server in order to request the contents of the user's inbox to be 'handed over'. I hope that we can appreciate the intelligence required to do this. Our user's request may require a lot of subtasks, such as figuring out who

[19] Most databases store information in tables that may well be linked to each other (i.e., relational). SQL is a way of stating how we would like to retrieve or deposit data into those tables.

[20] We can think of actors as having state and a state transition as indicating an event. An event could then be interpreted as an action. For example, the mobile email user goes from, say, state 'idle' to state 'waiting for inbox'. This is the same as saying that the action 'I want my email' has occurred. We do not wish to get pedantic with such things, but it is useful to understand this paradigm as it helps us to move to formal software thinking, like object orientation, when we need to (though we don't discuss that transition in this book).

the user is, where their email server is located, establishing credentials for accessing the server, knowing how to ask the server for the required information and then how to grab that information and hand it back to the email user in a usable format.

Furthermore, all of these intelligent subtasks are taking place while the user is moving, so another challenge is physically keeping track of the user and maintaining an open communications pathway – this is the job of the RF network. Perhaps the email server could not respond in a timely manner and the user has since moved location while travelling aboard a high-speed train. The interlocutor may have to switch to another part of the radio network to find the user again. As complex as this sounds (and it actually is), all of this is done automatically and, hopefully, transparently by the RF network.

In other services we can imagine that monitoring user movement may become even more critical, like knowing where the nearest hotel is, should the user wish to make such a request. This type of concern we can refer to as *mobility management*. We will look at this later in Chapter 12 when we consider the role of the RF network in our mobile services. Taking care of issues automatically, like managing mobility, is an essential function of the interlocutor and it has to take part transparently. We would like always to be able to ask, 'Where is the nearest hotel?' or 'what's in my inbox?', without having first to specify where we are in order for the process to be realizable. Such a constraint would contravene the usability criteria implied in our definition of a mobile service. Preferably, mobility should always work in our favour, never against us. At the very least, it should be transparent.

So, what is the interlocutor? Essentially, it is all of the technologies that are described in this book, all acting in unison. The nature of the interlocutor is a networked software service, which ultimately is a network of software applications colluding between themselves to carry out the required interaction, as shown in Figure 4.7. Some of the applications we can

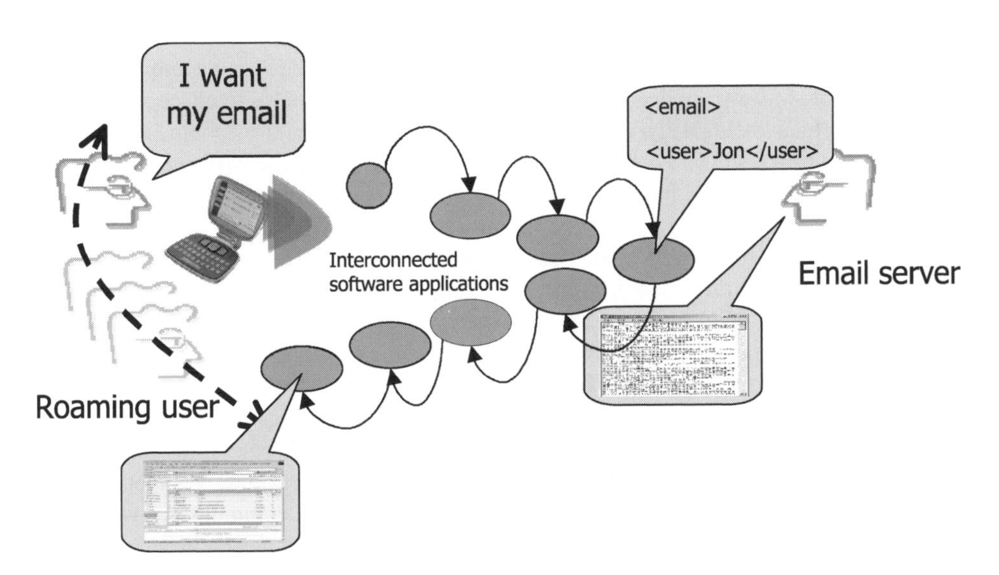

Figure 4.7 Interaction in a mobile service is courtesy of a network of interconnected software applications.

build for ourselves, while others already exist and are waiting for our custom applications to call on their services. It is important to realize that there is this network of software services underlying any mobile service – there is no one application that provides all the functionality. This will become clearer as we progress through the discussion and through the book.

Having conceptualized our mobile service and after scrutinizing its value in terms of fulfilling our mobile services definition and its display of the 5M attributes, we need to specify the interaction between each end: whether it's H2H, H2M, H2C or M2M. We first define who the actors are in our system, according to the exact nature of the mobile service. For example, in Figure 4.6 we were thinking of wireless email as a possible mobile service. In this case the two primary actors in our system are 'mobile email user' and 'fixed email server'. It is useful to add a prefix to actor names – either *mobile* or *fixed* – to make the mobility of the end points known, as it may be important to keep this visible in the subsequent analysis. This also allows for a differentiation between actors ostensibly doing the same thing, but under different circumstances, such as 'mobile IM user' and 'fixed IM user'. An application such as buddy tracking (discussed later), which gets used to detect where mobile users are at any one point, will involve an extra degree of complexity because all the users are mobile.

4.7 A LAYERED NETWORK CONTEXT FOR MOBILE SERVICES

We need to think in more detail about the nature and architecture of the networked software services acting as our interlocutor. We have identified that it has a networking role – the ability to connect our various actors, be they fixed or mobile – and it has a reactive and transformative role – the ability to interpret the states, actions and commands of the actors and present those to each other in a meaningful manner and format.

Let us first reflect on the networking aspects so that we can visualize a context in which the network of software applications must operate. To do this it helps to conceptualize the network as actually being a layer of networks, each with a distinct architecture and distinct attributes. Figure 4.8 shows a possible representation of the layered network viewpoint. Our mobile services operate within this perspective, and the figure clearly shows the actors at their end point locations.

Let's first examine the layers in the network, just at a superficial level, to satisfy ourselves that this model has validity, especially in terms of supporting our primary anticipated modes of interaction: H2H, H2C, H2M and M2M.

I should point out that, although the use of a layered model is convenient in our analysis, it happens to reflect quite accurately how the physical world of networked services software actually pans out in a mobile setting. Otherwise, I should clarify that the layering referred to here has nothing to do with formal communications or software paradigms, such as the OSI layer model.[21] Something like the OSI model would be a good guide to understanding how any two of the software applications in our network might communicate with each other. Our layered mobile network model is a much higher level of abstraction.

[21] For an introduction to the OSI (Open System Interconnection) model, see http://www.webopedia.com/quick_ref/OSI_Layers.asp

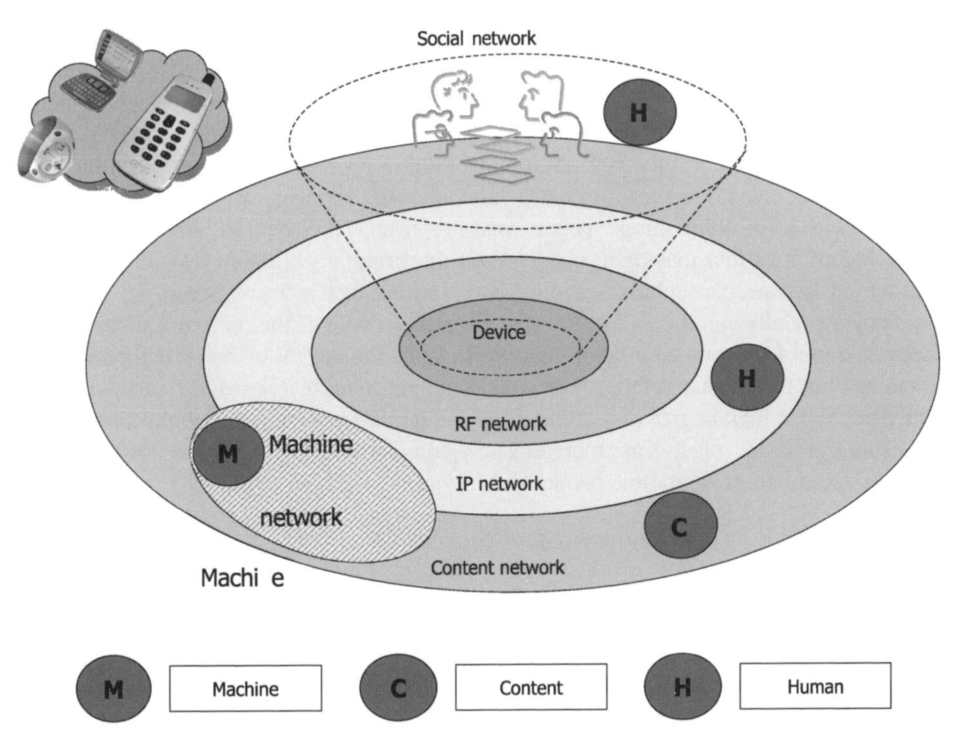

Figure 4.8 Layered network perspective of mobile service.

4.7.1 Social Network

The social network comprises the people who use the network. By explicitly including them in the model we are acknowledging and deliberately emphasizing that people who use mobile services exhibit social interaction as a primary form of expression. A social network is still a form of communications network, and we often forget this.

The social network is the highest layer in our network or its inner core, depending on how we view things; but, either way it is a sensible idea to give due reverence to the users! The inner core is a useful perspective to reinforce the notion that the users are at the centre of our mobile services cosmos. Users use (and abuse) the system and pay to do so. They are its lifeblood, so they deserve respect. In terms of software applications it would be more usual not to include the social network as part of the model, but we are concerned with the higher abstraction of mobile services where we expect services to be more intimately tied to social habits given the intimate relationship between a user and a device that they carry all of the time, perhaps even sleep with.[22]

The reason for including the social network is to underline that, ultimately, services are about people, except for the class of service M2M, where there are no people directly involved in the interaction. People will be the eventual beneficiaries, but not primary actors in an M2M service.

[22] I, for one, have often used my phone as an alarm clock to get me up in the morning.

We can think of mobile services as social-networking tools, as facilitators, magnifiers and multipliers of social interaction. It is important to reiterate our discussion from the book's introduction (Chapter 2) wherein we posited that mobile technologies have social-shaping powers.[23] New ways of communicating, interacting, socializing and living become possible and in ways that can become self-reinforcing. New ways of doing things will bring new, possibly unexpected demands for new technologies and services, which in turn will bring about new ways of doing things. This is what I refer to as the *virtuous circle of ubiquity*.

If we look at the text-messaging phenomenon that originally blossomed in GSM networks throughout Europe, particularly Scandinavian countries, then many observers recognize this as being a socially significant occurrence. Since the advent of the mechanical clock, time became a tool for social coordination, more so since the advent of portable time keeping (watches) and faster means of transport, which combine to make it possible to schedule social activities with a high degree of accuracy. The availability of mobile telephony, particularly text messaging, has made time more elastic again. As noted in Ling and Yttri's chapter 'Hyper-coordination via mobile phones in Norway':[24]

> *Owing to the recent yet explosive growth of mobiles, it is quite noticeable as a cultural phenomenon ... [one] version is the 'softening' of time; for example, sitting in a traffic jam and [texting] ahead to the meeting to let them know that you will be late.*

The other reason that we need to include the social network is to understand that it has its own ability to move information sideward without the other layers being involved. This may be useful to study and understand. People meet and exchange information, such as reading out text messages, showing each other picture messages, swapping telephone numbers, and reporting locations where 'air graffiti' notes[25] are hanging in space and so on. This P2P interaction is direct, and we can think of it as the most primary method of achieving a H2H service, albeit limited; but, it is something that occurs naturally in any case and involves mobile technology indirectly or in a particular fashion.

As we will consider in one of the following sections, the H2H interaction that occurs physically can then be assisted using devices, but still without the other layers, such as an IP network, being involved (e.g., using Bluetooth™ or Infrared).

4.7.2 Device Network

The device network comprises all the wireless devices – of any type – that interact with the network or each other directly. They are the primary access points into the mobile service world and represent the main interface between the actors and the available services.

Again, in a more conventional communications model or software system we may not have found it useful to go beyond thinking of a single-device access point; this is usually the case. However, here we are stressing that even the devices themselves have a sideward intra-networking capability.

[23] It is interesting to note that even the possession of a mobile phone has social impact, such as their iconic status in some post-Soviet bloc countries and in places like Hong Kong and China.

[24] Ling, R. and Yttri, B., 'Hyper-coordination via mobile phones in Norway'. *Perpetual Contact – Mobile Communication, Private Talk, Public Performance* (pp. 139–169). Cambridge University Press, Cambridge, UK (2002).

[25] We will examine the concept of spatial messaging later. 'Air graffiti' is a slang term that denotes how the idea of leaving a message pinned in mid-air (symbolically) is akin to daubing graffiti on a wall. Of course, this is a very narrow view of the potential of spatial messaging.

Device networking consists of three modes:

- *client domain* (CD) – between the device and the RF network to access other networks;
- *social domain* (SD) – between devices belonging to *different users*;
- *personal domain* (PD) – between devices all belonging to the *same user*.

The CD mode of networking will be commonplace in many mobile services. It uses the RF network to achieve nomadic or ubiquitous access to some other networked resource, such as another user (fixed or mobile), some content or a machine, all of which can be very remote from the user.

Within both the SD and the PD, devices will probably interconnect using similar technologies, such as Bluetooth™, Infrared and WiFi. SD networking would include such things as swapping contacts (e.g., phone numbers or other personal contact information), what has been historically referred to as 'beaming' in recognition of its light beam origins (infrared).

With the advent of low-power, cost-effective and short-range RF communications devices, it is no longer necessary to restrict inter-device communication to line-of-sight physics. This makes all kinds of exciting applications possible, such as multiplayer gaming or proximity sensing.[26] Moreover, the ability to connect without regard for device orientation and positioning means that it is easy to connect personal devices to each other without requiring clumsy effort that puts users off. This means it is also easier to interface several devices within a PD, as there is no need to line them all up nor be restricted to face-to-face alignment, which tends to preclude the use of more than one device pair. The book's introduction (Chapter 2) talked about PD communications and applications, as well as the idea of personal mobile gateways. Later sections in this book will expound on all these topics in more detail when we look at devices in more depth.

The fact that we may have devices talking directly to each other, whether in an SD or PD context, will have implications for our software architecture approach and which software technologies and interaction mechanisms we need. If a user has a sleek wireless watch, a keyboarded smart communicator and a tablet PC, then some degree of coordination between the devices may be necessary. For example, if the user is accessing email via their tablet PC, then there is no need to send email alerts to the watch. Similarly, all Session Initiation Protocol (SIP)-based calls could be routed through an SIP client on the tablet PC, perhaps itself on a WiFi link, with audio being handled via a Bluetooth™ headset. Another consideration is information synchronization. Perhaps a contact is entered onto the tablet PIM application, and this should be automatically available the next time a call is placed via the watch.

In the SD networking scenario our software solutions need to include appropriate protocols to allow devices to talk to each other directly, possibly without any intervention at all from another networked resource. This may have implications for such considerations as security, making sure we have an appropriate mechanism to authenticate a user from one device to talk to a software service on another device. If this is going to use the same

[26] For example, a Bluetooth™ network within a supermarket could be used to discover the presence of a mobile device.

authentication principles employed for network authentication onto the RF network (e.g., via a subscriber identity module, or SIM), then we need to ensure that we have the necessary provisions on the devices to access these authentication apparatuses. Under normal circumstances, such authentication mechanisms may not be accessible for localized inter-device communication; they may ordinarily rely on networked resources, such as an authentication server. This has implications for our software architecture, as it means that we need a wide-area RF connection to achieve authentication.

4.7.3 RF Network

In this book we will consider two types of RF network – *ad hoc* and permanent – this is to enable us to distinguish CD device communications from the other domains.

Two types of RF network:[27]

- *ad hoc* – exists for the duration of the communication only;
- *permanent* – always exists, thus implying the need for permanently installed infrastructure.

For devices that wish to talk to each other locally in the PD or SD configurations I refer to this as an *ad hoc*[28] RF network. This alludes to the nature of the network being transitory and only coming together for the short time it is needed and without the need for any permanent communications infrastructure, as indicated in Figure 4.9. The network arises for a period of use by virtue of devices talking directly to each other in sufficiently close proximity,[29] and then the network dissolves.

Ad hoc networks are continually set up and torn down as needed. In the case of social networks, devices cluster together to talk and then disperse again. With personal networks the devices may well be within range a lot of the time. The network is set up and negotiated between devices on an as-needed basis. We should not confuse this method of RF networking with the previously introduced method of software collaboration called P2P; this is a common confusion. As we will see later, P2P does not have to involve physical proximity and devices talking directly[30] to each other.

The other type of RF network we will refer to as permanent[31] and requires installed infrastructure. The network is always available, always accessible and is ordinarily a conduit to other networks, as indicated in Figure 4.11. In the early days of mobile telephony the RF network was a conduit to the Public Switched Telephony Network (PSTN). As we will see, with modern (second and third generation – 2G and 3G) cellular networks the permanent

[27] Note that we do mean network here, not connection (or session).
[28] Note that *ad hoc* networking can have particular meaning in certain RF networking discussions and contexts; but, I am not referring here to any specific technique or networking technology.
[29] The range may vary depending on the technology being used and the hardware implementation; but, this will be discussed later.
[30] By 'directly' I mean that the RF propagates physically from one device to the other (and vice versa).
[31] Although we recognize that an RF network infrastructure like GSM can be temporarily deployed, such as at sports events or conferences, while the need for infrastructure is, even in this case, permanent. If the infrastructure is removed the devices cannot communicate.

Figure 4.9 *Ad hoc* RF network.

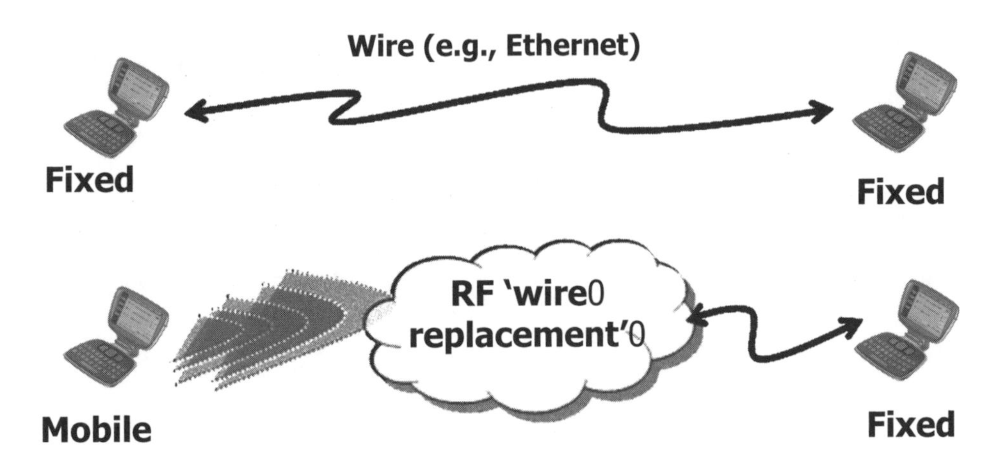

Figure 4.10 RF network as replacement for wires.

RF network is still a conduit to fixed telephone networks, but is also a conduit to IP-based computer networks, especially the Internet and the World Wide Web.

It is tempting to think of the RF network as just a conduit and nothing more, as if its design objectives were to be a faithful representation of a robust wire line connection (e.g., Ethernet CAT5 cable), but without there actually being a wire of course, as shown in Figure 4.10. However, as we will see when we discuss the characteristics of the RF network, it has other powerful capabilities that unintentionally (and recently intentionally) have become quite useful to third parties, such as the ability to locate a mobile device (and presumably its user) or to deliver a text message from a nonmobile application.

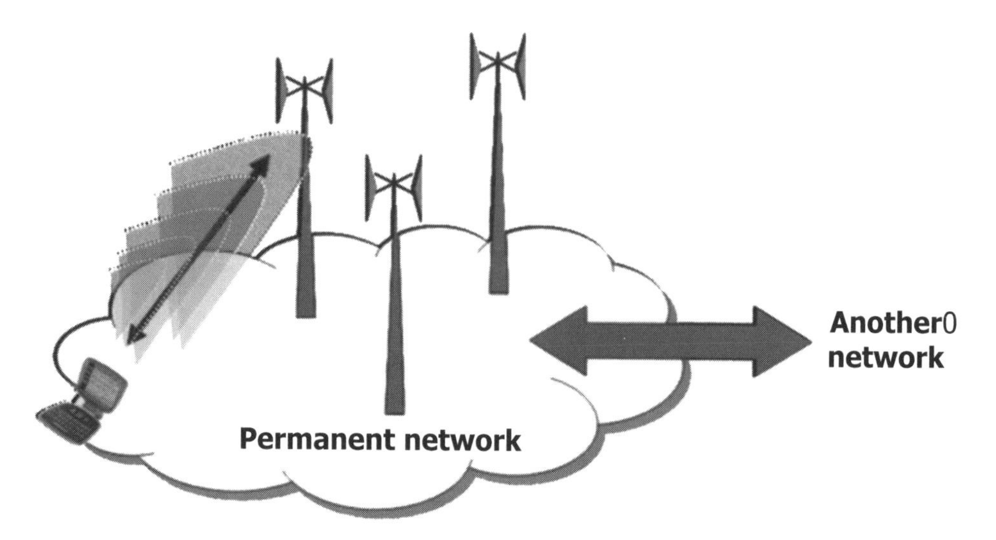

Figure 4.11 A permanent RF network.

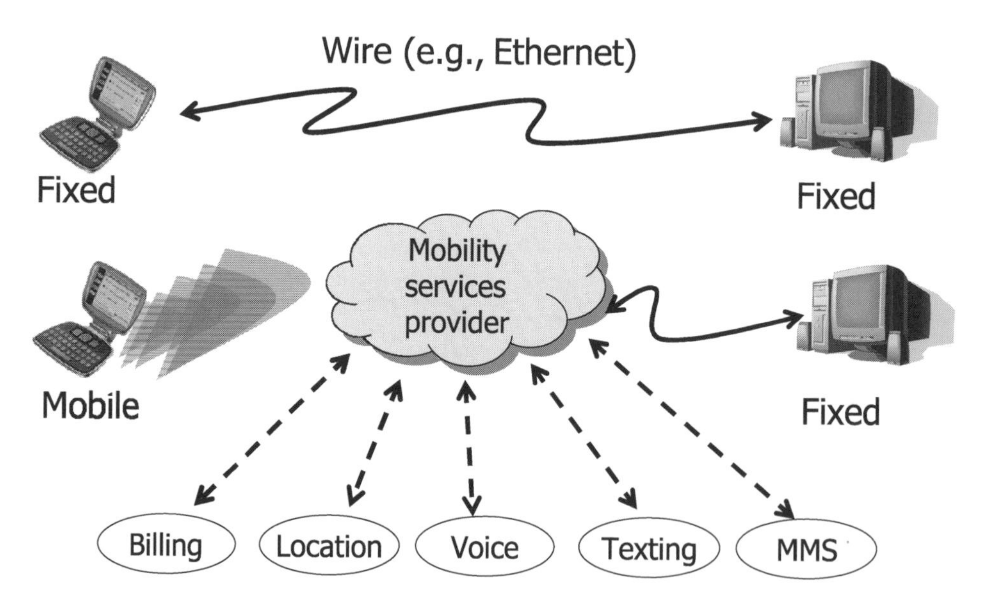

Figure 4.12 RF network can provide more than just a replacement for wires.

In fact, it is quite deceptive to think of an RF network as being a wire line replacement, the problem in perception probably coming from the use of the word RF, which tends to elicit a mode of thinking based on the 'wireless' physics of connectivity and nothing more. It turns out that the *cellular network* has many components in its infrastructure and is a substantial, intelligent network in its own right with powerful capabilities, some of which are listed below. The idea of the RF network as a powerhouse of assets is shown in Figure 4.12.

Some of these cellular network assets often go unnoticed, but are already important in mobile communications and will become more so for next generation services. Many of them relate to voice telephony, but voice has an important role to play in many applications, not just making calls.

Other features of the RF network:

- voicemail;
- text messaging;
- multimedia messaging;
- call handling (diversion, etc.);
- call conferencing;
- billing;
- fraud management;
- VPN connectivity;
- IP address assignment;
- charging;
- authentication;
- location finding;
- customer care.

It may seem strange to mention all the various cellular network features, especially items like customer care, which almost seem like incidental features of a cellular business rather than a feature of a cellular network. However, our minds should not become permanently focused on applications; rather, we need to think in terms of services. Taking the example of customer care then, it is perfectly feasible that we could develop a mobile application that took advantage of an existing customer care regime under the management of the cellular operator. If we think along these lines we can envisage being able to submit customer care reports electronically so that customer care agents are able to understand in what state our application was when a user experience difficulty (in the case of customer care manifested as technical help).

Let's not get bogged down at this stage with the operator-centric view of the world, which is very much driven by the exploitation of the infrastructure assets just listed; but, let's instead think of all these assets as having intrinsic value and capabilities that may be useful to exploit in any mobile service, whether it's offered by the mobile operator or through them by some other party. The powerbase or workhorse of our service may sit entirely outside the operator network, but symbiotically utilize the operator assets.

The relevance to our present discussion is that we are examining what it takes to build mobile services, working our way through the possible attributes and features of the *software services* network (our interlocutor). We should hold the idea in our minds that all the cellular

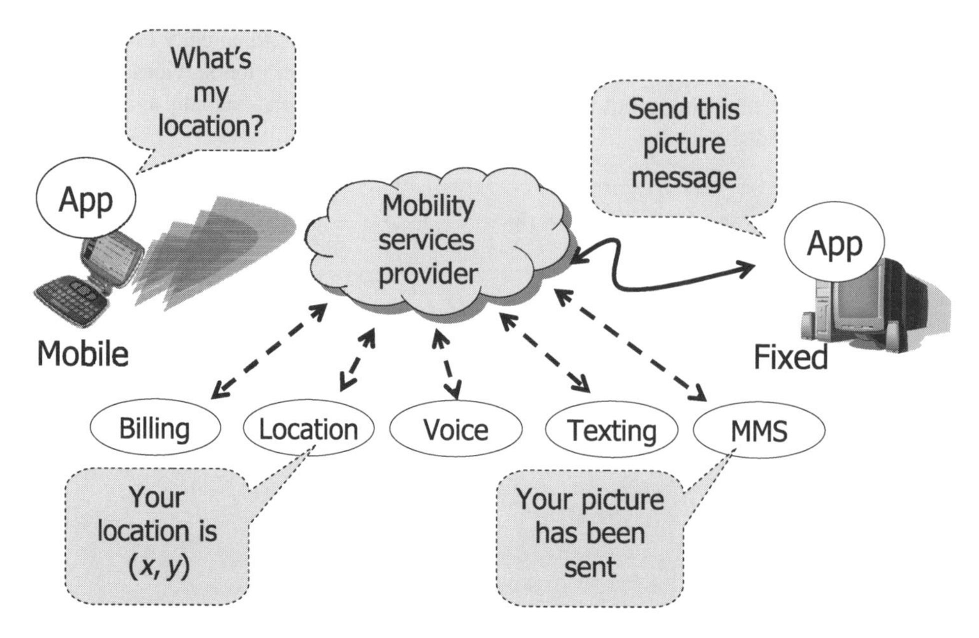

Figure 4.13 Application on the device and on the server can request network services from the RF network.

network features could be made accessible to our mobile services, whatever they happen to be or wherever they happen to reside once we get down to building them. We all will have our own ideas about services, but what I want to do is make it clear that the RF network is rich with capabilities, which, once we gain programmatic access to them (i.e., from our software applications), potentially enable us to deliver very powerful mobile services, as alluded to in Figure 4.13.

Let's briefly examine just a few examples of exploiting this kind of access to RF network assets. Let's imagine that we want to use our standard voicemail system to announce that a new email message has arrived. Better still, perhaps we want to have the new email messages read out using text-to-voice technology, but integrated audibly into the menu of our standard voicemail service. This is one possible way of unifying message collection, but it needs a means of integrating into the voicemail service, which usually is provided by a voicemail service platform hosted by the RF network operator.

If the voicemail system in the RF network was open to access from our software application, then possibly we could implement the proposed email-reading system. As an extension to this service, we may like to consider the possibility of inserting a new option on the voicemail help system – like 'press 5 for email announcements' or 'press 6 for important messages' – and this would take us through to an MP3 'jukebox' system on the Internet that we use to record messages from friends on our e-networking[32] bulletin board.

Historically, this level of integration into the RF network services has not been so commonly or easily available. The necessary service access points in the RF network were not available for all (or many) of the various network assets that might be useful. Furthermore,

[32] 'E-networking' is the term sometimes given to websites that facilitate business networking (i.e., meeting people).

the means of accessing these services was not conducive to popular programming and software paradigms, such as using Internet-centric models. On some networks only the messaging centres (text messaging and multimedia messaging) have interfaces accessible to third-party applications. Mostly, the network resources are for use only by their owners or in the way the network owner (operator) makes them available to select customers. They are not available as configurable software services to other service providers. However, this is going to change, especially with 3G and something called Open Services Architecture (OSA) and its close technological friend Parlay – topics that we will discuss in Chapter 12 when we look at the RF network in depth. It is therefore something we need to bear in mind as an architectural consideration when deciding how to build our mobile services.

Not all mobile services will need direct access to RF network resources. Some will simply make do with just the roaming connectivity power that the RF network provides. However, this is likely to be the exception. Straightforward access to a company Intranet might be such an example, although, as we discovered in Chapter 3, by applying the 5M analysis even such a seemingly innocuous application as corporate email had all kinds of exciting possibilities within the mobile context.

In summary, we have two ways of thinking about the RF network. One is as a wire line replacement and simply a means of roaming connectivity to another resource. The other view is of the RF network as a mobility services provider, able to take care of a wider range of concerns than just connectivity. Which view we take is not simply a matter of preference; it will be determined by what we are trying to achieve, but will also be influenced by which resources (features) the owners of the RF networks make available and the means by which they are accessible. However, nothing short of *total* access to the network's resources is going to suffice. As we will shortly argue in Section 4.8, mobile operators need to become adroit at converting their entire network into a services-hosting environment for third parties or else face extinction.

4.7.4 IP Network

In the case of CD connectivity via the RF network, typically we are attempting to connect as a client to some networked resource beyond the boundaries of the RF network. This resource will be a computer of some kind: be that a server, a desktop PC or some kind of embedded device – perhaps a dishwasher to turn it on, perhaps a remote-sensing station to measure rainfall, perhaps an on-board trip computer in a car. The possibilities are vast.

Whatever the resource we wish to interact with, clearly it has to be connected to a network. There are myriad data-networking solutions possible and available, but the prevailing solution is to base services and applications on the IP suite. Therefore, we will quite rightly focus on the IP network as the next natural layer in our network model beyond the RF layer. To reach beyond the RF network, there needs to be an IP-cognizant interface. Thereafter, there is a whole world of networks to connect with, especially the Internet. In any case, this is already the natural assumption of the operators who own the RF networks. The prevailing standards for wide-area RF solutions all assume inter-working with an IP network and that the boundary between the RF world and the fixed one is IP-based. Clearly, to ignore the Internet would be derangement in the extreme.

What this also leads to is the ability to assume certain approaches toward our service and software architectures: in particular, the adoption of the Web-centric paradigm for software services and related models. At the heart of Web-based architectures is the golden protocol of the modern computing era (namely, HTTP). This protocol had and still has a very simple design objective and, consequently, is able to provide an effective and highly scalable communications paradigm. This turns out to be useful for gathering support and momentum around Web-based solutions, HTTP being relatively simple to implement. This is even truer today with so many products and tools that make it easy for software providers to adopt HTTP. The plethora of HTTP-aware solutions made available in recent times is dazzling.

HTTP is so popular that it is often used as a default communications protocol in many solutions, even ones where the objective is no longer to deliver Web pages to browsers for visual consumption; this is increasingly so. For example, HTTP is at the heart of *Web Services*, which is an initiative to form a global IT consensus for how software services should talk to each other. Not surprisingly then, this model is what we see emerging as the preferred way for providing access to the mobility services in the RF network that we referred to earlier, such as the location-finding resources, text-messaging centres and lots more.

HTTP is also gaining ground as an interconnection solution for M2M connectivity modes. It turns out to be relatively simple and cheap to embed HTTP communications into embedded computers, it even being possible to implement an entire Web server on a single chip for a few euros. Other protocols may well make sense, but, given the prevalence of HTTP support in software tools, components and technologies, it is a sensible proposition to adopt it wherever possible (provided it is appropriate, as it may not be for certain types of communication modalities, but this will emerge in Chapter 5 on IP-based protocols).

As with the RF network, it is disingenuous to view the IP layer as just another data pipe through to the target resource, which ultimately is probably going to be a database of some form or other in the classical CS paradigm, although, as noted earlier, there are alternatives to CS that are gaining ground all the time, perhaps more so in the mobile arena than elsewhere. The IP layer has its own characteristics or capabilities, which we may well need to accommodate or utilize. For example, just as with the RF network, we need to consider security. There are ways of applying security models specifically to the HTTP protocol. Indeed, this has been an area of significant development ever since the Web became a widespread means of doing business and the need for secure financial transactions and secure authentication became apparent.

4.7.5 Content Network

This seems a strange name for a network, but we should remember that the network model that we are developing here is an abstract, high-level model that provides value in terms of thinking about how we build mobile services. For mobile services we need such a model so that we can take a holistic approach to our consideration of how to utilize various technologies to deliver the desired services.

The purpose of this book is to sketch out frameworks that we can use to think about mobile services, as well as to design them. As we have already seen, the RF network has two prime capabilities: not only can it support roaming connectivity, but it also has its own inherent resources and capabilities. In a more conventional approach to communications models, most likely we would omit this fact and that would become problematic. For example, we may postulate a service idea that includes voice or texting. Our network model along with an understanding of the capabilities of each layer will allow us to articulate that service idea into possible architectures, leading us in useful directions toward building the service.

Similarly, our consideration of the devices as forming a network in their own right also allows for a wider scope for mapping service ideas onto possible architectures. This also interplays with the networks that sit above and below the devices, such as the social patterns in the social network and the ability to form *ad hoc* networks in the RF layer. This view of devices as networks in their own right is valuable in aiding our thinking about how to build mobile services.

What I am highlighting is something that you may have already noticed: that the network model we have developed so far allows us to think of architectures that allow for interaction between layers as well as within layers, or any combination. This duality is important. This is also why I have elected to define the remaining layers in a similar vein, especially the content network, so that we maintain a consistency in our model, although this particular layer has its own attributes that are worthy of note.

The content network is really the apparatus that lies beyond the IP network; it hosts the information that the client device wishes to interact with (e.g., it is the user's email account sitting on an email server). It could also be the stock levels of soft drink cans in a vending machine. The types of content are quite diverse.

For any given mobile service the content could have a single source, such as an inbox or a household security system, or could come from many sources, such as aggregated news, portal services, a plethora of networked domestic devices or a hotels availability service. We can well imagine that a query to ask 'where's the nearest free hotel room' will require several databases of hotels to be queried. In such cases it is easy to see how multiple sources imply the need for powerful networking capabilities, the means to gather the information from different sources. However, there are other issues relating to service that the content network has to take into account, such as the need to provide redundancy and other types of resilience that ensure that any given mobile service is always available or able to degrade gracefully in terms of the quantity, quality and timeliness of the information it can produce.

In our model the content network is not just the content, it is also the infrastructure required to host it and deliver it. Here we may typically think of servers, such as Web servers, mail servers and database servers. These may require all manner of configurations depending on the nature of the servers and the implications of the service we are trying to deliver. Clearly, there may well be a need for some servers to network with others, possibly across organizational boundaries, such as a hotel information service needing to consult directly with various hotelier IT systems. This is all part of what we are calling the content network.

The content network also includes any embedded devices that provide information, such as vending machine sensors, remote diagnostics systems, household security systems, postal delivery systems and so on. However, these types of systems, where they are embedded into machines, we assign to a subdomain of the content network, called the machine network; we explain this in Section 4.8.

It is tempting to think of the content network as being passive. This comes from thinking in terms of the CS paradigm. We have already suggested that devices on the RF network will connect to the content network as clients, as indicated in our earlier introduction of the term 'CD'. A client usually requests services and gets a response. This is true in many cases and is a workable paradigm for many mobile services (we will address this topic shortly): namely, how our network model supports our different interactivity modes (H2H and H2C, etc.). However, it is important to stress at this point that it is perfectly possible for the content network to initiate the interaction unprompted by the client; this is a paradigm that is particularly unique to mobile with lots of potential to enable services to work in harmony with users' needs, habits and circumstances.

The mobile *content network* has three modes:

- *pull* – the RF device initiates information requests from the content network;
- *push* – the content network initiates information delivery to the RF device;
- *peer* – the content network initiates information requests from the RF device.

These communications modes may appear unconventional at first. 'Pull' is the most straightforward and probably fits with our expectations from the content network. We ask for an email message and we get it, we ask for a Web page and we get it, we ask for a stock price and we get it, very much a request and response cycle. These modes of interaction fit well with the CS architecture which we will discuss shortly.

However, 'push' is perhaps not so obvious to figure out, although the concept is simple, of course, and easy to appreciate. It is important to clarify that push is not the same as a polling technique: we may inadvertently think that some applications, like email, already use push technology, whereas they are actually based on pull technology. Email messages appear to be pushed to our mail client. However, this is not push. The email client establishes contact (polls) with the email server periodically to request (pull) the email messages to be sucked down for subsequent display. The RF network enables true push mechanisms wherein, *without any initiation from the device or user*, information can arrive at the device from some information source. If this information resides in the content network or is generated by the content network, then clearly an interface is required to allow the content network to initiate push messages via the RF network.

We can consider the push mechanism(s) as a key enabling technology for all kinds of interesting mobile services; so, we need to be aware of this aspect of our network model and enter it into our thought processes when we design our services.

There is yet another mode of networking to consider in the content layer: this is the *peer* mode. This is essentially like the pull mode in reverse. It is a mode where the RF device now becomes like a server, able to respond to requests for content that it is hosting itself. This could be useful in a number of services, but this approach is part of an emerging advance in software architectures called P2P computing,[33] not to be confused with the H2H mode of interaction we have already discussed, which appears to be related but is not because P2P computing can take place without human involvement in the information transactions.

An example of the peer mode of networking is a software service in the content layer requesting contact information from a phone's address book or searching a mobile device's picture memory for certain images.

[33] P2P computing is something more comprehensive than simply being able to assume either a client or a server role interchangeably; but, we wish to focus on this interchange here and not the broader P2P issues, some of which we examine later.

4.7.6 Machine Network

The machine network is an oddity in terms of our networking model because there is no unified view for considering interaction with machines; it does not fit neatly into the model. It is shown as a subset of the content network because, increasingly, machines will become Web-connected using tiny embedded Web servers. Although the electronics beyond those Web servers will be potentially sophisticated and certainly peculiar to each type of device involved, the Web server will mask the inner nature of the machine. If we think of a home automation system and its sensory output, then requesting the sensor measurements from the system via a Web server is no different to requesting a Web page built using information from a database. Indeed, one way to think of connected machines in an abstract sense is as databases of real time information.

Machine networking is becoming increasingly easier and can be done retrospectively at relatively low cost. Many machines have physical data ports, such as serial ports or similar. These can be easily converted to Ethernet ports using devices like terminal servers.[34] Once the device is on the Ethernet, then it becomes easier to integrate into a local area network where a Web server can reside and act as a Web interface to the machine. Alternatively, the terminal server itself could contain an embedded Web server. With solutions like these, many machines are potentially accessible via the Internet. Once that is possible, wireless access is achievable via the RF network's Internet gateway. Alternatively, an RF device could be attached to the machine itself, which is an especially attractive option for remote machines. However, with domestic appliances, these typically don't have any form of electronic interface other than the human one. For example, a central heating system has a temperature control, but not a serial port. Similarly, most alarm systems have access panels, but not a serial port. Networking domestic machines is a more challenging proposition which requires a deliberate attempt by manufacturers to incorporate networking options; this is currently a slow development due to an uncertain business model for home automation.

4.8 MODES OF COMMUNICATION ACROSS THE NETWORK LAYERS

Having examined the network layers, we need to justify this model in terms of its support for our primary modes of interaction: H2H, H2C, H2M and M2M. We should understand how these interaction modes work across and within the layers we have identified.

Figure 4.8 shows the location of the primary actors in our network model. Let's now see how each of the layers supports these modes.

4.8.1 Human-to-Human Interaction (H2H)

H2H interaction can take place in one of several ways. First, H2H can take place in the social network. We have already discussed this to remind ourselves of the potential impact of mobile services and devices as their influence extends into the social domain. We mentioned

[34] A terminal server is a small device that has a serial port on one side (e.g., RS232) and an Ethernet port on the other side. Using software running on a PC the serial port can be contacted via the terminal server via an Ethernet connection on the PC. As far as the PC is concerned it is talking directly to a serial port (COM port). See, for example, http://www.lantronix.com

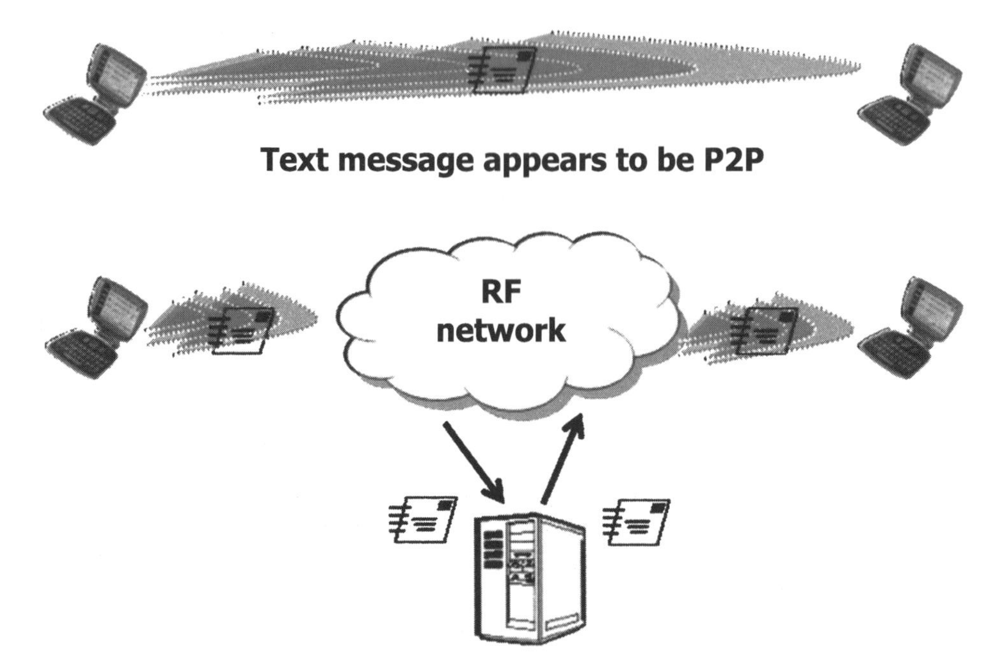
Text message appears to be P2P
RF
network
Text message actually goes via messaging centre

Figure 4.14 Despite appearances, text messaging is not a P2P application.

verbally swapping phone numbers as an example, showing friends a picture message or video message, and so on.

Second, H2H interaction can take place using an *ad hoc* RF network, such as made possible using Bluetooth™ or WiFi technology. Interactive 'chatting' is possible using Bluetooth-enabled personal digital assistants (PDAs), perhaps to allow colluding attendees of a meeting swapping surreptitious notes during the dull moments of the meeting. Note that, architecturally, this method of communication is P2P and will therefore require a specific software design approach. For example, just because we can form an *ad hoc* network using Bluetooth™ or WiFi-enabled[35] PDAs doesn't mean we can automatically carry out any familiar communications task, such as conducting an IM dialogue. In fact, using the popular IM services, like MSN Messenger, it is not possible to use such a service in P2P mode, as these services require an intermediary server, which in our model would sit in the content network or IP network. The requirement for an intermediary can go unnoticed. For example, text messaging, which at first seems to be device to device,[36] is not a P2P method of collaboration. All text messages go via a store-and-forward messaging centre as shown in Figure 4.14.

So, H2H communications across an *ad hoc* network implies that all the necessary mechanisms to establish and maintain the communications have to exist on the devices and

[35] As we will later find out, WiFi supports *ad hoc* networking as one of its modes of operation.
[36] Ignoring the obvious need for the RF infrastructure.

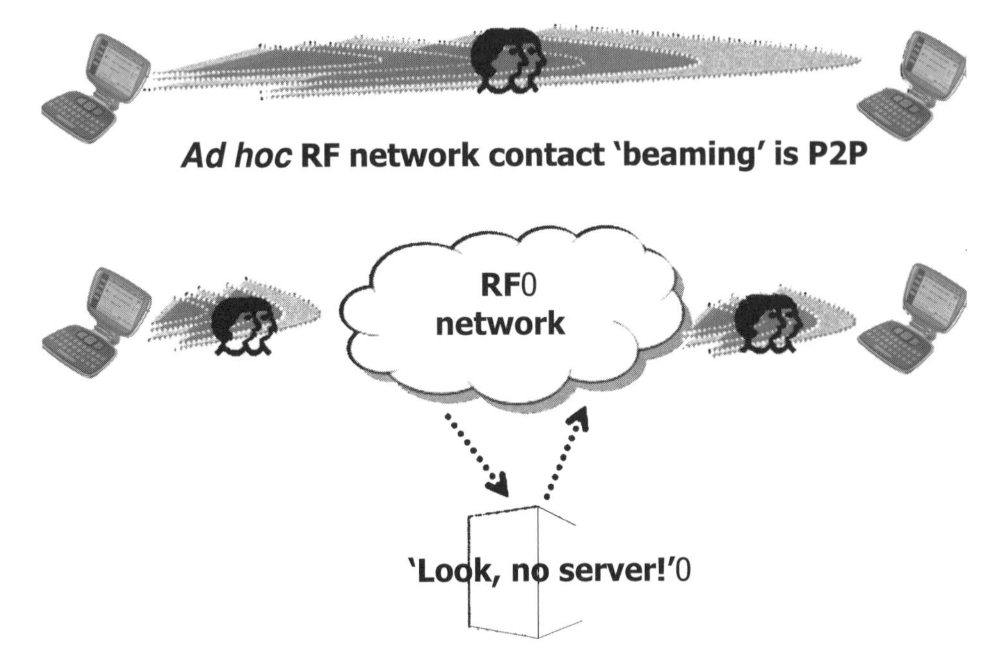

Figure 4.15 Contact 'beaming' H2H interaction using P2P.

nowhere else in the network. This is potentially a complex problem and is the challenge of designing P2P protocols like JXTA[37] (see Chapter 6).

H2H communications is viable only within the SD in the device network, when we are specifically thinking about *ad hoc* networks. Clearly, there is no such concept as H2H communications within a PD, unless we are prepared to generalize to the case of self to self as an instance of H2H interaction, but that's overly pedantic. It is also preferable to retain an unambiguous definition of H2H (namely, that more than one person is involved). We will tackle PD applications elsewhere.

It is tempting to think that the P2P computing paradigm only applies to *ad hoc* networks. This is not the case. The P2P paradigm can work with devices that are physically separated and unable to form an *ad hoc* network, thus implying the involvement of the permanent RF network and devices being connected in their CD mode. This is a perfectly viable option. As Figure 4.15 shows, two devices could interact in P2P mode across a wide-area permanent RF network, and there is no reason for both devices to be RF-connected, one of them could be fixed.[38] Actually, there is no inextricable link between P2P computing and RF connectivity. Any computing device could work in a P2P networking mode, even across ordinary wire line networks.

The key to the P2P computing paradigm within the mobile context is probably the absence of a server. This tends to allow for a more impromptu approach to software collaboration,

[37] JXTA is the open initiative by Sun to develop a set of workable P2P protocols. See http://www.jxta.org
[38] Technically, both devices could be fixed as P2P is a general paradigm and not specific to the mobile context.

such as H2H interaction might entail, irrespective of whether the RF connection is present or not, *ad hoc* or not. However, it may be too misleading to suggest that there is never a need for a server or some kind of intermediary computing platform. We might have noticed that in *ad hoc* networks there is automatically a degree of authentication as the device users (humans) can see each other (presumably) and there is also no need to find the mating device as physical proximity is both assumed and required. However, to mate with a peer device across a wide area network, such as will often be the case in the permanent RF context, then there is a need for mechanisms to locate the mate and quite possibly to authenticate it. Even with physically close devices the need for a proxy function performed by a server may be useful. For example, in a conference meeting hall with lots of attendees the prospect of discovering all the possible devices to talk with may be technically cumbersome and could be assisted by a central server that is already cognizant of the location of devices in the neighbourhood.

We can think of the need to locate as being a type of mobility management, a feature already present in cellular networks, which are required to keep track of where a device is at any one moment in time (while it is switched on). In a P2P session *mobility management* is not the usual P2P lexicon for finding a mate to engage with nor is the RF mobility management responsible for this function, although the similarities are interesting to ponder.[39] In P2P computing the concepts of service and device discovery are talked about, processes that have subtle differences from mobility management, but are essentially still providing a locating function (although it may not have anything to do with devices being mobile, just the information pathways to them being unknown). Our point of reflection is that it may be difficult to perform device discovery and any authentication process without an intervening server, so we should not assume that P2P computing is completely free of this need in general. Clearly, direct P2P communication *is possible* as this is the essence of device pairing in Bluetooth™.

Other modes of H2H interaction can involve users who are on a fixed network or can involve software services that provide the infrastructure for the H2H interaction, such as for IM and videoconferencing. In both cases we expect servers to be involved and we can imagine that, in effect, these servers are effectively the originators of content as far as our actors are concerned. Email is an example of an H2H interaction, although it is not real time and can also be used for H2C interaction, such as subscribing to email alerts that are generated by software services, like a stock market monitor. It can also be used for H2M interaction, such as a home security system sending out status alerts via email.

Any H2H software process that involves brokering, such as the store-and-forward capabilities of an email server or the store-and-indexing capabilities of a bulletin board, can be viewed from the mobile user's perspective as if the other actor is sitting in the content network. There is no cognizance of what is happening beyond the content network. Perhaps an alternative view of the defining characteristic of the content network is that the sources of information are usually going to reside in databases. We don't really care how the data got into the database, whether a human initiated the input (directly or otherwise), a machine or some other computer in the content network, like a syndicated news feed. By generalizing the content network to represent an information-hosting capability – no matter the source, human or automaton – we are able to take a unified view of the architecture of mobile

[39] In fact, in a wide-area RF network the issue of finding a P2P partner device on the RF network is related to the mobility management capabilities of the RF network; but, this should become more apparent in later discussions.

services, and this makes our network model quite useful. We will see later why this view makes perfect sense, when we look at CS software architectures and their prevalence in software solutions. We can think of the client component as being our social/device network and the server as being our content network. Between the client and server are the RF and IP networks, notwithstanding that either of these networks can end up hosting services in their own right to make the CS process more powerful.

In summary, all modes of possible H2H interaction have been shown to be supportable by our network model, and we now need to consider the H2C mode of interaction.

4.8.2 Human-to-Content Interaction (H2C)

As we traverse naturally from the device network to the content network the H2C mode of interaction is automatically suggested by the path we need to take, but we need to examine its nuances.

When we want to grab content from its source, as we have already remarked, it is most likely sitting in a database somewhere in our content network. This would imply CD connectivity for the device and the use of a permanent RF network infrastructure. Most often, this will be the case.

However, the possibility exists for content to be sitting on another device and for it to be accessible in P2P mode, so H2C interaction is not restricted to CS communications in the classical sense. Perhaps a user wants to access their friend's address book, having already gained permission to do so (i.e., a level of trust exists between the owner of the content and the requestor). In such cases there is no restriction on the type of RF connection; it could be *ad hoc* or permanent. I could be requesting access to a friend's address book, and it could be on a device that is located nearby, or anywhere within the wide-area coverage of the RF network or even somewhere on a fixed network. Possibly, it could be all three within one session of H2C communication, for a variety of reasons. For example, in accessing a friend's diary to book an appointment with them the initial session could be established in P2P mode with a nearby device – let's assume that we are both sitting in the same coffee shop. However, after some communications back and forth, my diary application is notified that my friend's 'master diary' is sitting on a Web-hosted calendar service, so my device changes mode and establishes a wide-area RF session to the content network. Of course, we could argue that the device should have gone straight to the central diary in any case, but why should that be the assumption? Also, we could argue that the friend's device would be in constant synchronization with their central diary, thus eliminating the need for the first (P2P) step, but that's also an assumption that needs qualifying. Perhaps someone else had only a few minutes ago added an appointment to my friend's central diary and it had not propagated to his device, simply because his device is out of range and thus synchronization is temporarily not possible and is suspended until coverage resumes. It could be that my friend's device is a WiFi/Bluetooth™ combination device and the coffee shop does not host a WiFi hot spot, so synchronization will not take place until the next hot spot is entered (or by some other means). My device has a 3G connection, and so it is possible to access the central diary (on the Internet) to get the most up-to-date appointments schedule for my friend.

This diary case is only one example of the possibilities of the device dynamically adjusting its communications mode to suit the desired task according to the current mode of interaction. To illustrate this further let us for a moment consider the applicability of multimode behaviour to the previous mode of interaction, which was H2H. In the same coffee shop scenario we may want to hold a three-way text chat session with another colleague or friend who is sitting in an office somewhere else. In this situation *ad hoc* Bluetooth or WiFi networking between my device and my friends is combined with permanent networking over a 3G link, with my device acting as a gateway for the communications between my friend and the other party. An interesting aspect of this scenario is that this configuration is not compatible with standard (current) IM architectures because my friend's device is hidden from the chat server. Either my device could act as a proxy for a standard IM architecture or, using P2P communications, it could act as a relay. In the latter case my device could even deliver the service via a Web interface and enable the entire session to be established irrespective of which IM protocols the other devices support or, possibly, even irrespective of my friend's device to engage in P2P communication.[40]

It is tempting to assume that the access of content requires a similar access mechanism whether it is sitting on a database in the content network or in a smaller database on a peer device. This could well be the case, but the essential difference is that P2P is generally a one-to-one association of user to content source, whereas a CS solution is likely to be a many-to-one association with possibly millions of users trying to gain access to an information source, possibly the same piece of information. Clearly, this dictates quite different considerations in terms of the information delivery architecture. Scalability, availability, security and other such concerns will be more crucial in the CS scenario. We will get to this when we discuss the CS technologies in Chapter 6.

4.8.3 Human-to-Machine (H2M)

The ability to interact with a machine has several possible modes:

- interrogating a machine for its operational data (e.g., status);
- receiving alerts from a machine based on operational status changes (state changes);
- controlling the operational status of a machine via remote control.

As we already noted when discussing the machine network, if the machines are networked using Web-centric software paradigms (i.e., HTTP), then the interrogation of information is akin to pulling content down (pull) from a Web server (content network). However, a machine itself may not be up to the job of providing a scalable or resilient Web interface directly, so one consideration is the provision of Web-based proxies in the content network that provide the direct interaction with the human (device), whether by CD or PD, as shown in Figure 4.16.

The advantages of going via a front end in the content network include:

- enabling a more scalable architecture to be realized;
- potential to use powerful Web-serving technologies (e.g., J2EE[41]) that are not easily integrated into a machine;

[40] For example, if my friend's device is able to surf the Web, then my device can 'become' the Web as far as his browser (device) is concerned and thereby allow for direct P2P communication to be established.
[41] Java 2 Enterprise Edition (J2EE) is a set of technologies useful for implementing massively scalable and robust applications, particularly those that support the CS mode of interaction.

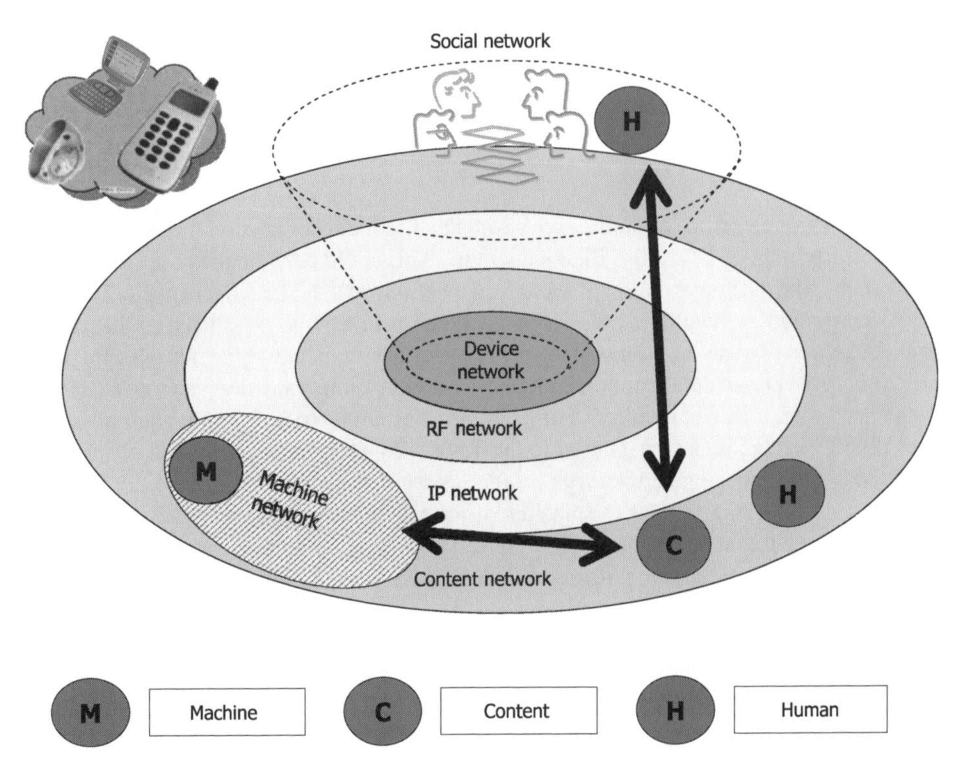

Figure 4.16 H2M interaction via proxy.

- easier to control security issues, such as authentication and encryption;
- possible to integrate access to the machine into a wider service portfolio (e.g., mobile portal);
- potentially lightweight protocols can be used for the machine-to-proxy interface, thus lowering the costs of implementation (including the link costs to get the information from the machine to the proxy).

In terms of linking the machine to the proxy the opportunity exists to use wireless networking, and often this will be a more convenient option, sometimes the only option. For wireless-enabled machines then, we have several possible modes:

- Remote connection via the content network using RF infrastructure. An example of this is an alarm that connects wirelessly back into a central control centre which itself sits in the content network.
- Remote connection via the RF network using RF infrastructure. An example of the ability to directly dial up an RF modem embedded in a vending machine and then connect to its embedded computer. This is the more standard means of remotely accessing machines like vending machines. This mode can also include the use of text messaging to either send control messages to the machine's embedded computer or to receive alerts, or even to pull status messages using auto-reply text messaging.

- Local connection via the device network in *ad hoc* RF mode. An example of this is using Bluetooth™ to connect a PDA to a local printer in order to print out an email. There are plenty of other examples of this 'remote control' method of networking, such as control of the television, automatic garage door, electronic fob entry to premises and so on.

H2M interaction need not be limited to CS mode. The local connection method just mentioned clearly implies a type of P2P connectivity. This could be widened to permanent RF networking, like the above option of remote connection via a modem, but using an IP-based networking solution. In other words, if the machine sits in the Internet, then a mobile device could talk directly to the machine using an IP protocol, without any server necessarily being involved. In the classical P2P architecture the communications pathways can move laterally to the next peer, if this is required. For example, a group of friends might each possess a hard-disk TV recoding device, similar to the Tivo.[42] If a program were missed, then one of the friends could interrogate his own recorder to see if it managed to record the program automatically. If it didn't, then the machine itself can contact a peer recorder in the group of friends to see if that device has the program. This in turn could interrogate a peer, and so on. This is the P2P technique witnessed in such file-sharing applications as Gnutella and the infamous music-sharing service Napster.

With H2M connections via the appropriate RF technology the possibility also exists for real time communication. For example, a nightwatchman for an accommodation block (e.g., apartments) could monitor a door entry system that has a camera and two-way audio intercom. Ordinarily, the watchman might be stationed at the guard's desk where the desktop intercom terminal is situated. However, if the watchman needs to leave his desk post in order to patrol the building, then the intercom function could be enacted as a mobile application on his mobile device. The presence of a visitor, announced by a push button, could cause the mobile device to be alerted and for the remote monitor application to be initiated. This would receive a live video feed from the door camera and connect with the intercom via a two-way audio channel. Speaking into the mobile device would cause the audio to be emitted from the intercom speaker. What's more, if the guard decides to allow entry, then a command issued from the mobile device could trigger the electromagnetic latch on the door, thus allowing it to open.

This type of application is a powerful example of what H2M interaction can achieve within the mobile context. What makes this possible is a variety of technologies. The key, as in all H2M or M2M solutions, is in the design of the machine interface itself (i.e., the interface to the network). This does tend to necessitate a novel type of machine hitherto not seen in the markets. But, the advent of ubiquitous mobile technology will surely promote these types of developments: the enmeshing of semi-intelligent and interactively responsive machines into 'the network'. The seamless transition from the RF network to the IP network seems a key to such services and applications. Matched with a seamless interface between embedded computers and the IP world then, it is all the more likely that we will see H2M applications emerge with increasing frequency. A key contributor to the trend will be the possibility to directly plumb these remote machines into the RF network using embedded RF technology, thus avoiding the grisly installation process and physical networking headaches associated with the plumbing of structured cabling and so forth.

[42] http://www.tivo.com

In summary, the network model we have established seems to support the various modes of H2M interaction, although there may well be more that we haven't explored – and some of these may emerge in further examples throughout the book.

4.8.4 Machine-to-Machine (M2M)

M2M communication is increasing viable as more and more goods leave manufacturing lines with computers embedded within them – computers able to engage with networks, should they be programmed to do so. The applications of embedded computing are constantly expanding, and it is easy to overlook the fact that most of the computing devices in the world are hidden from view, buried away in some microwave oven or hidden behind a panel in an elevator. Building control is an increasingly vital means of enabling sophisticated living and office quarters to function. Most buildings these days do not function without computers. Waterworks, elevators, fire systems, heating, windows, lighting and myriad other building utilities are pulsating with the hum of built-in computers amid a fabric of sensors, actuators and controllers.

The M2M component is frequently seen as the ability to tie all these embedded computers into a network so that they can connect to IT systems (i.e., other machines), housed locally or remotely, and that they have good reason to interact with the embedded machines.

So, what is the role of wireless in M2M systems? There are two considerations for the use of wireless:

- A cost-effective and convenient means to bring static machines into the reach of an IT system. Essentially, this is thinking of wireless as a wired line replacement, in order that a machine can be accessible to the nearest network infrastructure. The wireless link can be either short range, such as a WiFi connection, or long range, such as a GPRS or 3G modem – it depends on the physical location of the machine. Either way the use of wireless to integrate static (immobile) machines is the more conventional solution that is being seen with increasing frequency, especially now the price of embedding RF (of one form or another) is decreasing.
- The other consideration for wireless is looking at how to exploit the mobility (or roaming) aspect of RF networks. There is no reason why we should not, with today's ubiquitous technology, enable any machine to become network-enabled, even a moving machine, such as a car engine management system or a forklift truck. This reminds us of the 'movement' attribute in Ahonen's five Ms. We are now applying 'movement' to machines, whereas our previous emphasis was on the movement of users themselves (which is still very relevant of course). Within this category, we can apply 'movement' to previously static machines too, which may lead to interesting application possibilities. For example, a medical diagnostic ultrasound machine that is portable could be networked to allow previous patient images to be retrieved from an imaging archive in the content network.

A prime example of an M2M application is the tracking of assets through a delivery service, something that has become fairly commonplace with most courier companies. An asset is tagged at the beginning of the delivery process and assigned a unique ID. Usually, a barcode is attached to the packaging and the ID is encoded in the barcode or associated with it. At each stage of the delivery process the appropriate courier agent scans the barcode. The ID is given to the customer who can enter it into a Web-based application to track the whereabouts of the package. When the package moves through a transit warehouse it is easy to scan the barcode and for the location of the package subsequently to be ascertained, such as 'at Heathrow airport'. In the past the challenging part for the courier companies used to be tracking the final delivery to the destination address. However, using a wireless terminal (machine) the delivery agent is now able to get an acceptance signature and report the final delivery (and the signature) back to the courier's IT system (machine).

An alternative to barcode-based tracking is to use an RF device to constantly track the package by remaining itself with the package. This is made possible by the location-finding potential of the RF network, such that the RF device's whereabouts can be regularly interrogated. Clearly, it would not be cost-effective to put an RF device into each and every package in transit. For valuable assets this might not be a problem, and the customer could be charged extra for the service. Otherwise, the other approach is to ensure that the packages themselves are always travelling in vessels that have integrated RF-tracking technology. This could be the delivery van itself, for example, which could have an RF device mounted somewhere on its body. Parcel carrier palettes with integrated RF-tracking devices are an alternative. Tracking of goods need not be restricted to delivery services. Any asset that needs tracking could potentially be tracked using an M2M solution. Perhaps equipment hire companies would like to add RF tracking to their equipment, either directly or to the equipment cases. Even people can be tracked using this approach, and we talk about this when we look at location-based services in Chapter 13.

4.9 OPERATOR CHALLENGES

In the final section of this chapter we now turn to the challenges facing mobile operators in order that we can take these into consideration when constructing an approach toward building mobile services or toward building the underlying technologies that the services utilize.

Figure 4.17 shows us the 'old world' view of the network operator, which stemmed from a very simple business model. The operator builds a cellular network and sells mobiles that work on the network. The owners of the mobiles get billed for using it in a very straightforward manner, such as charged according to time spent using the network (i.e., 'talk time').

Figure 4.18 show us the 'new world' view that the operator has stumbled into. Suddenly, there are a lot more things to cope with. In terms of the key challenges facing the operator the following list is worth reflecting on:

- *give users something interesting to use and pay for*
  - not necessarily what they 'want' (as they don't know what it is they want yet! – creativity is required!),
  - compelling enough to generate revenue;
- *balance service delivery and subscriber demand with capital expenditure on the service delivery platform*;
- *cope with uncertainty in next generation*
  - content and service types,
  - pricing models,
  - consumption patterns;

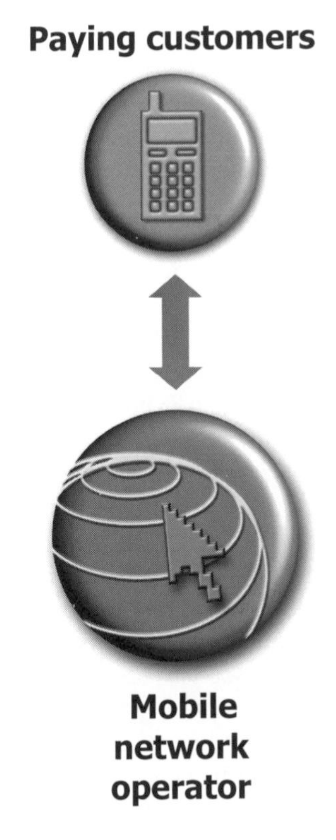

Figure 4.17 Operator 'old world' view.

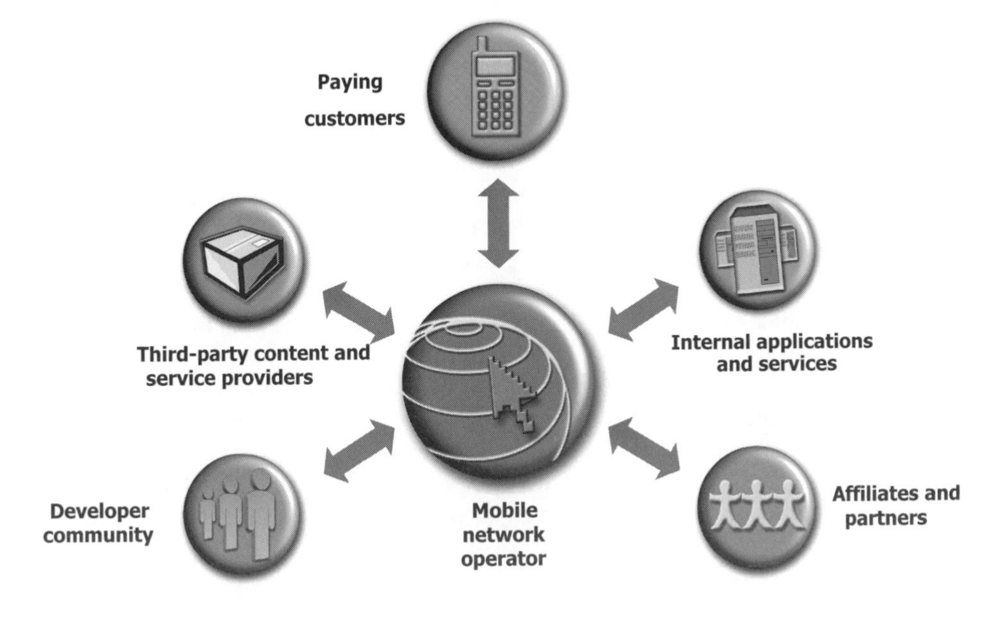

Figure 4.18 operator 'new world' view.

- *providing a rich, 'open', end-to-end service package*
  - building a third-party service platform that enables any application to be launched from the network, meeting both the operator needs and the third party's,
  - embrace a technological approach that the developer community will welcome,
  - reach the 'event horizon' of a leading IT company, not a utility company;
- *entering into 'networked business' models with*
  - affiliates,
  - third-party service providers, and
  - even competitors.

These challenges are reflected in the networked business diagram shown in Figure 4.18. Without the need to discuss the above points in detail we can summarize the essence of 'the problem' facing operators by focusing on the first point and the last two points, keeping in mind that if any of what we are trying to do involves ubiquitous service coverage, then more than likely these problems directly impact on our ambitions to succeed in this industry.

The first point is really another way of saying that we don't yet have a killer app (see Chapter 3 for a discussion on killer apps and killer cocktails). Granted, we did already admit in Chapter 3 that the very notion that a killer app exists is probably misleading, but, nonetheless, whatever service or services are going to transform the mobile market, currently there is no indication as to what they might be. The reason that they don't yet exist relates strongly to the last two points in the list of challenges, which also point the way to possible solutions. One of the major challenges is technical in nature, while the other is commercial and perhaps the most challenging of all, if not pivotal, in determining how this industry might move forward.

Given that we don't know what the killer app is and we don't yet have the killer cocktail either, the sensible approach is to ensure that we at least maximize the chances of either emerging. Contrary to what some of us might think about developers being the vanguards of moving this industry forward, this responsibility clearly lies on the shoulders of the mobile industry (operators and equipment vendors) and *not* the developer community or its entrepreneurial commanders. This inevitable conclusion is reflected in the last two points in the list of challenges. The first point really says that the operator has to provide a 'platform' on which it is extremely easy for competent developers to launch services. It is not a chicken-and-egg situation. The platform needs to be built, and this book is really in essence an exploration of a set of platform technologies that can create a fertile environment within which creativity can thrive. This platform has to be both rich in features and open. We should qualify these two qualities before moving on to explore the final link, which is the commercial basis for gaining access to the platform.

We should think of the entire operator's network as a 'service delivery platform', which is my preferred usage of this term, no matter its various emergent definitions or common usage[43] in the industry (as discussed later in the book). The richness should be reflected in the ability to gain programmatic access to a diverse set of the network's capabilities, such as text messaging, picture messaging, micro-payment, billing, location, call processing, games vending, voicemail and just about any useful asset that the operator has in its network, which

[43] The term 'service delivery platform' doesn't really have an accepted common usage yet anyhow, but is being pursued by some companies as a marketing tool to push their wares.

most definitely extends to include all devices running on the network. The openness should not be mistaken to mean 'open source', which might well happen. It really means *open access* to the platform. One aspect of openness is adherence to IT technologies that are widely used in the industry, avoiding as much as possible the temptation to introduce new incarnations or, worse still, entirely new approaches altogether, unless these are agreed on by the *entire mobile industry* by due and open process. It is tempting to go as far as saying that this restriction should be a tacit creed for all operators. Part of this openness means the use of open standards *as much as possible*, such as the current trend toward using Java-based platform technologies.

The last point in the list of operator challenges is a critical one, perhaps the most essential. No matter how easy it might become to develop, test and deploy a mobile service on the operator's network, unless the means exists to establish a meaningful and productive commercial relationship without friction, then the technological advantage of a feature-rich and open platform is likely to be of little consequence. Whichever part of the network platform that the service provider needs to access, there has to be an easy way of gaining *commercial access* to it that is economically viable for both parties.

On this point I return to the vision that Dick Snyder and I had many years ago when we first discussed the Zingo wireless portal that my company built for Lucent Technologies (see the mention of Zingo in Chapter 2). At the time our idea was rejected by operators and vendors alike who were still thinking in the 'old world' view of the classic value chain model that was regurgitated ad nauseam in analyst reports of the time. The reason for rejection was that our 'new world' view turned the classical value chain model on its head, which in itself was nothing radical because we were only aping what we saw emerging on the Internet.

While we may be tempted now to think that 'new economy' ideas sunk with the dot.com bust, the truth is probably far from it. It is my view that what Amazon has done with its marketplace affiliates scheme is what operators need to do with next generation networks. Just as Amazon gives access to its marketplace, which it heavily invested in, built, advertised and attracted the customers to in the first place, operators need to give access to their *marketplace*. Note the emphasis is on marketplace, not network. Giving access to the *marketplace* is completely different to giving access to the network, although both are clearly related. As a marketplace affiliate with Amazon, I am *one step* away from gaining access to a customer and from them gaining access to me or my 'shop'. Amazon brings the customer onto the network, but when it comes time to purchase the goods Amazon thrusts my wares and my shop to the fore. All this takes place at lightning speed and with dazzling efficiency and Amazon doesn't even care nor know what product I am selling, just that Amazon will receive their cut, whether I sell one product or thousands. The model is amazingly simple, but not without considerable technical investment on behalf of Amazon and a real commitment to *networked business*. Despite all the amazing Web technology used to build the Amazon platform, the most effective part of the implementation is the commercial process. If you have not already done it, I seriously suggest signing up to become an Amazon affiliate just to see how easy it is to become a business partner with Amazon. Try doing the same with any operator. Or, contrariwise, imagine Amazon asks its affiliates to spend months gaining commercial access to what amounts to a single software component to allow me to put items in an Amazon user's shopping cart, but that's all, and I have to build the rest of the business infrastructure and relationship myself! Try being an affiliate on that basis. But this is what operators currently expect of all but the paid-up-by-the-millions, strategic mega-partners.

Until and unless this changes the killer cocktail will not emerge. Granted, the platform and level of affiliation, even its nature, might be different for a network operator than for Amazon, but the essence is what has to be replicated, and this should be a major focus of network operators, the main areas of creative thought, consultancy, analysis and investment: *open networked business*! This book can only describe what lies underneath such a business model (namely, the technologies used to build the *operator affiliates platform*). Returning to the apparently aloof allusions of Chapter 2 – 'build it and they will come!' – what is it that should be built? Well, an operator affiliates platform or service delivery platform, call it what you will, is the answer. 'They' are the affiliates.

5

IP-Centric Mobile-Networking Power

This chapter is about the Internet and the role it plays in our mobile services model. Having introduced our high-level network model of the mobile services universe, the internet may seem a curious place to start. We could have addressed devices first, perhaps the wireless network, or maybe even the content network, having stressed its importance in previous chapters. We are going to start with the Internet, particularly the Web, due to its particular significance in influencing the shape of next generation mobile services technologies.

If adopting a more conventional approach, I would probably proceed with explanations of the Internet and then WAP (the Wireless Application Protocol). However, I reiterate that our approach in this book is to understand the connectedness of the mobile services cosmos. There are now many interesting things happening in relation to the Internet. It has moved quickly to provoke many new ideas to improve its networking power, which is at its heart. It is important to understand the source of this networking power and the new initiatives, so that we can transpose their significance to the mobile services context. In the end this will help us to understand how to enable interesting mobile services to emerge. In later chapters we will indeed examine some of the protocols and related technologies in depth, but for now we start at a high altitude and slowly descend into the critical details.

5.1 NETWORKING POWER

We have all heard the mantras about the Net – it is no less than the new creed – that in the Web-enabled 'new economy' the Internet is the new way of doing business: go online or die. All our affairs have to be 'netified' or else risk irrelevancy in the new epoch, or so we're told. The creed has as one of its tenets that the free flow of information on the Net is potentially the most pure form of economic and social interaction possible on

Next Generation Wireless Applications P. Golding

ISBN: 0-470-86986-0 (HB)

the planet, and, consequently, the informed and involved participants are empowered to engage in new levels of human interchange.[1] The Net can empower the expression of the individual. Some pundits commend the Internet as the most level playing field for presenting ideas and exchanging information – giving rise to all kinds of emergent, socio-technocratic themes like *netocracy*[2] to reflect a transition from capitalism and democracy to a new economic, sociological and political paradigm. Other theories like emergent democracy[3] are also attracting the attention of those who like to ponder on the wider societal-level significance of the Net. Irrespective of my view or yours some of the ideas that provoke such reactions are worthy of attention, especially if we are seeking to find hitherto untapped pre-echoes of a killer app. Our technical consideration of the Internet within the mobile context will not exclude these ideas, although we do not scrutinize them in this book.

In some ways the era of the Net marks the arrival of a golden age of IT, notwithstanding the IT crisis in 'the West'. I do not mean the fool's gold of the dot.com age. I mean a golden age in terms of what early computer enthusiasts once dreamt about the future of computing, the pervasive computing ideal, the future of science fiction books, everyone plugged in, switched on and able to command information wherever and whenever they want and about whatever they want. We dreamt of permanently connected wristwatch computers plumbed into a massively rich and intelligent[4] network. This no longer seems far-fetched. I think by the time you finish reading this book you will understand the potential for new computing paradigms in terms of new ways of benefiting from computing that we have not had access to before, mostly courtesy of the conjoining of Internet technologies with mobile technologies.

Despite being an area of ongoing and intense debate, we should not be in any doubt that the Web does herald a golden age of computing that has the potential to transform aspects of social and business interaction across many societies. The vanishing mirage of the dot.com shareholder frenzy should not detract from the core value proposition of the Web, which seems to be embedded in its networking power (see Box 5.1: 'The power of networks').

Box 5.1 The power of networks

Charles Boyd[5] identified three maxims that affect how businesses will have to manage their activities for competing in the Internet era, but these truisms are generally applicable to all aspects of networked life and apply to mobility. Boyd mentions Moore's Law, Metcalfe's Law and Coase's exposition on business transactions.

Moore's Law Gordon Moore is founder of Intel Corporation. He elucidated an observation that in its revised form became known as Moore's Law.[6] In it he states that every 18 months, processing power (of silicon microprocessors) doubles while cost

[1] Notwithstanding that a good deal of Internet currency still seems to be puerile spam, porn or Viagra.
[2] The netocracy, say pundits Alexander Bard and Jan Söderqvist, will be the new power elite, controlling networks – both social and digital – and displacing the bourgeoisie as the ruling class. See Bard, A. and Söderqvist, J., *Netocracy*. Pearson Education, London (2002).
[3] http://joi.ito.com/joiwiki/EmergentDemocracyPaper
[4] Although the Internet has nothing to do with computing intelligence and this largely remains an elusive goal.
[5] Charles Boyd is teacher of strategic management at Southwest Missouri University. I found his website had many interesting articles on digital economy issues. See http://www.mgt.smsu.edu/Boyd/index.htm
[6] http://www.intel.com/research/silicon/mooreslaw.htm

holds constant. Moore's perceptive insight has proved to be generally true and remains so for the foreseeable future. Telecommunications bandwidth and computer memory and storage capacity are experiencing a similar fate because they are heavily reliant on silicon processing devices.

Mobile devices are no exception. It is not widely appreciated how much processing power is required in digital wireless modems. Some people may think that wireless networking is improving because of advances in the underlying mathematical theory of radio transmission. Generally, that is not true. It is more to do with the ability of modern chips to affordably implement what wireless theory has predicted for many years. At the time that GSM (Global System for Mobile Communications) handsets were launched they had more processing power in their tiny packaging than was available in state-of-the-art PCs, which at the time saw the introduction of the newly designed and exciting Pentium chip (P60).

This makes it very affordable for individuals and businesses to be equipped with the electronic means to network easily. The price threshold for ubiquitous communications is within the reach of consumers, even very young consumers.

Metcalfe's Law Robert Metcalfe designed the Ethernet protocol for computer networks. Metcalfe's Law states that the usefulness or utility of a network equals the square of the number of users. This principle is a combination of a straightforward mathematical fact and theory. The factual part is simply noticing that in a network with N nodes that are able to connect to each other, then the number of connections possible per user is $N - 1$, thus making $N(N - 1)$ total connections:

$$\textbf{Total network connections} = N(N-1) = N^2 - N$$

The theoretical aspect is in assuming that some value can be given to a user by utilizing any of the possible connections open to them. It doesn't really matter what value in particular, but the benefit of the aphorism is in noting that the usefulness of the network increases in proportion to the square of the number of users.

The standards factor We can see that mobile telephony is a combination of Moore's Law and Metcalfe's Law, working to make mobile life possible. As already noted, it was Moore's Law that made mass consumption of mobile telephony possible. As we will see later in Chapter 12, when we discuss the RF (Radio Frequency) network, the ability to support large numbers of mobile users in a particular geographical area is only possible using digital communications techniques that require huge amounts of processing, both in the devices and the infrastructure (base stations). The realization of Metcalfe's Law in mobile telephony is through the adoption of standards, such as GSM, so that we end up with a large network rather than lots of smaller ones, thus N is able to scale quickly and we end up seeing the benefits. In GSM this has been driven by the decision of most operators in the world to adopt GSM, so Metcalfe's Law is given global momentum through the squaring of massive numbers.

It almost does not need stating that the Internet has clearly benefited from Metcalfe's Law – it is the archetypal example given when discussing Metcalfe's Law. So, clearly, if we have global wireless data standards for accessing the Internet, which we more or less do through 3G (third generation), then we have a very powerful network indeed.

This is our justification for putting the 'mobile Internet' as our centrepiece of the mobile network model (i.e., the Internet Protocol, or IP, layer).

We can sometimes see the effect of the square law as it becomes obvious when we pass some point of inflexion where the rate of increase has a 'snowballing' effect. In GSM this became obvious with the use of text messaging in many countries. Once the various operators within countries decided to allow inter-networking of the Short Message Service (SMS), its usage boomed.[7]

Transactions As we have already discussed earlier in the book, the ability to network can be utilized in several ways, reminding ourselves here of the human-to-content (H2C), human-to-human (H2H), human-to-machine (H2M) and machine-to-machine (M2M) modes of interaction. It may have been in our minds, during the above discussion of networking maxims, that networking is all (and only) about people connecting to other people. But, that's only part of it. We should not forget the other interaction modes identified in Chapter 4. Clearly, if there is value to be gained from interacting with a machine, then the addition of machines to the IP network also increases the value of the network. Some machines may already be networked, in which case we are combining the power of networks.

Connecting, say, M networked machines to N networked people we end up with a combined value $N^2 + M^2 + 2NM$. One example that immediately springs to mind, and is a network that we often overlook (as we take it for granted so much), is the Visa™ network, which is a network of machines. Combining this network with the Internet brings enormous pooled value.

Financial transactions are only one type of value attainable in a network, so we should not think exclusively about financial interaction, but nonetheless it is clearly an important consideration.

Ronald Coase won a Nobel Prize for his explanation of transaction costs in a 1937 article titled 'The nature of the firm', where he concluded that firms are created because the additional cost of organizing them is cheaper than the transaction costs involved when individuals conduct business with each other using the market.

Moore's Law and Metcalfe's Law have made many types of transactions cheaper. This trend is sure to hasten because digitization has cut transaction costs all the way to zero in some cases. Once a product or service is in place to be sold on the Internet, for example, there is zero cost for each additional transaction. The advent of marketplaces on the Internet, like Amazon and eBay, has brought amazing transactional power to the individual. This model is now appearing in the mobile operator landscape with similar marketplaces, such as the market for downloadable Java games. I-Mode in Japan has already harnessed the power of the marketplace by its revenue-sharing model, enabling thousands of producers to sell their wares into the I-Mode user base (network). We should recognize that in both cases the power of the network is the combined power of standardized devices and the billing network of the operator, enabling charges to be levied and shared.

[7] There were other factors lending a hand toward the success of text messaging, such as price factors and lower cost of access by the introduction of prepaid mobile phone accounts, but the power of inter-networking agreements soon became obvious.

To understand this and its significance for mobile services we should reflect a little on what networking is all about. Box 5.1 helps us to understand the general principles of how access to networks has the potential to add value. However, the analysis can leave us detached from the purpose of networks. After all, a network is not very powerful if it has millions of users who simply plug in and do nothing thereafter. This would be like the electricity grid (which technically we could call a network, at least physically).

Clearly, we are interested in *communications networks*, so the power of the network is in its ability to support communication or *information exchange*. Thereafter, the users can decide for what reason they want to communicate and what they want to say. But, what determines the value of communication? We can explore some ideas in order to see if we can gain insight that will help us to better understand the power of mobile networking.

5.2 THE 'ME' FACTOR IN NETWORKS

In our earlier discussion in the book on the success factors for mobile services we entertained Ahonen's five Ms as useful axes for identifying value: movement, moment, me, money and machines. A key attribute was the 'me' factor, which we said was the degree of personalization made possible by the service or the degree of personal relevance versus general relevance. Let's look at the role of personalization in determining how valuable networks really are.

First, let's take the example of business or social networking to illustrate some useful principles. We know that with business networking the adage 'It's not what you know', it's who you know' tends to be true. We understand that value often lies in knowing someone who can do something for us, which may involve introducing us to someone else (i.e., more networking).

Knowing people is the essence of social and business networking, but knowing particular people is even more essential. The realizable value of our network is more to do with the actual connections that are useful to us. For example, if we are trying to find a job in a particular line of work, then the only connections in our network that matter are the ones to people who can connect us to openings in our desired field of work. This tends to suggest that the power of the network is not related to its total size, but by the number of relevant connections. We may also find that, irrespective of what we are trying to achieve, certain people always figure more strongly in our network and we gravitate toward our connections with them.

How do people find out about jobs? Are relatives, close friends, acquaintances, employment agencies or newspaper advertisements the most useful sources about jobs? Mark Granovetter, a sociology graduate student, tried to answer this question in his PhD research. He found that, most often, people found out about jobs through acquaintances. The information people possess about job possibilities is affected by their placement within social networks, not by access to job advertisements.

Granovetter found that many of the paths between the job and the new employee were surprisingly indirect. Moreover, not only were the paths surprisingly long, the person who told them about the job was also often someone they did not know very well. Granovetter named this as *the principle of the strength of weak ties*.[8] This tends to suggest that if we want

[8] Granovetter, M., 'The strength of weak ties'. *American Journal of Sociology*, **78** (May 1973), 1360–1380.

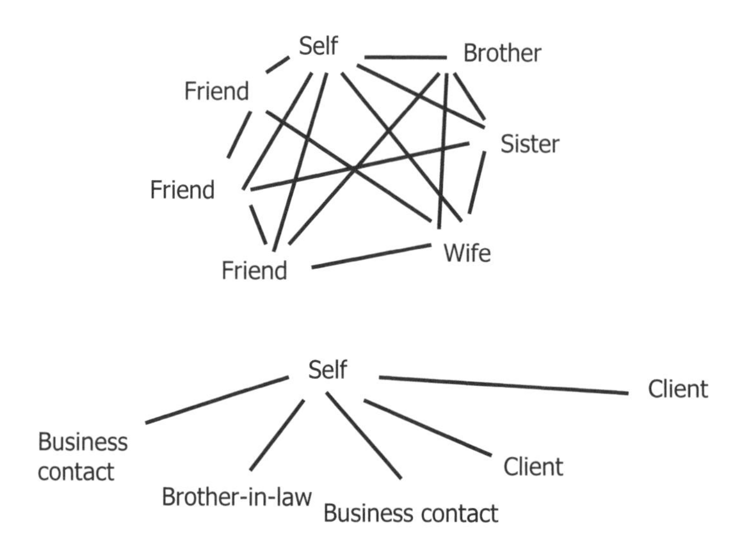

Figure 5.1 Weak ties and strong ties in social networks.

to find a job, then we are better off if we are well placed in large social networks *with people we don't know that well.* This reinforces the idea of 'whom you know', but demonstrates that our ability to participate in large networks in the first place should add value.

Granovetter was not able to explain conclusively the reason that weak ties had more value, but he offered an elucidation that has some merits and may be useful for our consideration of networks in mobile environments, as we will see later.

Granovetter posited that close ties, where we ordinarily expect more value due to their greater willingness to help (among other reasons), are also likely to be connected to each other. However, casual acquaintances are less likely to know each other. This is shown in Figure 5.1. The benefit of disassociated contacts is that they are likely to belong to their own networks and that information circulated in each network is more likely to be distinct within each network. In the close associates' network, information is more likely to be redundant as each contact has likely already heard it from someone else in the network. In other words, we will gain more information from loose associations as an aggregate than close associations.

The principle of the strength of weak ties suggests that it is going to be very useful for a user to be able to join networks easily and for conditions to be favourable for groups to form and to become manageably large. Clearly, any mechanism by which we can belong to a wider set of networks through weak ties, the better our chances it seems of gaining access to value. The Internet certainly offers this potential and ability. Currently, mobile networks do not offer this ability *intrinsically*, but they do offer the potential, as we will examine.

One aspect of weak ties that seems important is being able to discover who knows what or who knows whom. Discoverability, as we will see, is an important part of the networking process, and we will need to address this issue, especially within the wireless context where it possibly poses more challenges. We will return to this topic later when we look at how users are identified in mobile digital space. This is discussed in Chapter 13 on location-based services (LBSs), which not only discusses LBSs but also acts as a catalyst for fusing some of the ideas currently under review as well as other themes in the book.

One approach to discoverability is the ability to form or join networks that are appropriate for our particular needs. For example, if we are looking for a job within a particular branch of engineering, then perhaps we are better off connecting with a network that has a higher degree of association with that branch. Clearly, this should or could be a more valuable network, and having found it then the bigger the network the better, due to Metcalfe's Law (see Box 5.1: the power of networks), assuming we have suitable tools and mechanisms for tapping into the network's informational power.

A network like this, which reflects our specific interests, is as if we have our own *personalized* network and falls neatly into the 'me' attribute that we were looking for in defining useful mobile services. Therefore, in the provision of networking power via a mobile device we need to take into account the ability to form and join specific networks. This presents all manner of issues, not least of which is how to discover personal networks or even how to form them.

We should not confine our understanding of personal networks to the social domain or the H2H interaction as discussed in Chapter 4. The adage 'It's not what you know', it's who you know' may no longer apply. In the networked world it may well be possible to access other people's information without their *explicit* consent or involvement; something we will discuss in Chapter 13 on LBSs. Using peer-to-peer (P2P) architectures it may also be possible to traverse personal networks that are associated by friend-of-a-friend connections: 'It's not who you know, it's who they know.' Irrespective of the means of locating the information, in a connected world H2C interactions are also important. Here, the importance of the content is in its value to the task in hand, which in a social-networking context could be all kinds of information. This reinforces the reflection in Chapter 3 that the nature of content is far from the traditional view that emerged in the dot.com era, which is the broadcast view of content as information with wide appeal, like many forms of entertainment-related information. For many users of communications networks the potential may lie in the ability to discover a single piece of content that has maximum relevance to the seeker and has potentially life-impacting qualities.

'Content is not king'[9] is a paper by Andrew Odlyzko is which he argues that:

1. The entertainment industry is a small industry compared with other industries, notably the telecommunications industry.
2. People are more interested in *communication* than entertainment.
3. And, therefore, that entertainment 'content' is *not the killer app* for the Internet.

This observation would seem particularly relevant to the nascent mobile services industry. The main nonvoice applications for mobile phones are probably text messaging and ring tones. Text messaging is a form of communication and most of the content is personalized. Text alerts from information sources only accounts for a small percentage of text-messaging volumes, which are measured in the billions for the UK market alone.[10] Ring tones are a form of entertainment, but probably more accurately thought of as items of fashion and self-expression, albeit of a highly limited form. The most popular use of premium rate[11]

[9] http://www.firstmonday.dk/issues/issue6_2/odlyzko/
[10] http://www.text.it
[11] Premium rate messaging is also called reverse-billed messaging, which is where the user pays to receive a text message, usually for a much higher price than sending one, such as 100 pence reverse-billed versus 10 pence to send (or less).

messaging in the UK is dating. In other words, the mobile data market is already social network-centred, focused on the individual's need to communicate, even in terms of outward self-expression (e.g., ring tones, phone fascias, etc.). This has been verified by numerous studies.[12]

If we take two mobile phone users today, how do we know anything about them beyond their phone number? We probably don't, so it is not even possible to instigate a discovery mechanism, as there is no information to support discovery, just phone numbers and, at best, the physical addresses of users. However, what we do have is an extremely powerful resource indeed: the database of call records since the mobile networks started. This tells us who is talking to whom and can also be used to find groups with low transitivity.[13] This may or may not be useful as it is, but combined with other information and the power of networking capabilities deliverable via an IP network we might have the ingredients for exceptionally powerful networks that can provide the basis for useful marketplaces for all kinds of transactions.

What we are beginning to realize is that a network, to be powerful in adding value, should be capable of supporting personalized networking. It should allow groups of users with similar interests to cluster together and for their collective interests to be supported and magnified by the network (what Rheingold[14] called 'swarming'[15]).

To take another example, let's consider for a moment the apparently benign task of looking up train timetables. This is very much an H2C interaction, but it has the potential to benefit from social-networking power and become an H2H interaction. If all we are interested in is standard train times for the local train journey to the nearest city, then probably we can access networks with strong or weak ties to find this kind of information. However, if we are planning a train trip from London to Beijing and would like to know what is the best route, times, costs and related advice, then we are suddenly interested in accessing networks with weak ties. Actually, we need access to very specialized personal networks, such as people with real experience of the matter, people who love travel or, even better, people who have travelled to China or know of such people (via friend-of-a-friend association).

Let's say that we are already travelling through Europe, nearing its borders with Asia, and we want to know how which train to catch next. At that moment in time we would benefit from gaining access to a network that can facilitate this information for us. We are on the move and what we would dearly like is to be able to call someone who can advise us. But who should we call for advice? With a mobile device we have the ability to access networks wherever we are and whenever we like, possibly very powerful networks if we also have access to an IP network. If we have the necessary mechanism to discover an appropriate social network within our mobile network, then we could conceivably place a call (or possibly a few calls) and stand a high chance of ending up speaking to the right person.

[12] See Katz, J. and Aakhus, M. (eds), *Perpetual Contact – Mobile Communication, Private Talk, Public Performance*. Cambridge University Press Cambridge, UK (2002).
[13] These are groups with loose ties.
[14] Rheingold, H., *Smart Mobs – The Next Social Revolution* (pp. 174–182). Perseus Books, Cambridge, MA (2002).
[15] I don't much like the current obsession to use animalistic terms to describe human behaviour. Human qualities are often overlooked in the pursuit of technological progress, where human involvement is usually confined to consumption. Not much is said about other aspects of human need. Given the record numbers of text messages sent in the UK on Mothers' Day and Valentine's Day, perhaps we should refer to the collective messaging phenomenon as 'huddling' rather than the distinctly bestial term 'swarming'.

Taking this a step further, imagine that the person with the information has not only a means to share it but also an incentive to do so, such as a means to charge for their assistance. Is this the basis for a mobile marketplace for information? This is a passing thought, but one worth returning to because in the mobile world the micro-payment mechanisms are already in place to cope with such a model (unlike in the Internet world).

Regardless of the actual mechanisms used a mobile device is essentially and intrinsically a networking tool, and so it would be interesting to recast our mobile services definition to reflect the networking aspect more clearly, giving it special emphasis for the current discussion.

Revised mobile services definition

The ability to add value to our lives through networking while freely moving anywhere we are likely to go in conducting our usual day-to-day business and social lives.

Or (shortened version):

The ability to network and transact at any time and from any place.

5.3 THE INTERNET

The Internet is a universally accessible network of information, people and machines and represents a significant step forward in the wireless science fiction future; so, we need to understand it in more depth.

We should understand the essence of the Web. We should also understand and appreciate its impact on computing. This is because the Web will undoubtedly figure strongly in facilitating interesting and significant mobile services. First, Web protocols are prevalent and easy to integrate with, so it is relatively easy to build mobile services using a Web-centric model. However, probably more significant is the way in which the Web offers a huge amount of networking power (i.e., its appeal to mobile service developers). Many observers are coming to similar conclusions that the 'killer app' for mobile services is something to do with interpersonal networking or communications, not the consumption of content, which has been a traditional assumption that still echoes from the early dot.com implosion.

Early on in mobile service development there was a notion that operators should take control of matters and dish up their own à la carte menu of content, thus dictating the possibilities to the end users. This approach is based not only on a desire to control ownership of the customer as much as possible but also on an incorrect vision of the Internet – one that is possibly dominated by an outdated publishing model and one that overlooks the power of open networks when the membership is maximized, not confined in a 'walled garden'.

It seems certain that the 'walled garden' model will fail – should operators wish to pursue it. The one that will succeed is the model that most effectively enables mobile devices to access as many resources as possible on the Net and for personal networks to emerge as rapidly as possible. To do this we have to place the Net at the centre of the mobile services

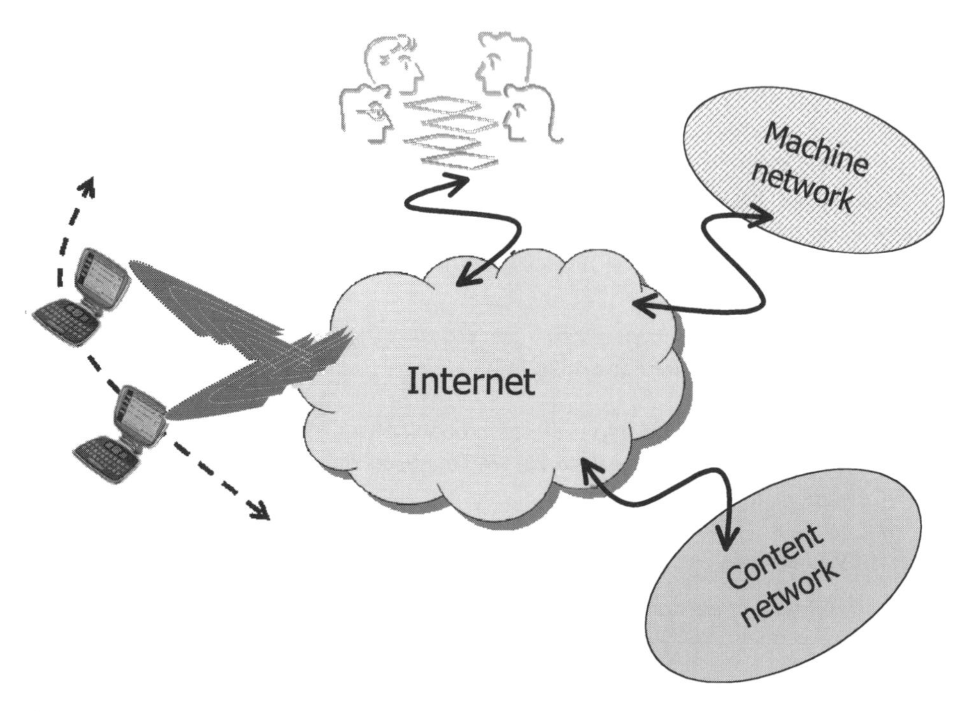

Figure 5.2 Internet at the heart of the mobile services universe.

model, as shown in Figure 5.2; this is probably the right approach given the huge networking power of the Internet. It may seem an obvious perspective or approach, but operators may be tempted to put their mobile network at the centre and work outward. This is understandable, given the investment in building RF network infrastructure and capacity, but may lead to an imbalanced approach.

Today we have truly powerful mobile networks that have already networked hundreds of millions of people globally and with global roaming potential. Yet, despite the power of the RF network (which is something we will examine when we look at the RF network in some depth in Chapter 12), the most powerful network is the Internet, and so it has the greatest potential to add value to users. Adding the RF network brings additional value, no doubt, and much of this book is concerned with how to bring about the multiplication of value in combining the potential of all the networks shown in our mobile network model, including the device network.

The emphasis we are placing in this part of the book is on the Internet, *especially the Web*. It is a valuable centrepiece to our mobile services universe, which is why I have chosen to tackle this part of the mobile services network model first (namely, the IP network layer), before looking at the other layers (such as RF and device layers).

Why has the Web become so widespread so quickly? What has it achieved exactly? I think we need a thorough understanding of what the Web does in order to nourish the creative thinking processes that I hope will send next generation mobile services into orbit, propelling us into the science fiction future.

In order to understand how we might benefit from a 'wireless web', we first need to examine what exactly we mean by the Web. While doing that in the following sections, we will look at what aspects of the Web provide challenges for wireless, but defer our detailed discussion on how to solve these challenges to the following chapters, especially Chapter 7, 'Content-sharing protocols vital to mobile services'.

5.4 THE CHALLENGES OF LIBERATING DATA

We are concerned in general computing with accessing information. Most useful information (content) resides in databases – that is Observation No. 1:

> **Observation No. 1** – most electronic information in the world sits in databases.

Databases are extremely efficient systems for storing collections of related information in a structured way that can be easily retrieved, organized and updated. The most widespread type of database is the relational database, where information is stored in tables. Each table will contain data that have a particular grouping, such as a table of customers, a table of products, a table of prices and so on.

Relationships can be made between tables in the form of relational links, such that a customer from the customer table can be identified with products in a product table (e.g., products they may have purchased). Database servers allow any meaningful subsets of linked data to be rapidly updated and accessed, serving a high number of users making different requests in a number of different ways. A common database example is to think of an inventory of products, such as books or DVDs. But there is no end to collections of data that have found their way into databases.

Clearly, we need to be able to access remotely – across a data network – the information held in databases. Let's be clear that when we talk of accessing the information our primary concern for now is to *view* it (visualize), although we should keep in mind that we may want to change it or add to it.

When viewing information we want to be able to see it in a format that makes sense. It should definitely be human-readable and preferably formatted in a way that is an aid to understanding and interaction. For example, we would like to be able to view data in tables where appropriate, with appropriate text rendering and so on.

We would also like to interact easily with the information if we need to alter our view of the current data set – say, from books on science to books about a particular topic in science by a particular author. In other words, we need navigability, a means to 'drill down' further into the information or look at other parts not currently visible to us.

In order to view the information and navigate its space we obviously need the ability to pull down information from the databases to our device in the first place. A generic architecture for our system is shown in Figure 5.3. We need to consider the possible mechanisms required in implementing this system and how we implement them.

If we consider the primary problems to be overcome in viewing information that comes from databases, this discussion will enable us to understand not just how the Web works but also why it works and what was the thinking behind it.[16] I have identified four main problems to overcome.

[16] The Web model was not originally designed for accessing databases, but to access collections of pre-formatted pages of information (i.e., documents) that sat on a hard disk on some remote PC. However, assuming these pages 'sit in a database' or will be populated with data from a database is a useful generalization and one that reflects a common occurrence in today's Web systems and does not deny the original design aims of the Web.

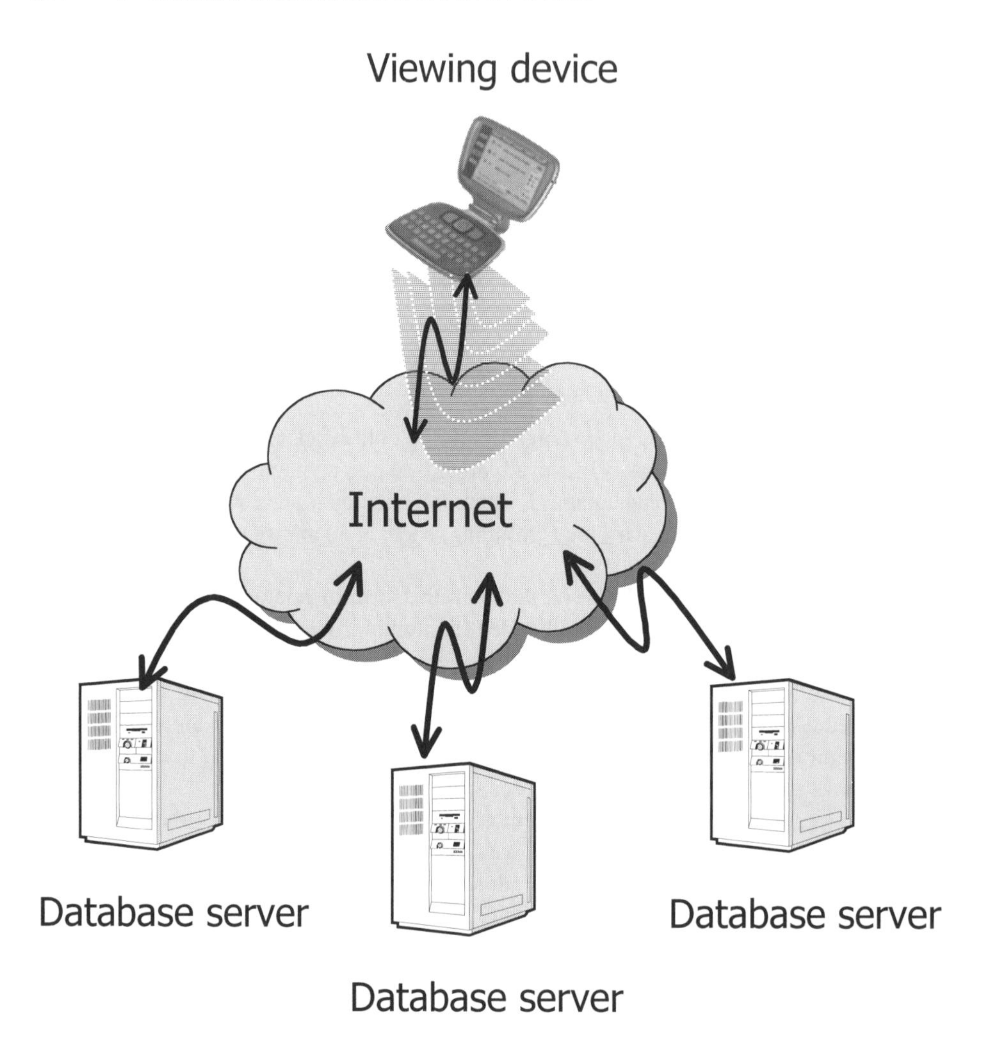

Figure 5.3 We want to view information stored in databases.

Challenges of visual interaction with databases:

1. *Formatting* – how to format the information retrieved from the databases into a viewable format and subsequently how to view it. Note that the information in the database would most likely *not* have been stored with visualization as a guiding factor in its structure (data types), but only with data representation in mind. For example, storing someone's name requires only a string of, say, 50 characters typically. Usually, we do not need to consider other information like what colour we wish to use to display the name in a particular application. Such display meta-information would typically come from the application that is responsible for generating the human interface.

2. *Visualization* – how to add the visual formatting to the information requested from the databases, once it has been retrieved.
3. *Protocol* – how to tell the database what information we want to view.
4. *Delivery* – how to get the retrieved information back to the user so that it can be viewed according to the formatting.

In discussing each of these issues in general and their solutions we will also examine the special challenges for mobile access as we go along. A full discussion of the particular solutions for wireless is given in Chapter 6.

We should first be aware that what we are about to discuss is the origins and workings of the Web, *not* the Internet. The Internet has been around for a long time and its underlying principle is that any machine that can 'talk' IP can talk to any other machine in the world that can talk IP. We should understand that IP has become the default and most widespread protocol set for computers to talk to each other – this is really the main thing to note about IP – and that there is a global data communications infrastructure which enables most PCs, irrespective of geography of locality, to access the Internet. In the computing cosmos, IP is just as English is in the international business world, the lingua franca of exchange. It is like the computing passport for global access. There is nothing inherently advanced about IP that makes it the obvious choice. Its prevalence and legacy make it an obvious candidate for network-enabling any computer application today. A detailed understanding of IP is not needed to understand this discussion (or the book), but an appreciation[17] would be useful.

Just like any language, what IP does is enable things to talk; it says how they should talk but not what they should talk about – the conversation. This requires higher layer protocols, and this is what we are going to look at now for the purposes of agreeing on how we can use IP to talk universally and generically to databases and display their contents on our device screens, both in the wired world and in the mobile one.

Observation No. 2 – IP is now the universal language for computers and devices to talk to each other.

Now let's look at how to address the four challenges that we identified above for gaining universal visual access to databases.

5.4.1 Challenge 1: Making Database Information Human-Readable

The information in the database does not contain (usually) anything about how the information should be presented to the user. It is just raw data. For example, customer records may describe customer name, address and so on, but nothing about where and how such information should appear on a screen, such as font, screen position, colour and the rest. In fact, as far as displaying information on the screen is concerned the native database format

[17] For a basic introduction see John R. Levine *et al.*, *The Internet for Dummies*. IDG Books, Foster City, CA (2002).

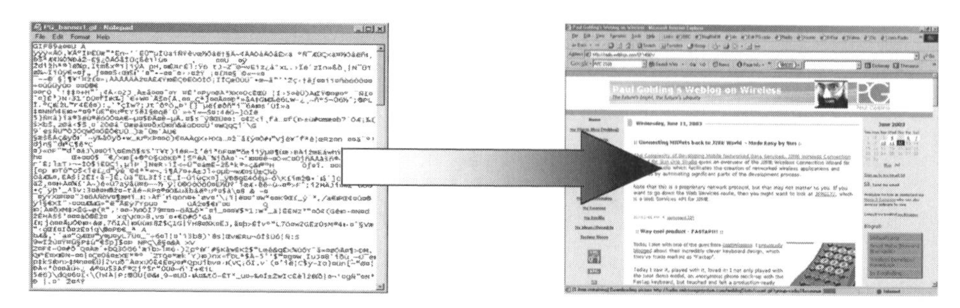

Figure 5.4 Information stored in database generally needs transformation to be visualized.

would probably appear as complete gibberish[18] and will definitely need transforming to a human-viewable format. This problem is shown in Figure 5.4.

Is there a *generic* means by which we can extract information from the databases in a format suitable for human viewing on computer screens, such that we don't need a different viewing program for each application that accesses the database or for each type of data? We don't want to have to install an accounts manager-viewing program, a customer management-viewing program, an inventory control program, a TV listings program and so on each time we have a different application that uses information in a database. This would be unwieldy to manage and most likely an expensive solution. In a mobile context it would in any case be untenable because most mobile devices cannot support installation of programs in the field (after production and sale of the phone). We are looking for a universal and generic means to view information sets from databases.

Solution 1 We use something called a *browser* that can accept information from our databases along with some kind of accompanying layout information that specifies fonts, text positioning, columnar layout, tabular views, colours, italicization, etc. The layout information is not really the data; it is called *metadata*, which means *data about the data*. What it tells us about the data is how to lay it out and how to display it. The layout metadata gets mixed up with the actual data in the same file and this gets sent up to the browser, as shown in Figure 5.5. Clearly, there must have been a process to extract information from the database and another process to add the mark-up.

We could standardize the layout data in terms of its format, but the actual data come straight from the database, be it the accounts, the inventory or the weather forecast.

The layout data are what we call *mark-up* as they 'mark up' the data, such as 'take this piece of data and make it bold', 'take this block of data and put it in a table, width 500 pixels, 4 columns', 'take this piece of data and make it into a large heading font'. The mark-up needs to conform to a standard method if we want to be sure that anyone using the appropriate browser (i.e., conforming to the mark-up standard) can access our database information, as the browsers will be expecting the layout data to be in a pre-agreed format or language.

These layout metadata are formalized into what is called a *mark-up language*, and in the case of the World Wide Web this language is known as *HyperText Markup Language*

[18] Gibberish: unintelligible or meaningless language **a** : a technical or esoteric language **b** : pretentious or needlessly obscure language (*Webster's Dictionary*).

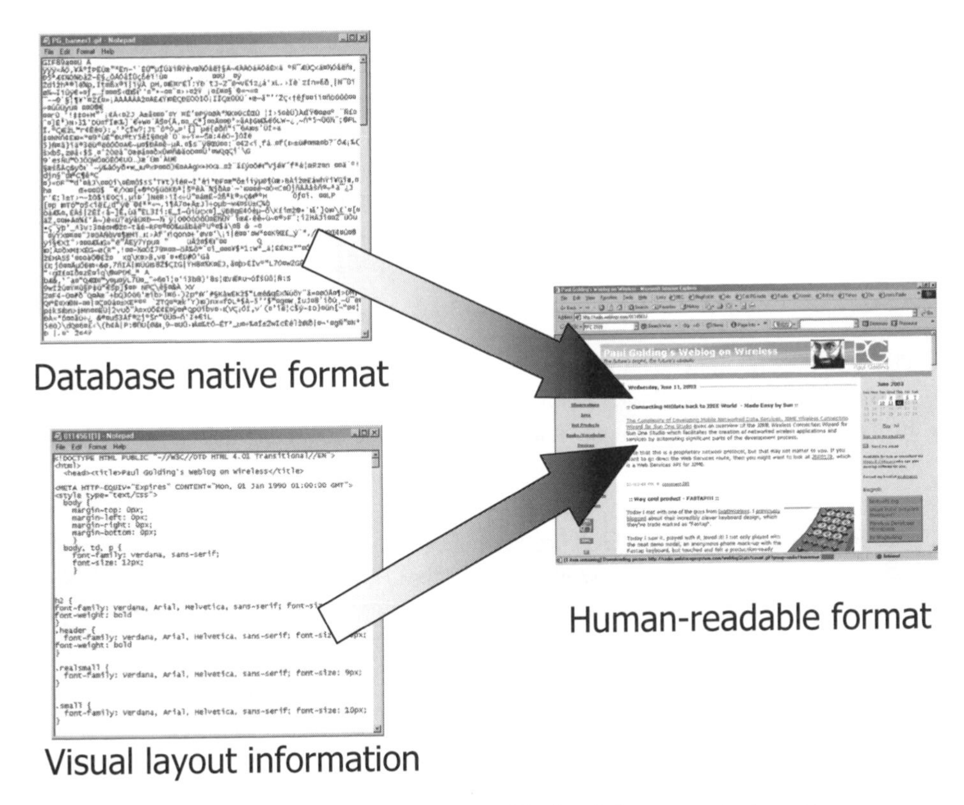

Figure 5.5 Combining database information with visual formatting information.

(HTML). Please see Box 5.2: 'A quick look at HTML' for an example of what HTML looks like. We will see why it is called *Hypertext* ML in a minute, and later in the book we will describe its mobile variants, like WML,[19] in more detail.

A Web server does the transformation process of adding the mark-up data to the actual data, as well as grabbing the data from the database. As mentioned earlier, clearly we need two things:

- a programmatic way of controlling the database extraction;
- a means of adding the mark-up metadata to the extracted information.

Box 5.2 A quick look at HTML

HTML uses the concept of tags to annotate data. Not being the actual data, but rather data about the data, this mark-up is referred to as *metadata*. 'Meta' is a Latin word indicating change, and that's exactly the effect it has on the marked-up information, transforming it into a structure that is capable of being displayed by a browser.

To take an example, if we want to indicate that a new paragraph has begun, so as to allow the browser to demarcate the paragraph with a suitable portion of white space,

[19] WML stands for WAP Markup Language, but there are several alternatives for mobile browsers, as we will discuss later in the book.

then we insert the tag <P> at the start of the paragraph and the tag </P> at the end. In general, the tags all take this form, using angle brackets '<>' to delimit each tag, and the forward slash '/' to indicate a closing tag in a tag pair, most tags coming in pairs to show the beginning and ending of a particular marking.

Another example would be <B> and </B> to indicate bold text. Tags can also have parameters, such as <font color=white>Description</font> where these font tags indicate that the colour[20] of the word 'Description' is white.

So if we take a data set from our database like this:

Deluxe Golf Trolley, 248, Burgundy

The program running on our Web server, having extracted these data from the database, can add general data, such as text to make it clear that the 248 means 248 British Pounds (Sterling), and the mark-up tags. So we could end up with something like:

```
<P>
<B>Product:</B> Deluxe Golf Trolley<BR>
<B>Price:</B> <font color=white>248 British Pounds (Sterling)</font><BR>
<B>Colour:</B> Burgundy
</P>
```

The tag we previously didn't consider is the <BR> tag which indicates a line break (i.e., a new line should start). Otherwise, we have already explained the other tags and the final display in a browser might look something like:

Product: Deluxe Golf Trolley
Price: 248 British Pounds (Sterling)
Colour: Burgundy

All of the available tags that we can use in HTML are already defined for us; we cannot change them. This is so information will appear the same in all browsers. The standard for HTML is maintained by the World Wide Web Consortium (W3C[21]).

So our Web server has to provide these functions. On demand, the user requests information from the database via the Web server, which accesses the database on behalf of the user and inserts the mark-up data and passes the combined information – called a page – to the user. The user's device receives the page in a browser, which knows how to interpret the mark-up in order to display the embedded information according to the mark-up. A page designer (i.e., a person who can design a suitable visual layout and design) determines how the information should be presented on the screen and uses appropriate mark-up chosen from within the available (fixed) HTML vocabulary. A programmer takes on the task of programming Web pages that can access the database and then merge the retrieved data

[20] Note that HTML tags use American spelling conventions, so the use of the spelling 'color' in the tag is not a mistake here.

[21] After the invention of the Web by Berners-Lee, the World Wide Web Consortium (W3C) was created (October 1994) to lead the World Wide Web to its full potential by developing common protocols that promote its evolution and ensure its interoperability. See http://www.w3.org/Consortium/

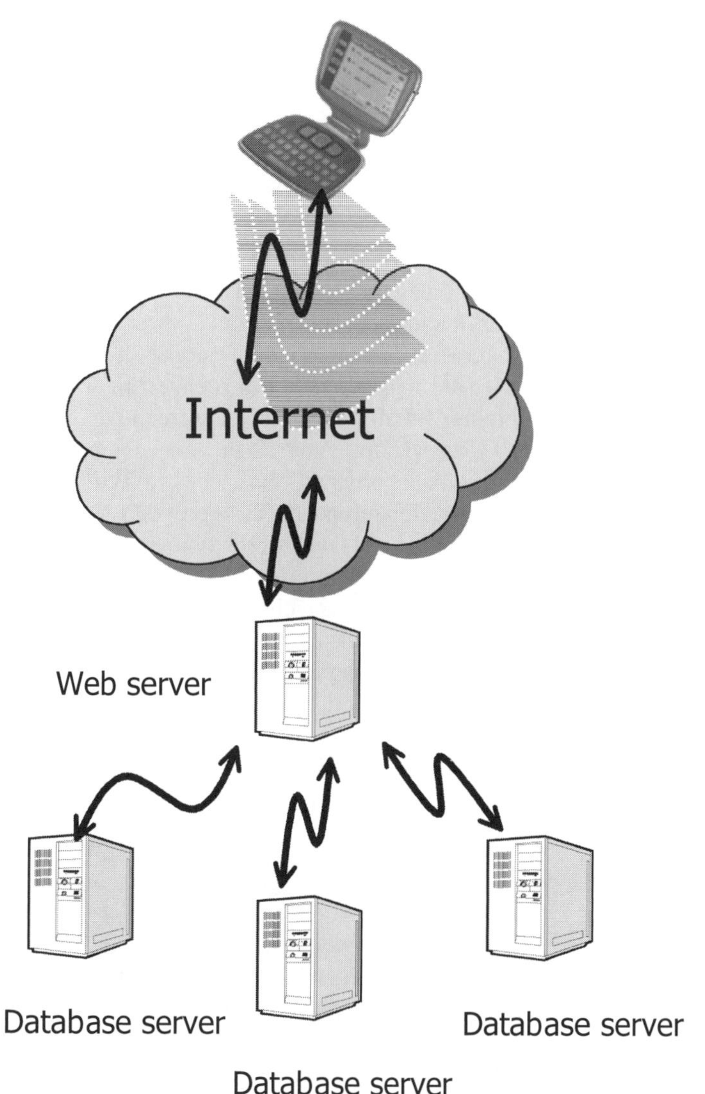

Figure 5.6 Web server in the middle.

with the mark-up, which itself is probably generated programmatically rather than stored as a template file, although this could be done and sometimes is.

5.4.2 Two Important Points to Note About the Browser Approach

Point 1 You may have noticed that we have now introduced a new networked computer element in our generic information delivery architecture that was not previously assumed or shown in Figure 5.3 (namely, a *Web server*). So we now have what is known as a *three-tier architecture* as shown in Figure 5.6. This is the so-called client/server (CS) architecture for software applications.

Observation No. 3 The Web paradigm is technically based on a CS architecture.

In our three-tier architecture both the Web server and the database server are sitting in the content network layer of the higher abstraction network model that we established and discussed earlier in the book.

Point 2 A browser is a generic display program capable of presenting any information that has been marked up using HTML. Therefore, it is a universal client not restricted to any one particular application or information set. This is its real power and is a reason for the popularity and significance of the Web. Because the browser can display *any information* (HTML formatted, of course), we are able to benefit from connecting with *any Web server* where the owner has decided to use HTML, which they are free to do because HTML is an *open standard and easy to learn*, even for a novice. Previous computer paradigms had often assumed that a proprietary software program was needed to facilitate each and every application with its associated information set and that these programs were confidential and proprietary in terms of their construction and how information was represented internally.

The shift from using proprietary information-processing software programs to a universal open information standard (HTML) and associated client (browser) constituted a simple yet significant shift in computing that heralded the explosion of the Web. This explosion had enough momentum that its impact is now being felt in the mobile world, the consequences of which are explored in detail throughout this book. As I hope will become clear, the growth of Web-enabled[22] mobile paradigms will be self-fuelling, as the power of the Web becomes magnified by mobility (ubiquity), which in turns magnifies the appeal of mobility: a virtuous upward spiral will emerge.

Wireless challenges

As we will explore in more depth later when we look at WAP in detail the problem with HTML for mark-up is that it has been designed and has evolved with large, colourful, rich display formats in mind. As is obvious, mobile phones do not have anywhere near the graphical display capabilities that the average desktop does. Hence, we can imagine that HTML is probably too verbose and over-specified for delivering mark-up information to browsers on phones. It is also the case that phones do not have the same human interface characteristics as computers, such as a mouse. Therefore, the means to interact with the data on a mobile device may not be consistent with assumptions intrinsic to HTML.

5.4.3 Challenge 2: Adding Visual Formatting to the Database Information

In general, even if we knew how we wanted to present the information to the user, database server products do not have any means of formatting the data for viewing. There is no concept of a Graphical User Interface (GUI) that the database server can apply to the information

[22] Web-enabled mobile paradigms are what we need, not mobile-enabled Web paradigms.

to make it viewable. This task is not the domain of the database server, which is optimized and designed just to serve raw data for consumption or formatting by another software program upstream. This suggests that the arrow in Figure 5.4 showing conversion from database format to human format is not a feature of the database server and, consequently, an external server of some kind is required to fulfil this function.

Solution 2 It has already become clear in the previous discussion that we need a new server – the Web server – to assist with extracting the information from the database and then adding the HTML mark-up before sending the requested information back to the browser. The way to think about the Web server is as a store of page templates representing the various views that the user may require of the target information. These templates only store the presentational information – the mark-up – but do not yet contain the actual information that the user requires from the database. There are placeholders in each place in a page template where the information from the database can be inserted.

The role of the Web server then is to handle requests from the browser for information, which comes in the form of a request for a page. The server extracts the required content from the relevant database and merges this with the appropriate mark-up template to produce a complete Web page for display in the browser. This process is shown here in Figure 5.7.

Wireless challenges

As we will explore in more depth later in the book, when we look at WAP, the challenge for wireless is to maintain the use of existing Web servers already being used by wired devices during the rising popularity of the Web. This is an obvious advantage. Having invested in Web servers and learnt how to use them and program for them, it would be a pity to have to introduce new types of servers to cope with wireless devices. We would like to be able to benefit from the quite extensive investment that has already been made in Web technology, infrastructure and – if we can – existing page design where we feel existing pages may in essence be useful to view while on the move (i.e., on our mobile phones). As we will find out later, the linchpin of the Web legacy is the HyperText Transfer Protocol (HTTP) protocol. We can think of software as becoming HTTP-centric, which is the essence of the networked computing revolution.

5.4.4 Challenge 3: The Need for a Protocol

We need a means to connect to the database via the Web server, including the ability to send in queries to the Web server, receive the resulting content and to locate the servers on the Internet in the first place. There might be thousands of databases (Web servers) accessible to us: how do we make sure that our device is talking to the right one? This is an addressing problem, just like calling someone on the phone. There are thousands (millions) of phones to call, and we solve the addressing problem by giving each phone a unique number (telephone number). We need a similar addressing system for our Web pages. Furthermore, just like phone directories, we need a means of looking up where on the Internet our server is in order to connect to it, to issue the request. Once connected to the right server, we need an

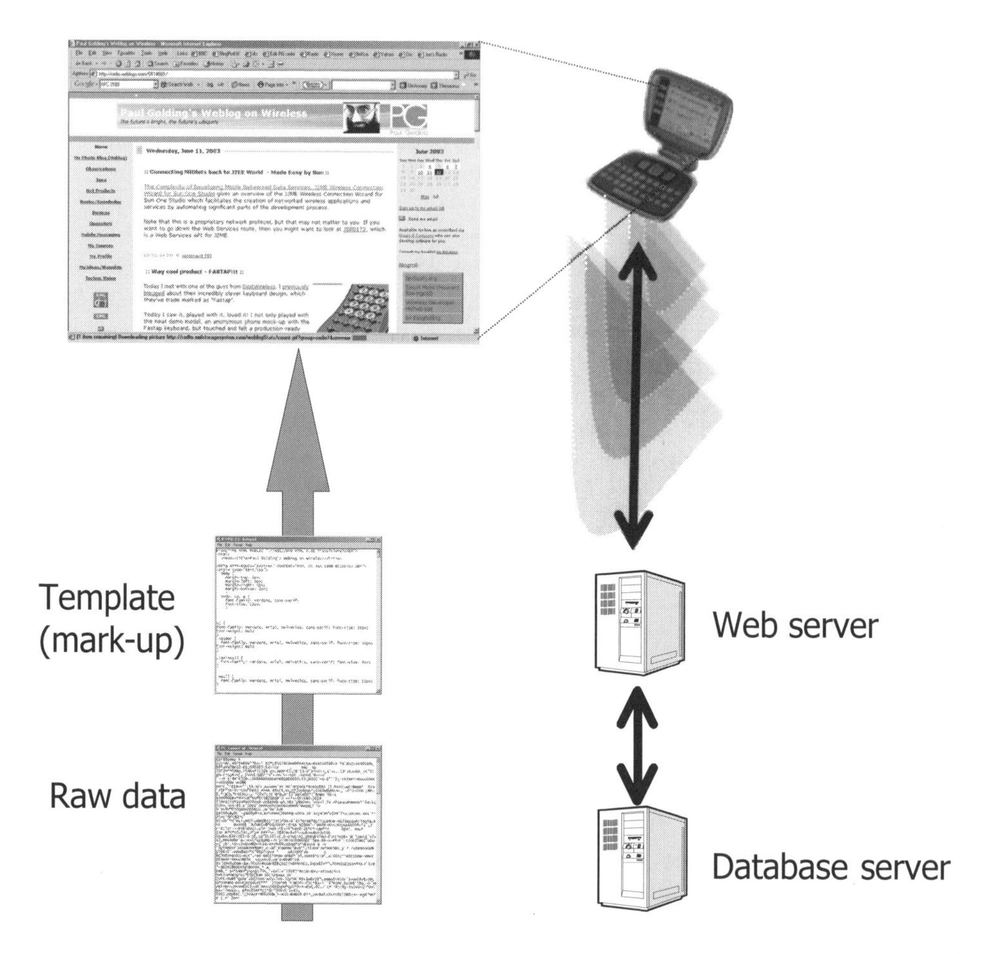

Figure 5.7 Web server presents database-bound information to the browser.

agreed language (protocol) in which to speak to the server to ask for the information set we require to view. A Web server may have thousands of page permutations, so it needs to know which one the browser requires from one request to another.

Solution 3 What we have is essentially a navigation and communication problem. How do we communicate from the browser to the Web server and how do we navigate from one page to another. We have HTML page templates waiting to be programmatically filled out with information from the database and then dished up to the browser, but how do we know which page is being requested by the user? This is the job of what has become known as the uniform resource locator (URL), which is a Web-specific addressing variant of the more generic uniform resource identifier (URI[23]). It is a means of addressing the page on a particular server, so that the server knows to use a specific page template in responding to a request.

[23] If we are just talking about web addresses, like 'http://www.web.com', then we can use the terms 'URI' and 'URL' interchangeably, as happens throughout this book. For a definition see http://www.wikipedia.org/wiki/Uniform_Resource_Identifier

We would like to alleviate the need for the user to specify in 'database language' exactly which data are required from the database to fulfil the request. This would be a very cumbersome approach and probably unworkable except by database gurus (who are a small minority, of course). Even then, this approach is not possible if the guru has no knowledge of the database structure or how the database servers expect to be queried for information.[24]

Databases are full of interlinked tables of the variety of information we are interested in viewing and manipulating. For example, for an online bookshop perhaps there is a table of books linked to a table of authors linked to a table of publishers linked to a table of prices, and so on. Let's say that the bookshop owner wants to view information about a particular customer and outstanding book orders against that customer. The information required to construct that view might be stored across several tables in the database. The way a database server works is that the information is requested by stating which tables we want, which columns in those tables and some qualifying condition to select the row or rows in these tables, such as 'customer number = 01728'. There is a special language to construct the request (query), called structured query language (SQL). It is supported by most database server products on the market.

The average Web user is not going to have a clue about SQL. If they had to specify the SQL each time they wanted to view a Web page, this would be unworkable. Here's a sample of SQL that shows why:

```
SELECT   [Equipment Inventory].SerialNumber, Products.ProductName,
Customers.CompanyName, [Equipment Inventory].DatePurchased
FROM   Products INNER JOIN
          Customers INNER JOIN
          [Equipment Inventory] ON Customers.CustomerID = [Equipment
Inventory].CustomerID ON Products.PartNo = [Equipment Inventory].PartNo
WHERE   ([Equipment Inventory].SerialNumber = '&num')
```

Despite its verbose appearance, this fragment of SQL is actually a very simple query against an uncomplicated database of hospital equipment, used to look up when a particular accessory (with serial number '&num') was purchased by the hospital. Imagine if the user had to type this query string into the browser in order to obtain the information: a daring proposition, even for someone who knows SQL!

What we need is a means of labelling a page with an address that we can enter easily into the browser such that the above SQL runs implicitly and the results are merged with the mark-up and sent back to our browser. This is exactly what happens on the Web. We assign URLs to pages and we simply request the URL from the Web server. For example, to obtain the information requested in the above SQL sample, we might simply type something like 'MyHospitalInventory' into the browser, which sends a command to the server, something like 'GET MyHospitalInventory' and the corresponding page *myhospitalinventory* runs on the server. Within the page template is the SQL needed to query the database and the mark-up to format the queried data. The query is run and the output formatted according to the template and sent back to the browser, as shown in Figure 5.8.

[24] Unless the guru designed the database in the first place, they would not know how to structure queries to extract the required information.

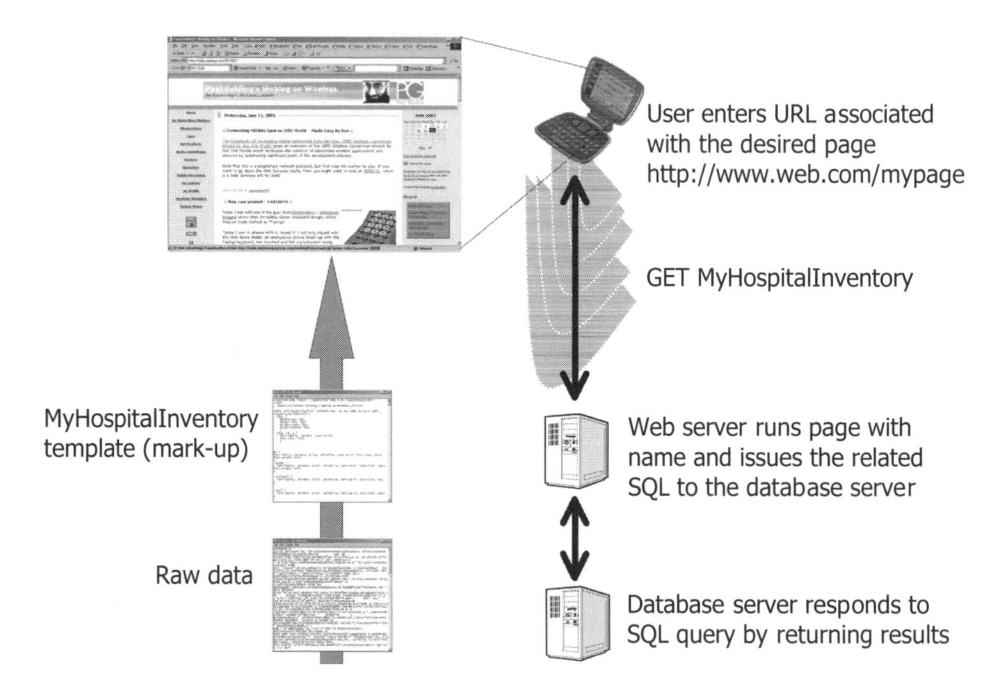
User enters URL associated
with the desired page
http://www.web.com/mypage
GET MyHospitalInventory
MyHospitalInventory
template (mark-up)
Web server runs page with
name and issues the related
SQL to the database server
Raw data
Database server responds to
SQL query by returning results

Figure 5.8 Converting Web page requests into database queries.

Let's say that the user now wants to change their current view of the data, perhaps to expand on a particular subset of the information for a more exploded view of the detail. There are probably countless ways that the user or users (as we have to contend with the entire user population for our application) may want to view and interact with the data. It would be extremely cumbersome and impractical to require each user to remember the entire set of different page names required to create these views. Even with simple page names like 'mypage1' and 'mypage2' or meaningful names like 'mycustomers' and 'myinventory', the task would be daunting and impractical. Thankfully, there is a way for the user to be alleviated of remembering the names of any pages beyond an initial entry page into the application (usually called the *home page*). This can be done using *hypertext*.

What the inventor of the Web Tim Berners-Lee came up with was the ability to embed the page names inside the Web pages in a way that the user only has to click on a visual link to invoke the browser to fetch the next desired page from the Web server. This saves having to type the name into the browser or from having to know or remember it in the first place. Furthermore, the method used by HTML to encode these links enables the page names to be masked behind clickable text or images that could say something meaningful like 'click here[25] for more information about this customer' as opposed to 'get MyCustomerInDetail' or some other obscure label (obfuscation being what software engineers traditionally do best). In this way we can continue to interact with the server without having to remember complex URLs or without having to know anything about the SQL behind the resultant queries that the Web server runs against the database server.

[25] Actually, using an explicit linking phrase like 'click here' is considered by the hypermedia aficionados to be against good design principles.

It is worth stating that even if there are thousands of different pages, there is no need to link to all of them on every single page as a means to allow the user to navigate easily to any page in the entire navigational space. The natural flow of tasks in any application usually entails transitioning logically from one related state to another, and the number of 'next state' options in any sequence is usually naturally limited to a few possibilities rather than the entire page set. Branching to another sequence in the application flow is usually done by returning to some fixed and known starting point (initial state), which invariably has come to be regarded as the function of the home page in websites. Web usability gurus talk about the desirability of designing a navigational model that allows a user to branch to any part of the application space in no less than three clicks on average. The sparseness of an application's navigational options from any one state to another is what makes hyperlinking an effective means to navigate through an application, even though the origins of hyperlinking are in branching off to separate information spaces, such as one might do using a footnote or external reference in an academic paper.

As just noted, although it works well within an application, hyperlinking is not limited to destinations on the same Web server; links can refer to other Web servers anywhere on the Internet. Therein lies one of the most powerful characteristics of the Web: the ability to refer to another completely independent resource *without the permission* of that resource (i.e., no authentication is required on the target server). This allows a massively connected information space to develop, which is how the World Wide Web grew so rapidly. Branching to any destination on the Web is intrinsically supported by both the URL format and the associated *GET method* that Berners-Lee came up with to fetch pages. This combination includes not only the ability to address a page on a server but also the Web server itself on which the page resides, as specified using a host name, such as those we are familiar with today, like 'www.myserver.com'. The default behaviour of HTTP and a Web server is *not* to authenticate a user, which means that it is easy to access any Web resource unfettered by such concerns.

Given that the Web server is programmatically generating the Web pages, it can dynamically embed the URLs beneath the user-friendly hyperlinks according to the user's current position within the particular navigational flow they have charted through the database. For example, if the user is paging through a sequence of customer records, a link can be included on each page that says 'next' under which is the URL for the next page, corresponding to the next chunk of records. On the server the Web software is keeping track of where the user is in the customer record set and creating the appropriate 'next' URL to ensure that it will link to the appropriate database query required to fetch the next record set in the flow. Of course, the user is not cognizant of such details. Their view of the process is visual and symbolic.

Berners-Lee did not invent hypertext or hyperlinking, but what he did was to include a powerful means of incorporating the technique into the HTTP/HTML paradigm, which is why the 'H' in both cases stands for *hypertext*. Implied in the specification is a requirement for the browser to render hyperlinks on the user's screen in a way that is actionable. Initially, the familiar underlined blue font was the default indication of a hyperlink used in conjunction with an altered-state cursor (e.g., pointing finger) when hovered over a link. Images can now be used for links, and there is no restriction on link text having to be underlined and blue, although the convention remains a useful one that aids usability by familiarity.

In order to understand the tie-up between HTTP, HTML and specifying database queries, we return to the example of the SQL query to look up a customer order for books. There is one element of the mechanism that we have not yet addressed: how to encode into the page name which SQL query we want to run on the server. Although we can implement a mechanism to associate page names with SQL queries, we also need a means to feed in any required parameters that a particular SQL query needs to operate, such as the customer name or customer number. This is one of the original design goals of the Web that Berners-Lee was interested in (namely, the ability to include search parameters in a page request). The method devised is to append the search parameters to the URL, so we end up with constructs like:

'GET MyCustomerInDetail ? CustID=123'

Here, the question mark is denoting that what follows are parameters; then we have a parameter name ('CustID') followed by its particular value ('123'). In this way we only need one page on the server that runs the corresponding SQL query, but we can feed it any parameters we like. As we will see in Chapter 7 the actual GET command is embedded in the protocol requests that are sent from the browser, so this is something the user never sees. What they do see is the URL – which contains the server address for specifying which server is referenced – then the page name and then the parameters, all prefixed with the protocol name (HTTP); so we end up with the familiar pattern:

http://www.myserver.com/MyCustomerInDetail?CustID=123

It is unlikely that the user would ever have to type this into their browser; instead, they would see a user-friendly hyperlink, such as 'view customer'. Clicking on this link would invoke a GET request to the above URL.

5.4.5 Challenge 4: The Need for a Delivery Mechanism

We have already indicated that we need a browser to view the formatted information. The browser has to communicate with a Web server to request the required information set. The Web server then in turn requests the required data from the database server (using HTTP), programmatically adds the formatting mark-up (using HTML) and now it needs a way to send this information back up to the browser in response to its request. HTTP provides the necessary mechanism to return data. The protocol is a two-way process. A command, like GET, is always met with a response, which leads us to think of this protocol as a 'fetch–response' mechanism, as shown in Figure 5.9.

Solution 4 Web servers sit on the network constantly listening for requests. A Web server is a software application that runs on a standard PC or workstation, particularly computers running the Windows and Unix-variant operating systems (and others too[26]). Web server programs are multi-threaded, which means they can listen to more than one request at the same time and handle more than one request at once. This is an essential part of their design, as they necessarily have to be capable of handling more than one user, otherwise they would be very limited for hosting *scalable* Web applications.

[26] Originally, Web servers were based on Unix; they still remain popular on variants, such as Linux, notably the open source server called Apache. Windows is also popular, but Web servers have migrated to high-end business platforms like IBM's AS400.

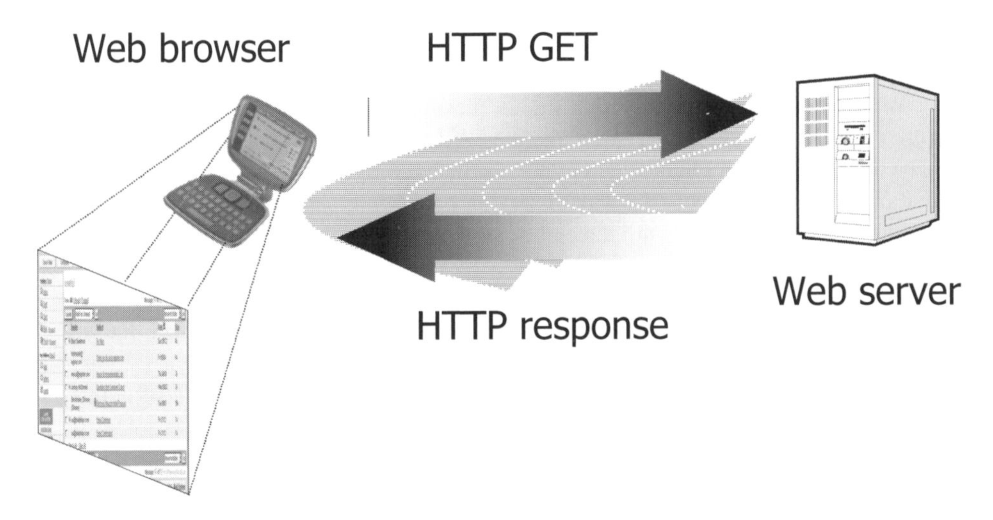

Figure 5.9 The GET–response nature of Web communications.

Requests are made to the server and responses are sent back, as shown in Figure 5.9. The requests are actually called GET methods, and HTTP specifies the exact structure of requests and responses, which is an Internet Engineering Task Force (IETF) protocol. In addition to the GET method Tim added another simple command called POST that is used to send text information along with the page request. This is a mechanism for uploading information to the server, such as could be fed into the database or used as parameters in the processing of a request. There is a mechanism for getting data from the user into the POST message, which is the familiar form on a Web page: when the user hits the SUBMIT button, the contents of the form are sent to the server using POST.

We should note again that the HTTP protocol does not require any authentication between the browser and the server. In effect, the browser acts anonymously and the Web server responds – see Box 5.3: 'Security (insecurity) of anonymous use'. All inbound requests that are valid (i.e., properly formatted requests for a page that exists on the server) will be responded to without asking for user credentials. Authenticating has to be added on as an extension to HTTP or at a different layer to the protocol: for example, either within the IP packets themselves or within the application, such as using the URL parameters to identify users, like 'GET mypage ? user=Tom' and then using forms to gather user credentials and subsequent server logic to filter out unauthorized users.

Box 5.3 Security (insecurity) of anonymous use

We have mentioned in the main text that the HTTP protocol, unlike other IP protocols, such as POP3 and FTP, or File Transfer Protocol, does not require authentication. We should be clear about what this means and whether or not it has any weaknesses. We should understand that HTTP can operate with a complete absence of any authentication mechanism, which is the so-called *anonymous user* mode of operation and is very much the default for the majority of Web servers, as well as the most widely used mode for serving pages. There is no authentication at all in anonymous user mode; it is not that authentication is simply being bypassed with default usernames and passwords – there really is no authentication at all.

We have already mentioned that by not placing any burden of authentication on either the client or the server we make the Web more agile, thus contributing to its growth. HTTP is a very lightweight protocol which removes much of the paraphernalia that might otherwise get in the way of serving or viewing pages. The lack of authentication has also helped toward the culture of openness on the Web, where sharing of information is encouraged, made easy and very much the norm. The potential downside to this is that protection of data, where required, is somewhat of an afterthought. While we should applaud the anonymity of the Web, we should recognize that, at times, some protection might still be required and useful.

Where authentication has proven to be required is in the establishment of online communities, many of which have naturally emerged from the grouping of interests as more and more users engage the Web. The other essential use of authentication is in membership of sites that offer additional services to subscribers, possibly for a fee, and so authentication becomes closely linked with the ability to make money. Generally, this is true for all financial transactions on the Web; authentication becomes important.

Some simple models for authentication have arisen which are easily supported by HTTP. There is a variety of ways to implement authentication, but in all cases we don't get very far in our consideration of authentication before the wider topic of security becomes a dominating anxiety, not least our natural concern for passing authentication information (usernames and passwords) securely. To do so we have to consider the related issue of encryption. These are topics we discuss in more depth in the coming chapters of the book.

The Internet is extremely easy to connect with, and this means that it is easy for unscrupulous people to plug in and get up to their antics, such as spying on authentication information in order to gain illegal access to Net-based resources, some of which could be quite sensitive. An increase in surreptitious activity can be expected and has been observed as more and more people join the Internet community.

Clearly, we need effective means of securing Internet financial transactions, especially Web-based ones. Such mechanisms need to be extendable into the RF and device networks or at least provide the means to interoperate successfully and without compromise.

A lot was said during the early WAP days about potential security issues. In fact, the marketplace was in confusion with some pundits saying that WAP was secure and others saying it wasn't. Some commentators even mentioned that WAP security didn't matter because in the case of GSM the RF network offered enough security of its own (meaning the bit of the RF network that is the actual RF connection, the so-called *air interface*, which is encrypted in GSM and other digital cellular standards).

Later in the book we will address these issues, and I hope to eliminate some of the confusion and myths about wireless security, an area that yet again seems in turmoil in the WiFi domain, this time also due to lack of understanding (and some poor system and service design, including lack of end user education and support). Security, by its very nature, continues to be a source of anxiety and can impede progress in data communications if it is poorly understood (or even if well understood).

We will look at the HTTP protocol in more depth later (see Chapter 7), as it is so vital to the infrastructure of modern mobile services, not only in the contexts we have been discussing so far but also in new and exciting areas, like gaining access to resources within the RF network, such as the location of the users.

There is one final aspect to the HTTP protocol that we should consider: the ability to support content format negotiation. Whenever a GET request is made to a Web server the browser is expected to issue information about the type of content that it is able to display. This enables the Web server to dish up content that is formatted accordingly. In mobile services this potential becomes increasingly important, as there is a greater diversity of device content-handling capabilities. This is why we have seen the mechanism for content negotiation become more sophisticated since the advent of HTTP, with working groups coming together solely to formulate standards initiatives in this area alone. It is such an important area that initiatives have grown out of both the Internet world and the mobile world, particularly the 3G Partnership Project (3GPP) and the WAP Forum.

5.5 DID WE NEED HTTP AND HTML?

People are often confused about why we needed or need a new protocol (HTTP) to facilitate the information-sharing process we have just been describing. After all, Internet protocols for information exchange already existed, including ones for passing text back and forth between two computers, like the email protocols (POP3, SMTP), FTP and so on (Newsnet, etc.). It is sensible to ask why these protocols were not adequate for the request–response processing of HTML files.

The answer is that HTTP and HTML are closely matched with the issues we have just been examining, whereas the other protocols are not. HTTP with HTML has the ability to:

1. Facilitate file requests from servers using a simple addressing scheme that can be easily mapped to a programmatic request on the target server – for example, to get information from a pre-written HTML file on a disk (*static HTML*) or from a database, subsequently manipulated into a HTML file (*dynamic HTML*).
2. Support anonymous information requests, which makes for an agile system.
3. Seamlessly refer users from one resource to another, on the same server or even on another server anywhere on the IP network, all made easier by anonymous access (lack of authentication).
4. Support content format negotiation.
5. Provide a foundation for universal data access – a common interface for disparate information sets and applications.

Within HTTP there is the implied idea that the information being requested is usually HTML. This is why we have, for example, the intimate tie-up between submitting an HTML form and the POST command, the form being a generic means of attaining information from a user. There is also the assumption that hyperlinks can be used to jump to the next resource request, and this is supported in the protocol, including, importantly, the ability to encode in hyperlinks a means to address resources anywhere on the Internet, as shown in Figure 5.10.

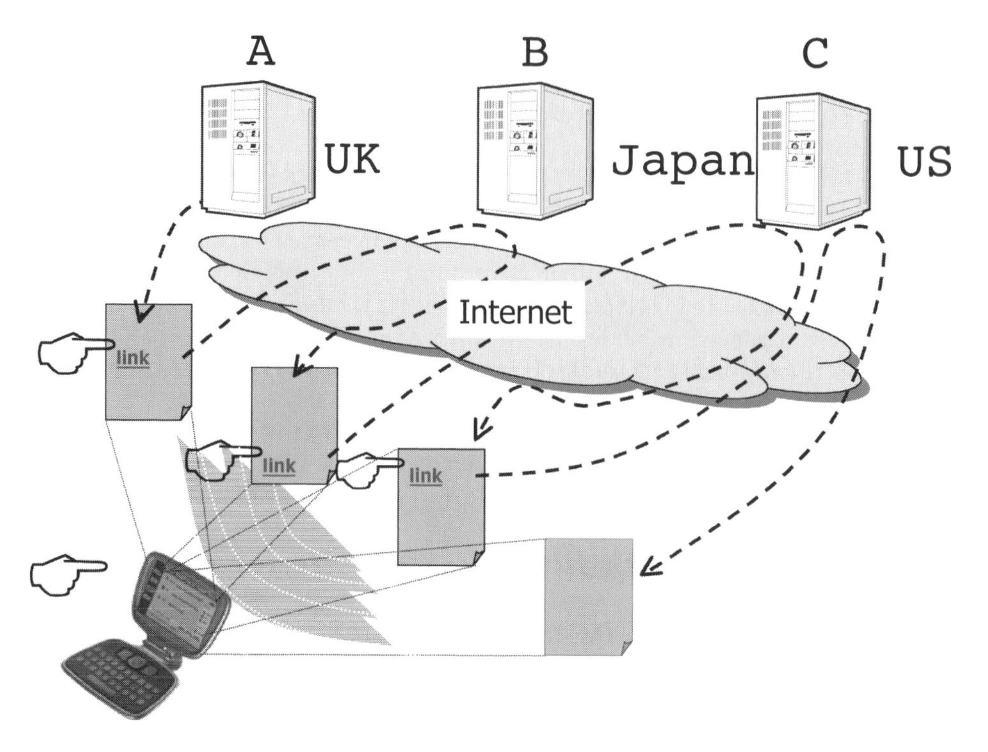

Figure 5.10 Hyperlinking supported by HTML/HTTP creates the World Wide Web (WWW).

However, it should be made clear that there is no other reason for the requested content to be limited to HTML. This will become significant later on in our discussion of *Web Services*.

The Web model is very simple – it is a constant stream of GET or POST commands resulting in pages being constructed and returned from the server. Since its release HTTP has remained a straightforward protocol that adequately supports this simple paradigm. Because the protocol is stateless,[27] does not require authentication and is text-based, it is relatively *easy* to implement in software and this has contributed to its popularity. The simple paradigm also turns out to be generally useful for a wide range of applications beyond the originally intended human-navigated journey through an electronic document landscape.

When we discuss CS software technologies in Chapter 6 we will see that Web servers have evolved into very powerful distributed software platforms. These are able to work together in clusters (server farms) to cater for huge demand, complex failure modes and to provide many other powerful and complex infrastructural functions that we do not wish to discuss here, but which are important. We will also see that the programmatic means of combining mark-up with database information is also a lot more powerful than perhaps indicated in the discussion so far and for many good reasons that we will also consider later.

[27] The meaning of 'stateless' in this context will become clearer in Chapter 7, where we look at HTTP in some depth.

More wireless challenges

On the surface, there appears to be no need to change the fundamental hyperlink paradigm when switching to wireless access for a hand-held device; so, we might wonder if we need a special version ('wireless Web') or will the standard one do? Of course, the driving factor toward wireless access is that there is already so much momentum behind the Web that it seems inevitable that we would want to access its benefits while on the move or generally from anywhere. It seems evidently a good prospect to base the wireless information access world, if there is a distinction from the wired one (something we will consider in Chapter 6), on a paradigm that has already proven to be effective for information sharing in the tethered desktop environment. However, if we proceed down this avenue of pursuing wireless access to the Web, then we need to think about the following potential problems:

a. Do the GET/POST commands running over IP involve any unnecessary overheads that would take up valuable bandwidth on a relatively slower wireless connection? We would not want to devote too much of our connection resource to merely issuing commands back and forth as opposed to sending the actual information we want to access. We will find out in Chapter 7 that HTTP/IP turns out to be a relatively inefficient protocol from this point of view (i.e., raw bandwidth requirements).

b. Is there any means by which we can compress the data being sent back and forth to improve speed on a slow link?

c. What about the URLs? Are these going to prove to be too long to have to enter by hand in a small device (possibly with a tiny alphanumeric keypad) during the times that we do have to enter them by hand? We will examine this issue later in the book.

d. What if our communications pathway gets disrupted, due to temporary loss of RF connectivity, which is common on some types of RF network? Will HTTP withstand such disruptions?

e. Is the visualization model for HTML usable for small devices? (For some allusions to the answer to this question see discussion about the WAP services model in Chapter 3.)

5.6 SIDESTEPPING THE WEB WITH P2P INTERACTION

Thus far, in our discussion we have assumed that the databases we are trying to access are not local to the devices, but are hosted on database servers in the content network, accessible via the IP network, using intervening Web servers somewhere (anywhere) on the Internet. In our high-level network model developed in Chapter 3 these network entities (servers) are both in the content layer of the network, as shown in Figure 5.11. The HTTP connection has been operating over the IP layer and is going to become our dominant focus when we scrutinize this layer in more depth throughout the book, especially in Chapter 7.

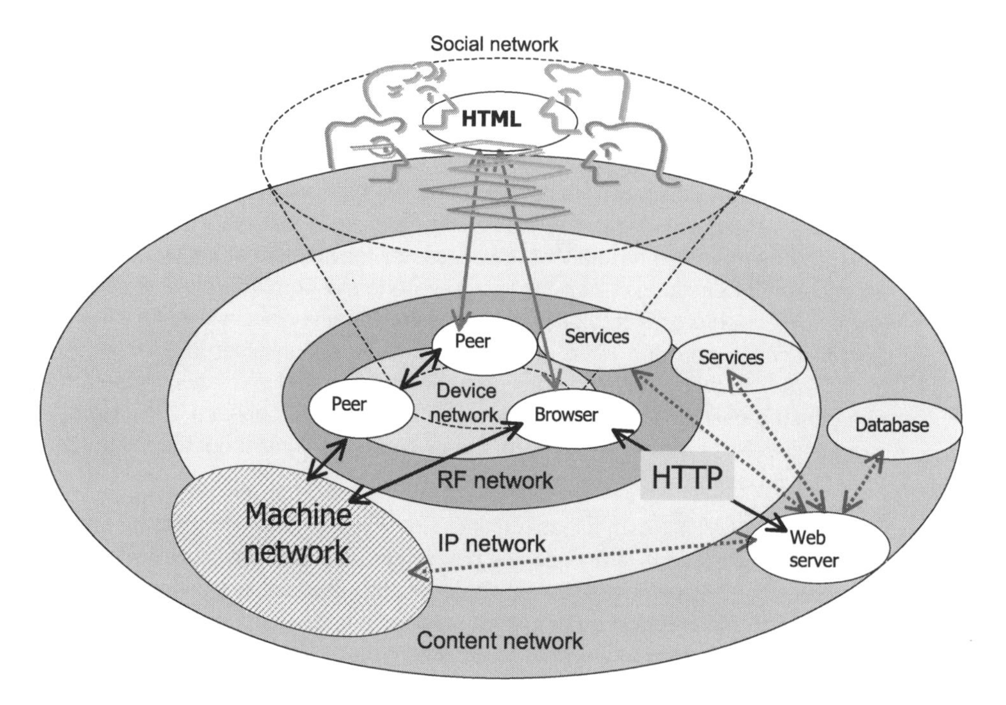

Figure 5.11 The Web server in the context of the mobile services network model.

As Figure 5.11 shows, P2P interaction does not appear to involve directly the content network. In all probability the information sets that we want to access in P2P mode are on the devices themselves, and not on database servers in the content network. We can still think of the information as sitting in databases, but these are device-bound and will be unlike the powerful database servers in the content network. It is tempting to think of the P2P mode as being like folding up the content network into the devices themselves, but this denies the entirely different characteristics of a P2P architecture compared with CS interaction.

With mobile P2P applications the content itself will almost certainly be restricted to certain types and forms. Users will share files, such as ring tones (see Box 5.4: 'Ring tone files'), MP3 files, pictures and video clips. File sharing is the original P2P application, as popularized by Gnutella for general file swapping and Napster for MP3 music files, until the music industry shut it down. There are specific types of information, such as address books and diaries, which are likely to figure strongly in P2P applications. We shall examine this in more detail when we look at LBSs in Chapter 13.

Box 5.4 Ring tone files

Ring tones have proven to be very popular with mobile phone users, part of the personalization of phones that has become an important element of mobile service design consideration. This is why Ahonen emphasized its importance with the 'me' attribute in the 5M paradigm.

One of the reasons that ring tones have been popular is that it has been relatively easy to get them. This is because ring tones can be described in very small files, small enough to allow them to be sent via a text message. Nokia, who have incorporated their support in their phones from very early on in the GSM market development, invented the most popular ring tone format. The format is the Ringing Tones Text Transfer Language (RTTTL) specification. An example of using RTTTL to describe musical notes is shown in the following diagram:

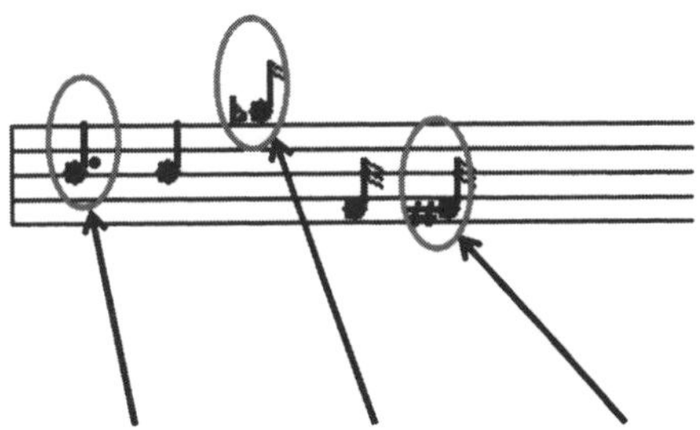

groovy toon:d=8,o=5,b=140:4b.,4b,16f#6,32f,32f#;

The diagram shows both the musical notes on a musical stave, and the textual description of these notes according to the RTTTL format.

It is easy to imagine the swapping of ring tones in a P2P network,[28] especially if a group of friends are in both a single peer group and in an adjoining peer group. A tone file could be searched for on immediate peers and, if not found there, then on the adjoining peers and so on.

For phones capable of sounding more than one note at a time (polyphonic) the Scalable Polyphony MIDI (SP-MIDI[29]) content format is a similar type of scheme for textually describing musical sequences.

The other aspect of P2P computing that makes it different from the CS approach is that the reference to other resources that hyperlinking achieves in the Web model is not so straightforward within a decentralized networking model. Let's say that we want to access a certain ring tone file we have heard about, called *Groovy Toon*. Our device would look for *Groovy Toon* on peers that it knows about (i.e., how to connect to). If it does not find the file on those peers, then it asks to be referred to their peers and so on. This peer hopping is a crucial part of the P2P computing paradigm. This decentralized networking arrangement is one where the processing and service provision is pushed to the edge of the network and away from the centre, as shown in Figure 5.12. For obvious reasons, centralization of resources has always offered efficiencies in any kind of system (not just computing ones), especially when the number of users gets large. Deliberately removing this advantage places a strain on the network that has to be accommodated by other means and may have deleterious consequences for the RF network.

It is interesting to consider at what level the peer communications take place. Systems like Gnutella use the IP network to establish peering. However, another possibility exists for devices operating solely on the mobile network, especially if they are on the same network

[28] Ignoring any licensing issues in swapping tones that represent legally protected musical property.
[29] http://www.midi.org/about-midi/abtspmidi.shtml

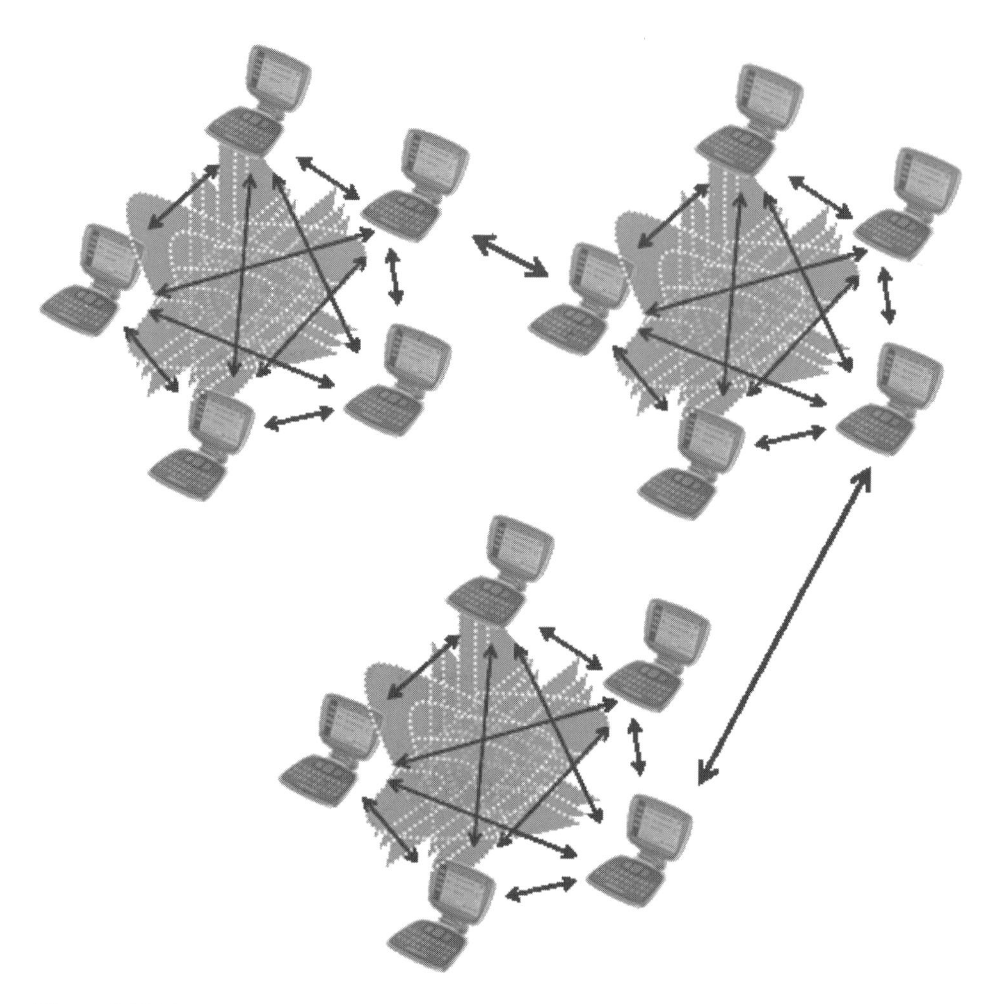

Figure 5.12 P2P networking.

(i.e., run by one of the operators). A P2P connection could be established in the RF network layer without leaving the RF network infrastructure. With P2P protocol suites like JXTA there is no reason why this can't be done as the protocols are network-agnostic.

Efficient P2P communication within the RF network is something the operator could offer to the developer community by offering device-programming interfaces that support such a mode. The relative pros and cons would be interesting to understand.

Purely as an aid to understanding P2P communications, it may be useful to imagine that much of the time P2P communication is a one-to-one, bi-directional, CS arrangement, as shown in Figure 5.13, though we should keep in mind that otherwise these two paradigms are very different. The differences will become obvious when we look at how to build mobile application servers for CS usage and later when we look at the architecture of devices. We also have to keep in mind that we are not confining our view of P2P to the publishing model that we described earlier when developing our understanding of the Web. While it is perfectly feasible for one device to browse content on another within a P2P context, the

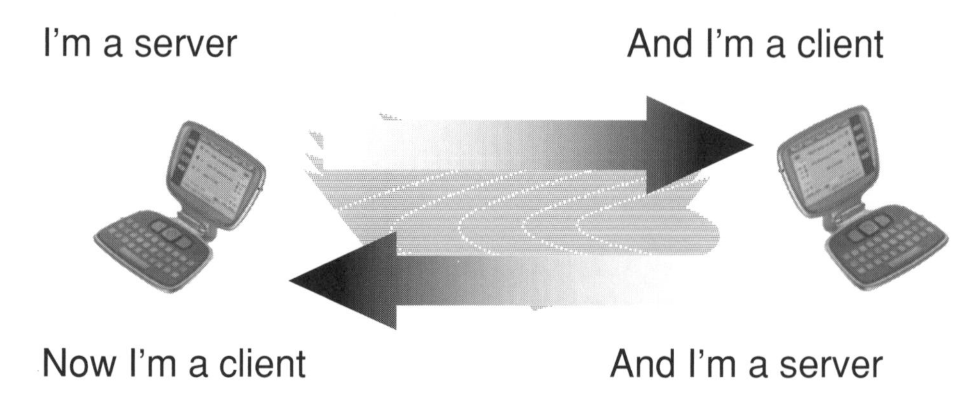
I'm a server
And I'm a client
Now I'm a client
And I'm a server

Figure 5.13 Thinking of P2P as CS back to back.

P2P paradigm has other objectives. These might include file swapping and other sharing paradigms to do with social interaction, whether for pleasure or business, such as variants of instant messaging and whiteboard sharing, where the lack of intervening central servers might facilitate an *ad hoc* approach to networking that makes it easier for new services to be adopted by interested user groups.

5.7 GOING BEYOND PUBLISHING WITH WEB SERVICES

There has been significant momentum concerning the adoption of the Web paradigm for all kinds of applications. It is so easy for anyone to publish information, particularly if the resulting Web pages do not require database queries and can be constructed as *static* pages using a Web-publishing tool, like CityDesk[30] from Fog Creek Software or the popular Microsoft FrontPage,[31] which is part of Microsoft Office™.

As of 14 September 2003 a glance at the homepage of Google,[32] the most popular search engine on the Web, shows the following line:

Searching 3,307,998,701 web pages.

That's a lot of pages on the Web, although this is not an authoritative count by any means, as many pages simply aren't searchable by the Web-crawling tendrils of search engines.

The popularity of the Web has resulted in widely available infrastructure, including Web servers, routers and firewalls configured (and optimized) for Web traffic. The other infrastructural momentum is that which has arisen around the number of software solutions with embedded Web access or HTTP awareness. This ranges from high-end enterprise systems, such as popular CRM[33] products like SAP (which now has a Web interface) to software development tools and libraries, of which there are a plethora offering built-in Web support. It is easy for a relatively novice software engineer to construct a program that is able to network with the Web.

[30] http://www.fogcreek.com/CityDesk/
[31] http://www.microsoft.com/frontpage/
[32] http://www.google.com
[33] CRM stands for Customer Relationship Management.

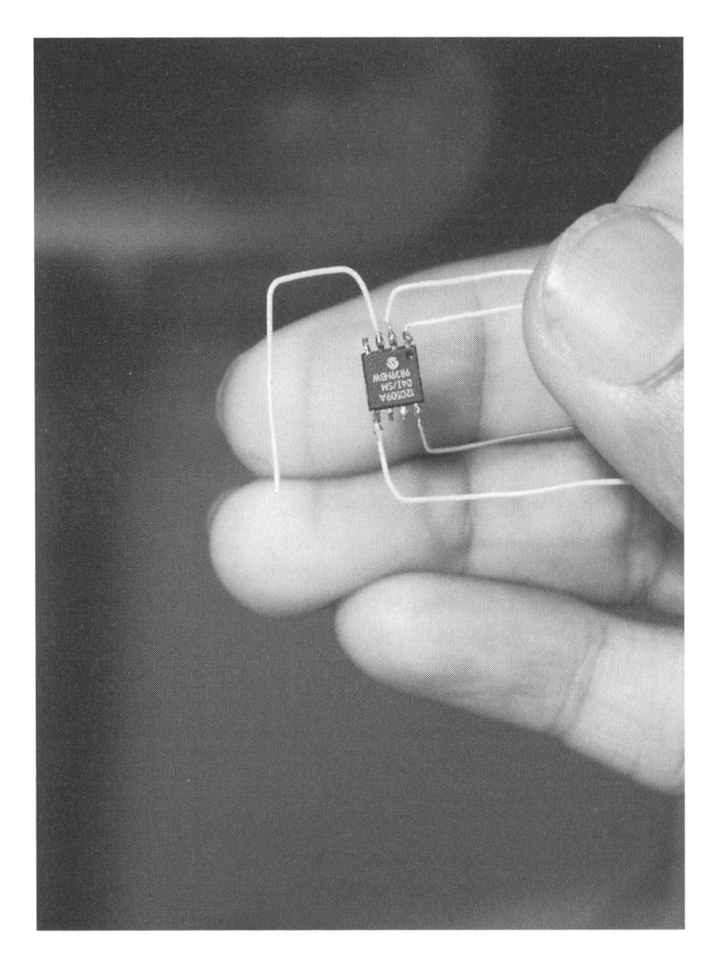

Figure 5.14 A very tiny Web server. Reproduced by permission of UMAS News Office.

The prevalence of the infrastructure also means it has an increasingly low cost base. It is very cheap to 'Web-enable' a product, and not just software products. It is increasingly common for machines to be networked using Web infrastructure. In fact, an entire Web server is now available as a chip. The single-chip computer in Figure 5.14 runs the iPic Web server, the world's tiniest implementation of a Transmission Control Protocol (TCP)/IP stack[34] and an HTTP Web server.

In M2M applications the trend is toward using a Web model to connect machines, rather than proprietary models that have been in use for quite a few years.

What we have been considering in the preceding discussion about the origins of the Web is a view of the Web as a human-centric document-publishing or information-sharing tool. The 'request–response' protocol paradigm of HTTP has been used to request chunks of HTML, the intent being to eventually render the output visually in a browser.

[34] Stack is just a name given to some software that can take care of how to shuffle packets to and from a communications channel, such as an IP connection. This concept is elaborated on in Chapter 10 when we discuss devices.

What is becoming apparent is that the Web can provide a useful backbone for any applications to swap information. For example, an inventory-checking system in one business could easily use the Web to talk to a stock-ordering system in another business, thus automatically ordering new stock.

For these embedded applications, HTML is no longer an ideal means of annotating data for this type of application interchange. The application requesting the information from another is more interested in the structure and meaning of the raw information, rather than how to display it (it may not be displayed at all). The motivation is to consume the information into the innards of an application, not send it to a browser – although that is where some of it may end up depending on the design and purpose of the system.

In the case of stock inventory, rather than embedding mark-up information to structure its visual display, it is preferable to embed mark-up to indicate the business-specific structure of the data, so that our application can identify interesting primitives within the information, such as item names, part numbers, current prices and so on.

The way of marking up the structure of data for application interchange is to define our own mark-up language or tag set, making it specific to the particular application being implemented. This is possible using something called eXtensible Markup Language, known by its acronym XML.

Within certain limits defined by the basic grammatical assumptions that XML imposes (such as the use of angle brackets '<>' and so on), we can have any tags we like, and it is up to our applications to make meaning of them. Therefore, for the example of inventory again, we could propose the use of custom tags like:

```
<StockItem>
  <ProductName>Deluxe Gold Trolley</ProductName>
  <ItemColour>Burgundy</ItemColour>
  <Price>248</Price>
  <Currency>GBP</Currency>
</StockItem>
```

This kind of data interchange using XML could be between any applications that are able to connect with the Web. Figure 5.15 shows a mobile device running a field sales application that is gathering updated stock information automatically (i.e., without user intervention). Any device connected to the Web can engage in Web Services dialogue: it does not have to be wireless. When we look at the RF network in Chapter 12 we will see examples of where Web Services are very important for mobile service delivery – even though the Web Services aspects do not involve wireless, but take place between servers in the content network and application gateways in the RF network.

We will look at XML later in the book, but we needed to introduce it now in order to understand how the Web has evolved since its inception. It has grown from being a purely human-centric Web to one that is also application-centric. Instead of pulling down HTML files from a server, Web Services request and post XML files. We may have realized from the above example that XML is a plain text format and has no special dependency on a particular operating system or programming language, which means it is another example of a universal model that means even more computers can connect to the application-centric Web, thereby increasing the value of the Web even further.

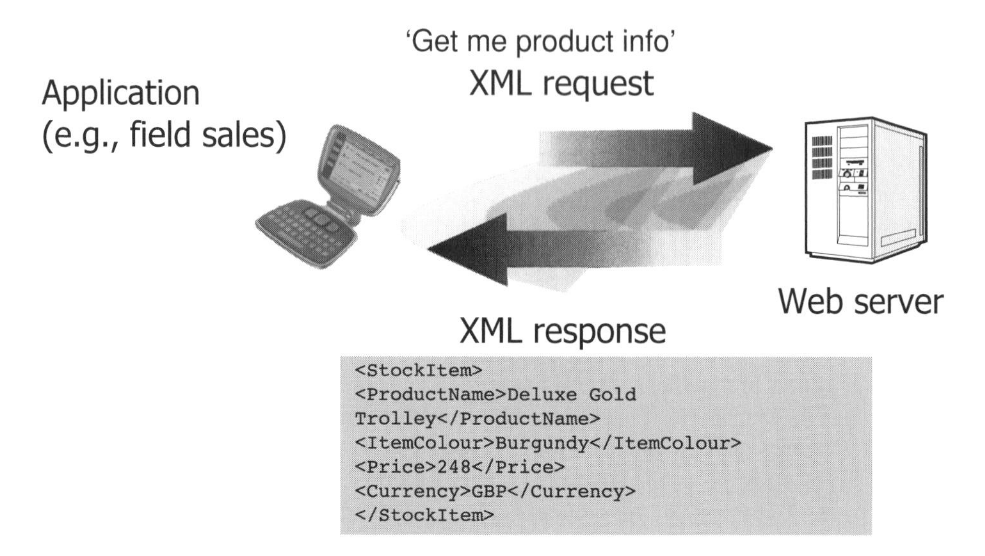

Figure 5.15 Applications talking across the Web using XML.

> A Web Service is any service that is available over the Web using XML to exchange messages, and not tied to any particular operating system or computer language.

5.8 SEMANTIC WEB

We have looked at how the Web is a major networking force today, not only for human-centric uses where visualization of information is important but also for application-centric uses where basic exchange of information enables physically separate applications to collaborate.

Most of the content on the Web today has human consumption in mind, and, therefore, the semantic meanings of the Web pages are what their visual form conveys implicitly. For example, a Web page displaying a photo album only means something to a person who knows what the pictures contain and to whom the content has relevance, and this information about the photo album is only available visually. An appointment in a Web-based events diary only has meaning to a person who knows what the event is about and to whom it is relevant. In other words, the meaning of information on the Web is usually conveyed by whatever is visibly evident on the page. Furthermore, these meanings are only accessible to human consumers, not to applications running on machines, because only humans, not machines, can see the pages.

It would be useful if applications could understand the meaning of information on the Web. For example, we might want to search the Web for news information relating to a particular field of interest. Probably, whenever we look at a news item on a page, we know that it is news. We can tell by certain formatting and context, even if the word 'news' is not itself displayed on the page. However, we need a way to enable an application to determine that a particular Web page contains a news story. We can do this using the *Semantic Web*.

> *Semantic (adjective)*:
> of or relating to meaning in language.

The Semantic Web is an attempt to add lots of metadata to the information contained in Web pages so that applications can recognize the meaning of the information described (i.e., comprehend it). Physically, the Semantic Web is not a different Web from the one we have so far introduced and discussed: it is an alternative context with semantic qualities accessible to applications, not just humans. To understand this concept better it is probably best that we introduce some of the ways that the Semantic Web uses to convey semantics.

First, it should come as no surprise that the manner of describing semantics is through XML. The beauty of XML is that it is both human and machine-readable. It is relatively easy to write a software program to recognize XML tags, find them in an XML message (called parsing) and then extract the information contained in the tagged fields. In fact, the use of XML has become so prevalent in general that there are lots of software tools and libraries to make XML processing easy and cost-effective.

Second, we need a framework to formalize an approach toward describing semantic information. This has emerged recently in the form of the Resource Description Framework (RDF). While XML allows us to add any structure we like to information, it doesn't actually tell us what the structure means. This is where the RDF comes in.

In just the same way that we describe things in human language, using constructs like sentences which have a subject, a verb and an object, RDF provides a similar ability that is accessible to machines. For example, we could use RDF to make an assertion about a particular Web page by giving it properties – such as 'this page contains news' and say things like 'this person is the author of this news item' (where 'this person' is someone's name found in another Web page). We can better understand these semantics by examining the concept pictorially, as shown in Figure 5.16.

In Figure 5.16 the handwritten labels represent annotations of the underlying Web-based information. These annotations enable us to understand the meaning of the content in the pages. Without these annotations we would not understand the meaning of the information implied by the symbols, nor would we understand the relationships between them that enable us to gain more insight into the meanings. This is exactly the same problem faced by computer applications unless we state the meanings. In the Semantic Web we use RDF to make these annotations (to add semantics).

The need for machine-digestible semantic annotations first arose because it became increasingly difficult to find information on the Web as it grew to be so huge. From Figure 5.16 we can see that the semantic labeling can take advantage of the interconnectedness of the Web. An RDF description can say things about one resource on the Web that links its meaning to another resource. As the figure shows, we can say that a certain page contains a photo of someone whose biography is contained in another page. When we get to that page, we can use semantic labelling to give meaning to the content, such as indicating that it is biographical in nature. Within the biography page, XML can identify biographical data, like first name, second name, date of birth, educational history and so on.

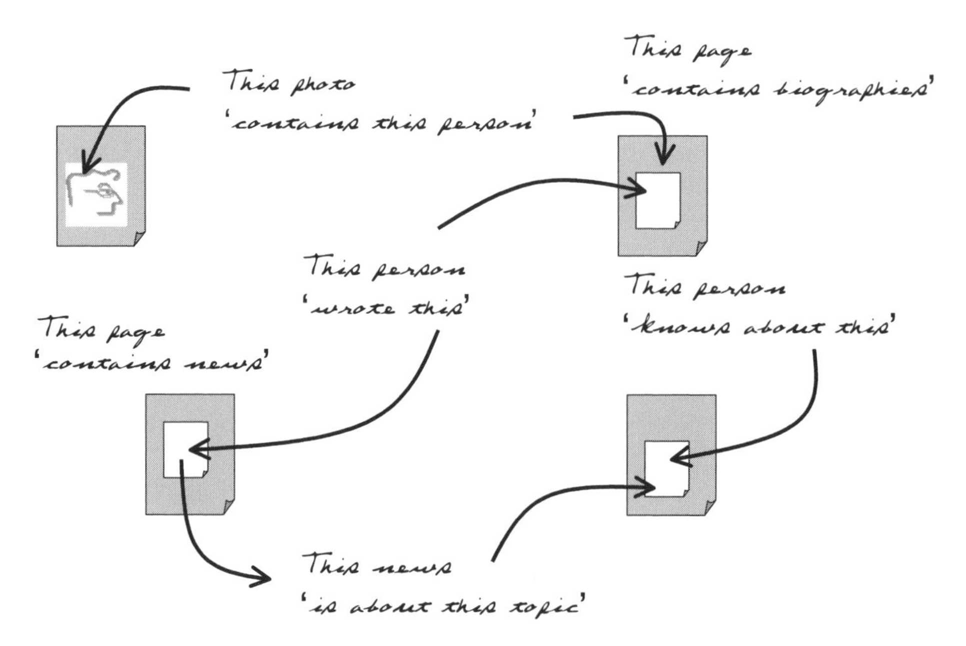

Figure 5.16 Adding meaning to content creates a Semantic Web.

```
<Biography>
  <FirstName>Joe</FirstName>
  <SecondName>Bloggs</SecondName>
  <BirthDay>24</BirthDay>
  <BirthMonth>8</BirthMonth>
  <BirthYear>58</BirthYear>
  <Education>
  ...
  </Education>
</Biography>
```

We will examine this topic in more depth later in the book. However, here we can briefly mention some uses of RDF already in common usage or beginning to emerge in the Semantic Web. Perhaps the most well known goes by the acronym of RSS, which stand for either *Really Simply Syndication* or *RDF Site Syndication*, the reason for the two names being that there have been two major and, sadly, separate efforts to define an RSS semantic framework.

Putting quibbles about acronyms aside, what semantic purpose does RSS fulfil? RSS is an attempt to allow any changes in website information to be summarized in order that applications can monitor a website for changes of interest. We might think that we do not need a particular semantic framework to do this, as we could possibly watch Web pages to see if their publication dates change, this information (i.e., publishing date) being a standard part of the HTTP protocol, as we will see later in the book. However, monitoring date stamps is not useful enough. We probably want to know what information on a page has changed. For example, in the case of news items there could be several additional news items on a page; so, we would rather find out about these in particular and not be satisfied just to know that the page has been updated. We may want to know what has changed and who has changed it. We could then look out for particular news items on a particular topic by a particular author, or, rather, we can use a software program to do the watching for us. RSS makes this possible.

Perhaps I have fallen into the trap of using news as an example of using RSS, the trap being that as news is such a common usage of RSS it is tempting to think that monitoring news is its only purpose. Actually, RSS is an umbrella semantic framework for tracking any website changes. For example, if we think about the possibility of posting picture messages to the Web according to where the picture was taken, then we could use RSS to add semantics for monitoring what is happening in a particular geographical location where pictures may often be posted.

If we think deeply about posting pictures from mobile phones, then there is a whole range of semantic information that might be useful and for which we would need a framework. We might need to know who's in the pictures, what the pictures are of, and where they were taken (i.e., ordinate information). We could use this information to gather pictures of certain people in particular places. We could also try to find people who were in the same place, but their pictures taken by different people. The variations are probably many and a discussion of this type of application can be read in Section 13.8: 'Getting in the zone', where we introduce other uses of RDF within a mobile services setting, placing emphasis on location and multimedia messaging.

The Semantic Web is an important development of the Web phenomenon and it adds incredible extra dimensions to its usefulness, many of which will become especially important in mobile services. What the Semantic Web does is enhance the networking power (Metcalfe's Law) we already discussed earlier in this chapter. It should enable us to think differently about an earlier topic that we mentioned: the discoverability of information in a network and the impact this has on its power. Discoverability is greatly enhanced by the addition of semantic information interweaved with our core data.

5.9 XML GLUE

Earlier in this chapter we looked briefly at XML. It is fast becoming a universal way of describing data that pass from one application to another. This technique is not bound to any particular protocol or network, nor is it bound to any particular network architecture, whether CS, P2P or any other. From an applications perspective, as long as applications can produce and consume XML, then we don't really care how the XML passes between collaborating agents, nor where they sit in the network, as shown in Figure 5.17.

Realizing the universality and network independence of XML, it is tempting to suggest the notion of a virtual XML highway. This is just an academic concept; but, as we progress through the book the idea that all software applications exchange XML messages becomes a powerful archetype by which to think about software collaboration. To implement such a highway globally would require an addressing and routing scheme. Of course, one already exists with IP, and with the advent of IPv6 we now have a big enough address space to

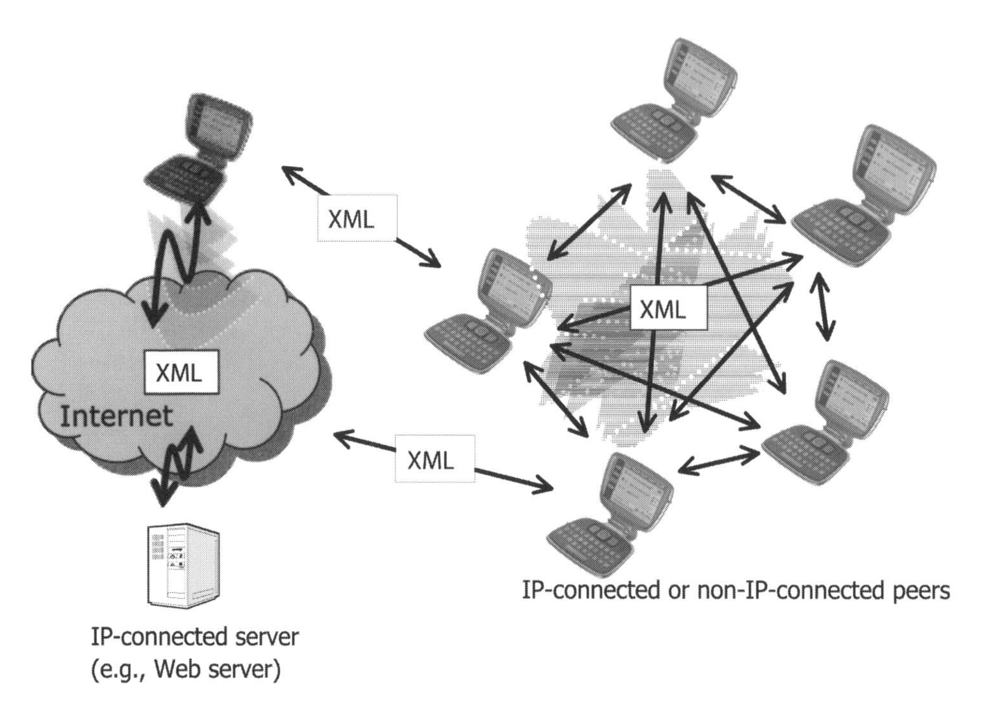

Figure 5.17 XML connects everything in the mobile network.

accommodate a unique address for every device on the planet. However, we have already alluded that IP is not always the best networking option, particularly for P2P transactions *within* the device network (more on this later).

It is likely that all wireless applications will eventually swap information using XML vocabularies. This is currently not the case. For example, we have already seen that the RTTTL tone format is not XML (nor is SP-MIDI). In fact, there are plenty of non-XML message vocabularies currently in use. XML is an open standard on which many other open software standards now rely and is used for all manner of internal and external information descriptions, not just message flows. We will be looking at many of these descriptions throughout the book, but for now it is worth highlighting that XML is likely to form the main backbone for passing data through our mobile network, as shown in our network diagram in Figure 5.18.

5.10 REAL TIME SERVICES

Thus far, our view of the Web has suggested a fixation with a page request model as a dominant application model, be the page constructed from HTML or as we have just discussed of XML.[35] Thus far, it seems that we are primarily interested in pulling down

[35] In fact, as we will see in the following chapters, HTML in its newest form XHTML is now an XML-based language – which previously it wasn't, despite appearances (i.e., stark similarities).

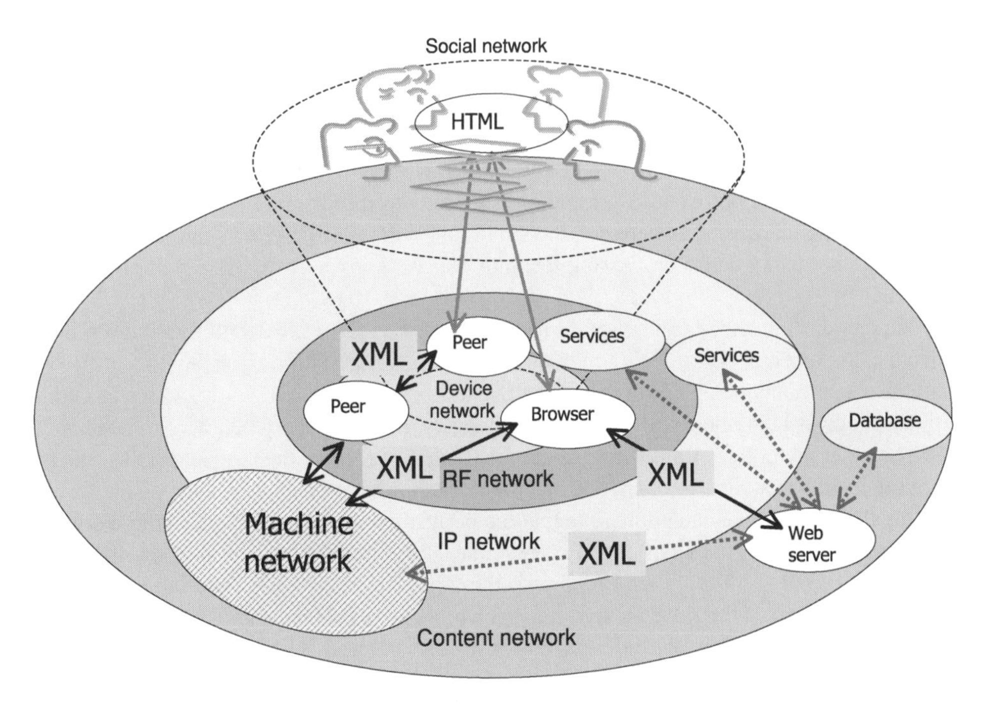

Figure 5.18 XML flows through the network.

chunks of information one page at a time, doing something with them (such as reading or further processing by our application) and then requesting some more.

Depending on the availability of resources within the entire Web infrastructure at any one point in time, we might expect variable performance in the infrastructure responding to page requests, most likely on a user-by-user basis: perhaps a server gets busy, or perhaps a communications pathway is heavily congested, or perhaps a back end database server is busy doing a computationally intensive task, such as indexing a database. During such times page responses might be slow. This will lead to variable performance. For example, we might get a page returned within two seconds on one try and then within twenty seconds on another.

> **Wireless challenges**
>
> In a mobile services context, users are generally less tolerant of delays; so, this is a matter for further consideration, suggesting that mechanisms that can ameliorate the variation in delay ('jitter') would be useful to implement in mobile networks.

With applications where the performance of the system manifested in time is not critical to the user's successful usage of it, we can think of these applications as being non-real time. However, there is a class of applications called real time applications. These are where the delivery of data within certain limits is essential for acceptable performance. These are time-critical applications, such as listening to audio files streaming across the Internet. In Chapter 12, when we discuss the RF network, we will explain the concept of voice digitization, as this topic underpins the design of digital cellular networks for real time speech transmission. For now, all we need to know is that it is possible to take digitally recorded voice and slice it into small audio files which when played successively can be heard as one contiguous audio presentation faithful to the original recording. We would not notice the interruptions between the small audio files, as they are seamlessly stitched back together using audio processing techniques (also known as digital signal processing, or DSP).

We can do this sliced transmission in two ways. We can wait for all the audio files to arrive at the receiver before being reassembled into an entire audio track. As long as we do not have a particular time limit for hearing the audio, this mode will work and we can think of this mode as being non-real time; when it's ready, it's ready, and then we can listen to it.

The other method for transmitting an audio track is to receive the mini audio files and try playing them *as we receive them*, one at a time, stitching them back together 'on the fly', as we go along. Here, we can appreciate that the performance of the network is more crucial, especially its delay characteristics. For example, if we receive the audio in two-second[36] chunks of sound at a time, then by playing these back in succession (without buffering[37], this means that from the time we start playback of a chunk) we have two seconds to receive the next file and stitch it on, before the user needs to hear it. Otherwise, if it takes more than two seconds to receive and cue up the next chunk, then we will have to endure a period of interruption in the audio. For example, if the network becomes loaded during playback and we have to wait eight seconds to receive the next chunk, then we have a six-second gap to fill, which for most audio applications is probably intolerable. Were this to happen regularly during playback, then the whole experience would become insufferable.

Using the most primitive forms of IP communication (i.e., the basic packet data mechanism), there is no guarantee that we will receive the audio files in order, especially due to routing differences.[38] The Internet is a network of networks and the ways of traversing those networks from A to B are many. If audio file N and audio file $N + 1$ take different routes, it is possible that $N + 1$ could arrive before N, especially if N has a particularly long route, dogged by such resource problems as congestion. This possibility is shown in Figure 5.19. Therefore, in addition to ensuring the timeliness of files arriving (i.e., mitigating excessive and variable delay), we also need to ensure that we have a method for reordering audio chunks (or any other real time media chunks) should they get out of order.

Fortunately, IP-based protocols have been designed to allow real time streaming of data for media applications, exactly to overcome the problems we have just been exploring. Protocols such as Real Time Streaming Protocol[39] (RTSP) and Real-time Transport Protocol[40] (RTP) are available for such applications.

The difference between these two protocols is that RTP is the actual means of enabling the streaming process to take place *vis-à-vis* the chunking, sequencing and buffering process

[36] In actuality, for various performance reasons the chunks will be in the order of milliseconds, not seconds, but for discussion purposes it seems easier to imagine audio breaks and the playback process quantified in seconds.
[37] Buffering is a data communications term meaning to store in a queue.
[38] The Internet is a vast labyrinth of physical network segments tied together with devices called routers. Routers do not necessarily send all the traffic from one source via the same route.
[39] See http://www.rtsp.org or consult the specification [RFC 2326] at http://www.faqs.org/rfcs/rfc2326.html
[40] Consult the specification [RFC 1889] at http://www.faqs.org/rfcs/rfc1889.html

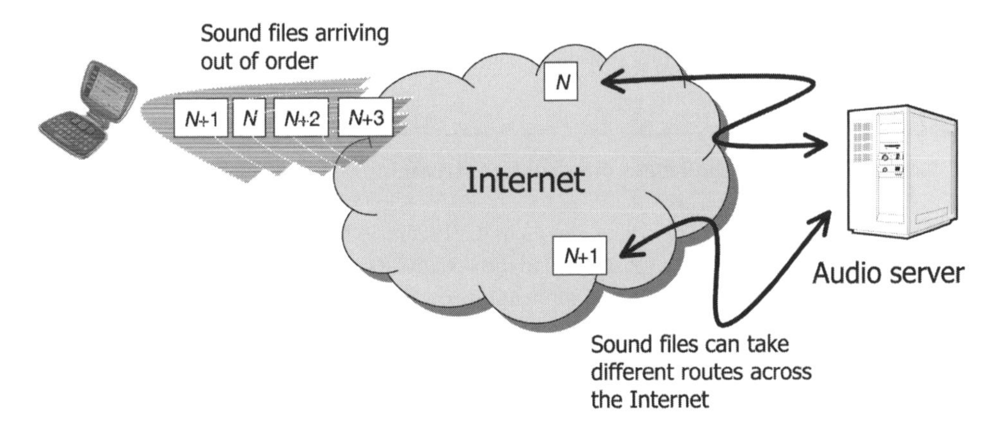

Figure 5.19 Streaming of audio files across the Net.

alluded to in the foregoing discussion of real time media data flows. To quote from the RTP specification:

> *RTP provides end-to-end network transport functions suitable for applications transmitting real-time data, such as audio, video or simulation data, over multicast or unicast network services. RTP does not address resource reservation and does not guarantee quality-of-service for real-time services.*

Multicast means to send out the media stream to more than one receiver at the same time, whereas unicast is to a sole receiver. The last sentence from the extract tells us that, although the RTP provides a means to implement the mechanics of 'chunked real-time transmission' (streaming) across the Internet, it does not provide any mechanisms for guaranteeing performance. That may seem a strange thing to testify given that our earlier discussion of the 'real time problem' seemed to suggest that performance guarantees (quality of service, or QoS) were exactly what we were looking for, such as might be required to avoid interruptions to the continuous playback of the media file at the receiver. However, the problem is with the essence of the Internet itself. At its core the foundational protocols did not include QoS techniques. Thereafter, the Internet infrastructure (e.g., routes and switches) has grown without these mechanisms in place and this inheritance remains, a problem only recently addressed by the newer version of IP – IPv6 – but this is not yet widespread. Wherever a QoS mechanism is available, this would need to be supportable by the RF network too (see Box 5.5: 'Wireless challenges for real time streaming').

What RTP does provide is a means to enable streaming of information where the streaming process is cognizant of the underlying information sources, be that audio or video. For example, it can cope with chunking according to the time base of the media, including synchronizing disparate sources that may be encoded using different compression techniques, either due to different source natures or as compelled by more stringent bandwidth availability for some sources.

RTSP is not a streaming protocol *per se*, despite its name. It does not facilitate the real time transport of media files. It is better to think of it as a 'remote control' solution. If we think of the media source server on the Internet as being like a CD player, then at the receiving (client, or 'media player') end we need a means to control the CD player, such as 'play', 'pause', 'rewind', 'select track 2' and so on. This is the function of the RTSP: to allow the exchange of these types of commands between the media player and the media server.

Box 5.5 Wireless challenges for real time streaming

In our mobile services network, as shown in its IP-centric form in Figure 5.2, information coming via the IP network still has to traverse the RF network and, potentially, the device network, before arriving at applications on the device, such as a digital audio player for the real time application we have just been considering.

Therefore, if we are able to utilize IP protocols, such as RTSP,[41] which accommodate real time services (e.g., media streaming), then we need to make sure that they are sustainable across our RF and device networks, such that these networks do not become the weak links. This indicates that the designers of the RF networks for next generation mobile services have to ensure that IP protocols can be supported, in general, and real time ones, in particular, without degradation. However, certain RF-related issues may present greater challenges, such as greater and more unpredictable contention for resources in an RF network, thereby adding greater burden on our network not to disrupt real time services.

Some of these issues relate to QoS, which is about providing network resources such that certain performance guarantees can be made on a user-by-user basis.

In Chapter 12, when we look at the RF network, we will examine the concept of content transmission techniques, which can adapt to the prevailing network conditions. Indeed, this is a particular challenge for designing mobile applications that remains with us despite massive improvements in RF network technologies.

[41] http://www.rtsp.org/

6

Client–Server Platforms for Mobile Services

We will now focus on how to deliver mobile services using the client–server (CS) architecture, which is the predominant approach for many mobile applications, mainly because of its widespread usage in Web applications, which are seen as a sensible basis for building wireless services. We have already discussed that this is not necessarily the case for all services and applications, but that the significance of the Web demands our attention.

At the end of Chapter 4 we discussed the challenges facing the mobile operators and said that in the absence of knowing in advance the killer app, or even the killer cocktail (tsunami), we should ensure that we have a platform that is ready to host applications in a manner appropriate to facilitating the emergence of killer apps. This is a multifaceted consideration, or it ought to be. One facet of the considered approach is to assume that CS architectures are a 'good bet' on the future of mobile services. This is a truism inasmuch as mobile devices will need to connect to other nodes on the Internet, and, therefore, we will need a means to establish connectivity. Moreover, as we have also stated, part of the consideration for mobile services is accessing content, and virtually all content is eventually found sitting in a database. As we have already concluded, the CS structure is well suited to accessing content in databases.

Accessing content was very much part of our mobile services definition, although we allowed content to be broadly understood as virtually anything that the user wanted to view *or produce* on their device, the deliberate breadth and vagueness of this definition allowing us to come out of the dot.com mindset of content as syndicated media or 'nuzak'[1] from corporate monoliths. This slant is not intended to be a nose thumbing at corporate information giants, but an attempt to place the strongest emphasis possible on the future individualized nature of mobile services and the probable emergence of a vast number of

[1] 'Nuzak' is a slang term coined in America to describe 'news-bite' infotainment.

Next Generation Wireless Applications P. Golding
 ISBN: 0-470-86986-0 (HB)

select markets, user groups, tastes and commensurate services, more 'cottage industry' than corporate dominance.

If the 'cottage industry' view is correct, then the underlying technology should be highly flexible and as much as possible allow the service creators to invest most of their available resources into only creating those parts of the service that are distinct and add value, rather than creating general infrastructural components. Even if the 'cottage industry' notion is incorrect, the desire to deliver a flexible platform is still relevant, especially given the need to provide a fertile soil in which to allow the seeds of killer apps to be planted. The infrastructure or much of it should be provided by the network operators. Currently, it seems that they do not really see this as part of their function or are not entirely comfortable with it, for understandable reasons to do with perceptions of 'self' and 'core competencies' as not being attuned to providing software platforms. However, if this is true, then it is a mistaken view. In our following discussion we suggest that the entire mobile network (the network of device, RF, IP and content networks) is really only a chain of software services. In essence, a good deal of the required services are intimately associated with assets that the operators own; therefore, the responsibility to make these accessible as a 'platform' lies on their shoulders and no one else's.

By adopting the CS approach to software services, the operators are able to take advantage of all that has already been done by software experts toward tuning this architectural approach to software implementation. As we will see, considerable achievements have already been made by these experts. Later in the book we will consider peer to peer (P2P) and other application paradigms relating to how software services materialize in the device network, such as the Java 2 Micro Edition (J2ME).

6.1 THE GREATER CHALLENGES

Figure 6.1 reminds us of the basic structure of our mobile network within which the CS architecture is utilized to deliver mobile services. We can see that the device communicates over the RF (Radio Frequency) and IP (Internet Protocol) network layers to the server in the content network. Figure 6.2 is an alternative view of this model taken by slicing through the networks. We can use this alternative view to highlight several observations about the CS approach.

In most mobile services we have a device at the top of the chain of events. We will look at devices in depth in Chapter 10. The next component in our chain is a set of protocols – a common language between the device and the server, represented symbolically by the arrow in Figure 6.2. The arrow represents two mechanisms. First, the method of communication between the two ends, which for the current discussion will focus on HyperText Transfer Protocol (HTTP) and its wireless associates like Wireless Session Protocol (WSP) from the WAP (Wireless Application Protocol) family of protocols. Second, we mean the format of the information passed by the protocol. As Chapter 5 discussed in some depth, the format of the data stored on the server or in a companion database server is not suitable for direct display on a device; it will be neither understandable by the device nor in a presentable format. As the figure shows, attempting to view the information in its native state will result in incomprehensible gibberish being displayed (the file snapshot on the far right). The alternative is to agree on a common formatting language, which in our case will focus on something like XHTML (eXtensible HyperText Mark-up Language), which is really

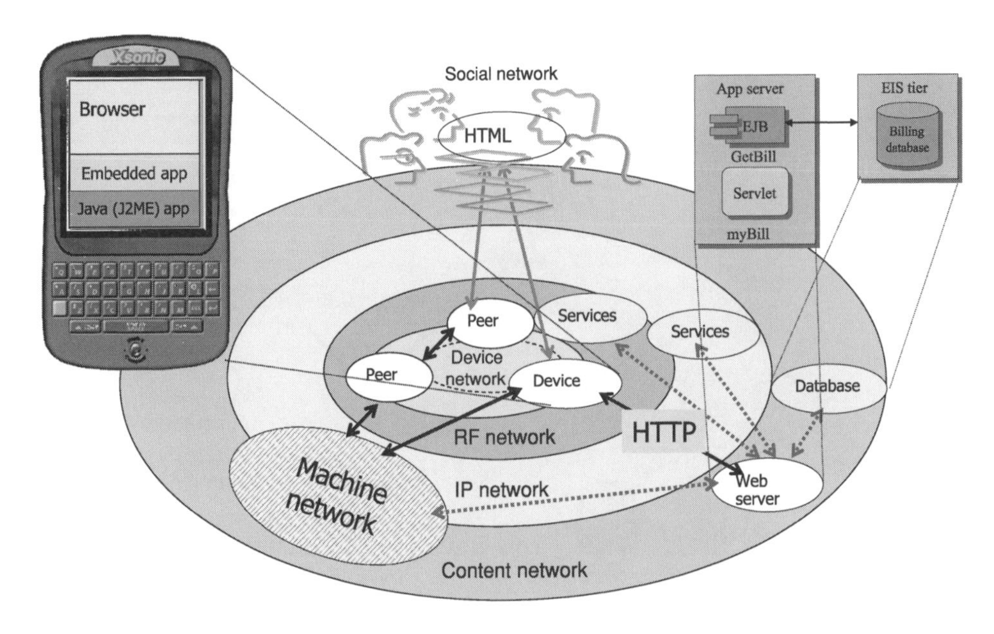

Figure 6.1 Mobile network topology.

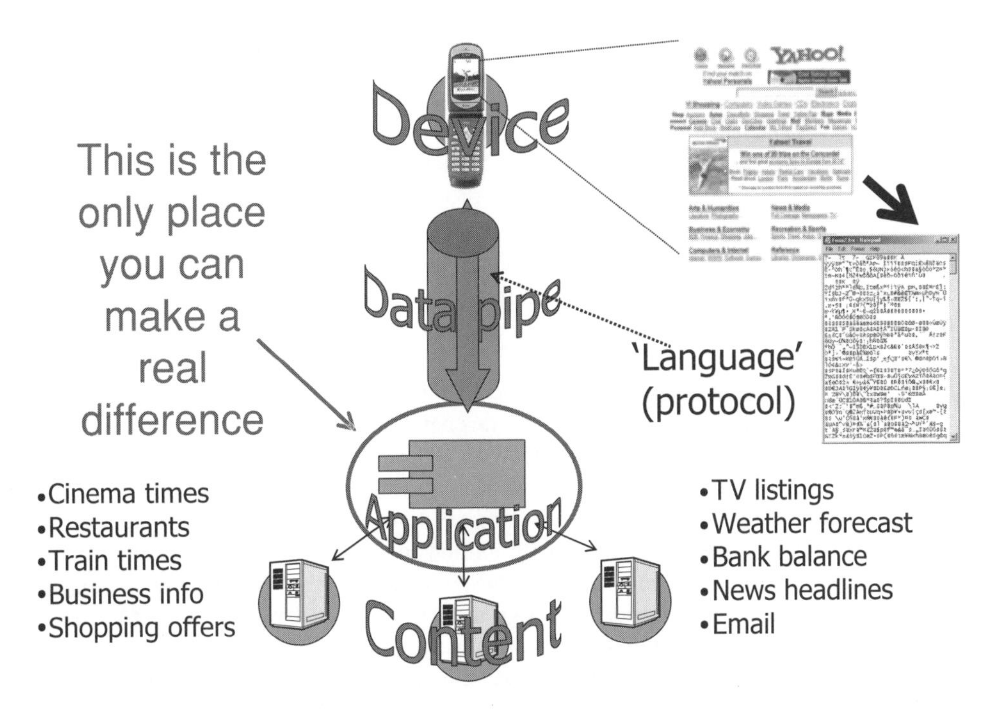

Figure 6.2 CS architecture.

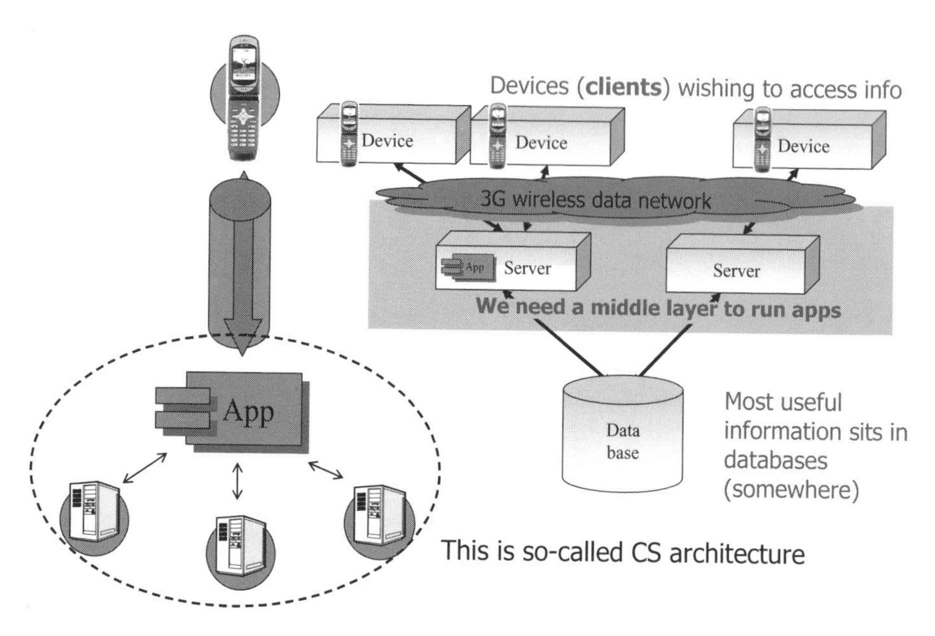

Figure 6.3 There's usually more than one client and more than one server.

the emerging standard for presenting information in browsers on next generation mobile devices.

At the bottom of the diagram we have the content, which here we represent as hosted by servers other than the ones we want to run our applications on, this most often being the case (as will become clear). The data-hosting servers are more than likely hefty commercial database products, such as Oracle or Microsoft™ SQL (structured query language) servers or other, industrial strength, scalable database platforms.

What we then need is the application itself: the 'brains' of our system. All other things being equal between competing systems, which invariably they will be,[2] the application is where we have the greatest scope to differentiate. It is up to us what we program and how we present the application to the user, and there is no need to settle for off-the-shelf solutions.

The final component in our mobile service is a bit pipe to carry the information from the application to the device and vice versa, this being assisted by common protocols like HTTP and a common data format like XHTML.

Figure 6.3 reminds us that we are not just dealing with one client, at least not if we want to offer a useful service with lots of users and resultant revenue; by targeting services at mobile operator customers we end up with millions of potential users. Furthermore, in a typical hosting environment we will need to host more than one application. Hence, we can appreciate that we have a formidable design task ahead of us if we are going to support reliably a large user base potentially accessing many applications.

The challenges of handling more than one client and more than one server is further complicated in mobile delivery platforms by the frequent need for different parts of the

[2] More often than not, all the mobile operators will have similar infrastructure equipment from similar vendors and they will all offer the same handsets or mobile devices. Applications are the biggest point of divergence.

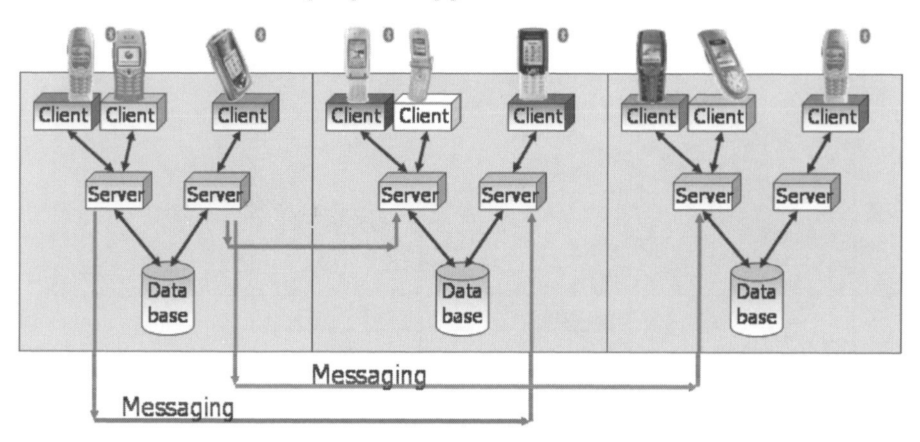
Lots of interrelated systems (e.g., data services, billing, etc.).
Lots and lots (hopefully) of different mobile users.
Client
Server
Data base
Messaging
Messaging

Figure 6.4 In a complex system there are many applications networked together.

system to collaborate by sharing common information and utility services. For example, a 'what's on?' guide for entertainment in a city may need to link to a mapping application on a separate server. If the user requires turn-by-turn, navigational directions, this may require reference to yet another application running on another server. Perhaps a user decides to visit a cinema first and wants to view the movie trailers on their mobile device. This may involve linking to another application on another server in order to download video clips, or even yet another server to download the latest video player before viewing the clips, or perhaps another server to allow video streaming to the device, especially for longer clips. Somewhere in all this flow of tasks we may have to charge the user for something, perhaps the 'what's on?' information, the map, the directions, the video clip or all of them. We may also have to authenticate the user across *all* the applications. Thus, we can see that a mobile service may require a high degree of interaction and collaboration between different systems, as shown in Figure 6.4.

Figure 6.4 shows the disparate systems exchanging messages with each other. As we will see in Chapter 13 when we consider location-based services (LBSs) in some depth, messaging is a process where disparate software entities can communicate with each other without the need to wait for the message recipient (other application) to be available there and then to receive or process the message; in other words, the process is *asynchronous*.

Figure 6.4 also illustrates an important point about the devices accessing our applications. There are likely to be many different devices with varying capabilities. Not only are there myriad device models at any one time being sold in the market but there is a continuous legacy of devices in circulation with fixed functionalities that cannot be upgraded in the field as well, thus further complicating matters.

The sophisticated software platform coping with multifarious devices in large numbers is complex enough, but as Figure 6.5 reminds us we still have a wireless data network to consider, which itself is extremely complex and perhaps not as transparent as we would like. For wide area network (WAN) services, like 3G (third generation) cellular networks, this system is not only complex in terms of its interfaces and behaviours but, as we noted

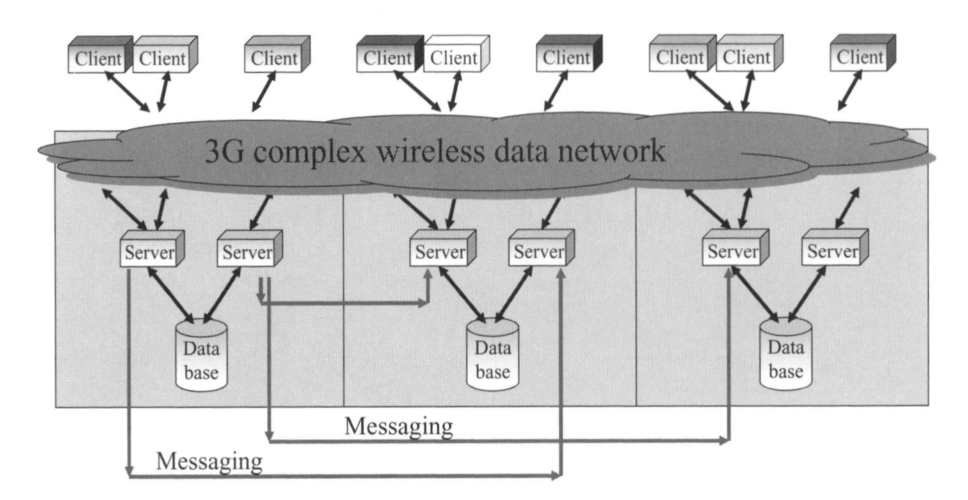

Figure 6.5 Complex system made more complex by the RF network.

in Chapter 5 (and we will examine in depth later), it also has its own internal resources that our applications may have to interface with. These include embedded services, such as a voicemail, location finding, text messaging – to name just a few. As we note in the figure, for many operators setting out along the next generation path they have no prior experience with a lot of the software technology, RF technology and device technology: everything is new! This has its own challenges, but plenty of opportunities too!

6.2 THE SPECIFIC CHALLENGES

Let's return to our earlier sliced view of the CS architecture and use Figure 6.6 to highlight some of the key challenges. We should reflect on these before we rush to build a CS system, as they highlight some useful points that could influence our design strategy for the better.

At the user interface end (namely, the device) the greatest challenge is for our applications to be *usable*. Of course, this is heavily influenced by the application design itself, but we should not forget that the device itself should not impede usability; in fact, it should offer opportunities to enhance usability. We will discuss this topic in more depth when we look at devices in Chapter 10.

At the very other end of the chain, the content itself should be *relevant*. This is perhaps obvious, but easily forgotten. Relevancy is not just about having the right content types and genres available it is also about making sure that the user's view of the content is relevant to them. For example, when viewing train times from a particular train station, clearly the user wants to view times that are relevant to that time and place. It would be no use giving information that is slightly out of date or for the wrong platform. With LBSs, ideally the user would not have to input which train station they are standing or sitting in – having been to the application before and getting the relevant train times via a programmable 'favourites'

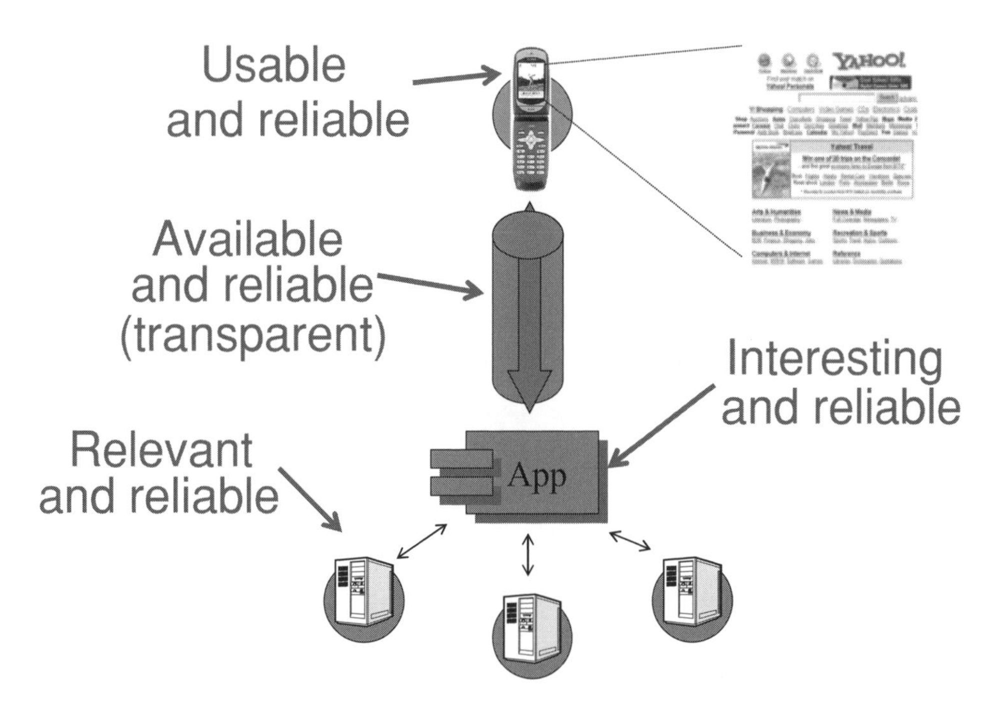

Figure 6.6 Some challenges for mobile services.

link configured in the application. This imperative of relevancy is a clear echo of Ahonen's 'me' attribute, plentifully discussed in Chapter 3.

For the physical link between the device and the back end the priority is that the link should be *available*. This is to the extent that as much as possible the RF nature of the link is practically *transparent* to the user. For example, if availability is limited to a narrower pipe than usual – due to high network congestion, say – then the user should still be able to use the services and benefit from them. The services should *adapt* to the available pipe, not merely expect the user to tolerate poorer performance than usual. Users may not have such a forgiving tolerance of the variability of mobile services, especially if they are relatively expensive to use, either in monetary times or expended effort (emotional and physical).

Finally, in considering the application layer itself, which is running on our back end servers and responsible for dishing up the relevant content and responding to users' requests, the most apt adjective that comes to mind when describing the key challenge is to make the applications *interesting* to the users. What this tells us is that the primary concern of the applications developers should be in putting as much of their effort as possible into making interesting things happen or allowing interesting things to happen: creativity is important.

There are many usage scenarios to consider and lots of interaction with other systems. Perhaps, just to provide a 'what's on?' application, it will have to work with a huge array of data sources just to grab the raw information. We then have to consider all the different presentation formats according to the vast array of devices that may want to access the content. We then have to think about how we are going to charge for the content – not just pricing, which is probably the accountant's concern – and how to make a financial transaction. Perhaps the accountants come up with some fancy scheme for banded billing,

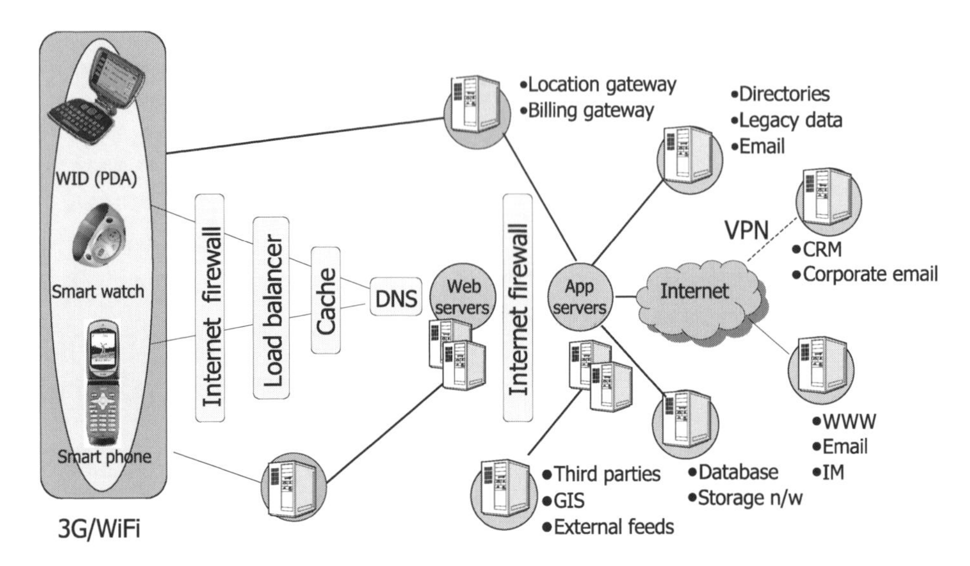

Figure 6.7 Example of the physical environment.

service bundles, add-on upgrades, pay-as-you-go plans and many other wonderful pricing schemes dreamt up in a huge spreadsheet. We probably don't want to have to keep updating our charging scheme once we have programmed the application, so we probably need to provide a pricing interface that enables our accountants' spreadsheets to prime the charging 'engine' directly; it could be updated as often as they liked. As soon as new schemes are dreamt up they are in operation the very same day, the same hour or the moment the accountant hits the 'commit' button on the wonderfully programmed Web interface provided for them.

This is all beginning to sound a bit daunting and we might be tempted to put the book down and go and do something else. Before doing that, let's first see where this is headed. Before we move on to look at how we are going to build our applications platform we should briefly reflect on its physical environment. Figure 6.7 gives us a very simplified view of the CS architecture within its physical domain.

On the left of the figure we have the device network, which has all the different device formats we have mentioned so far (we will thoroughly elaborate on these in Chapter 10). The devices can talk to each other in their own private (or shared) network, and this may at times complicate matters in the CS approach, but let us ignore such complications for now. The rounded rectangle around our device network represents the sea of RF connectivity in which the device network floats much of the time, which could be via a WAN (e.g., 3G) or local area network (e.g., WiFi). The RF network adds its own complexities too, but we will ignore these for now as we discuss them in Chapter 12 on the RF network.

The RF network interfaces with an IP network, which is the main infrastructure we use to reach our content network (back end). In this book we are not concerned so much with the physical nature of IP infrastructure, like routers, caches and the like, although we will look at some architectural concerns, like clustering of servers to enable a scalable mobile application environment. In this book our main concern with the IP network is in understanding the IP-related protocols that are needed to take advantage of the IP infrastructure in a wireless networked environment. This is principally HTTP and its wireless-optimized cousins, like WSP.

Once we get through the IP network we are now in the content network. As the figure shows, this is largely a farm of interconnected servers running our custom wireless applications, Web servers, databases and the like. We should also note how the content network could stretch out its tendrils into many other collaborative networks, including enterprise networks sitting behind firewalls somewhere else on the Internet (or via a private leased line).

6.3 SERVICE DELIVERY PLATFORMS

6.3.1 The Need for Software Services Infrastructure

What is probably obvious already is that many of the challenges for building useful mobile applications are going to be the same repeatedly. Let us list just a few of the common challenges (there are a lot more):

- handling different device types;
- coping with legacy devices;
- handling a huge user base with different account schemes, possibly with each scheme customizable by the user themselves;
- dealing with disparate data feeds including a mixture of content types, like pictures, ring tones, audio files and video clips;
- transforming data feeds to formats appropriate to our service needs;
- formatting user data to be appropriate for display on their device and according to their circumstances (Ahonen's 'me', 'moment' and 'movement');
- keeping track of users' usage patterns and preferences;
- providing current location information to enhance any mobile application that needs location enabling;
- providing a unified and openly accessible means for an application to charge the user;
- providing a means for pricing information and schemes to be kept current and to be automatically reflected in prices sent to users and added to their bills;
- providing payment mechanisms for users to engage in various commercial transactions;
- enabling users to be added to the system (and removed when necessary);
- providing discovery mechanisms so that mobile services can be discovered and subscribed to by users.

This list of potentially common requirements leads us generally to the idea of building a mobile service delivery platform (SDP) that has the capabilities to provide a lot of these infrastructural assets without having to explicitly include them (program them) into each application. Furthermore and perhaps most crucially, in trying to keep our developer community free to innovate and provide useful applications, by removing the concerns for these general housekeeping functions, they are freer to innovate.

It seems clear that SDPs as identifiable products are already emerging in the software marketplace from vendors providing these components to mobile network operators. Indeed,

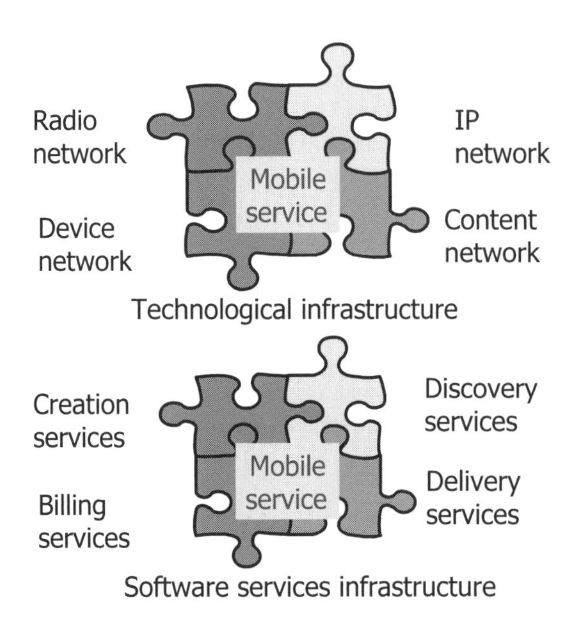

Figure 6.8 Two views of the mobile services environment.

this trend seems well under way, and the concept of an SDP is solidifying in the marketplace from vendors like Elata[3] and BEA Systems[4] coming together to form the *SDP Alliance*. In the words of the joint press release:

> *Leading software and service providers to the mobile industry – BEA, Elata, Incomit[5] and Volantis[6] – today announced the formation of the Service Delivery Platform (SDP) Alliance, an open forum to clarify issues surrounding SDP infrastructure and promote open standards and device platform independence. The SDP Alliance will help mobile operators and vendors to better understand infrastructure requirements and make informed strategic decisions to achieve their business plans for mobile data.*

In this book we will not only develop the theme of the SDP as an entity in its own right we will also explore the theme of identifying technologies suitable for building SDPs. Much of this latter discussion will centre on the Java 2 Enterprise Edition (J2EE) technology suite.

We will address these issues later in the book, but first we need to spend more time surveying the ingredients of a mobile service and application. Having done that it will become clearer what some of the issues are in managing software delivery and provisioning in a mobile network.

A useful approach when examining mobile services is to move away from a technological view of the infrastructure toward a software services-based view of the infrastructure, as illustrated in Figure 6.8. The top half of the figure shows our mobile service as something that utilizes the physical resources of the primary networks we have thus far identified as major parts of our mobile network model. The bottom half presents an alternative view as

[3] http://www.elata.com
[4] http://www.bea.com
[5] http://www.incomit.com
[6] http://www.volantis.com

far as our mobile service is concerned, which is that it only sees its environment in terms of the primary services available to it that enable it to deliver a useful service. As the figure shows, we can propose four main service areas:

1. *Creation services* – this is the availability of software services that enable us to create new services. For example, the ability to manage an appointment book in the user's diary. Whether this is a Web-based diary that sits in a networked database or a diary application that runs directly on the user's device, the implementation is not that important so long as it is possible to call on services somewhere 'in the network' to manage the diary. This enables us to create services based around diary events without having to program a diary application in the first instance. Combine this with other creation services, like a picture-messaging service, and we can glue the two together to enable a wedding anniversary reminder service by allowing pictures to be sent from one user to another on a particular anniversary occasion. A delivery florist service might be interested in such a capability. Each of these component services are potentially complex and time-consuming to develop. However, when made available as services, then our developer community is free to concentrate on interesting ways to combine these software services to create exciting mobile services. All of the housekeeping functions like handling different device types would be provided automatically. It is probably feasible to allow the users themselves to assemble their own services – 'services made to order' or 'roll your own service'. There are paradigms that could probably be developed to allow this level of self-construction or 'self-service' to take place.
2. *Discovery services* – after creating our service, which is then hosted by the SDP, we then need a way for our users to discover its existence. This is perhaps the next stage in the product (service) life cycle. To an extent, we are probably talking about a marketplace for services, and this is a concept that makes most sense within a portal framework, which is a place where users frequently go to access all manner of content and services, new and old. This is the most common discovery metaphor used by mobile network operators. In terms of a service for discovery, there is a need for mechanisms by which the application developer could make their service known to users. It would again automatically relieve the developer from developing certain housekeeping components. For example, a means for customers to register with the new service would be available as part of a common portal framework. If we wanted users to be able to personalize their service requirements within a range of possible options on offer, a common portal framework could manage this personalization. The same framework could be utilized to manage the selective offering of different service elements according to device types and to manage differential pricing and charging for the various service permutations. If the new service is tailored toward a particular demographic group or lifestyle category, then the framework could manage this process in terms of ensuring that only the target group are especially able to discover the new service. As a passing note, it is more than likely that XML (eXtensible Markup Language) would figure heavily as the means for creating service descriptors to control how the framework handles a service.
3. *Delivery services* – having created our service and made it discoverable to the target user community, our users need to be able to access the service itself. Delivery can take many forms: from enabling password access to the service to enabling content to

be downloaded to the device or both. However, another aspect of delivery is the ability of the service platform to allow inherent scalability. For example, if the new service turns out to be popular, then the platform should be able to apply sufficient resource to allow the service to scale. This should be intrinsic in the platform. There also needs to be mechanisms to facilitate any basis for a commercial agreement. For example, asking the end user to accept terms and conditions, specific licensing agreements and payment agreements should be part of a common services framework.

4. *Billing services* – once the service is made available to the user we will need to collect money from them according to our pricing strategy and charging model. This would vary in terms of complexity: from a simple one-time access fee to the service to a sophisticated event-based fee structure. For example, we could charge for downloading a game and then for accessing new levels and for high scores to be registered. Additionally, we could charge for multiplayer access and other variations, depending on what suits the gaming concept in hand. We don't want the developer community to be bothered with developing charging mechanisms, so this is provided as part of the common framework. The framework needs to provide the freedom to determine how the payments are made, such as various revenue share options via mobile operator-charging vehicles (like reverse-billed text messaging) and more straightforward credit card payment gateways, depending on what makes most sense.

In remodelling our view of the mobile services network to be a collection of software services that our application can access, we are reminded that a mobile service is ultimately a network of collaborating software programs acting in concert across the entire network, as shown in Figure 6.9, as a software services-centric model seems a useful and valid view to take.

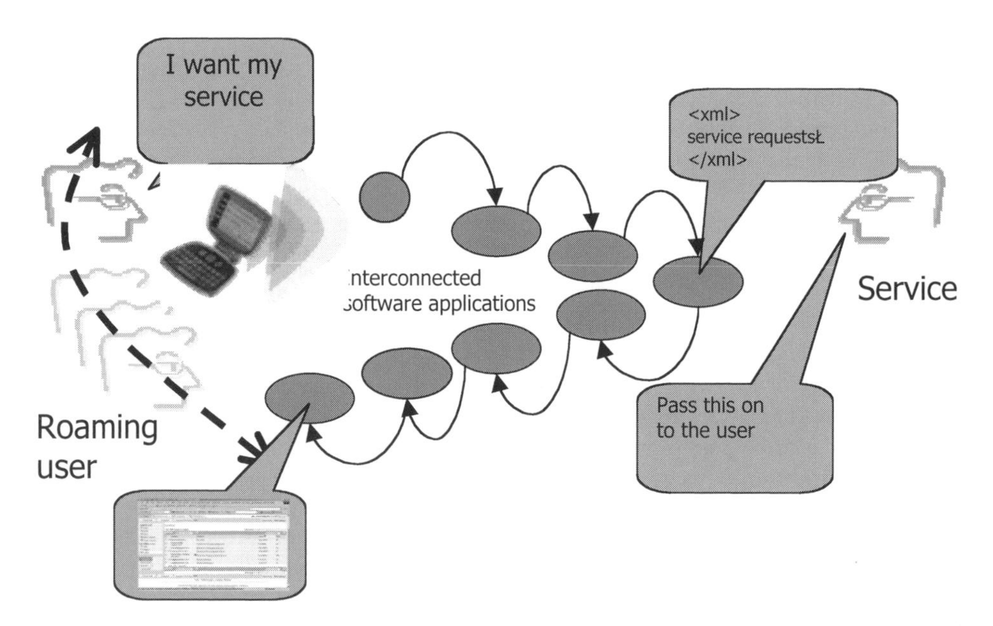

Figure 6.9 A mobile service is a network of software services.

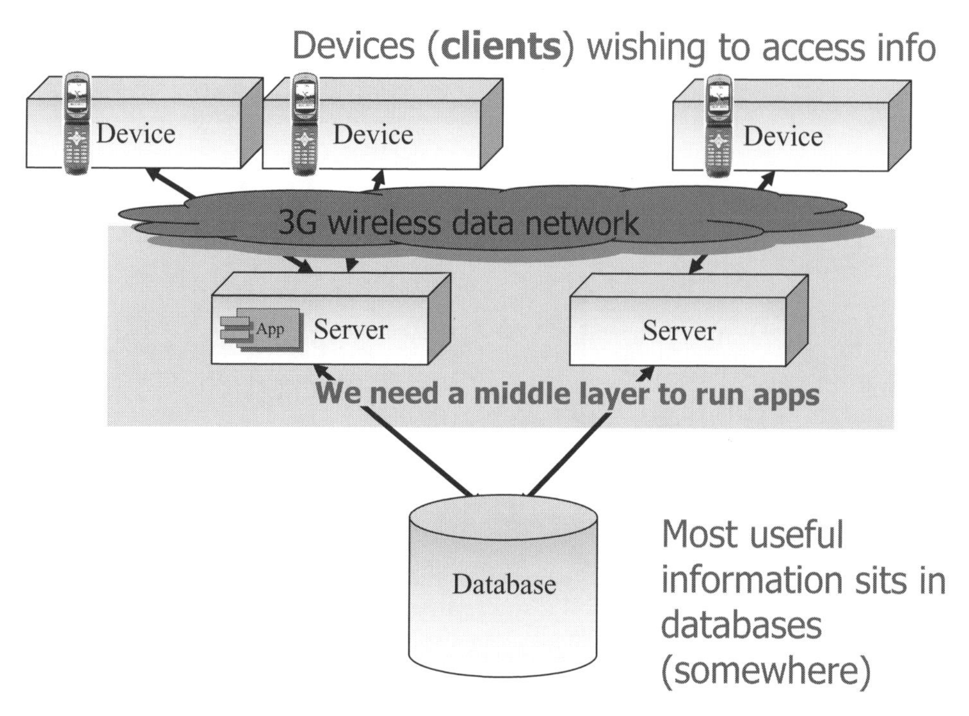

Figure 6.10 CS architecture at its simplest.

6.4 THE NEED FOR SOFTWARE SERVICES TECHNOLOGIES

If we are going to start building service delivery platforms for mobile service environments, then we need the software building blocks with which to construct such platforms. It's easy to talk about ideas like service delivery platforms and being able to provide common infrastructural services, like access to charging mechanisms and so on. But, how is this going to take place? What will be the actual implementation steps required both to realize and access such a service?

To understand what we might need let's return to our CS architecture for a moment and imagine that we are building such a system from scratch on which to assemble our mobile applications. We wish to understand the issues we may have to solve in order to build and deliver a CS system suitable for mobile services.

Figure 6.10 shows the classic CS architecture.[7] Let's try to think of all the design issues we would have to solve in order to make such a system work. You can try writing these down yourself on a piece of paper, but let me have a go first in order to convey the gist of the exercise.

[7] Why do we simply jump straight in with this architecture? The CS architecture is just a fancy name to say that we don't want to install applications either on every single device or on one server per device. Both of these are inefficient for many applications, especially those that imply a degree of centralized activity that is common to all users.

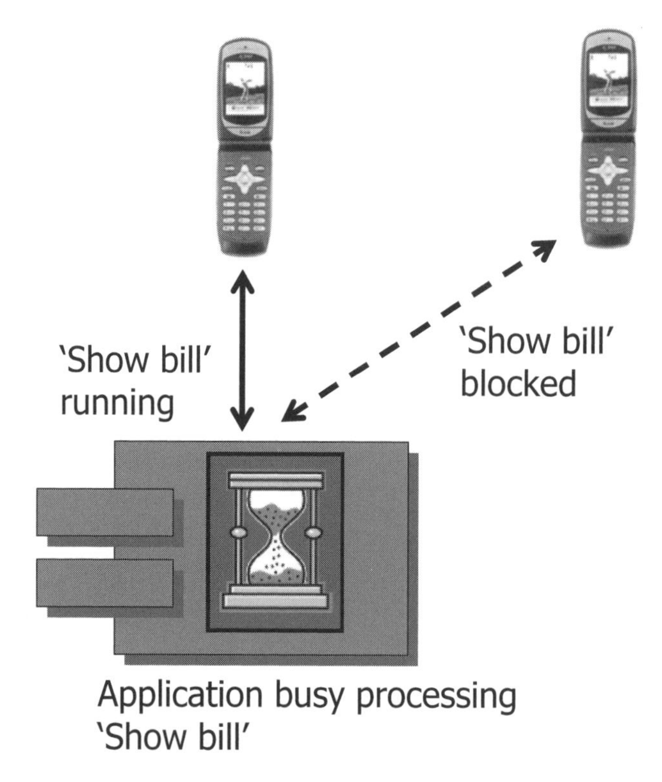

Figure 6.11 Application processing only one request at a time.

6.4.1 Example CS Design Issues

Let's take the server at the centre left of the diagram (Figure 6.10) and notice that it has two clients attempting to access the software running on it. For arguments sake, let's say that the application allows a user to see how much money they have been spending on their mobile phone bills (voice calls). We can imagine that the billing information is stored in the database at the bottom of the diagram.

Thinking carefully about the design issues and remembering that we have to program all of the above system from scratch (i.e., the server bit, not the database or the client software); some interesting challenges become apparent. These are challenges that many of us may not have pondered on before, unless we have stood back and examined the fundamentals of CS computing. For example, let's just think about the challenge of handling more than one client. We could probably write a program to present a table of billing items to the user (ignoring for now how we access the database and present the information, which are some of the other challenges). Let's say that this program receives an incoming message[8] from the device which says 'Show bill'. A naive implementation of the program will probably only support processing of one 'Show bill' message at a time.

Figure 6.11 shows what happens if our application can only process one message at a time from a client. While it's busy processing one message, other requests are unable to be

[8] By 'message' we mean something generic at this stage, not tied to any protocol.

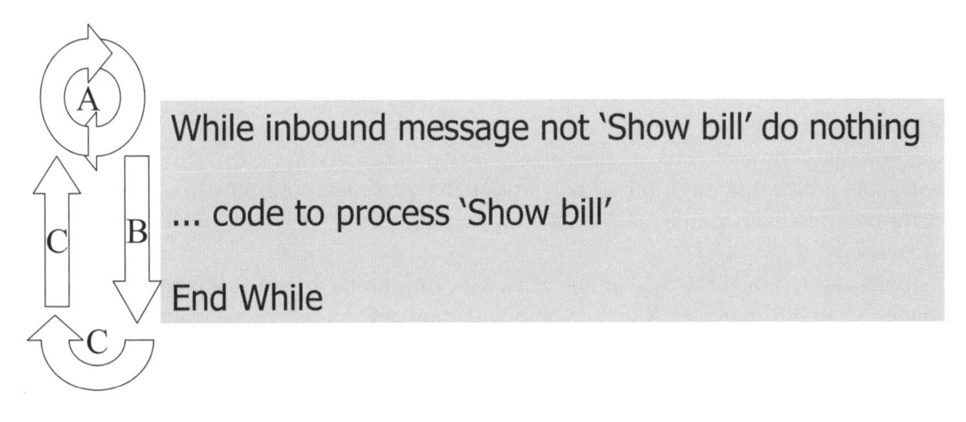

Figure 6.12 Loop code.

handled and so other users get *blocked*. Thinking about this in the crudest software terms, we can imagine that we have a processing loop in the software that sits idle waiting for inbound messages, as shown by the loop A against the piece of pseudo-code in Figure 6.12.

As soon as it gets a message the loop begins execution (path B). Thereafter, the application proceeds with execution of the code within the loop (still B) and so can no longer listen for new messages until it finishes (C) and can get back to the beginning of the loop where the listening process is reactivated (A again). In other words, during the processing part (path B) the code is busy. Consequently, during that time other requests are blocked.

This seems like we have gone back to computer science school for beginners to learn about loops; but, this simple example is rather powerful in illustrating our current enquiry into the CS architecture. Clearly, we have a problem with the implementation as shown, because it appears to imply that we can only handle one user at a time. A variety of solutions exists to solve this type of problem, but our challenge is not to suggest the solutions at this stage. We are merely attempting to identify key challenges, which have to be solved one way or another, so this is just one example of the design challenges posed by the CS architecture. The more important point is that we would have to address this problem if we were going to build such a multi-user system from scratch.

If you are feeling ambitious, then please go ahead and try to brainstorm other design issues. If you are not in the brainstorming mood, then continue reading while we uncover more of them.

If we carry on with the above example, then the next challenge we face is how to get the message to our application in the first place. For now, let's assume that we can build a Transmission Control Protocol (TCP)/IP link between the device and the server, so we don't have to worry about the really low-level data-networking details. However, we still have to think about how we trigger our processing loop. The first challenge is to agree on a format for the 'Show bill' message. Let's not complicate matters, so why not have just the string 'show bill' all in lower case letters in clear text. That should work; but, the next challenge is how to grab the message from the TCP/IP connection. For example, we could use TCP/IP to send the string in a datagram across the connection from the device;[9] but,

[9] Note that this implies that our device and the RF network can both support TCP/IP – we will come to this topic in Chapter 7.

how do we know the other end is ready to receive the message? For example, if we try sending the message while the application is busy processing loop B for another user, then the application may miss the message. If that happens, then how do we know that we missed the message and that we should initiate a retry? In other words, we need a protocol to send our message, a protocol that inherently solves all these issues. We are not going to identify a suitable protocol (though HTTP will work fine), as we are not trying to solve problems here, only highlight problems for the purposes of scoping the task of designing a CS architecture.

Let's say we found a way to solve the blocking problem and that we could service many 'Show bill' requests at once. There will come a point where we reach the resource limit of the computer running our application. At that point, the ability to handle 'Show bill' requests will saturate[10] and some users will experience degradation or even denial of service. The manifestation will likely be a delay in service as our network protocol will most likely engage in a series of retries on the user's behalf within a certain time limit. In a wireless environment that is particularly irksome the tolerance to delays is less, as usability guru Nielsen tells us:

How quickly?

0.1	seconds:	immediate
1	second:	uninterrupted flow
10	seconds:	limit of attention span
>10	seconds:	coffee break, do something else, ...

Nielsen, *Usability Engineering*,[11] chap. 5

We can see that if we block our users for a period longer than 10 seconds, then we have probably lost their attention, which means we have probably lost their business.

Previously, we had identified a blocking problem at the application level, now we have a problem at the server level. Just as we might envisage that we need more than one copy of the application attending to our user population, similarly it would seem to make sense to deploy more than one server running our application. In fact, we probably want to install a whole raft of servers if this is going to be a highly popular application with potentially millions of users. That may seem fanciful, but in a well-visited mobile portal hosted by a mobile network operator these numbers are realistic.

Using a *cluster* of servers instead of one server seems like a straightforward proposition. However, it has its own challenges. We have to think about how we assign users to servers. This clearly implies some kind of switching function to route traffic to the appropriate server in order to achieve what we call *load balancing*. This is so that one server does not end up saturated with requests while another is hardly breaking a sweat.

However, what if there is more than one message from a user? Perhaps we can envisage the 'Show bill' message followed by 'Show item', where the second message is used to

[10] The way a computer runs many loops 'at once' is to run them in succession, but very quickly such that each instance of the loop appears to be running adequately fast. Eventually, if too many loops are placed in a queue, the process will slow down.

[11] Nielsen, J., *Usability Engineering*. AP Professional, Cambridge, MA (1993).

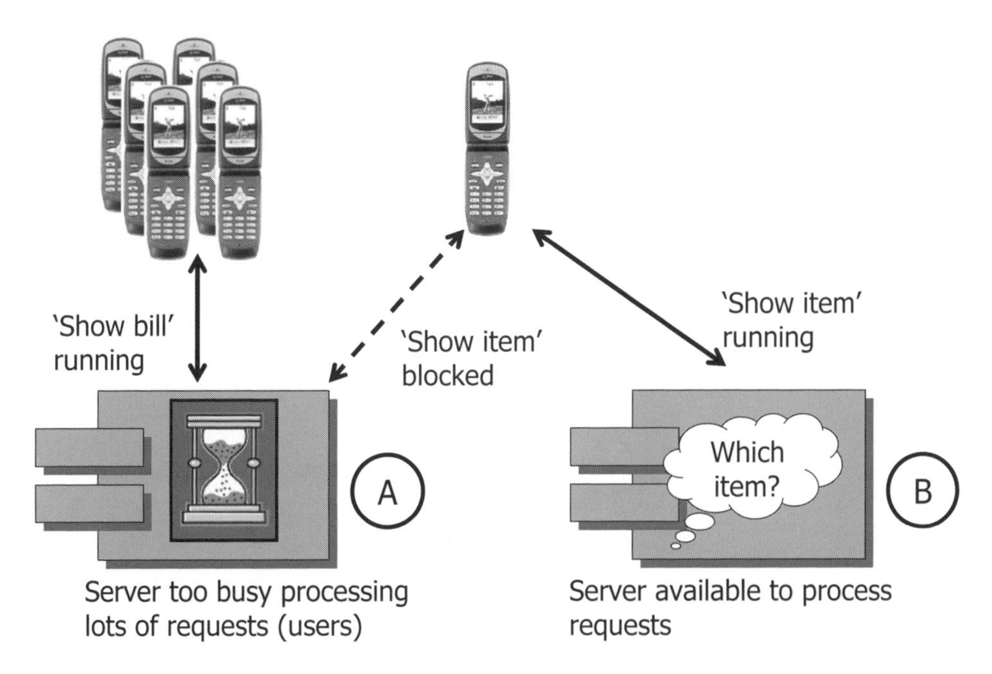

Figure 6.13 Being shifted from one server to another, the application loses track.

get information that is more detailed on a particular billing item in the bill. As Figure 6.13 shows, one of our users manages to be served by server A when the 'Show bill' message is sent. After a short while looking at the results on the user interface the user wants to see more detail, so the 'Show item' message gets sent for one of the items.[12] As the diagram shows, server A is now too busy to process the request, so somehow we manage to reroute our message to server B using a clustering process that is yet unknown (we will get to this later). However, this is where we run into a problem. The application on server B was not the original application that serviced our 'Show bill' request, and so it has no record of servicing such a request. Therefore, it cannot fulfil the 'Show item' request, as it does not know which bill is being referred to. Another possible problem is that server A may have opened the billing table for our user and has locked it (in the database server), so server B would not be able to gain access anyway.

Of course, this problem has many solutions, but we are deliberately highlighting naive implementation ideas for the purposes of pointing up design problems particular to the CS architecture.

We could continue to scrutinize the CS architecture and uncover many such problems. To save us the time of doing so, here is a list of challenges and some hints (in parens) as to the solutions. Don't worry if you don't understand the buzzwords or acronyms in the brackets – these are just here to whet your appetite and maybe to begin the process of connecting with solutions that you may have heard about, but don't yet know what they are.

[12] This is just a high-level thought exercise, so we ignore how the actual item is identified.

CS challenges:

- How do we handle lots of clients trying to access the same application? (*multithreading, resource pooling, load balancing*)
- How to implement more than one copy of the application across several servers and maintain contiguous processing and a framework for distributing load? (*low-level distributed software mechanisms, JMS, RMI, load balancing, clustering*)
- How do we optimize server access to our back end databases so that this link does not become the bottleneck? (*resource pooling, connection pooling, load balancing*)
- How do we operate a system where redundant database servers are being used to replicate data across different physical sites?
- How to handle different types of client – especially wireless?
- What protocol to use for the CS communications? (*HTTP, RMI, JMS*)
- How do we implement security at all levels of our system, such as authenticating users, setting access permissions on applications and preventing one application illegally accessing data from another?
- What protocol do we use to implement the server–database connection and how to we optimally write software to handle it, again without producing bottlenecks? (*JDBC*)
- What if any server fails – how do we swap over to another one without affecting service availability? (*fail-over*)
- What happens if something goes wrong with our wireless connection and we are unable to complete a critical sequence of tasks – how can we ensure we do not end up in an erroneous state? (*transaction monitoring*)

What we may notice about the above design challenges is that they are generic in nature, not specific to any particular type of service or application, mobile or otherwise. A pharmaceutical or financial application in a wired environment will have similar issues to contend with in a CS configuration. Therefore, we would really like not to devote our attention and resources to solving these problems, as they are *not adding any particular value to our mobile cause.*[13]

Fortunately, these problems are very well known and prevalent in many areas of the software industry. Therefore, solutions exist to tackle them for us, most notably in recent times the J2EE platform,[14] which is probably the most popular solution in use in the mobile solutions community.

[13] We are specifically talking about these low-level software process tasks, not about service delivery components common to mobile applications, such as a charging mechanism. The charging mechanism is a value-added software feature, and this would be built using the low-level software process principles we have just been discussing.
[14] http://java.sun.com/j2ee/

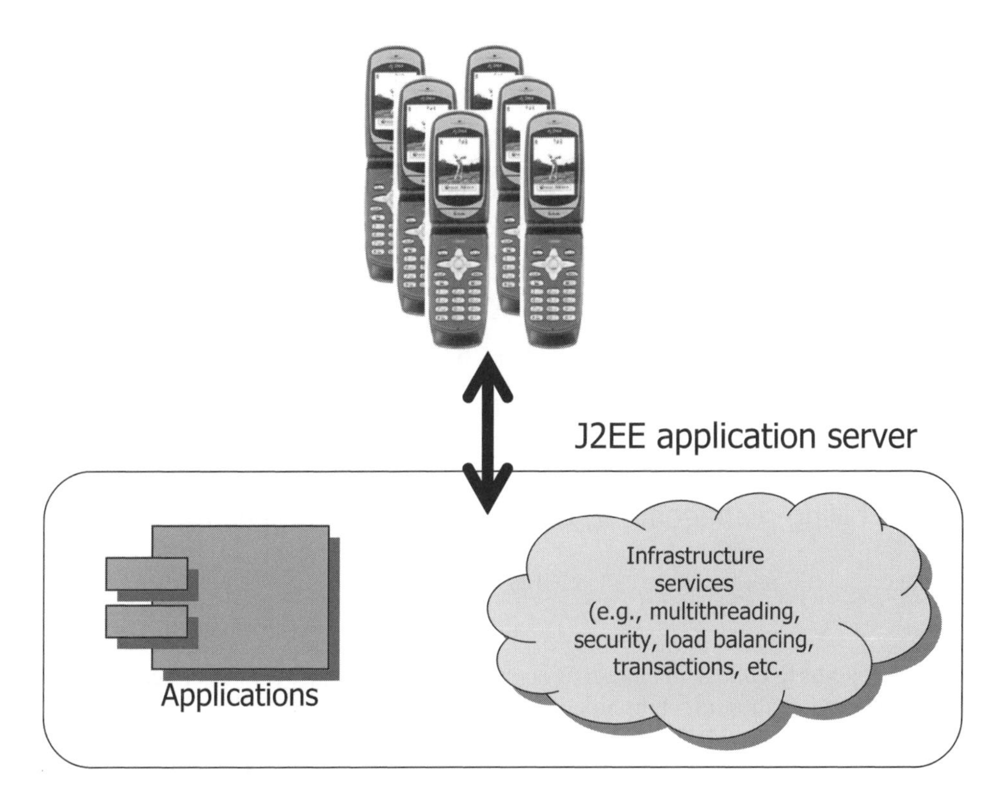

Figure 6.14 J2EE application server concept.

6.5 INTRODUCING J2EE – THE 'DIRTY STUFF' DONE FOR US!

Ideally, we want to be free to develop our mobile application software without having to develop software to take care of all the infrastructure issues we outlined in Section 6.4. The ideal solution would be to use a software platform that already provides these 'software infrastructure' services for us. This is exactly what a J2EE application server does, as shown in Figure 6.14. The J2EE platform comprises several software themes:

1. Java language support across a wide variety of underlying operating systems.
2. Infrastructure services provision for enterprise applications.
3. A programming model to enable infrastructure services to work.
4. Set of software services to provide powerful interfacing capabilities to enterprise information service tier (e.g., databases, mail servers, etc.).

Our interest for the current discussion is especially with items 2, 3 and 4. For now we are not interested in the Java programming paradigm. This is discussed in Chapter 10 when we look at using Java to program applications directly on the devices (using J2ME). For now we will just take the underlying J2EE mechanisms for granted (e.g., Java support) and not be so concerned with how the programmer writes code. We are more interested in how

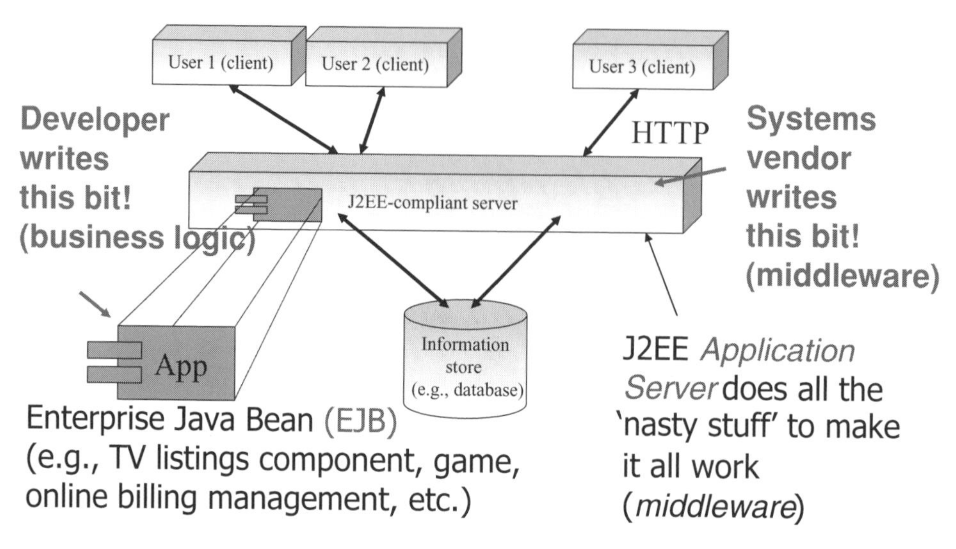

Figure 6.15 J2EE programs are called EJBs.

J2EE enables us to build a mobile services platform, so the infrastructure services concept is more relevant to the current discussion.

The essence of the J2EE approach is that developers are left to concentrate on writing software that is business-specific, while leaving all the 'dirty stuff', such as security management and distributed processing, to pre-programmed 'out of the box' infrastructure services. To do this we have to impose some rules on the software development process. First, we have to program using the Java language. Second, we have to package our custom applications as components called *Enterprise Java Beans*, or EJBs. Don't get worried about the jargon. An EJB is just a program, and for the most part we can think of EJB as just another word for 'program'. The good news is that as long as we stick to the packaging rules (known officially as the *EJB Developer Contract*), we can simply plug our EJB into any J2EE-compliant server (*application server*) and the server will take care of all the infrastructure stuff, which for the purposes of this discussion we will call *middleware*.[15]

As Figure 6.15 shows, we can envisage the EJB program as though it were a pluggable component in a J2EE system. The EJB plugs into the server and everything should work together in a harmonious fashion. We are free to focus on our EJB programming. The infrastructure services, like handling clustering and multiuser access, are all programmed by the system vendor who provides the *J2EE application server*. The beauty of this approach is that the system vendor is an expert at infrastructure services programming, and this is all they do – so we expect them to be good at it. That means we get solid and robust middleware to support our application, leaving us free to focus on the business-specific functions of our system, which we call *business logic*.

[15] The term 'middleware' tends to get used rather liberally to describe a host of ideas. In some cases an entire solution programmed to sit between our user device and some enterprise system (e.g., mail server or database) can be called middleware.

Because the J2EE specification is an open specification,[16] many vendors can enter the marketplace to provide J2EE application servers, and indeed this is what has happened. This means that wireless solution providers do not have to rely on one vendor for the underlying software platform on which to build mobile services delivery platforms. Thus, it is possible to avoid vendor lock-in, which is better for business for a number of reasons.[17] There are many J2EE vendors, such as:

1. BEA Incorporated, who provide *Web Logic Server* (WLS).
2. IBM, who provide *Websphere*.
3. Apache Jakarta (Open Source project), who provide *TomCat*.

It is up to the reader to confirm which solution is the best for their needs depending on a variety of factors. In this book we will talk about J2EE in general terms, but occasionally allude to specific features from BEA's WLS, not only because this is a highly popular solution already among mobile solution providers but also because I have more familiarity with this product than any other. Otherwise, no particular product is recommended or endorsed by this book.

To clarify a potential misunderstanding, we shouldn't jump to the erroneous conclusion that we lump all of our entire application code together into one EJB; that's not the case at all. Typically, we would divide our application design naturally into components that have functions grouped according to a common purpose. The Java programming language supports object-oriented design methodologies; so, typically, we would refer to these components as objects and they would usually reflect functional entities in the problem space we are working with. For example, returning to our earlier design problem of displaying a user's mobile phone bill on their device, previously retrieved from a database server, then we might expect an EJB called 'Bill'. This software object would respond to our messages like 'Show bill', 'Show item' and so on. If the application also allowed our use to configure their preferences for viewing billing information, we might expect an EJB called 'Preferences'. This might respond to messages like 'Show preferences', 'Set preferences' and so on. We can see how a design methodology emerges around the idea of identifying objects and the messages they respond to (and generate), but we do not discuss object-oriented design in this book.

6.6 WHY ALL THE FUSS ABOUT J2EE?

J2EE is certainly gaining a lot of momentum in the mobile services world, so we might well ask 'why all the fuss?' First, we should reflect on the circumstances we find ourselves in with next generation wireless services. As already ruminated the plethora of next generation mobile service opportunities and emerging technologies is already quite mind-boggling; we might even say perplexing (which is perhaps why you're reading this book). Therefore, as much effort as possible needs to be expended in creating new services that meet all the

[16] 'Open' means that it is publicly available for anyone to read and thereby implement a J2EE solution (application server).
[17] If we rely on one vendor and they give us poor service, we are stuck. Similarly, they may tie us into a very punitive vendor relationship that becomes costly over time due to monopoly of supply.

various criteria we have been suggesting throughout this book (e.g., Ahonen's five Ms, usability, scalability and so on). This means we do not want to spend effort on building anything that is not going to add direct value to the effort, such as optimized load-balancing algorithms and interfaces to databases, mail server interfaces or XML content feeds.

Recently, several problems have hit mobile operators at the same time. The average revenue per user (ARPU) has been trailing off as the voice markets begin to saturate and, consequently, pricing becomes more competitive in attempts to lure customers from one service provider to another (*churn*). At the same time, nonvoice services have become possible thanks to mobile data-networking technologies, new device technologies and a whole host of software technologies to choose from. This is a problem to the extent that operators are not sure what to do with the new technologies. With all this going on, there is pressure to produce new services and bring them to market quickly and cost-effectively.

In this environment of change and uncertainty, it makes good sense to invest in software platforms that provide a good deal of the software infrastructure services that we need to build reliable and scalable service delivery platforms. The existence of the J2EE open standard has already proven attractive to many mobile solution providers working in the mobile service markets. J2EE has therefore apparently emerged as the de facto standard for the mobile industry: the preferred technology with which to deliver powerful mobile services. J2EE is seemingly being adopted right across the range of systems required to support a mobile services business, not just the customer-facing services but internal business and operations support services too.

J2EE has rapidly become a widespread solution for applications delivery throughout the software industry; so, there is a growing network of support, including programmers, tools and other related solutions. An increasing number of independent software vendors supplying mobile solutions are building them using J2EE; so, the technology itself is already emerging as a preferred platform.

6.6.1 The Challenges of Integration

In the case of a mobile network operator the general ethos of service provision seems biased toward applications that include a good deal of ability to integrate with other applications and services, wherever that makes sense and at whatever level. That sounds vague, but it should become clearer as the extent of the software services horizon within the mobile context becomes more apparent as we progress through the book. To repeat an earlier example of possible integration, users would expect to be able to connect with a viewing application for cinema trailers (and movie reviews, ticket booking, etc.) from a general 'what's on?' application. In fact, probably the user doesn't know or need to know that these are two separate applications, but the integration may require a common context to be maintained. We may also require a diary application to store calendar events. After booking a cinema ticket the user may want to enter the time and date into their personal mobile diary. This will be yet another application. Again, the user should not have to leave (i.e., log off) the 'what's on?' application in order to enter (e.g., log on) the 'diary' application and then manually enter the event details. This would be a wholly unsatisfactory user experience. More than likely, all of these applications will appear via a common mobile portal, this too being an application in its own right. All the common services, such as the personal calendar, will be available via suitable *application programming interfaces* (APIs), and this will become clearer in the following chapters of the book.

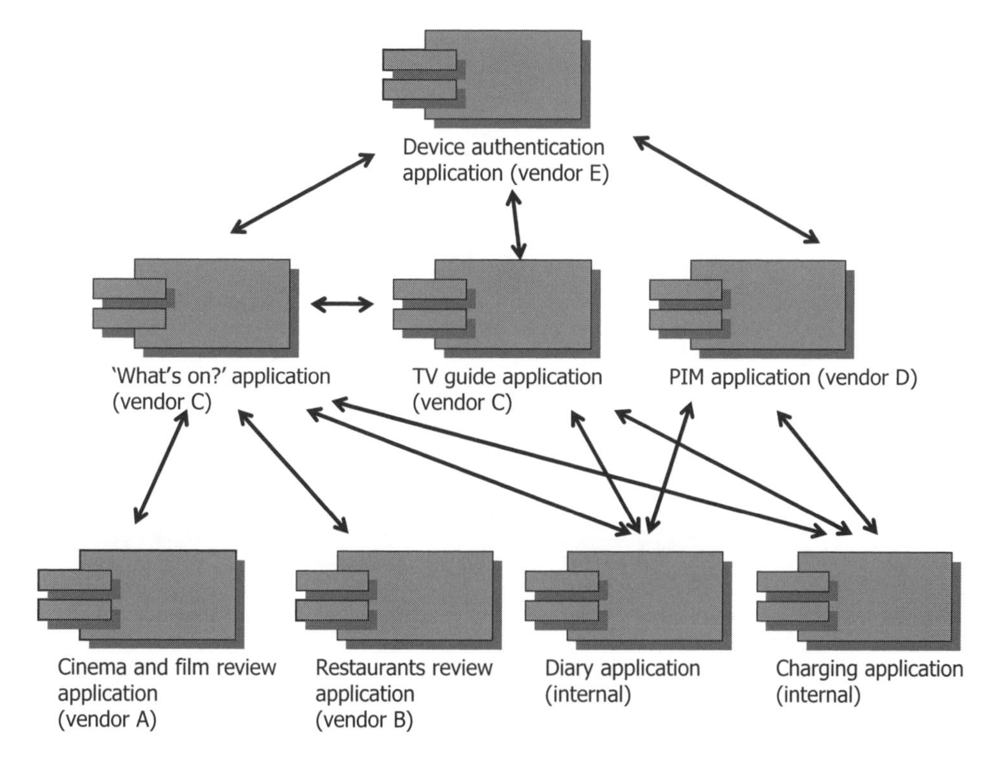

Figure 6.16 Interconnectedness of applications in mobile portfolio.

There are many challenges to integrating applications in the manner we have just suggested:

- we need a framework for integration;
- we need to decide how one application talks to another;
- if we want to apply certain constraints on applications, like support for a certain device set in a particular way, then we may need a framework to enable this to be updated without having to update each application;
- we need a means to provide unhindered access from one application to another, such that common applications, like a diary, do not get loaded to the point of breaking;
- we need a way for disparate software vendors to build applications to their internal specifications, but ones that still work within the integration schema in the hosting environment.

These challenges are each significant and collectively even more so. Figure 6.16 gives a feel for the levels of interconnectedness for just a few applications in the possible services portfolio. We should reflect on the implications of what the diagram suggests, keeping in mind that the number of applications shown is small and that perhaps a mature mobile portfolio will stretch eventually to hundreds of applications. The aggregated access to common 'house keeping' applications, like the diary or device authentication, will cause the load on these applications to be very concentrated. All the issues we have been discussing in this chapter will certainly come into play, not only for this application but also for all

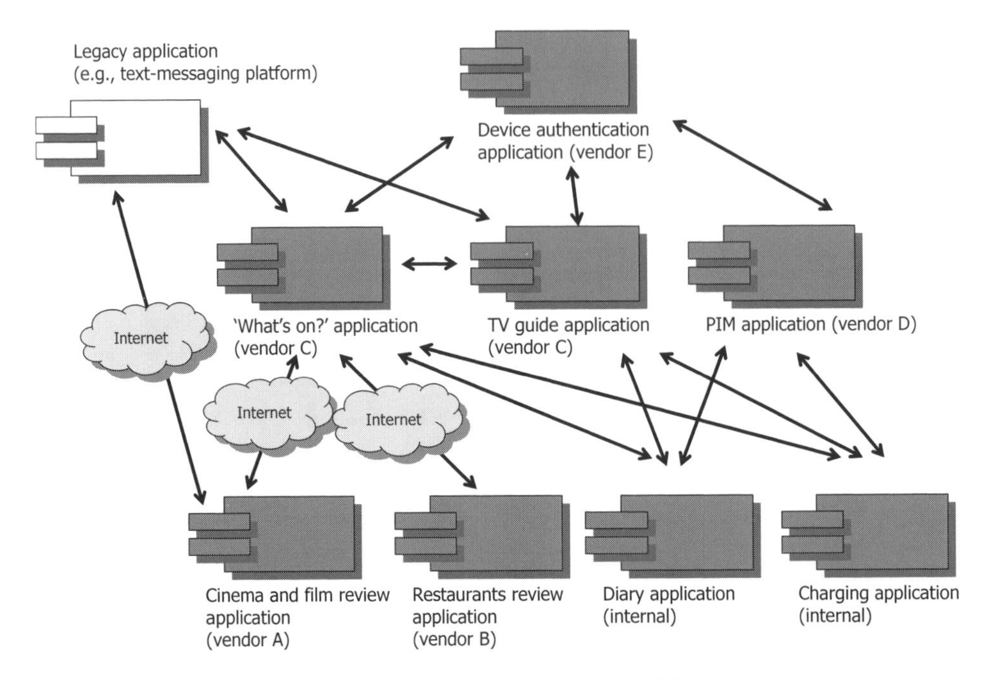

Figure 6.17 Interconnectedness compounded by Internet and legacy access.

of them. The entire application portfolio may have to handle millions of users 365 days a year, 24 hours a day, sometimes experiencing intensive peak loads.

The case has been made, at least conceptually, for requiring low-level software infrastructure services to assist with providing a robust applications environment. The need for a standard approach to integration becomes acutely apparent if we again review the diagram, this time to notice that each application is potentially coming from a different vendor. Were each vendor to adopt a unique approach toward interfacing with the service delivery platform, we can only begin to imagine the chaos that will ensue and the size of the integration task ahead. It would probably not be practical to cope with this problem in a cost-effective manner, if at all.

The adoption of J2EE as a common platform environment has made this type of integration effort much more manageable and resource-efficient. The problem is actually at two levels. As Figure 6.17 shows, it is not likely that all our applications will be physically co-hosted in the same place on the same set of servers. By necessity, some applications will be hosted elsewhere and will need to connect back into the applications framework, most likely over the Internet. However, there is also an added element to consider and cater for. We may have painted an idyllic picture wherein we deploy wonderful J2EE technology all over the place like magic dust; this is not the case. There will be a great number of legacy systems to accommodate. By 'legacy' we mean anything that already exists and is not running on the J2EE platform.

A mobile network has many resources already deployed in the network that will need exposing to the mobile services platform. Some of these legacy functions might perhaps be wrapped up as housekeeping applications, such as an existing user directory service, or some may be applications in their own right, like multimedia messaging. Therefore, we

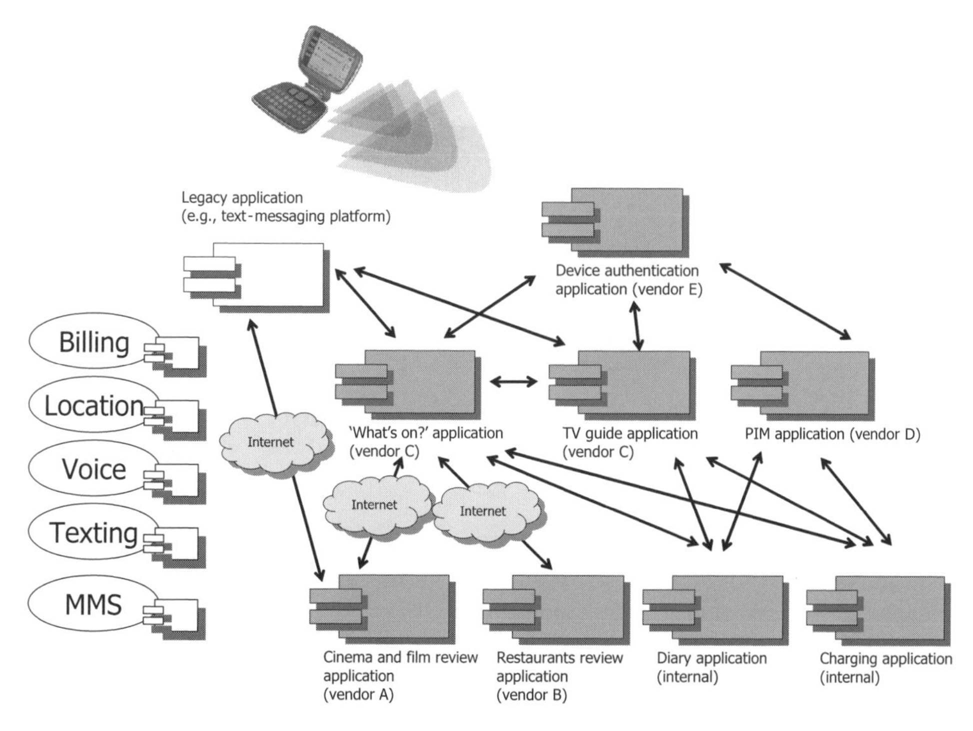

Figure 6.18 Many legacy systems in the RF network need to be accessed.

need to be able to cope with interfacing with legacy systems, not just in terms of primary connectivity but making sure that we can maintain security and scalability performance consistent with the rest of the platform as well.

We have already discussed how the RF network – in the case of a WAN such as a cellular network – has many network assets that the operator has invested in and can facilitate interesting services. It is essential that we can tap into these resources from our service delivery platform, and so support for interfacing with legacy systems becomes a powerful requirement in such a system, as shown in Figure 6.18.

The case for J2EE is even more compelling if we are able to support effective access to legacy systems. It turns out that this is exactly one of its strengths. What we need to implement the environment shown in Figure 6.18 consists predominantly in two things:

1. A set of software mechanisms to enable the interfacing.
2. A set of agreed protocols to pass meaningful messages between the collaborating software entities.

Our second requirement will be discussed in Chapter 12 on the RF network, when we look deeper into service delivery platforms – in particular, the *Open Services Architecture* (OSA) aspects of Universal Mobile Telecommunications System (UMTS) 3G networks – and the use of something called *Parlay X*. This is an initiative to make the mobile network resources available to external software entities, enabling these entities to access the mobile network and build services on top, as if the entire mobile network itself now becomes a platform for running external custom applications, like a giant 'mobile network operating

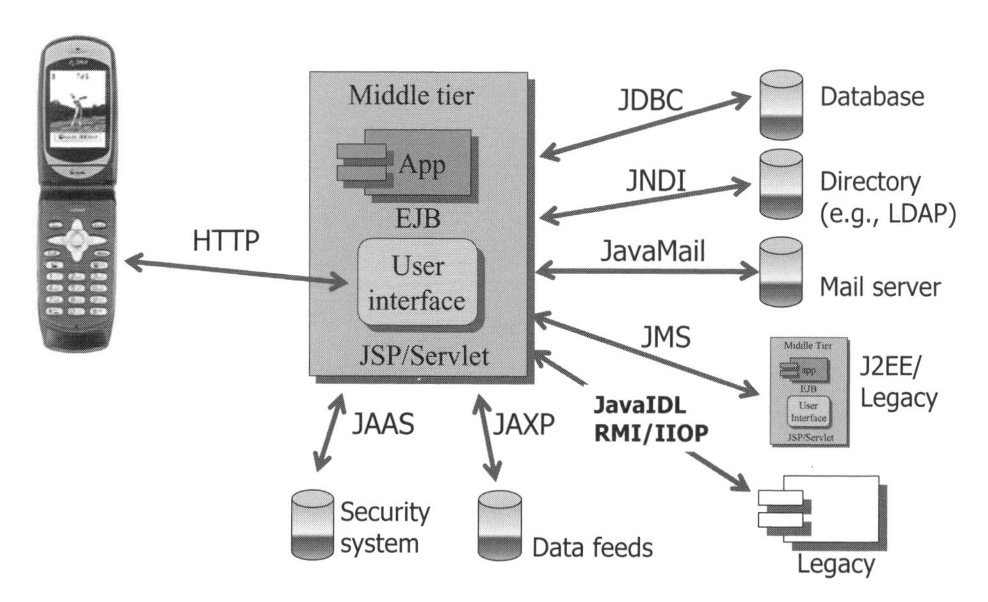

Figure 6.19 Rich set of interfaces available from J2EE services.

system'. This is clearly not an easy task. Just as with low-level software we need a set of mechanisms and interfaces to enable infrastructure services; similarly, we need suitable software mechanisms to enable OSA to take place. These have been defined already by a technology initiative (forum) called Parlay,[18] which, fortunately, has defined a particular instantiation of the interface that utilizes a loosely coupled *Web Services* interface.

Leaving aside the OSA/Parlay aspects for now, we still need to address the first issue highlighted above: providing the low-level software mechanisms to enable interconnectivity with disparate systems. Life would be hard if we had only one interface – called 'J2EE Speak', say. All our subsidiary systems, support systems, housekeeping applications and custom applications would have to be programmed to use 'J2EE Speak'; this is highly undesirable. However, this is where J2EE excels due to its rich set of system interfaces that come as standard as part of the J2EE infrastructure services, some of which are shown in Figure 6.19. Some of the interfaces supported by J2EE include:

1. *JDBC (Java DataBase Connector)* – this provides the programmer with a set of standard message formats and mechanisms to allow compliant database server products (of a wide variety) to be accessed. Due to the prevalence of database applications from the very beginning, J2EE has supported an effective database access model that has been present in Java for some time and builds on the principles of the widely successful Open DataBase Connector (ODBC) industry standard for accessing databases. From Sun's Java website, we read:

   JDBC technology is an API that lets you access virtually any tabular data source from the Java programming language. It provides cross-DBMS connectivity to a wide range of SQL databases, and now, with the new JDBC API, it also provides access to other tabular data sources, such as spreadsheets or flat files.

   JDBC is a highly developed database access solution and includes the ability to

[18] http://www.parlay.org

create, modify or delete all manner of data entities stored within a database server. The handling of data can include entire database manipulation in addition to low-level record access and manipulation within databases. Database servers from popular vendors, such as Oracle and Microsoft, are accessible via JDBC, which itself is an open specification, thus allowing any database server vendor to provide J2EE[19] access (drivers) to their database product. The JDBC API itself has a low-level form that enables these drivers to be programmed.

2. *JNDI (Java Naming and Directory Interface)* – this provides the programmer with a set of standard software message formats, interfaces and mechanisms to allow access to external entities that store information in directory structures, such as we might need to store the details of all subscribers in a mobile network, as found in GSM (Global System for Mobile Communications) and UMTS networks in the Home Location Register (HLR). The HLR is a central repository of key mobile subscriber information, such as equipment, SIM and STD numbers as well as other data, like encryption keys for user authentication. Often, vendors of HLR products provide directory protocol support, such as Lightweight Directory Access Protocol (LDAP). A product like Lucent Technologies Flexent Distributed HLR has such support, as mentioned in its product information:

   Use interfaces like Lightweight Directory Access Protocol (LDAP) to allow managed, two-way relationships between operators and third party vendors.

3. *JavaMail (Java eMail interface)* – this is self-explanatory; this interface provides software features to enable connection with any email system that supports the common Internet standards for mail transport (e.g., SMTP, POP3, IMAP4, etc). Clearly, a mobile user will more than likely have an email account, either their own account hosted by their chosen supplier or an operator account, or both (or many other possible options). The availability of a software service to handle the mechanics of email communication is clearly desirable within a mobile context. Moreover, protocols like SMTP, or Simple Mail Transfer Protocol, are also usable for communicating with nonemail systems, such as submitting push messages to a WAP Push Proxy Gateway (PPG).

4. *JMS (Java Messaging Service)* – this provides the programmer with a means to send asynchronous software messages to other software systems, like a kind of application-to-application text-messaging system (or email). Software messaging is a powerful means of enabling disparate software processes and systems to communicate, especially where the synchronous information pull mechanism of something like HTTP is not suitable for either submitting requests or fetching results. In fact, fetching information is probably not applicable to the software-messaging paradigm. We will discuss JMS in more depth in Chapter 13 when we look at LBSs. In this context, the nature of software being interrupted by location update events is very well matched to the software-messaging technique and provides a useful context within which to explain JMS. As we will see, messaging is a key consideration in the design of the J2EE platform.

[19] Other Java platforms, like the desktop standard edition (J2SE), also provide JDBC support; but, J2ME does not, as we will see later in the book.

5. *JavaIDL, RMI/IIOP (remote method invocation/Internet Inter-ORB Protocol)* – this is a low-level software mechanism for enabling one application to use the services of another as if they were both on the same machine (location transparency). It is a major underpinning technology in the J2EE platform as it enables applications to be distributed across a cluster of servers, for various reasons. This technique will be discussed later in the book.

6. *JAXP (Java API for XML Processing)* – as the name clearly suggests, this chunk of infrastructure software provides the programmer with powerful services for processing XML messages from other systems. JAXP enables applications to parse and transform XML documents independent of a particular XML-processing implementation. Mobile application and tools developers can rapidly and easily XML-enable their Java applications using JAXP. In a mobile service delivery platform, a good deal of information exchange is required to configure the applications within the common environment. Using JAXP the information exchange is implemented easily with XML.

7. *JAAS (Java Authentication and Authorization Service)* – this is a software service that enables our mobile applications to authenticate and to impose access controls on users. It implements a Java version of the standard Pluggable Authentication Module (PAM) framework, which enables new authentication technologies to be deployed without the need to change our applications to take advantage of them. For example, a message 'Authorize User' will still be supported, but underlying it will be a new (pluggable) method for carrying out the authentication process. The JAAS is flexible enough the allow integration of any legacy method of authentication that an operator or service provider might already have in place.

8. *HTTP (HyperText Transfer Protocol)* – already discussed in the book, the HTTP protocol stack is a powerful connection paradigm and the standard way by which applications will connect with client devices. The entire protocol-handling mechanism is available as a built-in software service on a J2EE platform. Many of its features will be discussed elsewhere in the book.

The existence of these powerful interfaces in the J2EE platform make it an ideal base on which to build powerful applications that can integrate with other systems, which is, and will be, the nature of any scalable and useful mobile services. The intrinsic capabilities of the J2EE platform to support these interfaces in a highly scalable and robust manner, while providing all manner of key, underpinning, software building blocks makes it an ideal choice on which to build mobile services. This will become much more evident, if it isn't already. We mentioned at the outset of the book that we aim to become 'smart integrators', using existing and proven technologies to build value-added services with great speed, moving rapidly with the tide of emerging next generation technologies and not against it. J2EE is a smart set of technologies that provides a worthy candidate to be a major part of the puzzle in building effective mobile services.

7

Content-Sharing Protocols Vital to Mobile Services

7.1 BROWSING

The browsing paradigm made possible by HTTP (HyperText Transfer Protocol) and HTML (HyperText Markup Language) has become a key mechanism for liberating information held in databases. A virtuous circle arose out of the universality of the browser, once it had reached critical mass. More browsers in circulation heightened the appeal of offering services that could be accessed via the browser. Many applications and services have since become 'Web-enabled'. Moreover, HTTP has become so widespread that it is no longer confined to browsers accessing content via Web servers. As we learnt in Chapter 5, HTTP can be used to gain access to any files on remote servers. If we combine these two capabilities, then we have an effective mechanism for 'surf-and-collect': the ability to discover and retrieve interesting content for mobile devices. This makes HTTP a useful backbone for downloading all kinds of content to a mobile device, such as games, pictures, logos, ring tones and so forth.

Many machine-to-machine solutions (M2M) have adopted HTTP as the means to exchange information, clearly demonstrating that this mode of exchange is no longer concerned solely with retrieval of visual information. In an M2M configuration we don't have a human at the other end wanting to view information: we have a machine wanting to consume it for some purpose or another. This is a trend that we also see emerging with the growing popularity of Web Services, a generalized means for any IT systems to exchange information using HTTP and the Web infrastructure.

There are yet other uses for HTTP. Over-the-air (OTA) synchronization of contact and diary information using the new Synchronization Markup Language (SyncML) can be done using HTTP. Querying a mobile location centre to get the coordinates of a mobile phone is done using HTTP. In fact, the list of uses for HTTP is growing daily and has extended

Next Generation Wireless Applications P. Golding
 ISBN: 0-470-86986-0 (HB)

way beyond its original scope (hyperlinked document retrieval). These days, even a lowly vending machine might be connected via HTTP.

This universal adoption of HTTP for many applications has taken place because software implementation folk are now very used to the paradigm, and there is plenty of know-how, tools, software and infrastructure available to implement and support it. Indeed, the availability of Web servers, Web server code, HTTP code and a whole host of related utilities often means there are very real cost and efficiency benefits in using HTTP. Coupled with its familiarity, this adds to the growing momentum behind the protocol.

HTTP and HTML (now XHTML, or eXtensible HTML) has become very widespread in general computing and now in mobile computing as well. It is therefore valuable to scrutinize its workings a little deeper than we did in the previous chapters where we have so far only discussed its essence, not its inner core. As we will learn, one problem with HTTP is that it was designed and has since evolved[1] very much with the existing Internet, data networks and abundant computing resources as the assumed ingredients at its disposal. This assumption does not hold for wireless networks.

In Chapter 6, when discussing HTTP and HTML at a very high level, we identified several challenges for the wireless environment. These were mainly to do with the limitations on connectivity (slower links, possibly intermittent in nature) and limitations or variations in user interface capabilities (screen sizes, keypad constraints, etc.). In Chapter 13 we will learn in some detail about device architectures and technologies, as well as how the RF (Radio Frequency) network functions. We will then better understand the nature of these limitations, although the superficial differences between the mobile computing world and the fixed one should already be obvious.

To address the limitations of the browsing model for wireless services, a collaborative industry effort arose that eventually spawned a new protocol family under the umbrella title Wireless Access Protocol, or WAP. In this chapter we will chart a course through the browser paradigm, explaining in some detail how HTTP and HTML work while explaining the WAP optimizations and other alternatives.

7.2 LINK–FETCH–RESPONSE ANATOMY (HTTP AND OTHERS)

This section is titled 'link–fetch–response' as this string of words reflects nicely the concept behind the browsing paradigm, as already briefly explained in Chapter 6. The idea is that we first link to a resource, usually via the user actually clicking on a link displayed in the browser or by entering an address that uniquely points to the resource. Linking to the resource causes the browser to initiate a *fetch* of the linked-to resource, which may be a Web page or some other content type, like a descriptor for a ring tone or a game. Later on we will look at the download mechanism for media files and games in more detail. The fetch request from a server hopefully causes the server to generate a response.

HTTP was very much designed with this cycle in mind and by assuming that an application called a *browser* was on the requesting end doing the fetching; so, the details of the mechanism have been designed to facilitate certain aspects of a browsing solution that arise when thinking about its specific design challenges. However, we should caution that the

[1] What very little that it has evolved, as its original design concept and specification has not needed much in the way of revision. It is also extensible in other ways outside its core protocol.

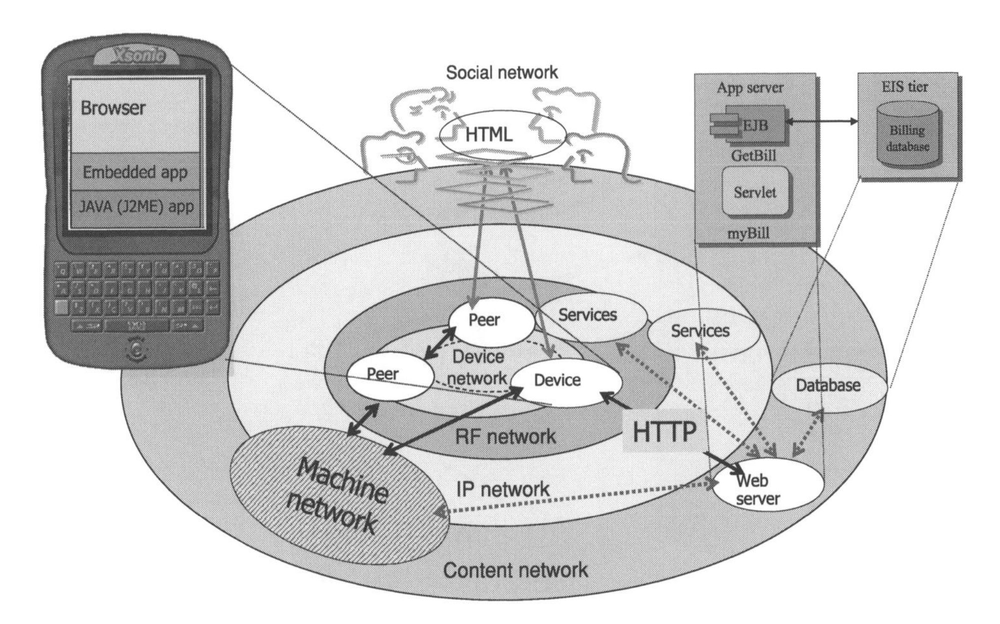

Figure 7.1 Device talking to server via the mobile network or networks.

requestor may not necessarily be a browser and, hence, the general term used is *user agent*, the terminology used in the HTTP specification.[2]

What we assume for HTTP is that the underlying transport mechanism used to shuttle fetch and response messages is the Internet Protocol (IP). This provides us with the essential means to connect one device with another or one software program with another running on physically separate devices. We do not describe IP here and the reader is referred elsewhere[1] to understand the details. All we need know is that IP provides the means to shift packets around, including a way of addressing those packets so that they can reach the required destination. Underlying the IP infrastructure is a bunch of devices called routers that ensure packets reach their destinations.

In Figure 7.1 we are reminded of our network topology and that our device layer interacts across a multitude of interfaces to get to the content layer. In this chapter the IP layer is the focus of our attention. We are not concerned with how the RF (Radio Frequency) network functions other than we assume for now that it is able to transport messages using the protocols that we are going to discuss here. On most 2.5 and 3G mobile networks, as well as alternatives like WiFi, it is reasonable to assume that the RF networks can support IP connectivity. In Chapter 13 we will examine how the RF network supports IP, but for now we won't concern ourselves with this issue. However, as we will come to learn, the IP connectivity offered by an RF network has unique characteristics that force us to consider better ways to implement the higher protocol layers, such as those under consideration for the link–fetch–response paradigm. This will soon become apparent when we look at WAP.

In the figure we can see the device network is able to run three primary types of application: browser-based, embedded or Java.[3] We examine each of these programming paradigms

[2] HTTP 1.1 found at http://www.w3.org/Protocols/rfc2068/rfc2068
[3] More specifically, we mean Java MIDlets, which we consider in Chapters 10 and 11.

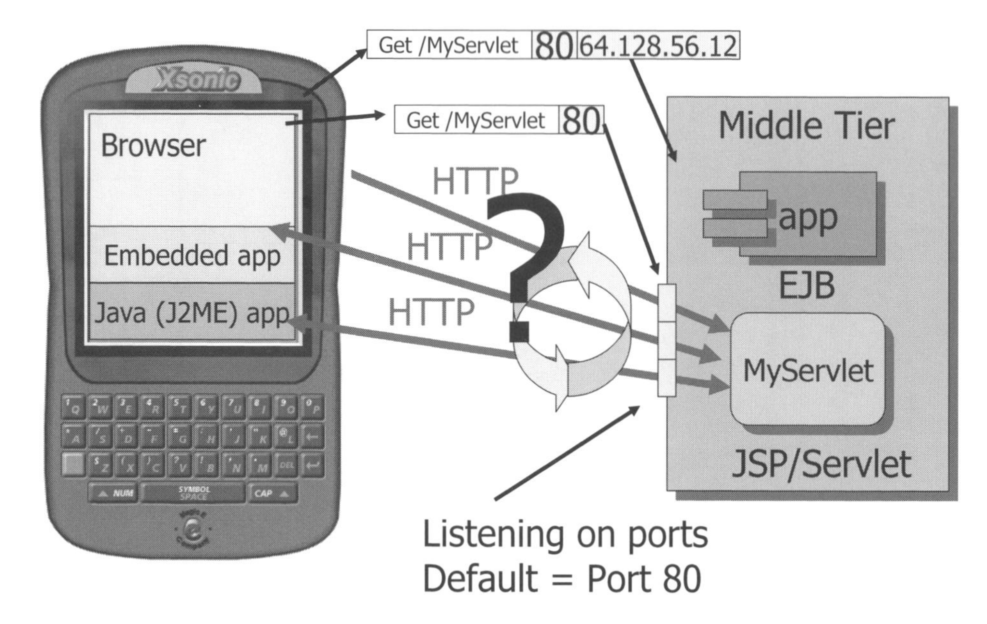

Figure 7.2 HTTP gets used in all the device programming paradigms.

in depth in a later chapter, but all we need to consider for the current discussion is that each of the paradigms will generally interact with the content network using HTTP or a similar protocol, like Wireless Session Protocol (WSP) from the WAP set of protocols.

In Figure 7.2 we see how all the main application paradigms can utilize HTTP to talk with the back end (server). The figure also shows how the back end has its own internal architecture, here reflecting the Java 2 Enterprise Edition (J2EE) method of delegating HTTP handling to programs called servlets and JSPs (Java Server Pages), which we describe in detail in Chapter 8.

In the figure there is a hint at how the HTTP protocol works. We can see messages called 'Get' requests being sent from the user agent on the device to the server. This is the name of the HTTP method that initiates requests for resources from the responding end of the HTTP dialogue, which is typically A web server, but is officially known as an *origin server* in the vernacular of the HTTP specification.

What is also shown in the figure is a reminder that the GET method requests for resources on the origin server do not reach the server by magic. They are routed via the IP network – the IP address being appended to the requests to enable routing to the appropriate server. The physical server that hosts our origin server not only has an IP address it also has some important additional addressing information called a port number. The default for HTTP is port No. 80. This means that any message sent on this port number will get processed by our origin server located on the corresponding physical server with the target IP address.

From now on we will not concern ourselves much with IP addressing and port numbers. We will simply assume that our user agent has the means available to it to enable HTTP messages to be carried to their desired location via an IP mechanism implemented on the device and within the RF and IP networks. What this usually means, as we will see when we look at devices in more detail in Chapter 10, is that the device will have another application

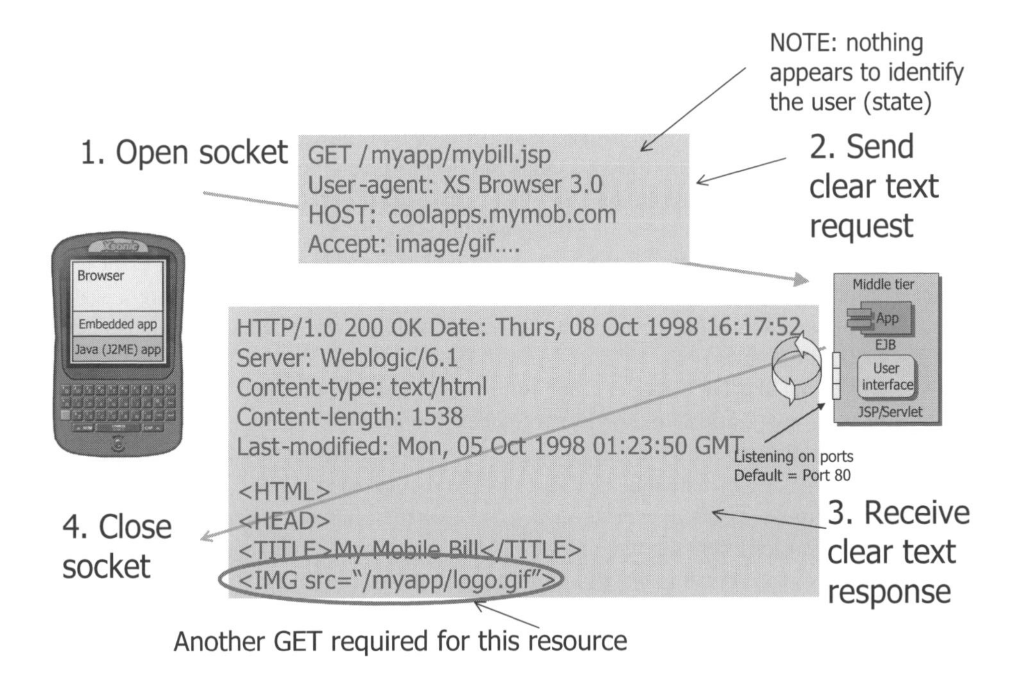

Figure 7.3 Basic HTTP request–response cycle.

running on it besides the browser, called an *IP stack*. The user agent actually makes its requests to send and receive messages to the IP stack, which in turn packages the messages up using the underlying IP protocols and routes these data packets to the RF modem included on the device. This mechanism will become clearer when we look at devices in Chapter 10; so, don't worry about it just now.

Our default or idle condition in the mobile network is to assume that our device and content network have an available means to communicate and that messaging pathways are open and ready to transport messages at the whim of the user agent. We also assume that our device will already have a valid IP address assigned to it, as will the physical target server (we will look at how IP addresses are assigned to devices when we look at the RF network in Chapter 12). Our server is now waiting and actively listening out for messages on its port (port 80); it is ready to respond as and when these messages come in. Let's now look in more detail at how this process works.

In Figure 7.3 we can see the very basic anatomy of a HTTP request–response cycle. Let's go through it step by step. An IP connection is established between the user agent and the origin server in readiness to pass messages – this is called opening a socket.[4]

To make a page request, what amounts to a small text file is sent to the origin server from the user agent. It contains a list of text lines, each one containing what we call a header field. We can see that the opening header starts with the keyword 'GET', thus denoting that

[4] Two software processes communicate via TCP (Transmission Control Protocol) sockets, which is originally a name given to the association of a Unix communications process with a TCP/IP link. It is really the idea of a system interface provided by the underlying platform (e.g., operating system) to applications, enabling them to bind a TCP/IP stream to a given port number and IP address. These bindings are uniquely identified so that several can be supported at once. The binding can be considered a socket.

this message is a GET method. Next to the keyword is a parameter. This is a pathname to a resource on the origin server. The resource – some kind of executable file – is required to generate content that the server subsequently sends back to the user agent. In this example we are showing a file name with a '.jsp' extension, which indicates that the resource is a type of Java program. In cases like this the program must execute on the origin server and programmatically produce the content. In other words, the program itself is not the content; we don't want the JSP file, we want what it produces for us, which typically will involve getting data from a database and formatting it for display in our browser interface to the user agent. If the file were marked with an .html extension (or similar), then the origin server opens that file and sends back its contents directly, which is the content that is displayed directly by the browser.

After receiving the GET request, our origin server either executes the file if it is an executable program (e.g., JSP or servlet) or fetches its contents directly (e.g., from a HTML file); it then attempts to send the output back as a text file message to our user agent. The response message begins with a header, just like the request message did, but this time with the resource output appended, everything sent in clear (i.e., human-readable) text. If this is the end of the session, then the two end points can dispense with the active IP connection by closing the socket. For this discussion it is not important to know what a socket is nor what it means in low-level software terms. Such concepts are of interest to programmers who implement IP stacks and of interest to the programmers who subsequently access these programs via an application programming interface (API).

7.3 IMPORTANT DETAIL IN THE HTTP HEADERS

Before going on to analyse the basics of the HTTP mechanism in the mobile context, let's make a few observations about the process just described.

In the request cycle (see Figure 7.3) we can see a header called *user agent*, which is a string that nominally informs the origin server what type of user agent is making the request. For example, it could indicate that the user agent is a particular browser vendor running on a particular mobile device. This information is potentially useful to the origin server, particularly in the mobile context. This is because there are many different types of device: from limited display mass-manufactured handsets, perhaps second generation (e.g., GSM without GPRS, like the Nokia 7110), to very feature-rich larger format colour devices (like the Sony-Ericsson P800). Even within the same device class there can be vast differences in display capabilities between different generations of device, as shown in Figure 7.4. Clearly, we would not want to send a response with detailed colour graphics to a device that is unable to display them. Thus, we can use the user agent header to assist our origin server with deciding what content to dish up according to device characteristics.

In theory this process is also aided by the *accept* header, which we can also see in the example. This is a list of content types that the user agent can accept. For example, if the device can display GIF images, then it says so in the accept header. If it can't, then this is omitted and the origin server can avoid sending such content. This is not the same as content *adaptation*, which we might perform using the user agent header. In the case of adapting the content dynamically, we may adjust the entire output to suit the user agent interface. This is what we might call *capability negotiation*. In the case of the acceptance of content we may still have the same content being generated by our application logic (i.e., the code within our JSP), but the origin server blindly does not bother sending any content

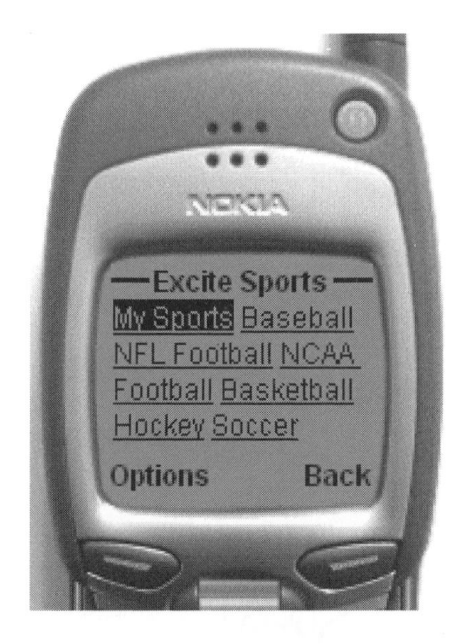

Figure 7.4 Differences in browser capabilities for mass market phones. Reproduced by permission of Nokia.

that the user agent has not said it can accept. So, in the case of embedded images that can't be accepted the server simply doesn't bother with sending them, whereas ideally what we need is for the page to be formatted differently in this case.

The original request may lead to a response that contains a Web page with embedded images. Images referred to in a Web page file (HTML file) are not stored in the file itself; they are referenced using image tags that point to the images stored somewhere else on the Web server (which could be on a different server from the origin server). Therefore, in order to display the entire page properly, including the images, the user agent issues a set of repeated requests to go get (literally 'GET') the remaining content, one image at a time, request by request, as shown in Figure 7.5. In other words, each page element results in a separate HTTP request. The results from these requests are visually aggregated by the browser.

Each embedded item is going to result in another request. Over a wireless connection this may not be a good idea because we might not have enough bandwidth to afford wasting some of it on the overhead of making several requests just to view a single page. On the other hand, we might deem this staggered fetch to be an advantage as it is possible that we could load the HTML into our browser first and be able to display the basic information on the page while waiting to load the images. In other words, we achieve a kind of multitasking or parallelism: while the reader starts reading we load the images in 'background mode'. Mind you, it is not a good idea to rely on this technique as a means to improve usability, because a user psychologically may feel the need to wait for the entire page to be loaded, thereby only adding to their frustration, not easing it. But, we will look at this issue in more depth as we proceed with the discussion, since there are other factors affecting page-loading and viewing time (latency).

We should not think that the overhead of making several fetches versus one fetch is trivial. Just by looking at the HTTP headers, perhaps we are tempted to think that the overhead is only a few lines of text, like three GET strings instead of one in our example and ditto for

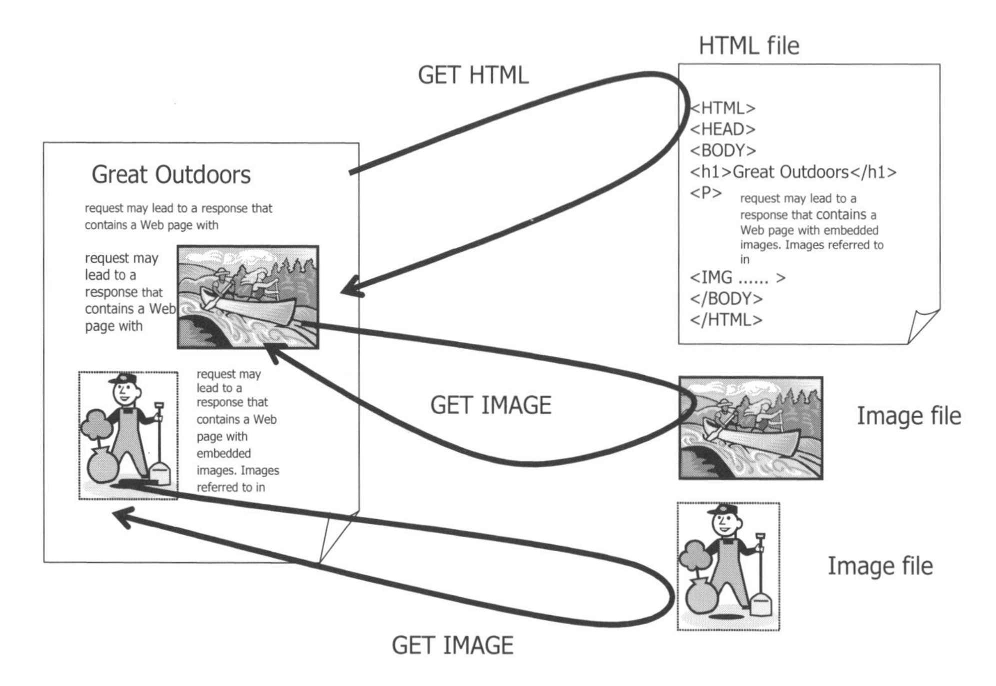

Figure 7.5 Browser page made from several files means several requests.

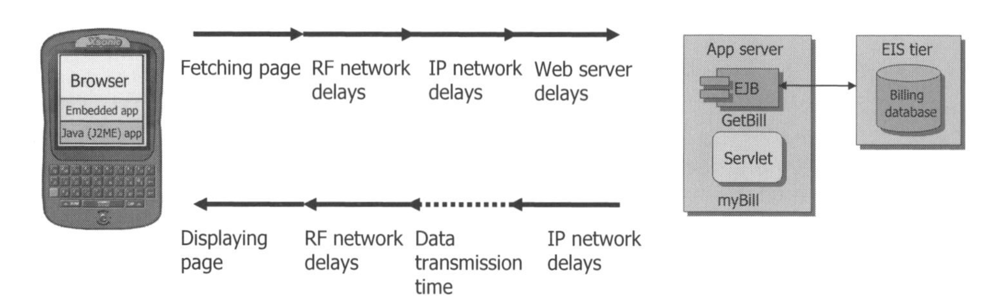

Figure 7.6 Example of delays in a network.

each HTTP header. This is true in terms of the actual amount of data fetched, but network latency is the problem, not necessarily the amount of data that need to be transported.

In any network there are delays, as shown in Figure 7.6. These are due to a variety of reasons; but, they are often due to buffering of data at different points, or network congestion, or waiting for resource allocation to occur, or possibly by virtue of the system design (such as unavoidably long intervals of interleaving[5] on an RF network). When we add up all these delays, they can become the dominant factor in network latency compared with the actual time spent sending the data (see Figure 7.6), which is why making unnecessary round trips in any type of network is not good design, particularly in a network that is prone to high

[5] Interleaving is the process of spreading out a chunk of data over many blocks of data, only using a small portion of each block to send the data. This is usually done to minimize the chances of a network error burst taking out an entire chunk of data.

latency (like many RF networks), where it should especially be avoided as it can have a severe impact on performance.

In browser-based communications using RF networks we really should aim to minimize the number of network requests to the minimum necessary to carry out the task in hand. There are a number of design strategies that can be deployed, as well as overhauls of the protocols to facilitate a more reliable method, which is what WAP is all about, as we will shortly see.

A key part of the header is the opening field of the response. This states the HTTP protocol version supported by the server followed by a status code. In the example given the status code is indicated by the string '200 OK', which means that the request was serviced without any problems. The format of the status is always a numeric code followed by a readable English summary of its meaning. We are perhaps more familiar with some of the other codes as we often see them cropping up when surfing the Web: a code like '404 File Not Found' is one that most of us are painfully familiar with. Later on, when we discuss authentication and security issues in Chapter 9, we will see that '401 Unauthorized' is an interesting status code that is used to initiate a secure session whereby the user agent (and in turn the user) must undergo authentication in order to verify that a request for content is legitimate. This authentication process is the kind we expect to find when using a Web-based email client for checking a private email account.

In the HTTP response shown in Figure 7.3 we can see the HTTP header contains a header *Last-Modified*. This enables the user agent to know how fresh the file is. This is useful for caching of information; this is a process where any resource we have previously fetched from a URL (uniform resource locator) is stored in local memory on the device. This process is illustrated in Figure 7.7. Step 1 is the ordinary GET process to retrieve all

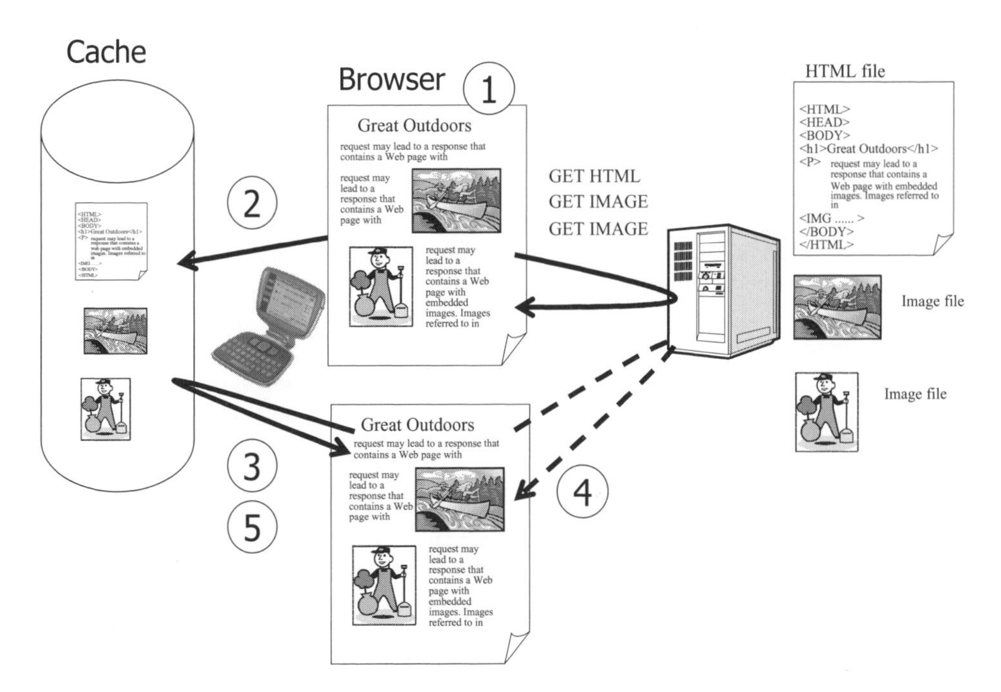

Figure 7.7 Browser checks its cache to see if content is already loaded.

the content from the URL in the headers. In step 2 the browser stores the content in an area of nonvolatile memory on a device called a cache. If the user later on requests the same resource, as in step 3, then the browser is able to detect that this content is already sitting in its cache. However, prior to loading the content the browser has to check that the content on the server has not been modified since it was last retrieved from the server, which is step 4. This step does not involve performing a fully blown retrieval of the content, as that would be self-defeating; in this step, aspects of the HTTP protocol are used to conditionally fetch the requested resource. The user agent adds an *If-Modified-Since* header, applying the timestamp from the *Last-Modified* header the last time the resource was actually fetched from the server. If this is not more recent than the timestamp for the same content in the cache, then the server issues a '304 Not Modified' status code and the browser loads the content from the cache instead, as shown in step 5.

This caching strategy saves unnecessary fetching of content, which not only saves the user power (battery consumption) and money (airtime consumption[6]) but also makes page loading much quicker because the browser is able to pull the objects from device memory very quickly indeed. There are other ways to implement tagging of content for the purposes of validating its freshness in the cache, but these are beyond the scope of this book.

We do not wish to ponder too much on the various HTTP headers and their usages, as we are more concerned with the specifics of using this protocol over a wireless link or with the possibility of alternative protocols that might be more efficient. To determine what we mean by efficiency, we defer this discussion until Section 7.5.4 when we look at *Wireless Profile HTTP* from the WAP family of protocols. Here, we will also look at some of the other headers and how they are especially relevant to wireless communications, even though this was not the original intention of the protocol designers. We will also examine how the most recent version of HTTP – 1.1 – is better suited to wireless communications than its predecessor 1.0, although most Web servers and Web clients these days use HTTP 1.1.

7.4 THE CHALLENGES FOR HTTP OVER A WIRELESS LINK

Having discovered how HTTP enables chunks of information to be retrieved from an origin server, we might well ask what impact a wireless link has on the process. This is a sensible question and, as this is our main concern in this book, it makes sense to spend time looking into such issues rather than scrutinizing the standard HTTP configuration in more depth. As we will see, to overcome significant problems due to poor wireless link performance we have to resort to alternative protocols, although still very similar in nature to HTTP.

First, let's look at what the characteristics of an RF connection might be with respect to establishing HTTP-like communications over the link. During a communications session between a mobile device and its communications companion (e.g., server), the data to be passed are sporadic: sometimes there is something to be sent, sometimes there isn't. But when there are data to be sent, such as a picture file, then the two ends have to synchronize to ensure that the communication is successful.

It is generally the case that, even in relatively reliable wired transmissions, it is not possible to transmit data without experiencing some errors in the transmission, which manifest

[6] More accurately, we mean money spent on transferring data, which is more likely billed by data amount, not transmission time (though the two are related, of course).

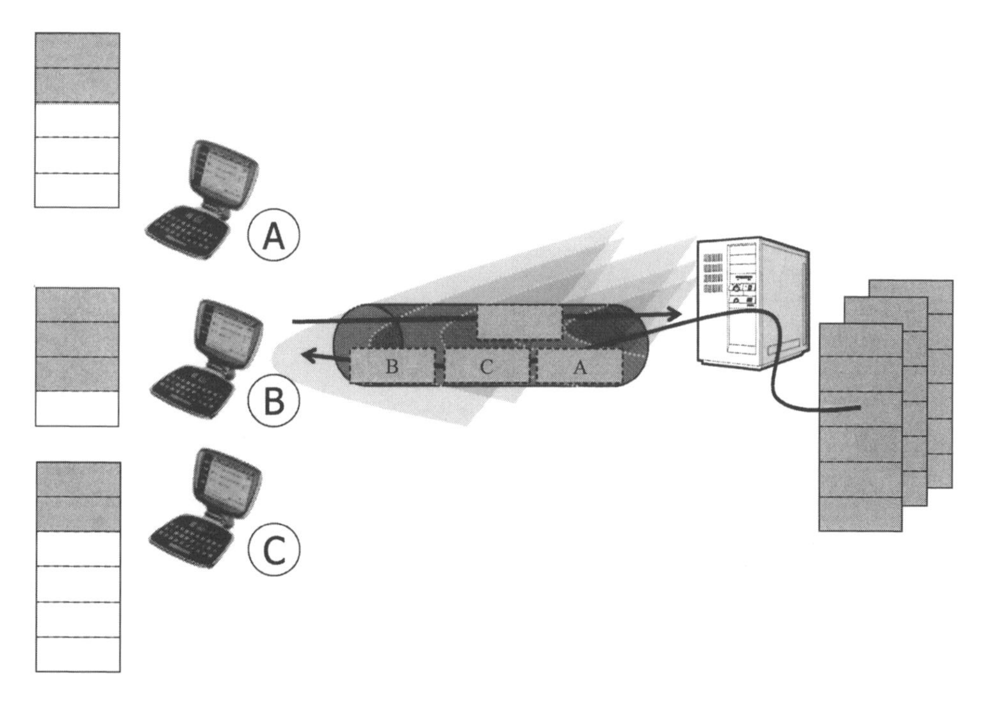

Figure 7.8 Links can be shared using a packet approach.

themselves as small parts of corrupted data. Also, there are times when the link is not available at all, and these times vary in length from very transitory to lengthy outages. These two effects combined are a problem, but one that can be ameliorated by sending data in chunks (packets) and not in one go. This is convenient anyway because of the sporadic nature of most data communications, as just mentioned. By sending data in packets we can make sure that the link is not tied up between two parties while others are also waiting to use the link. When a packet has finished its transmission the link is relinquished for someone else to have a go. In this way many devices can share the same data connection, as shown in Figure 7.8 by devices A, B and C all receiving packets over the same link. This segmentation of data into packets is taking place at a level below our HTTP communications plane using a technique called Transmission Control Protocol over Internet Protocol (TCP/IP). We can imagine that the three files shown to the right of the server in Figure 7.8 are each requested by a single GET method request from the corresponding device. It is the underlying TCP/IP software that is doing the segmentation for us, and this is quite transparent to our user agent and origin server software. As far as our user agent is concerned, it issues a 'GET' request to the server and expects an atomic response. In other words, the user agent receives the entire response in one go and is ignorant of the low-level 'packetization' of the response. We will see later when we look at this process in the wireless context exactly how it is that the user agent does not need to know about the segmentation.

The benefit of sending data in packets is in being able to implement some kind of flow control to take care of intermittent link performance. We can ask for packets to be re-sent if too many errors were experienced. We can also increase packet size to accommodate conditions when the chances of error are less or decrease if we think that resending is

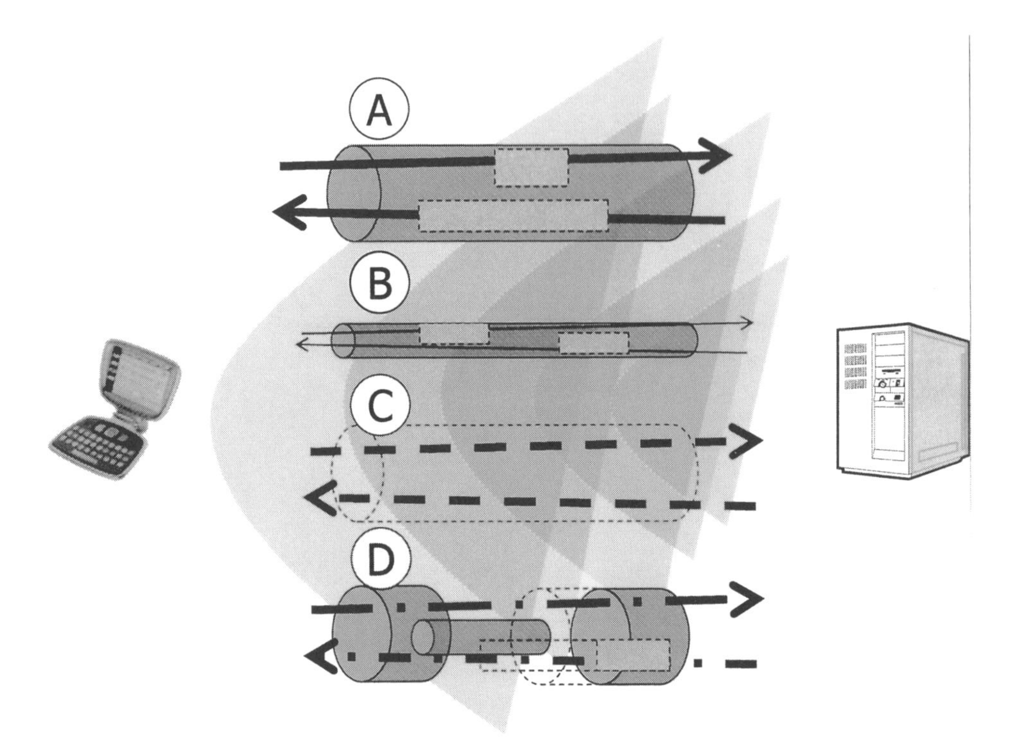

Figure 7.9 Variable packet size and delays due to variations in RF link performance.

likely; we would become very inefficient if we had to keep sending large packets as we end up wasting time sending the same data over and over again, even though only some of them might be corrupted. This point is best illustrated by considering the limitations due to RF variability and showing how packet-based transmission can be used to overcome them.

What we find with RF communications is that, due to the nature of the RF link, the transmission characteristics for data are very variable, even though powerful signal-processing techniques are used in attempts to maintain the best performance possible on the air interface.[7] The variation in the link and the effects this has on the data communications can best be shown diagrammatically, as shown in Figure 7.9.

Sometimes we can achieve very good link performance, as shown in A by the thickness of the data pipe. The link is reliable and available for a good period of time and the link quality is good enough to send data at high data rates. In this situation we can send data in large packets because we are able to sustain good communications for longer periods of time and with a lower chance of error.

In the case of channel condition B we see that the link is very slow, as indicated by the narrowness of the data pipe. This may be a temporary condition or possibly even a permanent feature of some RF links, like GSM (Global System for Mobile Communications) data calls. In this situation we tend to send data in small packets so that we do not have to resend too

[7] It is tempting to think that the error correction techniques used in wireless communications entirely overcome the hostile conditions of the RF channel; this is not the case. There are always bit errors on an RF link. Error rates vary according to the link conditions and how much error correction is used. However, high error correction means fewer data get through due to the overhead of inserting redundant bits into the data stream which enable the correction process to function.

much data if a packet gets corrupted, thereby attempting to maintain a reasonable throughput on the link.

In situation C we can see that the channel is simply not available. This delay in availability may be variable. This is not a problem *per se* for packet-based communications as we generally have a handshaking procedure that checks that the recipient is ready to receive before we attempt transmission. This flow control means that data do not get lost; they just get queued up at the transmitter. What we do not want, however, is for the transmitter to switch off because the link is down. The link absence may be for only a short while. Therefore, part of the flow control mechanism is to keep retrying the handshake. Clearly, this interval should not be too long; otherwise, we may end up not bothering to transmit even though the link has become available again, thereby inadvertently adding to the latency problem.

Finally, condition D reminds us that we may well experience any combination of the above scenarios during a particular packet transmission attempt. The link may slow down or even disappear for an interval while we are transmitting. This may occur within the context of trying to send a file or within the context of sending a packet. In the former case we can appreciate that the ability to dynamically adjust packet size and things like retry attempts (and intervals) may be useful. In the latter case the ability to cope with link variations during a packet transmission should not cause an undue failure in the communications progress. Our protocol should be able to cope with both these situations.

We have seen that RF communications suffers from variable link performance. However, similar problems, but with different causes, already existed in ordinary data communications networks – the reason that IP-based communications already has the necessary flow control procedures in place. We may be tempted for a moment to think, therefore, that we don't need to do anything different for RF channels as opposed to wired channels (link Ethernet over twisted-pair cables); this is not the case, primarily for two reasons.

First, some RF links can become so hostile that the standard IP flow control algorithms are no longer suitable for coping efficiently with the punitive conditions. As an example, if the link becomes very prone to errors, then the optimal packet size may be smaller than the IP protocols have allowed for; so, we end up with packets that are too big and data are repeatedly and needlessly resent. Another example is in the segmentation of large files into packets. If that process is such that the corruption of a packet requires the entire file to be re-sent, then that process is going to become extremely cumbersome for a relatively slow RF link.

Second, RF links are invariably a lot slower than their wired cousins, so it is essential to make best use of the link. If the IP mechanisms require a lot of data to be sent just to manage the flow control, then that is eating up valuable bandwidth that the user would rather be spent sending the actual data, like the picture or game being downloaded.

All of these issues have been thoroughly scrutinized by the WAP Forum, which is now under the auspices of the Open Mobile Alliance (http://www.openmobilealliance.org). Subsequently, a set of protocol optimizations have been designed to accommodate packet-based transmissions over RF links. Due to its prevalence, HTTP has been used as the guiding communications paradigm on which to base the WAP protocols.

7.5 WAP DATA TRANSMISSION PROTOCOLS

In this section we will look at the WAP protocols that enable us to implement a link–fetch–response communications paradigm, like HTTP, but in a manner that works well on an RF

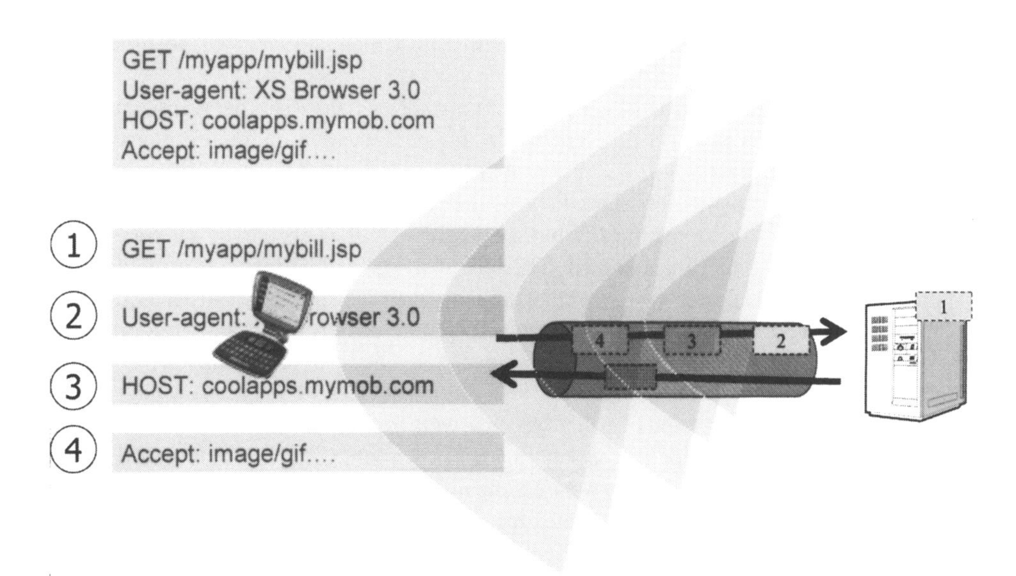
GET /myapp/mybill.jsp
User-agent: XS Browser 3.0
HOST: coolapps.mymob.com
Accept: image/gif....
1
GET /myapp/mybill.jsp
2
User-agent: rowser 3.0
3
HOST: coolapps.mymob.com
4
Accept: image/gif....
4
3
2
1

Figure 7.10 Segmenting an HTTP GET request into packets.

link. Before we look at the details of the protocols, let's first talk a little about protocol stacks and how data communications actually gets implemented in software terms. This enables us to better understand the concept of service layers within a communications protocol, as this may be useful.

7.5.1 Protocol Stack Paradigm

In our discussion of HTTP we saw that the protocol consisted of a set of communications primitives based on request methods, such as GET. However, we have just learnt that the underlying mechanism of TCP/IP is a bit more complicated. Perhaps and most probably, a single GET request may not make it in one attempt to our origin server: it may need to be divided up into segments and sent one piece at a time, allowing for retries and handshaking, etc., as shown in Figure 7.10.

The exact size of the packets may not be as we have shown them, nor is it likely that the packets will be delineated neatly along the boundaries of HTTP headers as shown; this is just to illustrate the principle. But, the key point is that the user agent is not at all responsible for this segmentation process and the underlying flow control. The user agent is a software application that knows about HTTP methods and in this case has issued a GET method. So, we might ask to whom the request has been issued. While logically it has been issued to the origin server, physically the user agent has actually sent the GET request to another piece of software running on the device, and it is this process that handles the communications across the link, as shown in Figure 7.11.

The TCP protocol uses an even finer grain underlying protocol: the IP itself. The IP level is really about how the packets get constructed, whereas the TCP level is more about how they get used to establish a reliable communications pathway using flow control. Therefore, there could well be a set of software to handle TCP, which itself utilizes a separate software service to carry out the IP packet processing.

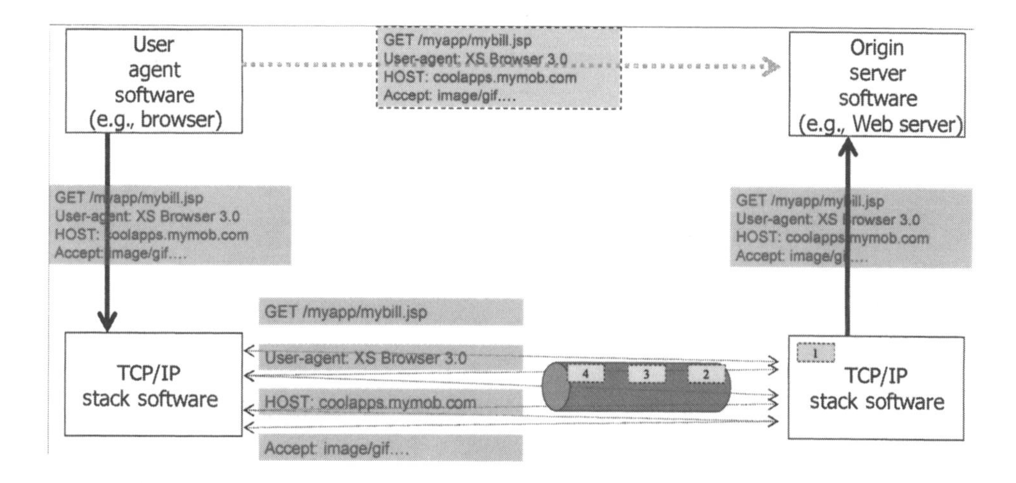

Figure 7.11 User agent talks HTTP to another software process, but it 'thinks' it's talking to the origin server.

We can see that this arrangement is a layer of responsibilities and functions within the communication process, with one functional layer stacked on the top of another, which is why the set of software gets referred to as a *protocol stack*.

In Figure 7.11 we can see this stack – although we are only showing the division between the HTTP part and the TCP/IP part – which groups the TCP and IP processing together (which is common in any case). As we will see in Chapter 10 when we look at devices the TCP/IP stack software is probably part of the generic device system software and can be accessed by any user application running on the device. In this way our user agent programmer only needs to implement the HTTP protocol handler in the software and use an appropriate software-messaging technique to pass the HTTP requests to the TCP/IP stack software.

As far as the user agent is concerned, logically it is talking directly to the origin server, as shown by the dotted pathway across the top of the diagram. This is a useful abstraction for designing and thinking about communications protocols and is well established in data communications.[8]

7.5.2 The WAP Stack

Having examined the idea of a layered protocol model, with the various functional layers taken care of by components in our software stack, we can now look at the optimizations to this process proposed by WAP.

It is fairly obvious that if, as we claimed earlier, the mechanics of the TCP/IP process does not work well for RF links (to varying degrees, depending on the link technology), then we could propose removing the TCP/IP stack and replacing it with something else. This is certainly one approach that has been advocated, as we will see. However, there is a design limitation that we should think about placing on how we replace layers in the protocol stack, apart from the actual requirement of RF friendliness. This limitation is related to the pre-existence and prevalence of HTTP.

[8] It is so well established that it has been formalized into a standard model called the Open System Interconnection, or OSI, layered protocol model.

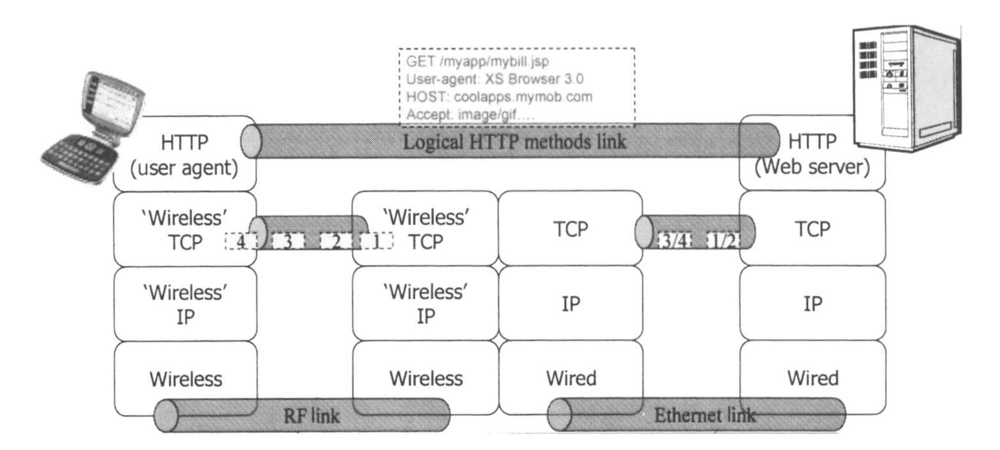

Figure 7.12 WAP proxy 'talks' TCP/IP by proxy for the wireless stack.

The advent of mobile services is something fairly new, and so we have the luxury of developing and implementing new protocols in devices without fear of remaining compatible with old ones, simply because there aren't any to be compatible with. However, on the other end of the connection we have already understood that the prevalence of HTTP is extensive and, consequently, a vast Web-based infrastructure has arisen to cater for all manner of services built on the HTTP and browser paradigm, such as the ubiquitous Web server. Therefore, to build our mobile services we would rather stick to the use of *existing* Web infrastructure and not have to deploy something new to cope with modified protocol stacks. Certainly, if we take the J2EE platform, as discussed in the last chapter, then we want to be able to tap into all its benefits without deviating from the standards on which it is built, especially HTTP.

This was exactly the design limitation that the WAP inventors placed on their protocol design: that WAP-compatible devices should still be able to communicate with standard (i.e., HTTP-compatible) Web servers so that existing applications developers could use existing and familiar design approaches, infrastructure and tools; this, no doubt, was a wise decision.

However, immediately we see that we might have a problem. On the one hand, we need to do away with HTTP running over TCP/IP as it is too cumbersome for hostile RF connections and, on the other hand, we want to stick with HTTP running over TCP/IP as that's how Web servers work; these requirements seem incompatible. Fortunately, there is a solution to this problem: the use of a gateway or proxy. As the name suggests, this device maintains a standard HTTP session with a Web server by proxy on behalf of a device that does not then need to communicate using HTTP over TCP/IP, as shown in Figure 7.12.

What we end up with is the user agent able to communicate with the Web server in the most optimal fashion possible. Thanks to the software stack method the user agent software still thinks it is talking to the origin server, which due to the proxy can still be an ordinary Web server. Web application designers can continue to use their well-honed skills and the available infrastructure and Web-related products to deliver services to our mobile user. The origin server could well be a J2EE application server as discussed in Chapter 6, so we maintain all of the benefits we identified with going with the J2EE approach – we have not had to compromise on any part of it.

The diagram shows how the 'wireless' TCP and IP layers are maintained between the device and the proxy and then converted to the standard TCP/IP protocols. This would tend to indicate that these protocols have to be fairly similar in design concept in order for the conversion process to be possible in a meaningful and efficient manner. However, clearly there will be differences, which is why the diagram deliberately illustrates a different packet flow on either side of the proxy; this is what we would expect. The flow on the left-hand side is optimized for a wireless connection, whereas the flow on the right-hand side is optimized for a wired connection.

WAP has gone through a series of design evolutions and we have now two major releases: WAP 1 and WAP 2.[9] In WAP 1 the main design criteria were link optimizations within a context of exceptionally slow RF services available at the time. For example, with early WAP devices on cellular networks the only connection available was to establish a circuit-switched data call. On GSM devices this was theoretically a maximum data rate of 9,600 bps, which often meant no more than 2–3 kbps in practice. For a network like GSM, the data bearer itself could support an IP connection; so, at the low levels this could be utilized for data communications. However, many alternative RF data systems at the time, particularly in the USA, were unable to support IP communications (e.g., the two-way paging networks).

Box 7.1 Long thin networks (LTNs)[10]

Cellular networks are characterized by high bit error rates, relatively long delays and variable bandwidth and delays. TCP performance in such environments degrades on account of the following reasons:

- packet losses on account of corruption are treated as congestion losses and lead to reduction of the congestion window[11] and slow recovery;
- TCP window sizes tend to stay small for long periods of time in high BER (Bit Error Rate) environments;
- the use of exponential back-off retransmission mechanisms increases the retransmission timeout resulting in long periods of silence or connection loss;
- independent timers in the link and transport layer may trigger redundant retransmissions;
- periods of disconnection because of handoffs or the absence of coverage.

Research in optimizing TCP has resulted in a number of mechanisms to improve performance. Some of these mechanisms are documented in Standards Track RFCs (request for comments) and have been accepted by the Internet community as useful and technically stable. The Internet Engineering Task Force (IETF) PILC[12] group has recommended the use of some of these mechanisms for TCP implementations in LTNs [RFC 2757[13]].

[9] There are minor version numbers, like WAP 1.3, WAP 1.3 and so on, but for simplicity I prefer to stick to the major version numbers to highlight that we are dealing with two quite different sets of design recommendations.
[10] Extracted from http://www.faqs.org/rfcs/rfc2757.html
[11] Windowing is discussed in the main text when we look at Wireless-profiled TCP (W-TCP).
[12] Performance Implications of Link Characteristics (PILC). See http://www.ietf.org/html.charters/pilc-charter.html
[13] http://www.faqs.org/rfcs/rfc2757.html

This led to a WAP 1 protocol stack that was highly optimized for wireless communications and that did not assume anything about the underlying network, other than it could support datagram-based (i.e., packet, but not IP) communications. It was the responsibility of the WAP implementer for a particular wireless solution to implement the necessary adaptation software (and possibly hardware) for the WAP protocol stack to work over the underlying data bearer. Because there is no universally supported datagram protocol (e.g., IP-based UDP) for all wireless solutions, there was a need in the WAP specifications to define a datagram protocol from scratch, which was denoted Wireless Datagram Protocol, or WDP.

The IP-based User Datagram Protocol (UDP) is adopted as the WDP protocol definition for any wireless bearer network where IP is available as a routing protocol, such as GSM, CDMA and UMTS. UDP provides port-based addressing, and IP provides the segmentation and reassembly in a connectionless datagram service. It does not make sense to use WDP over IP when the ever-present UDP will suffice and is already very widely implemented and understood. Therefore, in all cases where the IP protocol is available over a bearer service the WDP datagram service offered for that bearer will be UDP. UDP is fully specified in [RFC 768[14]], while the IP networking layer is defined in [RFC 791[15]] and [RFC 2460[16]].

The bearers defined in this specification which adopt UDP as the WDP protocol definition are:

- GSM circuit-switched data;
- GSM GPRS;
- ANSI-136 R-data;
- ANSI-136 circuit-switched data;
- GPRS-136;
- CDPD;
- CDMA circuit-switched data;
- CDMA packet data;
- PDC circuit-switched data;
- PDC packet data;
- iDEN circuit-switched data;
- iDEN packet data;
- PHS circuit-switched data;
- TETRA packet data;
- DECT packet/circuit-switched services;
- UMTS circuit-switched data;
- UMTS packet data.

[14] http://www.faqs.org/rfcs/rfc768.html
[15] http://www.faqs.org/rfcs/rfc791.html
[16] http://www.faqs.org/rfcs/rfc2460.html

Our main concern then for the current discussion is the higher layers and, in particular, how the HTTP/browser paradigm ('link–fetch–response') is achieved in the wireless environment.

WAP 1 specifies an alternative to TCP, called Wireless Transport Protocol (WTP). The good thing about WTP is that it was designed from scratch with the fetch–response paradigm in mind for browser-like applications; so, it is appropriately structured. However, it is still intended as a generic transport protocol for wireless applications, not just for browsing; so, browser-specific ideas are left out and are instead incorporated into a higher layer protocol called the Wireless Session Protocol (WSP).

As the capabilities of wireless carriers have evolved, particularly with the advent of 3G (third generation) solutions like UMTS (Universal Mobile Telecommunications System), the need for highly optimized protocols lessens. The extent to which they are no longer needed is debatable, but it is worth reminding ourselves that even with 3G the reality of realizable data rates is still far below what has been prevalent for years within cabled networks in the Internet arena, which is why there is still a need to think about optimizing the protocol stack for mobile applications. However, there is always an appeal to utilize standard protocols from the IP family as these are so well supported and understood that the benefits of using them are desirable. Therefore, with WAP 2 the evolution of WAP has been toward using the standard HTTP/TCP approach, but with some suggested modifications – notably ones that will not take the implementations outside of what the existing specifications can support. The analogy that comes to mind is the use of higher octane unleaded fuels in cars. These fuels allow for an optimized performance where a car can take advantage of it, but they will still work in standard cars. This approach has led to Wireless-profiled TCP (W-TCP) and Wireless-profiled HTTP (W-HTTP).

We will begin with the WAP 2 protocols, first, as these closely resemble what we have already been discussing thus far with HTTP and, so, enable us to build on these ideas while uncovering some of the design challenges that will subsequently enable us to understand the design ethos of the WAP 1 protocols for completeness (given that WAP 1 is currently a widely used protocol).

7.5.3 Wireless-Profiled TCP

TCP itself is really about adding error handling and congestion handling on top of the crude, underlying packet transport provided by IP. With TCP we gain the ability to get feedback from the receiver about how well the data are being received, so that we can resend any lost information. Also, in a packet-based network where lots of users attempt to share the same link, hoping that there are always enough gaps in the transmission of others to send our own packets, there is the danger of congestion. If this occurs, then we need the ability to detect it and to back off transmission to ease the congestion.

Now, with TCP we take the data we want to send and we divide them into segments. A key mechanism for TCP is something called *windowing*. This is an important concept to grasp in order to understand wireless profile optimizations or the need for them.

Whenever a segment is sent from the TCP sender, the sender wants to know whether or not it was received correctly so that, if required, a segment can be re-sent. This error control is a key function of the TCP protocol layer. To implement this feedback the TCP receiver is required to acknowledge (called an 'ACK' message) that segments have been received correctly, as shown in Figure 7.13.

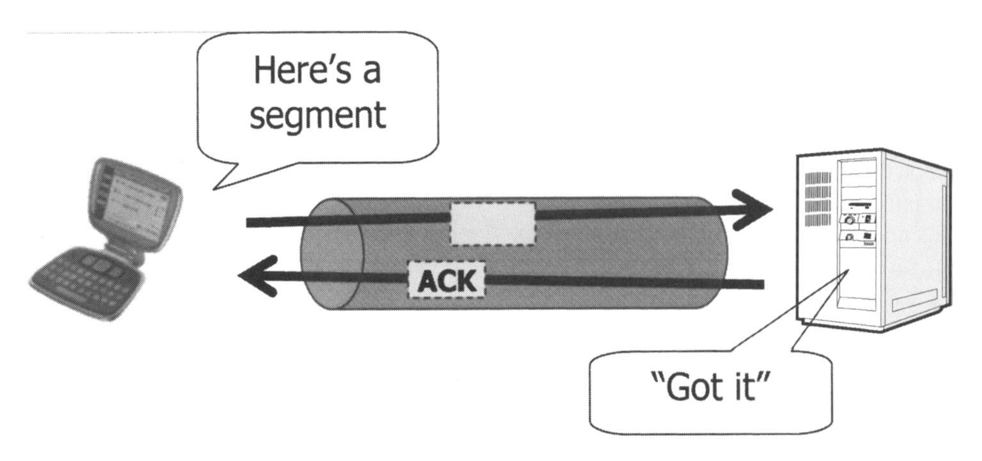

Figure 7.13 TCP send and acknowledgement (ACK).

The process of send and acknowledge constitutes a feedback loop. An important feature of the loop is its response time. There will be a certain delay in the segment getting across the network to the receiver and another delay in the ACK getting back. In wireless networks this delay can be particularly long for a variety of reasons, mostly to do with how the RF resources get shared between users. This is a problem in itself for wireless networks, as discussed in Box 7.1: 'long thin networks'.[17]

Regardless of the delay length, it would be a very inefficient process to have to wait for an ACK before the next segment gets sent. This is because the transmitter will not be transmitting anything and, so, the available bandwidth is not being used. This is particularly problematic for wireless networks where the bandwidth is a scarce resource. It would seem silly if we only have a slow link, to spend half the time (say) just waiting around for acknowledgements instead of sending data. For the user it would mean that loading browser pages takes longer than is theoretically possible, and this would detract from the user's enjoyment of a WAP service.

To solve the problem of waiting for the ACK the transmitter simply carries on transmitting without waiting for a response, segment after segment, up to a controllable amount of segments called a *window*. This is the windowing process, as shown in Figure 7.14. Basically, the TCP sender is hoping (assuming) that there is not going to be a problem with the segments that get sent, so it just carries on transmitting to make best use of the bandwidth while waiting for an ACK. What the receiver does under this arrangement is wait to receive all the segments in the window before sending an ACK, and the ACK is now used cumulatively to acknowledge all the segments rather than one at a time. This process gets pipelined so that the sender slides the window along and sends the next window. It checks for the ACK from the previous window and keeps going as long as the ACK gets received before a timer expires that gets set for each window.

There are several problems with this approach in the wireless network, as well as other problems with TCP that are to do with the subtleties of the actual mechanism and related

[17] Note that some of the terminology used in Box 7.1 might not be accessible until you complete reading the current section of the main text.

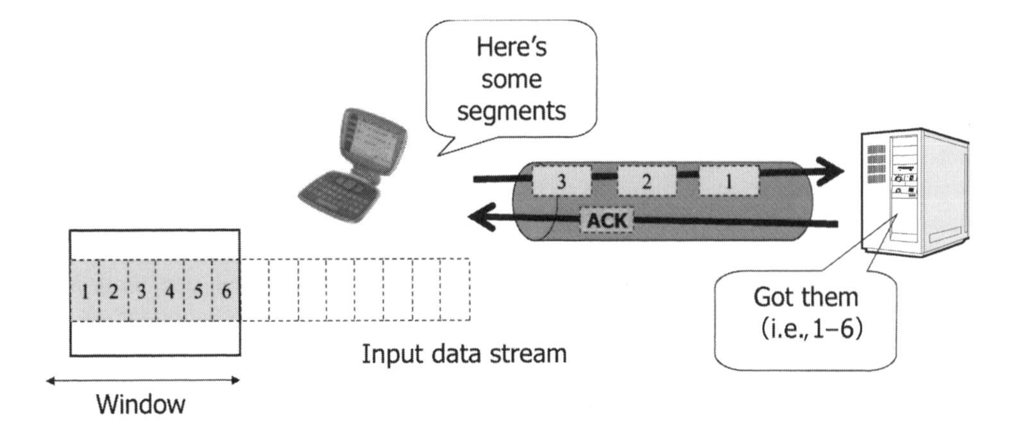

Figure 7.14 TCP segments get sent several at a time in a window.

mechanisms (e.g., congestion control). We have painted a slightly simplified picture in our explanation in order to convey the essence of WAP optimizations.

For an LTN the delay in receiving the first ACK may be much longer than in ordinary wired networks. Since most TCP implementations assume a wired network, of course, the sizing of the *initial window* is too small for the wireless case. In physical terms this means that the transmitter will stop transmitting at the end of the window and wait for the cumulative ACK. It will still have to wait because the wireless connection may involve considerable delays. The windowing process was invented to avoid waiting; so, the waiting due to a long delay is undesirable. This problem can be overcome by making the window longer; this is one of the mandatory changes to TCP implementation for WAP 2, turning TCP into W-TCP.

What becomes apparent from pondering on the windowing process is that the cumulative ACK implies that if something goes wrong in the window transmission (and reception), then the entire window of segments will get re-sent as part of the error avoidance strategy. The problem with this approach is that not all of the segments may have been corrupted, but they will all get re-sent; so, this results in needless (redundant) data transmission. This problem is now compounded by the increased window length suggested by WAP 2; we will have to re-send a lot more data if things go wrong! Hence, for WAP 2 the TCP profile utilizes *Selective ACK* (SACK) [RFC 2018[18]] where each segment is acknowledged explicitly. Therefore, only the corrupted segments need be re-sent: a process known as *selective retransmission*. We will return to this topic when we look at the WTP from WAP 1.

Clearly, the size of the window is an issue that affects link performance. In an ordinary cabled network it may be necessary to adjust the window if the network gets congested. In a congested network there will simply be no point in transmitting lots of segments in the hope they will get through. The amount sent should be backed off. In TCP there is actually no mechanism for discerning if a packet is lost due to corruption or congestion. This can be taken advantage of in the wireless case to reduce the initial window size, should it prove

[18] http://www.faqs.org/rfcs/rfc2018.html

too big for sustained link performance. In other words, we set the initial window to be large and we rely on the congestion control to back it off if it proves too large.

On the other hand, congestion control can be a problem for a wireless link if it gets used to correct for packet loss rather than congestion. If we decide to back off our transmission rate, either by adjusting the window or by deliberately imposing a waiting time before transmitting, then we need to attempt to recover from this situation as soon as conditions improve. TCP has its own algorithms for recovery. The problem is that the conditions under which congestion occur versus packet loss in an RF network are different, so recovery strategies should be adjusted accordingly. Packet loss can be very transitory in an RF network compared with congestion scenarios. If several packet losses occur during an RF performance episode, the TCP link can drastically curtail performance under the assumption that severe congestion is present on the link. This is made worse by the slow recovery attempts that are appropriate for mitigating congestion. In other words, an interruption to the RF link may cause undue performance losses as an artefact of TCP rather than actual link performance; this is clearly undesirable.

The performance of TCP under wireless conditions has been studied on many occasions (see Box 7.1: 'long thin networks'). The results from several of these studies have been incorporated into the specification for W-TCP. There are many possible optimizations and we have only discussed two of them: large initial window and selective ACK. These are the only two that are mandatory in the W-TCP specification. There are plenty of other optimizations that are recommended, and their effect on performance will depend on a variety of factors that are beyond the scope of this book.

7.5.4 Wireless-Profiled HTTP (W-HTTP)

In discussing W-HTTP we need to resume our explanation of HTTP in order to understand some of the benefits that HTTP 1.1 offers over its predecessor 1.0, especially as W-HTTP is modelled on HTTP 1.1.

Pipelining of Requests on a Single TCP Link We have briefly examined TCP in the preceding discussion. What we did not examine is the process to establish a TCP connection, other than creating a socket, which is the software binding to the TCP/IP protocol stack according to the desired destination address for the packets and so forth. To establish a TCP connection between two end points, there is some initial handshaking that goes on to make sure both ends are ready to exchange data. Part of the start-up process in TCP is what we refer to as a *slow start*, wherein the performance parameters that affect TCP communications are set to pessimistic values in case the network is congested from the start, not wanting to exacerbate the problem. The effect of the slow start algorithm is exactly as its name suggests: the TCP link performance is initially not optimal.

As discussed when we looked at the browser paradigm, with a Web page there are often objects (called *inline objects*), such as images, that are stored as separate files on the Web server and need to be fetched via separate HTTP requests, each object requiring a separate HTTP GET request. This creates quite an overhead in response time due to the slow start nature of TCP. With HTTP 1.0, as we saw in Figure 7.3, after a successful response from the origin server, the connection gets closed by the server. Therefore, to fetch a referenced object, like an inline image, the client has to open a new TCP connection with the server.

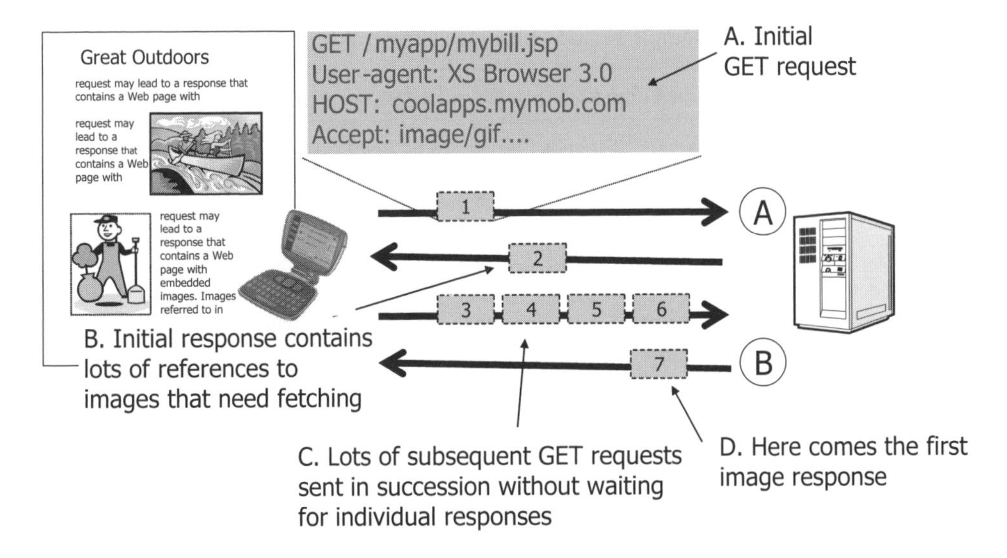

Figure 7.15 Pipelining of HTTP requests over a second TCP connection.

This takes time, not only in the TCP handshaking, which is a lot of control packets being sent back and forth, but also in the overhead of the slow start every time a new TCP connection is established. As we will see later, WTP does not have this 'start-up' problem.

HTTP 1.0's solution to improve performance is for the user agent to open concurrent TCP connections and request the inline objects via independent HTTP requests. However, this causes congestion and processing overhead. To improve matters, HTTP 1.1 has the ability to keep the same TCP connection open and to pipeline the HTTP requests, which means many GET requests can be sent in a row without waiting for the responses, as shown in Figure 7.15. This makes maximum use of the open TCP connection. This aspect of HTTP 1.1 is carried forward into W-HTTP; in fact, all of the core features in HTTP 1.1 are specified as required by W-HTTP.

The pipelining aspect of HTTP 1.1 is carried into W-HTTP by default. It is particularly beneficial for wireless devices as this requires less processing overhead than opening many concurrent sockets, as would be necessary with HTTP 1.0 to achieve a better user experience.

The improvement of HTTP 1.1 over 1.0 is shown in example measurements taken from an experiment to investigate HTTP 1.1 optimizations.[19] These are shown in Table 7.1. On the left-hand side are the measurements taken from retrieving a sample file for the first time in which there are plenty of inline image references. We can see that the performance of HTTP 1.1 is clearly better than 1.0 when pipelining is used. Note that these results are for HTTP 1.0 using multiple concurrent connections (up to 6), whereas all the HTTP 1.1 results are for a single TCP connection. The results on the right-hand side are for a subsequent fetch of the same page, but this time verifying if each resource is newer than the cached one in the user agent's cache. Here we can see an even better performance of HTTP 1.1, which is because other features of 1.1 can be utilized to improve performance (e.g., using *byte ranging*, which is not discussed here but gets discussed later when we

[19] http://www.w3.org/Protocols/HTTP/Performance

Table 7.1 Example performance measurements with HTTP 1.1.

	First time retrieval				Cache validation			
	Pages	Bytes	Seconds	%overhead	Pages	Bytes	Seconds	%overhead
HTTP 1.0	559.6	248,655.2	4.09	8.3	370.0	61,887	2.64	19.3
HTTP 1.1	309.4	191,436.0	6.14	6.1	104.2	14,255	4.43	22.6
HTTP 1.1 pipelined	221.4	191,180.6	2.23	4.4	29.8	15,352	0.86	7.2
HTTP 1.1 pipelined with compression	182.0	159,170.0	2.11	4.4	29.0	15,088	0.83	7.2

look at segmentation and reassembly [SAR] in downloading Java applications to mobile devices).

In terms of optimizing HTTP 1.1 performance, we still may need to use more than one TCP connection after all, but we avoid the many connections that HTTP 1.0 implementations found necessary to spawn. It is useful to open an initial TCP connection to grab the XHTML content from the origin server, as shown by connection A in Figure 7.15. As discussed in Chapter 6, in a J2EE environment we generate the XHTML content dynamically using Enterprise Java Beans to assemble the content from databases or from other information sources. The process of writing the dynamic content onto the TCP link back to the device takes time, but we can still receive some of the content before the entire XHTML file arrives. By opening another TCP link (connection B) we can begin to pipeline requests for any of the inline content that we can already see referenced in the initial portion of the XHTML. In this way we can achieve an even higher degree of parallelism and thereby improve the overall response time. The net result is a better user experience for the end users, which is an important consideration in the design of any mobile service.

In terms of pipelining of requests or even multiple connections via HTTP 1.0, it is important to realize that the browser is normally capable of displaying the initial XHTML content prior to receiving the inline images (or other embedded content), so the user gets to see the partial page displayed. This improves the apparent response time as far as the user is concerned and, so, is a desirable implementation feature.

With WAP 2 the specifications dictate that caching of fetched content should be supported and be compatible with caching methods for HTTP 1.1, as already discussed. Additionally, the WAP gateway, if used, should be capable of caching data itself, and there are some extra headers that WAP agents can use to ensure cache maintenance on the gateway can be controlled.

Using Compression with W-HTTP (HTTP 1.1) In terms of improving user experience there are yet more things we can do to improve underlying performance, such as the use of compression – an optional feature in HTTP 1.1 that should be implemented in W-HTTP user agents wherever possible. The advantage of compression is that the content in the responses is compacted. The HTTP headers themselves do not get compressed, but the XHTML content does. Even where link-level compression is available, such as V42bis, by using something like *zlib* compression[20] on the HTTP payloads, performance can be improved and without relying on the wireless infrastructure to implement link-level compression (which cannot be relied on[21]).

With HTTP 1.1 there is an HTTP header called *Accept-Encoding: deflate*. If this header is included in the initial GET request, then the origin server knows that it can compress the response payload. In other words, the user agent is indicating that it is able to accept compressed content as it has the capability to uncompress it. Compression is not required for image fetches as the image file formats themselves, such as PNG[22] and JPEG[23], support internal compression by default.

Compression in itself can reduce the size of the XHTML content, and on a slow wireless link this will improve response times. But, there are some subtleties with this approach that are interesting to note. First, compression requires more processing time prior to displaying the XHTML, as we need to run the compression algorithms to extract the content. This is an added delay not only for displaying the initial XHTML content but also in our user agent being able to extract inline references and initiate the request pipeline to get the rest of the content (although on pages for very small device displays we expect the number of inline images to be small). On a very slow device this may prove problematic and the gains of compression can be lost, but there is no hard and fast rule. As with all these protocol optimizations the actual implementation affects overall performance. It is not a straightforward matter to be able to get these things right, especially for all scenarios.

There is another reason that compression is more important than it might seem at first: the low-level operation of TCP and how it interacts with the pipelining process. The details of the problem are beyond the scope of this book, but we can briefly mention the principles. It is more efficient in the pipelining process if we can buffer up several requests and send them at once. This reduces the amount of control packets that need to be sent. The buffering process will incur a delay if we have to wait for the buffer to fill before sending out (flushing) its contents. The problem with this is that we may have an ACK sitting in the user agent buffer that the server needs before it can send out more responses. In which case we may incur an extra delay in waiting for a timer to expire, indicating that the buffer should now be flushed (in the absence of enough content to fill it and cause a flush automatically). By compressing the XHTML the receiver gets more of the page in one go and, consequently (on average), more inline image references. These references result in more GET requests, which will fill up the outgoing buffer and cause it to flush sooner. It has been shown by experiment[24] that this can improve performance.

Using Cascading Style Sheets with XHTML-MP We will look at the technology of mark-up languages for WAP a bit later in the book, but for now, while it seems pertinent, we should mention something about the use of cascading style sheets (CSSs) to improve performance.

[20] http://www.gzip.org/zlib/
[21] Just because a particular technique is referred to in the technical specifications for a cellular network does not mean that it will be implemented on all networks, nor on all devices.
[22] Portable Network Graphics – see http://www.libpng.org/pub/png/
[23] Joint Picture Experts Group – see http://www.jpeg.org/
[24] http://www.w3.org/Protocols/HTTP/Performance

CSSs allow the browser to understand how it should layout text and what formatting to apply. By supporting CSSs, it is possible for rich layouts to be achieved without resorting to the use of image files to achieve similar results. CSSs are textual descriptions of layout criteria and, as such, they can represent certain visual effects in a more compact fashion than using image alternatives.

For example, the image shown in Figure 7.16 with the caption 'solutions' requires 682 bytes to be stored as a GIF image. Using XHTML and a CSS, the same content can be represented by:

```
P.banner {
color: white;
background: #FC0;
font: bold oblique 20px sans-serif;
padding: 0.2em 10em 0.2em 1em;
}

<P CLASS=banner> solutions
```

solutions

Figure 7.16 Typical Web page image.

In WAP 2 the mark-up language recommended is called XHTML-MP – a derivative of XHTML-Basic – which, as we will see, is a specification for using XHTML for display-limited devices. However, XHTML-Basic does not include the use of CSSs. The WAP Forum has specified support for a limited version of CSS in XHTML-MP, so we should be able to take advantage of its features by enabling us to make visually interesting pages that load faster. The style information itself can be stored separately in a file called a style sheet, which has to be fetched using a GET request. This is apparently an added overhead. In the pipelining scheme the overhead will be reduced. However, the real advantage in performance comes by using a single style sheet across many pages in our application. This is because the style sheet itself will get cached by the user agent, so there is no need to keep fetching it on every subsequent page request. Combining this with its powerful display enhancements, to be used wherever possible in preference to images, the use of CSSs is a potentially powerful performance-enhancing feature for WAP 2.

7.6 WIRELESS PROTOCOLS – WTP AND WSP

7.6.1 Introduction

Having discussed the wireless profiles for TCP and HTTP, we have covered the main direction for WAP support in the future. However, WAP 1 protocols are already in wide circulation in millions of devices and will continue to be used for some time to come as they are extremely efficient, especially on slower wide area network (WAN) links, like GPRS (General Packet Radio Service). Many newer devices will support both WAP 1 and WAP 2 stacks; these are not backward-compatible, so they have to be implemented as separate stacks on the device – WAP 2 will not handle a connection with a WAP 1 proxy (and vice versa). This does not mean that we need two browsers! The trend is for a single-browser implementation that can handle all the various mark-up languages that have evolved within (and alongside) WAP. The browser will use either a WAP 1 or a WAP 2 stack to access the corresponding stack on the other end of the link, but it cannot use a WAP 1 stack to access a WAP 2 source (and vice versa).

WTP and WSP are not compatible in any way with TCP and HTTP – they are separate protocols, although similar in many respects. The design ethos for both these protocols is to facilitate the fetch–response paradigm in the most optimal manner. This is not like TCP, which is a generic protocol that has been designed to support many varied higher layer applications, not just Web browsing (HTTP). TCP can support protocols like FTP for file transfer, telnet for remote terminal access, SMTP for email and so on. WTP has been designed to support WSP (Wireless Session Protocol), so it in some respects already mimics the fetch–response structure that WSP (and HTTP) has. WTP is built on a messaging principle rather than the streaming[25] principle that TCP/IP supports.

Both WTP and WSP have been designed from scratch to cope with LTNs, so they have overcome some of the limitations of TCP and HTTP, as we will now explore.

7.6.2 Wireless Transport Protocol (WTP)

WTP is a transaction-based protocol; it has been designed with the assumption that the wireless device and its peer (e.g., server) will engage in an ongoing conversation that can be broken down into distinct transactions, as shown symbolically in Figure 7.17.

The transactional nature of WTP makes it ideally suited to fetch–response applications, like browsing. WTP supports a variety of features that make the protocol ideally suited to the greater challenges of the potentially hostile RF network.

The protocol actually supports three modes of transaction, two of them primarily concerned with one-way data flow (push) and the other concerned with requesting data in the fetch–request manner (pull). We will primarily focus on examining the characteristics of WTP via the latter transactional mode, as this is the one used to support WSP, the WAP equivalent of HTTP.

WTP does not require the three-way handshaking that TCP uses to establish a link. WTP simply sends an invoke message as the start of a communications cycle. This invoke message is always used to initiate a transaction on the link, whether for the first transaction or the 100th; so, the inefficiencies of the TCP slow start are done away with in WTP.

WTP allows for a much stronger verification regime for ensuring that messages have been both received *and processed* by the application that is using WTP, which might be a

[25] Here, we do not mean the real time streaming that multimedia applications require, we mean that the structure of the protocol is agnostic of the content, viewing the input as a stream of bytes to be sent in a reliable manner; whereas WTP views the content as being transactional or conversational (message-based), like HTTP, and groups protocol primitives around the completion of transactions.

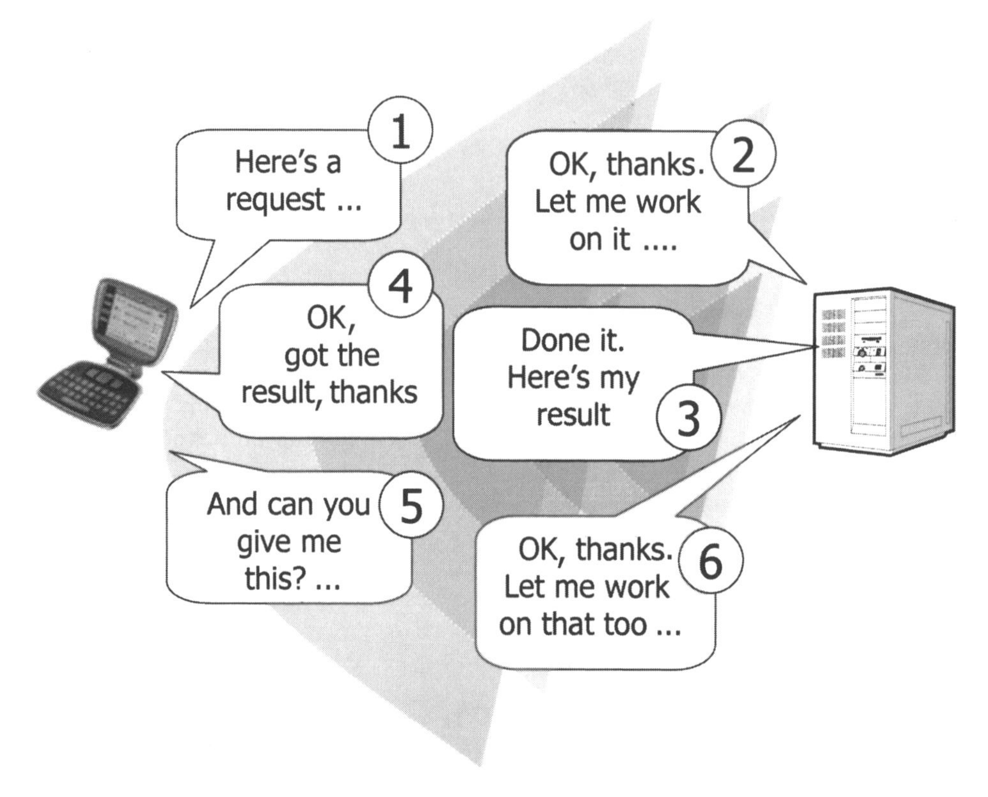

Figure 7.17 WTP supports dialogue at the application level.

WSP stack program, for example. This is unlike TCP, which only allows for verification of reception, but does not verify that the higher layer application using the protocol has successfully consumed the message; this can be compared with text messaging from one mobile to another. With the Short Message Service (SMS) in GSM, the sender can ask for a message delivery receipt to indicate that their original message to the recipient has actually been delivered to the device. This is like a delivery acknowledgement and is paralleled in WTP by the delivery of an ACK packet (packet data unit, or PDU). However, WTP goes a step further; it has another level of acknowledgement that allows the higher layer application to say that it has consumed the message. The nearest parallel is to imagine a mechanism in text messaging whereby the sender could be notified that the recipient has actually opened and read the message, just like some recipients will by their own volition send an 'OK' message back.

Figure 7.18 shows us the WTP transaction sequence in some detail (though some of the finer details are ignored here for brevity). Step 1 is the user application, which could be another layer in the stack (e.g., WSP), invoking a transaction. The transaction is initiated over the link (step 2) and passed to the user application at the other end (step 3). Note that we do not wait for an acknowledgement (ACK) before sending data, we have included it with the invoke payload going across the wire to save time. The WTP stack software will send an ACK (step 4) to acknowledge successful receipt of the data. There is then another

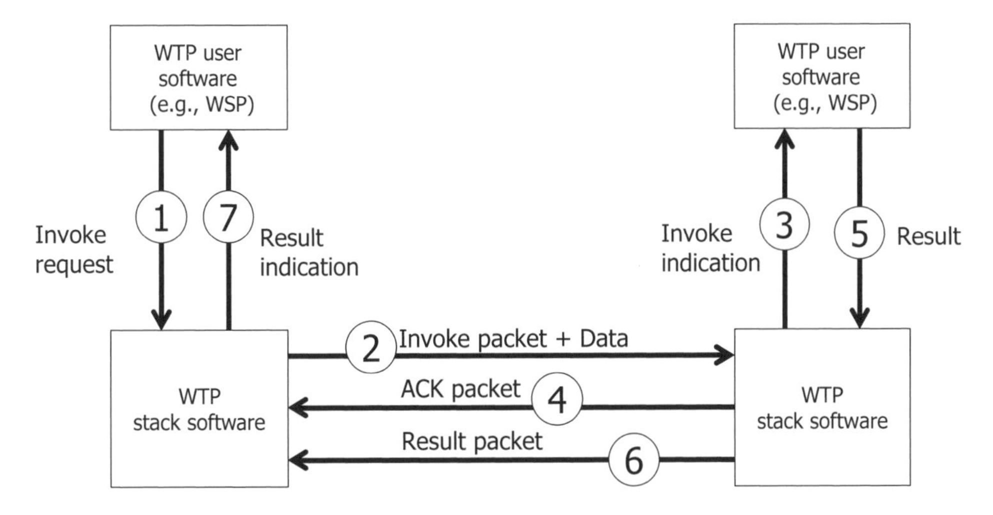

Figure 7.18 WTP provides strong verification of message transaction.

level of acknowledgement, which is referred to as a *result* within the WTP nomenclature. In step 5 the very application that has consumed the received data is sending a result request back to the WTP stack to indicate that all is well with the data. This gets sent back to the WTP provider (step 6) and emerges as a *result indication* (step 7) to the originating application.

Note that the original invoke was to initiate a complete transaction and, subsequently, more data packets could have been sent by the WTP user application as part of a single transaction. The receiver is aware from the start that more packets are expected and so will begin a process of tracking where it is in the progression of the transaction. All packets are marked with a *Transaction ID (TID)* to commonly associate them with a single transaction wherever that may be so.

There are various details of the WTP protocol that make it ideally suited to wireless transmission channels. In the above example we could dispense with explicit ACK packets wherever we expect a result packet to be sent, as this implies that the data were received, thereby making explicit ACK redundant. Result and ACK packets also allow data to be appended to their tail ends – data that are not part of the transaction. This is referred to as 'out of band',[26] as it is not really part of the main dialogue. This allows other processes at either end of the link to swap information on the back of the WTP transaction. For example, this could be used to send RF measurement information or some other link quality indicator; this might be useful for the higher layer application to adjust its behaviour accordingly. Perhaps during episodes of poor link performance a browser could automatically switch to accessing a sparser set of Web pages, such as a text-only version of a site.

[26] Strictly speaking, this is not out-of-band communication, but it is similar. The point is that data can be sent riding on the back of an existing transmission; so, no specific connection was required to be set up to carry the supplementary data, which is also quite independent of the main transaction.

7.6.3 Concatenation and Segmentation

The transaction capabilities of WTP are quite advanced and support two methods of dynamic adjustment to the underlying RF bearer capabilities. First, WTP supports a concatenation technique that allows several WTP packets to be aggregated before being sent, so as to fit them into a single (larger) RF bearer transmission unit. This will allow maximal efficiency to be achieved on certain types of bearer, especially those that are like very slow datagram services (e.g., some paging networks).

Second, WTP supports a SAR technique that allows very large transactions to be distributed over successive WTP packets. The WTP stack itself can reassemble the segments before passing them to the upper layer (WTP user). There is often some confusion that arises when dealing with SAR techniques as this can be done at different layers in the entire protocol stack. As we have learnt, WTP packets flow over a WDP link, which is the WAP-defined low-level datagram protocol. Wherever possible, WDP defaults to UDP – its equivalent – and uses IP to structure the packet flow. IP has SAR capabilities. Later in the book, when we look at downloading Java Archive (JAR[27]) files to mobile devices, we will see that HTTP 1.1 also supports SAR by using its *Range Requests* HTTP headers to ask only for segments of a resource rather than the whole thing.

Given that all these levels can support SAR, which one should we use? This is application-dependent to a degree, but a key feature of WTP SAR is its selective retransmission capability, which is something we have seen before with the SACK option in HTTP 1.1 (which is made mandatory for W-HTTP).

7.6.4 Segmentation and Reassembly in Action

It is worth exploring the SAR capabilities of WTP for a moment as they are a very useful feature of the WAP stack, especially now that transactions have become bigger thanks to richer content, such as pictures and games.

Let's say we have a 30k picture file that we want to download to our device to use as wallpaper on the display. Using WTP SAR we can divide the file into 20 units of 1.5k and send them as a single transaction, notifying the receiving WTP stack that a segmented transaction is taking place. It is perfectly valid to have a multi-packet transaction that is not a segmented higher layer message, so we should not think that transactions bigger than one packet are automatically using SAR.

The WTP provider on the receiving end will buffer up the segments in readiness to reassemble them. With the basic SAR capabilities of WTP only 256 segments can be accommodated, so this places a limit on the SAR process.[28] In terms of implementation the WTP stack must have enough memory allocated to allow for successful buffering.

If some of the 20 packets in our image file get lost during the transmission, then the WTP stack issues a *Negative ACK (NACK)* to specify which packets have not been received properly. The WTP provider at the transmitting end can then retransmit the lost packets, but not those that were already successfully sent. In contrast to this, any loss of the IP packets

[27] A JAR file is how a Java application gets packaged up for distribution. Here, we are specifically referring to packaged J2ME applications that are designed for small devices and can be loaded OTA.
[28] This limit can be overcome using *Extended SAR*, but this is an optional feature of the WTP protocol.

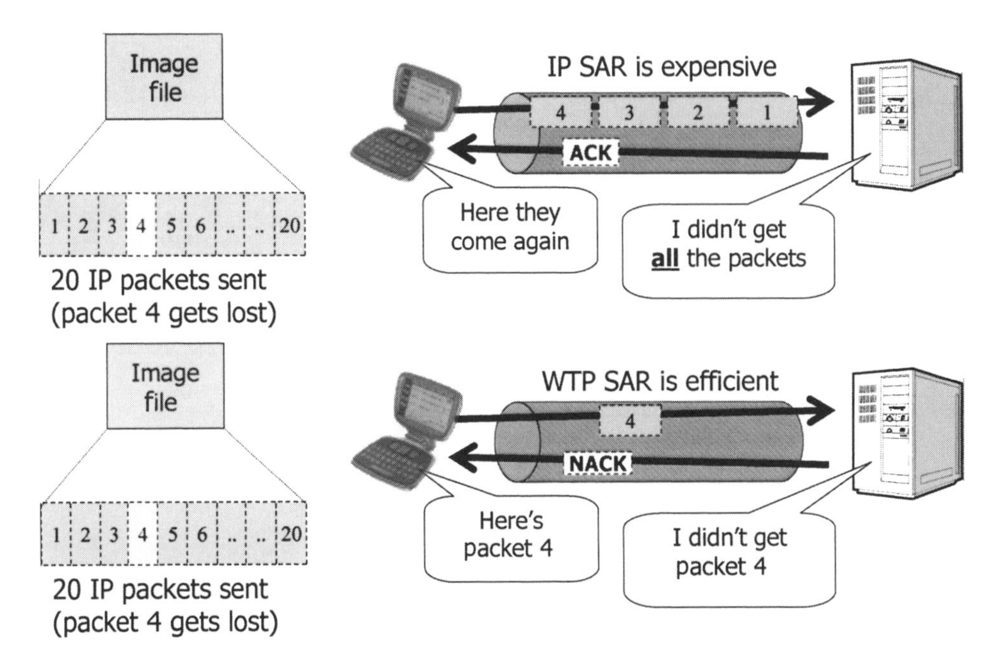

Figure 7.19 Selective re-sending of WTP SAR is more efficient.

with IP segmentation cannot be selectively recovered; all the packets are required to be re-sent. This contrast in behaviour and performance is illustrated in Figure 7.19.

7.6.5 Wireless Session Protocol (WSP)

WSP is very similar to HTTP in concept and much of its implementation; this is what we expect, given that WAP devices are expected ultimately to fetch content from standard Web servers, which, we as know well, only support HTTP (mainly 1.1).

Similar to HTTP methods, WSP supports all of the HTTP 1.1 headers, except that there is a requirement to ensure that headers are listed in a certain order in the request. This is to allow the use of *code pages*, which is a grouping of different headers into collections. With WSP, commonly used (well-known) headers are replaced with a shorthand form to condense the overall HTTP request for transmission across the RF link. For example, instead of using the string 'GET' in the header, the hexadecimal code '40' is used and, instead of the header string 'Last Modified', the hexadecimal code '1D' is used. With HTTP requests, it is common that a lot of the headers remain the same from one request to another, especially if the requests are within the same domain (e.g., a WAP portal). By grouping the static (unchanging) part of the request into code pages, these can effectively be cached and substituted for a short-form version that indicates a particular page is being used. When received by the device or the proxy the code page is looked up from the cache and its values can be extracted and used.

Most of the HTTP header region is assigned binary token equivalents to achieve the shorthand form as a means of compression; this applies in both directions. For example, the status codes of responses are also encoded. Some of them are listed below, along with their hexadecimal tokens ($0 \times xy$, where x and y are hex numbers):

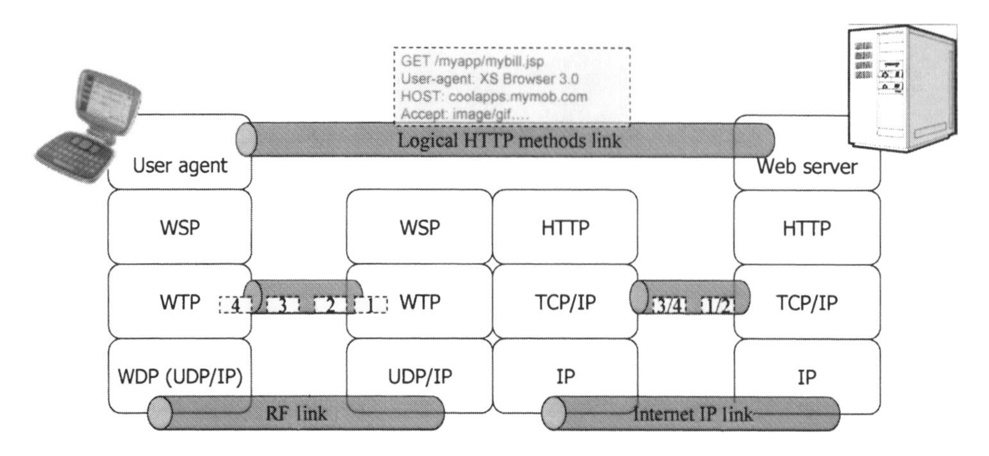

Figure 7.20 WSP requires a proxy.

'200 OK, Success' = 0 × 20
'201 Created' = 0 × 21
'202 Accepted' = 0 × 22
'203 Non-Authoritative Information' = 0 × 23
'204 No Content' = 0 × 24
'205 Reset Content' = 0 × 25
'206 Partial Content' = 0 × 26
'300 Multiple Choices' = 0 × 30
'301 Moved Permanently' = 0 × 31
'302 Moved Temporarily' = 0 × 32
'303 See Other' = 0 × 33
'304 Not Modified' = 0 × 34
'305 Use Proxy' = 0 × 35
'307 Temporary Redirect' = 0 × 37
'400 Bad Request – server could not understand request' = 0 × 40
'401 Unauthorized' = 0 × 41
'402 Payment Required' = 0 × 42
'403 Forbidden – operation is understood but refused' = 0 × 43
'404 Not Found' = 0 × 44

Clearly, the use of tokenized headers is not directly compatible with HTTP 1.1, even though semantically WSP and HTTP headers are identical, except for some WAP-specific extensions. Therefore, different from W-HTTP, we have to rely on a WAP proxy to convert our WSP sessions to HTTP sessions, as shown in Figure 7.20.

WSP relies on the reliable transport mechanism of WTP; namely, the strongly verified transaction capabilities discussed earlier. As noted in our discussion of WTP, it is possible not only to acknowledge messages but also to verify that the consumer has successfully attained a meaningful result from the transaction, as shown earlier in Figure 7.18. This method of transaction is used by WSP in its *connection-mode*. An alternative mode – *connectionless-mode* – provides less reliable transport, but can be used without the need to establish session credentials and, so, is more efficient for applications that require one-off interactions on an infrequent basis. This is useful for applications that require data to be pushed to a client, such as an email alert; we will discuss push separately in Section 7.6.6.

Other aspects of WSP that are notable are its ability to support asynchronous requests, similar to the pipelining of HTTP; so, it would be possible to achieve overall efficiencies of pipelining end to end. WSP supports this queuing technique and does not require the responses to be in order (i.e., in step with the requests or synchronous).

WSP also implements multipart responses where the content can consist of several objects grouped together into the body of a single response. This uses a technique called multipart MIME, which we will discuss later when we look at the Multimedia Messaging Service (MMS). It should be noted that multipart transfers are used specifically for media types that are by their nature multipart and can be handled properly by the user application sitting on top of WSP. We should not mistakenly think that this is a technique for grouping together several requests from an origin server into one 'concatenated' response.

7.6.6 WAP Push

So far we have been looking at methods for a client communicating with a server on a transactional basis where the user is actively requesting content during some kind of browsing experience. The content could be varied: from pages to display in a browser to ring tones and other downloadable objects that could be subsequently utilized on the device.

The underpinning paradigm is the link–fetch–response, where the link aspect is some means of addressing a remotely stored resource that we subsequently attempt to fetch from a hosting entity that can respond with the content. Protocols like TCP/HTTP can be utilized to do this, although they have inefficiencies that become apparent over LTNs. These can be mitigated by utilizing optimizations within the scope of the protocol specifications, albeit via some largely optional implementation features.

WTP and WSP enable us to do better. They are highly optimized protocols for LTNs. The efficiencies offered are useful for implementing usable mobile applications and greatly mitigate the constraints of the RF link, which ideally we would like not to be there at all in terms of its limiting effects. The WAP protocols are a step closer to making the link transparent to the user, to making the constraints less intrusive than they might otherwise be within the context of browsing and downloading content.

However, WAP goes even further in enabling the delivery of useful mobile services by offering an alternative communications paradigm called *push*. This is the ability to send content to the user without them first requesting it. As we will see, this paradigm is also useful going in the other direction to enable users to push unsolicited content to other users. The advent of devices such as camera phones makes the utilization of this paradigm all the more likely. Some applications of push messaging to the client include:

- email notification – 'You have new mail from Fred – click to read';
- location-sensitive advertising – 'SuperMart's 10%-off everything – click for map';
- news alerts – 'Budget headlines now available – click to read';
- travel alerts – 'Paddington 5.20 cancelled – click for timetable'.

Here, we are showing the use of push as a means to display alerts to the user, but we should not be fooled into thinking this is the only use of WAP Push. The push mechanism is a protocol feature; it is part of the WAP stack – it is not an end user application. However,

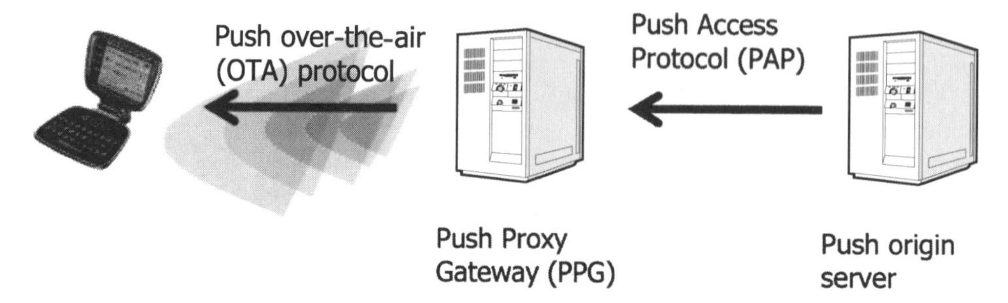

Figure 7.21 WAP Push architecture.

using push to send displayable alerts is an application. It is probably the most likely use of WAP Push and is definitely worthy of consideration in a wide variety of contexts, especially as WSP enables content that can include Web pages to be pushed. This is why our examples above are actionable; they all contain links that can be clicked as a response to the alert. This would cause a browsing session to be initiated, whether by HTTP or WSP. This is one of the powerful features of the WAP Push mechanism.

The push mechanism is inherent in WSP/WTP, so there is no need to implement any special technique to support it. A WAP-compliant device should have a WDP port dedicated to the reception of WAP Push messages and always be ready to receive WAP Push messages. The messages can be identified by the sender as destined for a particular device application using an application ID field in the message. The advantage WAP Push has over WSP/WTP is that it can take place over any bearer that supports WDP, which includes SMS. Recall that with RF bearers that support IP, WDP defaults to UDP in those cases and then WTP runs over UDP. However, the way that most RF networks work is that an active IP session[29] has to be established before IP communications can take place. Hence, if a device has no active IP connection on its RF network interface, then we are not able to utilize a UDP connection to support push messaging. However, we can run WDP over SMS, and devices that support SMS are always ready to receive text messages; there is no need to establish a messaging context (session) first, other than the device being within RF coverage, of course.

WAP over SMS enables us to send WAP Push messages without an IP session. A special type of WAP Push message can be sent to ask the device to establish an IP session for subsequent push communications where more substantial amounts of content could be pushed. We are able to detect that an IP session has been established by asking the device to register with a Push Proxy Gateway (PPG), which is the network element that is responsible for exchanging push messages with the device, as shown in the WAP Push architecture diagram in Figure 7.21.

As Figure 7.21 shows, the PPG is the network element that sends the push messages to the device. Messages are submitted to the PPG using HTTP 1.1 wherein the HTTP requests use a method called POST; with the POST method, content can be attached to an HTTP request. This is different from the GET method, where requests consist of header information only (HTTP headers), but do not have a body. With GET methods we are only expecting to see a body in the HTTP responses from the origin server, the body being used

[29] We will look at what an active IP session means later; but, we can imagine it is like a mobile device needing to plug into the IP network before it can talk over IP, even though there are no plugs in a wireless network, of course!

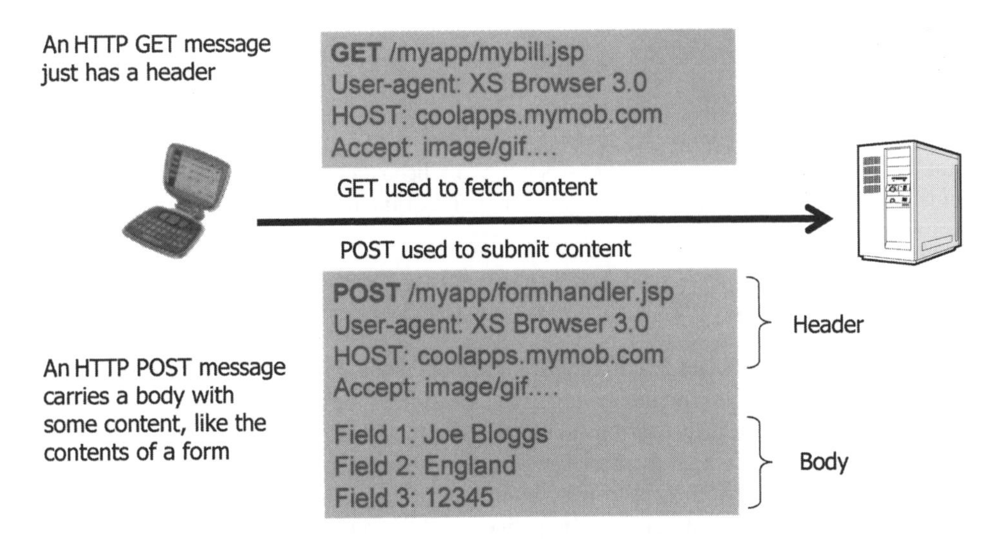

Figure 7.22 HTTP POST method used to submit content.

to carry the requested content from the resource identified in the requesting URI (uniform resource identifier) header field (part of the GET method). The actual message contents submitted to the PPG using POST must conform to the PAP.

The PPG reminds us of the WAP gateway we saw earlier and in some ways it has similar functionality. However, it has the added capability of being able to store messages to forward them later to a device that is unreachable (i.e., out of RF coverage). Because we are using WSP, even in connectionless push mode, we are still able to receive an acknowledgement from the device that the push message has been delivered to it (but not to the higher layer application). This enables a retry mechanism to be implemented.

If we know that the objective of a particular WAP Push message is to visually notify the user of an actionable item (e.g., link to another server), then we can send a particular message type called a *Service Indication*. This contains an embedded URI that points to a resource on an origin server. The objective of the push message is to ask the user to visit this resource, usually via a browser. The user can elect to click on the link, and this will automatically initiate an IP session and subsequent WSP or HTTP (or W-HTTP) session to the referenced resource.

For WAP 1-compatible devices, WSP works fine for asynchronously pushing messages to mobile devices. However, with WAP 2 both the standard or wireless-profiled forms of HTTP and TCP do not support push mechanisms; therefore, we have to implement a push channel using an alternative approach. The way this is done is by turning HTTP on its head and asking the mobile device to act like an HTTP origin server so that it can accept POST method requests (Figure 7.22). If this is done, then we are able to send it content over a TCP/IP link.

It is fairly obvious that this would require an active IP session between the device and the network. Therefore, in the absence of an IP session we can still revert to the WAP Push over SMS; the use of more than one protocol to achieve a task is perfectly plausible. It is just a matter of implementing both protocol stacks in software on the device. It is perfectly acceptable to use WAP Push over WTP/SMS to initiate a subsequent browsing session with an origin server via HTTP/TCP (or W-HTTP/TCP).

7.7 PEER TO PEER (P2P)

P2P is a new and exciting area of computing and due to its infancy clearly established protocols have yet to emerge that are as firmly established as HTTP is for the Web. P2P is not that widespread anyhow, so there has not really been a chance for a set of protocols to emerge, but one of the efforts that is attracting a lot of interest is the JXTA project, which we will discuss shortly.

7.7.1 Defining P2P

We have already introduced P2P in the early chapters of this book; so, it seems a bit late to be defining it now, but it is an interesting topic, nonetheless. We can dispense with some of the more meaty academic discussions that might surround this topic and concentrate on P2P in the mobile context, which is after all where our interest should lie, although we should always have in mind that mobile-to-fixed communication can still take place in P2P mode. The philosophical origins of P2P don't quite fit the mobile context anyhow. There is a thought process that, taking into account the increasing prevalence of broadband connectivity and the equally widespread availability of massively powerful desktop computing, concludes that reliance on centralized processing might be outdated or unnecessary in some cases: not in all cases, but certainly some. With mobile networks the same abundance of resources does not exist, although under certain conditions it could approach some wire line configurations, such as a powerful laptop networked via WiFi.

The key concept that underpins our consideration of P2P is that mobile devices can talk to each other, without the need to establish any link with the content network or, possibly, without having to involve the IP network either. The most obvious case is the direct association of two or more devices in a personal area network sustained by Bluetooth™ connectivity. This seems to be the more interesting case to apply research and development effort in the short term. The reason for this is that it seems obvious that dual-mode or even tri-mode devices will become commonplace: dual-mode[30] being Bluetooth™ and 3G, and tri-mode being the addition of WiFi to these two. It therefore seems obvious that, increasingly, devices will talk to each other directly or will be able to network without the need to go through a server (although the absence of a server is not a necessary and sufficient definition of P2P[31]).

The important part of the P2P concept in the wireless space is that devices may want to connect with each other without going through a mobile cellular network, the reason being that the other modes of RF connection would provide potentially greater speed and be at lower cost, possibly no cost at all. This does not mean that we assume an either/or networking situation, where we use either the WAN or a localized network. Oddly, both could be used in a P2P application. For example, a connection between two devices could be established using a WAN and the switch-over to a local connection in order to swap a file; the obvious example is music files.[32] What's interesting about this type of application

[30] It is recognized that there is a trend in the PDA (personal digital assistant) market toward WiFi with Bluetooth dual-mode devices, but here we are talking about the wide-area connection being one of the required modes.
[31] For a more in-depth discussion of what is P2P see Brookshier, D. et al., *JXTA: Java P2P Programming*. Sams Publishing, Indianapolis, IN (2002).
[32] Note that despite the failure of Napster, music file swapping is still a possibility. *Digital Rights Management* schemes could make file swapping possible within a controllable framework.

is that the ability to determine the location of a device makes it all the more possible to find a neighbouring device to swap files with, assuming that this isn't already obvious by virtue of direct association with a nearby colleague or friend.

7.7.2 Some P2P Concepts

It is a common misconception to think that P2P networking is based on a single protocol; this is not necessarily the case. It really depends on a number of factors, but in the first instance it should be borne in mind that P2P networking can work in several different modes:

- *Point-to-point* – this is a direct communications pathway between two devices, although the devices could be co-located or separated in terms of a direct RF connection. A variety of protocols could be envisaged for point-to-point, and there is no need for it to be limited to one, even within the context of a single communications session. For example, in a P2P session, devices may need to talk to each other to exchange control information. This could be done using a protocol separate from the one used to exchange session information, such as music files. In the latter case it could be that it makes more sense to stream information from one device to another, thus requiring a suitable streaming protocol, whereas control information would be exchanged using some other means.
- *Neighbour-linking* – within a P2P network, devices may need to talk to each other to establish an interlinked mesh so that information and people can be found via the mesh. We will discuss this idea in Chapter 12 when we look at *spatial messaging*, which is a type of application built on the location-finding capabilities of the RF network.
- *Broadcast* – this is the ability for one device to send the same information simultaneously to other devices in a defined group. It need not be restricted to a one-way process. All devices within a defined group could broadcast to every other device in the group. An obvious example of this is multiplayer gaming where the moves of a player need to be made known to all other players in real time.
- *Multipoint* – this is similar to broadcast, except that the means to get the information to each device group is via hopping from one device to another. This should not be confused with the neighbour-linking protocol(s) mentioned above, which are to do with discovering other devices ('out of session'), not communicating with them 'in session'.

All of these modes of networking are required to implement a P2P system. In the mobile context there are also implications relating to the RF network and, in particular, the locality of the peer devices. For co-located devices that are able to communicate physically with each other directly (such as via Bluetooth™), there are different considerations from those that are indirectly networked, realizing that either mode can still support P2P communications.

Direct P2P connectivity exhibits unique challenges compared with indirect P2P, most notably the challenge of potentially having to support the entire communications session without any intermediary server functions whatsoever. For example, this would imply that the P2P clients would have to be able to find means to authenticate each other. This may not be a problem because, as already noted, authentication in one sense is implied if the communicating parties are visually within contact and visual identification is possible. However, we should not ignore the cases where visual identification is not possible simply because the two parties hitherto had never met.

There are many issues, like authentication, that need to be considered when designing a P2P-based application in the mobile context. No one set of protocols has emerged yet to offer a comprehensive and satisfactory conclusion, and there is nothing stopping the adventurous among us from designing their own P2P protocols, which need not be IP-based. Indeed, one of the contenders for P2P protocols is JXTA, which is transport-agnostic in terms of the protocol set.

7.7.3 JXTA

JXTA started as a research project at Sun Microsystems, Inc. under the guidance of Bill Joy and Mike Clary, to address the growing interest in P2P architectures as viable solutions to certain types of computing problems. JXTA is a set of open and generic P2P protocols that allow *any* connected device (mobile or fixed) on the network to *communicate and collaborate*.

Despite is origins at Sun and its use of the letter 'J' in the project name,[33] JXTA protocols and possible implementations are intended to be computer language-agnostic – not solely or especially Java – though one naturally expects a bias toward Java within the open developer community attracted to this project.

JXTA proposes a set of protocols that address the common needs of a P2P solution, in the same way that TCP/IP proposes a set of protocols for hardware networking. However, JXTA technology connects peer nodes with each other, not hardware nodes. TCP/IP is platform-independent by virtue of being a set of protocols; so is JXTA. Moreover, JXTA technology is transport-autonomous and can exploit TCP/IP as well as any other transport standards, which technically could include the native data protocols for UMTS within a 3G environment. JXTA offers the following protocols:

- Peer Discovery Protocol;
- Peer Resolver Protocol;
- Peer Information Protocol;
- Rendezvous Protocol;
- Pipe Binding Protocol;
- Endpoint Routing Protocol.

These protocols offer various solutions to the P2P networking problem, and it is not intended to describe them in any detail here, only to introduce them and to identify some interesting aspects pertaining to mobility.

Before explaining the basis of the protocols we need to introduce some other concepts from the JXTA P2P technology suite:

- *Advertisements* – an advertisement is an XML document that names, describes and publishes the existence of a resource on the P2P network, such as a peer (a P2P participant), a peer group, a pipe (see below) or a service. We can see that advertisements enable peers to find out what exists on the P2P network, but no mention has yet been made of how to find these adverts (i.e., a discovery mechanism).

[33] In fact, JXTA is a play on words, influenced by the word 'juxtapose' (to place side by side); hence, it is not an acronym at all, more a sort of onomatopoeia.

- *Peers* – a peer is any entity that can communicate within the P2P network using the JXTA protocols. This is analogous to the Internet, where an Internet node is any entity that can talk using the IP protocols (some or all of them). As such, a peer can be any device, including embedded devices like domestic appliances (e.g., burglar alarm).
- *Messages* – messages are designed to be usable on top of asynchronous, unreliable and unidirectional transport. Therefore, a message is designed as a datagram, containing an envelope and a stack of protocol headers with bodies. A message destination is given in the form of a URI on any networking transport capable of sending and receiving datagram-style messages, which could include wireless or wire line, and on any type of RF network. Messages can include credentials used to authenticate the sender and/or the receiver, but no set way of doing this is specified; this needs to be negotiated (as with the Secure Sockets Layer, or SSL).
- *Peer groups* – typically, a peer group is a collection of cooperating peers providing a common set of services. The specification does not dictate when, where or why to create a peer group, or the type of the group, or the membership of the group; it does not even define how to create a group. Very little is defined with respect to peer groups except how to discover peer groups using the *Peer Discovery Protocol*.
- *Pipes* – pipes are virtual communication channels for sending and receiving messages and are asynchronous. They are also unidirectional; so, there are input pipes and output pipes. A pipe's end point can be bound to one or more peers. A pipe is usually dynamically bound to a peer at runtime via the *Pipe Binding Protocol*.

Let's return to the protocols for a brief explanation; they do the following:

- *Peer Discovery Protocol* – this protocol enables a peer to find advertisements on other peers and can be used to find a peer, peer group or advertisements. This protocol is the default discovery protocol for all peer groups and, so, has to be included in the implementation of any peer software. Peer discovery can be done with or without specifying a name for either the peer to be located or the group to which peers belong. When no name is specified all advertisements are returned, but clearly this is a costly option within the mobile context.
- *Peer Resolver Protocol* – this protocol enables a peer to send and receive generic queries to find or search for peers, peer groups, pipes and other information. Typically, this protocol is implemented only by those peers that have access to data repositories and offer advanced search capabilities.
- *Peer Information Protocol* – this protocol allows a peer to learn about other peers' capabilities and status. For example, one can send a *ping* message to see if a peer is alive. One can also query a peer's properties where each property has a name and a value string.
- *Rendezvous Protocol* – this protocol allows a peer to propagate a message within the scope of a peer group.
- *Pipe Binding Protocol* – this protocol allows a peer to bind a pipe advertisement to a pipe end point, thus indicating where messages actually go over the pipe. In some sense a pipe can be viewed as an abstract, named message queue that supports a number of abstract operations, such as create, open, close, delete, send and receive.

- *Endpoint Routing Protocol* – this protocol allows a peer to ask a peer router for available routes for sending a message to a destination peer. Often, two communicating peers may not be directly connected to each other, as discussed above when we talked about the general concept of *neighbour linking*. Any peer can decide to become a peer router by implementing the Endpoint Routing Protocol.

The use of JXTA for wireless is being looked at by some members of the JXTA community and has led some researchers to find interesting ways to combine JXTA P2P concepts with the emerging and evolving wireless technologies,[34] such as the mobile version of Java, called Java 2 Mobile Edition (J2ME), which is a software technology we discuss in a later chapter in the book.

The main challenge with JXTA in a mobile environment is ensuring that the protocols enable the most efficient use of resources within the resource-limited RF network. There are other concerns relating to the robustness of JXTA within a temperamental mobile environment. Neighbour linking needs to be resilient to any of the neighbouring peers finding themselves in a hostile RF situation, including complete signal loss. The rate at which the network can heal and pipes be re-formed will be critical to system performance. There seems a need to research the efficiencies gained by mapping P2P protocols onto the RF network in a way that is cognizant of network characteristics.

[34] Arora, A., Haywood, C. and Pabla, K., *JXTA for J2ME – Extending the Reach of Wireless with JXTA Technology.* Sun Microsystems Inc. (March 2002) – http://www.jxta.org/project/www/docs/JXTA4J2ME.pdf

8

J2EE Presentation Layer

In Chapter 6 we have already looked at how scalable and robust mobile applications can be built using a software platform technology called J2EE (Java 2 Enterprise Edition). The essence of J2EE was the division of software into infrastructure services and business-specific services, as shown in Figure 8.1. The infrastructure services is inherent in the J2EE platform and is what enables our business-specific application to be multi-user, scalable, secure, able to support redundancy, able to connect with databases and so on.

In Chapter 7 we looked at protocols to support the link between the devices and the platform. Our main focus was on link–fetch–response paradigms supported by Internet Protocol (IP) protocols like HTTP/TCP and by made-for-wireless protocols from the Wireless Application Protocol (WAP) family, like WSP/WTP.

In this chapter we will examine in more detail how J2EE supports these protocols and how we might structure our business software to take advantage of the power of J2EE while simultaneously supporting protocols like HTTP (HyperText Transfer Protocol).

We also need to look at exactly what happens on the device in terms of subsequent processing of the content fetched to the browser from the server. We have mentioned the browser application as a key component in many mobile services, but we have not yet looked at how it works in any detail or how it ties back to the J2EE programming paradigm. After all, with browser-based solutions the application logic and guts is implemented and hosted by the J2EE platform, not by the device. This detachment has its own challenges that we have to take care of in the J2EE world, and it is not always obvious how to do this. There are also some challenges that are very specific to the mobile environment, such as how to cope with many different device types. This is a challenge for our application designers and one that needs to be addressed. Further questions are: what if we design an application to access an email server and how are we going to make sure that it works equally well or satisfactorily well, on all devices? When we look at mark-up languages and their relationship with device capabilities, we will need to address some particularly thorny problems.

Next Generation Wireless Applications P. Golding
 ISBN: 0-470-86986-0 (HB)

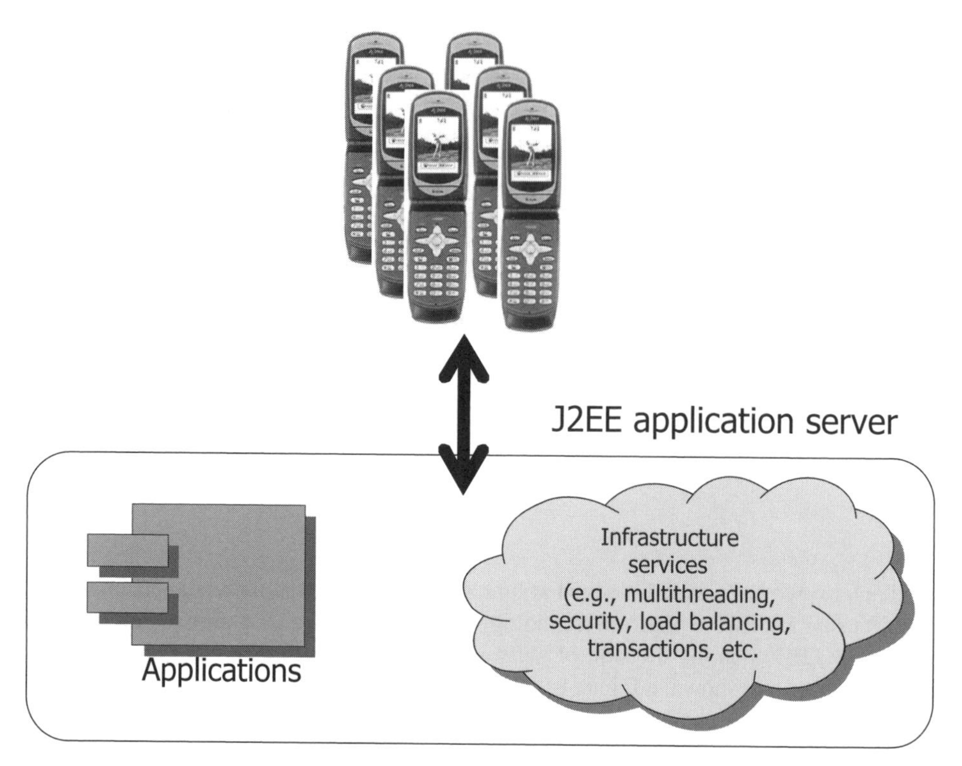

Figure 8.1 Division of software into infrastructure and business services.

8.1 SEPARATING PRESENTATION FROM BUSINESS LOGIC

When we introduced J2EE we discussed the concept of encapsulating our business-specific software into distinct software units called Enterprise Java Beans (EJBs). We explained how, if we put all of our required business functions into EJBs and then planted these in a J2EE application server, we could rely on the server doing all the challenging and complicated middleware stuff, such as enabling the bean to be accessed by many clients at once without danger of it collapsing under the strain. In Chapter 9 we will look in detail at exactly how this is able to take place by examining the underpinning technologies of J2EE. We will also find that, to support a wide range of possible solution scenarios, there are actually three types of EJB. However, for the time being we just assume that, irrespective of the type of bean, by using them we get all the goodness of the infrastructure services automatically, as if by magic!

Most of the time, when developing an application that runs on a J2EE application server, the entire set of software functions we end up needing will fall neatly into two functional areas. The first area is concerned with implementing the user interface (UI) and the other is everything else, or the 'brains' of the application, as shown in Figure 8.2.

The software concerned with the UI is generally referred to as the presentation logic, or presentation layer. The rest, or the 'brains', we refer to as the business logic, or layer.

This neat division of effort is something that can be exploited by our J2EE architecture. It is possible to provide the programmer with a means to focus on one area or the other by offering them dedicated J2EE support mechanisms that work particularly well for each case.

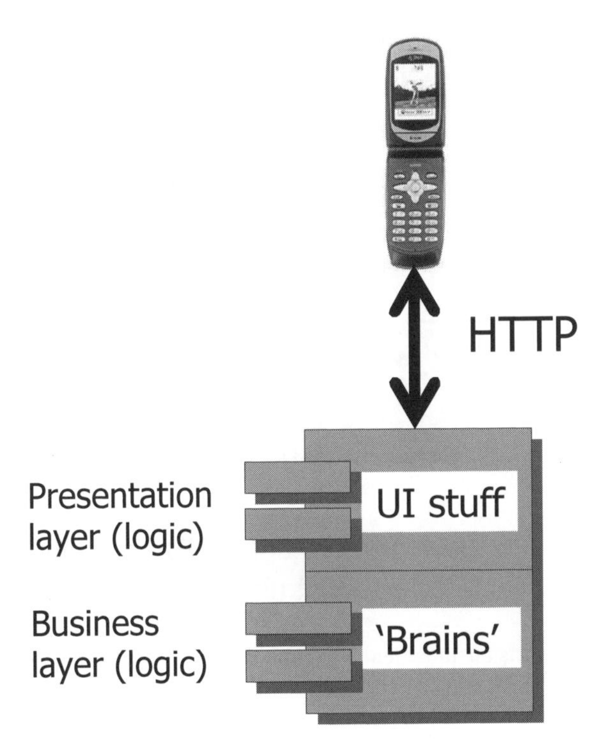

Figure 8.2 Presentation and business layers.

For example, in the case of the presentation logic the powerful built-in HTTP capabilities of the J2EE platform can be offered to the programmer in a model that makes programming for HTTP connectivity easier. By separating out presentation logic from business logic it also means that we can maintain these portions of software independently of each other, which is good for managing the software development process. It is especially useful for mobile services where the evolution of device types tends to be quite rapid. Consequently, we need to constantly revisit our UI design to tweak it for new devices as they come onto the market, although we will look at strategies for making this particular process as painless as possible.

J2EE has three types of software program that we can use to create an entire J2EE application: *servlets*, *Java Server Pages* (JSPs) and *EJBs*. We already know something about EJBs; in Chapter 9 we will look at them in more depth. Let's look at these J2EE software entities to see what the essential differences are and how to use them.

8.1.1 Servlets and JSPs – 'HTTP Programs'

As shown in Figure 8.3 there is a simple distinction between EJBs and the UI entities (servlets and JSPs). The EJBs are always associated with accessing what is referred to as the Enterprise Information Services (EIS) tier. The EIS tier is where all the heavy database servers reside: directories, legacy systems, mail servers and so on. We can access a good deal of the EIS tier via the powerful J2EE application programming interfaces (APIs), like Java DataBase Connector (JDBC), but it is essential that we do this from EJB code if we

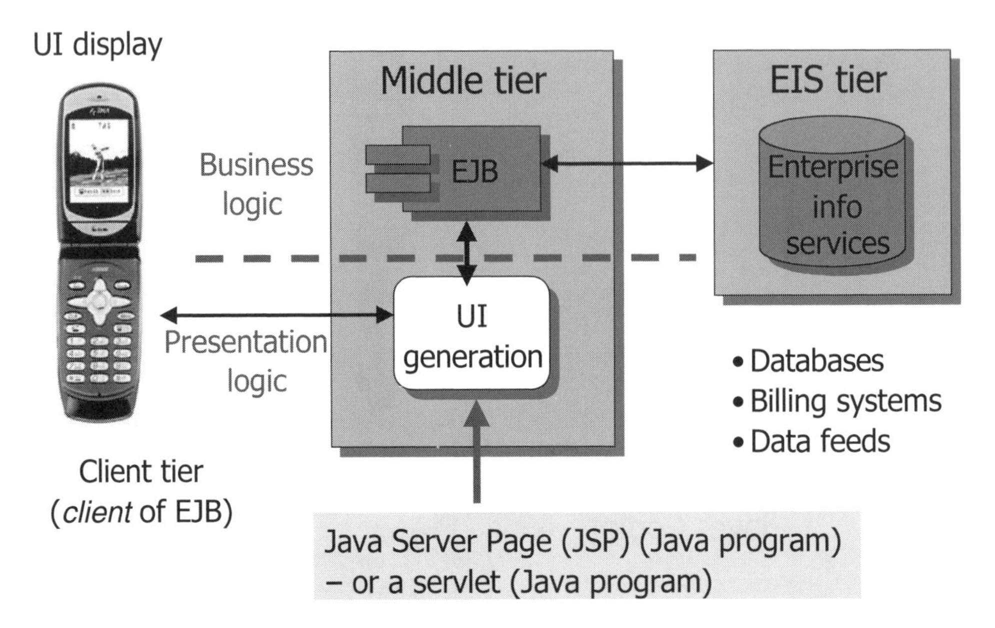

Figure 8.3 J2EE supports business and presentation tiers via different program types (EJB, servlet or JSP).

want to build a system that truly benefits from the J2EE infrastructure services approach. In Chapter 9 we will explain that EJBs have particular characteristics which enable the J2EE infrastructure services concept to operate.

The servlets and JSPs act independently of the EJBs, but can call on the services of the EJB to carry out useful work under the command of the user via the UI. The servlet and JSP program types are specifically designed to be used for facilitating browser-based applications accessing the J2EE platform via HTTP. In particular, it is assumed that any software written to run inside a servlet or JSP is going to produce a text-based output that will be streamed over an HTTP link back to a browser that initiated the program's execution in the first place by sending an HTTP request to the J2EE server. This tells us that one of the roles of a J2EE application server is as a Web server (or *origin server* in the HTTP vernacular). In other words, servlets and JSPs are programs that sit on the J2EE server ready to run whenever an HTTP request is routed to them by the application server's Web engine, as shown in Figure 8.4.

If we look at the sequence of events in Figure 8.4, we can better understand the role of servlets and JSPs and how they function in a J2EE environment. We will also be able to appreciate the intimate connection these program types have with the HTTP protocol.

Step 1 is the user initiating a request via a link to one of the servlets on the server. Of course, our user doesn't know anything about servlets or JSPs, so they are not at all cognizant that their Web request has anything to do with running these entities on a J2EE server. All the user knows is that they have either entered a Web address in their browser (via the address entry box) or clicked on a link in a page that has this request resource address beneath it.

Step 2 is the *user agent* responding to the user's request by generating an HTTP request using the GET method. We can see in the header for the request that the resource required

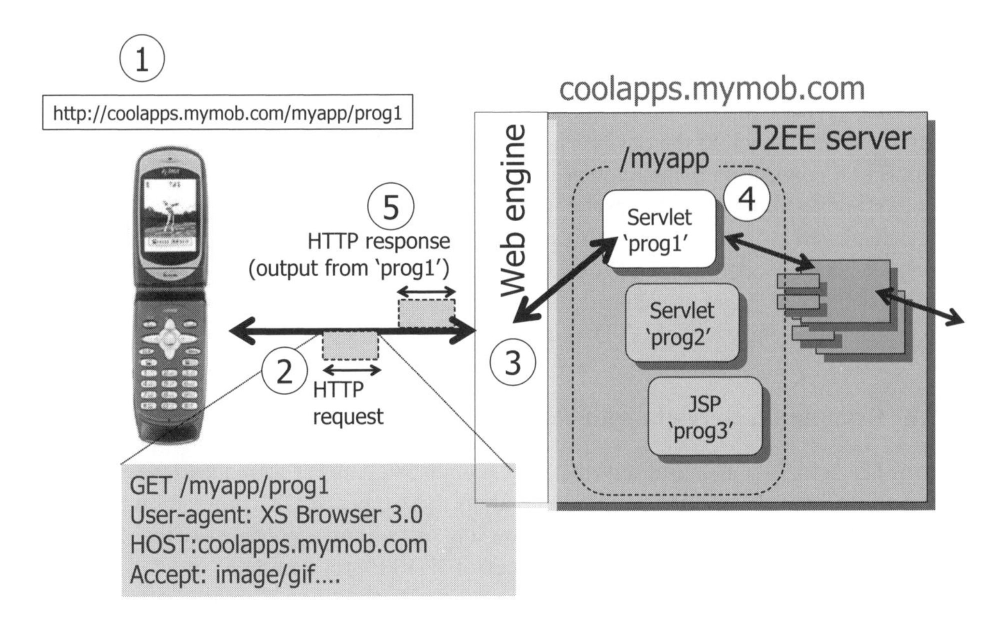

Figure 8.4 Servlets and JSPs are 'HTTP programs'.

on the server has the pathname of '/myapp/prog1'. The uniform resource locator (URL) for the request has the domain name 'coolapps.mymob.com', and the *user agent* uses this information to route the HTTP request to the corresponding server. When the request reaches the server via the RF (Radio Frequency) and IP network layers it is routed by the server's operating system to the J2EE server application. The operating system or its Transmission Control Protocol (TCP)/IP stack component knows to do this because the J2EE server has registered itself (bound) with the appropriate port (port 80). Communications to port 80 are processed by the Web engine component of the J2EE engine. The Web engine extracts the pathname from the request – '/myapp/prog1' – and 'looks' in the Web folder '/myapp' to see if there is a program registered there with the name 'prog1'. On finding that there is indeed a servlet with this name at that path location the Web engine proceeds to schedule the servlet for execution on the J2EE platform, as in step 3.

All of the HTTP header information is made available to the servlet, and it is the job of the Web engine to pass this information to the servlet (or JSP). This is done by making the information available via a Java API, which in this case happens to be called *servlet*; this is a Java program that has methods that can be accessed by the servlet when it executes. One of these methods can be used to extract HTTP headers from the request that spawned the servlet's execution. If there were any parameters in the HTTP header that the servlet needed, then it has full access to them via the API.

In step 4 the servlet is now executing on the server and has full access to all the J2EE APIs, including the ability to call EJBs to carry out business logic tasks. For example, this servlet could be responsible for creating a Web page to list ring tones on a particular theme. The list of available ring tones is probably sitting in a database along with their descriptors (e.g., tone name, theme and device compatibility). An EJB may have been programmed to fetch tone names from the database. The returned descriptors may be in XML, or eXtensible Markup Language, so the EJB also extracts the appropriate XML fields, such as the title of the tones (e.g., from a <Title> tag pair).

On completing the bulk of its execution, including any calls to EJBs (which could be on the same server or another one), the servlet has now generated a page of XHTML, or eXtensible HyperText Markup Language (or other appropriate browser language) and is ready to stream it back to the device via HTTP. The servlet does this by writing the page to the HTTP connection using another method within the *servlet* API; this is step 5. The HTTP response output from the servlet is now received by the *user agent* on the device and it gets displayed in the browser interface. The user can now see a list of ring tones. These may have been encoded in the page as links, probably linking to a related servlet or JSP (e.g., 'prog2'), which can reveal the detailed information about a tone, like its cost and how to order it; hence, the cycle of HTTP requests and responses continues.

8.1.2 Comparing Servlets with JSPs

Having just described in outline how the process of dispatching HTTP requests to servlets or JSPs notionally works, the first thing to know about these types of J2EE programs is that they are optimized and designed for the HTTP (browsing) paradigm. A good deal of their features were originally conceived with the challenges of 'HTTP programming' in mind. Hence, wherever possible we should aim to use these entities for building Web user interfaces for our Web-based applications. Technically, there is actually nothing preventing EJBs from servicing HTTP connections, but they are not ideally suited for this task. When programming for Web-based interaction there are some common challenges that programmers face. In keeping with the J2EE infrastructure ethos, some of these challenges have already been addressed by built-in facilities within the J2EE platform; so, we should think twice about circumventing these in order to implement our own. These built-in features, pertaining to HTTP, are part of the JSP and servlet programming environment.

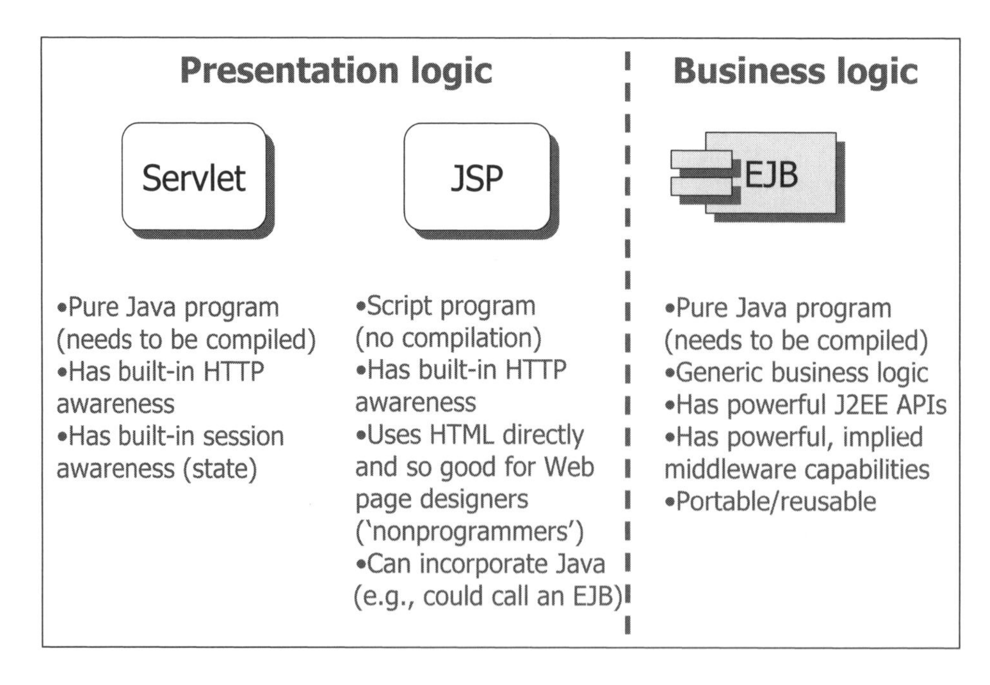

Figure 8.5 J2EE program types.

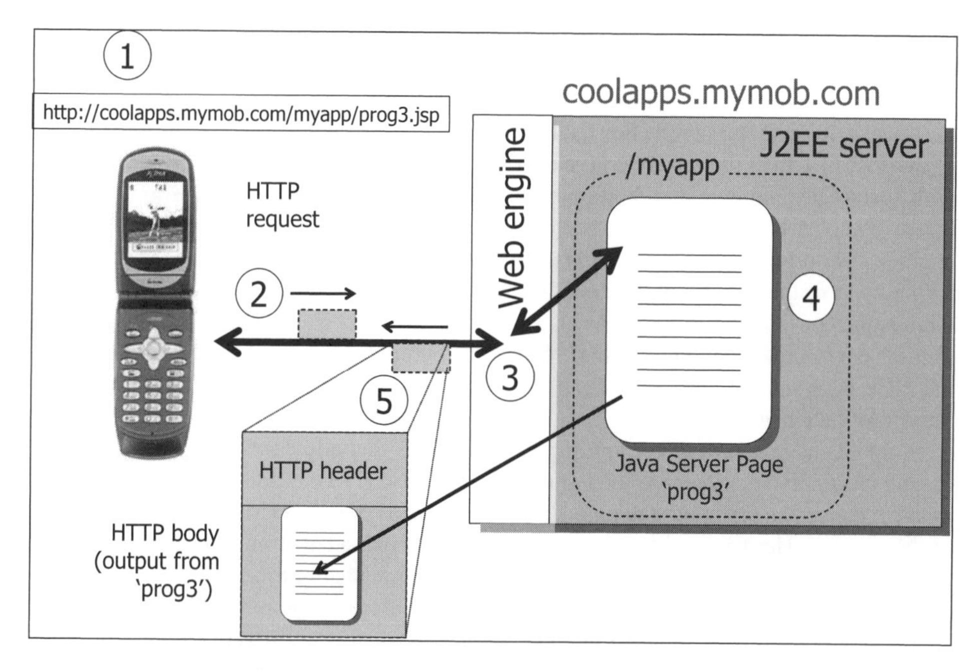

Figure 8.6 JSP execution.

The essential difference between a JSP and a servlet is shown in Figure 8.5, where we also make a rudimentary comparison with EJBs. We can see from the figure that what JSPs and servlets have in common is they're built in 'HTTP awareness'. As we saw in Section 8.1.1 these program types can be executed in direct response to an inbound HTTP request from a user agent. More than this, these programs are designed so that their outputs are fed back to the requesting user agent automatically. The programmer does not have to make any special effort to track inbound requests and match outgoing responses, no matter how many users are concurrently accessing the program. J2EE automatically supports both concurrency and making sure that the responses get back to the requesting agent.

The essential difference between a JSP and a servlet is that the JSP is constructed like a script and can include the direct use of XHTML (or HTML, etc.) in its construction, whereas a servlet is a pure Java program.

The JSP is written, as are servlets, as a text file. The JSP subsequently gets stored on the server as a text file. When the J2EE server runs a JSP it will simply attempt to open the file and output all of the JSP's contents as the body of the response to the inbound request, as shown in Figure 8.6.

As we can see in Figure 8.6 the user agent processes a link (step 1) with the appropriate address (URL) to eventually create an inbound HTTP request (step 2) that gets routed (step 3) to the corresponding JSP on the server. This JSP is run as a script on the server (step 4). What happens is that the JSP file is opened and then each line is stepped through, line by line,[1] until the end of the file. Whatever the J2EE Web engine finds on each line is written straight back out to the HTTP body of a corresponding response to the inbound request, as seen in step 5. The entire response, its header and its overall construction are all

[1] Just like any kind of script, even a film script, the script is followed one line at a time from the beginning to the end; hence the reason a JSP is called a scripted program. It has a linear flow from top to bottom.

taken care of by the J2EE server – the JSP programmer does not have to bother with any of these infrastructure tasks. The running of a JSP works just as if our device's user agent simply picks up the contents of the text file and pulls them back up to the requesting device.

The exception to this simple flow is if the web engine encounters special scripting inserts in the JSP, of which there are two types. The first type (called *scriptlets*) are demarcated with special beginning and ending markers: '<%' to begin an insert and '%>' to end an insert. What lies between these markers is Java code and will get executed by the J2EE Web engine. The output *from the execution of the code* is what gets sent out to the HTTP body for that part of the script, not the code itself. In Java, as with all programming languages, there is the concept of a program writing its output as a text stream that the programming runtime environment handles, such as routing the output to the screen. In the case of a JSP the J2EE environment takes the output stream and inserts it into the HTTP output flow at the appropriate places.

Scriptlets have access to certain implicit Java APIs that allow the code in the scriptlet to gain programmatic access to the request and response HTTP headers. Also, for inbound POST requests the scriptlets can gain access to the fields of any HTML form that is posted in the request. The values from the form are already extracted from the input stream by the J2EE Web engine and are made available programmatically to scriptlets, to make the programmer's life easier and the overall processing of pages more efficient.

The other type of special insert in a JSP is something called a JSP tag. Essentially, these are directives to the Web engine to substitute information wherever one of these tags is found. We will return to explain these in more detail later when we look at how to develop JSPs that can adapt their output to suit the display characteristics of the device that is requesting the page to be run.

A servlet is similar to a JSP; it is invoked in a similar way, as per steps 1–3 for the JSP. However, at the execution stage the entire servlet is run as a compiled Java program (or *class file*, which we explain in Chapter 10 when we look at devices). The file itself is not returned in the body of the HTTP response. The HTTP body is constructed by the program itself using whatever programming logic it chooses, including any number of references to other Java programs, like EJBs, to get the job done.

Essential Differences Between Servlets and JSPs Clearly, a servlet can only be constructed by a Java programmer and, probably, a fairly experienced one at that. However, a JSP can be constructed by someone with a quite different skill set, someone who knows about mark-up languages and how they get displayed in browsers (especially as a JSP script is primarily constructed using mark-up languages). It is likely that some JSP programmers will be able to learn enough about JSP to have a go at handling the different inserts, like scriptlets. It is often the case that scriptlets are passed around from programmer to programmer like tiny off-the-shelf utilities, almost like components (but at a much smaller and less sophisticated level than EJB components). Just to confuse us a little, but for the sake of completeness and in order to eradicate a common misunderstanding, it is possible to pass around more substantial components that make JSP programming more powerful; these are called *beans*. They are not the same as EJBs! We can think of a bean in a JSP as being like a portion of scripted code, but without the need to actually include the code in the JSP, just a reference to it (i.e., the bean).

To understand beans a little better, we could imagine a routine in software that checks the HTTP *user agent* header and looks the string up in a look-up table somewhere (e.g., could be a flat file or a database). We could use the output from the routine as a condition on which to decide how to format the rest of the JSP file, so that an appropriately formatted page is sent back in the HTTP response. For example, if the device is not capable of handling colour images, then we will send monochrome ones. This is the technique of adaptation that we alluded to earlier and we will investigate in more depth later in this chapter.

Perhaps another programmer has already figured out how to program this function. Having spent the time to program it, they can encapsulate the code as a bean. Instead of the JSP programmer having to include all of the code from the bean (cutting and pasting the code itself), they can simply include a special JSP tag that points to the code (bean) elsewhere on the system. This is a very useful approach as it means that JSP programmers can insert ready-made components into scripts, components that might otherwise be too difficult to program without this kind of reuse. It also means that the component can be maintained independently of the JSP (and vice versa). If the bean programmer decides on a more efficient coding approach or needs to check for new headers, then these changes can be made without having to change the JSP.

8.2 MARK-UP LANGUAGES FOR MOBILE DEVICES

So far we have addressed the process of a mobile device fetching content from an J2EE server using an HTTP link. This model applies to many mobile service scenarios. As we discussed earlier when we looked at WAP, WAP protocols will also work with this model, as they have been designed to work with HTTP very closely. As far as our J2EE environment is concerned, it will receive HTTP requests from WAP devices just like any other type of request. Hence, we do not need to resort to specific design approaches in these cases; we can still use JSP and servlet programs for our presentation logic.

What is unique to the mobile context is the nature of the information that the JSPs and servlets will produce. Clearly, the information is targeted at devices that have various display characteristics, so we expect the need to adapt our designs to match each device accordingly. Ordinarily, Web pages for the desktop world are very information-rich and can have very complex layouts. This is what we expect from resolutions that are typically 1024×768 pixels and rarely less than 800×600 these days.

We have seen the complexity of browser display capabilities evolve from something akin to a word processor document to something capable of supporting arcade game levels of complexity in both display and interaction, all within the Web browser. Contrast that with a popular mobile device from the current Nokia series 30 family of phones, like the 3410, where we still have a monochrome display and the resolution is 96×65 pixels. Figure 8.7 shows us what happens if we try to fit a standard Web page into the screen of a 3410. Clearly, even if this were possible, the resolution is insufficient to display the page, and it is simply too small to view. Going the other way, if we were to display the information at its real size and scroll around to see the page, then we can see how impractical that would be. The page image is shown in the figure with some sample screen locations overlaid on top, each sized to the 3410 resolution. Asking the user to scroll horizontally and vertically to see the page would be too much to ask for.

The other limitation is the ability to interact with the Web page using a pointing device (e.g., mouse) and a keyboard. On devices like the 3410 there is no pointing device to speak of and the keypad is a poor substitute for a QWERTY keyboard. We will be looking at

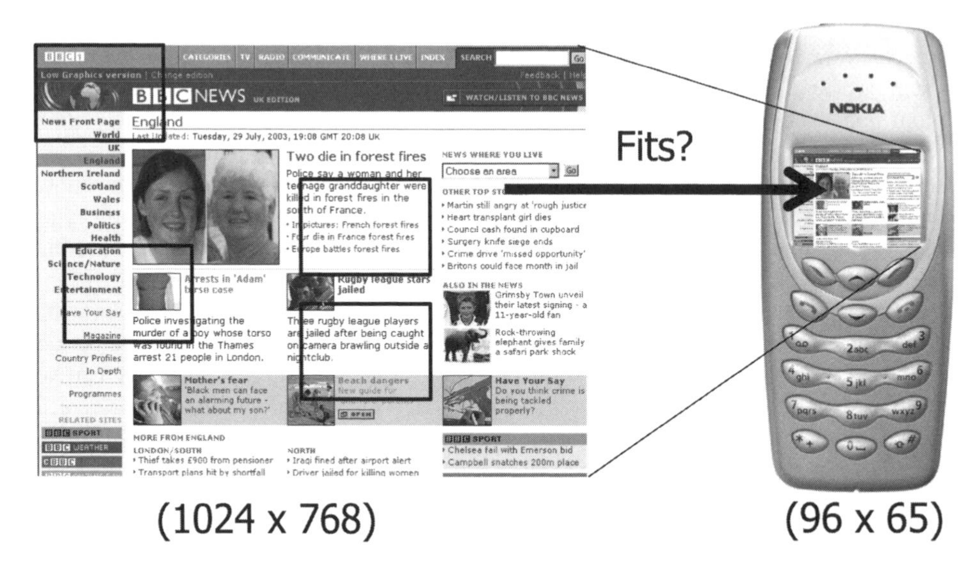

Figure 8.7 Which fits which? Reproduced by permission of Nokia and the BBC.

keypad solutions in Chapter 10 when we look at devices, but for now we can appreciate that there are potentially some major obstacles to overcome when it comes to user interaction. Not surprisingly, more and more innovations are emerging that deal with this problem, but for the time being we still need to accept this comparatively major limitation.

The people who brought us the WAP protocols that we looked at earlier in the book have proposed various techniques to adapt the Web browser display technologies to mobile devices. In this section we will look at mark-up languages aimed at mobile devices, notably those proposed by WAP 1 and WAP 2 specifications.

8.2.1 The Foundation of Mark-Up Languages is HTML

As discussed earlier in the book, the Web originated with the combined emergence of HTTP and HTML HyperText Markup Language. These two technical concepts were thoroughly intertwined from the start. HTML is a language that enables text and graphical information to be formatted for viewing in the browser. It is a means to control how the information gets displayed. For example, using HTML we can display text in columns, formatting some of it as headers, some of it as paragraphs and enlivening it with italics, boldface and underlining. We can insert pictures into the text and control where the pictures get displayed and even what size they get displayed at. We are so used to viewing Web pages that it hardly seems worth mentioning what they look like, but the point to grasp is that the formatting of any given page is controlled by describing to the browser how we want the page to look. Therefore, in a typical HTTP request, two types of information are returned in the body of the response: content for the user to look at and instructions to the browser on how to present it to the user. Often, we do not distinguish between the content and mark-up language when discussing Web solutions. We may refer to both of them as ‘the HTML’ or ‘the HTML file’.

As most of us are probably aware, HTML is a language that consists of tags. With any paginated information we can always treat the page as a linear list of content and work through it from top to bottom. Similarly, a browser opens an HTML file (or accepts the HTML stream via HTTP) and interprets it from top to bottom, or beginning to end. As the browser flows through the page we can think of turning formatting instructions on and off to control the layout, such as 'start boldface' and then a while later 'end boldface'. Those of us familiar with old-generation word processors are probably familiar with this concept, as these control codes were actually visible in the word processor file before the advent of WYSIWYG[2] solutions where the formatting itself became visible, not the codes.

With HTML we similarly insert control codes into the page flow, but they are referred to as *tags*. In the case of boldface text we use the tag <b> to turn on the boldface and </b> to turn it off:

HTML:

```
This is <b> bold text </b>
```

Displays as:

This is **bold text**

We can use a variety of tags to affect the text, including sizing, italics and so on. The <p></p> tags mark paragraphs, <i></i> mark italics, <u></u> mark underlining, <font size = "4"></font> controls font size:

HTML:

```
<p> This is <b> bold text </b>, getting <font size="4">bigger</font>, getting
<font size="5">bigger</font>.</p>
<p>Next paragraph, say something <u><i>profound</i></u>.</p>
```

Displays as:

This is **bold text**, getting bigger, getting bigger.

Next paragraph, say something *profound*.

Our JSP or servlet programmer has to know HTML in order to insert the appropriate tags into the output stream of their program in order to achieve the desired layout in the browser on the requesting device. The challenge then is how to support the same concept for mobile devices. So far, the tags to format text do not seem to present any problems for a mobile device. We could easily support this type of formatting in a small display device, as shown in Figure 8.8, although we can see that already we are filling the entire screen area with just a few words.

In addition to font-formatting tags there are a range of housekeeping tags in an HTML file that are used to indicate other information that may be of use, such as the title of the page. Page title may not be necessary, but they are often used by search engines to index a search, so judicial usage of titles may be a benefit. In the example shown in

[2] WYSIWYG = What You See Is What You Get.

HTML:
```
<p>This is <b>bold text</b>, getting
<font size="4">bigger</font>, getting
<font size="5">bigger</font>.</p>
<p>Next paragraph, say something
<u><i>profound</i></u>.
</p>
```

Displays as:
This is **bold text**, getting bigger, getting bigger
Next paragraph, say something <u>*profound*</u>.

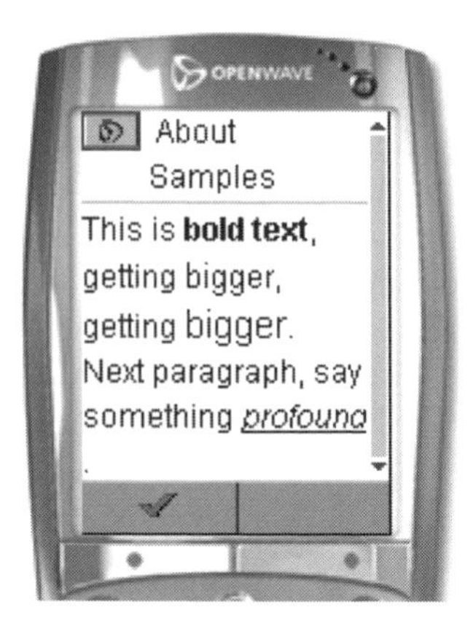

Figure 8.8 Text formatting in a mobile browser. Reproduced with permission from Openwave.

Figure 8.8 the title is ‘About Samples’, which is indicated in the HTML as <title>About Samples</title> and gets displayed in this case at the top of the screen. We can see that this information may be useful to give context to the user, so that they know what the page is referring to. For example, if this were a TV listings application and we were viewing information about programmes on BBC1, then we might think to use ‘BBC1’ as a title. In this way the user will see a visual indication that the times being viewed belong to BBC1 and not another channel. This is probably a good example of how designing pages for mobile devices has different considerations than when doing so with desktop browsers. Something that seems a relatively insignificant design detail in a desktop situation can have greater significance in the mobile environment and seriously affect application usability.

So far, so good: our HTML model seems to be working in the mobile environment. However, we have only touched on a very small subset of the HTML tag set. Originally, HTML was a very sparse language and the options for formatting were limited. This was in keeping with the original precept of the Web: as an information oracle for scientific information. Anyone who has studied science knows that scientific papers are columns of text with the occasional diagram[3] inserted. As profound as the contents might be the layout is not that stunning, not like the page of a youth culture magazine or a car brochure, for example.

Not surprisingly, the evolution of HTML since version 1.0 has progressed toward supporting a wider variety of purposes, well beyond the journal paper layout. Originally, desktop browser manufacturers, in particular Netscape, started to introduce their own tags in order to meet the demands of more ambitious webpage authors. This initially led to a fragmentation of HTML, with different tags supported by different browsers. However, in 1994, the World Wide Web Consortium (www.w3.org) was formed, and this eventually led to a major overhaul of the language to HTML 3.2, which is the most widely supported version of the language.

With HTML 3.2 (and its successor 4.0) the richness of formatting has increased considerably. For example, we are able to support tables using the <table></table> tag pair

[3] In fact, HTML 1.0 did not support images, but this arrived in HTML 2.0.

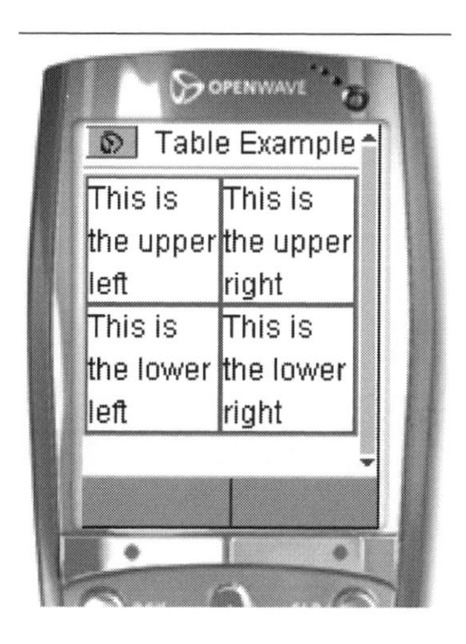

Figure 8.9 Tables in a mobile browser. Reproduced by permission of Openwave.

to indicate the table itself, and the <tr> </tr> to start and end a row, with <td> </td> to indicate a column cell within the row:

```
HTML:
<table border="1">
  <tr>
    <td width="50%">This is the upper left</td>
    <td width="50%">This is the upper right</td>
  </tr>
  <tr>
    <td width="50%">This is the lower left</td>
    <td width="50%">This is the lower right</td>
  </tr>
</table>
```

Displays as:

This is the upper left	This is the upper right
This is the lower left	This is the lower right

Tables like this can also be supported in modern mobile browsers, as shown in Figure 8.9. However, tables can be used to support complex layout arrangements rather than to display tables in their own right. This is where things may start to get tricky for a small display.

8.2.2 The Mobile Evolution (WML)

Originally, when WAP was first being proposed, there were no colour phones on the market. Most phones were monochrome and had limited display capabilities, mostly character-based

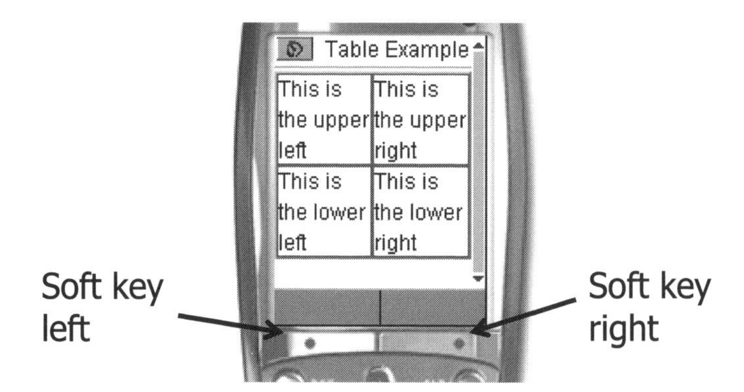

Figure 8.10 Two-button interface.

and with a correspondingly poor pixel resolution and greyscale depth. Because of these obvious limitations, much of the complex formatting possible with HTML 3.2 (as it was at the time) would simply not have worked on mobile devices. With hindsight, what may have made most sense was to have gone back to an older version, like HTML 2.0, and tried to support this. In this way Web programmers would have still been able to use a familiar language and there wouldn't have been any concerns about formatting compatibilities, as the options were limited to mostly text formatting and insertion of images.

However, there was the need for some optimizations, and this led to a new language being developed called Wireless Markup Language (WML), which actually ended up being different from HTML.[4] Despite its departure from HTML, many of the features of WML made a lot of sense and are worth exploring because the principles are still valid and because some of the features have ended up in what we have today (which is no longer WML, but we will get to that in a moment).

The main ergonomic challenges for mark-up language applications on mobile devices are:

- no or limited pointing device;
- small display;
- tiny alphanumeric keypad.

In addressing these issues the WAP architecture proposes a standard interface model: the presence of a two-button interface to which soft keys can be defined in the WML code, as shown in Figure 8.10.

In addition to the soft keys the user can select options from the keypad. In both cases – soft keys or keypad – the user is actually selecting a standard anchor tag in WML that is used to link to another page. We can see the links displayed along with the numeric shortcuts (access keys) in Figure 8.11. In the figure we see that link 3 'Another Link' is highlighted; the user could scroll down the list of links using the arrow keys or could select and activate the link in one step simply by hitting the '3' key.

[4] We are avoiding a discussion of the historical development of WML and its merits versus other approaches, as this is largely irrelevant now.

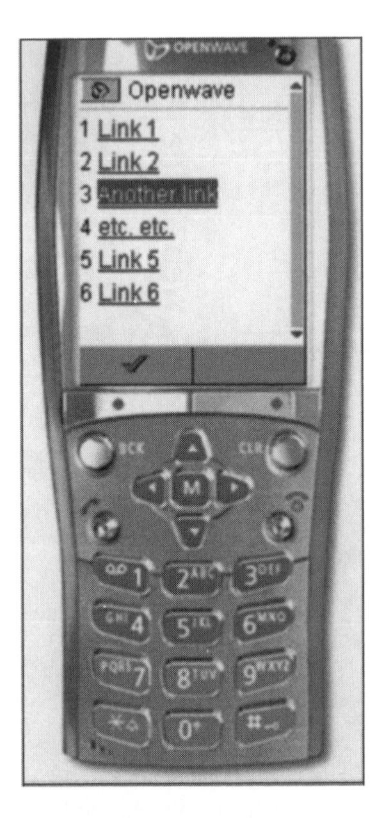

Figure 8.11 Access keys to make link selection quicker via numeric keys. Reproduced by permission of Openwave.

The other optimization that WML incorporates which makes it quite different from HTML is the concept of decks and cards; perhaps this was not obvious until we realized that the HTTP/HTML paradigm involves single-page requests per HTTP cycle. In other words, when the browser requests the resource from a single URI (uniform resource identifier) it is expecting to display the returned content in its entirety in one go. If the user subsequently clicks on a link on the fetched page, another HTTP cycle will be initiated to fetch the linked page. Clearly, we could foresee an arrangement of pre-fetching of pages to speed up user response, but there is no such model in HTML, even though there is no reason why HTTP couldn't support it; with WML, however, this provision is allowed for.

To distinguish between requested pages and pages that are fetched in the same cycle the WML model is based on decks and cards. A deck is like a page, but contains several cards (or only one) and these are the basic unit of content for rendering in the browser interface window. The reason for this mechanism is the small display size of some devices; it may be so small that the reader can digest the contents very quickly and proceed to the next linked resource. If this is stored on the origin server, then another WSP/HTTP fetch cycle is required and this has an obvious overhead. If the display buffer to render content is very small, then the user will request new pages from the server with a high frequency and delays will become a dominant factor in the overall experience.

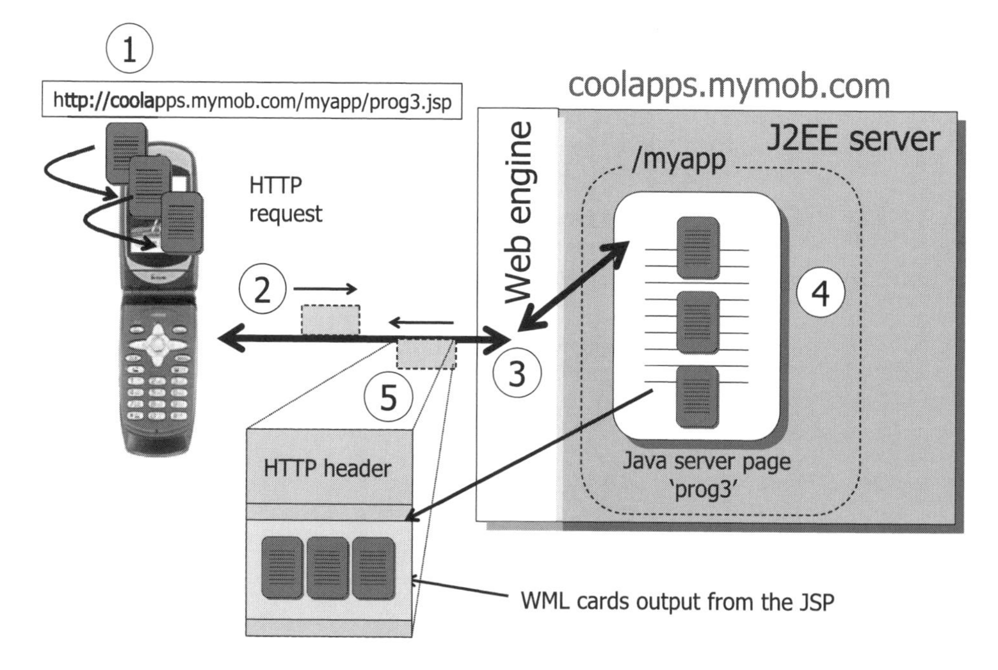

Figure 8.12 Fetching WML deck and cards from a JSP.

The way to think of decks and cards is that cards are like HTML pages and several of them can be fetched at once. Once fetched, only one card is displayable at a time, but the user can link from one page to the other and, because a card is already stored in device memory, it is relatively quick to go from one card to another.

Figure 8.12 shows us what is going on with decks and cards: as usual, the browser makes a request (step 1/2) for a particular resource on the J2EE server to run (step 3); in this case it is for a JSP. The JSP is run by the Web engine, and it outputs code that is WML and forms a deck of cards. The cards are part of one contiguous file (deck). Because the WML file (deck) is one file, it gets passed in one go via the HTTP response (step 5) and via a WAP proxy, which we do not show here. Eventually, the cards end up at the device and the user can switch between them without the need for any further HTTP requests, as indicated in the figure by the cascading arrows between cards on the device.

WAP Binary XML (WBXML) The potentially slow link speed of the RF network poses additional challenges. In Chapter 7 we examined how WAP protocols can be used to optimize communications over an RF link, but we did not discuss issues related to the actual information being passed over the link; this does make a difference. In the case of HTML a large proportion of the body of an HTTP response can be the HTML tags as opposed to the actual content.

If we consider our table example again, then we can see how much of the data are simply metadata rather than actual data. If we separate out the two types of data, then we can see the difference more clearly:

Metadata (tags)	Content
<table border="1"> <tr><td width="50%"></td> <td width="50%"></td></tr> <tr><td width="50%"></td> <td width="50%"></td></tr></table>	This is the upper leftThis is the upper rightThis is the lower leftThis is the lower right
128 characters (including spaces)	90 characters (including spaces)

In this short sample of HTML code 59% of the information is metadata. When transmitted over an RF link, we spend over half the time just transmitting metadata that the user never gets to see (and doesn't want or need to see) – it is simply baggage as far as the user is concerned and its inclusion results in a worsening of the overall user experience in terms of tardiness in the responsiveness of the application.

With WML the actual WML stream that gets transmitted instead of a HTML stream is binary-encoded to achieve compression of the metadata. This is done in a manner similar to the compression of HTTP headers in WSP, whereby a hexadecimal code (token) is substituted for a well-known string in the WML.

The tokenizing of the data steam is a relatively straightforward process. Fortunately, we do not have to ask our JSP or servlet programmer to implement this. Tokenizing takes places at the WAP proxy. Our JSP or servlet outputs the WML in the body of an HTTP response stream back to the proxy. The proxy performs the tokenizing and places the stream back out on the link, using WSP, and the WAP browser is required to de-tokenize before it can display the file. Strictly speaking, because the WML compression is a tokenized data stream it is possible to extract the WML tags and content without the need to have the entire file available beforehand. This means that the browser display could be written to in parallel with receiving the WML stream. This may improve display response as far as the user is concerned, though this depends on the exact implementation of the browser on the device. One concern in doing this might be that the file turns out to be incorrect WML syntax; this might prevent the page from displaying depending on how forgiving the browser is of bad syntax. However, this should be circumvented by strict validation of the file at the time of tokenization.

The encoded WML output is referred to as WAP Binary XML, or WBXML, and the following table gives an idea of the tokens that would get used in our table example:

Tag	Hexadecimal token
'<table>'	1F
'<tr>'	1E
'<td>'	1D
'<p>'	20

In WML the table tags are slightly different from HTML in terms of the allowable parameters. For example, we cannot specify the border parameter, as we did in our HTML example, which indicates to show visibly the borders of the table. The correct syntax for a similar table in WML would be:

```
<table columns="2">
  <tr>
    <td>This is the upper left</td>
    <td>This is the upper right</td>
  </tr>
  <tr>
    <td>This is the lower left</td>
    <td>This is the lower right</td>
  </tr>
</table>
```

	Metadata (tags)	Content
WML	<table columns="2"> <tr><td></td><td></td></tr> <tr><td></td><td></td></tr> </table> (81 characters = 81 bytes[5])	This is the upper leftThis is the upper rightThis is the lower leftThis is the lower right (90 characters = 90 bytes)
WBXML	1F 53 03 02 00 1E 1D 01 1D 01 01 1E 1D 01 1D 01 01 01 (18 bytes)	03 This is the upper left 00 03 This is the upper right 00 03 This is the lower left 00 03 This is the lower right 00 (98 bytes)

With the WBXML coding scheme we have reduced our metadata overhead to only 16% of the overall response stream across the RF journey of the HTTP response. This is a significant saving, albeit for a solitary example.

Comparing WML with HTML and cHTML The main issue with WML has been its digression from HTML, as WML is not compatible with HTML, although there is a degree of similarity between the tag sets in each language. The optimizations for WML are useful, especially the deck construct and the binary encoding, although the tokenized encoding is something that could be implemented for HTML if there were ever a need to do so.

The programmable soft keys are actually a feature developed specifically as part of the WAP concept; hence, WML contains specific tags for gaining control over these keys. This can make applications altogether more usable, as one-touch operation of key commands is possible. With HTML, or its ilk, there is no model for soft key interaction. The paradigm

[5] Assuming UTF-8 coding, which means 1 byte per English character.

assumes the existence of a pointing device. In the absence of such a device the user has to scroll down lists of links and select the one they want to visit.

The main concern with WML has always been its deviation from HTML, as Web programmers are forced to learn a new language. This has been considered an obstacle to the take-up of mobile application programming using the browser model. How much this is true is difficult to say, but a comparison has often been made with the relatively greater success of a competing solution called iMode – originally a Japanese 'standard' developed by NTT DoCoMo.

The iMode system uses a language called cHTML, where the 'c' stands for 'compact'. It is literally a cut-down version of HTML, removing many of the tags, especially ones that would be unlikely to be useful in mobile browsers. The iMode service comprises more than just mobile Web pages, screen savers, email and so on: it is more akin to a portal service, but the first thing DoCoMo did was invent the method of delivery. Japan has a long-established record of doing her own thing, so the adoption (creation) of iMode in place of WAP is not at all surprising and is not necessarily a rejection of WAP on technical merits.

cHTML is designed to meet the requirements of small information appliances with limited memory and processing power. Its design is based on the following four principles:

1. Current HTML W3C recommendations – cHTML is defined as a subset of the HTML 2.0, HTML 3.2 and HTML 4.0 specifications; this means that cHTML inherits the flexibility and portability of the standard HTML.

2. Lite specification – cHTML has to be implemented with a small-memory and low-power CPU. Frames and tables that require large memory are excluded from cHTML.

3. Viewable on a small monochrome display – cHTML assumes a small display space of black and white. However, it does not assume a fixed display space, but it is flexible for the display screen size. It also assumes a single-character font.

4. User-friendly operation – cHTML is defined so that all the basic operations can be done by a combination of four buttons: *cursor forward*, *cursor backward*, *Select* and *Back/Stop* (i.e., Return to the previous page). The functions that require two-dimensional focus pointing like 'image map' and 'table' are excluded from cHTML.

The major features that are excluded from cHTML are as follows:

- *JPEG (Joint Picture Experts Group) image* – an image format that utilizes a high degree of processing to display (due to the handling of the compression method, which implies a mathematical process to display the image rather than simply reading the pixel values from a file).

- *Table* – the handling of tables requires potentially a lot of memory due to the nested nature of the tables and the implication this has on software data structures.

- *Image maps* – an image map is the ability to turn certain parts of an image into clickable links. This potentially requires complex data structures to maintain the clickable areas as well as a lot of processing power to dynamically detect the 'mouse'[6] cursor position on the image and compare it with the image map.

[6] Most mobile devices do not have a mouse, and many lack any kind of pointing device at all, especially one that will work effectively in two dimensions.

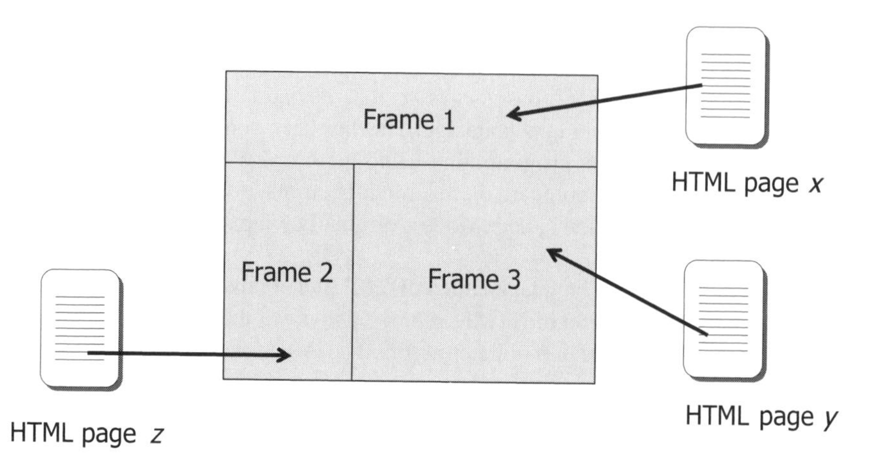

Figure 8.13 Frames in the browser used to display more than one HTML page.

- *Multiple character fonts and styles* – again, this will take up too much memory and the advantages of different fonts for a small device are arguable.
- *Background colour and image* – in early mobile devices colour support was irrelevant anyhow.
- *Frames* – with desktop browsers it is possible to display several HTML pages at the same time by dividing the screen into frames, as shown in Figure 8.13. Again, maintenance of this construct in the mobile browser would require potentially excessive memory and processor resources.
- *Style sheet* – a style sheet is a means of defining styles – like font colour and font type – but in a block of code that is separate from the main HTML. This makes the maintenance of look and feel easier, as it is separated out from the content. For example, we would still use a header tag in our file, like <h1></h1> to tell the browser to display oversized text as a means of demarcating a section heading. However, we may decide that all such headings should be in Garamond font and burgundy in colour, perhaps to match a corporate identity colour scheme. We can control such appearances using a style sheet. For early mobile devices the scope for style was limited, so style sheets were not supported by cHTML. We will return to style sheets when we look at XHTML-MP.

Despite repeated claims to the contrary, the use of cHTML is probably not the main reason for the apparent success of iMode in Japan. Other factors figure heavily, including the revenue share that developers can gain from providing their content via iMode. Social factors in Japan are quite different from other countries[7] and have meant an altogether different reliance on mobile devices than seen in most other regions. However, cHTML

[7] For example, the Japanese spend a lot of their time out of their homes, because often homes are very small. This 'out of home' living tends to engender a different social dynamic. See Rheingold, H., *Smart Mobs: The Next Social Revolution*. Perseus Books, Cambridge, MA (2002).

became a source of inspiration for how to choose a suitable mark-up language for the WAP 2 era, as we will soon discuss.

8.3 FULL CIRCLE – WML 'BECOMES' XHTML

As we discovered in Chapter 7, when looking at mobile protocols, WAP progressed from WAP 1 to WAP 2. In doing so it moved a step closer to the fully blown desktop browser model, especially with respect to the protocols and the wireless profiles of HTTP and TCP, which are completely compatible with standard HTTP and TCP.

It would be folly to suggest that we dispense with wireless optimizations altogether; this is especially true when considering which mark-up language to use. Clearly, mobile devices have severe display constraints compared with desktop devices. This would suggest that utilizing an identical approach is not going to work – a topic we critically examined in Chapter 2. Certainly, the idea of viewing standard Web pages on mobile devices is anathema to giving the mobile user a compelling experience. All our notions of usability are contravened by asking a user to wade around a page that is up to 20 times too big for the display.

From WAP 1 to WAP 2, the era has seen the advent of many more devices capable of displaying colour and a higher density of pixels. Screen sizes have also increased on many devices. At the same time, as we have already noted when looking at protocols, the ability to connect a wireless device to the Internet has become easier. As we will see when we look at RF network protocols in Chapter 12 mobile networks have evolved from circuit-switched technology to packet-switched.

The main advantage of packet switching is not actually the greater network speeds that are theoretically achievable.[8] With packet-switched connections there is no need to initiate and maintain a data call, like we used to do[9] with modems at home when accessing our favourite ISP (Internet Service Provider). In effect, modern devices are already connected and ready to transfer data, like sending a GET request to a server and receiving the HTTP response. The 'line' only becomes active for that period of time during which we pay for the data transferred.

For the end user the act of accessing data sources, like a WAP site, is a convenient process, as the line is always connected and so delays with connection set-up are avoided. Moreover, the process is affordable as there is no need to hog an expensive resource (a switched circuit).

Given the improvements in devices, we could posit technical arguments why switching back to HTML from WML might be a good idea. However, the reality is probably that the likeliness of users being prepared to use mobile services is increasing all the time, due to the improvements we have just been discussing. Therefore, if we want to attract more widespread development of mobile websites, then it is probably a good idea to make this as easy as possible for existing Web designers; this necessitates the adoption of a language that is compatible with existing approaches, such as HTML.

As we will now examine, HTML itself has evolved during the WAP 1 to WAP 2 time frame. Any attempt to adopt HTML for mobile applications would need to be synchronized

[8] Packet switching in itself does not improve data transfer rates. The higher speed referred to is a result of other improvements in implementation.

[9] Some of us still use telephone line modems rather than broadband connections.

with the evolution of HTML itself. However, just as we did with the wireless profiling of HTTP and TCP, any performance improvements we can find that are within the confines of HTML technology would probably be worth pursuing, especially if they advance usability.

XHTML is Modular HTML is a language for formatting the display of information. We have seen that the language itself comprises of tags; tags are used to indicate formatting states within a stream of data that is going to be displayed in a browser, like <bold> to turn boldface on and then </bold> to turn it off some time later in the information stream.

Earlier in the book we briefly looked at a generic way of annotating data using tags, called XML, or eXtensible Markup Language. The use of XML has mushroomed of late. Many tools and APIs have emerged to make processing of XML easier for us. It is relatively simple now to adopt XML as the means of describing any information set that we care to exchange between two devices, no matter their application. Because of the increased momentum behind XML and its obvious similarity to HTML, a decision was taken by the World Wide Web Consortium (W3C) to make sure that HTML was fully XML-compliant.

XML uses tags to describe data. The tags are mostly undefined, which is the whole point and why it is called 'extensible'. However, the rules for defining the tags are very much fixed. Sticking to the rules is like obeying the principles of grammar and punctuation, or syntax. Defining our own tags within the rules is like formulating our own vocabulary and doing this in conformance with XML syntax constitutes making a well-formed XML document. To achieve conformance with XML, HTML has become XHTML, but the features of the language (i.e. the vocabulary itself) remain unchanged.

By moving to XML, there is an opportunity to modularize the language; this is a slightly strange concept at first, but makes perfect sense – as we will see in a moment when we reflect on why we bother with standards at all.

With any XML document we need a means to determine what the allowable tags actually are. This is done by writing a special document called a Document Type Definition (DTD) or an XML schema, which is an alternative approach. For example, it is in the DTD for XHTML 1.0 that we will find that the <bold> and </bold> tags are actually allowable. DTDs also tell us about tag relationships, or nesting. If we take an example XML vocabulary to describe anything to do with films (movies), we could imagine a file snippet as shown in Figure 8.14. We can see the tags in pairs and the information contained within them, noting that some tag pairs (e.g., <song></song>) contain yet more tag pairs, whereas some tag pairs (e.g. <title></title>) contain the annotated information itself. This nesting of XML tags leads to a tree-like structure in the file, as shown in Figure 8.15, which is a useful data structure to accommodate as it reflects the implicit organization of many data sets.

Within an XML document there might be the need to refer to several DTDs, each one taking care of a particular subset of the vocabulary that is allowable within our file. Taking our fictitious Film XML (FML) as an example the topic of films is clearly quite broad. We could envisage being able to describe so many detailed topics related to a film: for example, we could include information about where a film was shot, each location, the details of the props used, the extras in the film taken from the location and so on.

Were we to include every conceivable facet of film-making into our XML definition, it might become a bit unwieldy, especially if we wanted to extend it at a later date. An alternative approach is to use a DTD for each distinct topic subgroup within the overall information space concerning films; this is modularization.

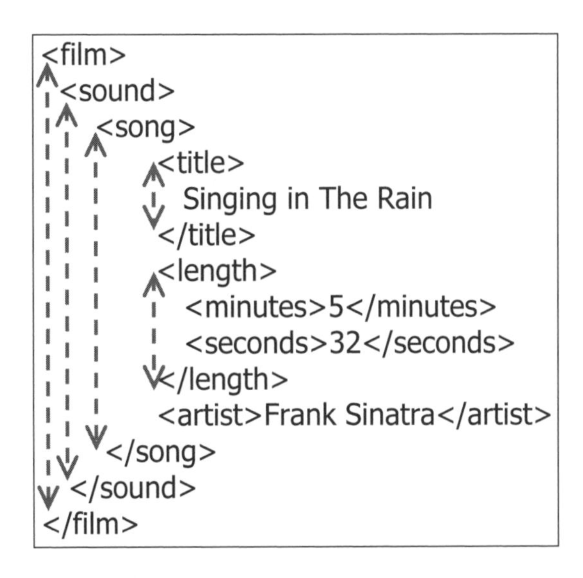

Figure 8.14 Example of using XML to describe films.

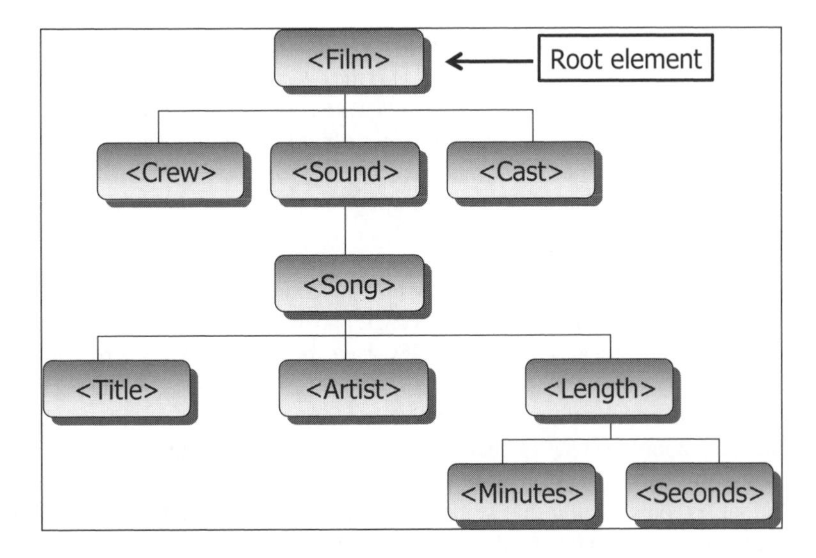

Figure 8.15 Tree structure resulting from nested tags in XML.

Modularization allows for compatibility to be maintained between systems while allowing for extensibility within an application space: for example, various software solutions could be envisaged that relate to the film-making industry. If we want to standardize the information exchange between these applications, then we could define and adopt our FML, similar to the example we have just been looking at. However, a particular group of applications might only be concerned with the audio aspects of film-making. These applications would want to exchange information about soundtracks; so, they could use the openly published DTD for soundtracks and thereby achieve a guaranteed level of compatibility by sticking to the DTD and the resultant XML vocabulary and structure.

However, perhaps a particular part of the industry recognizes a need to go further within the realms of describing soundtracks, wishing to include information about the recording equipment and sound formats. Instead of adopting a new standard altogether the interested parties could agree on an extension to the existing FML simply by adding a new module to include the means to describe the data they are interested in exchanging. They could publish the DTD for this extension back to the public domain and, for the sake of clarity, name their hybrid slightly differently, something like FML-A, with the 'A' standing for 'Audio'. It is also possible that within FML-A the agreeing parties could have jettisoned some of the existing modules within FML that are not especially relevant to the applications catering for audio; this is the benefit of the modular approach – it allows both subsetting and supersetting of the base (host) language.

XHTML-Basic Having just looked at the modular approach adopted by XHTML we can now move on to look at a particular incarnation of XHTML, getting back to our mobile network and away from the world of films, interesting though they might be (perhaps we should make a film based on this book ☺).

XHTML-Basic is a version of XHTML aimed at small information appliances, particularly mobile devices (though not exclusively). To quote from the W3C document defining XHTML-Basic:[10]

> *HTML 4 [foundation of XHTML] is a powerful language for authoring Web content, but its design does not take into consideration issues pertinent to small devices, including the implementation cost (in power, memory, etc.) of the full feature set. Consumer devices with limited resources cannot generally afford to implement the full feature set of HTML 4. Requiring a full-fledged computer for access to the World Wide Web excludes a large portion of the population from consumer device access of online information and services.*

In our foregoing discussions we have already mentioned at least three approaches to markup languages for mobile devices: HTML, cHTML and WML. We could be selective about how we use HTML but, because there are many ways to subset HTML, there is a danger of ending up with many subsets, possibly as many as are defined by the various organizations and companies with competing or disparate interests in this matter.

Without a common base set of features, developing applications for a wide range of Web clients is difficult; this is because we might design a user interface (UI) assuming we can use tables, only to find that some of our users do not have table support in their browsers. Clearly, such a fragmented approach will ultimately work against all of us: users and developers alike.

The impetus for XHTML-Basic is to provide an XHTML document type that can be shared across communities (e.g., desktop, TV and mobile phones) and that is substantial enough to be used for content authoring that works well enough within the limitations we may find applicable to our application. However, by choosing XHTML as a basis for XHTML-Basic, we automatically inherit the modular approach; thus, new community-wide (i.e., shared common 'standards') document types can be defined by extending XHTML-Basic in such a way that XHTML-Basic documents are in the set of valid documents of the

[10] http://www.w3.org/TR/xhtml-basic

new document type. Thus, an XHTML-Basic document can be presented on the maximum number of browser clients.

In our earlier discussion about cHTML we listed some of the absent features dropped from HTML and briefly discussed the possible reasons for the omission. Certain types of mark-up were not predisposed to implementation in a mobile browser running in a resource-constrained environment. Most of the limitations are related to memory and processor constraints. We treat the over-the-air constraints separately by applying compression techniques afterward, just as we did with WBXML and its tokenized compression method. In other words, the need to overcome device limitations is uppermost, whereas the need to overcome file size problems is treated separately and is not a major consideration for the language definition.

From our consideration of resource constraints and their impact on the design of cHTML we can appreciate the similar need to limit XHTML for mobile devices. Many of the design decisions behind XHTML-Basic mimic the considerations for cHTML, not surprisingly since cHTML was submitted for consideration by the W3C.[11]

Let's briefly examine the official W3C comments on the rationale for XHTML-Basic before going on to look at how the WAP community used this as a foundation for XHTML-MP, for good reason as we will see. The design rationale for XHTML-Basic covers the following points:

- *Style sheets* – these are recommended, but limited in scope. We will defer our discussion of these until we reach XHTML-MP, where a more comprehensive treatment is required (and relevant).
- *Form handling* – basic form handling (*Basic Form* XHTML Module) is supported with some limitations.
- *Presentation* – many simple Web clients cannot display fonts other than monospace. Bidirectional text, boldface font and other text extension elements are not supported. It is recommended that style sheets be used to create a presentation that is appropriate for the device.
- *Tables* – basic table handling is supported (*Basic Tables* XHTML Module). Note that in the Basic Tables Module, nesting of tables is illegal.
- *Frames* – these are not supported.
- *Scripts and events* – something we have not discussed so far in relation to mark-up languages is the concept of scripts and events. We can see how XHTML tags can be used to drive the display of content. However, perhaps we want to do something more complicated. In desktop browsers we can insert scripts into our pages to enable us to programmatically manipulate the mark-up content once it is on the device. We can think of this as being like being able to run the JSP or servlet on the device, simply to allow some degree of processing of the content (though clearly this analogy does not extend to accessing the J2EE platform and its APIs, etc.). The scripts can programmatically gain access to any of the content in the XHTML page; for example, we could move an image around the screen.

[11] http://www.w3.org/TR/1998/NOTE-compactHTML-19980209

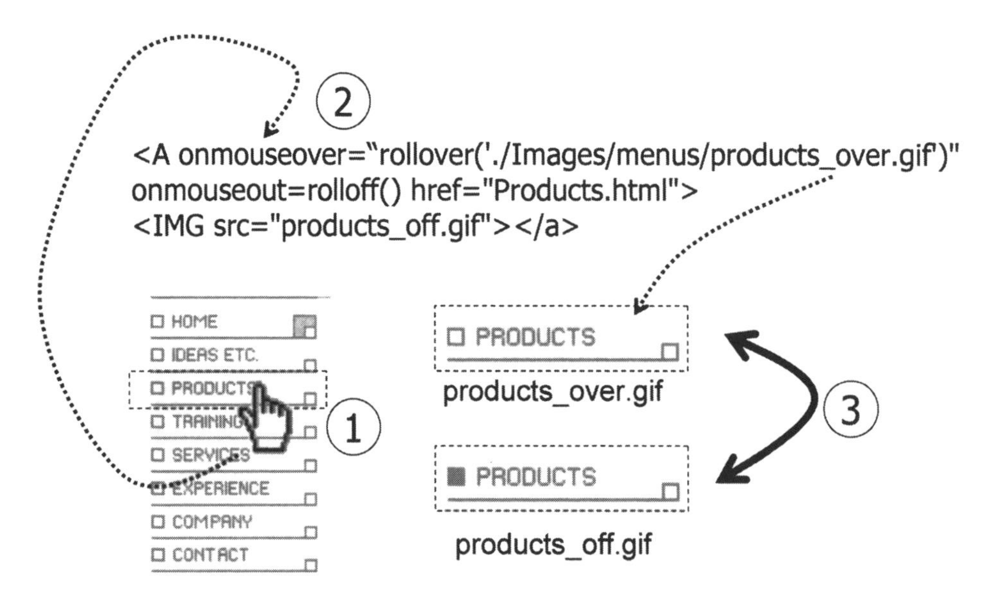

Figure 8.16 Rollover effect is achieved by events and scripting.

Perhaps the most classic example of widely used scripting is the now famous 'rollover' effect, as shown in Figure 8.16; This is where the image under the cursor changes when we place the cursor above it, a technique typically used to give a degree of positive feedback that we have selected an option. The figure shows the mouse cursor over the image (step 1) causes the *onmouseover* event to be fired in the associated container for this image element, which in this case is an anchor <A> tag (hyperlink). Inside the opening <A> tag we have specified the *onmouseover* attribute, thus letting the browser know that when this event occurs it then calls our scripting function (*rollover*), as indicated in step 2. This function is not shown, but is written using another language called Javascript and is listed at the top of the HTML page where the browser knows to find it in order to run it when this event fires. Consequently, the original image gets swapped for another image (step 3); this is how the rollover effect gets accomplished.

The event model and scripting capability that we have just mentioned implies that our Web browser is a lot more complicated than perhaps envisaged. It does a lot more than simply render HTML to the screen. Figure 8.17 gives us a better idea of what a browser might consist of. We have five main elements:[12]

- *Browser display* – the software that handles the actual UI and can render XHTML (or others) into a meaningful graphical representation on the screen, most likely using the graphics' APIs of the device (which we will discuss in Chapter 10 when we look at devices).

- *HTTP/WAP stack* – if not already present as an accessible API on the device, then the browser needs its own HTTP/WAP stack to enable interaction with the origin server. This

[12] Note that these are somewhat arbitrary for the sake of illustrating the detailed operation of a modern browser. Actual browser implementations may follow alternative architectures, although the functions would be similar.

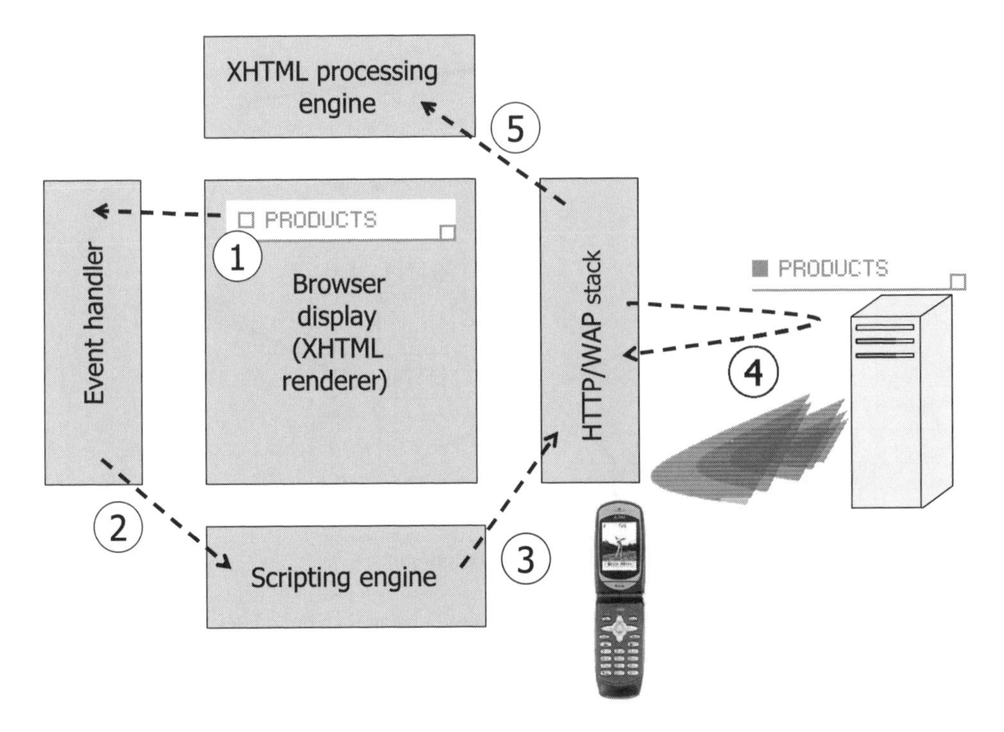

Figure 8.17 Browsers are a bit more complicated than we thought.

part of the browser may well also include the cache-to-cache pages, images and other fetched objects.

- *Event handler* – having defined the possible events that the browser could respond to or generate (e.g., like *onmouseover*, *onmouseoff*, *onpageload*, etc.), we need a software process to detect events and then pipe these to the scripting engine.
- *Scripting engine* – we need an API that fully exposes the mark-up document to access via a scripting language, which itself needs an execution environment. The scripting engine provides this facility; we typically refer to tags or events from the script using a *Document Object Model (DOM)*. We can think of this as modelling the entire document as a class and then exposing its structure via properties and methods that our script can call.
- *XHTML (HTML/WML) processing engine* – this component actually parses and validates the inbound mark-up document and converts it to a form to pass to the browser display module.

In Figure 8.17 the sequence of a typical rollover effect is shown. In step 1 the image within an <A> tag is pointed to by the pointing device. This causes the event handler to detect the event and to pipe this (notify) to the scripting engine (step 2). The scripting engine receives the event and executes the appropriately tagged code segment that is bound to the event within the mark-up (e.g., via an event attribute in the <A> tag), as shown in step 3. In this example we imagine that the executing script accesses the pointed-to image in the document

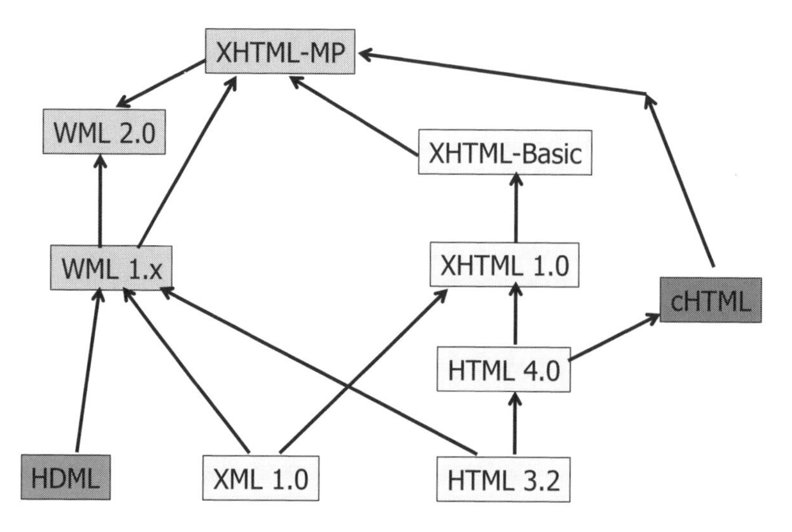

Figure 8.18 Relationship of different mark-up languages.

and replaces it with an alternative image (the 'rollover image') from a URL specified within the script. This causes the scripting engine to fetch the image from the URL (step 4). The final step is to process the content for display (step 5).

We have just described the complexity of a browser that supports scripting and events. Because of the implications in terms of device resources, in XHTML-Basic there is no scripting support, so the browser architecture is less complex.

The problem with scripting and events is not just with heavy resource implications but also with defining a suitable event model and scripting API for the document (i.e., defining the DOM). Events and API possibilities in certain devices may not make sense in others. For example, a mobile device with voice-calling capabilities would benefit from the ability to link an embedded number to the device address book. We would need an event to indicate that we have clicked on content tagged as a phone number.[13] However, this would imply a particular scripting API that is not necessary on devices that do not support voice, such as dedicated email devices.

However, richer mobile devices which support full XHTML will generally include scripting. Wherever scripting is supported, the object model for it will most likely be simplified, as there are still plenty of things that a mobile device can't do compared with a desktop environment – something that we will consider when we look at devices in Chapter 10.

XHTML-MP (Mobile Profile) – the Final Frontier Before discussing the essential elements of XHTML-MP, let's recap the relationship and evolution of mark-up languages thus far, something that we can show graphically (see Figure 8.18).

As we have discussed on several occasions, the advent of the browser paradigm began with HTML, which became widespread and feature-rich at version 3.2. This version migrated to version 4.0 and was converted to XHTML 1.0, using XML and modularization to develop a host language that could be used in its own right or as a basis (or container)

[13] This feature is supported in WML and on some WAP 1 devices.

for other languages. The HTML standards community (W3C) took it on itself to define a subset of XHTML, called XHTML-Basic, aimed at resource-limited devices.

As the figure shows, in parallel with HTML progression the WAP community was busy with its own activities, initially inspired by several vendors' technologies, most significantly the Handheld Device Markup Language (HDML) from Unwired Planet. To promote a more vigorous standardization effort, eventually the WAP Forum coalesced efforts with the W3C. At the same time, NTT DoCoMo had already submitted i-Mode's cHTML as a technical note for the W3C to put into consideration for lightweight applications.

What we have ended up with is XHTML-MP; this is a specification owned by the WAP community, which is now part of the Open Mobile Alliance (OMA). While the specification is standards-based in terms of its dependence on XHTML-Basic, the specification is not a W3C specification. This in itself does not pose a problem, especially as the whole point of XHTML-Basic was to provide a host language that other communities could extend to suit their needs, which is what the OMA has done.

Of course, XHTML-MP is an open specification, not an edict. As such, there is no need for anyone to conform. However, while the aim of the language is to support browser applications on very resource-limited devices, the aim of the specification is to engender sufficiently widespread support that a large population of devices can display XHTML-MP pages; this will make it easier for developers to target applications at mobile users. The language is easy to use, due to its reliance on the prevalent XHTML (HTML) foundation, and using it pays off as a large number of users will be able to access content authored with it. In this regard the 3G Partnership Project has adopted it as part of the standards for UMTS (Universal Mobile Telecommunications System), the global 3G (third generation) standard; it is already a recognized standard for most 2G and 2.5G standards, such as GSM.

Before considering the details of XHTML-MP, we should discuss WML 2.0. This specification uses XHTML but with the unique semantics of WML 1 and is used for *backwards compatibility only*. In other words, new applications based on WAP technology should not be using WML any longer. WAP 2 applications should only be created using XHTML-MP. We therefore do not discuss WML 2.0, although we will look at browser-specific extensions to XHTML-MP that are motivated by some of the features of WML that did not make it into the XHTML-MP specification. There are also similar features from cHTML that need noting.

Using XHTML-MP XHTML should be very familiar to Web designers and authors; many of the familiar design features are present. XHTML-MP inherits the following abilities to format content from XHTML Basic:

- structure;
- text;
- hypertext;
- list;
- basic forms;
- basic tables;

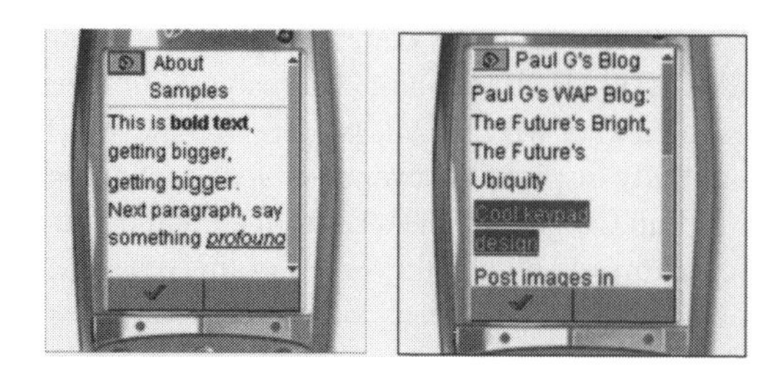

Figure 8.19 Basic XHTML site. Reproduced by permission of Openwave.

- images;
- meta-information.

Additionally, it supports:

- subset of the fully blown XHTML forms module;
- subset of the XHTML presentation module;
- style sheets and style attributes.

Except for structure, it is easiest to explain these by example. The structure support is the ability to divide XHTML content into components, consisting of a title block, header block and the body itself. Most of the content that gets displayed in the browser interface is contained in the body block and comes from the rest of the XHTML elements we have listed above. The title information can be displayed in the browser, and the header block can contain important meta-information. This is a way of specifying HTTP headers that should be contained in the response when the Web engine writes out the XHTML stream in a HTTP response: for example, we can use a meta tag to explicitly set a validity period for the purposes of caching. This is particularly useful for information that changes regularly, like news pages.

With the basic text display capabilities of XHTML-MP, previously shown in Figure 8.8 and repeated on the left of Figure 8.19, we could have a basic and navigable WAP site, such as the one shown in Figure 8.19, which is a WAP version of a weblog.

In Figure 8.19 we demonstrate the basic features of text formatting (e.g., plain, **bold**, *italics*, large fonts, etc.). We can also see the use of the all-important hyperlink, such as the highlighted 'Cool keypad design' link shown, which would take us to the next page in the weblog.

The greatest feature of XHTML-MP is its richer formatting and graphics support as compared with earlier WAP browsers: for example, we can take an image (b) and insert it into the page (see Figure 8.20). Moreover, the image types supported can be GIF, JPEG and PNG, which are all standard formats that Web authors are used to. We can even take an image (a) and use it for the background of the page.

Both these ideas are amalgamated in Figure 8.21.

A major feature of XHTML-MP is that we can add the formatting that affects style by defining styles in what is called a style sheet, or a cascading style sheet (CSS). The real beauty of the style sheet is that it is a separate file from the XHTML page. Initially, this is a problem because it means our browser has to initiate a separate HTTP fetch cycle to go

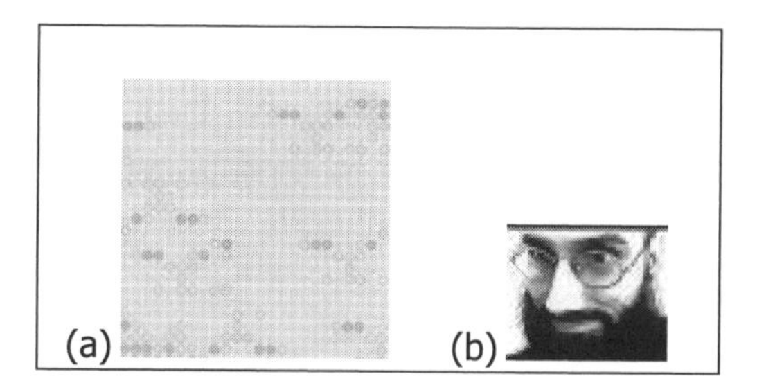

Figure 8.20 JPEG images supportable by XHTML-MP.

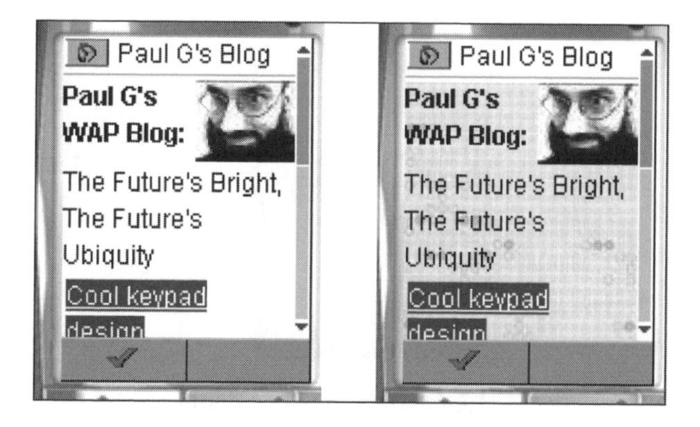

Figure 8.21 Adding images to XHTML-MP pages.

grab the style sheet. What's more, it has to do this before we can start to see how the initial page should look in the browser; so, it is difficult to preload the page or carry out these tasks in parallel (using the pipelining of requests that we discussed earlier).

We use the style sheet to tell the browser about the look and feel of any pages that reference it, as shown in Figure 8.22. We can define style information for more or less every display element that the browser understands. For example, to make the title of the page as shown – 'Paul G's WAP Blog:' – we use the following tag in the XHTML-MP:

```
<p>Paul G's WAP Blog:</p>
```

This tells the browser that this is a paragraph. This is not quite the same as the grammatical idea of a paragraph in English. It is really a block element that contains text, inline images and related content. The XHTML-MP rules (from the associated DTD) determine what tags can be nested within the paragraph tag. As we mentioned inline images, then this must be one of the permissible nested tags, which is how we insert the image seen in the top right of the screen shot. To do this, we use the image <img> tag,[14] as below:

[14] Notice that the tags are in lower case; this is not a strict requirement of XML's syntax rules, but case sensitivity is. Hence, once defined as lower case in the DTD, we have to stick to lower case. Otherwise, in theory, our browser should reject the XHTML-MP file as it is not *valid* XML.

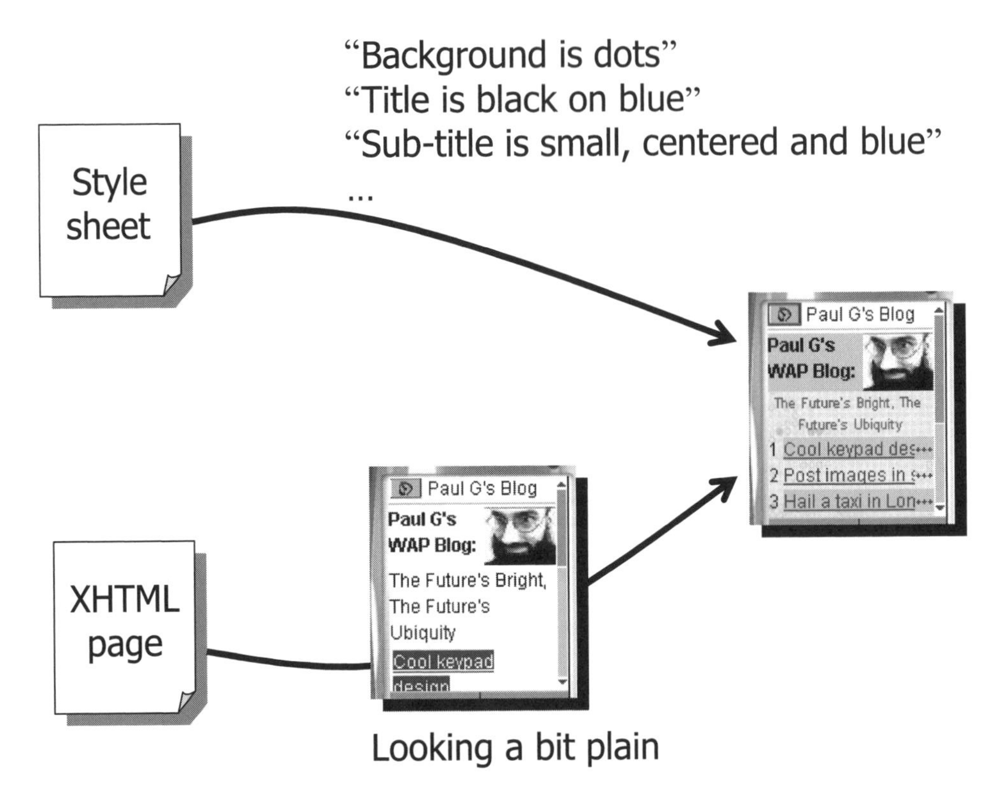

Figure 8.22 Separate style sheet tells browser about the look and feel of pages.

```
<p>
<img src="pgtiny.jpg" align="right" />
Paul G's WAP Blog:
</p>
```

The <img> tag is using a couple of attributes. The first is the source attribute that tells the browser where to find the image. Here we have the value – actually a URL – 'pgtiny.jpg'. Because there is no domain name or other path information in the URL, the browser assumes it to be in the same path from where the XHTML-MP page was fetched, so the browser initiates another fetch to go and get the image. Don't forget that the browser can issue this fetch as soon as it finds this tag when processing the input HTTP stream containing this XHTML-MP content. In other words, it can use pipelining to speed up fetching the image before waiting for the entire XHTML-MP page to be loaded (streamed in). The other attribute is the *align* attribute that controls alignment on the page with respect to the browser window – the options being to align to the left, centre (*center*)[15] or right.

To apply a look-and-feel style to our pages, we can define different styles in the style sheet and then reference them in our pages. For example, let's say that for page titles – such

[15] XHTML, like many similar specifications, is produced in American English, so common words like 'centre' are spelt with American spelling (e.g., center).

as our 'Paul G's WAP Weblog:' – we want to display them in bold font on a blue background. We can define a style called 'Pagetitle' (the name is up to us within the syntactical rules of CSS). This would look something like (the line numbers would not appear in the CSS file):

```
1  .pagetitle {
2     color: black;
3     background-color: #99CCFF
4     font-weight: bold
5  }
```

This information would appear somewhere in our CSS file. Line 1 is the name of the style. The name begins with a period ('.') to indicate that this is a custom style. Line 2 indicates that the font colour is black.[16] Line 3 specifies the background colour for the paragraph, and line 4 indicates we want the text to be boldface. Line 1 and line 5 contain the opening and closing parentheses (curly brackets) to demarcate the style.

To use this style in our page we use the attribute *class*, which can be used in most tags to define the applicable style, so we end up with our paragraph tag looking like this:

```
<p class="pagetitle">
<img src="pgtiny.jpg" align="right" />
Paul G's WAP Blog:
</p>
```

So, when the browser displays this paragraph block, it knows to use the style class called 'pagetitle', which it finds in the associated CSS, which is linked to the XHTML-MP page using the link tag:

```
<link href="style.css" rel="stylesheet" type="text/css" />
```

The key attribute here is the href that contains the URL to the style sheet, which the browser again assumes as relative to the current page and fetches it.

The benefit of this approach is that every page in our browser-based application can have a common look and feel by referencing the same CSS; each file simply links to the same URL where the CSS is to be found, as shown in Figure 8.23.

With a common style sheet, whenever we want to update or change the look and feel of the application, we only have to change one file. This makes our application easier to maintain.

The style sheet can be stored in the browser's cache. This means that once loaded, we don't have to fetch it again, speeding up the application. Not only is the style information available locally on the device, so we don't have to fetch it, but the XHTML-MP pages themselves are also sparser because on average they contain less information. On a relatively slow RF network connection this can make a noticeable difference.

[16] There is a slight inconsistency in the syntax here as we do not have to write 'font-color', just 'color'.

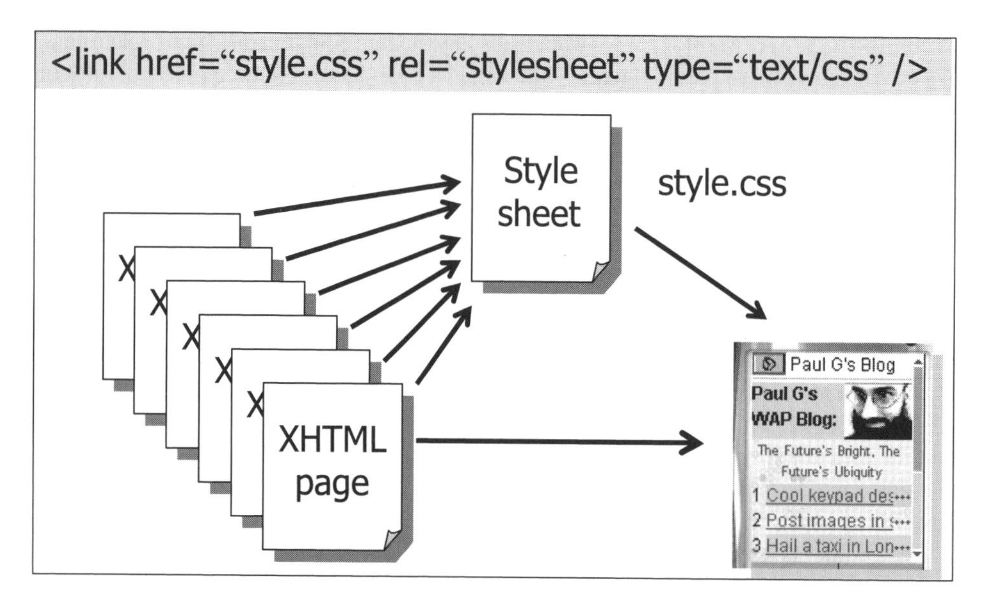

Figure 8.23 Achieving the same look and feel by linking to the same style sheet.

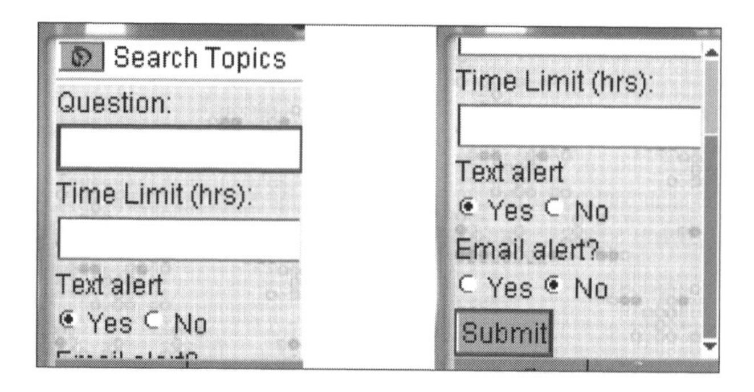

Figure 8.24 XHTML-MP supports forms.

A key feature of WAP or any browser application is the ability to gather input from the user. This is done using forms, recalling from earlier discussions that the content of the form gets sent in the HTTP stream using the POST method. XHTML-MP supports forms, as shown in Figure 8.24. This shows a page to gather input from the user who wishes to submit a question (see form field 'Question') to be answered by a consultant within a certain time limit (see form field 'Time Limit (hrs)'). In addition to the text boxes, forms can support the usual form widgets – such as radio buttons (the 'yes/no' options, as shown), select boxes and all the rest.

Browser-Specific Enhancements to XHTML-MP As with all technologies, competing companies always attempt to bring their own interpretation of what's needed in the marketplace, despite the existence of standards. This is also true of WAP and XHTML-MP.

#	Icon	Name	#	Icon	Name	#	Icon	Name
1		exclamation1	2		exclamation2	3		question1
4		question2	5		lefttri1	6		righttri1
7		lefttri2	8		righttri2	9		littlesquare1
9		littlesquare1	10		littlesquare2	11		isymbol
12		wineglass	13		speaker	14		dollarsign
15		moon1	16		bolt	17		medsquare1

Figure 8.25 Some of the Openwave browser icons.

Perhaps the world's leading supplier of mobile browsers is Openwave, which is no surprise, given the origins of this company: it was originally Unwired Planet, the inventors of HDML, which was the inspiration and technical basis for WML.

If we look at the Openwave Browser (Universal Edition), then we can examine its support for XHTML-MP and become aware of some of its interesting features. The extensions to XHTML-MP supported by Openwave Mobile Browser fall into the following categories:

- legacy HTML elements (tags) and attributes (for their counterpart XHTML tags);
- useful elements and attributes from XHTML 1.1 (i.e., fully blown XHTML);
- useful elements and attributes that are taken from WML;
- Openwave's proprietary extensions (only two of them).

When looking at some of these extensions, it is difficult to understand the rationale behind including them. Possibly, as with so many technical 'enhancements' to products, particular application scenarios arose that could not be adequately addressed with the basic technology set. Hence, someone decides to graft an extra capability onto the side of an existing standard just to accommodate these exceptions.

One of the most elegant extensions can be found in iMode's extension to cHTML: the ability to reference local source images. These are images that are already stored on the device in its physical memory during product assembly. These images are actually icons, some of which are used to convey emotions, just like the emoticons used so prevalently now in text messages, emails and Instant Messaging. Because these images are built into the device, there is no need to fetch them from a Web server using HTTP, so they load very quickly. The Openwave Mobile Browser includes hundreds of icons that can be used in XHTML-MP documents. In fact, the WAP Forum itself specified a set of WAP *pictograms* that are also supported by the Openwave Browser, although Openwave has a larger set of proprietary icons. Similarly, iMode supports a custom set of icons called *emoji*.

WAP browser icons can be used wherever an image can be used within an XHTML-MP document and can be used as inline monikers for lists. A sample of the icons built in to the Openwave Browser is shown in Figure 8.25.

Using these built-in graphics enables the Web page author to add a touch of colour or graphical reinforcement to information, which when used diligently could enhance

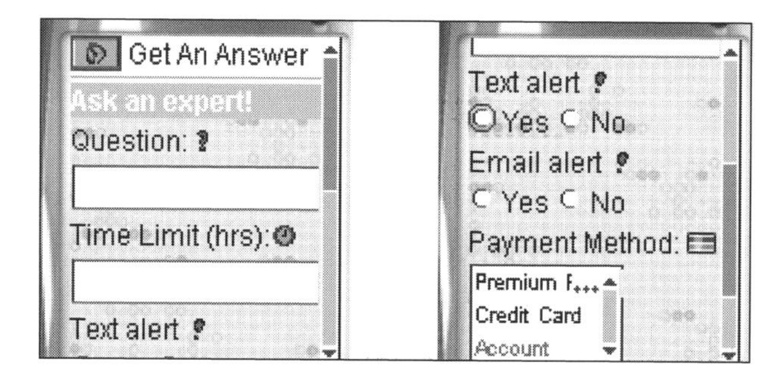

Figure 8.26 Using icons on mobile Web pages.

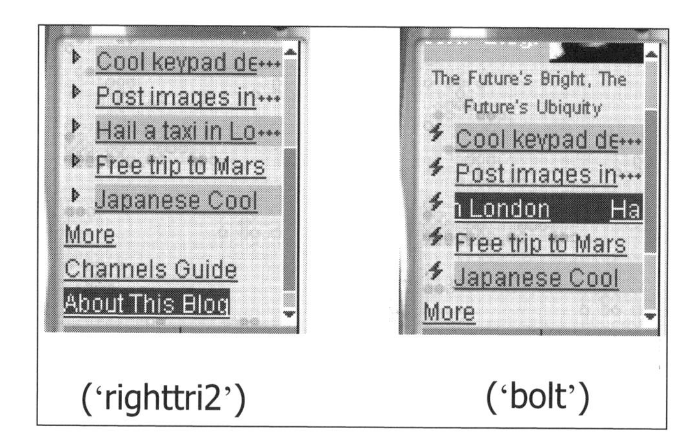

Figure 8.27 Using browser icons for bullet points.

usability – a key consideration for mobile applications. We have used some icons to redo some of our sample weblog application, as shown in Figure 8.26.

Icons can also be used inline as bullet points in lists, as shown in Figure 8.27, where we have taken the 'righttri2' and the 'bolt' icons from Figure 8.25. We insert these into the bullet lists via the style sheet of course, thus enabling us to change all the bullets in one easy step across all pages using the list element in XHTML-MP:

```
UL
{
  list-style-image: localsrc("bolt")
}
```

This style is different from the custom-defined style we used earlier, where we made up our own style name – 'pagetitle'. Here we are specifying the style that is to be used for an existing XHTML element – an *unstructured list* element – demarcated by the <UL> tag (see footnote 17). We can see how we build this list by looking at the following XHTML-MP, which illustrates several other features:

```
<ul>
<li class="blue">
<p class="nowrap"><a href="story1.html">Cool keypad design</a></p></li>
<li>
<p class="nowrap"><a href="story2.html">Post images in space</a></p></li>
<li class="blue">
<p class="nowrap"><a href="story2.html">Hail a taxi in London</a></p></li>
<li>
<p class="nowrap"><a href="story4.html">Free trip to Mars</a></p></li>
<li class="blue">
<p class="nowrap"><a href="story5.html">Japanese Cool</a></p></li>
</ul>
```

Notice how we begin and end the entire list using the <ul></ul> tag pair, which is how our browser picked up the reference in the CSS for the <UL> style.[17] To place an item in the list, we use the <li> tag element; sometimes we have added the class attribute to this tag, using our own custom style – 'blue'. This is placed in alternate list items in order to give the striped effect seen in Figure 8.27, which makes it easier to read the items on a small display. The other style worthy of note is in the embedded text within the list items, which uses a paragraph <p> tag to allow a style to be applied that specifies no wrapping of text onto the next line. This is why the list items appear to run off the side of the screen. However, if we look carefully at the left and right image in Figure 8.27, we can see how placing the focus on a list item causes it to scroll sideward, like one of those LED signs in a bank. This is called the marquee effect, which is supported by the Openwave Browser, though not all browsers.

As mentioned, iMode also supports browser icons called *emoji*. In Japanese, *ji* means character. Thus, *kanji* are characters originally borrowed from the Han Chinese repertoire and *gaiji* are 'foreign characters'. *Emoji* are characters invented by NTT DoCoMo for people to use in text messages on their mobiles. The most obvious example is the well-known 'smiley face', often encoded in ASCII as **:)** and called an 'emoticon': thus, 'emotion' + *ji* gives *emoji*. We show some of these in Figure 8.28.

Since DoCoMo uses standard Web infrastructure, including basic HTML and HTTP, we need to know how these icons are encoded. They use Unicode's 'Private Use Area', a built-in range of character codes that are there for people who want to use their own nonstandardized characters.

To my relief, as Figure 8.29 shows, at least *emoji* include some characters related to mobile telephony, whereas WAP and Openwave icons do not (unless I have missed them)!

Guidelines for Mobile Web Page Authoring Having looked at the evolution of mobile Web browser mark-up languages, we can briefly examine some of the issues facing us when designing mobile applications using the browser paradigm.

[17] Note that CSS syntax is not case-sensitive, which is why we end up with UL in capital letters; thus still works even though the element in XHTML must be lower case.

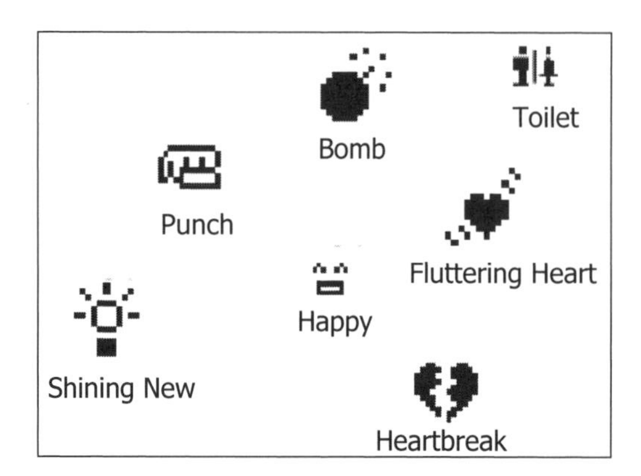

Figure 8.28 Some *emoji* characters.

Figure 8.29 Some *emoji* icons for mobile telephony.

We have already noted on many occasions throughout the book that usability is a key consideration for mobile design, more so than with the desktop scenario where there is a greater threshold for tolerating poor design. The golden rule is to make life easy for the user, which generally means several things:

- Do not deviate from accepted UI norms.
- Minimize user scrolling and 'deciphering' of content to find what they are looking for.
- Think 'task-oriented', not 'style-oriented': that is, let the user easily achieve what they set out to do, not simply be impressed by the visuals (looks).
- Make the page set as sparse as possible to achieve the task in hand and be prepared to cut things right back to the bone.
- Avoid the use of jargon wherever possible. Don't start using words like 'central' instead of 'home'. Only use jargon if it is a well-understood part of the user's vernacular.

When looking at the specifics of design, here are some things to think about:

Beware of the gateway – don't forget that a WAP gateway may still be present in your system, even if you didn't put it there. This is especially true if you are writing applications to work with, in or alongside a mobile operator's portal. Some gateways are very strict in checking for valid XML; so, if you're sloppy with the XHTML-MP coding, then pages that work fine in emulators may not work when published live. The best approach is to use a validating editor or a validation tool, which should be a standard design practice. Some gateways also behave strangely and may start doing things like transforming content, especially images.

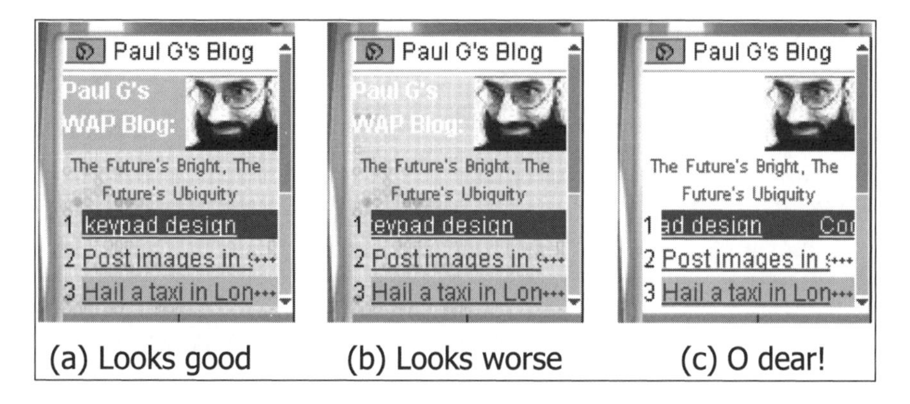

Figure 8.30 Lack of browser features wiped out the title.

Be prepared for this and the impact it may have on your design. Hopefully, you will use sparse and easily discernible images anyway (see below).

Think task list, not brochure – in designing page layouts, think about the vertical flow and make it as efficient as possible for the user to complete the task presented for each page(s). As with newspaper writing, keep all the important information at the very top of the news item or at least write a summary there of the rest of the article. Use similar principles in your design not just for news articles but also for every content type. Think about partitioning the content so that not too many important ideas or chunks of information are present on any one page. Try to direct the user toward being decisive. For example, don't give them the option to see information that you could easily have included anyway on the current page. It is often tempting to sculpture the design around what is easy to code (especially in the JSP/servlet sense) rather than what is easy for the user to accommodate. Asking a user to jump to another page to see the prices of items is not a very good idea if you could have easily included the prices on the current page, just with a little more coding effort at the back end.

Beware of app killers[18] – unfortunately, some browsers will not support a feature that you may have relied on in your design, such as background fills or background image. So, if you used a white for font colour, then you may end up not getting what you expected, as shown in Figure 8.30. Image (a) looks fine; it is exactly as we intended it. In image (b) we see that style sheet support is limited and we are unable to display the background fill on the title text, resulting in poor readability against the light background. Image (c) has really done it for us; there is no background support and the background image has also dropped out, leaving us with an invisible title – Oh dear!

We could test our design in every single browser, which sometimes may be necessary (and we tackle this problem later), but we should design for graceful degradation. This is different from lowest common denominator design where we don't use any of the 'advanced' features just to be on the safe side. Graceful degradation is taking an approach that stands the greatest chance of still being displayable, even in browsers with many features missing.

[18] Not 'killer apps', of course.

Figure 8.31 Lack of browser features gracefully handled.

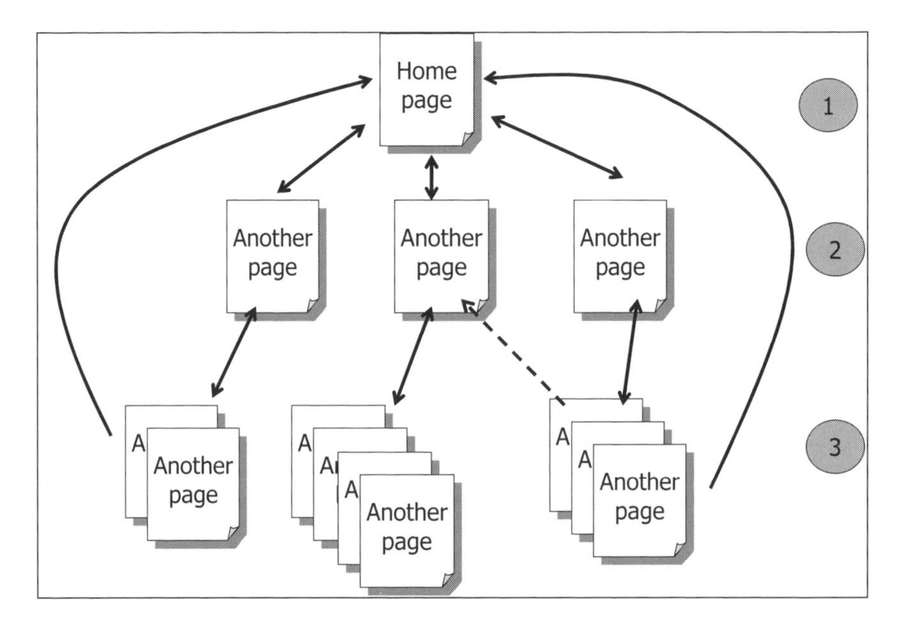

Figure 8.32 Applications should be shallow and easily navigated.

An example of this is shown in Figure 8.31, which shows the same design as shown in Figure 8.30, but this time with a few subtle changes.

First, as seen in (a), we also specified a background colour in addition to an image. If the image is not displayed, perhaps we can still benefit from a more appropriate background colour than the (presumably) default white. However, if we also stick with black colour font for the title (as in b), then it still looks fine under all cases, even if we regain our background image support, as shown in (c): in all cases the page is still readable.

Navigational efficiency – designs should be navigationally shallow with typically three links of depth and a consistent navigation model on all pages, as shown in Figure 8.32. We can see that every page should have a single click to get back home, an obvious point but surprisingly often overlooked. At any sublevel of the application, we should consider having the ability to easily move up and sideward (dotted line) in one click, where this makes sense – perhaps to view very similar and related topics: for example, in a TV listings application when drilling down to a view of what's on a particular channel and then clicking on a programme to see the detail. Having got there, we could well imagine that our user didn't find anything they were interested in and may well like to see the details of what else is on at the same time, but on a different channel. Rather than make the user go up and then back down, we could consider an immediate sideward link, such as 'what's on channel 2 at this time?'

We should remember to use meaningful titles in the XHTML-MP title element, as this generally appears in the top of the browser and is valuable navigation context information, helping us to know where we are. We should make good use of this and change the titles to match the content and refrain from repeating the application or company name over and over again – an all-too-common mistake. We should then make page titles really count and not merely repeat what the XHTML-MP file title already states, which is redundant and a waste of valuable screen space. Use text links only and avoid clickable images unless it is obvious that we are expected to select an image, such as a group of icons on a home page. Beware of browsers that can't support any colouring of hyperlinks, like the Sharp GX10 implementation of the Openwave Browser which only displays black links and ignores all style information. Also, beware of browsers that put links on a new line, even where one wasn't intended.

Design elegance – we should aim to keep colours simple and images simple, clear and relevant. As already noted, we should beware of using background colours and images when they are not supported. Our colour schemes should aim to use high-contrast design to aid visibility on small devices, especially with poor backlighting or under poor light conditions (which may mean overly bright as well as dim). We discuss device displays in Chapter 10 and discover that there are several predominant types of display technology, some much better than others. We should avoid developing and testing on a feature-rich device, only to discover later that on other devices (perhaps a popular device) the colours look washed out or that colour guides (e.g., striped lists) are hard to discern. Wherever possible we should use standard link colours and be consistent across all pages. If we can, then it is better to use CSS in place of images that are simply text blocks, as this will typically speed things up; although we should still be careful of variations in CSS support that might lead to excessive or unintended variation in our look and feel. The deeper we go in the hierarchy the more we should consider simplifying page design for speed, such as eliminating graphics altogether. By the time the user has drilled down the design to get somewhere they want to go, they are no longer interested in what pages look like – they want to get 'in and out' speedily in order to get the job done.

It is likely to be safer if we try to use lists wherever possible, avoiding the use of tables to construct similar layout flows. This is due to the variance in table support that exists across browsers, with some browsers unable to draw a table without displaying borders, even if these are not described in the mark-up code. To display a table, rather than a list, a device probably has to perform floating point calculations. On a very slow device this will take time, and thus the page will be slow to display, perhaps adding to the lag that the user experiences when fetching the page and decreasing user satisfaction. Tables do not render consistently on all devices, even some with the same browser. Some devices will display all columns with equal width, overriding the style information in the page. This can lead to particularly chronic layout failures that might even lead the user to conclude that an application failure has taken place. Of course, where data are naturally tabulated, tables are ideal for formatting the information; so, the idea is not to avoid them altogether, but generally to avoid relying on them for content formatting.

Browsers are not all the same – this is the bugbear of browser application design for mobile environments: usually no two browsers are alike and they don't all fully support XHTML-MP, even if they claim to be 'XHTML-MP-compatible'. This can make life difficult for the application designer who ultimately may have to consider designing an application that can dynamically adjust its presentation layer to suit the device in hand. Apart from some of the display differences we have already highlighted above, we need to pay particularly close attention to:

- cookies;
- level of language support;
- external CSS support;
- cache size.

Cookie support can be very limited on some devices, perhaps with the ability to store only a few cookies at best. This can be mitigated by using the gateway to support cookies. For an application that is being specifically designed to run through a known gateway, cookie support at the gateway can be exploited (if available). However, for applications where we are unsure about the level of cookie support we should aim to minimize their use or, possibly, avoid them altogether. The usage of cookies and possible alternatives is discussed in Section 8.5, when we look at something called *session management*.

In terms of language support, it is always worth consulting with a browser user guide first to understand which features are supported and which are not. We should not only be careful about understanding how many of the core XHTML-MP elements are supported but we should also pay close attention to extensions, particularly if these involve HTML. We may deliberately or inadvertently revert back to using HTML elements, some of which are not supported by XHTML. In particular, the way attributes, especially relating to style (e.g., 'width'), are supported in XHTML is now different from HTML. A design that uses HTML might work in a backward-compatible browser, but perhaps not in a strict XHTML implementation. This will give us problems later when we try supporting users with XHTML-only browsers. They will get errors on the pages that use HTML wherever it is incompatible with XHTML.

Language support goes side by side with CSS support, or WAP CSS (WCSS) support to be more precise. Not only should a designer be careful to check the browser guidelines on support for WCSS they should also look out for chronic deficiencies altogether, like the inability to support an external CSS, where the designer is expected to embed style information in the header block of the XHTML-MP file instead.

Cache size is an issue if the user is positively relying on the cache for improving performance in order to get away with using rich graphical content. There is also an issue with caching the external CSS: the designer should not assume that this support is automatically available, despite this being the intended wisdom behind it.

8.4 MANAGING DIFFERENT DEVICES

8.4.1 The Device Variation Problem

In our previous discussion it became apparent that not all devices are the same. This is a rather obvious point given the broad range of form factors and something we discuss in Chapter 10, when we look at devices. However, we have also just seen that even on devices that may ostensibly be similar there could be variations in browser implementation. This may preclude using a common design approach; were we to go down that route, we may end up gravitating to the lower common denominator. This may lead to design compromises that we are unwilling to accept.

There are copious mobile device parameters to take into account: browser, language support, operating system, user preferences, screen size, colour support, connection speed, graphics capabilities, audio capabilities, etc. The catalogue of features keeps expanding. It is obvious that not every application design will work for every user in all situations.

When we look at a device like the Sony Ericsson P800 mobile phone, which is really a wireless personal digital assistant (PDA), we see that the integrated browser can access the Web using HTML, XHTML, cHTML and WAP. The screen, as shown in Figure 8.33, is a $\frac{1}{4}$ VGA (videographics array) touch screen with 4,096 colour depth. This means that when fully opened (i.e., the keypad is flipped out), the screen size is 208×320 pixels.

By contrast, one of the leading Java games devices (at the time of writing) is the Nokia 3510i, which has a comparatively small screen at 96 by 65 pixels (see Figure 8.34), but an impressive 4,096 colour depth. However, the 3510i has a WAP 1 browser, which means we

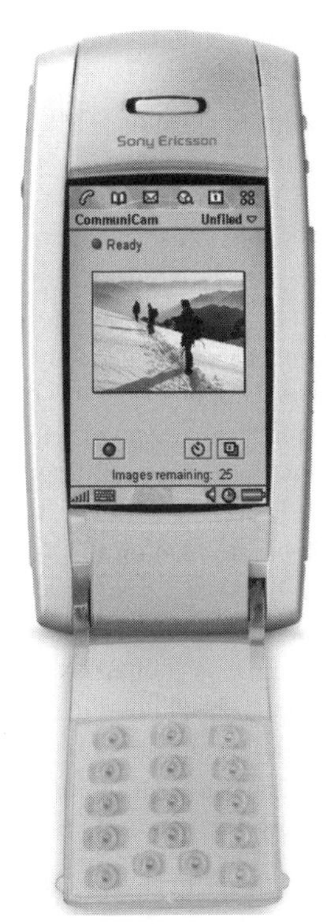

Figure 8.33. Sony Ericsson P800. Reproduced by permission of Sony Ericsson.

Figure 8.34 Nokia 3510i. Reproduced by permission of Nokia.

have to code in WML. This is not too much of a problem if we wanted to use WML for our design, since the P800 supports it. However, a quick glance at the relative screen sizes in terms of resolution shows us that a common design approach may be problematic, as demonstrated in Figure 8.35. It hardly needs explaining, but common formatting is going to be a problem. As an example of formatting problems, we might decide to use a header image like the picture portrait seen in our previous examples of the WAP weblog. This image, which is 53 by 43 pixels, would appear quite large on the 3510i, taking most of the display space, as shown in Figure 8.35. On the P800 the image would perhaps be a bit small, tucked away as it is in the top right corner.[19] It seems we might benefit from using two different page designs here or at least two different image sizes. The latter option would still necessitate two different WML pages, as the images would have to be stored as two separate files on the Web server, thus requiring two separate file names. This would have to be reflected in the WML, so we would need two different WML pages to dish up these two different images.

This is perhaps a trivial example of formatting problems. If this were an application involving the publication of articles, like the WAP weblog example, then the length of the articles (number of words) may have to be changed to adapt to the displays. Perhaps we could fit an entire article on the P800 display, which with a bit of scrolling is still manageable. However, on the 3510i we probably need to think about chopping the story up into smaller

[19] The figure is not quite a true representation, as the P800 browser window does not occupy the entire screen space available.

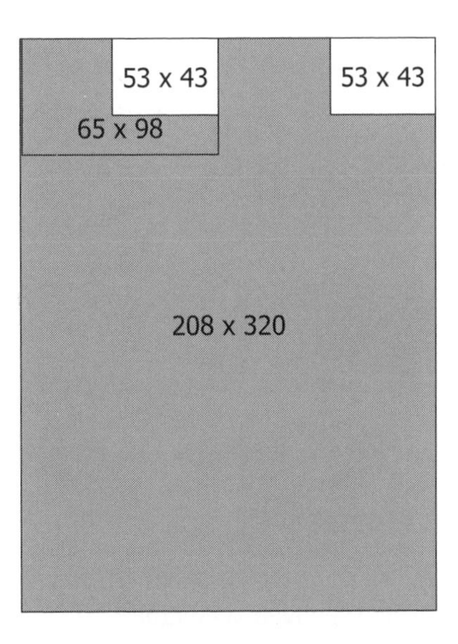

Figure 8.35 Pixel (not physical) size comparison between the 3510i and the P800.

chunks and asking the reader to download one chunk at a time. Such an approach could be made adjustable to any device, simply by altering the 'chunk' size to suit the device's display size.

Depending on the application, we may decide that WML (WAP 1) is just too limited a language to express the rich content we would like to support. The P800 can support other languages, so it is likely that we would want to exploit one of these. Looking critically at the P800's capabilities, we notice the very advantageous pointing device courtesy of the touch screen and stylus. This means that we can implement exact positioning if we choose. The higher processor speed of the device also means that we are not too fussed about the processing overhead in displaying tables. Moreover, with full XHTML support, we could even consider image maps, should we feel that the size of images involved are not too problematic given whatever limitations the RF connection might impose. With a more conventional PDA, like a Palm Tungsten T, we may have a high-speed WiFi connection available and don't really mind about image sizes very much in terms of their loading times.

The problem we face is how to handle different device types if we choose to design different user interfaces to match device capabilities. One solution is simply to build *N* applications for *N* device types, where *N* could be in the worst case scenario the total number of device types we are designing for or a group of device families with similar display capabilities. From our consideration of J2EE, this would mean that while our business logic (EJBs on the back end) could definitely remain the same, we would have to implement different presentation logic for each application. In the worst case scenario we end up with *N* different JSPs to control our view of the interface. Clearly, there is potentially a significant overhead in maintaining all these different applications, even if we are still able to use some component technologies within the servlets and JSPs (e.g., with Java beans[20]).

[20] We didn't discuss Java beans, but they are not the same as EJBs. Java beans are repeatable chunks of Java that we can reuse as plug-and-play components in our JSPs and servlets. They don't necessarily have anything to do with the J2EE platform and all its amazing capabilities that we have been discussing in this book.

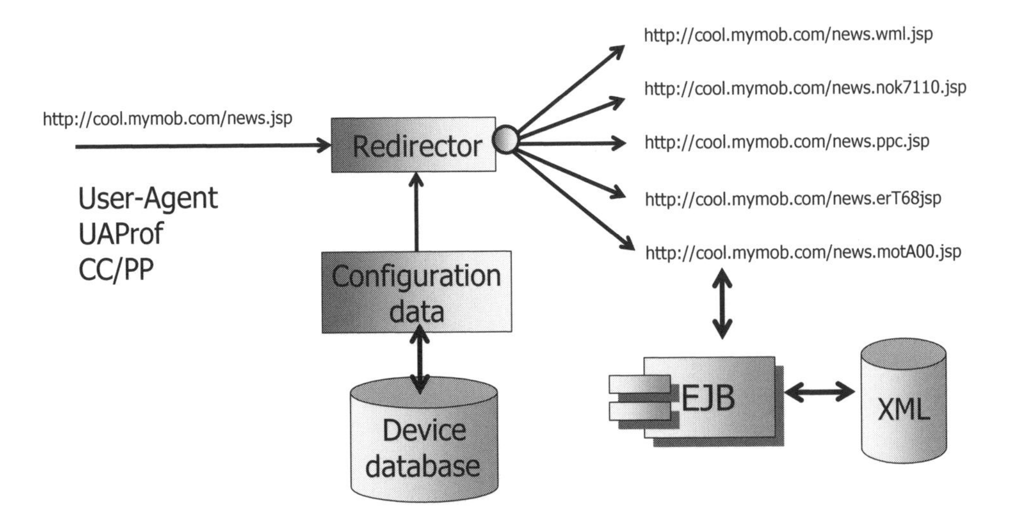

Figure 8.36 Device detection and application control.

Another solution, which we will examine in the remainder of this section, is to build a framework that lets as much commonality as possible be implemented and allows dynamic content adjustment to the different device types.

8.4.2 Building Device-Independent Applications

The general approach toward the problem of designing for different devices can be seen in Figure 8.36. We can use a software element called a redirector: our entry point into the application. There should be only one entry point, as this is easier to publish and link to as the starting point. The redirector somehow senses what type of device is accessing it. Based on this information, it consults with a device database and extracts configuration data that get used to subsequently redirect the user's browser to another resource that can better handle the display characteristics of the device they are using. This entire process is ideally automatic. The user does not have to specify what device they're using nor do they have to manually link to the redirected page – somehow, as we will find out, the browser can 'take them to it' automatically. The redirection process is all taking place at the presentation layer level. The underlying business logic remains the same and is probably none the wiser for the process.

The use of a redirection scheme does not save us from being smart about how we design our pages. Earlier we discussed guidelines for XHTML-MP design. These guidelines were largely general and apply to other mark-up schemes. Moreover, these guidelines apply even when using a redirection approach. The reason is that we do not want to test for each and every browser/device combination, which in all likelihood is not feasible anyway, especially if we are required to gain access to new devices to conduct testing. This is very often not that easy to do, especially if we are designing applications that support devices as they come on to the market. Gaining access to pre-release devices for testing purposes is notoriously difficult, even for some bigger companies that have a modicum of influence on device manufacturers. It is often engineering limitations that dictate that very few test devices are available before volume manufacturing commences, and this problem is passed on to the application designers.

We should always aim to design pages to degrade gracefully in the absence of certain browser features, as discussed earlier. This is good design practice, which will pay off in the end, as it means, ultimately, we will stand the best chance of attaining a favourable user experience, which is essential to the success of a mobile application.

If we wish to deploy the redirection (or 'resourcification') approach, then we need to consider how to achieve several mechanisms:

- detecting device type (or browser type);
- capturing device (or browser) capabilities;
- dynamically adjusting the presentation logic based on the above two information sources.

Let's look at each of these areas in turn, beginning with recognizing the device type.

Detecting and Capturing Device or Browser Information When discussing the details of HTTP, we understood that the request for service from the server consists of issuing a method, such as GET, along with associated fields (header fields) in the header of the request. One of those fields is called *user-agent*.

The *user-agent* request header field contains information about the user agent originating the request; this has several uses. First, it enables the Web server to amass statistics about the devices and browsers accessing the server. Log files collect this information, which can then be analysed offline. The *user-agent* field can also be used for tracing protocol breaches by 'rogue' devices; but, probably most importantly for our current purposes, it can be used for recognition of user agents for the purpose of automatically tailoring responses to match the capabilities of the user agent's display and other content-handling functions.

User agents are expected to include this field with requests, although there is no particular convention for doing so. The field can contain multiple product tokens as well as comments identifying the agent and any subproducts, which form a significant part of the user agent. For example, in addition to identifying the browser itself, it could be used to identify that the host device has a particular audio codec available for decompressing sound. Perhaps the codec was downloaded by the user after purchasing the device and so is 'nonstandard' for the browser; thus, we would not expect that our presentation layer would know of its existence unless we specifically informed it. Support for a particular codec may be useful for the server to know, so that it can include sound files optimized for the available codec. This will involve the presentation layer rewriting its output stream to ensure that the appropriate links (<a> tags) are included to reference the relevant sound files on disk (or on the audio-streaming server). Possibly, the codec is the latest one available and offers an especially high degree of compression, making it ideal for conveying sound files across a relatively slow RF network.

It is useful to include information about the particular capabilities of the device beyond its standard means. We can regard this as preference information. There are perhaps many other preferences that a user could specify, even regarding the user interface itself; this is not a problem as such. Once we have a mechanism in place to adapt our HTTP output stream per user agent, then there is no reason why we can't adapt it on a user-by-user basis – we could easily offer that degree of customization of the output stream.

Within the WAP environment, the ability to determine device capabilities and preferences is not limited just to browsing, it also extends to WAP Push. Later on, we will see how WAP Push is used to send multimedia messages that contain pictures, sounds and mark-up to glue them together. Clearly, capability information is useful in this context: there is little point in pushing an image to a device that cannot display it!

Our 'resourcification' problem has now extended to include not just capabilities but preferences as well. This had already been highlighted by the standards community as an important feature in the desktop browser world and is the subject of a W3C project to define a more substantial means of identifying capabilities and preferences, a technique formally known as conveying *Capability and Preference Information* (CPI).

The W3C initiated a project called Composite Capability/Preference Profile (CC/PP), and the WAP Forum adopted this and extended it into the WAP environment, sticking to the original project title of *User Agent Profile*, or *UAProf*.

The W3C has since gone on to expand its work in this area, recognizing its importance across a wide array of browser-based applications and situations. In September 2001 the W3C forwarded a draft proposal[21] for the formation of the Device Independence Activity (DIA), wherein they mentioned that the draft document:

> *... celebrates the vision of a device independent Web. It describes device independence principles that can lead towards the achievement of greater device independence for Web content and applications.*

The document is worth reading as it highlights a much wider vision than the confines of our current discussion, which is still related to the more conventional browser paradigm. However, the W3C DIA project takes into account all possible access methods, such as accessing content via an audio-only connection. A mobile devices working group was established for a time and its output was fed into the DIA project; so, the work of this group is fully expected to take into account mobile devices, our current topic of interest.

As we can imagine, the method of specifying capabilities and preferences needs to take into account a wide variety of device types and application scenarios; so, inevitably, it has moved way beyond simply identifying a suitable string to encode in the *user-agent* header field.

The CC/PP project has worked to produce a definitive vehicle and framework for delivering capability information within the context of the Web. We will briefly look at the CC/PP profiling method itself before considering its use within the WAP context. The latter is important as it tells us how the CC/PP information gets conveyed, handled and processed within a network where WAP protocols are being used.

CC/PP is based on RDF, the Resource Description Framework, which was designed by the W3C as a general purpose metadata description language. The origins and ethos of RDF are interesting. For a while, the Web has been attracting vast amounts of information to its cloud. However, the origins of the Web are in human interaction with information, notably the visual interaction paradigm of the browser we know quite well. However, it has long been recognized that machines themselves have difficulty in processing information on the Web. If we want to write a program to go and look for information on a particular topic, then how do we identify the target information in a meaningful manner? For example, there

[21] http://www.w3.org/TR/2001/WD-di-princ-20010918/

may be lots of pages about cooking, but how would a computer know that it is looking at a recipe say? Moreover, even if it could find a recipe (perhaps searching for the keywords 'cooking recipe'), how would we know what the ingredients are? In other words, we are articulating a problem to do with metadata, or semantics. We are not just interested in how to visually format information, we are also now interested in how to describe or annotate information – to add semantics.[22]

RDF can be expressed using XML; no surprise there. But we are not as interested in the details of RDF itself as we are in how to use it to support our CC/PP objectives, coming eventually to its use in WAP environments. A complete sample RDF file is given in Box 8.1 at the end of this chapter ('Sample CC/PP RDF file'). Just to illustrate its usage, we can look at a few snippets of XML, especially those relating to screen size, as this was an example of a capability we were looking at earlier.

```
<prf:ScreenSize>121×87</prf:ScreenSize>
<prf:Model>R999</prf:Model>
<prf:ScreenSizeChar>15×6</prf:ScreenSizeChar>
<prf:BitsPerPixel>2</prf:BitsPerPixel>
<prf:ColorCapable>No</prf:ColorCapable>
<prf:TextInputCapable>Yes</prf:TextInputCapable>
<prf:ImageCapable>Yes</prf:ImageCapable>
```

We should be used to the familiar XML tags by now, but the 'prf:' bit seems new: this is referring to a namespace, which is an advanced topic. Simply put, although we can define our own tags in XML, as we did when we were defining our tags for FML earlier in the chapter, the problem is how to avoid treading on someone else's toes. Perhaps the name we use is already taken by another vocabulary, especially if the name is somewhat generic, such as 'title'.[23] There is nothing wrong with mixing different XML vocabularies within one (host) document; however, this is when things are likely to conflict. To get around this problem, we refer to the namespace of a vocabulary, giving that space a moniker, like 'prf' in our case. We can then prefix all tag names from that namespace, so that they are uniquely identifiable with the namespace, thus avoiding naming conflicts. In the case of CC/PP, 'prf' is used here to refer to 'profile', which is the schema[24] used to define the allowable tags within the CC/PP profile.

Getting back to the XML snippet, we can see that the tags are quite self-explanatory as we would expect, because the objective of using XML is that it is also human-readable. The issue of screen size we discussed earlier is covered by the tag <prf:ScreenSize>; so, on discovering the CC/PP profile for this device, we can extract its screen size by searching for this tag. Other tags in the snippet also relate to screen capabilities, including colour support and image display capability. There are many tags in the UAProf that relate to various components of the device capability set. A sample of the structure of the UAProf, along with a few of the tags, is shown in Figure 8.37.

[22] Text missing.

[23] It's always a good idea not to use names that are too generic. If we mean the title of a film, then perhaps 'FilmTitle' is better, as it is more descriptive. However, there are arguments for sticking with the word 'title' on its own, as namespaces can solve such conflicts for us.

[24] A schema is like a DTD, see earlier in this chapter.

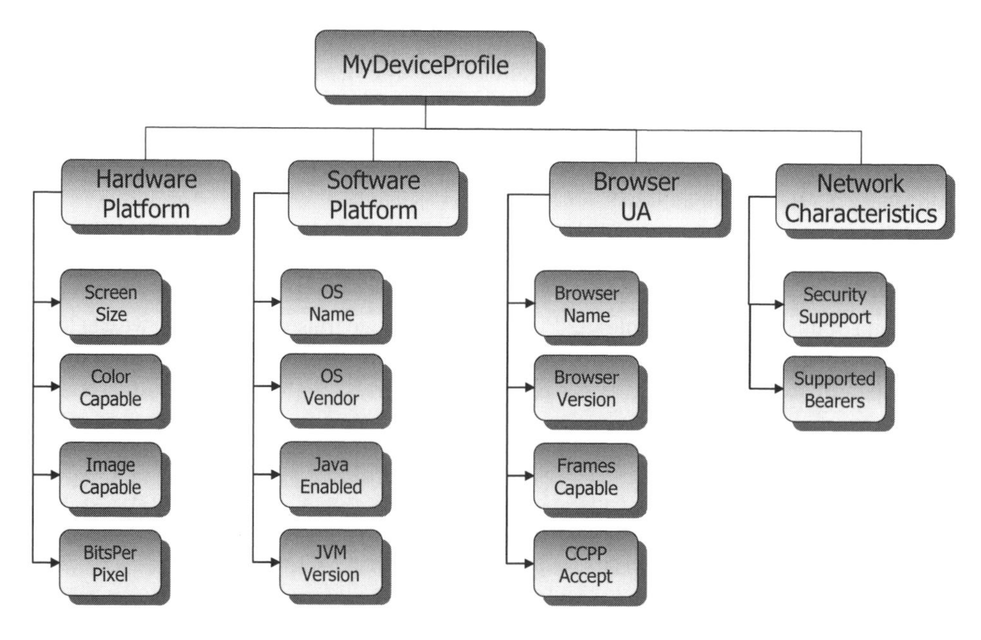

Figure 8.37 Structure of UAProf (partial).

The issue still remains as to how this information gets conveyed from the client to the server. We have already mentioned that it gets used in WAP Push, which is puzzling at first. If we are going to push information to a device, how do we know its capabilities in advance? Let's now look at these issues.

Conveying CC/PP Information The suggested method for exchanging CC/PP information is captured in the W3C technical note[25] defining the CC/PP exchange protocol. This is a protocol that is based on the HTTP extension framework, which is a means to convey information via HTTP headers. The HTTP extension framework is a generic extension mechanism for HTTP 1.1 and is designed to interoperate with existing HTTP applications.

There are two ways of conveying the profile, both of which are supported by the WAP UAProf specification, though in a slightly modified form than the one we will discuss here.

The first method is to send a URI specifying where the CC/PP RDF file can be found on the Web. As we can imagine, this is a particularly efficient technique, because it means the amount of information we have to transmit from the user agent to the origin server is very little, which is particularly compatible with sending such information over relatively slow RF networks. The second method, as expected, is to insert the RDF file into the HTTP request header itself, thereby conveying in one go the entire capability set of the device. That being said, it is possible to modularize the CC/PP (the specification allows for the description of capabilities and preferences in subunits called components). We can, if appropriate, just send a portion of the overall capabilities. This has several uses: not only

[25] http://www.w3.org/1999/06/NOTE-CCPPexchange-19990624

can it save on the amount of data to be transmitted but it also allows the user agent to convey subsets of information that may be pertinent, like the codec update we mentioned earlier.

The first method uses the *Profile* header field, the contents of which are a URI pointing to the RDF file. Examples of profile headers are:

```
Profile: "http://www.aaa.com/hw","http://www.bbb.com/sw"
Profile: "http://www.aaa.com/hw","1-uKhJE/AEeeMzFSejsYshHg==",
  "http://www.bbb.com/sw"
```

We can observe several features of the header. First, there is no need to stick to one URI. Several can be used in conjunction with each other. This might be a good idea: perhaps with one pointing to a repository of device capabilities and another pointing to a repository of separately maintained browser capabilities. The profiles are listed in order of priority, with the first one listed taking precedence.

The next thing we may observe is that in the second example the second entry in the field is somewhat strange; it certainly doesn't look like a URI that we are used to. It is called a *Profile-Diff-Name*: this is actually a pointer to a profile contained within the HTTP request itself, which is found under a separate header field called *Profile-Diff*. The contents of a *Profile-Diff* header consist of the RDF itself. The strange-looking string in the above example (second entry in the *Profile* header field) is actually a base 64-encoded MD5 digest of the RDF file. That sounds like a complicated process, but we will look at the actual mechanics of MD5 later in Chapter 9, when we discuss how to authenticate HTTP sessions. All we need to know here is that this process takes the RDF file and produces a very compact fingerprint of it. The reason this is done is to allow any caching process (such as at a gateway or the user agent itself) to perform look-ups using the digest which is short in extent, rather than comparing entire RDF files to check for changes.

The *Profile-Diff-Name* entry in the *Profile* header is inserted into the list in a position commensurate with its priority with respect to the other entries. There can also be more than one *Profile-Diff-Name* corresponding to more than one *Profile-Diff* header, which is why there is a single-digit prefix (in this case '1') to indicate which embedded RDF file is being referred to. *Profile-Diff* headers themselves are postfixed with the corresponding digit, like *Profile-Diff-1*, *Profile-Diff-2*, etc.

More than one *Profile-Diff* (i.e., RDF file) can be supported, as this may be a legitimate way of indicating capabilities. For example, the device itself may have an RDF, but subcomponents on the device (e.g., audio codec) may have their own RDF, in which case we probably want to convey all of them or those that are appropriate.

In the WAP UAProf specification[26] these headers are supported in both Wireless-profiled HTTP (W-HTTP) and Wireless Session Protocol (WSP), but in the former case they are renamed to *x-wap-profile* and *x-wap-profile-diff*.

There is no need to refer to external profiles, particularly if they are not available or they cannot be relied on. If this is the case the entire profile is embedded in the request, in the manner we have just described; an example follows:

[26] http://www.openmobilealliance.org/wapdocs/wap-248-uaprof-20011020-a.pdf

```
Profile: "1-P1GRkSjKK50aTWXXndFcSQ=="
Profile-Diff-1: <?xml version="1.0"?>
<RDF xmlns="http://www.w3.org/TR/1999/PR-rdf-syntax-19990105#"
      xmlns:PRF="http://www.w3.org/TR/WD-profile-vocabulary#">
  <Bag>
    <Description about="HardwarePlatform">
      <Defaults>
        <Description PRF:Vendor="Nokia"
                     PRF:Model="2160"
                     PRF:Type="PDA"
                     PRF:ScreenSize="800x600x24"
                     PRF:CPU="PPC"
                     PRF:Keyboard="Yes"
                     PRF:Memory="16mB"
                     PRF:Bluetooth="YES"
                     PRF:Speaker="Yes" />
      </Defaults>
        <Modifications>
          <Description PRF:Memory="32mB" />
        </Modifications>
        </Description>
    <Description about="SoftwarePlatform">
  ...
</rdf>
```

There is also no reason why a *Profile-Diff* header field cannot be added by an intermediary network element like a gateway. Perhaps the gateway has additional information about the subscriber which may be useful for the origin server to know. Such a technique also allows a consistent approach to be adopted with devices that do not support the CC/PP exchange protocol and only support the *user-agent* field. The gateway could add the profile information itself, based on what it knows about *user-agent* fields, device capabilities and profile information, even about the user (e.g., from information stored in a mobile portal's personalization database).

If we embed the profile in a WAP session using W-HTTP, then we keep the text description as it is; however, when using WSP we can apply tokenized compression to render the header into WBXML, which condenses it considerably, just as we saw with WML in our earlier discussion.

Within the CC/PP exchange protocol we also have the ability to acknowledge that the profile has been accepted by the server. This is done by passing information in a header field called *Profile-Warning*. This header field contains a code number that indicates the success of the exchange, similar to the way status codes are used in HTTP. There is one code for each RDF file that the server processed or attempted to process, hopefully corresponding to the number that was referenced in the request. To make each warning clear, the name of the target RDF file (*warn* target) is mentioned along with each code:

```
Profile-Warning: 102 http://www.aaa.com/hw "Not used profile",
        202 www.w3.org "Content generation applied"

Profile-Warning: 101 http://www.aaa.com/hw "Used stale profile",
        102 http://www.bbb.com/sw "Not used profile",
        200 18.23.0.23:80 "Not applied" "Wed, 31 Mar 1999 08:49:37 GMT"
```

There now follows a list of the currently defined warn codes, each with a recommended warn text in plain English, and a description of its meaning:

- *100 OK* – MAY[27] be included if the CC/PP repository replies with first-hand or fresh information. The warn target indicates the absolute URI that addresses the CC/PP descriptions in the CC/PP repository.
- *101 Used stale profile* – MUST be included if the CC/PP repository replies with stale information. Whether the CC/PP description is stale or not is decided in accordance with the HTTP header information with which the CC/PP repository responds (i.e., when the HTTP 1.1 header includes the *Warning* header field whose warn code is 110 or 111). The warn target indicates the absolute URI that addresses the CC/PP description in the CC/PP repository.
- *102 Not used profile* – MUST be included if the CC/PP description could not be obtained (e.g., the CC/PP repository is not available.). The warn target indicates the absolute URI that addresses the CC/PP description in the CC/PP repository.
- *200 Not applied* – MUST be included if the server replies with nontailored content that is the sole representation in the server. The warn target indicates the host that addresses the server.
- *201 Content selection applied* – MUST be included if the server replies with content that is selected from one of the representations in the server. The warn target indicates the host that addresses the server.
- *202 Content generation applied* – MUST be included if the server replies with tailored content that is generated by the server. The warn target indicates the host that addresses the server.
- *203 Transformation applied* – MUST be added by an intermediate proxy if it applies any transformation changing the content coding (as specified in the *Content-Encoding* header) or media type (as specified in the *Content-Type* header) of the response or the entity body of the response. The warn target indicates the host that addresses the proxy.

By analysing these codes the user agent can take corrective action if necessary. A similar heading is supported in WAP – *x-wap-profile-warning*.

When using CPI for WAP Push, there is no request from the user agent; so, there needs to be an extra step in the protocol in order for the push server to know the target device's

[27] These codes are taken straight from the W3C document where the words MAY, SHOULD and MUST are used to indicate the desired level of conformance required from an implementer of the specification.

capabilities before pushing any content to it. This is provided by a standard HTTP request that gets made from the Push Proxy Gateway (PPG) to the device, remembering that in WAP the device itself is capable of responding to HTTP requests, this being the mechanism by which data get "pushed" to the device (i.e., via HTTP POST requests). To first get profile information, the PPG issues an HTTP OPTIONS method, which is like a GET request, but without any resource actually being requested. As such, the device responds by sending back a response that only contains header information, but no body (as body wasn't requested). The HTTP header can contain the relevant profile headers (e.g., *x-wap-profile*) and, so, the PPG now has the CPI it needs before attempting to push content to the device, content which can now be suitably adapted to device capabilities and to user preferences.

A profile repository typically would be an HTTP server that provides *UAProf* CPI elements on request. An origin server would go to the HTTP server to fetch the corresponding profile it found referenced in the *Profile* header field. *UAProf* profiles may reference data stored in repositories provided and operated by the subscriber, network operator, gateway operator, device manufacturer, service provider or any valid host on the Web. Specialist companies or agents may choose to track device capabilities and generate profiles on behalf of interested parties. Different ways of tracking profiles may emerge, some of which are as follows:

- Manufacturers and software vendors may provide static profile resources that describe the client devices and applications (e.g., browsers) that they produce. Such profiles will describe the hardware and standard software elements that exist on the client devices, as well as optional software elements that get loaded later.
- Network operators may provide additional profile information that includes critical information about network characteristics. Additionally, they may choose to operate gateways in such a way that they provide profile information that defines permitted service levels through the communications infrastructure. This facility may also be used to allow extra preference information to be stored and presented on behalf of the user, based on service preferences they have made known to the operator, most likely via personal profile information associated with mobile portals.
- Subscribers may look to other entities to store preference information if they wish, perhaps to gain their own control over how they manage profiles. However, this would have to be a mechanism that the user agent supports. Just because a user agent can insert any profile URI in the header doesn't mean to say it will. Possibly, some browsers will use constant header information preset at the time of installation. Even so, this could still be overcome on devices that offer sufficient programming access for a local proxy to be installed (on the device) to transform headers under the user's control.
- Service providers, including enterprise IT managers, could choose to manage their own profile repositories. Perhaps this is done with security in mind where a particular authentication technique is used, in which case the origin server needs to be made aware of it.

Independent of the content of the stored profile, policy controls may be imposed that limit what profile information is delivered to a particular requester. Any such limits or methods employed to ensure compliance are beyond the scope of this discussion. Also, whether or not it is a good idea to use profiles for the purposes of controlling policies and controlling application performance will need some reflection.

8.4.3 Dynamic Page Generation Schemes

Although we now have a mechanism for understanding device capabilities, we still need to solve the problem of adapting our presentation layer's output to suit each device according to its profile. Before looking into the detail of any mechanisms, we need to recall our discussion at the start of this chapter, which dealt with the J2EE platform and its presentation layer capabilities. The platform supports several types of Java program (namely, JSPs, servlets and EJBs). We noticed that JSPs and servlets were fantastically well optimized for writing output streams over HTTP. However, it was not made clear exactly how we should combine these elements to make best use of their combined capabilities. It is pretty clear that EJBs always sit in the background and do the crunching and grinding of our business logic, talking to databases, XML streams, legacy systems and so on. But what about the servlets and JSPs? Is there a particular usage of these program types that could make our lives easier, particularly now that we understand that what we want is to produce a framework that adapts content dynamically?

In software engineering we have the concept of *patterns*, which basically involves a way of configuring software components to achieve a particular task. There is no real rocket science in this process. Largely, we take something that appears to work, refine it and then publish the idea as a pattern.

One J2EE pattern stands out from the crowd when producing mobile browser-based applications: this is called the 'Model View Controller' pattern, or sometimes MVC (and other names too).

The MVC pattern consists of three kinds of main classes: 'model', 'view' and 'controller'. The model represents the application's objective (core tasks) or its essential data; the view is the display of the model; and the controller takes care of the UI interaction with the user input.

We can see the MVC pattern on a J2EE platform shown in Figure 8.38. The EJB, the stalwart of business logic processing, is clearly the model in the pattern. The JSP, because

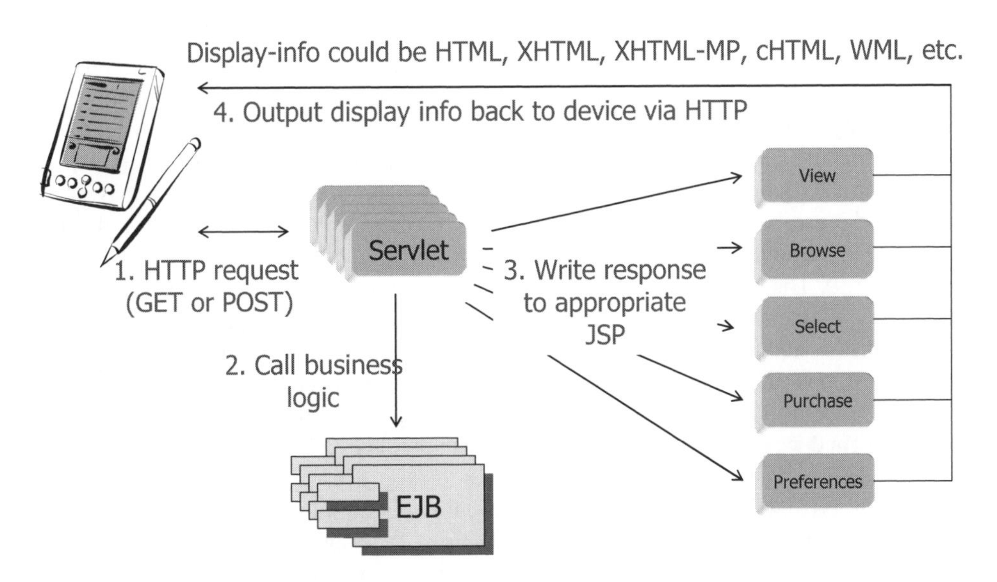

Figure 8.38 MVC design pattern.

of its superior support for page design (as noted earlier), is used to control the view – this is what produces the final output stream back to the user via HTTP. This leaves the servlet – which must be the controller. This also seems to be a good fit, as it naturally can handle HTTP input streams due to its intimate connection with the Web engine (via the servlet API); but, it also offers us the entire range of Java programming power to control how we want to handle the input stream. Comparing this with Figure 8.36, we can begin to sense that the 'redirector' in the figure is going to be supported by a servlet; but, let's dig a little deeper into the pattern and the adaptation problem we are trying to address.

Before continuing, we should clarify that the MVC pattern is generally applicable to J2EE application design for the Web. It is not confined to using servlets to select views based on device capability. In fact, this is not its origin at all. The MVC pattern can be used to develop scalable Web applications that are relatively easy to maintain. The controller actually is used to determine which page to serve next according to where the user is in the page or navigational flow of the entire application. In other words, we might want to progress from the 'login' page to the 'check my account' page. There may be a JSP to handle each function – LoginJSP and CheckAccountJSP. We can use the servlet to determine where in the design the user is or what page they are requesting and then divert to the appropriate view, accordingly. This is the primary function of the model: to react to user interaction. However, we are currently considering extending this control function to achieve a finer level of control, which is to select the appropriate view based on user request and user device type (and capabilities, as per CC/PP).

We will look at several methods for achieving the adaptation based on the MVC pattern. These methods are by no means exhaustive of all possibilities. As with all software, there is more than one way to develop a solution, depending on what we are trying to achieve, what we know, what restrictions may be in place and a plethora of other considerations, not always logical ones.

Reviewing Figure 8.38 for a moment, we can reflect on the possible means to achieve adaptation, introducing some of the programming techniques as we go along.

In step 1 our servlet gets called by the user agent via an HTTP GET request. Within this GET request we have three methods of passing parameters to the servlet. These methods are discussed in Section 8.5, when we look at state management (or session management). But for now, we just assume that we are able to receive parameters from the user into our servlet.

These parameters may be used to determine what the user wants to do next. For example, if the user is currently examining their mobile airtime account details, then we may need to know the user's unique ID (account number). Let's assume that the user has already logged in to the application using a secure authentication technique (which we will discuss in Chapter 9). The user would have entered some username to log in, and this is what we use to identify them in our system. Once the user logs in with this username, we may look up information about them, like their account number; this is what we need in order to look up any information from the accounts database. We therefore need to keep track of it for the duration of the user's session. The user may be looking at a particular part of their current phone bill, and we have programmed a JSP that can produce a phone bill view, as shown in Figure 8.39.

The process of writing an output stream to produce a phone bill view is clearly a repeatable one, and we end up using the same JSP to produce the view for each user and for each instance of each user's bill. It doesn't matter if we are looking at today's or yesterday's bill, the output will be same for a given device; just the data will change. Let's say we have a JSP called

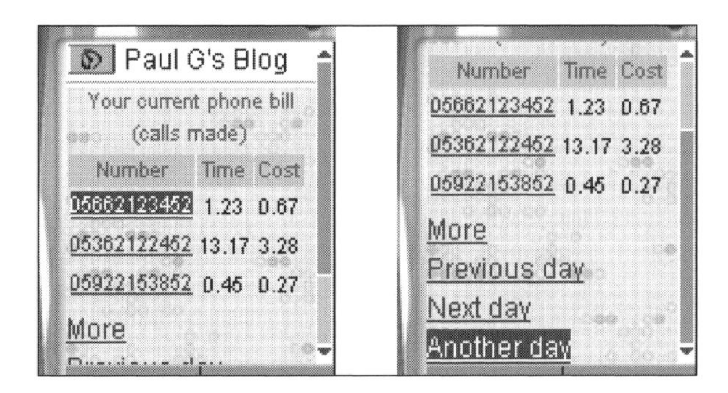

Figure 8.39 Viewing a phone bill via WAP.

MyBill.jsp to produce this view and recall that HTTP itself is a stateless protocol. We saw that there was a single instruction to fetch data, called GET; we could have issued that same method for the same page 50 times and the server is none the wiser. There is nothing inherent in the protocol that enables discernment between the first and the one hundred and first requests. This is why we need to feed in parameters to the servlet, so that in handing over control to the JSP it knows that on this occasion we are actually interested in viewing the bill from last Tuesday and that the bill in question is the one associated with Account ID 1234 (or whatever).

Let's imagine that somehow our servlet has access to the parameters that we can use to determine what to do next (i.e., state information). We can envisage a servlet called 'GetBill' that understands this is a billing enquiry. The servlet first extracts the parameters we have just been discussing. From these (however they are formatted) it understands that the billing request is for a particular date and account holder. If this were the first time that this particular user/device has accessed the application, then the servlet (actually, the 'login' servlet) would extract the CC/PP header field information. This information could then be processed by some business logic to check the CC/PP RDF file. This process would consist of two parts. First, we need to identify that we have valid profile information and that we can subsequently handle this device. To simplify our discussion, let's assume that we have divided our output views into three categories:

- small devices with XHTML support;
- small devices with WML support;
- large devices with XHTML support

Our business logic that processes the CC/PP information will attempt to put our user into one of these categories. If this is not possible based on exact profile information, then a default category is chosen – one that gives the best chance of graceful degradation. Once our user's device is categorized, this category information is stored somewhere so that it is accessible on subsequent page requests. On these requests the profile headers are ignored by the servlet as the categorization has already taken place. Either we find a way of subsequently passing the category information in the inbound requests or we look the category up using a look-up table that stores categories against account ID, which will work because we have already stated that we are able to extract the account ID as a parameter in each request (even though we have not yet discussed how to do that – we will do so in Section 8.5).

Of course, we may well have lots more categories; it is after all a design decision based on the nature of the UI and how flexible it is across different devices. If it is very flexible, then we probably only need a few categories, as in our current example – with each category applicable to many devices with similar attributes. For very specific layout control and usability enhancements, we may have many categories, possibly with only one device per category.

The second function of the servlet is to actually call the relevant business logic to process the billing request. In this example, business logic is likely to involve talking to a billing database server via an EJB. The EJB would handle a request to the server via the JDBC API and extract the relevant billing data. At this point we need a place to store the data, as they will be needed to build the output stream via the JSP. The actual mechanics of storing (or persisting) session data on a J2EE platform are beyond the scope of this book, but let's just assume that there are ways of passing data from one program to another, not forgetting that servlets and JSPs are just two types of program, of course. Here, we are assuming that, having handed the call over to the EJB, the EJB is now left out of the process and control is handed back to the servlet, which subsequently invokes the JSP. In other words, the EJB does not itself act as an interlocutor between the servlet and JSP.

We shouldn't forget that all of this process is taking place within the confines of the powerful J2EE platform; so, we are automatically gaining all of its benefits, as outlined earlier in the book, like scalability across clusters thanks to Remote Method Invocation (RMI) and so on.

Moving on and ignoring how data get passed from the servlet to the JSP, we are interested now in how the JSP adapts its output to the device. First, given that we only have three categories in our example, we could have actually considered having three very similar JSPs, something like 'MyBill_SmallXHTML.jsp', 'MyBill_SmallWML.jsp' and 'MyBill_LargeXHTML.jsp'. The servlet would then act as a redirector to the appropriate output page according to the category information that it extracted in the first instance, whether directly from the request or via a look-up table referenced by Account ID. We will not discuss this approach because it is easy to understand without further explanation and, furthermore, because it is an unlikely approach due to its cumbersome nature, especially when large numbers of device categories are involved. We probably will have too many categories for this to be a viable option, as we would end up with too many variants of each JSP and the maintainability issue will become a dominant and inhibiting factor.

We will consider here two ways that can be used to adapt the billing information:

- *JSP tags* – this is where we write conditional JSP code such that the appropriate tags eventually get written to the output stream based on a filter that picks each tag according to what the designer wants and what the capabilities of the device are.
- *XSLT* – this is where the billing information from the business logic gets stored as XML and we use the XSLT language to transform the XML to the particular vocabulary and structure we want for our output.

Adaptation Using JSP Tags The best way to discuss JSP tag libraries is via an actual example. Here, we examine the WURFL project, which stands for Wireless Universal Resource File. This is an open source initiative[28] that involves two concepts. First, the

[28] http://wurfl.sourceforge.net

resource file itself is an XML configuration file that contains information about all the capabilities and features of a wide range of devices. This is in apparent conflict with the CC/PP idea where this configuration information is held in separate RDF files. For now, however, this doesn't affect our illustration of the second concept – the focus of our enquiry – which is the support of JSP tags.

As we have previously discussed, it is possible to write mark-up output directly into a JSP file; so, we could write our XHTML, XHTML-Basic, XHTML-MP or WML directly into the JSP. We also know that we can mix Java code in with the mark-up, but that starts to get messy and is also not easy for page designers who don't know Java; it is also not very maintainable. But with JSP tags, we can use such tags as the ones defined for us by WURFL:

```
    :
<wurfl:if capability="xhtml_display_accesskey">
        :
    billing table mark up goes here without numbering
        :
</wurfl:if>
    :
```

This piece of code outputs the mark-up between the <wurf:if> tag pair, provided that the device accessing this page is XHTML-compliant and automatically displays numbers alongside <a> tags if they contain the 'accesskey' attribute. The 'accesskey' attribute can be used to allow a user to select a link on the page by pressing one of the digit keys on the keypad as a shortcut, rather than selecting the link with an arrow key and then activating the link. If the browser automatically displays the shortcut numbers for us, then we don't have to output them ourselves. To avoid having two sets of numbers, we detect this level of support and use the tags to select the piece of mark-up that either displays the numbers explicitly (e.g., using the image element <img> and possibly locally sourced WAP icons) or implicitly (the browser does it for us, removing the need for the <img> tags).

What is actually happening on the J2EE server is that the tags are being mapped to Java classes that execute when each tag gets processed. In the case of the WURFL tag library the code behind the tags is checking the *user-agent* header field and then trying to match this against the XML resource file. Thanks to the powerful inbuilt XML-processing capabilities of the J2EE platform, this process is efficient and scalable across our application. The types of capabilities that WURFL identifies are grouped into feature categories:

- general product info (e.g., device brand name, device model name);
- WML UI features (e.g., table support, soft key support, etc.);
- cHTML UI features (e.g., displays access keys, *emoji* support, etc.);
- XHTML UI features (e.g., accurately represents background colours, supports forms within tables, etc.);
- mark-up support (e.g., WML 1.1, 1.2, 1.3; XHTML Basic; XHTML-MP; cHTML 1.0, 2.0, 3.0, etc.);

- cache features (e.g., cache can be disabled, cached items' lifetime, etc.);
- display features (e.g., resolution, columns, image sizes, etc.);
- image formats (e.g., wbmp, gif, jpeg, png, etc.);
- Miscellaneous (e.g., phone supports POST method, etc.);
- WAP Wireless Telephony Application support (e.g., supports voice call from browser, supports access to phone book from browser, etc.);
- security features (e.g., supports the HTTP Secure Socket Layer, or SSL);
- storage attributes (e.g., maximum URL length, number of bookmarks supported, etc.);
- download fun support (e.g., ring tones, types of ring tones, wallpaper support, screen saver support, etc.);
- WAP Push attributes (e.g., various parameters to do with WAP Push support);
- Multimedia Messaging Service (MMS) attributes (e.g., can send MMS, can receive MMS, built-in camera support, etc.);
- J2ME support (e.g., MIDP version, memory size limits, colour depth, API support, etc.);
- sound formats supported (e.g., WAV files, midi-monophonic, midi-polyphonic, etc.).

The structure of WURFL is interesting because it defines device features in such a way that they automatically inherit from a parent feature where applicable. This will let a new device inherit all the capabilities of the family of devices from which it appears to descend. For example, a new device that mentions the Openwave Browser (UP.Browser) in its *user-agent* field will automatically inherit the capabilities of the Openwave Browser, even though the device is not currently entered in the resource file. This means that new devices do not default to the lowest level of support (not to do so would be undesirable and unrealistic): newer devices should support the newer features, not the oldest ones.

For historical reasons[29] the WURFL itself is not CC/PP-based, although it could be made so, depending on the need. What might be considered as value-added for the WURFL project is its comprehensive repository of configuration information that is constantly being updated via an active and vigilant developer community. The fact that this is happening via the open source community is also positive, as it means that there is strong support that the accuracy and quality of the resource information is reliable.

Let's put the particular strengths or weaknesses of WURFL aside for a moment, as the principle of JSP tags is the important feature to grasp here. The tags are XML, so they are easy to learn and incorporate into the JSP file. They avoid the need to insert Java code, although we still need a way to get our actual user information into the JSP file – such as the billing records for the example we have been discussing. This will involve some coding.

If we didn't have WURFL tags, we could use another tag library, including the possibility of developing our own. Let's return to our earlier example, where we only had three device categories. We mentioned how our servlet could identify the actual CC/PP information and then categorize our device:

[29] WURFL existed before CC/PP had fully emerged.

- small devices with XHTML support;
- small devices with WML support;
- large devices with XHTML support.

In this case the JSP could use an inbuilt mechanism to accept the category parameter directly from the servlet, thus avoiding the need for JSP tags; we would use snippets of Java code to conditionally write out the chunks of mark-up we require for each category, but still all within the one JSP file.

The point about using JSP tags is that they are easy to understand and insert into our JSP file. The WURFL example shows us that it is possible to control mark-up output based on attributes rather than device types, which is a subtle but important point. Rather than have constructs like 'if p800 ... ' or 'if 3510i ... ', we would much rather have constructs like 'if screen > 300 ... ' or 'if Java supported ... '. With CC/PP and JSP tags we could do this by mimicking the semantics of the RDF in our JSP. So, for example, we could envisage tags like the following made-up examples:

```
    :
<ccpp:if wtai.address="true">
      :
    Encode billed numbers to allow adding to address book
      :
</ccpp:if>
    :
```

When the JSP page runs, the Java class behind the <ccpp:if> tag will use a J2EE XML API like the Java API for XML Processing (JAXP) to search the embedded CC/PP RDF file for the following XML snippet (looking for the <prf:WtaiLibraries> tag):

```
1. <prf:WtaiLibraries>
2.   <rdf:Bag>
3.       <rdf:li>WTA.Public.makeCall</rdf:li>
4.       <rdf:li>WTA.Public.sendDTMF</rdf:li>
5.       <rdf:li>WTA.Public.addPBEntry</rdf:li>
6.   </rdf:Bag>
7. </prf:WtaiLibraries>
```

On finding the <rdf:li>WTA.Public.addPBEntry</rdf:li> entry the class can return true for the condition, and this in turn will cause the mark-up between the tag pair to be included in the output stream.

Provided the necessary semantics are available in the CC/PP RDF file to identify the level of control we are seeking in the output stream (JSP file), then there is no need to resort to resource files. All the information we need about a device's attributes is actually sent to the servlet or JSP via the HTTP header (*Profile* header field). This means that a highly maintainable process of device adaptation could be implemented, which is a useful achievement, especially given the complex sea of device possibilities that is expanding all the time.

Adaptation Using XSL It is possible to transform an XML document into another format using the *XML Stylesheet Language* (XSL). This is a very powerful capability and can be used in a variety of ways to address the current problem of adapting content to a particular device, obviously assuming that the content is described using XML in the first place.

We will introduce the concepts here, but a detailed exposition of XSL is beyond the scope of this book, as XSL is a complex set of programming technologies in its own right. We will focus on the application of XSL to our problem, rather than its internal mechanics, recognizing that there are actually several ways to apply XSL to the adaptation problem.

We have already learnt how XHTML can be delivered to a browser via the powerful J2EE presentation technologies, such as JSP and servlets. The browser can display XHTML because it understands the XHTML tags and can interpret them in terms of the implied layout semantics. For example, the existence of a <table> tag pair in XHTML tells the browser that a portion of screen space is going to be occupied by content in a tabular fashion. The browser's graphical rendering engine will interpret how to transpose the table contents to the screen. In terms of any style information, such as the colour of table cells, fonts and so forth, we already know that a separate file sent to the browser will contain the style information – the CSS.

However, some of our devices may not support XHTML at all or support it, but with limited capabilities in terms of screen size, colour depth and so on. The general proposition in this case, bearing in mind the availability of XSL, is to produce the page content in XML and then transform it to the desired final output form using XSL. From this we can see where XSL gets its name. Just as CSS adds style information to XHTML, we can think of XSL as adding style information to XML. Clearly, the main difference is that XHTML has a defined set of tags and hence a constrained style sheet format (i.e., CSS), whereas the tags we use in our XML page descriptions could be anything; so, we need a distinct style sheet format to match the chosen XML vocabulary. The general principle is shown in Figure 8.40.

The transformation process shown in Figure 8.40 gives us several problems to think about. First, where is the XML coming from in the first place? Second, is the XML file

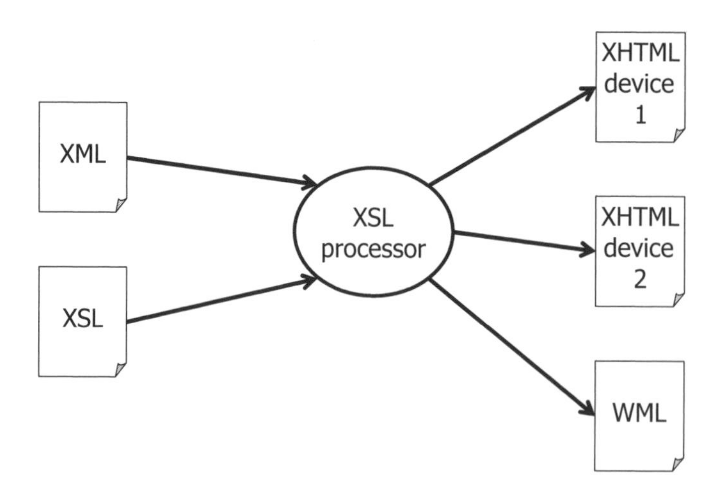

Figure 8.40 XSL transformation process.

purely the content, or does it contain the basis for driving the layout, or is the layout implied or contained in the XSL? Third, what is driving the overall process in terms of selecting which of the final output forms is required? We will examine some possible answers to these questions and keeping in mind that there is no one approach to this solution and that the final approach will depend on a number of factors that should become apparent as we proceed.

In terms of the initial XML input, we are probably better off limiting the XML to describe purely the content without any inference to its intended layout. We do not have to produce XML natively from the database, although some XML database products are making their way into the marketplace. Adding XML tags to the native data can be done using the EJB that is responsible for collecting the data from the database. If we imagine an XML file to describe downloadable games, we might have something like:

```
<?xml version="1.0" encoding="UTF-8">
<gamelist>
<game>
  <title>Space Invaders</title>
  <type>arcade</type>
  <description>Classic space invaders game from the 70s</description>
  <screenshot>images/spa.gif</screenshot>
  <jad-url>jads/spai1.jad</jad-url>
  <price>3.40</price>
</game>
:
</gamelist>
```

We can see from this XML listing that the content is purely descriptive about the games; there is no layout information at all. It is up to the page designer how this content gets laid out, but this needs to be tempered by the layout restrictions of the device. The fact that the XML contains a reference to a screenshot image of the game clearly has display or layout implications. Not all devices will be able to display an image. Those that do may need to display the image differently, depending on such factors as the width of the screen size. For narrow displays the image may need to be included in a vertical flow of information. For wider displays it may be possible to show the image in a horizontal flow, such as to the right-hand side of the description.

What we need now is the means to go from the raw XML content to the final output form. What follows is a simplified explanation so that we can avoid having to go into any details that would require a working knowledge of XSL.

What we can imagine is that the page designer comes up with a set of layout templates to cover a range of devices. These layouts could themselves be described in XML, which we will call *page profiles*. These would be generic 'families' of layouts, similar to our previous example when we had the notion of device categories, such as:

- small devices with XHTML support;
- small devices with WML support;
- large devices with XHTML support.

We can think of these page profiles as being mark-up-independent descriptions of the page layout: for example, we could have a tag called <navigation> to indicate the navigation area for the page or a tag called <header> to indicate a page header area, such as might be used for logo and page titles. Within each file there can be variations on the basic layout according to the finer capabilities of a particular device, such as its ability to support a background image in a table. Information is also added into the profiles to indicate what type of device capabilities are required in order to support the associated layout variations.

Now, what we are working toward is having an XSL file that can be applied to the XML content file in order to produce the desired output. The fascinating realization is that XSL itself is an XML-based language (a language written in XML) and, so, could potentially be used to produce XSL or vice versa. We can take the CC/PP RDF files, which are XML, to identify the capabilities of the device requesting a page. Using the page profiles as a basis for the output, we can transform the RDF files into XSL files that are tuned for the device. This XSL file in turn is what gets applied to the XML content in order to produce the required output for the requesting (target) device.

This may sound a tad complicated and the detailed implementation probably is, but its power is in its flexibility to handle all types of devices directly from the CC/PP semantics, which is ideally what we want. We can think of the process as follows:

1. We process the CC/PP file to produce an XSL file that in its structure will have code that looks for the custom layout elements defined in the page profiles: for example, part of the XSL output will be code that produces a navigation area. To do this the XSL, when processing the page profile in the next step, will be looking for a tag pair called <navigation>. Now, because our XSL has already been created by transforming the CC/PP file, in the case of a device that can support background images this facility will be reflected in the output XSL. If there is any tag element in the page profile that specifies a background image, then this will be picked up by the XSL and propagated through to the output.

2. The XSL that was produced from the first step is now applied to the page profile to transform it to an XSL file that can be finally applied to the content. The XSL at this stage now contains the code necessary to produce the final output, such as the XHTML tags in the case of an XHTML-capable device. These tags are now suitably formatted in terms of, say, any inline style information required for the target device. All that's missing is the content, which comes via the next step.

3. We can think of the final XSL file as the final output file, such as an XHTML file – properly crafted for the target device, but just missing the content; this is now inserted by transforming the content XML into the final output, which really amounts to a merging of the content with the mark-up contained in the XSL template.

The above process sounds a bit convoluted. However, the reason for using XSL in the first place is that it is a powerful language for processing XML, unlike Java or any other general purpose language. XSL has been specifically designed for handling XML and has special constructs and subcomponents to assist with the process. We have not gone into these; for completeness, however, XSL is actually made up of several components, including XSLT which is the transformation language itself and XPath which is another language specifically for referring to elements within an XML file and recognizing the treelike structure of XML

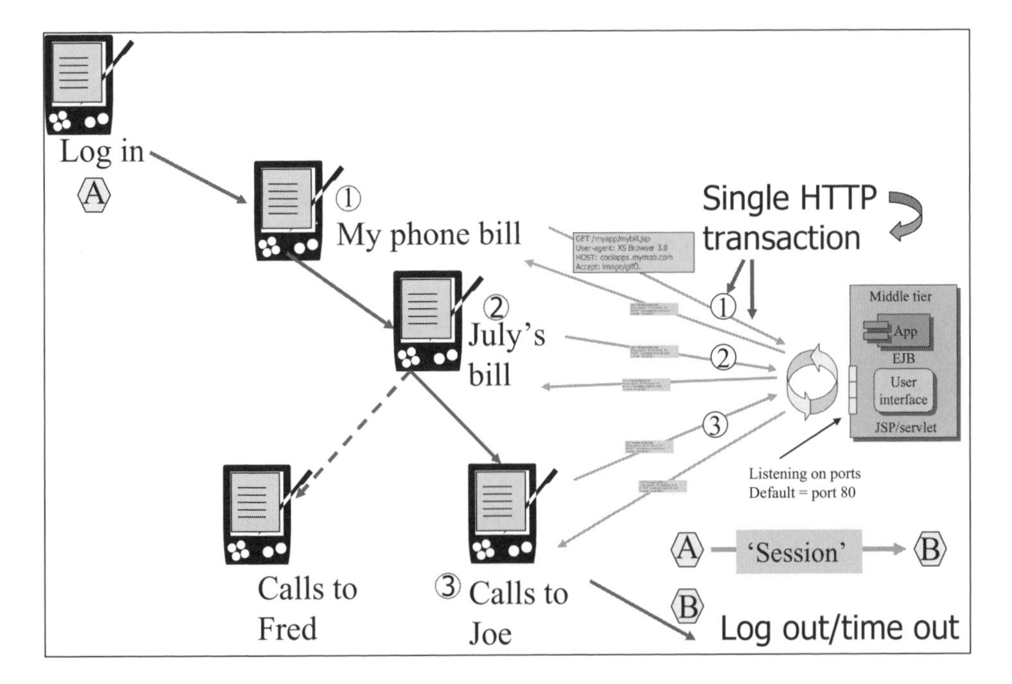

Figure 8.41 Tracking where we are in a Web application is a challenge.

files (see Figure 8.15). The point is that the CC/PP input driving our device capabilities is an XML file. The content is increasingly going to be in XML. Therefore, in an XML-rich process or pipeline the ability to handle the XML using an optimized language built for the job makes sense.

8.5 MANAGING SESSIONS

We have been looking at JSP and servlets for handling and generating HTTP streams. In Section 8.4 we identified a design pattern called MVC that enables a unified approach to browser applications, allowing us to handle requests, call business logic, and then hand off to the appropriate JSP to generate the output stream (the UI).

The previous discussion omitted any detail about how to support state information in our application; this is known as *session management*. This problem is easiest to identify by using an example.

Figure 8.41 illustrates the problem well. Here we have a sequence of page requests within our example billing application. Each page is requested by the user, initially by entering the application's home page and then drilling down (and back up) the application by successive link selection. As we have been discussing, the billing application is most likely handled by one servlet or JSP on the server (ignore the issues relating to MVC patterns for now). A 'session' can be defined as the traversal in time from the start of the application through to the end, either by logging out (or leaving) or timing out. This is shown as the journey from A to B in the figure.

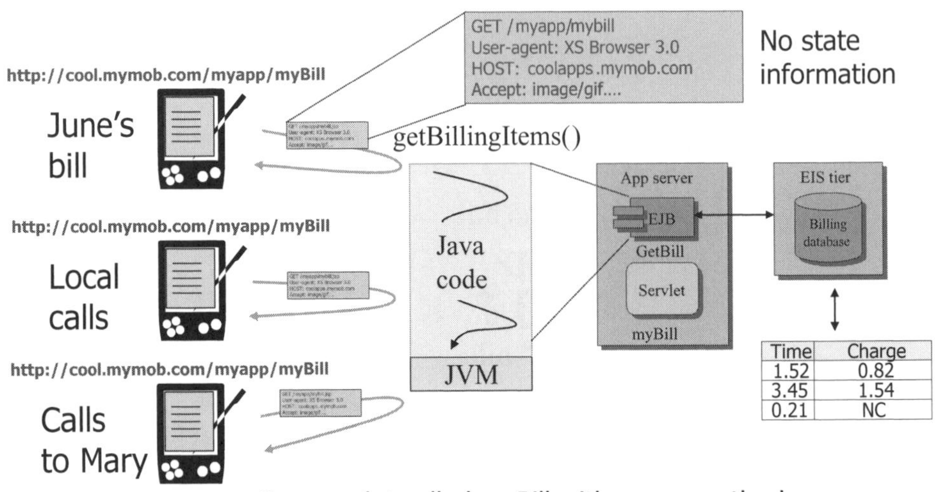

Figure 8.42 Session management.

It may seem an odd reflection, but how do we know where we are in the application? Or, more accurately, how does the servlet know whereabouts we are in the application in order to invoke the correct business logic and redirection to the appropriate JSP? Figure 8.42 perhaps illustrates the point better. Keep in mind that to invoke the servlet our user agent has to issue a GET method on its URI, which we show here to be http://cool.mymob.com/myapp/myBill. This ultimately points to the myBill servlet that will get run by the J2EE Web engine when it receives this request. However, within the application we have several possibilities that the user can enjoy, such as showing billing items according to the month we made the calls (i.e., June in the figure) or the type of calls (i.e., local calls) and even to whom we made them (i.e., to Mary).

Each time, we end up calling the same servlet. For argument's sake, the servlet is also calling the same EJB ('GetBill') and actually accessing the method 'getBillingItems()'. The problem lies in the empty parentheses, as presumably we need to tell the EJB which billing items we are interested in, so that it can construct the appropriate query to run against the billing database using JDBC.

Before we elaborate on how to get these parameters from the user, let's look at the problem in a slightly different way, as shown in Figure 8.43.

In this example, we are showing more or less the same problem, except we are trying to access billing items for the month of June for lots of different users (at the same time, of course). We are not bothered about how many users, as we know that our J2EE platform can handle it, but how do we know, for each servlet request, who is whom? Unless we know who is accessing the servlet, we will not know which billing records to query in the billing database; this is a big problem.

The point being made is that on each occasion, whether it's between users or between different views of the same billing records, the user agent issues the same GET request to the same URI (servlet). Apparently, there is no state information telling us either where we are in the application or who is using it. Clearly, we need a mechanism for managing sessions on a per user basis, otherwise our application will not work.

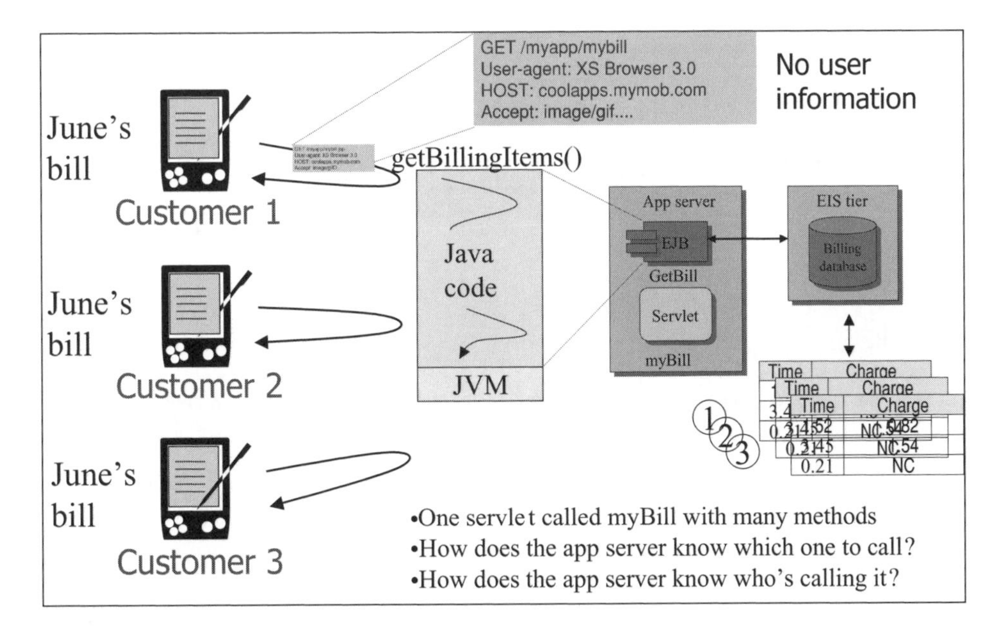

Figure 8.43 Session management between users.

This session management problem was not seen as an issue with the early origins of the Web. As we discussed earlier in the book, the aim of the browser paradigm was to publish documents and allow embedded hyperlinking to other documents. Within this paradigm there is no apparent need for session management. A scientific paper on particle physics can be served up no matter who asks for it, or what they did before, or what they want to do next. In other words, the need to maintain state was not required and not apparent, hence HTTP remained a stateless protocol until the kinds of problems we have just been discussing eventually became apparent, especially with the dynamic (programmatic) generation of output, as opposed to static documents sitting on a Web server hard disk.

8.5.1 Cookies to the Rescue

Fortunately, those clever people at Netscape many moons ago realized that session management was required, so they solved it. Their idea has subsequently become a public standard[30] [RFC 2109].

The solution is elegant and is important to understand if we want to go deeply into mobile application design using the browser paradigm or, possibly, any mobile application that uses HTTP, whether it is browser-based or not (as we can use HTTP from nonbrowser applications, as noted on several occasions in this book).

The cookie is simply a value that gets sent back and forth between the origin server and the user agent on a per-user, per-session, per-application (or path/URI) basis. In other words, we add a label to the incoming requests so that we can uniquely identify them.

The cookie process is shown clearly in Figure 8.44, which shows a response after the user agent has made a request from the server (GET method). Let's assume that this is the

[30] http://www.ietf.org/rfc/rfc2109.txt

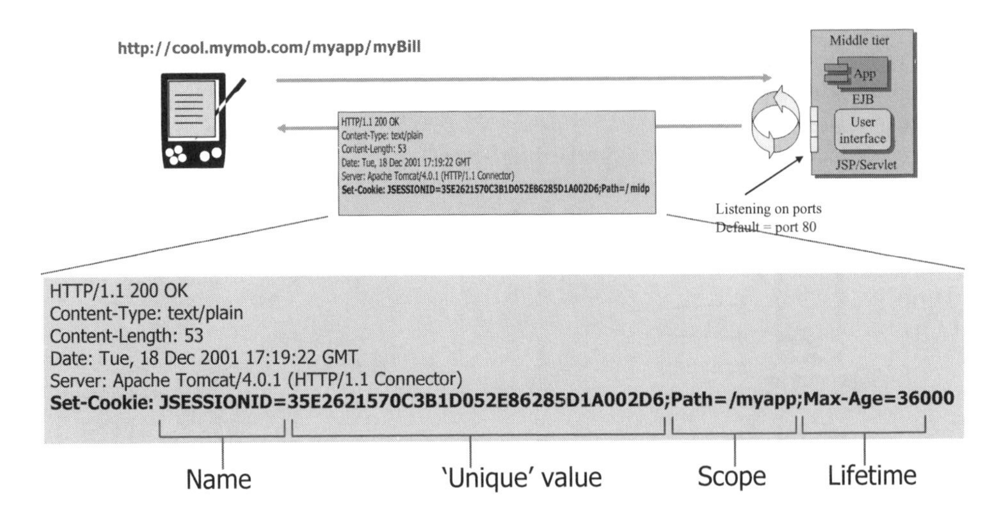
http://cool.mymob.com/myapp/myBill
HTTP/1.1 200 OK
Content-Type: text/plain
Content-Length: 53
Date: Tue, 18 Dec 2001 17:19:22 GMT
Server: Apache Tomcat/4.0.1 (HTTP/1.1 Connector)
Set-Cookie: JSESSIONID=35E2621570C3B1D052E86285D1A002D6;Path=/midp
Middle tier
App
EJB
User interface
JSP/Servlet
Listening on ports
Default = port 80
HTTP/1.1 200 OK
Content-Type: text/plain
Content-Length: 53
Date: Tue, 18 Dec 2001 17:19:22 GMT
Server: Apache Tomcat/4.0.1 (HTTP/1.1 Connector)
Set-Cookie: JSESSIONID=35E2621570C3B1D052E86285D1A002D6;Path=/myapp;Max-Age=36000
Name
'Unique' value
Scope
Lifetime

Figure 8.44 Adding cookies to the HTTP dialogue.

first request in a session and we are using our billing example again. The server, noticing that this is the first request, issues a cookie to the user agent. This is a unique alphanumeric value added to the HTTP response as a header field called *Set-Cookie*.

As the figure shows, cookies have three parts. The first is the cookie name (e.g., 'JDSESSION'); the next is the value of the cookie, which can be any string of characters. Then there is the scope of the cookie, or the path on the origin server which this cookie applies to. Finally, there is the age of the cookie, which is the length of time (in seconds) that it is valid.

What the user agent is required to do is to store the cookie and its parameters in a cookie table (memory). As long as the cookie is still valid (*Max-Age* has not been reached), then the user agent must append the cookie to any requests for resources on the path stipulated in the cookie. By passing a cookie back and forth, the server has a means of uniquely identifying a session, provided that a unique value was used for each unique user (session). On a J2EE server the Web engine creates a session object that is referenced using the same value as the cookie, as shown in Figure 8.45. This is done transparently on behalf of the JSPs and servlets being accessed by users. For example – using our billing application again – for each unique user that accesses the application (which we assume is only accessible via the path '/myapp') the Web engine sets a cookie and creates an associated session object. Whenever a JSP or servlet is subsequently accessed by a user, the relevant session object is made available to the JSP or servlet that is executing on the Web engine. This is done without requiring the JSP or servlet programmer to track the cookie (and its reference) in software; there is a J2EE API available that the programmer can access to set or get values from the session object.

Anything can be stored within the session object, as long as it's a valid Java programming entity; even objects can be stored. In this way we can keep track of a user's state. As shown in Figure 8.45, we could use the session object to store such user information as a username and for our billing example we can store state information about the current filtered view of the bill itself, such as 'month=July'. There really is no limitation on the way we structure information in the session object.

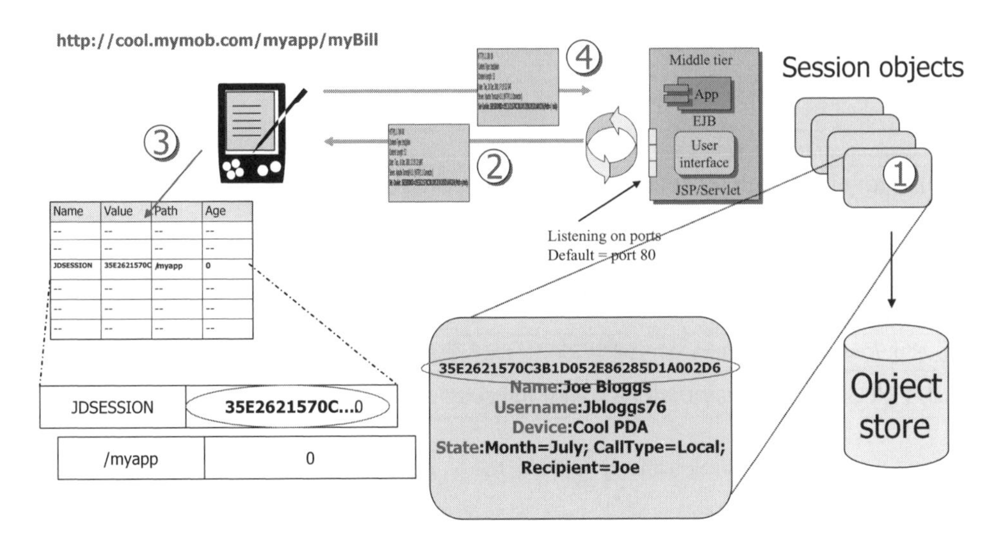

Figure 8.45 Session persistence using cookies.

Supporting Sessions and Parameters Via the URI What remains is to figure out how the user agent indicates to the server exactly what information it requires from a given servlet or JSP. For example, we know that the servlet 'MyBill' will produce a bill listing, but its URL is:

http://cool.mymob.com/myapp/mybill

How do we tell the server that what we actually want is the listing for July? There are two main ways of sending this parameter to the servlet, both of which apply equally to WSP (WAP 1) and W-HTTP (WAP 2):

1. By embedding the parameters within the URL itself and appending them to the servlet or JSP address (URI) as a query, like:

   http://cool.mymob.com/myapp/mybill?parameter1=this¶meter=that

2. By sending the parameters as fields in a form, so they get passed in the body as part of a POST request.

Let's consider the two methods and their implications for mobile application design and the foregoing discussion on cookies.

The first method is very straightforward and widely used. If, for example, we want to tell our MyBill servlet to get us the billing information for July, then we could construct a parameter called 'month' and pass this as part of the URL:

http://cool.mymob.com/myapp/mybill?month=july

This URI causes the user agent to send a GET request to the primary resource, which is located at the address to the left of the question mark. Everything to the right of the question mark is ignored as far as the GET request itself is concerned, except that the JSP and servlet are able to access this part of the string. As we might expect, the J2EE API for servlets and JSPs enables these parameters to be accessible in the code on the page and can be used

http://cool.mymob.com/myapp/mybill?SESSION=ab34sje93nslk3&month=july&calls=local&filter=long

(a) Full URI with query

http://cool.mymob.com/myapp/mybill?SESSION=ab34sje93nslk3&month=july&calls=local&filter=long

(b) Query truncated by limited URI memory on device

http://cool.mymob.com/myapp/mybill?SESSION=ab34sj&month=july&calls=local&filter=long

(c) Full URI with query restored by shortening the session ID

http://cool.mymob.com/myapp/mybill?SESSION=ab34sje93nslk3j&m=7&cl=loc&f=lg

(d) Full URI with query restored, with full session ID, by abbreviating parameters

Figure 8.46 URI-encoding truncation problems on mobile devices.

thereafter however the programmer chooses. For example, we can pull the parameter called 'month' from the request string and ask the J2EE API for its value, which in our example would return 'July'. We can then take this parameter to run the appropriate query on the billing database.

Using URI strings to encode request parameters is also useful for sending session values instead of using cookies. For example, we can just as easily have the following URI:

http://cool.mymob.com/myapp/mybill?SESSION=ab34sje93nslk3&month=july

This has the same effect as using cookies, except that the value does not persist in a cookie memory or via HTTP headers, but does persist via the links in the pages returned by the server. If each link in the page has the session ID encoded into it, then whenever the user selects a link the server will automatically get passed the session ID. Identifying a session is a large part of the session management problem, and we now have an alternative method to using cookies.

Using the URI to encode session information, rather than cookies, has some important implications. First, a user might have cookie support disabled. This might be because they perceive that cookies are evidence of having visited a particular site, which the user may not want to have used against them at a later date.[31] Second, cookies simply might not be supported or only in a limited fashion. This is particularly true of mobile devices. Some devices do not support cookies at all, while others support only a few cookies. It is therefore desirable to either avoid their usage altogether or to initiate a Web application with URI session encoding and then move to cookies if a cookie gets returned from the browser.

There are some further problems with this approach on mobile devices. In particular, many mobile devices have very limited memory capacity, and so very stringent memory-saving techniques tend to get deployed. One of these seen on some devices is the limited amount of memory available to process a URI. If the URI is too long, then it can end up getting truncated by the user agent, as shown in Figure 8.46. This can have catastrophic implications, as seen in (b), perhaps causing the application to stop functioning correctly.

[31] Cookies can be used for more than just session management; they can be used to store anything. This can have surreptitious applications, but this is beyond the scope of this book – see Cookie Central (http:// www.cookiecentral.com/).

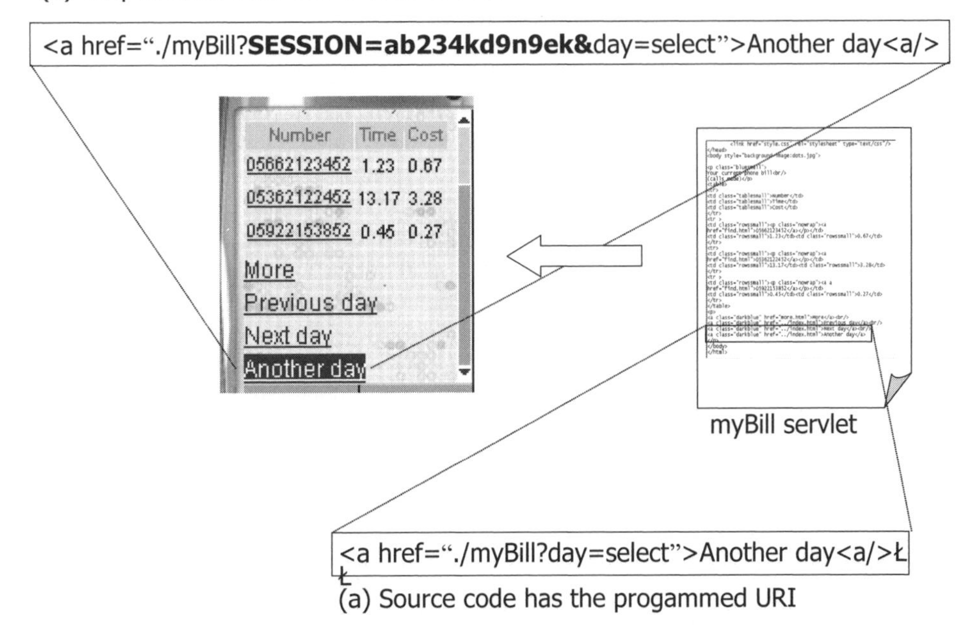

Figure 8.47 URL rewriting by the J2EE server adds the session ID automatically.

We can well imagine that if some parameters get chopped from the request, like 'filter=long' in (b), then the application will not be able to determine correctly what the user is trying to do (i.e., look at long-duration calls).

There are several approaches to fixing this problem. Before looking at these, we should understand that the process of adding session IDs to URI encoding is done automatically by the server itself, not by the programmer, as shown in Figure 8.47. The J2EE Web engine simply rewrites URLs in the output stream to include the required session information.

To mitigate URI truncation issues by mobile devices, the first thing we can try is to shorten the session ID, which is typically a very long string on most J2EE servers. This may recover the situation some of the time, as shown in (c). However, most J2EE servers generate the session ID automatically and its length is not configurable. A notable exception is Weblogic Server, where 'WAP support' is a configurable option for Web applications. If selected, this causes the session ID to be truncated,[32] which is probably not a problem most of the time (unless it becomes ridiculously short, of course).

Shortening the session ID may help us some of the time, but often we find that the rest of the URL dominates the string length anyway, so we may still have problems. One solution is to abbreviate the parameters, as shown in Figure 8.46(d). Abbreviating parameters can make the URL much shorter. In fact, with a set of very sparse abbreviations, combined with a degree of encoding, we can make the parameters very short indeed. However, the problem with

[32] In fact, the shortening of the session ID by Weblogic Server is via the removal of information that is needed to make session IDs work across a redundant cluster of servers; so, it would seem that using this option with WAP devices means that seamless fail-over from a primary to a secondary server is not supported within a mobile browsing session.

this approach is that it can make matters very difficult when it comes to programming, testing and debugging, because the URI strings can become very obscure and make it difficult for the programmer to understand what's going on: good for obfuscating, but bad for maintenance!

An alternative to using the URI to encode values is to embed them in the page requests by appending them to the request body. This entails using the POST method, so that form values can be submitted to the server. This can be done either by explicitly displaying a form or by hiding it and simply using it to convey parameters, whether or not the user knows it. With an explicit form we are limited to using the form interface as the navigation technique. For example, in the case of our billing application we could ask the user to select a month from a dropdown menu on a form and then submit the form. The selection from the drop-down menu would get sent to the server in the POST message. However, often we do not want to use a form for navigation as it can become too cumbersome for the user – as a form presentation is limited to certain widgets, like drop-down menus and selection boxes, which can sometimes be hard to include in a page while maintaining a usable layout.

8.6 MMS AND SMIL

MMS uses its own method of presentation, which can include the use of WML, but more commonly involves the use of an alternative mark-up language called Synchronised Multimedia Instruction Language, which is abbreviated as SMIL and pronounced as 'smile'. Discussion of this technology is deferred until Chapter 13 on location-based services (LBSs), wherein we discuss the use of MMS within the LBS context (although the explanation of MMS technology is general and not specific to LBS.)

Box 8.1 Sample CC/PP RDF file

```
<?xml version="1.0"?>
<RDF xmlns="http://www.w3.org/1999/02/22-rdf-syntax-ns#"
xmlns:rdf="http://www.w3.org/1999/02/22-rdf-syntax-ns#"
xmlns:prf="http://www.wapforum.org/profiles/UAPROF/ccppschema-20010430#">
<rdf:Description ID="MyDeviceProfile">
<prf:component>
<rdf:Description ID="HardwarePlatform">
<rdf:type resource="http://www.wapforum.org/profiles/UAPROF/ccppschema-
20010430#HardwarePlatform"/>
<prf:BluetoothProfile>
<rdf:Bag>
<rdf:li>headset</rdf:li>
<rdf:li>dialup</rdf:li>
<rdf:li>lanaccess</rdf:li>
</rdf:Bag>
</prf:BluetoothProfile>
<prf:ScreenSize>121×87</prf:ScreenSize>
<prf:Model>R999</prf:Model>
```

```
<prf:InputCharSet>
<rdf:Bag>
<rdf:li>ISO-8859-1</rdf:li>
<rdf:li>US-ASCII</rdf:li>
<rdf:li>UTF-8</rdf:li>
<rdf:li>ISO-10646-UCS-2</rdf:li>
</rdf:Bag>
</prf:InputCharSet>
<prf:ScreenSizeChar>15×6</prf:ScreenSizeChar>
<prf:BitsPerPixel>2</prf:BitsPerPixel>
<prf:ColorCapable>No</prf:ColorCapable>
<prf:TextInputCapable>Yes</prf:TextInputCapable>
<prf:ImageCapable>Yes</prf:ImageCapable>
<prf:Keyboard>PhoneKeypad</prf:Keyboard>
<prf:NumberOfSoftKeys>0</prf:NumberOfSoftKeys>
<prf:Vendor>myprofileprovider</prf:Vendor>
<prf:OutputCharSet>
<rdf:Bag>
<rdf:li>ISO-8859-1</rdf:li>
<rdf:li>US-ASCII</rdf:li>
<rdf:li>UTF-8</rdf:li>
<rdf:li>ISO-10646-UCS-2</rdf:li>
</rdf:Bag>
</prf:OutputCharSet>
<prf:SoundOutputCapable>Yes</prf:SoundOutputCapable>
<prf:StandardFontProportional>Yes</prf:StandardFontProportional>
</rdf:Description>
</prf:component>
<prf:component>
<rdf:Description ID="SoftwarePlatform">
<rdf:type resource="http://www.wapforum.org/profiles/UAPROF/ccppschema-
20010430#SoftwarePlatform"/>
<prf:AcceptDownloadableSoftware>No</prf:AcceptDownloadableSoftware>
</rdf:Description>
</prf:component>
<prf:component>
<rdf:Description ID="NetworkCharacteristics">
<rdf:type resource="http://www.wapforum.org/profiles/UAPROF/ccppschema-
20010430#NetworkCharacteristics"/>
<prf:SecuritySupport>
<rdf:Bag>
<rdf:li>WTLS-1</rdf:li>
<rdf:li>WTLS-2</rdf:li>
<rdf:li>WTLS-3</rdf:li>
<rdf:li>signText</rdf:li>
</rdf:Bag>
```

```
</prf:SecuritySupport>
<prf:SupportedBearers>
<rdf:Bag>
<rdf:li>TwoWaySMS</rdf:li>
<rdf:li>CSD</rdf:li>
<rdf:li>GPRS</rdf:li>
</rdf:Bag>
</prf:SupportedBearers>
<prf:SupportedBluetoothVersion>1.1</prf:SupportedBluetoothVersion>
</rdf:Description>
</prf:component>
<prf:component>
<rdf:Description ID="BrowserUA">
<rdf:type resource="http://www.wapforum.org/profiles/UAPROF/ccppschema-
20010430#BrowserUA"/>
<prf:BrowserName>Ericsson</prf:BrowserName>
<prf:CcppAccept>
<rdf:Bag>
<rdf:li>application/vnd.wap.wmlc</rdf:li>
<rdf:li>application/vnd.wap.wbxml</rdf:li>
<rdf:li>application/vnd.wap.wmlscriptc</rdf:li>
<rdf:li>application/vnd.wap.multipart.mixed</rdf:li>
<rdf:li>application/vnd.wap.multipart.form-data</rdf:li>
<rdf:li>text/vnd.wap.wml</rdf:li>
<rdf:li>text/vnd.wap.wmlscript</rdf:li>
<rdf:li>text/x-vCard</rdf:li>
<rdf:li>text/x-vCalendar</rdf:li>
<rdf:li>text/x-vMel</rdf:li>
<rdf:li>text/x-eMelody</rdf:li>
<rdf:li>image/vnd.wap.wbmp</rdf:li>
<rdf:li>image/gif</rdf:li>
</rdf:Bag>
</prf:CcppAccept>
<prf:CcppAccept-Charset>
<rdf:Bag>
<rdf:li>US-ASCII</rdf:li>
<rdf:li>ISO-8859-1</rdf:li>
<rdf:li>UTF-8</rdf:li>
<rdf:li>ISO-10646-UCS-2</rdf:li>
</rdf:Bag>
</prf:CcppAccept-Charset>
<prf:CcppAccept-Encoding>
<rdf:Bag>
<rdf:li>base64</rdf:li>
</rdf:Bag>
</prf:CcppAccept-Encoding>
```

```
<prf:FramesCapable>No</prf:FramesCapable>
<prf:TablesCapable>Yes</prf:TablesCapable>
</rdf:Description>
</prf:component>
<prf:component>
<rdf:Description ID="WapCharacteristics">
<rdf:type resource="http://www.wapforum.org/profiles/UAPROF/ccppschema-
20010430#WapCharacteristics"/>
<prf:WapDeviceClass>C</prf:WapDeviceClass>
<prf:WapVersion>2.0</prf:WapVersion>
<prf:WmlVersion>
<rdf:Bag>
<rdf:li>2.0</rdf:li>
</rdf:Bag>
</prf:WmlVersion>
<prf:WmlDeckSize>3000</prf:WmlDeckSize>
<prf:WmlScriptVersion>
<rdf:Bag>
<rdf:li>1.2.1</rdf:li>
</rdf:Bag>
</prf:WmlScriptVersion>
<prf:WmlScriptLibraries>
<rdf:Bag>
<rdf:li>Lang</rdf:li>
<rdf:li>Float</rdf:li>
<rdf:li>String</rdf:li>
<rdf:li>URL</rdf:li>
<rdf:li>WMLBrowser</rdf:li>
<rdf:li>Dialogs</rdf:li>
</rdf:Bag>
</prf:WmlScriptLibraries>
<prf:WtaiLibraries>
<rdf:Bag>
<rdf:li>WTA.Public.makeCall</rdf:li>
<rdf:li>WTA.Public.sendDTMF</rdf:li>
<rdf:li>WTA.Public.addPBEntry</rdf:li>
</rdf:Bag>
</prf:WtaiLibraries>
</rdf:Description>
</prf:component>
<prf:component>
<rdf:Description ID="PushCharacteristics">
<rdf:type resource="http://www.wapforum.org/profiles/UAPROF/ccppschema-
20010430#PushCharacteristics"/>
<prf:Push-Accept>
<rdf:Bag>
```

```
<rdf:li>application/wml+xml</rdf:li>
<rdf:li>text/html</rdf:li>
</rdf:Bag>
</prf:Push-Accept>
<prf:Push-Accept-Encoding>
<rdf:Bag>
<rdf:li>base64</rdf:li>
<rdf:li>quoted-printable</rdf:li>
</rdf:Bag>
</prf:Push-Accept-Encoding>
<prf:Push-Accept-AppID>
<rdf:Bag>
<rdf:li>x-wap-application:wml.ua</rdf:li>
<rdf:li>*</rdf:li>
</rdf:Bag>
</prf:Push-Accept-AppID>
<prf:Push-MsgSize>1400</prf:Push-MsgSize>
</rdf:Description>
</prf:component>
</rdf:Description>
</RDF>
```

9

Using J2EE for Mobile Services

In the previous chapters we introduced the Java 2 Enterprise Edition (J2EE) as a platform for building mobile applications and then went on to establish the dominant communications paradigms for establishing interaction between our mobile device and the J2EE platform: namely, HTTP (HyperText Transfer Protocol) and its wireless companions in the WAP (Wireless Application Protocol) family. We now want to look at how J2EE works in more detail and to understand exactly how its powerful infrastructure software services support our proposed communications paradigms. We will proceed to look at devices in some detail in the following two chapters.

In this chapter we will also look at how we make sure that the platform can be accessed securely, both in terms of implementing security within the J2EE environment and then ensuring that we can extend security policies to the devices and, ultimately, the users. It would be no use implementing a fancy means of making sure that only authorized users access a particular mobile service on the server if we then couldn't enforce this via the device, so we need to examine security within the entire mobile network context, revisiting our mobile network model.

In building our platform we will look at how harnessing the power of J2EE to support mobile devices is not enough; we also need to be able to support access to our mobile services platform by other participants who want to offer their services to the end users while taking advantage of any common infrastructure built on the J2EE platform. We will take a look at ways in which this might be achieved and how such an arrangement would fit with an integrated services model, able to cope with all the key challenges of providing services that we have elucidated so far throughout the book. By 'integrated' we mean that, irrespectively of whether an application is hosted locally by the operator or remotely via a third party, the software infrastructure services (e.g., scalability, security, availability, adaptation, etc.) still function and do so in a consistent manner across all services.

Next Generation Wireless Applications P. Golding
 ISBN: 0-470-86986-0 (HB)

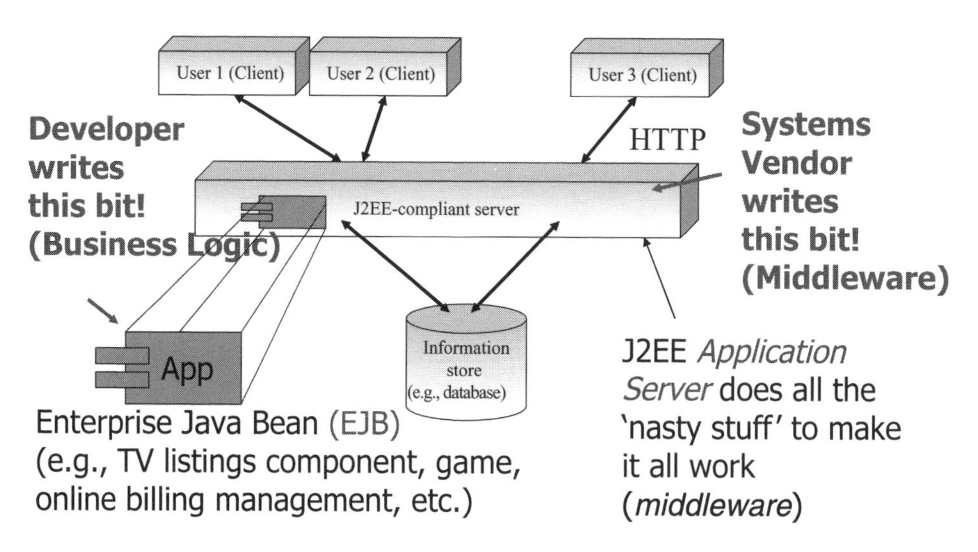

Figure 9.1 J2EE approach.

9.1 TECHNOLOGIES UNDERPINNING J2EE

As Figure 9.1 reminds us, J2EE is a platform that divides the gamut of software needed to implement a scalable, robust solution into two parts. The first is infrastructure services that are likely to be commonly required for all applications. We call this infrastructure 'middleware'; it provides a host of powerful functions like distributed processing, transaction management, security management, multithreading and so on.

The other half of our software is the custom application itself, which we implement in units called Enterprise Java Beans (EJBs) where we construct the value-added business services that make the overall application do something useful. With J2EE we have an approach that enables us to focus on 'our' bit while the application server does the rest (middleware). When introducing J2EE, we also learnt how the J2EE platform comes with a wide range of powerful system interfaces, such as the capability to connect with industrial strength database servers or mail servers. There is no need for us to implement any of these services. We utilize these services in our application by accessing the platform services via application programming interfaces (APIs). These interfaces are common to all J2EE implementations. Similarly, the low-level interfaces to enable the infrastructure services to function on behalf of our application are also common to all J2EE applications. This commonality is thanks to the J2EE specification being an open specification, accessible to anyone who would like to implement a J2EE platform, whether as a free-of-charge open source initiative (e.g., Tomcat) or as part of an expensive enterprise solution portfolio (e.g., BEA's Weblogic Server and its companion products).

9.1.1 Containers – The J2EE 'Glue'

It is obvious that there needs to be some connecting mechanism between our EJB and the infrastructure services. If in actuality the EJB really only contains our application software and nothing else, then the infrastructure services will have a hard time enacting the middleware function because there is no relationship between it and the EJB. If we think about an example, then perhaps we can understand why this won't work.

Let's say our EJB provides a look-up function that provides a list of URIs (uniform resource identifiers) where our mobile device can download ring tones. The bean will take an input value as a search parameter and then look for ring tones that match this topic, such as 'James Bonds[1]'. Probably, our ring tones are already organized as part of a content management system (CMS) for mobile content, which our bean is part of. Each ring tone has a description associated with it to identify its theme and title. More than likely these descriptions will be in descriptors that are structured as XML (eXtensible Markup Language) documents and probably stored in a database underlying the CMS. To access the database, we can use a J2EE API, which is a block of software already written to take care of the low-level interface tasks associated with connecting to a database server. This piece of software is called the Java DataBase Connector, or JDBC. We could imagine a command (method[2]) like 'gettones("James Bond")' that will return a record set from the database containing all the descriptors for ring tones that match the *James Bond* theme. We probably want to extract the exact titles of the matching tones in order to send them back to our user who made the request. We could use the J2EE API for processing XML (JAXP, or Java API for XML Processing) to extract the content of the XML tags called <Title>. So far so good – all of the difficult code to process database requests and XML parsing is provided for us by the J2EE platform in the JDBC and JAXP programs.

Now, what happens if another user wants to do the same thing at the same time, but for a different theme, like 'Hip Hop'. Our bean is still running on the server to handle the first request: this is problem No. 1, because we want the bean to process the second request. Now, let's make things a bit more challenging and imagine that actually we have 10,000 concurrent requests (this is a popular site for ring tones) and that the physical server (computer) we are using runs out of steam at 8,000 requests; this is problem No. 2 (scalability), which needs to be solved after we have solved No. 1, of course (concurrency). We already know that J2EE has the capability to solve these problems – that's what it was designed for. But we are trying to understand conceptually how that process might work.

In our EJB all we have written is the code to take the input topic, make the query and gather a collection of titles from the XML returned from the database. In other words, we have stuck to our 'business' and only written code to carry out the tasks we want it to do (hence 'business logic'). We know that computers run exceptionally fast and that via their operating systems they can appear to run many tasks simultaneously by swapping processes on and off at the central processor (e.g., Pentium). However, our EJB is not an operating system-level task. It is a piece of software that runs within the J2EE run-time environment.[3] Creating EJBs and setting them running is the job of the J2EE application server. Furthermore, the J2EE platform can arrange for EJBs to be created[4] across a bank of servers, so that the processing load can be distributed. In some cases this will involve the EJB having to talk to another EJB on a different server. This is a mechanism we will look at in some detail in just a minute.

[1] Ian Fleming's 007 secret service agent.

[2] In Java, software commands or functions (what used to be called subroutines) are called methods.

[3] We discuss what a Java run-time environment (JRE) is in more depth when we look at Java running on devices in Chapter 10.

[4] An EJB is just a program. By creating an EJB we bring the program to life, so that it is up and running on the system, ready to do its stuff, not just sitting on a disk.

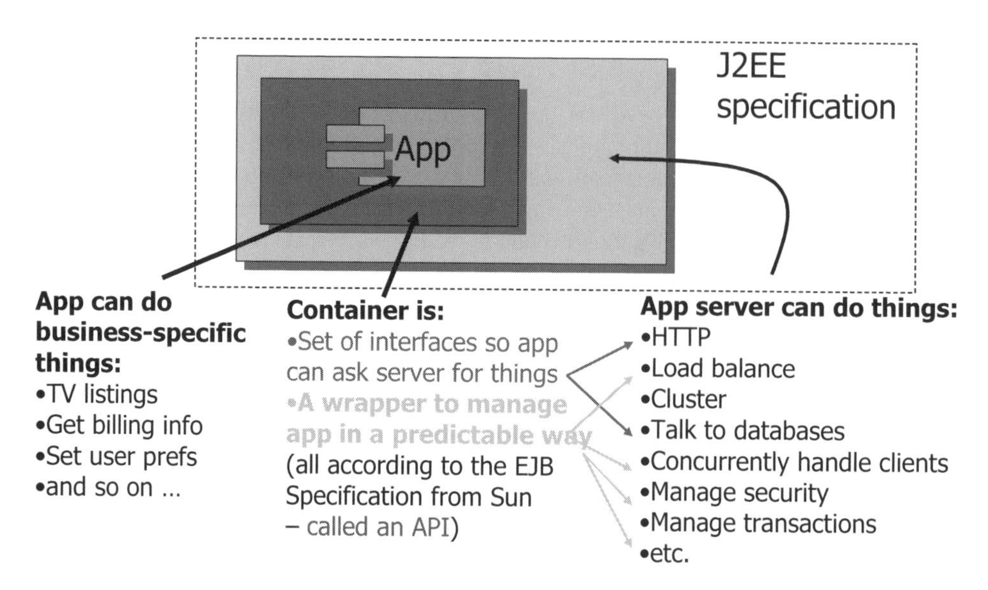
J2EE
specification
App
App can do business-specific things:
•TV listings
•Get billing info
•Set user prefs
•and so on ...
Container is:
•Set of interfaces so app can ask server for things
•A wrapper to manage app in a predictable way
(all according to the EJB Specification from Sun – called an API)
App server can do things:
•HTTP
•Load balance
•Cluster
•Talk to databases
•Concurrently handle clients
•Manage security
•Manage transactions
•etc.

Figure 9.2 EJB sits inside a wrapper.

Because we have not written any code to help our EJB be created or to enable it to talk to other EJBs on remote servers, we need to make sure this code is available. Some of the code is system code that is part of the J2EE platform, but some of it actually needs to be bolted on to our program like a wrapper, so that the EJB is able to interact with the platform in a predictable manner and carry out these system tasks on our behalf, usually in concert with the application server's own processes.

The wrapper function is shown in Figure 9.2. We can see our EJB sitting in the middle; it contains, as explicitly programmed by the programmer, the business logic – the actual code we want to run to carry out the task required of this EJB, which in our example provides the ring tone look-up function. The EJB sits inside a wrapper, which is housekeeping code that runs with our business logic as part of the EJB software process running within the J2EE platform. The wrapper provides the critical link between the EJB and the application server. At one level, it enables the EJB to access various interfaces, but another critical function is to enable the low-level infrastructure mechanics to operate, such as concurrency, program life cycle management (e.g., creation, suspension, destruction), transaction monitoring and distributed processing (inter-EJB communication across a network).

The wrapper function, while representative of how EJBs work, is not a function that is rigidly defined in the J2EE specification in terms of its physical representation on the server. Much of its implementation detail is left to the platform vendor to decide how best to put into action. However, we shouldn't forget that the wrapper function is enabling a component model to be achieved whereby we can program to a certain specification and then rely on being able to plug our EJB into a J2EE platform and expect it to work. In other words, the wrapper, while proprietary in implementation detail, is still used to realize an open interface, so at the black box level it is consistent with what a J2EE server expects.

In Section 9.1.2 we look at one of the critical functions of the wrapper code and a key underpinning technology for distributed processing within a J2EE environment (i.e., many J2EE application servers running in concert).

9.1.2 RMI – The EJB 'Glue'

One of the key design goals of the J2EE platform was to provide robust support for distributed processing; this is the ability to run an application across many servers. There are potentially several reasons why this is useful, perhaps foremost of which is the ability to achieve a scalable architecture. This is essential if we want to build a mobile services platform that can handle as many users as we want using a simple and consistent programming approach. We do not want a system that can only work effectively on one server, perhaps handling thousands of users, only to find that we have to throw it away and implement a new system to handle tens or hundreds of thousands of users across several servers. We would like to be able to take a single design approach to all these cases, an approach that simply scales when we need it to. This is a function that we would like the infrastructure services to support, not the EJB programmer.

Fortunately, this ambition has been fulfilled in the J2EE platform architecture. One of the key underpinning mechanisms for enabling this is something called *remote method invocation*, or RMI. The fundamental building block for Java programming is something called a *method*; this is a unit of Java code that does a specific task. Associated tasks, common to one functional objective within our software, may be collected together into a collection of methods encapsulated into a class. For example, we could have a class called 'MyBill' that represents a user's mobile phone bill. It is common for classes to model real world artefacts in this manner.

Within the class, we might have methods like 'UpgradeAccount', 'ChangeMyDetails', 'FetchCalls' and so on; all performing functions that are operations we are likely to want to carry out on our bill. These may not represent one-for-one correlations with tasks the end user has in mind, like 'Pay my bill please', but will more likely represent tasks that the programmer perceives will enable appropriate end user services to be realized, like paying the bill. We could argue that in a well-designed system a lot of these methods and end user tasks might well correlate directly, but that is not our concern here. A class is like a template – it is a definition of software behaviour achieved by specifying methods and then stating how they actually behave, which is done using the Java code that resides within the methods themselves. When a class is loaded, initialized and executed, it becomes an object. There could be many objects alive at any one time from one class. We could imagine lots of users accessing the bills, and for each user we have an object dedicated to the task.

An EJB is a particular instance of a class. It is a special type of class which implies that along with the user-defined class comes all of the wrapper code to enable J2EE infrastructure services to be accessed. The methods in a bean can be called from another bean or, as we discussed in Chapter 8, from servlets and Java Server Pages (JSPs). The RMI mechanism is a means to do this between servers, as shown in Figure 9.3, which shows a servlet (or JSP, or another EJB) on one physical server (server 1) apparently calling the method from a bean on the same physical server. This co-located software collaboration is the normal process in Java. However, what is actually happening is that somehow the server is conspiring with another server where the target EJB ('MyBill') actually resides (server 2). As far as the servlet is concerned, the bean is on the same machine. The servlet programmer does not have to implement any software to go find the EJB elsewhere on the network and then figure out how to talk to it over the network: this is done thanks to RMI, an underpinning mechanism used by J2EE.

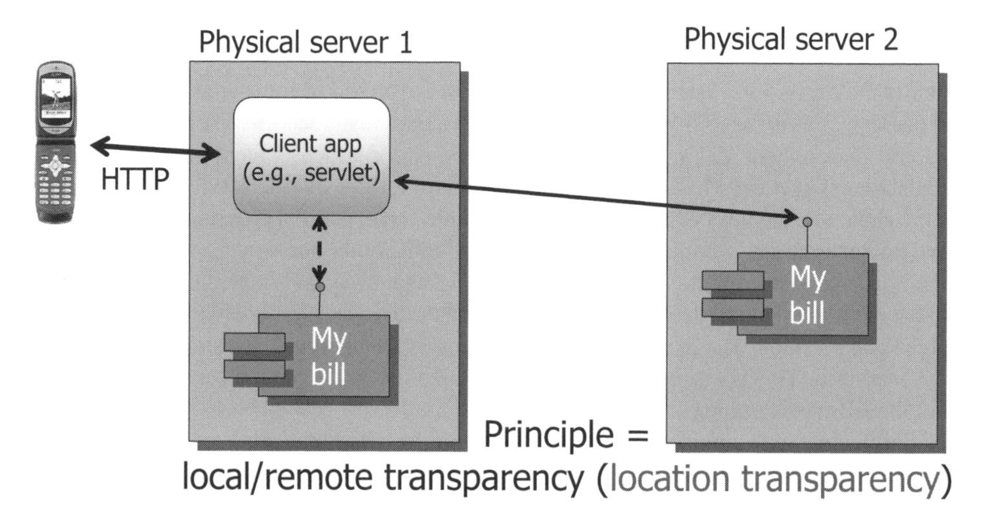

Figure 9.3 RMI.

The mechanism of RMI enables what we call *location transparency* to be achieved. This is where the location of the software object being referenced does not actually matter nor is it known in advance by the referring object. Location transparency is built into the J2EE platform by default. In fact, this technique is not new and is not unique to the J2EE platform. An earlier attempt to create an open standard for managing objects in this way had already been undertaken and is called CORBA, which stands for Component Object Request Broker Architecture. Some vendors had already implemented systems using this standard, although J2EE has in many ways supplanted CORBA due to wider acceptance in the software and IT industry. To achieve compatibility with CORBA, especially under pressure from some vendors who participated in the J2EE specification process, the RMI mechanism is compatible with parts of CORBA. The full name for the mechanism is actually RMI-IIOP, where IIOP stands for Internet Inter-ORB Protocol, which is an Internet Protocol (IP)-based protocol to talk to calls methods on any system that implements CORBA.

9.1.3 Stubs and Skeletons – The Inner Workings of RMI

Having introduced such strange sounding names, like Inter-ORB, the inner workings of RMI introduces us to yet more interesting names, this time called *stubs* and *skeletons*. While this sounds like a horror movie, it is actually something quite elegant and magical from the wonderful world of J2EE.

As Figure 9.4 shows, the reality of location transparency is that our software receives help from a pair of software mechanisms that act as interlocutors for our two Java programs that wish to collaborate in a distributed manner. The idea is rather simple, and a determined programmer could probably implement a rudimentary system from scratch,[5] although this is not necessary thanks to J2EE.

[5] For an illuminating example of how this might be done see Monson-Haefel, R., *Enterprise Java Beans*. O'Reilly (2001).

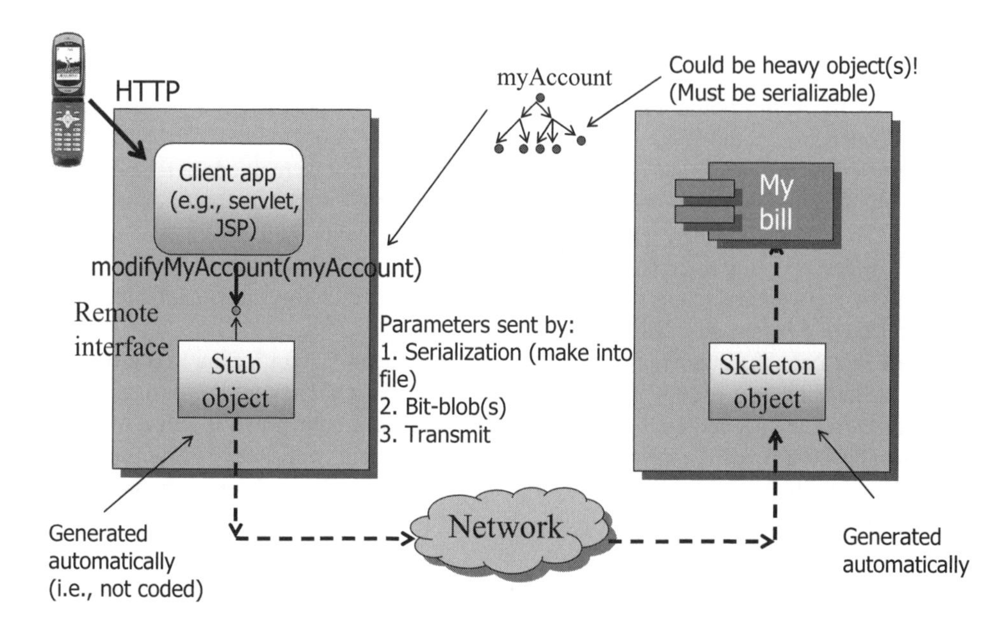

Figure 9.4 Stubs and skeletons collude.

Let's imagine again that we have a servlet that wishes to call an EJB, but, unknown to our servlet (and the programmer), the target EJB resides on a machine somewhere else on the network or possibly on lots of machines on the network – we will focus on just one. For our servlet to think it is communicating with the EJB locally, as per our original claim for location transparency, clearly there has to be an entity on the same machine as the servlet which is 'standing in' for the EJB; this is what the *stub object* does. It pretends to be the EJB 'MyBill' as far as the servlet is concerned, so that the servlet is able to pass its requests for accessing methods, as usual (the servlet doesn't really know that anything different from usual is happening).

The stub object rather cleverly 'mimics' the interface of the EJB and can happily accept the appropriate method calls on its behalf. However, the stub is just a shell – it doesn't actually have the inner workings of any of these methods. If it did, then the whole exercise would be pointless, of course, because effectively it then becomes the EJB anyway.

The stub object knows where to find a real 'MyBill' EJB on the network and uses a network protocol to route the method request from the servlet over to the EJB on the target machine. For now we will ignore the issue of how the stub 'knows' the methods of the EJB in order to successfully mimic them, and we also ignore how it knew where to find an actual implementation of the EJB in order to route the requests to the appropriate remote server.

In keeping with the J2EE ethos, it is not a requirement that the EJB explicitly program any special communications handlers to process the requests from a stub object. Remember that we don't require our EJB programmer to do anything special in order to benefit from the infrastructure services of the J2EE platform. The EJB has the user-defined methods implemented by its programmer, and these can be called in the usual way via a co-located piece of software, which is a mechanism that is inherent in the way the Java programming language works. To complete the process that our stub object initiated, we have a mirrorlike object on the target server called a *skeleton object*, courtesy of the J2EE server (i.e., not programmed by the EJB programmer). This accepts from the network the incoming method requests from the stub and then passes those method calls from the servlet by proxy on to the actual EJB. At this point, it is as if the stub and skeleton do not exist and the servlet has a direct connection to the EJB.

Once the EJB has received a networked request to run one of its methods, it takes its turn in execution on the server's processor and carries out its inner workings for the appropriate method. The results are then ready to be returned to the caller, and this is done using a reverse process, passing them back to the skeleton who contacts the stub who passes them up to the servlet which remains none the wiser about where and how the job got done. This is the magic of RMI on the J2EE platform!

There are some characteristics of the RMI process that are useful for the EJB programmer to understand as a user of the technology. One of these relates to the passing of any parameters that are needed by the EJB to carry out the required method. Because this process is happening by proxy, we need to send the parameters over the wire from the stub to the skeleton. If these parameters are large in size, then this can become a performance bottleneck, particularly if there is a high volume of RMI requests between lots of colluding objects, remembering that the mobile context of this process probably results in large numbers of clients generating plenty of requests. In some cases, parameters can be software objects in themselves, which can reference yet further objects. Moving objects around poses several challenges. The technique employed is a process called *serialization*, which involves converting an object to a self-contained flat file representation that will pass over a communications pathway. Not all objects will serialize, but for those that do we have to keep in mind that choices made in the design in our software solution may have unforeseen consequences on performance when running on a distributed platform.

An interesting observation about the stub–skeleton collusion is that it is an entirely automatic process. The code to produce the stubs and skeletons is generated by the J2EE platform. The actual details of their construction and the low-level integration of the infrastructure services with our own user-defined (EJB) code is a matter of interpretation for the J2EE vendors. This is an issue beyond the scope of this book. For now, we should be happy that we have a means to build scalable mobile service delivery platforms based on a reliable and robust software infrastructure, like J2EE.

Before we move on from this topic to look at other facets of the J2EE platform in relation to mobile service platforms, let's briefly address some of the issues that we glossed over in our cursory glance at RMI. First, when the stub object got involved in the process, it already seemed to know something about the EJB it was masquerading as, in order to mimic its interface. This would be exceptionally clever if that were possible without some prior reference to it (real magic). In any particular implementation of a distributed system using J2EE the collaborating servers are not acting aloof of each other. When a J2EE application is installed, it is effectively installed 'onto the platform', which means on all of the collaborating machines at once, not necessarily a particular machine. Installation processes are complex: in effect, we can think of each machine as receiving at least the templates[6] for all the EJBs, even if they don't intend to run them – this is the minimum required for our servlets and JSPs to gain meaningful references into the EJBs.

[6] In Java parlance these templates are called *interfaces*, but that is not really important to grasp to understand the general principles.

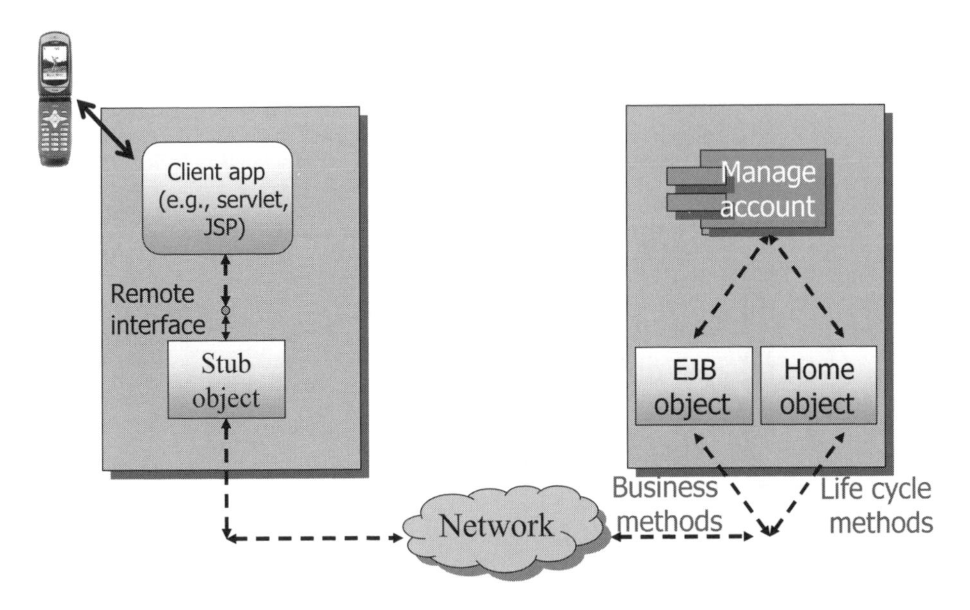

Figure 9.5 Skeleton a little more complicated than we thought.

Another challenge is the invocation of an EJB. Just as with a desktop application, such as a word processor (even though it is installed on a particular PC), that doesn't mean to say it is running and ready to do work – we have to start it; the same applies to EJBs. Furthermore, if, as we have portrayed matters, we need a skeleton hanging around for each EJB that we are running, we also need to start these, not forgetting that in a large multi-user system we are talking about large numbers of EJBs and their interlocutors running at the same time.[7] However, we can't sensibly have a system where all of our EJBs have been started, along with their proxy skeletons, for as many users as we are likely to encounter. This would take up resource irrespective of the actual demand and dynamics of the working system, which doesn't make sense.

To get around these problems, we have *life cycle methods* for an EJB that enable us to start them up (create) and shut them down (suspend or destroy). A special object exists to carry out life cycle processes called a *home object*, as shown in Figure 9.5. Each EJB has at least one home object on the server where the EJB is going to run. We can assume that the home object is always running and available to receive requests, unlike the skeleton object or its associated EJB. All we have to do is first of all locate the home object and then ask it to create an EJB for us. We can think of the home object as having a locatable address on the network where our servlet can easily find it. This is the only time that the servlet programmer has to be aware of what is actually going on in the J2EE distributed environment. A small piece of code needs to be included at the start of the servlet code and then gets called to find the home object by its name, something like 'findHomeObject("MyBill")'. This is a slightly simplified explanation, but fairly accurately represents the process.

[7] A technique called *instance pooling* can be used to avoid having one EJB and skeleton for each user, but that is an advanced topic.

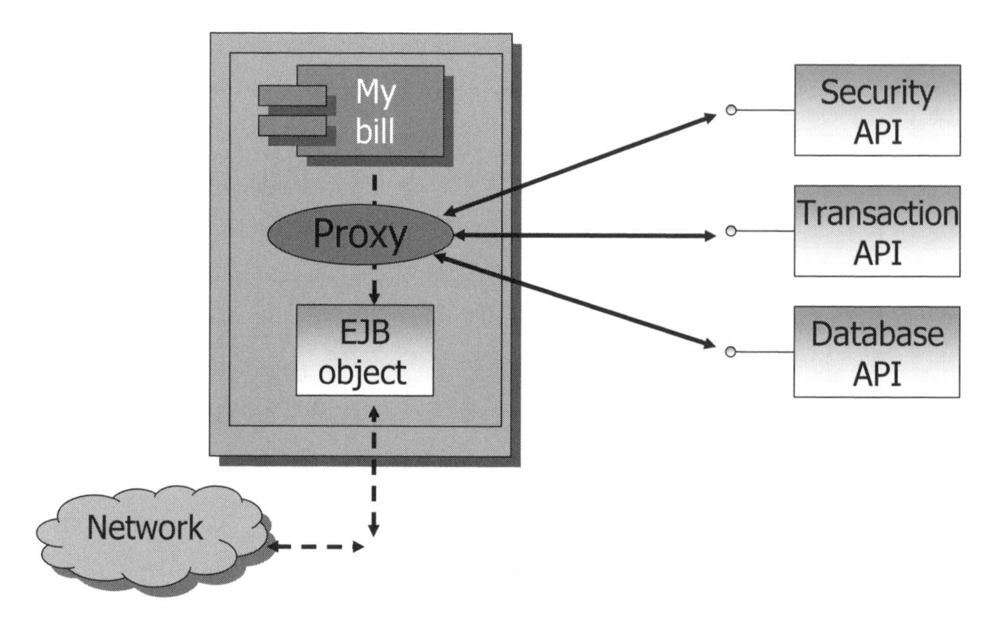

Figure 9.6 EJB object as a proxy.

Ignoring the details of the software process executed to find the home object – once the servlet has found it, requests can be made for the home object to create an EJB and its skeleton. The home object then tells the servlet the address of the skeleton so that its stub knows where to look for it. What actually gets created on the target server is not officially called a skeleton (even though it has that function), rather it is called the *EJB object*. The reason it is so called is that it actually is the object we end up talking to once things are under way. We never ever get to talk to the actual EJB directly.[8]

The fact that we do not talk to the EJB object directly actually offers the opportunity to implement a neat mechanism by which the J2EE server can now do all of the 'nasty middleware stuff' that we so desperately didn't want to program ourselves in our software, leaving it to the J2EE platform.

As shown in Figure 9.6, because the EJB object intercedes between the servlet (or any calling program in general) and the EJB being called, this gives the EJB object the chance to carry out infrastructure services for us, such as checking the security settings of the bean (i.e., are we allowed to use it?), restoring the state of the bean from a previous use (i.e., persisting state between uses) and, if necessary, keeping track of all our activities as part of a transaction sequence that must be monitored. The alternative is that we could have programmed all this ourselves in the EJB, but that is getting away from the ethos of J2EE. Moreover, it is not just a case of programming these proxy functions, we still need all the infrastructure services to go with it, like the security management system, transaction monitor and so on. These are all built in to the J2EE platform.

[8] To an extent, this is all semantics. The EJB object and the EJB may be hard to distinguish in reality depending on how the J2EE server has been implemented, but the abstraction is a useful way to specify J2EE and to understand it. Implementation of any technical mechanism can always deviate from the model by which we designed it, usually by virtue of optimizations that make sense when building it.

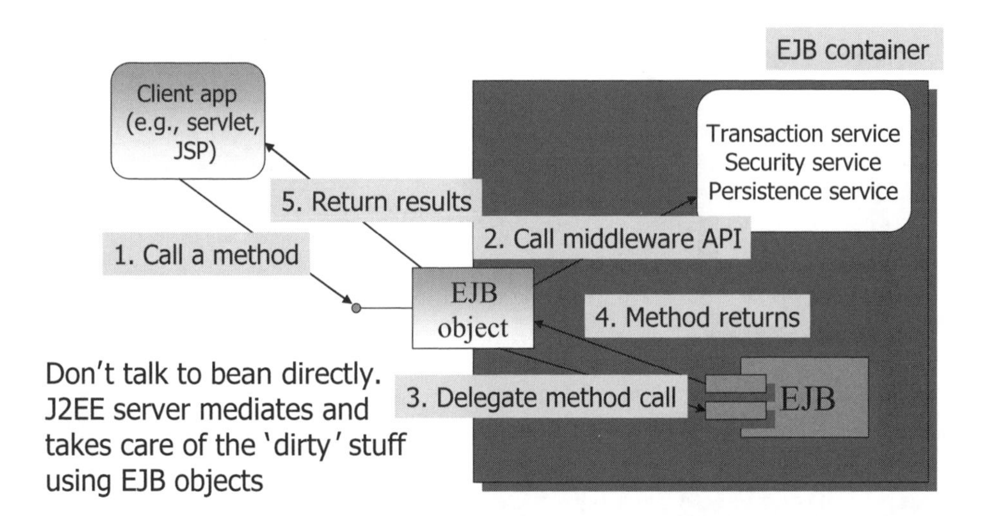

Figure 9.7 Proxy function of an EJB object as part of the process flow.

In Figure 9.7 we can see the proxy process in a bit more detail. In step 1 our servlet makes the method call to our EJB, which it does by referencing the local stub object. This reference was achieved during the 'start-up' process when the servlet first located the home object and invoked the EJB and its EJB object. Part of that process involves the J2EE platform passing a reference to the servlet that it can use as if the referred object were the EJB, whereas it is actually the local stub – which we are not showing in the diagram for the sake of simplicity. The EJB object, acting as our skeleton and proxy, receives the method call. It then accesses the infrastructure services for the EJB on our behalf. This could involve checking security permissions on the bean, monitoring transactions and many other housekeeping functions. The key point is that the EJB object is doing this for us and we don't have to program. After infrastructure services have been satisfactorily completed, only then does the method call get passed to the EJB itself, as in step 3. The method, if successful, returns the results (step 4) and these get marshalled back to the servlet (step 5) using the reverse process to step 1.

9.2 MANAGING SECURITY

We have looked at some of the fundamentals of how the J2EE platform functions. In Chapter 8 we looked at the mechanisms for connecting mobile devices making use of HTTP, using JSPs or servlets to handle the inbound HTTP streams and to generate the outbound ones.

Returning to our goal of using J2EE technology to build a mobile services platform, we need to consider some of the implications of handling lots of users and lots of applications within a commercial context. We can envisage that different users will have different applications requirements and will therefore only subscribe to the ones of interest. It soon becomes apparent that we need a method for securing our applications. Some of our requirements might include:

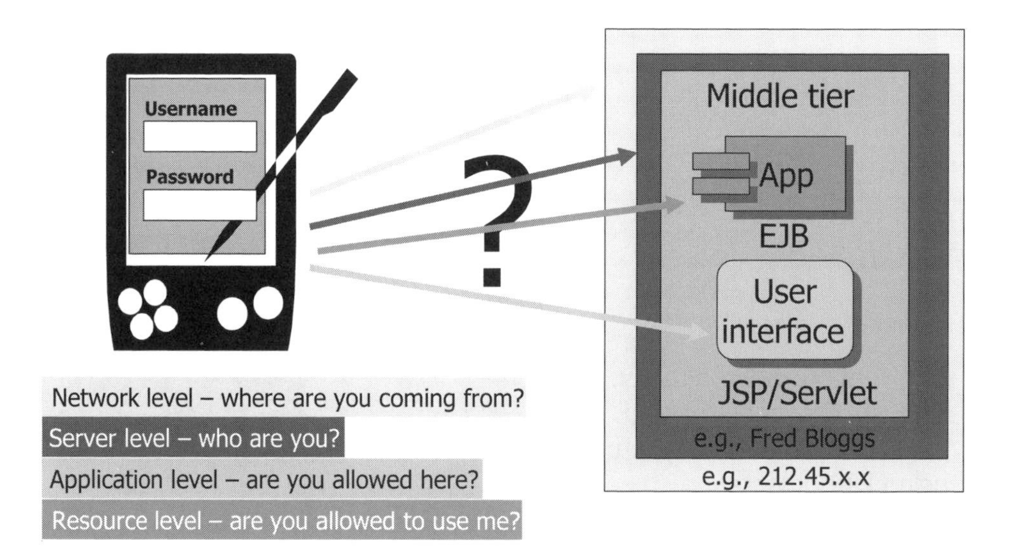

Figure 9.8 Different points to secure a service.

- being able to authenticate a user onto the services platform (i.e., that the user is a valid subscriber to the platform services);
- being able to control subscriber access to particular applications (i.e., that a subscriber can access the applications they have paid to access and no others);
- being able to protect one user's data from being illegally accessed by another user;
- being able to centrally and easily manage security mechanisms in a maintainable and scalable fashion.

We will now discuss how to address these issues.

9.2.1 Securely Connecting the User

The first challenge in the security mechanism is finding a way to authenticate the user who is accessing the platform. All authentication systems follow a similar challenge–response pattern. The user is challenged for their credentials and has to respond correctly in order to be allowed onto the system. In our J2EE world our users connect using HTTP, so the mechanism has to work over HTTP as this is the first point of contact that the user has with the system. Also, as HTTP is the common access method for all applications, it means that any security methods implemented at this level can be used across all applications.

As Figure 9.8 shows, authentication can take place at several levels in our system. We can authenticate at the network level, quite irrespective of whom the user is and what the application is. This is typically done at the IP level by allowing certain IP addresses and preventing others, as well as using related techniques; this is outside the scope of the J2EE platform and more related to physical networking and how the local area network is set up, including the Transmission Control Protocol (TCP)/IP networking on the physical servers.

The next level of authentication is at the server level, which can still be done using IP address blocking, but at a different service access point on the machine. The way to think of this is that, in the absence of network-level authentication, the server itself still has the ability to prevent HTTP packets from reaching the Web engine on the server. Of course, both of these levels can be used conjointly: a general level of filtering out of disallowed IP addresses can be done at the network level and then something more fine grain by the server. It is good practice to use both these methods.

The next level of authentication is at the application level, where individual applications can be assigned authentication access controls. The resolution of access control can be anything from restricting access to an entire domain (e.g., www.mymobileapps.com) right down to an individual JSP or servlet. Irrespective of access control resolution, the best approach to achieve application control is by adding authentication mechanisms to HTTP, as we will discuss.

If we are going to use HTTP to authenticate our users, then we need to achieve two things: the first is implementing a method to add an authentication mechanism to HTTP; and, second, we need a method to connect this layer of authentication to the security mechanisms within the J2EE platform itself. There is no point in enabling HTTP requests to be authenticated if we cannot subsequently pass user credentials to our JSP or servlet in order for it to run and fulfil the successfully authenticated request. We also need to ensure that these mechanisms are compatible with WAP (e.g., Wireless Session Protocol, or WSP), so that WAP 1 and WAP 2 devices can securely access our services platform via HTTP, either directly (e.g., Wireless-profiled HTTP W-HTTP) or via a proxy (e.g., WSP).

9.2.2 HTTP Authentication – Basic

The principle of authenticating an HTTP connection is quite straightforward. The first step in authenticating a user is to challenge them, which means that the server has to ask the user agent to identify itself or, more likely, its user.

When a servlet (or JSP) is requested to be run in response to an inbound HTTP request, the J2EE platform first checks if the resource is protected or not. It is considered protected if particular permissions are assigned to the resource other than allowing *everyone* to access it. Allowing everyone to access a resource amounts to no protection at all and, therefore, no authentication is applied, which is the so-called *anonymous user* mode. Assigning permissions to a resource in the first place is done using the J2EE security subsystem and has nothing to do with HTTP. The link between the two is how the J2EE server Web engine responds to an HTTP request for a protected resource.

If a resource is protected, then the Web engine must first return an HTTP response to the user agent without running the required resource. This makes sense as we are not yet ready to send the content anyway, not until we have authenticated the user. In the HTTP response, instead of the usual '200 OK' status code in the header followed by the output from the requested resource, the status code '401 Unauthorized[9]' is sent back to the user agent, which tells the browser application that it must first authenticate if it wants to access the requested resource. On a desktop browser, this would cause the familiar 'logon' window

[9] Note that the spelling of 'authorized' with a 'z' is the Oxford English American English spelling that is faithful to the relevant technical specifications.

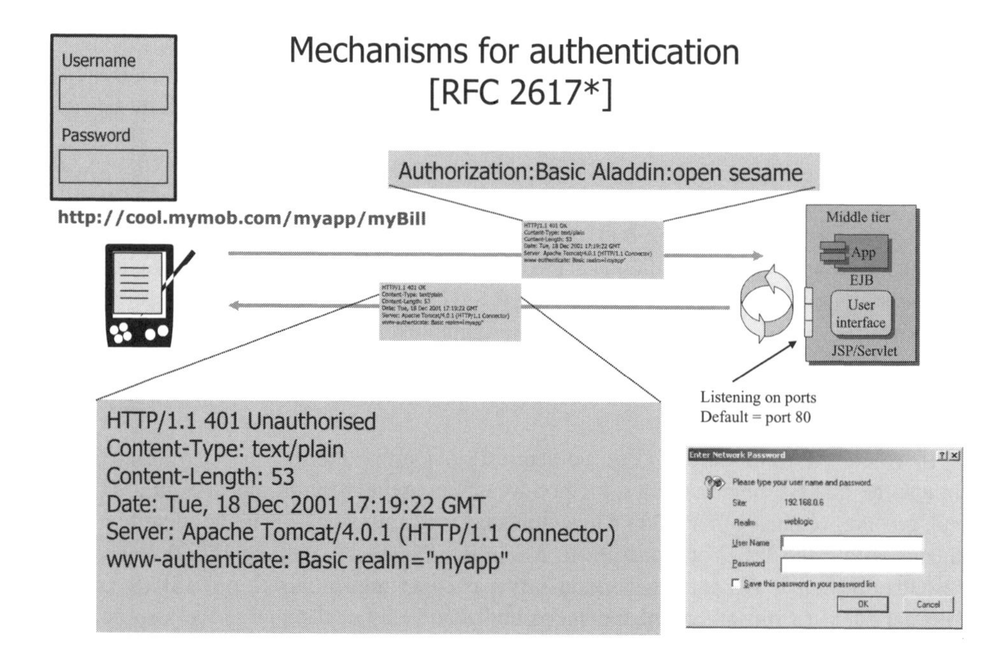

Figure 9.9 Authenticating an HTTP connection.

to pop up, as shown in the lower right-hand side of the figure. It is up to the browser on the mobile device to display a similar logon prompt on the device user interface (UI). However, before doing this, it checks the HTTP response header for another field that indicates how the server would like the user agent to authenticate, as there is more than one available method on most Web servers.

The main two methods supported by HTTP 1.1 are called *basic authentication* and *digest authentication*. Both of these methods of authentication should be supported by WAP devices, either using W-HTTP or WSP. In the latter case it is the duty of the WAP proxy to forward the authentication information in the header. The particular field used to identify the type of authentication is labelled thus:

- "www-authenticate:basic" for basic authentication;
- "www-authenticate:digest" for digest authentication.

We will look at both cases, beginning with basic, which is shown in Figure 9.9. The response to a basic authentication challenge is very straightforward. After eliciting the username and password tokens from the user, the user agent writes out the same resource request (GET method) to its output HTTP stream, but includes an authentication header – 'Authorization:Basic *username:password*' – where the appropriate username and password information is inserted. For example, if the username is 'Daredevil' and the password is 'Maximus', then the header will contain the string 'Authorization:Basic Daredevil:Maximus'. If the username is 'Aladdin' and the password is 'open sesame', then the string would read 'Authorization:Basic Aladdin:open sesame', as shown in Figure 9.9.

Figure 9.10 shows what happens at the J2EE server when we get a request with the 'authorization' header. In this example, we are considering a request to the resource with

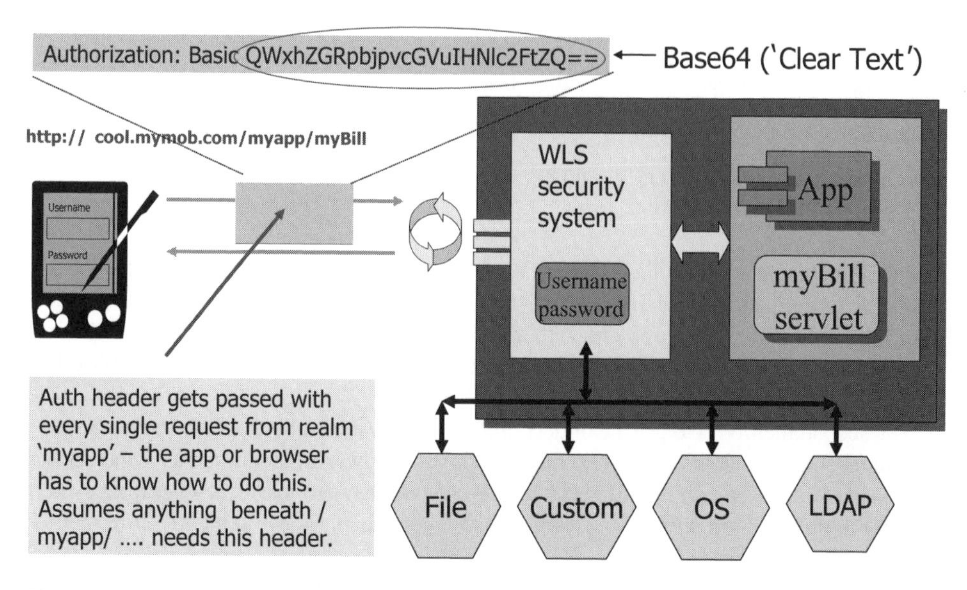

Figure 9.10 Processing a basic authentication request.

address http://cool.mymob.com/myapp/myBill: the address of a servlet on the server. HTTP is a stateless protocol; so, the J2EE Web engine doesn't know or is not required to know that this request is in response to an authentication challenge. It simply assumes this to be the case by detecting the 'authorization' header in the request. Because the servlet 'myBill' or possibly the entire application path '/myapp' is protected (i.e., requires authentication), the Web engine will check for the presence of the 'authorization' header. In our example the Web engine detects the appropriate header and, so, proceeds to extract the username and password token. However, notice that the token looks rather different from our 'Aladdin:open sesame'. It has changed to something altogether more obscure. This is a result of the user agent encoding the token string using a technique called *Base64 encoding*. This is *not* an encryption technique; it is merely used for obfuscation of the string so that it is made harder for a human to read casually, were someone able to intercept it. This is not at all secure enough to defeat a determined and well-equipped interceptor, but we will look at this weakness and its significance in a moment.

What happens to the token is generally the same on all J2EE implementations. The token is passed from the Web server engine to the server's security subsystem and checked against a database of valid and authorized users. If the credentials match those on the server and the resource is tagged as being accessible by the authenticated user, then the Web engine will run the servlet as normal and return its output stream in a standard HTTP response along with a '200 OK' status header. If the user does not check out or they are authentic but don't have permission to access the 'myBill' servlet, then the 401 challenge is resent.

The method that the J2EE server uses to authenticate the user on the J2EE server is implementation-dependent. The example in Figure 9.10 portrays several methods that are supported by BEA's Web Logic Server (WLS), so we will mention them here as examples of possible authentication mechanisms. The WLS security mechanism can check user credentials against a variety of possible user directories:

- flat file stored on the J2EE servers;
- the underlying security system used by the host operating system;
- an external directory system;
- a customized approach via the J2EE Java Authentication and Authorization Service (JAAS) API.

The first method is to consult a flat file that is stored on the server file system that is local to the J2EE server installation. This file is locally encrypted so that a casual observer cannot interpret its contents. The problem with a flat file method is that it is neither scalable nor very maintainable. If we have a large mobile service portfolio with a substantial subscriber base, then this method will not suffice.

The second method is to use the underlying security system inherent in the host operating system, such as Windows or a Unix variant. This is a powerful technique as it means that a consistent approach can be utilized where we do not have to leave the operating system security administration should we want to use it for our platform authentication method across all applications and servers.

The third available method is to use the Lightweight Directory Access Protocol (LDAP) to send our authentication requests to an external directory somewhere in our network. This approach is likely to be a common one, because the opportunity of maintaining a central directory of all subscriber information is highly attractive and more than likely already undertaken by many network operators. A centralized directory can be used to store and associate any information with a hierarchy of users, not just authentication details. In a mobile services scenario we may want to use a directory to store subscriber information, like user address, account type, device type, services tariff and so on.

The final option is to utilize a security API within the J2EE platform in order to implement a custom solution or at least a custom interface to an existing (legacy) solution that is not catered for by any other access technique. We might well have our own favourite security system or directory equivalent which is not accessible via any of the other methods thus far mentioned. In order to continue using the available system and make our J2EE applications compatible with it, we can utilize the security API to create a custom authentication interface. This mechanism can be as elaborate as we like as we are able to utilize the full power of the Java programming language and J2EE APIs to implement it, perhaps involving an API like JDBC to talk to an external database that underlies the external user directory.

9.2.3 HTTP Authentication – Digest

The problem with basic authentication is that the username and password are both sent in clear text over the link, albeit obfuscated, but weakly so. To appreciate the vulnerabilities of this approach we need to digress for a moment into the world of hacking (computer spying). As shown in Figure 9.11, we can think of our hacker as the 'man in the middle', a person who is fully able to access the message flows across our network going in both directions. We shouldn't concern ourselves with how this is possible. It is a well-established assumption in the computer security world that not only is this possible but also our default frame of mind should be to assume that it is likely or definitely going to take place. A word of caution from the security gurus is not to assume that the hacker is some 'evil spy' with a fantastic set of

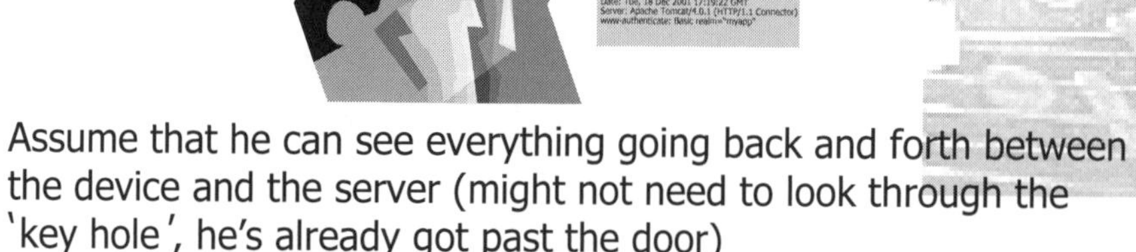

Assume that he can see everything going back and forth between
the device and the server (might not need to look through the
'key hole', he's already got past the door)

Figure 9.11 The despicable hacker.

spying apparatus and gadgets from a Hollywood movie; It could well be the friendly office cleaner or an employee. There are plenty of places that the hacker could be accessing the link, so anything is possible. Once we are comfortable (uncomfortable) with these gruesome facts of life, then we can go about the business of deploying techniques to address them.

It should be reasonably clear that if our hacker can snoop on our HTTP requests, then immediately they might gain access to critical user information by copying the 'authorization' header information from the inbound requests. Base64 encoding can be easily undone with the most rudimentary of computer programs. It is an openly available and widely known algorithm that does not pretend to be an encryption algorithm; so, there is no need to make it difficult to decode anyway – in fact, the reverse is preferable: it should be easy to decode for the purposes of computational efficiency.

With the 'man in the middle' threat firmly planted in our minds, let's proceed to explain an alternative to basic authentication that may deter or overcome this threat. We will first explain a technique called *digest authentication* and then see how it addresses our concerns with basic authentication. However, we will then proceed to see if there are cunning ways to defeat even this improved scheme, as impressive as it seems at first glance.

In general, there are really only two ways we can approach the problem of securing our authentication method. We can encrypt the password information or we can encrypt the entire link so that nothing can be read going across it. Digest authentication is an attempt to encrypt the password.

In Figure 9.12 we can see the process of digest authentication; but, it needs a little explaining to figure out what's going on. As with basic authentication – because our resource is protected – this results in our J2EE Web engine issuing a '401 Unauthorized' header in the response. However, the 'www-authenticate' header has an altogether different structure than we saw with the basic authentication. We need to focus on the field called *nonce*. At

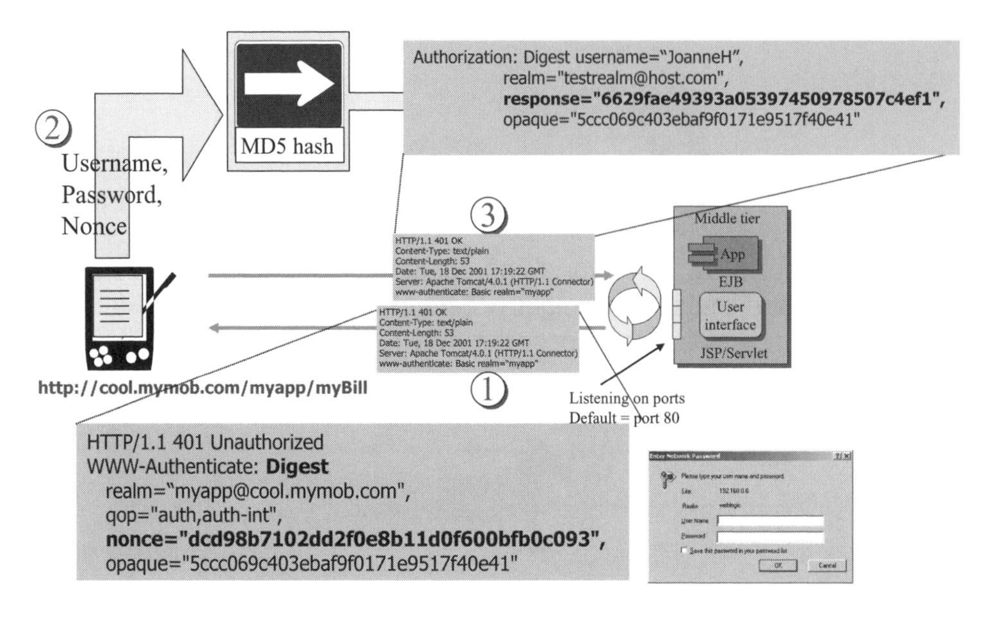

Figure 9.12 Digest authentication.

first, this sounds like a rude word[10] (well, it does if you're British judging by the sniggers I get from training classes). However, it refers to a unique string of characters to be used as the basis of our authentication process.

What the user agent must do with the nonce is to put it through a mathematical process called an *MD5 hash*. If that sounds complicated, that's because it is – at least at the deep mathematical level. However, the concept is rather more straightforward. The function is what mathematicians call a 'one-way' (or 'trapdoor') function. This means that once we have submitted the inputs and generated an output (*hash*), it is extremely hard to reverse the process, even if we know all of the input parameters except the one we are looking for (i.e., the password).

Having calculated the hash, the user agent embeds it in a repeat request for the resource, using the 'Authorization:Digest' header, which is similar to basic authentication, but has different fields. One of the fields is the username, which the server clearly needs to know in order to authenticate our user. Unlike the basic authentication method, we do not attach the password. Instead, we send the hash in the 'response' field (see Figure 9.12).

At the server end the Web engine detects a request for a protected resource, so the first thing it does is look for an 'Authorization' header, which it duly finds. On detecting that the header suggests a digest authentication response to a challenge, the Web engine processes the fields by performing the same hash function that the user agent carried out. The Web engine has the username, which is what it needs to look up the password from the underlying security apparatus, whatever it happens to be. Because the Web engine generated the nonce, it has all the same inputs available that the user agent had to generate the digest response field (the hash itself). Therefore, the Web engine repeats the calculation and then compares

[10] *Webster's Dictionary* definition of 'nonce': date: 13th century **1 :** the one, particular, or present occasion, purpose, or use <for the *nonce*> **2 :** the time being.

the locally generated hash with the one embedded in the 'Authorization' header. If they are the same, then it is safe to assume that this request has been generated using the same password on file for the user, hence the user is now successfully authenticated.

In the digest process, notice that we have avoided transmitting the password altogether, unlike the basic authentication method. This is a much better proposition and evidently more secure. According to the proponents of digest authentication, the rationale behind the design is:

> *An improved, minimal security scheme for HTTP is proposed. This scheme does not require the use of patented or export restricted technology and is believed to provide the best effective security possible within those constraints.*[11]

At first glance it seems that the digest process is remarkably effective. It is certainly an elegant idea and the mathematical basis for it gives a sense of robustness. However, those constraints mentioned in the justification above turn out to be its downfall. On careful scrutiny of the digest process it is actually very vulnerable after all. Let's now take a look at its weaknesses by putting on our hacker's caps.

To uncover the weaknesses of the digest process, let's not forget that as a hacker we may have full access privileges to what's happening 'on the wire'. We are the 'man in the middle' with full visibility of the message flows. Moreover, what we should have stated earlier is that we are able to change messages to suit our purposes; this is what it means to have 'man in the middle' privileges (the life of a hacker[12] sounds so exciting, doesn't it?; but I hasten to add that I do not condone it in any way, but we do have to be aware of the nature of real world threats, especially if we need to design against them).

Figure 9.13 lists the major problems with digest authentication. We will briefly mention them, but we do not need to ponder on the solutions too deeply. Soon we will discuss a different approach that makes these problems irrelevant.

The first method of attack is to simply take the request packet and replay it to the server. A naive implementation of a Web engine will not notice any difference from a previous request for the resource. Indeed, a repeat request may well be a legitimate one. We can imagine that our spy has stolen the message and then, whenever they want to access the protected resource (which could be any kind of sensitive material, like financial information), they simply send the request in to the server and siphon the response.

The second method of attack is to edit the authentication challenge from the server and simply change the string 'digest' to 'basic'. The user agent gets fooled into thinking a basic authentication request has been made and gladly (sadly) offers up the username and password token in the subsequent request. Note in this case that the user agent must have actively requested the resource in the first place, otherwise it will not be expecting to process a response from the server; so, there is a restriction to when our spy can carry out this attack. However, let's not be lulled into thinking that this constraint deters the hacker. The hacker is not going to be sitting on the wire with a 'stethoscope', actively 'listening' to our messages; they will most likely deploy a piece of electronic apparatus somewhere in the link to do all this automatically, so timing is of no consequence to them.

[11] Taken from the specification, which can be found at http://www.faqs.org/rfcs/rfc2617.htm

[12] The life of a hacker usually ends up abruptly terminated by a prison sentence, so this is not one to try at home, folks. For the curious, I recommend reading Mitnick's interesting book *The Art of Deception*. John Wiley & Sons, Chichester, UK (200 ●).

Problems with digest authentication

HTTP/1.1 401 OK
Content-Type: text/plain
Content-Length: 53
Date: Tue, 18 Dec 2001 17:19:22 GMT
Server: Apache Tomcat/4.0.1 (HTTP/1.1 Connector)
www-authenticate: Basic realm="myapp"

1. Can replay the response and then gain access to the URI
2. Can intercept www-authenticate:Digest and replace with www-authenticate:Basic and then do the hash
3. Can make hash responses for common passwords from the dictionary
4. Spoof the server credentials and offer Basic challenge

Figure 9.13 Digest is perhaps not that good after all.

An alternative to this is the similar approach of offering a basic challenge, but using deception. The spy sets up a server that masquerades as the original server of interest. This is done by mimicking the UI. The spy lures the user into accessing the protected resource, presumably because the user thinks that doing so is legitimate, possibly having been enticed onto it via an email message (or WAP Push message). Seeing that the UI is no different from usual, the user is none the wiser and happily attempts to log in to the system, thereby giving their username and password to the hacker. This type of attack is called spoofing. It seems that mobile users are particularly vulnerable to this type of attack and should be vigilant (see Box 9.1: 'Mobile browsing security threats'[13]).

Box 9.1 Mobile browsing security threats

Mobile users should be careful about responding to messages (i.e., WAP Push) that take them to WAP sites. They might be fake messages.

When I started thinking about the vulnerabilities of authentication, it occurred to me that mobile sites are perhaps a lot less safe than their desktop counterparts.

One way to gain a password from a user is to spoof a server. Pretend to be a particular website and then ask your users to log in. *Voilà!* If they bite, then you have their username and password. This has been done on numerous occasions with famous websites and is an ongoing threat. In fact, regular users of sites like Ebay should learn to become vigilant against these types of spook attacks. With mobile sites it appears to be even easier.

First, on most mobile browsers, to save real estate, the URL (uniform resource locator) display box is not displayed or there simply isn't one. This means a user

[13] Taken from the author's weblog: http://radio.weblogs.com/0114561/

typically has *no* idea what website they are actually on in terms of its Web address – surfing on mobile sites is sometimes an eerie experience, like walking in the dark. If a user is directed to a spoof website, they would have little or no idea.

Second, due to sparse interfaces, it takes little effort to mimic a mobile website, perhaps just by copying a logo at the top of the screen.

WAP Push will soon start to become more widespread. It has been slow to catch on, but with the increasing number of picture-messaging phones the necessary inclusion of the WAP browser means that more and more mobile users can access mobile sites, whether they know it or not. However, they don't need to know it to respond to a WAP Push message – the phone itself will take care of accessing the embedded URL in the message if the user chooses to respond.

The other weakness is with the WAP Push mechanism itself. It uses text messaging as the transport mechanism. With text messaging it is easy to change the sender's address in the message. In fact, many text-messaging bureaus offer this service to their bulk-messaging customers. This has some interesting and legitimate uses, but can also be malignant. It is easy to spoof the sender in order to make the message look legitimate, thus adding to the bait used to lure an unsuspecting user onto a spoof site.

Mobile users should be educated in the dangers of responding to WAP Push messages.

The final approach in outsmarting digest authentication is to perform a dictionary attack. This simply involves repetitive entry of different password options from the dictionary until a matching hash is produced to the one that the user agent generated. This type of attack should theoretically be easy to thwart, simply by enforcing users to utilize nondictionary strings for passwords. However, the reality of sensible password policy enforcement is often too embarrassing to mention because it frequently doesn't exist.

From what we can tell about the vulnerabilities of digest authentication, we can see that despite its apparent sophistication it is not that secure. It is still worth implementing or deploying instead of basic authentication, as there are different levels of determination and sophistication that we might expect from our hacker. We shouldn't assume that we are inevitably going to confront the world's best hacker as our foe and, therefore, in the absence of a sufficiently strong technique we should simply not bother; this would be folly and it is better to always implement the stronger of any security option if we are able to do so. Casual hackers may well be deterred.

9.3 END-TO-END ENCRYPTION

Ultimately, the best solution is to secure the entire pipe and make sure that nothing can be gleaned from our communications in the first place. If this is achievable, then any amount of hacking sophistication by being the 'man in the middle' should be to little or no avail.[14] A protected link defeats all hacking attempts that have thus far identified, except for some types of spoofing, which, as Box 9.1 identifies, may be particularly troublesome for mobile users.

[14] In the sense that any technical means will be hard to implement. However, the greatest sources of risk are often found elsewhere, such as the social engineering methods described in fascinating detail in Mitnick, K., *The Art of Deception*. John Wiley & Sons, Chichester, UK (2002).

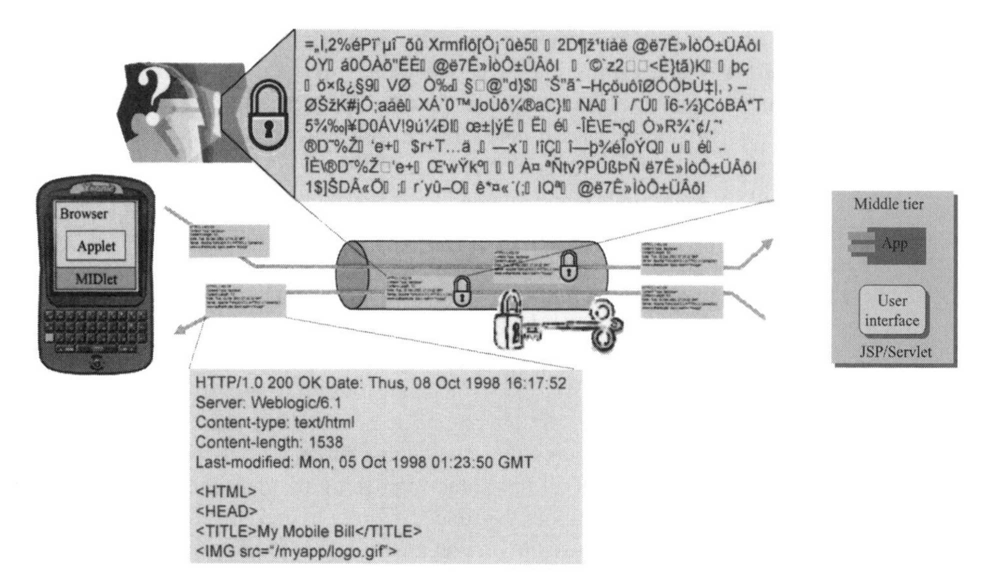
Browser
Applet
MIDlet
Middle tier
App
User interface
JSP/Servlet
HTTP/1.0 200 OK Date: Thus, 08 Oct 1998 16:17:52
Server: Weblogic/6.1
Content-type: text/html
Content-length: 1538
Last-modified: Mon, 05 Oct 1998 01:23:50 GMT
<HTML>
<HEAD>
<TITLE>My Mobile Bill</TITLE>
<IMG src="/myapp/logo.gif">

Figure 9.14 Securing the pipe.

Figure 9.14 shows us what securing the pipe is about. Using encryption, we cipher all the information flowing between our device and server, with these two end points being the only the true end points of the ciphering. In a moment we will clarify what we mean by 'end-to-end encryption', as there are common misconceptions about this concept.

Ciphering results in the data becoming completely obfuscated to everyone, unless they have the means to decipher it, although that is a capability that we can user clever means to guard against, as we will see presently. Only the intended recipient has the required deciphering ability. Because the information is obscured from everyone except the intended recipient, we don't really mind about a 'man in the middle' or even 'men in the middle'. Any amount of sucking data from the pipe in order to attempt reading it is no longer a concern. In fact, we could say that this is really the goal of an effective ciphering mechanism: we shouldn't really care about unintended recipients gaining our protected information, maliciously or otherwise. Our ciphering will take care of this threat for us.

As shown in Figure 9.14, we can think of ciphering as putting data into virtual caskets using a lock and key. We lock the data up in the casket and send it. We hope that the protection is sufficiently robust that it cannot be compromised. The only way to unprotect it is to unlock it with a lawfully acquired key. The lock-and-key concept is a useful and obvious metaphor for understanding encryption, so we shall stick with it for our current discussion.

With the lock-and-key process, the mechanism for secure transmission is obvious. The sender locks the data with the key, sends the locked data in the casket and then requires the recipient to unlock it with their own key. A wonderfully simple process, but with a subtle, yet significant, defect. The challenge is *key management*.

In order for the recipient to unlock the data, they must have a copy of the key. This presents some challenges. The first one is distribution. How do we safely get a copy of the key to the intended recipient? If this were a house and two people shared the house, then giving them a copy of the key is easy. However, what about if someone were coming to

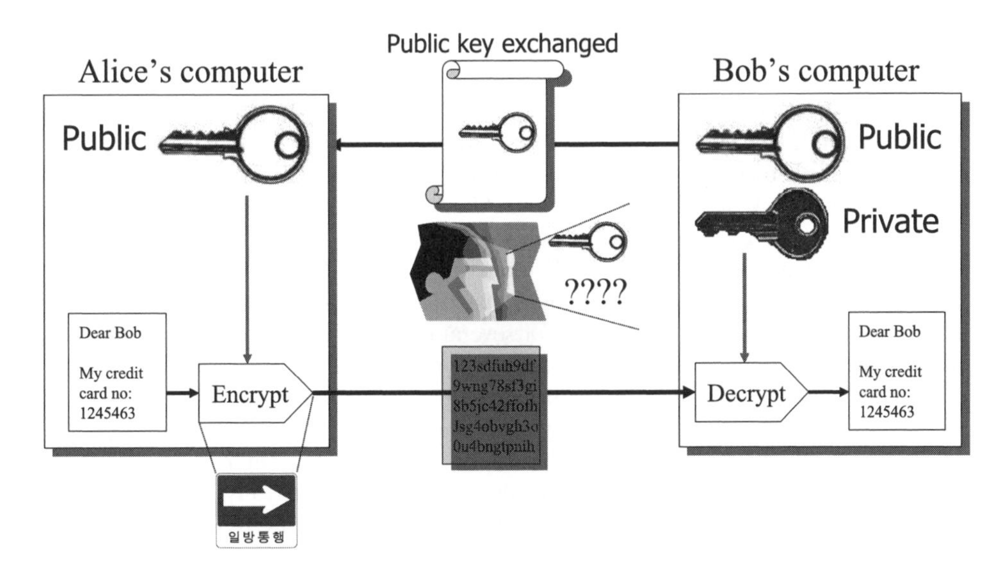

Figure 9.15 Public key cryptography.

stay in your house and they were traveling from the other side of the world to get there? How would we safely get the key to them? We could post it to them, but it might get lost or intercepted. In the latter case it could be copied without us knowing. Alternatively, we could send the key via a courier, but we would have to know that we could trust the courier.

Let's say we could find a means to securely distribute a copy of the key: what about making this available to more than one recipient? If we had lots of houses that we rented out to overseas visitors, then we would have to take care of lots of keys. It would not be a good idea to have identical locks and issue everyone with the same key. This would mean that one visitor could open the house of another. Similarly, with a large website, we don't want to give a copy of the same key to each user, so we prefer to have unique keys. This poses a potentially significant key management problem, possibly involving thousands of keys. Not only does this make distribution that much more risky, but we now have to ensure that for each transaction with a user we end up using the right copy of the key to cipher the conversation.

Fortunately, some rather clever mathematicians have worked on these problems and have come up with a genius solution called *public key cryptography*.

9.3.1 Public Key Cryptography

Perhaps the most famous example of public key cryptography (PKC) is the *RSA*[15]*method*. The principles of this technique are shown in Figure 9.15, which we will now describe.

With PKC, the genius idea is to have two keys. Both of these are generated (created) by the sender and are unique to the sender. In computer science terms a key is actually a long number, long enough that it would be too difficult to guess, but we will return to this point later.

[15] RSA is the most common and well known PKC method and is named after its inventors: Rivest, Shamir and Adelman. Their patent on the method has now expired, so the algorithm is free for anyone to use without paying royalties.

When the sender generates two keys (using a computer program on their device), one is designated the *private key* and the other the *public key*. Under no circumstances is the private key to be revealed to anyone – it is to be kept safely hidden away. However, the uniquely associated public key can be shown to anyone, which is actually what it is intended for.

As the figure shows, let's imagine that Alice wants to send her credit card details to Bob in a secure manner. What Bob does is to give Alice his public key. Don't forget that this is not a problem as long as Bob gives his public key and not his private key. Alice then uses Bob's public key to lock the data. This is where the power of PKC is exploited. A special mathematical one-way function is deployed that allows data to be locked with the public key. It is known as a one-way function for several reasons. First, in the case of PKC the public key itself cannot be used to unlock the data once locked. This seems remarkable and quite unlike our metaphor of the house key. Imagine a lock that we can secure with a key, but then cannot reopen with the same key: quite mind-boggling.

The one-way public key is the magic of PKC. It means that Bob can freely distribute his public key without any fear of it falling into the 'wrong' hands. He really doesn't care who gets hold of it. With this approach the key distribution problem is solved. The only way that Bob can open his encrypted message from Alice is to use his private key, which only he has possession of. Bob should never distribute his private key nor does he have any reason to disclose it.

There is another clever trick that can be done with public and private key pairs. Securing a message with Bob's private key can only be undone using his public key. This may seem a strange thing to do. It seems as if everyone could open the message; however, that's the point. If we can open a message and read it, then we know that the message in question must have come from Bob and so we can use this process to authenticate Bob.

9.3.2 Using PKC to Secure Web Connections

Using PKC we can secure Web-based transactions using HTTP and WAP (see Figure 9.16). We will now look at this process in general and then look at some of the issues concerning WAP, especially securing a WAP 1 connection.

The process of securing an HTTP session takes place at the transport level, which means that the entire HTTP communications link gets encrypted. In other words, every single HTTP message is encrypted from top to bottom and both the *user agent* and *origin server* are largely ignorant of the encryption process.

Recalling our description of the Transmission Control Protocol (TCP) in Chapter 5 on IP protocols, we may remember the idea of a socket: a socket is really a programming entity, like a handle used by a higher layer application to pipe its data through (or receive) with the notion that it will squirt out somewhere on the other side, this being identified by the IP address and port number bound to the socket, as shown in Figure 9.17.

Ideally, what we want is the ability to secure the socket connection, so that information goes in one end (source socket), gets encrypted and comes out the other end decrypted (at the destination socket). This is what transport-layer encryption is all about, as shown in Figure 9.18.

In Figure 9.18 we can see that each socket has a key for ciphering the data that flow through it. Consequently, the connection is locked and secure all the way from one application to the other. Later in the book we discuss where and how this cryptography gets carried out on

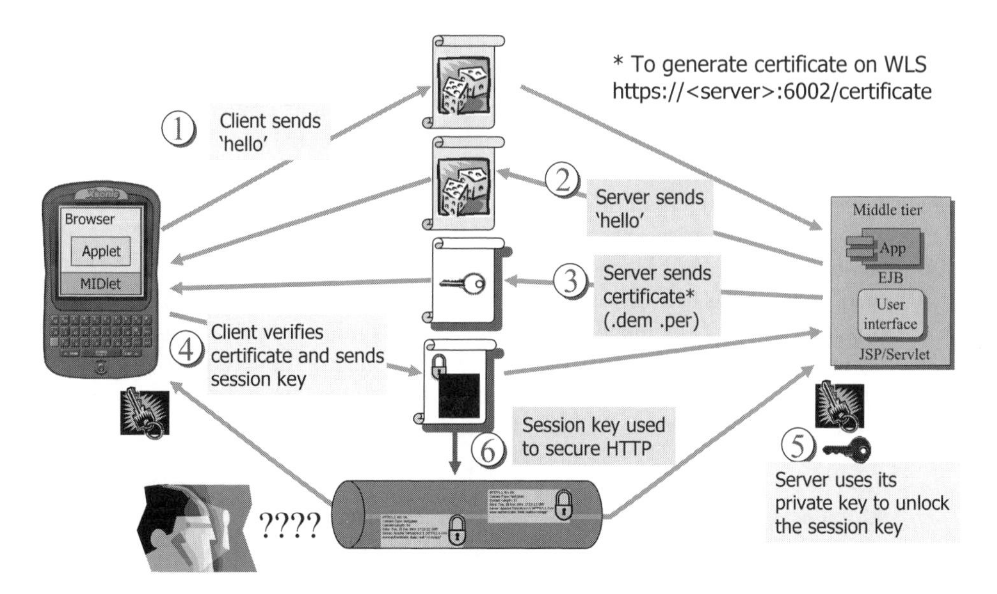

Figure 9.16 Using PKC to secure the Web.

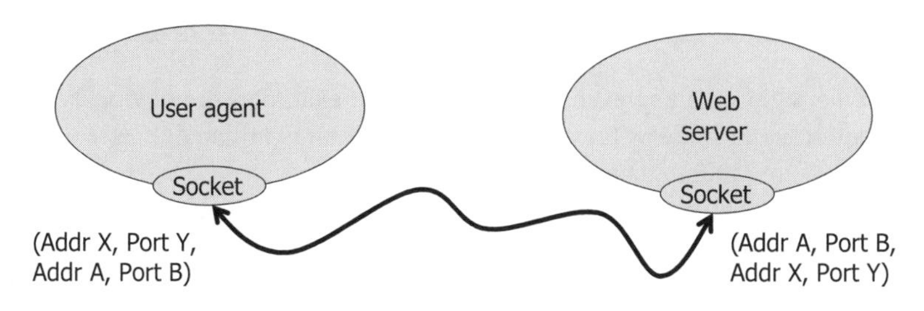

Figure 9.17 Sockets.

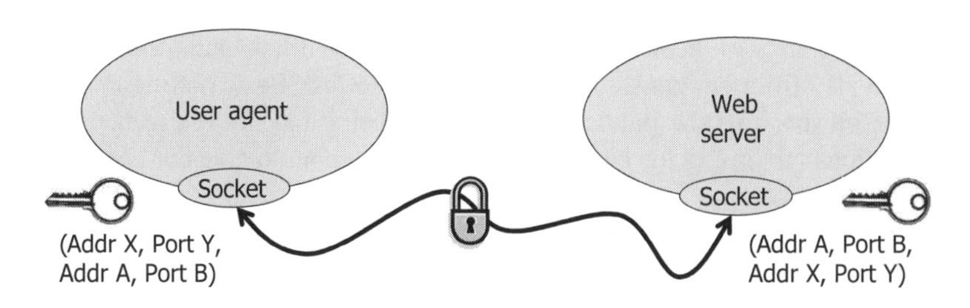

Figure 9.18 Secure sockets.

the device and make the point that a library of cryptography software routines is probably a standard inclusion these days on any mobile device, such is the importance of cryptography to an increasing number of mobile services.[16]

[16] And the good thing about having cryptography libraries is that programmers do not have to worry about implementing these tricky techniques themselves.

The method of securing sockets that we will look at here is shown in Figure 9.16 and is referred to by two names: the most apt for our discussion being the Secure Sockets Layer, or SSL. This was a method of securing HTTP connections originally included into browser software made available by Netscape. It has now become a standard, although it has migrated to a new specification and name, known as Transport Layer Security (TLS), which has a wireless counterpart (in the WAP architecture) called Wireless TLS, or WTLS.

Stepping through the process shown in Figure 9.16, we can find out how SSL/TLS works, as they are both essentially the same in principle.

The first undertaking between both ends is a handshaking protocol. Like handshakes between people, in data communications they serve a similar function to greeting each other. Our mobile device and server exchange 'hello' messages. These contain a request for an encrypted conversation to take place, laying down some parameters of the conversation before proceeding with it. We can think of this as follows:

> **Mobile:** 'Hi there – I want to talk with you over a secure link. I prefer to use RSA to exchange keys, but can do Diffie-Helman too – what say you? My opening phrase is "skipjack tuna".'
>
> **Server:** 'Thanks for the request. I can talk to you securely, but prefer we use Diffie-Hellman key exchange. My opening phrase is "wiggly bigly".'

Apart from the legitimate key exchange names, these examples are obviously contrived; however, they serve to show us how the SSL conversation is initiated. In a real system the opening phrases are random data.

The juicy bit of the process is what happens next. The mobile receives the server's public key. This clearly allows the mobile to securely send information (i.e., HTTP requests) to the server, which can decode using its private key. However, there are a couple of subtleties that add a degree of complexity to this process, although the principle itself remains a sound one.

Authenticating the key is our first challenge: how can we check that the key we receive actually belongs to the server purporting to send it? If we think that we are about to transfer money from our bank account using MobileCash Bank (fictitious name), then how can we be sure that the public key is indeed theirs and not one generated by Joe Hacker (fictitious villain). In the latter case we could be about to give our bank details to a thief.

The standards community have agreed on a technique for embedding keys into a file called a digital certificate. We can think of these as having an electronic stamp on them from a trusted authority (certificate authority, or CA). The principle is that the certificate is not easy to obtain unless the applicant really is who they say they are. The CA should carry out checks to verify credentials and the legitimacy of the applicant.

Once we have a certificate containing our public key, the mobile can check the certificate first, before extracting the public key. The certificate will claim to belong to MobileCash Bank and we can check for a digital stamp from a trusted CA. This digital 'stamping' itself is based on PKC. This is something we touched on at the end of Section 9.3.1. The CA 'signs' the certificate by encrypting it using the CA's own private key. The CA's public key is widely available.[17] If we can read the certificate by decrypting with the CA's public key, then we can trust that it was indeed issued by the relevant CA. Ultimately, it is up to the users of the system to trust (or not) that the certificate has been issued with due diligence.

It is not enough that we rely on the certificate to authenticate the server. The certificate simply authenticates the public key, but we still don't know who we are talking to until we ask for cast iron proof. The way this is done is for the mobile to send a random code to the server and ask for it to return the code encrypted using the server's private key. If we are able to receive the same code by unlocking it with the corresponding public key, then we know that we are talking to the right server. This assumes that the server has never shared its private key and that the certificate it sent the mobile checked out. To protect the random code from being spoofed by someone else, it is sent encrypted using the server's public key.

The details of this authentication process are more complicated still. As with the *digest authentication* technique used for HTTP session authentication, the server actually sends an encrypted hash of the original random code back to the mobile. The mobile can compute the same hash locally and compare. The hashing is really to protect the server from sending data that it doesn't want to send, the problem being that by using its private key the data are effectively being 'signed' by the server. The server doesn't want to 'add its signature' to any old data and send them out, as this could be used surreptitiously elsewhere. With a strong hash (like MD5 used for digest authentication) the worst that a third party can do is to decrypt a message from the server that is garbled once decrypted.

The hashing involves the secret data exchange and the original public 'hello' messages, which we said were in fact random. This combined approach achieves several objectives, including making it difficult for the hacker to carry out replay attacks, which may be useful even if the contents of the requests remain unknown (encrypted). But this is an advanced topic and is not covered here.

Thus far, we have only covered steps 1 to 3 in Figure 9.16, which is how we set up the secure conversation. This is all taking place at the socket level and, eventually, our sockets have a reliable, trustworthy and secure link. To indicate to higher layer applications at either end that this is a secure channel, a special port number is actually reserved for SSL connections for Web traffic: port 443. By assigning a separate port from the unencrypted connections (port 80), we can dedicate resource just to handling this port, independent of any software that may already be serving port 80.

Having gone through all the elaborate procedures and fancy mathematics to establish a link using PKC, the next step (step 4) in the SSL process seems a bit odd: the mobile generates a session key for a symmetric encryption session. This is back to the original idea where both sender and receiver have the same key. A moment of reflection reveals that this will work safely, given that the mobile can send the session key securely to the server using the server's public key to encrypt the session key. In other words, not only does PKC solve the key distribution problem in its own right, it also solves the key distribution problem for any other encryption technique. The only question that remains is why bother to revert back to symmetric processing for the remainder of the HTTP session? The answer is that symmetric algorithms are a lot less resource-intensive than asymmetric ones.

[17] In fact, in certain cases the public keys for various CAs are already included in our browser software. In effect, this is a delegation of trust. By using these browsers we are agreeing that we trust the CAs whose keys we are going to use.

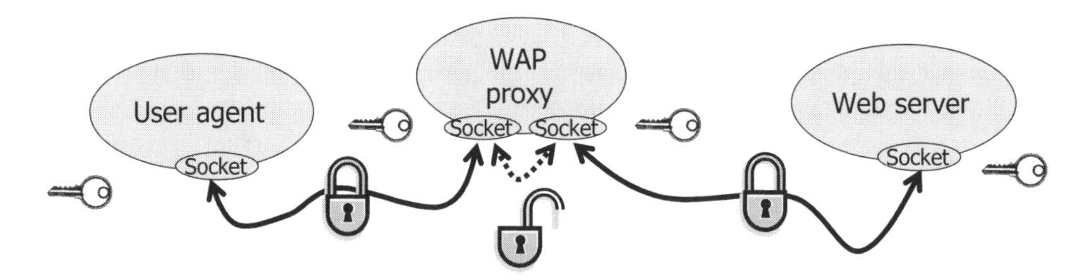

Figure 9.19 WAP proxy can't maintain a transparent connection.

The processing requirements for PKC are not really a concern for desktop systems. After all, they have oodles of processing power and memory to throw at problems, cost-effectively too. Nonetheless, even in the immobile world the processing requirement is still a problem – not for the client, but for the server handling thousands of sessions concurrently; therefore, a less intensive process is still worth it. We should not be misled into thinking that the lower resource requirements mean a weaker cipher; this is not the case at all. Symmetric ciphers are still very powerful and practically unbreakable with even an exceptional amount of applied resource.

For mobile devices, the use of asymmetric encryption for the entire session would definitely be out of the question, especially on devices with very limited processing resources. In fact, even the initial exchange using public keys can be a very slow process on a low-end device, so much so that the cipher experts have devised new ways of handling the maths involved in the calculations. Elliptic curve cryptography is one such example and is now a widely used technique for securing mobile browser sessions.

9.4 APPLYING SSL TO WIRELESS

In essence, the process of applying ciphering to a wireless browser session is the same as just described in Section 9.3. In fact, the security layer between the mobile and the WAP gateway is the same process exactly. The problem is that the gateway introduces a discontinuity in the ciphering process that cannot easily be circumvented. As we noted when looking at IP protocols for mobile applications, the Wireless Transport Protocol (WTP) and WSP are not the same as their Internet counterparts TCP and HTTP.

The difference between WSP and HTTP is what causes the bigger problem. If nothing else, the contents of many HTTP messages are converted from WAP Markup Language (WML) to WAP Binary eXtensible Markup Language (WBXML) at the gateway.[18] This means that we cannot transparently pass data through our gateway; the WAP gateway process necessitates conversion. This implies that we have to decrypt the data at the gateway in order to carry out the conversion process. We then need to encrypt the data again onto the opposing connection on the other side of the gateway. TLS only works if the higher layers require a transparent connection end to end (socket to socket). This is not the case with a WAP gateway, as shown in Figure 9.19.

[18] With WSP, the headers are also encoded, as we found out earlier in the book.

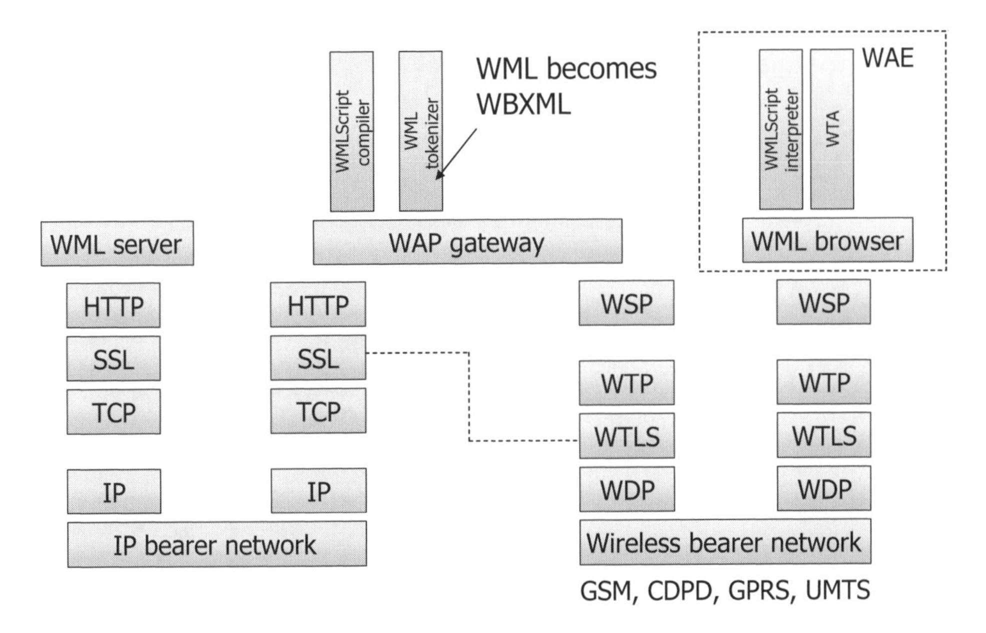

Figure 9.20 WAP stack not the same as HTTP stack.

For WAP 1, there is no way of getting around this encryption discontinuity at the WAP proxy. WSP and HTTP are not compatible, even though they are semantically similar and practically identical for the most part. The WAP stack and the HTTP stack are not the same for WAP 1, as Figure 9.20 clearly shows us.

This problem has caused a lot of controversy about WAP and was a contributory factor to its lack of uptake in the early part of its life. However, it is probably safe to say that the performance of early WAP devices was a much bigger problem and that, for the most part, the supposed 'insecurity' of WAP was more of a perceived problem than a real one for most applications. But, how should we approach this problem?

First, for applications that need to use WAP 1, the real impact of the discontinuity needs to be assessed. The point of vulnerability is the WAP gateway; this is the only place in the end that an attack can be successfully carried out. Therefore, if we are confident that our WAP gateway is well secured physically, then we may not really have much of a problem. If we still have doubts, then for applications that warrant it we could deploy a private WAP gateway and ensure that its integrity is maintained without recourse to any third party, whom we may not fully trust. In many enterprise applications it was very easy to deploy a WAP gateway internally within the company firewall, either through the firewall itself or by private dial-in or data access connectivity. Circuit-switched access from a WAP device is possible over standard dial-in servers which an enterprise may already have for remote workers. These can be configured to support fast connect times with mobiles on digital cellular networks when the dial-in lines are digital themselves (e.g., ISDN). The benefit of using digital connections is that they are faster to establish and they are usually more reliable. Accessing ISDN modems requires ISDN v. 110 or ISDN v. 120 rate adaption protocol support from the local Global System for Mobile Communications (GSM) operator and the dial-in system; this is widely supported.

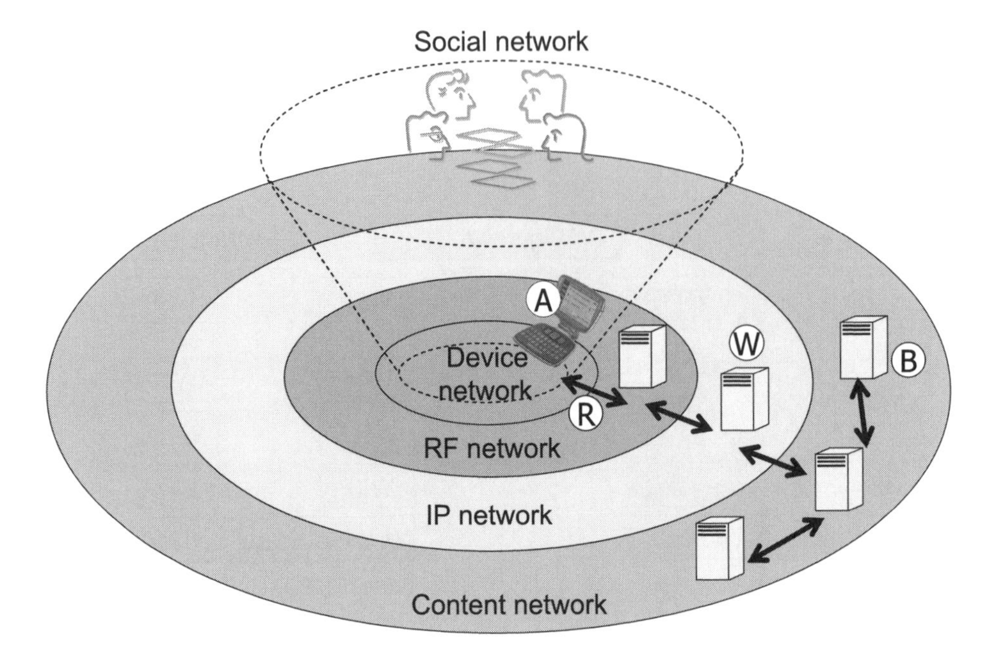

Figure 9.21 Parts of a mobile connection, not all them secure necessarily.

The discontinuity of the WAP proxy introduces us to the concept of end-to-end encryption. A lot of the confusion with the security of WAP once arose due to a misunderstanding of what constitutes a secure connection. In digital cellular systems like GSM and UMTS, the air interface (RF link) is encrypted. As it happens the encryption is quite strong (in most cases[19]). If we return to our mobile network diagram established earlier in the book, then we can see what the problem is with determining whether a connection is secure or not.

Figure 9.21 illustrates the problem. We are interested in passing data securely from our user all the way to the application server that needs the secure information. This could be password information, like that needed to complete a '401 Unauthorized' authentication cycle in a HTTP session. It could also be information submitted in a XHTML-MP (Mobile Profile) form which needs to be sent securely using the HTTP POST method to the server, perhaps containing credit card details.

If we are sending this information to server B in the diagram, then we need the link to be secure for the entire pathway from device A to server B. There should be no momentary break in the encryption nor should there be any portion of the pathway that is unencrypted. The confusion concerning securing WAP connections over digital cellular networks came from the idea that air interface encryption provided the necessary protection. Some mobile operators even mistakenly advised that using WTLS was not necessary because air interface encryption was sufficiently robust. This is the equivalent to just encrypting portion R in the diagram, leaving everywhere else unencrypted and, consequently, vulnerable.

[19] GSM air interface encryption comes in two types: one is strong and the other is weak. GSM operators in certain designated countries are only allowed to use the weak one so that they can be 'spied on by everyone else' (use of language is my own).

Shown as it is, the notion that encryption of link R is sufficient protection for a secure application is clearly absurd. If the data are unencrypted everywhere else, then that's where our hacker will attack and with very little effort required. Ideally, the encryption must begin on the device A itself and end at server B and vice versa going the other way. The next best thing is to deploy WTLS, which means we achieve encryption at least from device A to gateway W. If configured correctly, we can also encrypt the path from gateway W to server B, but the gateway itself becomes a point of vulnerability, as previously discussed; however, we might be able to live with that for many situations in practice.

The ideal solution is true end-to-end encryption. Don't forget that in our network model the content network can include going out to an enterprise network from the mobile operator network or, as we will soon discuss, going out to a third-party network where our application is physically hosted. In these cases we still need to maintain the encryption to the source (destination) server, which is what our diagram indicates, but we may need to examine what this entails physically when we look at topics like *Web Services*, a method to extend our mobile delivery platform via the Internet to other platforms hosted elsewhere.

9.5 END-TO-END ENCRYPTION IN A MOBILE NETWORK

We have established that end-to-end encryption is the optimal solution for protecting confidential user information. Moreover, the use of encryption helps us to provide a secure channel for authentication; some of the encryption techniques even lend themselves to authenticating in a strong manner. With WAP 1, end-to-end encryption is simply not possible, though we can get quite close; matters can be made relatively secure if we can properly secure the gateway itself.

An important realization is that both with digital cellular networks and other wireless networks like 802.11, despite the ability to achieve over-the-air encryption, this does not allow us to be lax about application-level or transport-level encryption end to end. Air interface protection is only one small part of the problem. In fact, if application-level or transport-level encryption is in place, then arguably we can dispense with air interface encryption altogether – at least for data transfers; although we still need the encryption to protect other types of information such as voice traffic.

Given the issues with WAP 1: how can we achieve end-to-end encryption in a mobile application? Well, first, we should point out that our current security consideration has been within the context of established software architecture paradigms, particularly the client–server approach based on the link–fetch–response or browser modality using HTTP or HTTP-like protocols.

As we will find out in Chapter 10 when we look at devices in detail, there are several ways to develop an application for a mobile device. Browser-based presentation of a server application is only one method, but we can develop stand-alone applications too, which sometimes will want to communicate with a back end somewhere. However, even with these embedded applications, we may well end up choosing to use HTTP as the method of communicating with the back end, for all the reasons one would expect to do with the prevalence of the protocol and the infrastructure and tools supporting it, not least of which is the powerful J2EE platform itself.

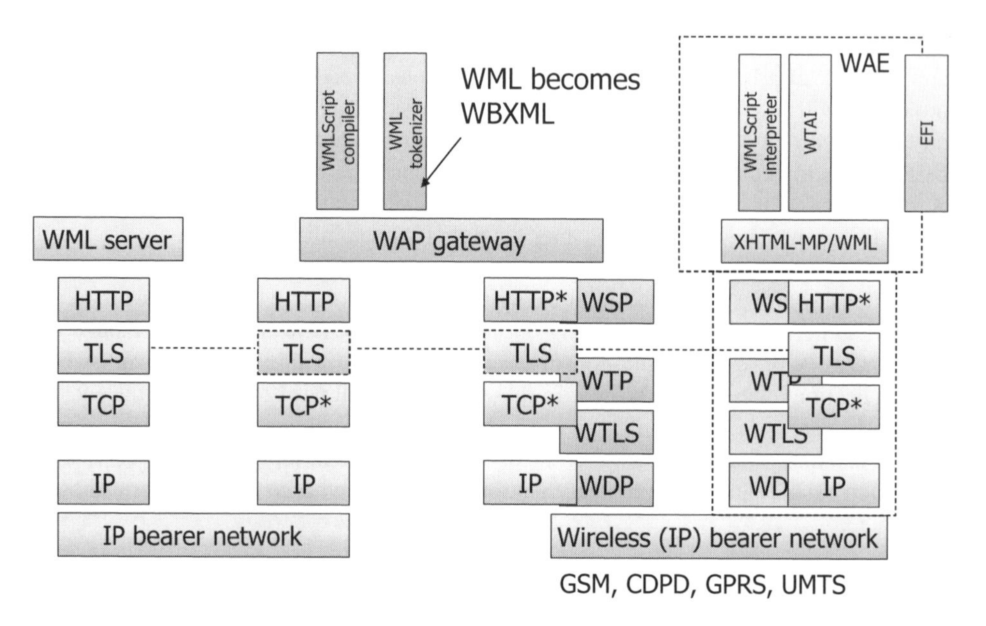

Figure 9.22 WAP proxy can still be present for WAP 2.

Due to its prevalence, we can expect to find HTTP software libraries on many devices. Wherever there is HTTP, there is usually HTTPS (secured) which is the naming convention for using HTTP over an SSL (or TLS) link.

For embedded applications, there is no need to do anything that precludes connecting directly with an origin server. Similarly, as we have seen with WAP 2, our user agent is capable of connecting directly to the origin server using native HTTP and TCP protocols, without a proxy. However, in both cases we may still choose to use or may have to use a proxy.

In the configuration of local area networks hosting our origin servers the network may be set up such that a proxy server is deployed as necessitated by the administration policy for HTTP access to networked resources. There are all kinds of reasons for this, which make it a common practice on many networks. We will not discuss the reasons here, but we will look at the implications for implementing end-to-end security using HTTP.

Even though the use of HTTP/TCP protocols is specified for WAP 2 browsers, we may still wish to utilize a WAP proxy. There are various reasons for this: some are general as found in standard HTTP networks and some are specific to the nature of optimizing and supporting wireless access.

First, the wireless profile of TCP is not going to be a standard configuration on most Web servers, even though we noted that the profile consists of extensions, or options, to the standard TCP implementation. This means that a particularly comprehensive implementation might be compatible with features of Wireless-profiled TCP (W-TCP), like *selective ACK* (SACK). However, if we want to be sure about gaining the benefits from its optimized performance, then this is one reason to deploy a WAP proxy, as shown in Figure 9.22. Here we can see the two WAP stacks side by side; this is how they would exist in actual products, should we choose to support both – which is what most WAP browsers will do for some time to come. WAP 1 and WAP 2 stacks are not interoperable, so the only way to support both of them is to run a dual-stack solution.

We can see from the figure that when we are running a WAP 2 device it can go via the proxy using TCP or W-TCP. The proxy can handle this and proxy the connection over a standard TCP link on the server side. If we are running with encryption, then the encrypted data are still pushed around using TCP, so there is no problem in going from W-TCP to TCP across the proxy. We can think of this simply as the way in which encrypted packets get queued up and sent and the way in which they get resent if necessary. In a moment we will consider what happens to the encrypted data at the proxy. We have already learnt that with WAP 1 protocols the data cannot continue their journey from here without being transformed, which means being unencrypted and re-encrypted so that transformation can take place. This is not required with WAP 2, as we will shortly learn.

There are other reasons why using a WAP proxy makes sense, even in a WAP 2 environment.[20] We discovered in Chapter 8 that part of the mechanism of managing Web sessions is to use the state maintenance features of cookies. Yet, not all devices support cookies to the extent that we might like. Some devices have a very limited cookie memory, and so extensive use of cookies across our applications portfolio could be problematic. Although there are solutions to mitigate this problem, like URL rewriting, we may still be better off sticking with cookies, especially as the use of cookies is part of the standard session management model for JSPS and servlets in the J2EE platform.

As well as these issues, the WAP gateway/proxy is often a more fully fledged product than simply a proxy: it may include WAP Push features, for example. Some gateway products also include mechanisms for keeping records for billing and can even offer so-called 'real time billing' capabilities, which is really a way of assigning different access metrics to different resources so that these can be billed later; this is also known as *content-based billing* or *event-based billing*. If this feature is required, then a mechanism for it needs to be provided somewhere in our content network. We could implement it ourselves as part of our J2EE application, but this type of feature is going to be a common requirement across all applications that need content-based billing in place. Therefore, we need a mechanism for enacting this across our applications, as a platform-wide feature accessible to all applications. While we could envisage a variety of solutions to this problem, why not use an existing network entity, like the WAP gateway, if it can do it for us?

Fortunately, when it comes to securing our applications, WAP 2 has the capability to support end-to-end security, even with a proxy in the way.

WAP 2 clients support TLS 1.0, as do most Web servers and J2EE application server Web engines. In order to support end-to-end security we use a technique called *tunnelling*; the name is fairly self-explanatory. What happens is this: the proxy that is under the instruction of the client passes all its data unhindered to the server and vice versa.

To establish a TLS tunnel, the WAP client uses an HTTP method called CONNECT, as defined in RFC 2817. It is required that the HTTP proxy server should support the HTTP CONNECT method in the same manner.[21] From the CONNECT method, the proxy knows simply to pass on the HTTP traffic to the server without touching it.

Any successful (i.e., 2*xx*) response to a CONNECT request indicates that the proxy has established a connection to the requested host and port and has switched to tunneling the current connection to that server connection. It may be the case that the proxy itself can only reach the requested origin server through another proxy. This may sound strange, but

[20] Our expectation is to run mixed WAP 1/WAP 2 environments for some time to come.
[21] http://www.faqs.org/rfcs/rfc2817.htm

this could well be the case if we use a WAP proxy for all the reasons just noted, but end up accessing a content network with its own standard HTTP proxy (for reasons we haven't discussed, but usually related to caching, security and access control, etc.).

In this case the first proxy makes a CONNECT request of the next (cascaded) proxy for a tunnel to the target server. Our WAP proxy should not respond to the WAP client with any $2xx$ status code unless or until it has either a direct or tunnel connection established to the requested resource.

We can see that the HTTP CONNECT method and the support for TLS at both end points of our service imply that end-to-end security is achievable.

10

Devices

10.1 INTRODUCTION

Recalling our collage of mobile ideas that dotted the prelude to this book, our imaginary lunchtime journey was highly connected with myriad information spaces merged with physical space via mobile technology. The access points to the information spaces were mobile devices. These appeared, even in our limited excursion, to come in all kinds of guises, from wristwatches snuggled on our wrists to cigarette box mobile gateways that sat snugly in our pockets.

By now we understand that a user can be mobile-networked by a variety of means and that there are apparently endless possibilities for engaging with useful mobile services. We now wish to look at the immediate human interface to the mobile network: electronic mobile devices of one form or another.

We wish to explore the possibilities for mobile devices in this chapter, investigating how they might work at a functional level. For the most part, we will not concern ourselves too much with device implementation issues involving the underlying physical electronics. Where relevant, it may be valuable to consider certain issues concerning the electronics, such as power consumption, memory capacity and so on, but still only at a general level.

For good reason, from the start of our exploration of mobile services we did not relegate the device considerations to any form factor or incarnation that might be suggested by the history of the mobile phone, especially the notion that we will only have or require one device to enable us to become mobile-networked. Of course, it is tempting to think in this mould, because all of us have mobile phones and the movement to mobile services is often construed as an extension or evolution of mobile telephony, courtesy of the addition of mobile data technologies.

Our notion here is different. We have made various attempts to define and examine the mobile services landscape in general without confining how mobile connectivity might be

Next Generation Wireless Applications P. Golding
 ISBN: 0-470-86986-0 (HB)

realized. For example, it is just as feasible for a mobile service to be realized through WiFi as it is through a 3G (Third Generation) connection, or both. It is just as feasible for mobile devices to talk directly to each other peer to peer (P2P) as it is for them to connect to some remote server, most likely, as we have suggested and reinforced, via an Internet Protocol (IP) network.

However, we have not restricted ourselves to IP connectivity either. A P2P connection could use any low-level protocol to shift the bits physically from one application to another. Along the way, we have also understood that the mobile network is not just about people injecting and gathering meaning from the clouds of information that exist in our virtual information space. Software too, especially with the help of the Semantic Web, can participate in more ways than first anticipated, actively seeking out opportunities and information on the users' behalf or, as strange as it currently seems, on their own behalf.

The myriad ways of handling information in the mobile network and the rapidly developing device market (still thanks to Moore's Law) leads us to think, as already suggested, that an appropriate model for the device end of the service delivery chain is to think in terms of a device network. This can be inter or intra-personal. Attached to that network can be devices of various ergonomic propositions: a Bluetooth™ pen has no graphical user interface whatsoever, neither does a personal mobile gateway, but a wireless-enabled personal digital assistant (PDA) or a wireless-enabled laptop both have powerful graphical interface capabilities, the laptop especially so.

In developing our understanding of the device network, we need to take into account the possible device ergonomics, the various networking propositions and the primary functions that each device or a combination of devices is trying to provide to the user in an effort to support mobile services as per our definition in Chapter 3.

To approach this topic, a useful starting point is to construct an abstract high-level view of the device by representing its functions as a set of layers, as indicated in Figure 10.1.

We can think of the software paradigm for a device as being centred on a simple process: that is, to respond to either interface events or network events, then react to those events with some kind of processing in order to carry out a useful *task* and then send output back to the network or to the interface. As a set of tasks this process can be coordinated to make a useful mobile service. The granularity of this process may vary: from a large number of tasks being required to carry out a useful service to perhaps just a single task. There is another type of event that is independent of the interface and the network: a time-based event driven by the clock.[1]

This software paradigm can be summarized visually, as shown in Figure 10.2. This simple model is the most basic operational view of a mobile device. Of course, we will expand on this throughout the remainder of this chapter, talking about how this model actually works and the various ways that it could be implemented.

10.2 INTERFACE ELEMENTS

Somewhere in the device network there needs to be a sensory interface between the device and the user. We will not spend too long considering this layer, but it is useful to summarize

[1] A mobile device will usually include a real time clock on board which knows the time, keeps 'ticking' away and is able to notify the software services that a particular time has been reached.

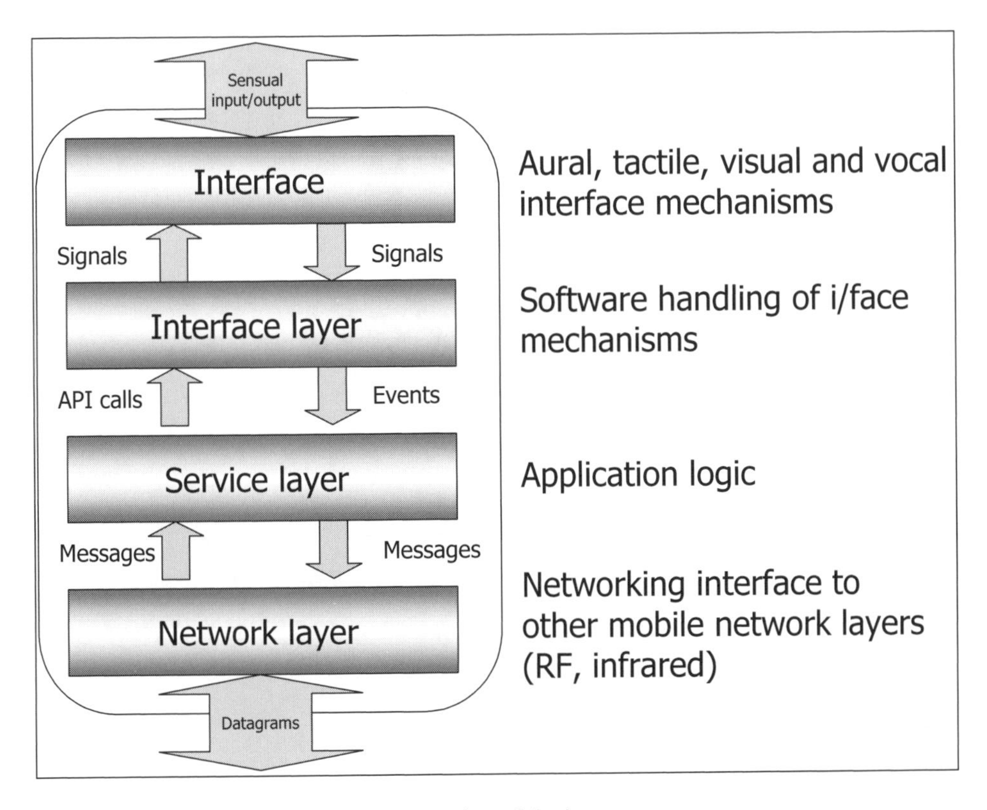

Figure 10.1 Abstract, layered representation of device.

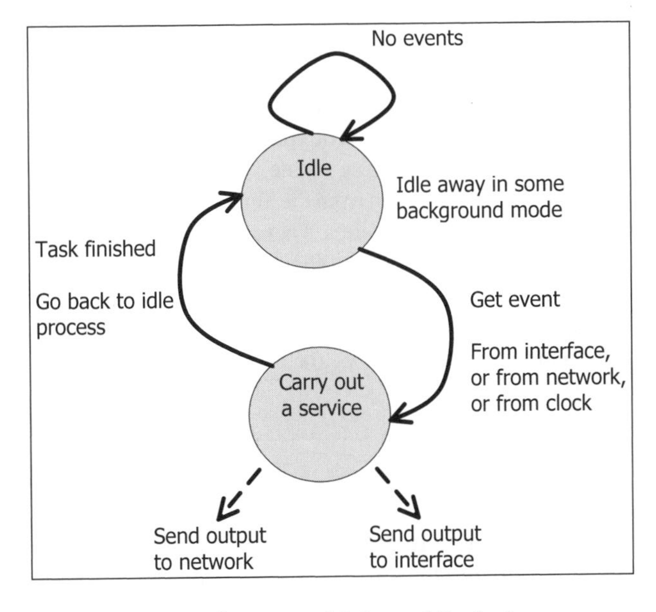

Figure 10.2 Basic (simplified) software model for mobile device.

Figure 10.3 PDA showing built-in miniaturized keyboard.

its characteristics. The most common types of interface can be divided into the following categories:

- tactile (pertaining to touch);
- aural (pertaining to hearing);
- vocal (pertaining to voice);
- visual (pertaining to sight).

10.2.1 Tactile Interface Elements

This is the ability to interact with the device via the hands and fingers, such as through a keyboard or pointing device (stylus) in conjunction with a reactive touch display. These interface modes are designed for enacting commands and for inputting text, which could be in any language or some kind of drawing input.

Keyboards can come in various modes: from the alphanumeric 12-key keypad common to mobile phones to full-blown qwerty keyboards. There are also many versions of miniaturized keypads that are added to PDAs, as shown in Figure 10.3.

For smaller devices, such as standard mobile phones, new and innovative keyboard designs, such as Fastap™ by Digit Wireless,[2] enable a greater density of keys to be achieved than was previously thought possible, thanks to clever ergonomic design concepts (see Box 10.1: 'Fastap to a quicker message'[3]).

Box 10.1 Fastap to a quicker message[3]

Today I met with one of the guys from DigitWireless. I previously blogged about their incredibly clever keyboard design, which they've trademarked as 'Fastap'.

[2] http://www.digitwireless.com/
[3] Taken directly from my weblog: http://www.paulgolding.info

Today I saw it, played with it, loved it! I not only played with the neat demo model – an anonymous phone mock-up with the Fastap keyboard – but also touched and felt a production-ready unit under test by the operators (can't say who – sworn to secrecy).

Only when playing with the keys did I at last appreciate the genius of the design (see image below). Its simplicity was shaking and its inventor sure has ergonomic insight. The principle is that the letter keys are raised above the phone face, like hills, and the digit keys are sunk in-between, like valleys; that bit I understood already. Each raised key has enough space around it to more or less (a little less) occupy the space of a key on a qwerty keyboard – believe it or not! The finger does not clumsily hit the numeric keys by mistake, as these are sunk, creating the impression there's no key there at all. I loved that magic bit! By placing the finger on the sunken valley, over the desired digit key, the finger can't help but depress some combination of the surrounding raised hill keys and the software picks up the combination and knows how to map it to the desired digit. Genius!

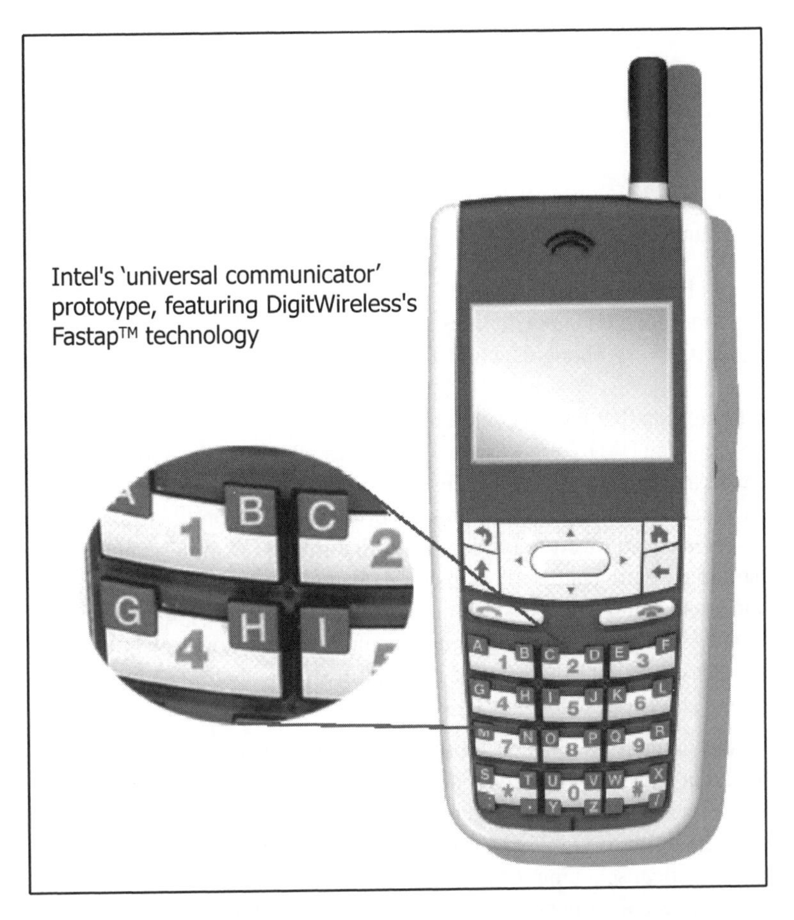

Intel's 'universal communicator' prototype, featuring DigitWireless's Fastap™ technology

So what's my reaction? Is this another tech idea that has no real application? Absolutely not! This idea must surely come to fruition and I can feel that it has the potential to become commonplace. The reason is that it simply is a better way to text, especially for those of us who cannot get to grips with predictive text (PT) input. I am a good user

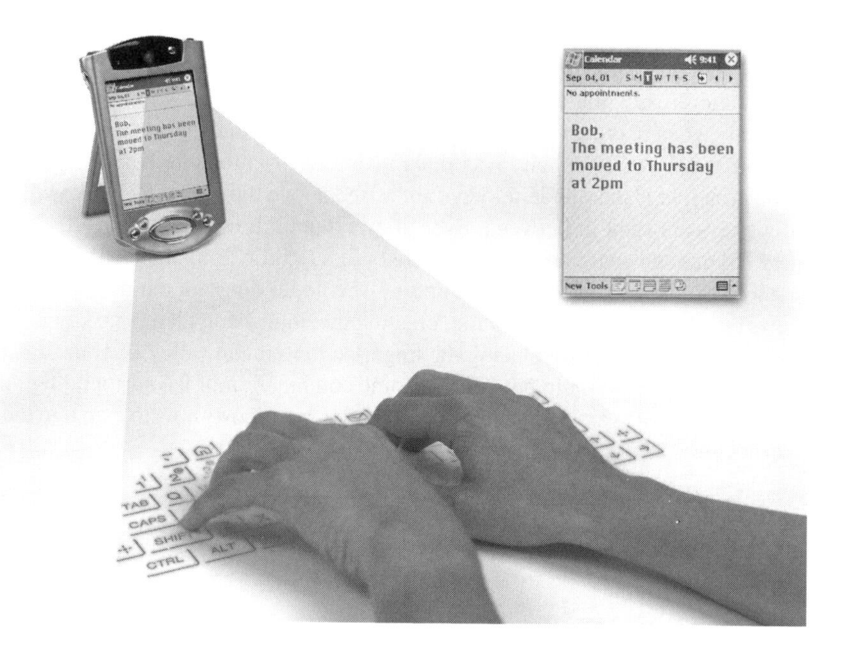

Figure 10.4 Light-projected keyboard attached to a PDA. Reproduced with permission from Cannesta.

of PT, until I hit a word that's not in the dictionary or I want to deliberately abbreviate. However, most people I know don't use PT or find it somehow clumsy; although, to be fair, I think it is very phone-dependent (T68i is one of the best implementations I've seen).

Another reason that Fastap could be in demand is the gathering interest in wireless Instant Messaging (IM). Certainly, new initiatives, like Wireless Village (and the PAM Forum-entering Parlay) are generating momentum toward a big push for IM into the wireless space, but this push overlooks the not-so-small problem of boringly slow typists – referring to qwerty users! I would like to use IM on my phone, but it's just too frustratingly cumbersome: Fastap could change that. If it does, then I think it will be a winner and could well assist IM take-off along with email, of course, which is the other mobile service still waiting to gather momentum, though its problems are not just character entry, but often poor usability, full stop.

As probably already familiar to most of us, keyboards do not have to be mechanical devices at all. On most PDA devices available a small virtual keyboard can be displayed on the screen. This virtual keyboard simulates both a numeric keypad and a qwerty keyboard.

The other emerging possibility is the use of light-projected keyboards that use clever projection and optical scanning techniques to project a keyboard image onto a flat surface and then sense where the fingers are on that image, as shown in Figure 10.4.

The technologies used by Canesta,[4] the inventors of the projection keyboard, are able to sense where objects are in 3-D space. This technology has wider applications for mobile device input mechanisms, as Canesta mentions on its website:

[4] http://www.canesta.com

Electronic perception technology permits machines, consumer and electronic devices, or virtually any other class of modern product to perceive and react to objects and individuals in the nearby environment in real time, particularly through the medium of 'sight,' utilizing low-cost, high-performance, embedded sensors and software.

This technology works by attempting to assign to each pixel of visual information a distance metric from the camera so that objects within the image can be assigned spatial depth. This is done using radio waves in a similar fashion to how radar signals are used to detect the distance of remote objects, such as aircraft, although clearly at a different range altogether.

In modern computing devices we are now used to the familiar point-and-click interface paradigm, as popularised by the windowing style of interface, an idea that actually grew out of intent to provide an interface method that would correlate to an object-oriented programming paradigm. This idea was first commercialized by the labs of Rank Xerox.[5]

The point-and-click interface has a variety of tactile possibilities, such as dragging, bounding, resizing and other visually guided tactile moves that come from point-and-clicking in a Windowed world. These originally arose from the use of a mouse as the primary point-and-click device. A standard desktop mouse is not a useful solution for most mobile devices, so other options have arisen, most notably stylus pointers used with PDAs and tiny joysticks now available on some mobile phones, including 'smart phones'.[6]

Tactile Feedback It is also worth mentioning that tactile feedback or alerting is not only possible, but is already widely used in mobile phones in the form of vibrating alerts. Such tactile feedback systems can be used in other ways, such as giving the sense of collision (e.g., racing games) or explosion (shoot-'em-up games).

10.2.2 Aural Interface Elements

The use of audio output is an obvious feature in a mobile phone where we need to be able to hear the other person or people (e.g., in a conference call). It is possible to either have an audio output connector for a headset or to incorporate a speaker. Speakers can be low power, such as the tiny speakers used in a handset for placing next to one's ear or could be higher powered, such as a hands-free solution might predicate. Higher powered speakers will place a greater drain on the battery, so they are not common in devices with small battery capacity.

Of course, on laptops, speakers are common. However, on smaller form factor devices, such as PDAs or phones, then speakers may or may not be included. We have already noted that for certain applications, such as push to talk (PTT) mentioned in the book's prelude (Chapter 1), the use of a built-in speaker may improve the overall user experience.

There is no need to confine aural output to voice-generated signals. We can deploy aural output to support text-to-speech applications. This would be especially useful for accessibility for sight-impaired users, but can also be deployed for convenience or safety, such as reading of emails while travelling in a car. With new technologies, such as AT&T's

[5] The Xerox 8010 'Star' Workstation was the first commercial realization of the ideas on object-oriented user interfaces developed at Xerox Parc in the early 1970s.

[6] There is no standard definition of a 'smart phone', but originally the term arose to explain a device with ostensibly a mobile phone form factor, but with PDA-like capabilities. However, the use of the term has become blurred and we only mention it for completeness. Our consideration of devices incorporates all possibilities, no matter their marketing terms.

Natural Voice^{TM}, it is possible to achieve text-to-speech output that is increasingly natural sounding, as if read by a human rather than produced by a synthesis of artificial vocal sounds (phonemes).

Aural output can also be fed via a Bluetooth^{TM}-connected earpieces, which are becoming increasingly commonplace.

10.2.3 Vocal Interface Elements

A microphone on a mobile device is necessary due to the telephony requirement, but there are other applications that might benefit from vocal input, such as recording voice notes, voice control and speech-to-text conversion.

Increasingly, digital signal-processing (DSP) techniques can be used to assist with voice input. One possibility is noise reduction: to remove the unwanted background noise that might be encountered in a travelling vehicle or a crowded place. All of us know the difficulties in trying to be heard (or in hearing) while using a mobile device in a crowded room. DSP can be used to address such problems, but it always comes at a price: the hardware and power consumption differentials implied by the computationally intensive algorithms that the smart guys have invented to ameliorate such interface problems.

DSP can be applied in ways that are already used in immobile applications, but have not yet penetrated into the mobile device family. One such application is audioconferencing, where a mobile device could be placed in the centre of a group of people to allow a fully duplex audioconference call to be placed. This would imply the need for a speakerphone and circumferential microphone pick-up technology. It also requires DSP to eliminate echo and to enhance audio perception and the quality of signals generated by a group of talkers.

We could well imagine a useful combination of conferencing technology with the electronic perception technology mentioned earlier, as deployed in the light projection keyboard. For example, a group of users could sit around a device and all take part in a simultaneous chat session alongside an associated audio call, courtesy of obliquely projected keyboards spread across an arc on the desk. This idea could also be used for games, such as multiple choice quizzes and so forth.

10.2.4 Visual Interface Elements

We are familiar with ever-richer graphical displays courtesy of liquid crystal display (LCD) technologies that are improving all the time. It is commonplace now for new mobile devices with the smallest screens (e.g., mobile phones) to have colour displays. PDAs have increasingly better displays with better contrast[7] capabilities and higher resolution, all deliverable with achievable power consumption that enables portability.

Newer LCD technologies, such as the reflective variants, also enable better power consumption due to their ability to harness ambient light to enhance the display, rather than backlighting. Reflective LCD technology makes use of only the surrounding ambient light to illuminate the display. Beneath the surface of the display a rear polarizer is combined with a reflector assembly to bounce back ambient light to aid screen visibility.

[7] Contrast is the ability to discern between different colours, especially in the presence of high ambient light levels. Clearly, displays should aim for high-contrast capabilities so that images can be clearly seen, including by visually impaired users.

Transmissive LCD display technology is how a normal display operates. To view the screen, there must be a continuous backlight, and this must be brighter than ambient light to save the screen from looking washed out and subsequently becoming difficult to read. These displays are naturally very power-hungry and operate best in darker lighting conditions.

Transflective LCD technology is a combination of reflective and transmissive types. The rear of the LCD's polarizer is partially reflective and combined with a backlight for use in all types of lighting conditions. The benefit here is the backlight can be switched on only when there is insufficient outside lighting. Conversely, when there is enough ambient light it can shut off to conserve power. Transflective takes the best of both worlds and enables viewing in dark environments. Additionally, the display won't 'wash out' when viewed in direct sunlight. However the contrast rating is not as high as a purely transmissive display.

There is a high degree of association between display technology and tactile input methods – the most obvious being the manner in which a windowed graphics environment requires point-and-click navigation, manipulation and data entry. Therefore, a pointing device has to be able to move a cursor on the screen or be able to click active areas in the parts of the windowing display responsive to tactile control.

With stylus pointing devices, the LCD screen needs to be sensitive to stylus position and pressure. This is achieved by a translucent overlay membrane that is able to act as a transducer from mechanical pressure to electronic signals which indicate whereabouts on the membrane the stylus is applying pressure.

There are a variety of types of touch technology available, but the five major ones include analog-resistive, capacitive, infrared, acoustic wave and near-field imaging. Of these, analog-resistive are commonly used for small displays, such as PDAs.

Analog-resistive touch technology is comprised of a glass overlay that fits exactly to the shape of a flat panel display. The exterior face of the glass is coated with a conductive, transparent layer. A clear, hard-coated plastic sheet is then suspended over the glass overlay. The interior face of the plastic sheet is also coated with a conductive layer. Between the glass and the plastic sheet there are thousands of tiny separator dots about one-thousandth of an inch thick. When a stylus applies pressure to the surface of the display, the two layers make contact and a controller (i.e., silicon chip processing device) instantly calculates the x and y coordinates. This accounts for analog-resistive overlays' very high positional precision (i.e., fine resolution).

The combination of affordable, low-power touch screen technology with affordable LCD displays has enabled the mobile device market to become established. The transition from monochrome to colour displays is now fully under way, and only devices that are very much budget-priced will continue to have monochrome displays, although for some manufacturers of mobile phones the inventory costs are altogether reduced by standardizing on colour displays across an entire product range.

Display size is still an issue, especially where portability is a key concern. Clearly, the more portable something has to be the smaller and lighter it also has to be. This invariably means a small display and immediately presents usability challenges, ones that impose a different set of considerations when applying the windows design metaphor. Usability is made all the more challenging by the obvious tactile differences between a stylus and a mouse, plus the ergonomic differences, particularly the ability of a desktop mouse to accommodate more than one button and even more interface devices, such as a scroll wheel.

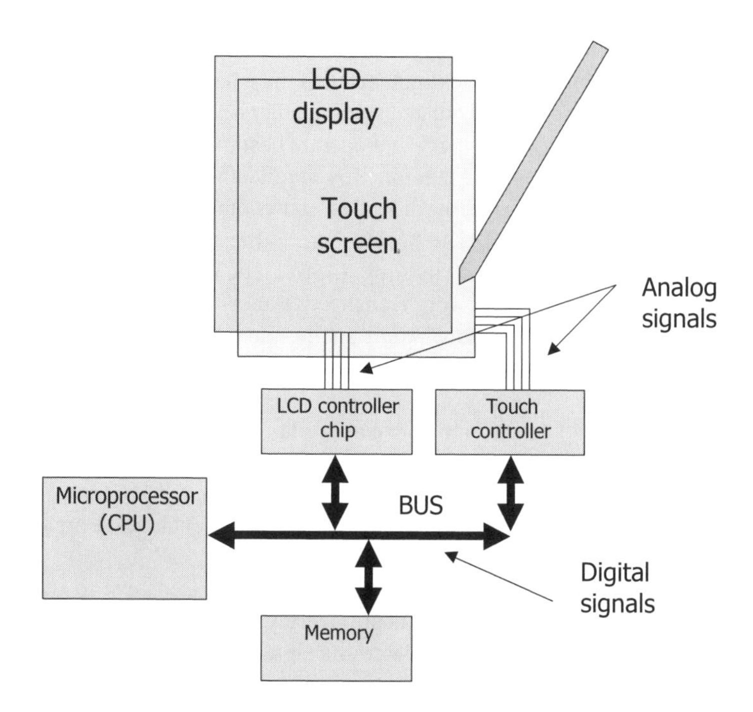

Figure 10.5 Microprocessor interface to interface peripherals.

10.3 INTERFACE LAYER

The interface layer is what enables the electronic signals in our interface devices to be understood by the software services layer. Any interface device is supported by a silicon chip device that handles electronic signals – usually analog; to and from transducers – and electromechanical or electro-physical components.

What we eventually want is for our services software to be able to drive interface devices or to respond to their outputs in response to stimuli from the outside world, usually due to human interaction. The controller chips for interface devices can be controlled using low-level digital commands issued from a microprocessor or similar, such as a microcontroller or general purpose DSP device. This would usually be the same processor that runs our services layer. Each controller chip, such as a controller for an LCD display, has its own specific set of commands that it will respond to via its microprocessor interface; this interface is usually referred to as a bus, as shown in Figure 10.5.

The microprocessor will run all or most of the custom software on our mobile device. There may well be an additional processor to carry out any DSP functions, such as speech processing. We will be looking at possible mobile device architectures later, but for now our concern is with the handling of the interface peripherals. We have posited a layer in the software that is called the 'interface layer'. This consists of a set of device drivers. These are specialized, low-level pieces of software that contain all the subroutines required to control the peripherals and to pass data to and from them.

The role of the device driver is not only to handle its associated peripheral but also to present a simplified software interface to the service layer in order that the programmer of the service layer does not have to concern themselves with the details of peripheral operation. For example, the command to display a blue dot (pixel) on the LCD display may require a whole sequence of binary words to be written to the appropriate memory location on the LCD controller chip. However, the service layer would just like to issue a command like 'dot (120, 160, blue)' where 120 is the x coordinate and 160 is the y coordinate.

The device driver enables this abstraction of the interface to be achieved. However, it is usually the job of yet more software to make an even more powerful and abstract interface for the service layer. Perhaps the service layer would like to issue commands like 'rectangle', 'button' or 'menu' and for the appropriate display widgets to be enacted on the LCD. We can imagine an entire chunk of software dedicated to providing a windowing interface paradigm to the higher software service layer.

We should remind ourselves that the set of possible peripherals for interfacing is increasing all the time, so we should not get overly fixated about displays. The inclusion of cameras is increasingly common on mobile devices of all form factors. A camera is based on a light-sensitive matrix and is like an LCD in reverse: instead of writing to the controller chip to energize various pixel locations on the screen for subsequent lamination (reflective, transmissive, etc.), we grab values from the camera matrix to understand input light values.

A camera could involve us having to handle a continuous stream of images, just as with a video, so that we can subsequently grab images – either stills or sequences. We can imagine that to keep processing an inbound stream of digital light values is a computationally intensive task for a microprocessor. Therefore, the controller might be designed to grab images and write them directly to an area of memory on the bus without processor intervention: a technique called direct memory access (DMA).

We do not want our service layer to be concerned with the low-level mechanics of controlling a camera or any peripheral; so, we would expect the device driver to do a lot of the work for us. Just as with the windowing display, we would also hope that we can find at our disposal a piece of software that can provide high-level interface primitives for the service layer to call on, such as 'grab image', 'rotate image' and so on.

What we are describing here is the availability of toolkits or libraries of pre-written software that we can use to take care of housekeeping functions, such as windowing and image manipulation, leaving our service software to concentrate on delivering or facilitating the actual mobile service we have in mind, like picture postcards (photo messaging) or videoconferencing.

10.3.1 Interfacing via the Network Layer

It is important to note that sometimes the human interface is not on the mobile device that runs the service layer. There could well be a separate interface device that communicates with the main device, or more precisely the software services layer, but does so via a networked connection. An example of this is the use of Bluetooth™ (BT) interface devices.

The most common device is probably the BT headset, in which the audio input (vocal input) is streamed across a BT connection to the processing device. Such a configuration could also be used for the voice control of a device, not just for telephony.

Another example of a BT-connected interface is the BT pen, an input device that enables ordinary handwriting on either normal or special paper which can be used to capture textual or graphical input for our services layer. This penned input could be used in raw form as vector or bitmapped graphical input, such as for taking free-form notes or making sketches,

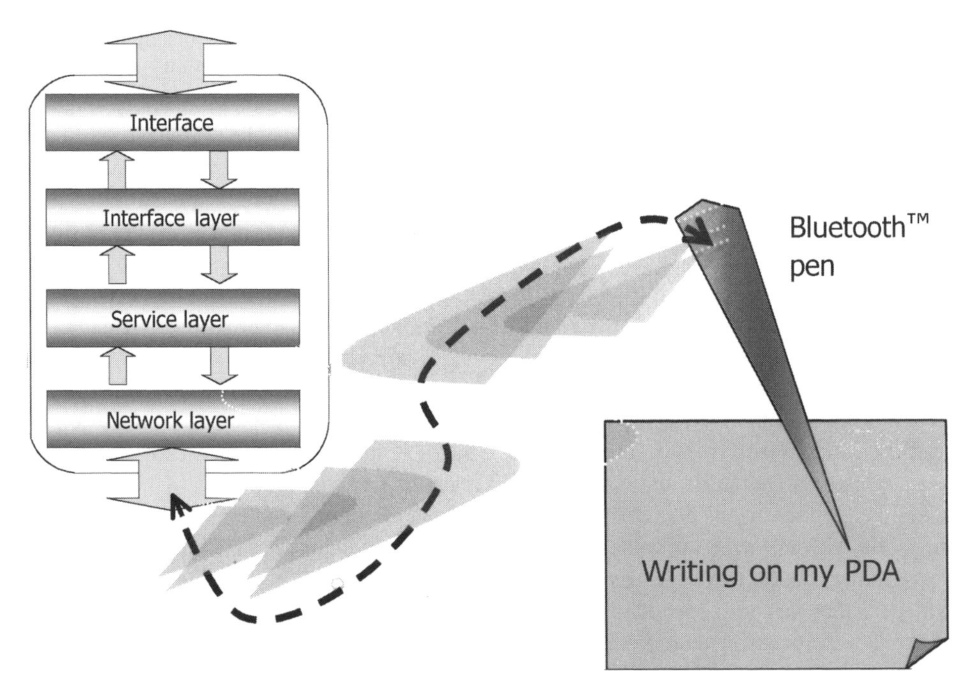

Figure 10.6 Textual and graphical input via a Bluetooth™-connected pen.

or can be used in conjunction with a handwriting recognition process to convert the pen marks to character form, such as ASCII or UTF (two different standards for digitally encoding characters).

With any of these usages for computerized input peripherals, interface handling is no longer carried out locally and the services layer does not talk to a local software process that handles interfaces. Instead, the interface dialogue is taking place via the networking layer, such as through a BT protocol stack in the case of the examples we have just considered. This configuration is shown in Figure 10.6.

10.4 SERVICE LAYER

The service layer is the 'business logic' of our mobile device application. A mobile service in its entirety can involve many interlocking functions, services and software spread across the entire mobile network. There are three modes of implementation:

1. Client/server.
2. Peer to peer.
3. Stand-alone.

These are purely modes of implementation, mostly reflecting the mode of collaboration with other software services in the network; they are not categories of application. For example, an application like a calendar tool may well be able to operate in all three modes of operation depending on the context of use.

A user may wish to consult with their calendar, just to read appointments and survey the coming day's events, etc. This would seem not to require access to other services in the network, which makes it a stand-alone application. However, even this apparently innocuous requirement requires a lot of processing on the device, so we should not think of it as a trivial task just because we are not involving any networked activity at this stage. For a moment, let's pause to consider what the processing requirements might be:

- accessing a particular area of memory to read calendar values;
- processing the calendar values to transform them from an internal digital representation to a human-readable one;
- creating a whole set of display widgets to display a calendar on the screen;
- rendering the graphics widgets on the LCD display, pixel by pixel;
- reacting to point-and-click stimuli on the touch screen;
- converting the point-and-click stimuli to appropriate calendar-handling functions, such as 'get next month';
- converting such functions as 'get next month' to a meaningful set of internal instructions to gather the appropriate data from memory and assemble them into a particular order;
- generating textual (character-based) output on the LCD screen and in the right place.

We can see that there is a lot of processing potentially taking place just to display a calendar in stand-alone mode.

The calendar application can also operate in other modes. We could find other peers to interact with and swap appointments. This would involve using the network layer to communicate with them; we will describe the network layer in a moment. We could also interface with software services running on the Internet: for example, to share calendar spaces or to download appointments from event calendars, such as sports events, exhibitions, shows and so on. We may also want to connect with a calendar in our enterprise network and synchronize with it.

A great deal of what we want to do with a mobile device may involve common functions across a wide variety of applications. For example, accessing a calendar might be a common requirement, as might accessing an address book or placing a call. There are many such functions that are potentially common to some or all applications. It would be advantageous to provide a set of common software building blocks to take care of these functions and for them to be available for the core software to call on. This concept can be achieved in a number of ways, and we will be examining them throughout the rest of this chapter and in Chapter 11, when we look at particular device software techniques, such as Java 2 Micro Edition (J2ME).

Later in this chapter we will propose a possible generic device architecture that will take into account some of these common elements.

10.5 NETWORK LAYER

Finally, in our layered abstraction of a mobile device that we have been discussing (see Figure 10.1), we have the network layer to consider. This layer is concerned with getting data on and off the device, so it is a key enabler of our mobile service world. Part of the network layer is the actual hardware to achieve wireless networking. We have already introduced some wireless-networking technologies, particularly those that enable us to connect to the IP network layer in our mobile network model (namely, digital cellular technologies and wireless LAN technologies). In Chapter 12 we will describe in detail the Radio Frequency (RF) network and explain its functions and characteristics.

In terms of the implementation of a wireless connection on a mobile device, we can think of the wireless hardware as a separate entity or module, usually referred to as a modem (stands for modulator/demodulator).

An RF modem is a peripheral on the mobile device, just as the interfaces usually are (e.g., LCD display driver chip); so, we can generalize the previous discussion about device drivers and software abstraction and apply the same principles to the task of interfacing to the modem. Typically, the modem will have a controller that facilitates the low-level software messaging interface with the microprocessor in our mobile device via the bus. This controller chip will have its own command set and protocols for receiving data via the bus to drive the functions of the modem. Therefore, on the main device processor we will need a device driver (software program) to handle the low-level control commands and data flows on the bus, while presenting these in more abstract terms to our service layer software. For example, the process to initiate a call via the modem might involve the writing of an elaborate software message to the modem's controller input on the bus. Rather than have to know how to do this and the formatting of the low-level message, the service layer programmer would rather just be given a simple software command, like 'MakeCallTo(this number)'. It is the job of the device programmer to translate this higher level command to the low-level message format and to take care of using the appropriate mechanisms to get the message to the controller's input on the bus.

Very often, a mobile device manufacturer will use a modem chipset from a third-party provider who specializes in modem design and manufacture. The chipset will come with data sheets that specify how to interface with the controller functions within the chipset. However, it is common these days for the chipset provider to also supply device drivers and libraries of software that make interfacing to the set even easier. Later on we will examine the various ways and levels that these libraries can be integrated into the device and how this eventually impacts on the development process for service layer software.

The software interfaces that our service software has to deal with may vary from device to device depending on the particular underlying software architectures that the device manufacturer has chosen, such as the operating system and the associated support for programming languages like Java. One can argue that much of the value added and skill in device manufacturing these days is in how effective the underlying software platform is in making it easy for the developer to quickly develop powerful applications for the device and being spared the low-level operational and implementation details. This will affect how much the device gets used to support various mobile services. In turn, this will affect device popularity and hence sales.

The software interfaces to the modem relate to how the modem is physically integrated into the mobile device. The modems may be fixed and part of the device circuitry or they may be pluggable into expansion slots. In the latter case the expansion slots will have their own controllers on the bus, such as a Personal Computer Memory Card International Association (PCMCIA) controller for PCMCIA peripherals. Industry standard expansion slot mechanisms have their own associated signalling protocols on the expansion bus. This is usually implemented by incorporating a controller device on the main processor bus that bridges to the expansion bus. This implies the need for a software handler to control the bridge.

As we will see, the way that peripherals are handled by the service layer software depends ultimately on how the device manufacturer chooses to integrate these peripherals and make them available to the upper software layers (service layer). Most often, all the peripheral handling in a device is glued together by the operating system, which might be custom-designed and proprietary to a particular device or family of devices, or it may be an industry-wide operating system licensed from a specialist supplier of mobile operating systems.

It may seem strange to the uninitiated, but, typically, an RF (Radio Frequency) modem is like a computer within a computer as far as our mobile device is concerned. The inner workings of the modem may well involve its own microprocessor, memory, bus, peripheral connections and so on. Most digital wireless transmission techniques involve huge amounts of processing at the air interface level, just to achieve modulation and demodulation, RF interference mitigation and so on. Then, on top of the raw processing of bits in and out of the antenna, we have the many protocol layers dictated by the particular RF standard in consideration, such as GPRS General Packet Radio Service, UMTS Universal Mobile Telecommunications System or WiFi.

Due to the amazing achievements of silicon design, entire systems are implemented on a single chip; so, many of the modem elements just described may well all be on the same piece of silicon, packaged into a single chip carrier.[8] This contributes enormously to lowering costs and power consumption. It is worth noting that there is an order of magnitude difference in power consumption between driving digital signals (by switching transistors on and off) across a silicon chip rather than across a printed circuit bus (metallic tracks on some kind of insulating substrate). The electrical properties of printed circuit tracks (e.g., capacitance and inductance) are such that electronic components have to drive output signals with a much greater push (energy) to get the electric currents flowing across the circuit tracks. Systems on silicon, therefore, usually offer a saving in power consumption.

It is useful to take a brief look at what these modems might look like at the functional level; so, let's take an example of a wide-area RF modem that is both UMTS and GRPS-capable (as we would commonly expect for wide-area devices, until such time as UMTS subscriber penetration completely outstrips the need to maintain GPRS networks). Figure 10.7 shows a reference design from Texas Instruments which shows various components required to implement a UMTS/GPRS modem. Let's take a brief look at the components in the reference design:

- *UMTS (WCDMA) RF transceiver* – takes care of the actual RF interface, processing the modulated signals in accordance with WCDMA principles as discussed in Chapter 12.
- *GSM/GPRS RF transceiver* – takes care of the GSM/GPRS RF interface, processing the modulated signals in accordance with the GSM TDMA principles as discussed in Chapter 12.

[8] A chip carrier is the physical packaging that bonds the silicon chip to the input and output pins that are then soldered to the printed circuit board hosting the device.

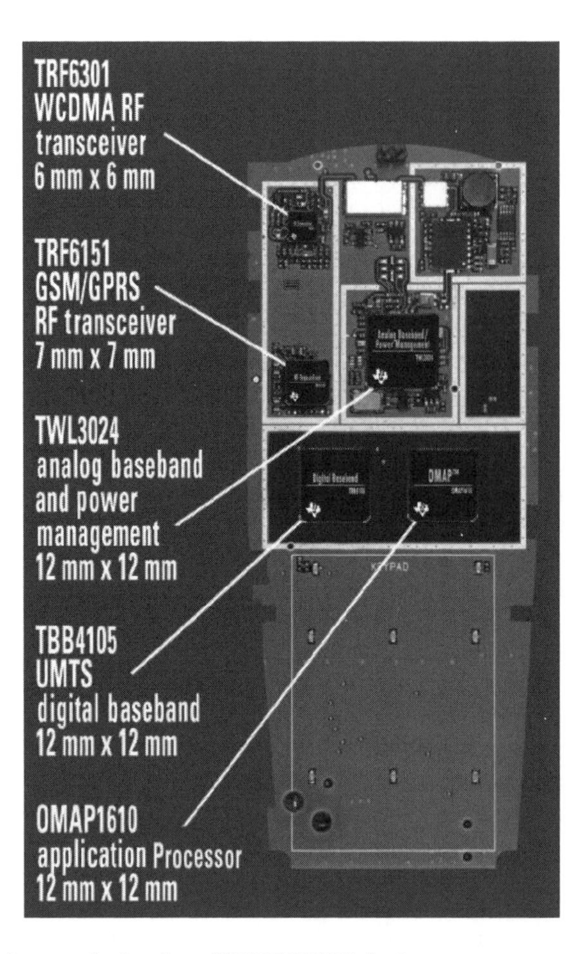

Figure 10.7 A reference design for a UMTS/GPRS device.

- *Analog baseband and power management* – this chip brings the RF signals down to their unmodulated frequencies and handles audio-processing aspects in the case of voice and digital data stream interfaces in the case of data. It also includes power management functions for the entire chipset, including intelligent battery-charging supervision.
- *UMTS digital baseband* – a good deal of DSP is required to enable digital cellular transimission to reliably take place, such as error coding (convolution coding) that enables the hostile effects of the RF path to be mitigated. Ciphering is also carried out on this chip with the help of acceleration processors that offload some of the intense cipher calculations from the core processor on this chip.
- *Application processor* – this is an entire mobile device processor on a chip, as shown in Figure 10.8, which includes many of the peripherals we have discussed so far. Where some interfaces seem to be missing, this is because they are present on other devices in the chipset. For example, the touch screen interface is on the analog baseband chip.

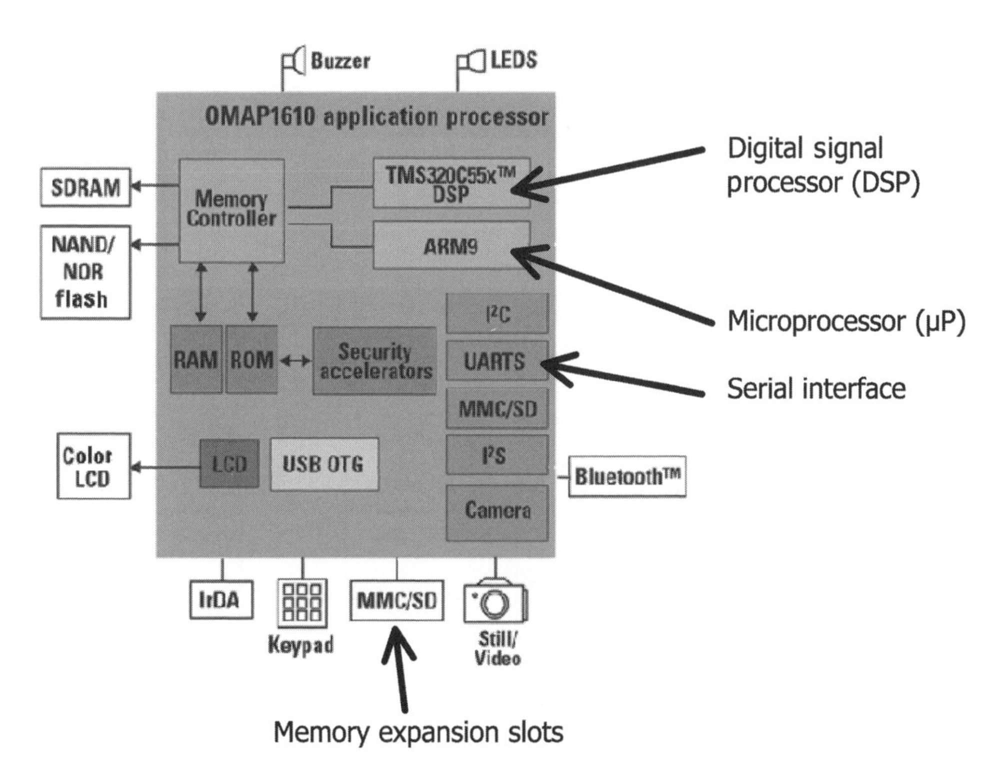

Figure 10.8 Application processor architecture from Texas Instruments' reference design.

In this section we only wished to focus on the major characteristics of the network layer, and hence the issue of what software is required to support the reference design shall be deferred to our later discussion on generic device architectures (Chapter 11).

In principle, the support of other wireless interfaces, such as Bluetooth™ and WiFi, is a similar affair. We expect to incorporate a similar modem chipset that presents an application interface to the core of our mobile device platform. The actual nature of the interface, how it is implemented and how it is presented to the core is a matter for further elaboration later in this chapter.

Throughout the book we have seen that there is usually an intimate relationship between mobile applications and networking protocols like HTTP (HyperText Transfer Protocol); hence, we would expect there to be a means of providing a way for our service Layer on the mobile device to easily gain access to such protocols. What we require is a software library on the device that can support HTTP without that much care about how this is implemented in the network Layer. Such a requirement seems sensible, and, as we will see, it turns out to be practical and increasingly available on many mobile device platforms.

10.6 ROLE OF DSP IN DIGITAL WIRELESS DEVICES

It is useful to review the basic architecture of a 2G (Second Generation) mobile telephone, such as a GSM (Global System for Mobile Communications) phone. This will enable

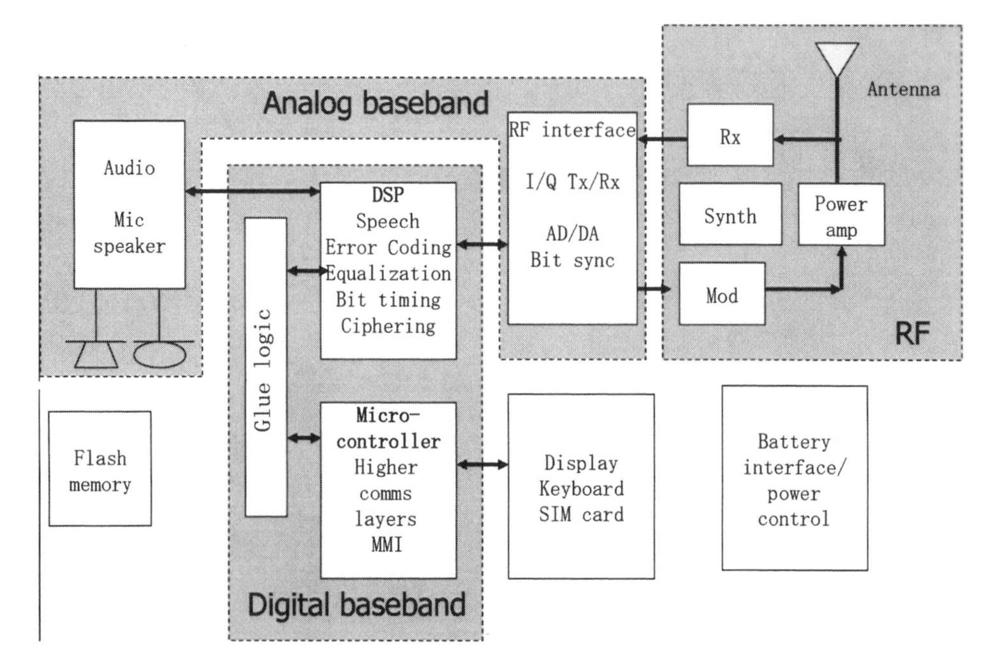

Figure 10.9 2G digital cellular phone architecture.

us to understand what the basic building blocks are of a digital communications device, particularly the modem aspects such as the RF and baseband modules, and to see how DSP processors are really at the heart of digital cellular (and wireless) devices.

We can see the basic components of a digital cellular device, such as a GSM phone, in Figure 10.9. We will briefly explain the functional blocks, just for completeness, though it is not really our intention to examine digital wireless modems in any more depth than this in this book.

10.6.1 Radio Frequency

Modulator/Power Amplifier The raw bit stream that we wish to transmit is used to create a corresponding pattern in the RF transmission from the antenna, which can be detected by a receiver at the base station. This pattern in the RF gets created by varying the frequency slightly about a central value such that at certain points in time, corresponding to the bit rate, the RF signal has a certain characteristic that enables the receiver to detect the underlying bit values at corresponding bit intervals. The modulated signal is fed into a power amplifier that drives the signal into the antenna with sufficient power to radiate the required distance to the base station.

Synthesizer To generate the underlying RF carrier that gets modulated we need to generate a signal at the required carrier frequency. We probably also need other signals at fractions of the RF carrier frequency to drive other parts of the signal processing chain in synchronicity with the modulation process. It is the job of the synthesizer to produce these reference fractional signals.

Receiver (Rx) We need to be able to detect input RF signals from our antenna which are being received from the nearby base stations. It is the job of the receiver to detect signals within the required frequency band. It does this by applying filters to block out extraneous signal frequencies and by using a tuner to resonate with the required signal and detect it for further processing and eventual bit extraction.

Antenna RF signals need antennas to convert from electrical signals to electromagnetic waves that can propagate through air. The antenna also acts as a pick-up for signals that impinge on it and resonate at the frequency to which it is sensitive. Antenna lengths are precisely chosen to facilitate the pick-up process at the desired frequency band.

10.6.2 Analog Baseband

IF Staging We cannot handle the signal processing directly from the raw bit rate straight to the RF or vice versa. The problem is mainly in the receive chain where the RF signal is oscillating too fast for it to be converted into a digital signal that can be processed by a computing device at the raw rate. We therefore mix the signal down to an intermediate frequency (IF) where we can then deploy conversion. Generally, whenever signals get processed in electronics circuits, there is some kind of degradation in quality. There is a compromise between the chosen IF frequency and the reliability of conversion in terms of key metrics like signal quality, component costs and power consumption.

Analog–Digital Conversion The signal from the RF chain eventually needs to be converted into a digital representation whereby we use integer values to represent signal amplitudes. To do this conversion we use a device called an analog to digital (AD) converter. Such devices are selected according to how fast we want them to sample the analog signal and the accuracy of representing each sample as an integer. Particularly important is the range of integer values we can accommodate, which is called resolution, or dynamic range (though not exactly). With AD converters, cost is related to resolution and sample speed.

Generally, AD converters are difficult to implement on the same silicon substrate as a large digital circuit, such as a microprocessor or DSP. This is because mixing analog circuit technology with digital usually requires stringent silicon chip-manufacturing conditions – silicon not always being the ideal substrate for analog electronics in any case.

10.6.3 Digital Baseband

Bit Synchronization and Extraction Having retrieved our input signal from the AD converter, we then need to extract the actual raw bit values. This may sound strange as the AD converter is presenting us with raw bit values; so, what do we mean by bit extraction? Usually, the AD 'oversamples' the input data stream. In other words, it is not just taking a digital reading of the analog signal at each point where it thinks the bit is, but at many

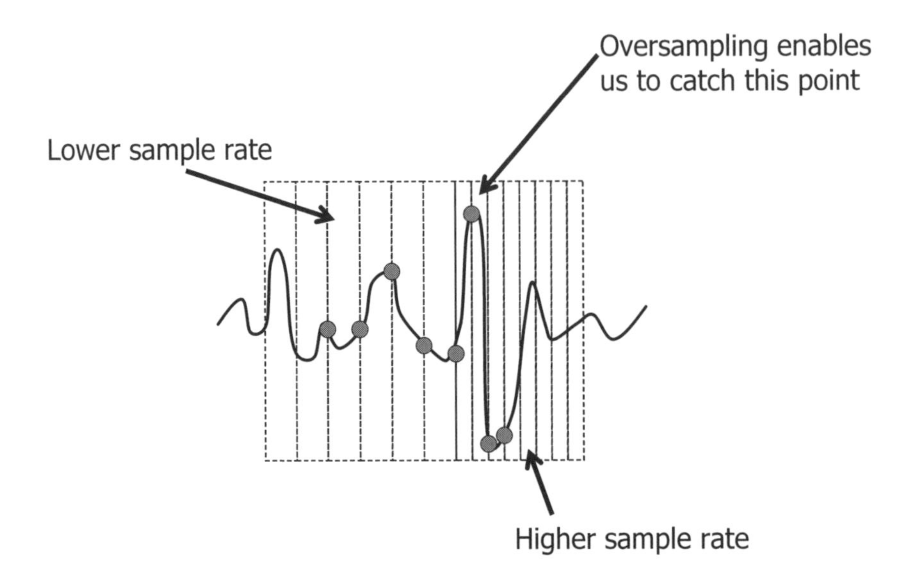

Figure 10.10 How sample rate affects signal quality.

points in-between. This enables our receive process to use interpolation techniques in order to stand a better chance of finding the actual bit values that were transmitted.

We can see this in Figure 10.10 where a higher sampling rate enables us to better determine a point where there is a sharp inflexion in the signal. There is a fundamental minimum rate (called the *Nyquist rate*) at which we have to sample to catch the basic detail in the signal at the rate we are expecting it to change; but, if we exceed this sample rate, then we can do better at detecting the inflexions, particularly if our sampling is not ideally synchronized with the transmitted bit rate. Even so, we will see that, due to adverse propagation effects, even a perfectly synchronized system can still benefit from oversampling.

Equalization As just mentioned, adverse propagation effects can affect our ability to extract the correct bits from the sampled data stream. This is because the RF signal seldom reaches our antenna directly from the base station transmitter. If we pause to think for a moment, we should be able to confirm that most of the time that we use a mobile phone we cannot see the base station – assuming we were close enough to do so with the naked eye and that we knew what one looked like (and it is not disguised as a tree, as shown in Chapter 12).

As Figure 10.11 shows, the RF signal bounces off objects on its way to the mobile device. However, because RF signals spread out as they propagate, we can end up with several instances of the signal bouncing around and arriving at the antenna. Under normal conditions, we can assume that these signals are basically copies of each other. However, they will actually have different power levels according to how far they have travelled and any path-specific attenuation experienced along the way, such as penetrating a building versus travelling unhindered over a field or over water.

Because these different signals take different paths, they are shifted in time with respect to each other. We can appreciate that the signal that has travelled a little further is a bit delayed.

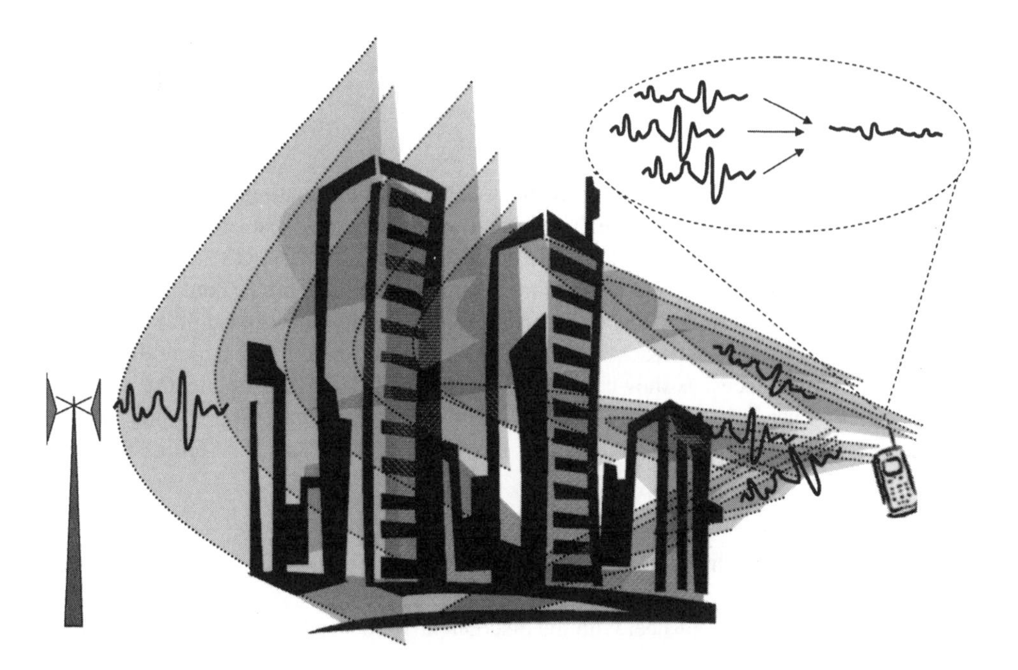

Figure 10.11 The effects of multi-path propagation on the RF signal.

These multi-path signals combine ('add' together) in our antenna and receiver. Due to phase differences, they can end up combining in ways that no longer resemble the original signal. Even without understanding the actual mathematics of it, we can perhaps appreciate what would happen if we listened to an audio signal, like a radio, on several radio sets, each running slightly behind or ahead of the other, the phase difference constantly changing: it would sound garbled at times and at other times not. This is what happens to the RF signal.

We can mitigate this multi-path effect using a technique called equalization. In fact, without equalization, GSM would not be possible; it is that crucial to digital RF transmission. The way it works is simple in principle, but complex in practice and implementation; this is why it accounts for a significant proportion of the computational complexity, and therefore cost and power consumption, in a mobile phone receiver.

We can imagine the sum path between the transmitter and the receiver as being modelled by several dominant pathways. If we were able to detect somehow the exact nature of these pathways, than we could apply some kind of reverse process at the receiver. It turns out that this is possible so long as we have a means to measure the accuracy of our model each time we want to equalize a stream of bits in the receiver. We can do this by making sure that for a certain period of time we transmit a sequence of bits that is known in advance. When we receive these bits they will come out all scrambled due to the multi-path effect. What we can do is to take the known bits and keep putting them through our model, adjusting the model through all its permutations until the output stream matches the messed-up bits we actually received. We then know that we have a good model, and we can use it in reverse to undo multi-path interference.

This sounds like a brute force approach, and in many ways it is; it therefore requires a lot of processing power. The actual process is slightly more complicated than I have explained here, but not quite so brute force in implementation. We can use elegant mathematical tricks to streamline the process, although overall it still remains a computationally intensive and power-hungry task.

Error Correction Equalization removes the effects of multi-path propagation as much as possible, but not enough to ensure that we have an uncorrupted bit stream. The challenge when transmitting digital data streams is to achieve a very high level of data integrity that enables us to preserve what we are trying to transmit. Depending on the content, it may or may not be resilient to errors in the transmission. But, let's say we are transmitting financial information contained in a spreadsheet and even the corruption of a single bit of data could have grave consequences, as shown in Box 10.2: 'Effect of bit changes on text'.

We therefore need very aggressive and robust error correction capabilities when sending bits across the RF network. We can achieve this using a technique called error correction correcting (ECC).

Box 10.2 Effect of bit changes on text

Imagine we have a spreadsheet with the discount at 1000. As ASCII code, this is a bit stream of:

`0011 0001 0011 0000 0011 0000 0011 0000`

Imagine that a single bit gets changed (shown underlined):

`0011 1001 0011 0000 0011 0000 0011 0000`

Then we end up with a price of 9000 instead. A simple bit flip has cost us 8000!!!

If we were considering HyperText Markup Language (HTML) codes instead of ASCII ones, imagine we have a price of £79, which in HTML is:

`£ 7 9`

which in Unicode Latin hexadecimal is:

`26 23 31 36 33 26 23 35 35 26 23 35 37`

which in binary is:

`0010 0110 0010 0011 0011 0001 0011 0110 0011 0011 0010 0110 ...`

Again, if we just flip one bit, as shown underlined:

`0010 0110 0010 0011 0011 0001 0011 0110 0011 0010 0010 0110 ...`

when rendered in a web browser our £79 now becomes 79 pence. Were this the latest price uploaded to our server from our mobile device – subsequently used in hundreds or thousands of online sales – then we have lost a lot of money.

The way that error coding works is by adding additional (redundant) bits to the bitstream which we can use to tell us something about the other bits. Actually, that is probably a fairly loose way of describing the process, the problem being that error correction principles are somewhat mathematical in nature and so difficult to describe here. They all rely on a theorem called Shannon's Theorem which says that basically we should always be able to devise a coding scheme to correct errors.

The error detecting and correcting capabilities of a particular coding scheme are correlated with its code rate and complexity. The code rate is the ratio of data bits to total bits transmitted in the code words, once we have added the error protection bits. A high code rate means information content is high and coding overhead is low. However, the fewer bits used for coding redundancy the less error protection is achieved. A trade-off must be made between bandwidth availability and the amount of error protection required for the communication.

The most common type of coding involves the use of convolutional codes. We will not look at what these are other than to note that the complexity lies in the decoding process, not the encoding. This is an extra burden in the receiver processing chain in addition to the considerable burden already added by the need for equalization. However, it turns out that decoding convolutional codes and performing equalization can both use the same algorithm, called the Viterbi Algorithm. Therefore, we can benefit from using the same process for both parts of the chain. This becomes particularly agreeable if we use a special co-processor to speed up Viterbi processing, as we save on power consumption and silicon costs by not needing such a fast main DSP processing core.

We need robust error-correcting techniques, especially for data applications, as we have already seen (see Box 10.2: 'Effect of bit changes on text'). What we have just discovered, however, is that the more coding we use the less bandwidth we have for the actual bitstream we want to protect. Previously, we had understood that the presence of errors in the first place has something to do with the quality of the RF 'channel' (i.e., the sum propagation path). What we can therefore postulate is a variable rate scheme that adapts to the channel conditions. When the channel is good the amount of error correction coding is reduced (i.e., the code rate goes up), and when it is bad the error coding is increased (i.e., the code rate goes down).

Adaptation of the error correction code rate is something we may already have heard about, perhaps unwittingly, in the GPRS world, where the coding schemes CS1, CS2, CS3 and CS4 get talked about a lot in the specifications (e.g., base station and handset specifications). We should clarify that we are not talking about classes – such as Class A and Class B – which are used to denote the level of GPRS service that can be achieved simultaneously with a GSM phone call; this is an altogether different concept. Coding schemes 1 to 4 refer to the amount of coding going on to protect the data stream. The handset can negotiate different coding schemes for use in talking to a base station depending on perceived channel conditions.

Voice Transcoding The underlying reason that digital cellular voice networks became possible when they did was due in no small part to the newly possible ability to digitize the voice and then *compress* it for transmission and decompress it at the receiving end. We discuss the significance of this technique with respect to its impact on digital cellular in Chapter 12 when we discuss the RF network. One of the most significant contributory factors in making the compression and decompression process possible was the availability of cost-effective DSP technology with low-power consumption. This continues to be a major factor in the success story of cellular technology.

Voice compression is referred to in digital cellular as voice transcoding. This is because the voice is first digitized into one format and then it is converted from that format to the compressed one; hence, from *transformation* and *coding* we get the term 'transcoding'. This should not be confused with an identical word that is used in the wireless developer world to refer to transforming Web pages from one mark-up language to another automatically, so that they can be viewed successfully on mobile devices. The term *vocoder* is often used to refer to the apparatus (software and/or hardware) that carries out the transcoding function.

The underlying premise for voice compression is that within the range of sounds that the voice box (vocal chords) makes during speech, many of them are not critical in the hearing process in order for the voice to retain fidelity and still remain intelligible to the listener. We can think of the voice as being made of various components at different frequencies. We can throw away or diminish some of these components and it won't affect fidelity, although the more aggressively we dispense with some or all of the components the greater the degradation in the speech quality.

Originally, voice compression was no different from any other type of audio compression. It was understood that the voice would remain intelligible if we filtered out the high-frequency components, as most of the vocal energy is in the lower frequencies. As we remarked earlier in our consideration of sampling RF signals: in order to catch faster inflexions in the signal we have to sample at a greater frequency, notwithstanding the requirement to sample at least as fast as the Nyquist rate, which is basically twice the highest frequency component we want to reconstruct in our digital representation of the analog world. Hence, if we have voice components as high as 15 kHz (components moving as fast as 15,000 inflexions per second), then we need to sample at 30 kHz. Let's imagine that for each sample we use up to 8 bits to represent the level of the signal coming into our AD converter: for this to happen we potentially have to transmit 240,000 ($8 \times 30{,}000$) bits per second (bps).

If we ignore the higher frequency components, as they are not required to make the speech intelligible, then we could sample at say 8 kHz and end up with 64,000 bps in our bitstream. This is already a substantial reduction in the number of bits we have to transmit. At this stage, we have not done any compression – just filtering or subsampling of the signal.

What is interesting to note is that most of the time the signal will not be changing rapidly, so we can deploy a technique to transmit the changes from one sample to the next, rather than the entire sample. We refer to this as adaptive differential pulse-coded modulation (ADPCM), which in itself is a form of compression. It is also a form of compression that does not sacrifice quality, as we are not throwing anything away by sending the difference values. We can reconstruct entirely the absolute values at the other end. Of course, we notice that this process is very susceptible to bit errors. If we failed to transmit the original reference value on which we subsequently reconstruct the original samples, then we can imagine that we could easily change the signal altogether; this might cause loss of fidelity or even loss of intelligibility. Yet again, we see why error coding is so important – as it was for nonvoice data sources (see Box 10.2: 'Effect of bit changes on text').

ADPCM is actually the usual way of representing audio signals digitally as a result of its compactness. However, using modern DSP techniques, we can do a lot better than this for voice. ADPCM is a technique that is generally applicable to audio. However, the voice has its own unique aural characteristics that enable it to be modelled in a way that we can attempt to synthesize.

We understand that vocalized words can be deconstructed into such constituents as phonemes. Phonemes can be thought of as instructions for articulating speech sounds, and so a phoneme can be described in terms of the behaviour of the vocal apparatus that occurs when a speaker articulates their particular representation of the phoneme.

Phonemes can be divided into consonants and vowels. In the articulation of consonants the flow of air from the lungs through the vocal apparatus is cut off or impeded. In the articulation of vowels the flow of air from the lungs is not impeded, but the vocal organs are used to change the shape of the oral cavity and thus make different sounds for different vowels.

Amazing as it may seem, using DSP and mathematical models of how phonemes get translated into measurable signals, we can attempt to model a person's vocal apparatus in software. The reason this is useful is that, instead of sending sound samples, we can transmit codes that represent phoneme production. It turns out that this is a much more efficient way of compressing speech signals. This type of approach is called parametric encoding, as we are now transmitting vocal parameters rather than sound samples.

Speech compression technology is changing all the time, as is the ability to deploy large amounts of intensive DSP processing at lower cost and lower power. This is why we have seen an evolution in speech compression offerings in the 2G cellular systems. The advent of multimedia communications in the Internet world has also resulted in greater interest in this area.

Speech compression is lossy, which means that some of the vocal excitement of the speaker gets lost in transmission. The more we compress the lower the fidelity and the lower the bandwidth required to carry the signal. Newer algorithms have sought to improve fidelity at lower bandwidths, rather than attempt to lower the bandwidth yet further. They are also adaptive, which means that – should we choose to – we can dynamically adjust quality and/or bandwidth to meet various requirements when running our voice network.

Encryption It was understood early in the development of cellular radio that eavesdropping would be an impediment to commercial success. First-generation analog cellular systems were notorious for this problem, with many high-profile cases of surreptitious behaviour, often involving famous people. Therefore, encryption of the signal travelling across the RF channel was deemed an essential requirement for 2G systems, and it remains so for 3G.

Encryption of the signal is a sophisticated task that involves intensive bit-level manipulation of the bitstream. This process is best carried out by DSPs, as these devices have excellent bit manipulation capabilities, as one would expect. Nevertheless, it is common to see the availability of ciphering accelerators. These are small dedicated digital circuits that are companions to DSP core processors. They get used to offload the processing overhead from the DSP, thus often lowering implementation costs and power consumption.

10.6.4 The Digital Signal Processor (DSP)

As we have seen, the heart of all digital cellular phones is a DSP: it plays the key role in digital baseband processing. The dominant DSP in this market is Texas Instruments' TMS320C54x, which can be found in many modern cellphones. This processor is responsible for modulating and demodulating the data stream, coding and decoding to maintain the robustness of the transmission in the face of transmission bit errors, encrypting and decrypting for security, and compressing and decompressing the speech signal. In other words, it carries out all the baseband processing on one chip.

Carrying out lots of signal-processing tasks on one chip means that it has to operate very fast; this is very much a real time operation. Voice samples or data packets are arriving at a periodic rate, so each of the baseband tasks has to be carried out before the next set of samples or packets arrive for processing, not forgetting that this has to be done for both the transmit and receive processing chains concurrently.

A DSP carries out instructions that tell its internal processing core and mathematical engine how to process the incoming bitstream. The sophistication of the digital transmission algorithms and the hard real time constraints mean that a DSP has to carry out millions of instructions per second (MIPS) in order to meat its deadlines.

In early 2G TDMA (Time Division Multiple Access) phones these functions could be accomplished with 30–50 DSP MIPS. As vocoders have become more complex and as data rates have risen in 2.5G phones, the total DSP load has risen past 100 MIPS. DSP technology has kept up with this requirement while amazingly also getting more efficient in terms of power consumption due to improved silicon fabrication techniques, such as the ability to operate the internal transistors at a lower voltage.

The CDMA (Code Division Multiple Access) standards require a somewhat different functional partitioning, because data rates generated by spreading are too high for direct processing by a general purpose DSP device. While the DSP can still be used to process at the basic data rate (functions like forward error correction, encryption or voice compression), custom silicon hardware operating under the control of the DSP must be used to process and modulate/demodulate the spread–spectrum signal.

10.6.5 Summary

Our discussion of basic mobile phone architectures has enlightened us about the sophistication of mobile devices as a necessary response to sophisticated RF-networking requirements. Here we have only discussed digital cellular standards and implied architectures, the most basic of which we have introduced and examined their basic components. We discovered that a large amount of intensive processing is required to move data across an RF network, mostly due to the harsh conditions that our RF signals experience and our desire to maximize RF spectrum capacity by compressing signals into their smallest form without compromising integrity and quality too far.

We have not yet discussed other DSP functions that require even greater processing power than already demanded by an RF modem. Applications like voice recognition, voice to text, speech synthesis and video compression all require large amounts of DSP processing. We will discuss these later in relation to their impact on device design and how we develop applications that require a DSP element. As we will see, it is not necessarily the case that a maths-intensive application requires a DSP: some newer microprocessor designs incorporate certain features to accelerate maths functions in a DSP-like manner. The distinction between a DSP and a general purpose processor is unclear in these cases.

10.7 SUGGESTING A GENERIC DEVICE ARCHITECTURE

Let's now develop our previous discussion to start expanding our understanding of device architectures and explore what possibilities are open to us in terms of developing software

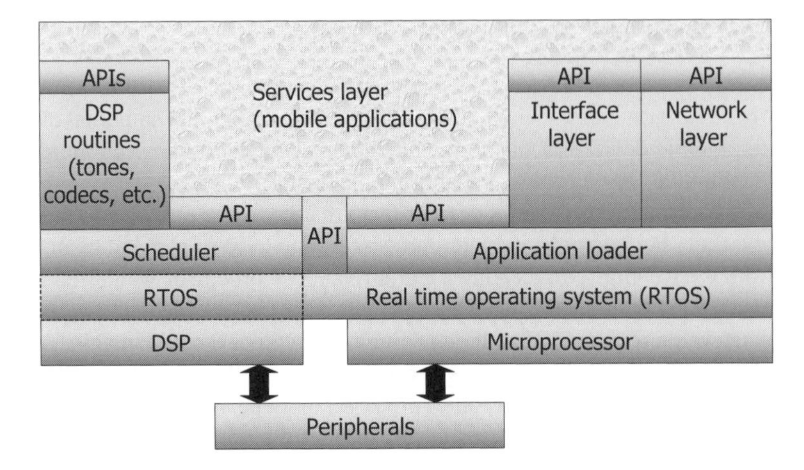

Figure 10.12 High-level generic architecture for mobile device.

services to run on mobile devices as part of any mobile service offering. We reiterate that there are three basic modes of application on a mobile device:

1. Client/server.
2. Peer to peer.
3. Stand-alone.

We have posited that, no matter the mode, a basic layered model for a mobile device can be constructed that consists of interfaces, interface software, services software and network connections. The bulk of our service is enacted in the services layer and we refer to this as the 'business logic' of the mobile application to draw a parallel with the software approach we witnessed with J2EE (see Chapter 8).

Thus far we have only considered the layered model as a notional concept, but we have tried to convince ourselves that it is a sound and complete model. But, how can we actually deploy services using such a model (or architectural basis).

We can begin to approach the basis for software and applications deployment by adding some sound software patterns to our model. We will tackle this process from two sides. First, we will examine what framework we need to deploy software on a device, regardless of its mode or function. In other words, we will introduce basic software practices for embedded devices. Second, we will propose common functions that we would expect to find on any mobile device or across a range of devices, things that are commonplace with little need to justify, such as a man–machine interface (MMI) for telephony (i.e., being able to dial a number, place a call, etc.).

Let's take the first approach and look at a possible generic architecture for pulling our device together based on what we know so far from the layer model and its implications. A refined block diagram for our mobile device is shown in Figure 10.12.

We can still see our four layers in the diagram, but they are now set within a context of other companion or support software elements and hardware entities needed to support the layers and to glue them all together. We will now discuss this refinement of our layered model in more detail.

10.7.1 Core Processor and Operating System

The workhorses of a mobile device are the microprocessor and DSP. Ultimately, our application software runs on these devices, although, as noted earlier, the sum total of software running on a mobile device extends to include those elements running on the processors embedded within the various chipsets that bring the device to life.

Microprocessors are mostly serial devices that are able to run only one logical unit[9] of software at a time. For example, the microprocessor may have to respond to an incoming message from the RF chipset indicating a call arrival. The processor would run a piece of software to activate the ring tone, display a call alert icon, possibly actuate a vibration alert motor or whatever is required to handle this event: to do this we could imagine a software process like 'handle call'. If the processor is running the 'handle call' process, then it is not running any other process. However, other processes may be vying for attention, such as a battery charge meter indicator process, an email alert process and so on. We don't have to think too hard to bring a long list of potentially useful and meaningful processes for our mobile device processor to run.

Due to its serial nature, the processor has to swap continually between vying processes, so that they all get a turn at being executed. If this process is done quickly enough, then swapping is not noticeable just as if there are *N* virtual processors for *N* processes waiting to be executed.

When each process gets control of the processor, we may have to implement some degree of resource control to make sure that processes do not violate the resources of other processes. For example, we wouldn't want the vibrate alert process to start writing rubbish to the LCD display or to start sending messages via the RF data connection. We need some kind of process containment.

These types of challenge, such as scheduling of processes, swapping them in and out of the processor, protecting resources and other mechanisms, are all the function of the operating system. In our case we have the added challenge that certain processes need a deterministic means of gaining access to the processor and completing their designated tasks. For example, we would not want an email-sending process to be able to hog the processor to the extent that the call-handling ('handle call') process is stalled and we end up missing a call. This is why we have the need for what is called a real time operating system (RTOS), the idea being that certain tasks have a very definite time limit for initiation and completion and this is what we mean when using the term 'real time'.

In real time systems the success of a process running on the processor is not just measured by its computational output it is also related to how long it took to do it. Time really is 'of the essence', unlike in standard desktop operating systems which generally deploy a 'best efforts' approach; processes finish when they finish and this is deemed acceptable most of the time. 'Best effort' is often turned into 'acceptable effort' by the application of brute force – plenty of processor speed and memory.

Clearly, what is meant by 'real time' is a matter of degree. Even on a standard Windows™ desktop, an email-polling process may well block our ability to type a letter, but not to such a degree that it becomes significant. However, more critical to our discussion is that in such a system there is no means to control deterministically the scheduling and completion of

[9] What constitutes a unit of software is not something we want to get into here, but I am thinking of software processes as units rather than lines of assembler code (or higher languages for that matter).

processes, so the more real time that things become the less likely they are to work on such a platform.

The concept of 'hard real time' perhaps needs to be introduced. This is the notion that unless a process completes by a certain time, its output is deemed a catastrophic (or complete) failure. Catastrophic need not mean that a nuclear power station is about to superheat into oblivion and irradiate everyone around it. In our case we would probably deem that the inability to receive calls due to call-handling processes being blocked is catastrophic. Were it to happen often enough, the mobile service may be deemed to be a failure and would result in collapse of the service or even the business; that would be pretty catastrophic.

'Soft real time' is where the success of the process diminishes the more postponed it gets. We may consider that the reception of picture messages (via multimedia messaging service, or MMS) has a softer real time constraint than call handling. We might receive an MMS message indication while listening to our favourite MP3 track or while watching an MPEG-4 video clip. If our device is too busy to download the picture message payload from an MMS centre, then it can try again a bit later. Depending on whether we have already been notified of the availability of a message and on its importance, the delay in being able to view it may or may not be problematic to us, but probably not catastrophic.

An RTOS is a specially written operating system that attempts to handle process-timing issues as just described. There is no single, agreeable and sufficiently comprehensive definition of an RTOS, but we can elaborate on its characteristics in order to understand better what we require on our mobile device.

A hard RTOS must guarantee that a feasible schedule can be executed, given sufficient computational capacity and once external factors are accounted for. In this case external factors are devices that may generate interrupts, including network interfaces that generate interrupts in response to network traffic, which in our case can be due to data or voice. In other words, if a system designer controls the environment of the system such that events are within known bounds the operating system itself will not be the cause of any delayed computations.

The types of mechanisms used to implement an RTOS are beyond the scope of this book, but the basic principle is the ability of the RTOS to gain a high degree of control over scheduling of software process threads.[10] The degree of control available enables one thread to be suspended and pre-empted by another thread so that, with sufficient processing resources and judicial scheduling, real time constraints can be accommodated.

This concept is shown in Figure 10.13. Here we show two software processes. Let's imagine one is an MPEG-4 video media player and the other a Bluetooth™ pen handler. We can see that each software process has various threads of execution through its code. Thread X is the media player thread that fetches more encoded packets from an Internet-hosted server via an RF network connection. Ordinarily, thread X would run to conclusion, but at time t1 the Bluetooth™ pen gets lifted from the page and wants to send a stream of vector pen strokes to the mobile device, to be handled by thread Y in software process B. Usability tests have shown that an upper limit of 500 milliseconds is tolerable for displaying the pen strokes. We therefore have until time t2 to complete the handling and processing of the pen input.

Because thread X was going to fetch a few more seconds of video from the server, we would not have been able to meet the deadline t2, so we use our RTOS to suspend thread X

[10] A thread being a particular path of execution through the software process that is active.

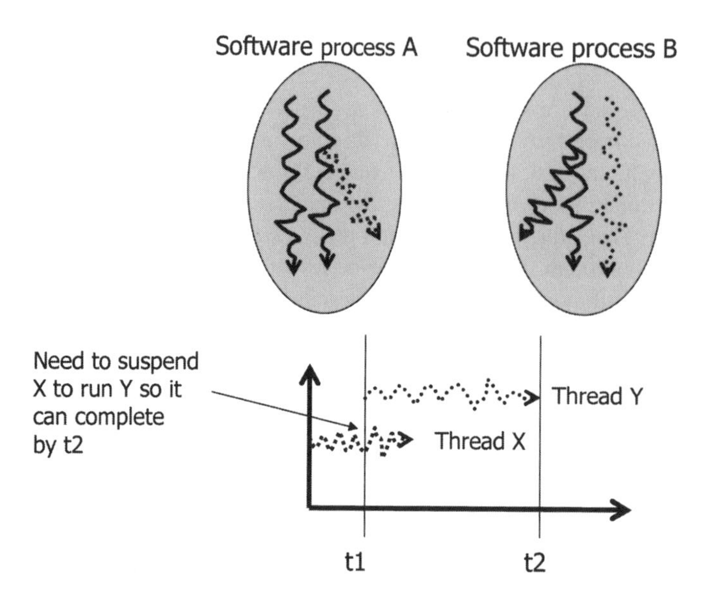

Figure 10.13 Concept of thread scheduling to achieve real time deadlines.

in favour of giving greater priority to thread Y. This is not a problem for our system as the delay in buffering the video will not affect playback, unless we did it so much that thread X never completed or was severely retarded.

Priority manipulation of program threads can benefit a variety of applications: for example, media players that are increasingly common on mobile devices (MP3, WAV, MPEG-2, etc.). The operation of a media player can be tied to the media rate required for proper playback (i.e., 44 kHz audio, 30 frames per second video). So, within this constraint a reader thread and a rendering thread can both be designed to wake up on a programmable timer to buffer or render a single frame and then go to sleep until the next timer trigger.

Timer-based pre-emption provides the tempo of execution that allows the priority to be assigned above normal user activities, but below critical system functions that must not be allowed to become starved of processor time. With well-chosen priorities, playback will occur consistently at the given media rate. A well-written media player will also take into account quality of service, so that if it doesn't receive adequate CPU time it can reduce its requirements by selectively dropping samples or follow an appropriate fallback strategy to attempt masking the problem from the user. This will then prevent it from starving other processes as well.

We may also wish to treat certain user events preferentially within the system. An example might be a continuous data stream from a Bluetooth™ pen as we attempt to jot down notes during a meeting. We would prefer that the mobile device appears responsive to our note taking, even while doing other tasks, such as possibly audio-recoding the meeting directly into a memory card (e.g., MMC) or playing back a video stream as indicated in the earlier example.

This works well when we increase the amount of concurrency within the application and when the event can always be handled in a small, predictable amount of time. The key concern here is the frequency at which these events can be generated. If they can't occur too frequently, then it is safe to raise the priority of the thread responding to them. If they can occur too frequently, as when writing a particularly long sentence fairly quickly, then other activities will be starved under overload conditions if we keep assigning to the pen handler thread all the available CPU time.

There are different programming strategies that can be employed to address these sorts of issue, such as dividing responsibility for events into different handling threads with different priorities, using timers to force suspension of priority threads and so on. In our earlier example of suspending the video-buffering thread to enable a more responsive handling of our BT pen stokes we could imagine that we might better serve the video process by having two threads that fetch video: one that is a housekeeping thread to keep the buffer fed under normal circumstances and the other a higher priority remedial thread that runs if the buffer is nearing underflow or has already emptied. Under this condition, we can suffer the infrequent delay in pen response in favour of keeping our video stream running.

The reader should be aware that these challenges exist and that different mobile devices will have different mechanisms available to assist our application with achieving the desired real time performance.

We should also appreciate that under certain conditions it may not be possible to have sufficient control over how the RTOS treats our application. Some of these issues will arise further in the ongoing discussion in this chapter and in Chapter 11, particularly when we look at the pros and cons of different approaches to developing applications.

10.7.2 Digital Signal Processor

We previously discussed the importance of a DSP for the RF modem aspects of a mobile device; but, we are also concerned with general DSP functions and how they might get integrated into our mobile devices and made accessible to the services layer. With the ever-increasing need for computationally intensive processes, like video compression, voice synthesis and so on, the inclusion of a DSP on a mobile device is an obvious idea. We should caution that there is an alternative approach, as we will see shortly, but for now the inclusion of a DSP does seem sensible.

Let's consider a DSP in isolation: usually, it is deployed to run a very narrow set of functions that are highly mathematical in nature, involving a good deal of repetitive tasks, most likely vector-based (e.g., matrix manipulation[11]). It is not intended to provide general purpose computation, although in theory at least there is no reason why it shouldn't. However, this would generally not be a good idea because for each unit of processing power, measured in some general processing sense, a general purpose device is cheaper than a DSP. Also, programming tools for general purpose processors are more plentiful and more advanced, including, significantly, the existence of available operating systems to support the software processes.

Operating systems for DSPs are not that widespread and have limited features. For example, as a rule, we would not expect any support for graphical output devices, never mind support for graphical interface paradigms like windowing. On a DSP we usually

[11] Many processes involving audio-visual streams can be reduced to matrix maths problems. Hence, a processor that is optimized for processing matrices would be useful. Matrix manipulation typically involves repetitive arithmetic operations, and a DSP is highly optimized for such processes.

expect some kind of basic task loader and scheduler, but nothing like the sophistication of a full-blown commercial RTOS.

If a mobile device has a DSP available for running software, then it would usually be expected that the developer would develop it almost independently of the general purpose microprocessor device. Any collaboration between the two would usually have to be managed by judicious programming on both devices, most likely involving passing of messages between the two. There may or may not be existing support for this messaging process built in to the RTOS or any utility code that may come with the DSP.

The concept of a unified RTOS abstraction for a multi-processor environment, supporting both DSP and general purpose microprocessors, is not widespread. There are a variety of challenges, not least of which is devising a programming paradigm that would make sense for such an environment. Among other challenges, this would include grappling with the scheduling of tasks running on two processors. It is hard enough devising a sensible real time scheduling scheme for tasks running on one processor; but, when we have to manage parallelism, matters become significantly more challenging.

We should not be dismissive of the dual-processor approach. After all, it works well in the case of the typical RF modem where there are generally two processors – one DSP and one microcontroller. A common paradigm, or metaphor, is to set things up in a way that the microcontroller is viewed as the master and the DSP as the slave. In other words, the microcontroller dictates the pace and, probably, is the instigator of what gets run by the DSP at any particular time. Many applications will fit the metaphor well, especially as DSP functions are usually very deterministic, thoroughly enshrined in a rigid real time framework that is repetitive and predictable: data are injected into the input of the DSP, tasks get run one at a time and then data get squirted out – a predictable processing pipeline.

There is an alternative approach to the dual-processor architecture. It is one that mitigates many of the problems we have just discussed, particularly how to provide a unified programming metaphor. Modern microprocessors are becoming more and more DSP-like by the inclusion of modified or additional circuits that are optimized for certain DSP (mathematical) operations. This idea has been pioneered on the desktop environment where DSP co-processors were once expected to become popular, but never quite made it to the motherboard. They do exist of course, but in the form of embedded solutions (like a modem), inaccessible to the general programming environment.

When examining the idea of including DSP structures within general purpose processors, the decision by Intel to introduce Multimedia Extensions (MMX) technology is what comes to mind. This was initially an attempt to add some processing circuitry that could speed up graphics programming or any vector-based processing (see Box 10.3: 'Vectors and matrices'). For a while, Intel gave one of their Pentium chips the MMX moniker, but later dropped any special reference to it after the circuitry became commonplace in subsequent processors in the Pentium family. From the Pentium III onward, the enhancements got even better and became known as Streaming SIMD Extensions (SSE and later SSE2 for the Pentium IV).

The technology used for MMX and SSE is the well-known processor architecture called Single Instruction Multiple Data (SIMD). Usually, processors process one data element with one software instruction, a processing style called Single Instruction Single Data, or SISD. In contrast, processors having the SIMD capability process more than one data element in one instruction. Let's say that we want to take an image using a camera phone to be sent via

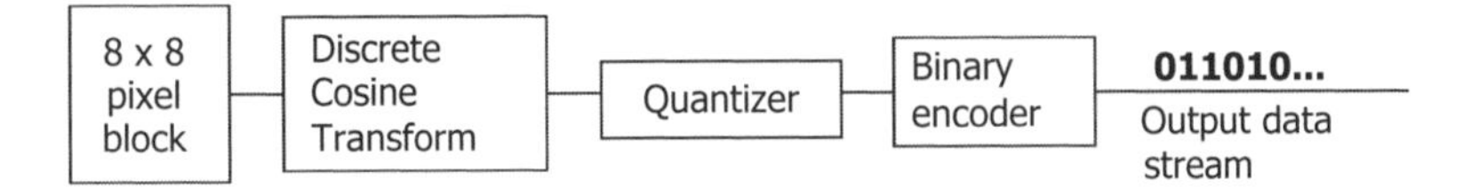

Figure 10.14 JPEG image processing.

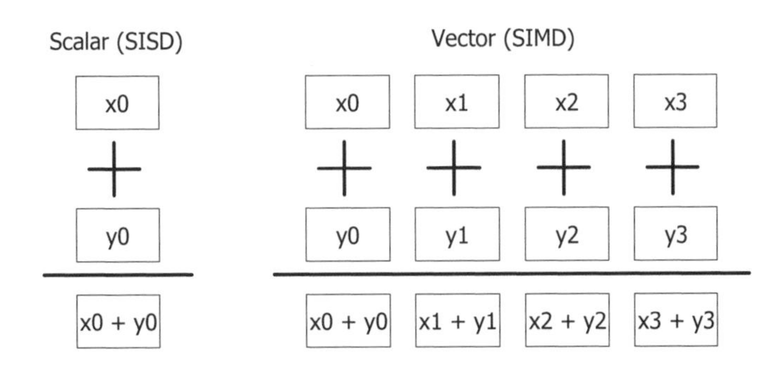

Figure 10.15 Parallel processing (SIMD) speeds things up.

MMS. We first wish to apply compression to the image using the familiar JPEG approach. The processing chain for JPEG is shown in Figure 10.14.

To implement JPEG compression, we have to take the entire set of pixels in the image and apply mathematical processes, such as the Discrete Cosine Transform (DCT). The DCT is applied to 8×8 blocks of pixels at a time. This does not mean we have 64 operations to perform; on the contrary, the actual processing for each of the 64 pixels is more complex, entailing several operations per pixel, mainly successive multiplications and additions.

The resulting operation to compute the DCT for a single 8×8 block of data is 970 operations on a standard processor without MMX or SSE. When we talk about 'operations on a processor' we really mean clock cycles, which is the basic unit of time on which all operations are synchronized. In the case of MMX architecture, according to Intel,[12] this can be dramatically reduced to 280 clock cycles and still further reduced to 250 clock cycles using SSE technology.

This four times reduction in time (or fourfold increase in speed) is what we expect, as the SSE circuitry deploys four arithmetic units in parallel where previously we would have only had one, as shown in Figure 10.15. Therefore, we can construe that once a mathematical process is under way, involving repetitive arithmetic manipulations, we would expect to be running four times as fast using MMX or SSE.

Intel has developed a wireless version of MMX called Wireless MMX™, and this is part of the new breed of devices aimed at mobile applications or Internet appliances. These devices are part of the X-Scale™ architecture. The Wireless MMX™ technology gives a large performance boost to many multimedia applications, such as motion video, combined graphics with video, image processing, audio synthesis, speech synthesis and compression, telephony, conferencing and 2-D and 3-D graphics. Even without MMX, the Intel XScale™ core provides the ARM V5T Thumb instruction set and the ARM*

[12] http://cedar.intel.com/media/pdf/appnotes/ap922/ap922.pdf

V5E DSP extensions. To further enhance multimedia applications, the Intel XScale™ core includes additional Multiply-Accumulate functionality as the first generation of what Intel calls 'media processing technology'.

It is fair to say that the distinction between a separate DSP processing element and an integrated one is blurred in terms of hardware instantiation. For example, the Wireless MMX™ technology is actually implemented as a co-processor in the Intel X-Scale processor core, so it is like a separate processor in some regards. However, the degree of integration with the main processor is such that the software developer does not really have to concern themselves with a disjointed programming model. This is especially true with some of the newer software-compiling technologies that are automatically able to map suitable code segments onto the appropriate instructions in the MMX or media co-processor.

10.7.3 Application Loader

In our basic device architecture we have proposed that applications run on top of the RTOS. However, we did not discuss how the applications get loaded in the first place and how they get registered with the RTOS so that it knows where to find them and how to launch them. This is the job of the application loader. We will not go into this in much depth; however, in a mobile context we are going to be interested in being able to download applications to our device via the RF network – be that from the IP network or from a friend or colleague in a peer group, using P2P interaction. The ability to load over the air (OTA) has become increasingly important, although historically it has not been present on that many devices, even though it is not a particularly challenging technical feat to achieve. Some of the new mobile-aware operating systems and related programming environments have started to address this issue.

We will be looking at OTA provision in more depth in Chapter 11, particularly in relation to the use of Java (J2ME) on a mobile device, although we will discuss the general issues of OTA later in this chapter.

10.7.4 Application Programming Interfaces (APIs)

In our proposed basic device architecture we have elaborated on some of the key attributes of a useful mobile device:

- core general purpose microprocessor to run our software on;
- an operating system, preferably real time-capable (RTOS), to run on the target processor;
- provision of a capacity to handle complex mathematical tasks, such as image processing, either by dedicated DSP devices or via 'DSP acceleration' techniques on the core processor;
- a means to install and load our software onto the RTOS and start executing it.

This just leaves us now with the job of writing our software to run in the services layer and being able to take advantage of the provisions just mentioned. The software might need to access a server (client/server), access another peer (P2P) or simply perform computations on the device for the user's benefit, without the need to network at all (i.e., stand-alone mode). To write the device software, we mainly need two things:

- software development tools to write and test the code and then produce an installable application to run in the services layer;
- a known means to access the device resources from our code, like the RF modem.

The first of these is mainly catered for by the existence of a compiler that can take our high-level language and convert it to the instruction set understood by the target processor (whether core processor, DSP or both).

The second matter is particularly important. This is about how to make our software work with the software already bundled with the device, like the software that controls the telephony features. Usually, the bundled software will come with published APIs that tell us how to access and use them from our app.

We often think of computer programming as being about learning programming languages, data structures and data-handling algorithms. This is only one aspect of the discipline of programming. Among other requirements, a programmer needs to know how to use the APIs they are expected to make their program work with. In fact, it is often a more useful measure of a programmer's ability than what languages they know.

An API is a strange name, as it tends to imply something insubstantial: an interface doesn't sound like much. However, in the case of an API this is not so. It is better to think of APIs as software programs that come as pre-installed libraries of software functions on the operating system. The pre-installed routines can do things on behalf of our application, without us having to concern ourselves with the details. To access these library routines, the programmer has to know how to point to where they are on the device (somewhere in memory) and to know the names of the routines and their expected parameters. The supplier of the API usually publishes documentation saying how to use it. It is then just a matter of writing our application code and calling these API functions at the appropriate point in our code.

For example, let's take a look at the Pocket PC 2003 mobile device operating system. This operating system can run on an Intel X-Scale processor, which reminds us that we can probably do powerful mathematical processing if we need to, utilizing Wireless MMX™. In reconsidering our block diagram (see Figure 10.12), we are now venturing that the network layer is accessible to us via an API or possibly several APIs, perhaps one for each wireless networking device that we need to access. For example, Pocket PC provides an API to interrogate or configure the status of any Bluetooth™ modem that might be present on the mobile device. In considering the use of Microsoft™ APIs relating to Bluetooth™, consider the following function:

BthSetMode

Use the *BthSetMode* function to set the Bluetooth mode of operation and reflect it in the control panel. This function is also used so that that state persists irrespective of hardware insertion and reboot.

Syntax

```
int BthSetMode(
  DWORD dwMode
);
```

Parameters
dwMode [in] Indicates the mode of operation that the Bluetooth radio should be set to. See BTH RADIO MODE for possible values.

Return values
Returns ERROR_SUCCESS on success or an error code describing the error on failure.

Requirements

Pocket PC Platforms:	Pocket PC 2003 and later
OS Versions:	Windows CE .NET 4.2 and later
Header:	Declared in bthutil.h
Library:	Use bthutil.lib

What this piece of documentation tells us is that there is a software routine called BthSetMode that enables us to set the operating mode of the Bluetooth™ hardware. For our current discussion, the key piece of information here is the bthutil.h and btutil.lib at the end. What this basically tells us is that there is a piece of software that already exists on a Pocket PC device called *bthutil*, this being the software realization of the API. This software, if we call it from our program, will take care of configuring the Bluetooth™ modem for us at our application's behest. It knows how to talk to the Bluetooth™ modem to get things done. We simply ask it, and it will make the configuration happen on our behalf. There is no need to concern ourselves with how this happens at the lowest level – this is the beauty of the API approach. In the case of the API example above we can access its benefits by inserting the piece of code indicated into our application, thus:

```
int BthSetMode(
  DWORD dwMode
);
```

For this discussion we don't need to know about the anatomy of this function: the principle is what we are trying to illustrate. In the case of the Pocket PC 2003 and its sister the Smart Phone OS, this API is provided courtesy of the operating system itself. The operating system has a built-in awareness of Bluetooth™ peripherals. For this to be so, clearly the Bluetooth™ peripheral must conform to an agreed interface specification for it to be accessible like this via the operating system in an expected and standard manner. This would mean that any Bluetooth™ peripheral being inserted into a Pocket PC 2003 device would subsequently be accessible via the associated API.

In the case of the Bluetooth™ protocol set it is quite diverse and complex. The API concept allows us to focus on what we want to do with Bluetooth™ at the application level, rather than how a Bluetooth™ modem works at the low-level interface. Let's say we want to write an application to implement a Bluetooth™ chat application for use between a group of peers (a P2P application) – we want to focus on designing and programming the peer-chatting paradigm and coming up with a nice user interface, etc. We do not want to bother with the details of how we find other Bluetooth™ devices and make a connection with them. Using the API approach, we could simply issue a 'command' to sniff for other Bluetooth™ devices within range and, then, using the results subsequently establish a paired communication pathway.

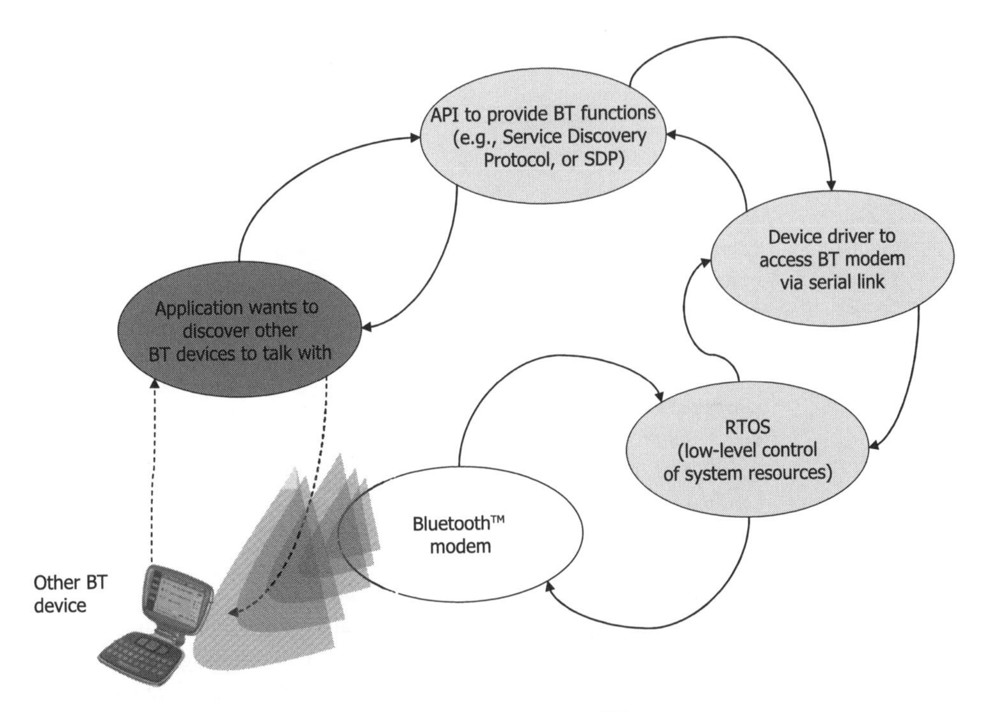

Figure 10.16 Using an API to sniff for other Bluetooth™ devices in the region.

Looking at Figure 10.16, we can see how the provision of a Bluetooth™ API works. Our software is the blob on the left and is what we write and focus our creative energies on. We can imagine that we are talking direct to the Bluetooth™ modem or, even better, transparently to other Bluetooth™ devices, as shown. The other blobs are the API, device driver and RTOS which are working to make this happen.

What we can conclude from our discussion on APIs is that they are a useful way to enable the services layer to gain access to underlying resources on the mobile device. Different APIs can exist to take care of different functions. We have just looked briefly at an API for a Bluetooth™ peripheral running on a Pocket PC device. This still fits with our layered model. We have a layer of network services – which can be any type of wireless standard – as long as an API is presented to the services layer so that mobile applications can be developed for the wireless modem concerned. In some cases the API will be part of the operating system, as we have just examined, and in other cases the provider of the peripheral chipset or device will provide the API themselves.

If the API comes from the peripheral vendor (either directly or via a third party), then it must be written to run on the host RTOS. Currently, the most portable way of doing this is to provide APIs written in C, this being the most common language for embedded systems. However, other languages for APIs are becoming increasingly popular, especially Java; but, we defer this consideration until later.

Using this approach the applications programmer responsible for programming a mobile application can rely on being able to access some powerful features of the mobile device without having to concern themselves with the interface details. For example, if a GSM modem is present on the device and the services layer wishes to send a text message (via the Short Message Service, or SMS), then in the case of Pocket PC an API exists for this:

```
HRESULT SmsSendMessage (

  const SMS_HANDLE smshHandle ,
  const SMS_ADDRESS * const psmsaSMSCAddress ,
  const SMS_ADDRESS * const psmsaDestinationAddress ,
  const SYSTEMTIME * const pstValidityPeriod ,
  const BYTE * const pbData ,
  const DWORD dwDataSize ,
  const BYTE * const pbProviderSpecificData ,
  const DWORD dwProviderSpecificDataSize ,
  const SMS_DATA_ENCODING smsdeDataEncoding ,
  const DWORD dwOptions ,
  SMS_MESSAGE_ID * psmsmidMessageID );
```

Again, we need not be concerned with the anatomy of this function call, but those who have tried it will tell us that interfacing to a GSM modem to perform text messaging is actually quite a messy affair at the low level, one that involves numerous serial interface messages. All that messiness has been removed from us courtesy of the API, and we are left with one simple function call to take care of sending a message.

This principle is not unique to Pocket PC; we have simply taken examples from this particular operating system to illustrate the API concept. The API approach will exist on all mobile devices and is how we expect our services layer application to interface with the rest of the mobile device. All the common mobile device operating systems support this approach. The issue that stands is whether or not there are common formats for the APIs themselves. Certainly, in the case of the different RTOS solutions this is not so – even for common functions like sending a text message.

10.8 MOVING TOWARD A COMMERCIAL MOBILE PLATFORM

Thus far we have been building our view of what a mobile device consists of. We have started out with a simplified layered model that captures the essence of what a mobile device does. Our first level of refinement was to propose an infrastructure for the layers (namely, an operating system) and a means of engaging with peripherals and the networking layer via defined software entities called APIs.

What we would now like to do is to further refine our architecture by the inclusion of certain utility applications that we might find useful on a mobile device and that are a sufficiently common requirement that we might expect to see them on most devices. Our further refined architecture is shown in Figure 10.17.

The figure shows more detail added to the services layer. Essentially, we have added programs that will run in the services layer, some of which are useful utilities in their own right and accessible to the end user, and others provide an additional framework for

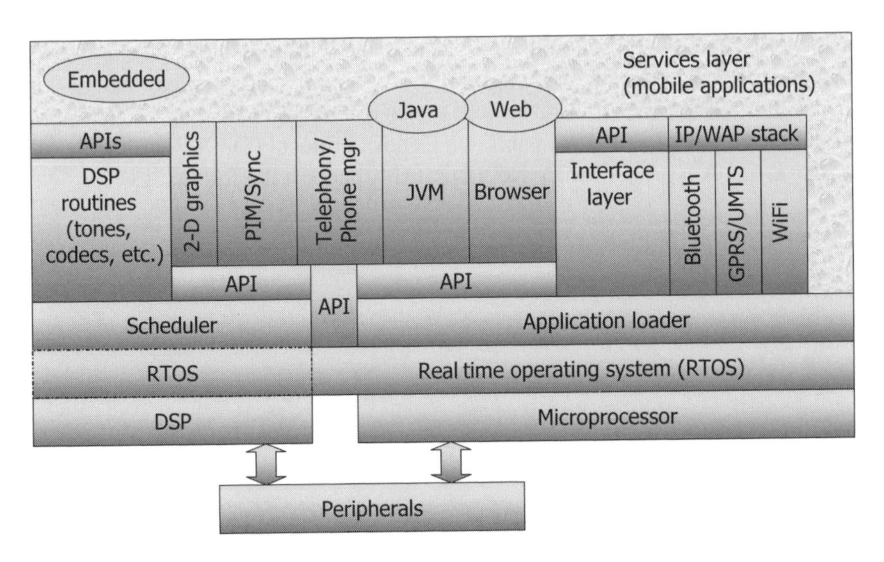

Figure 10.17 Typical mobile device functions.

supporting and deploying our own service applications on the device. In this section we will examine what these applications and utilities are. We will move toward demonstrating how this model is reflected in most commercial solutions available and for various form factors of device. This will give us confidence that we have established a viable model and will enable us to better grasp the various mobile platform solutions being offered. The utilities that we will be considering fall into four categories:

- communications utilities;
- interface utilities;
- personal information management (PIM) utilities;
- application support utilities.

This is a good point to state that each commercial mobile platform is different, despite having common traits, and it is these are what we are trying to identify here. Particular approaches to each commercial solution can only be weighed up in light of real application requirements and associated commercial concerns.

10.9 COMMUNICATIONS UTILITIES

10.9.1 IP/WAP Stack

At this stage of the book we should already understand the importance of being able to connect with an IP-based network, either using IP itself or the Wireless Application Protocol (WAP) refinement (1.x or 2.x). Therefore, it is useful to add a software library to the mobile device, which enables any application to exchange IP or WAP messages without having to be concerned with the low-level programming required to do this.

Low-level IP and WAP programming is concerned with message packet construction, error handling, re-transmission, packet assembly and so on. All this functionality is already included in the library, which is called a *stack*, as discussed in Chapter 5. The stack will support all the commonly used Transmission Control Protocol (TCP)/IP protocols, such as DNS, DHCP, etc. A typical IP stack would probably support the following:

- Transmission Control Protocol (TCP);
- User Datagram Protocol (UDP);
- IPv4/v6 support (IPv6 addressing as well as IPv4 addressing);
- Internet Control Message Protocol (ICMP);
- Point-to-Point Protocol (PPP);
- Domain Name System (DNS);
- Dynamic Host Control Protocol (DHCP) for IP address assignment;
- HTTP 1.1-compliant client stack.

This is likely to be a minimum set, although IPv6-addressing support may or may not be present. Additionally, we might expect the support for at least some of the secure protocol layers to support encryption and VPN connections, such as:

- Transport Layer Security (TLS) and Secure Sockets Layer (SSL) (the TLS module is effectively an enhancement of the SSL protocol);
- IPSec is an IP layer protocol used to secure host-to-host or firewall-to-firewall communication and provides tunneling, authentication and encryption for both IPv4 and IPv6 so that secure connections can be made over the Internet, primarily for mobile enterprise access.

The stack enables the mobile application developer to concentrate on developing the core of their application without having to be concerned with implementing the IP protocols that it may require to connect to the outside world.

Accessing the stack from the service layer is via an API that is produced by the device manufacturer or, if applicable, by the operating system provider. Although TCP/IP is strictly speaking not an operating system function, it is common for such functionality to be bundled with the operating system and be optimized to run on it. Often, no distinction is made between the actual operating system – or the RTOS – and the utility libraries that get bundled with it: they all get referred to as the operating system. These days, more and more utilities are included in the bundle; this is how an operating system vendor seeks to gain support (customers) for their particular product offering.

It is likely that, in addition to providing an API to access the stack, a toolkit of some kind will also be provided, most likely as part of a software developer's kit (SDK). At a minimum, this would provide documentation on how to access the API, how to program for it and utilize its services effectively. There may also be sample code demonstrating how to utilize some or all the API services (such issues as the programming language used to access the API will be discussed shortly).

Figure 10.18 Phone dialler application on a Pocket PC. Reproduced with permission from Microsoft ®.

10.9.2 Telephony Control/Phone Manager

At least one of the devices in our possible device network will have integrated telephony; therefore, we expect some basic telephony control software to be made available on the device, as well as software to control the phone settings, such as the ring tone. The telephony manager will have its own user interface to allow calls to be made (or similar tasks) and will have the ability to respond to key presses, whether from a mechanical keypad or via a touch screen, as shown in Figure 10.18. The capability of driving a display will also be included, showing the number being dialled, the numbers being received and so on.

On devices that incorporate telephony an array of telephony control and call management functions are to be expected; most of these should be familiar to us. Certain telephony functions are expected to be present according to the requirements of digital cellular standards like GSM and UMTS, features like phonebook, call diversion, called ID display and so on. Typical telephony features include:

- dialling management (the ability to dial numbers);
- phonebook store and manipulation;
- call barring;
- call records (last numbers dialled, received, missed, etc.);
- call diversion management (divert always, on busy, etc.);

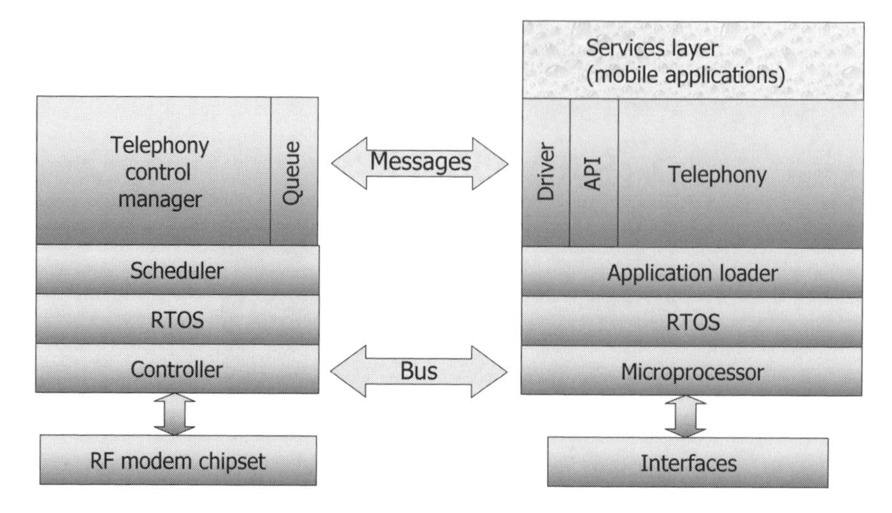

Figure 10.19 Telephony management architecture.

- ring tone selection and volume control, including vibration alert where applicable;
- cellular network selection;
- call timers and possibly cost calculators;
- call-waiting control.

There may be other control features available, but this will depend on the appropriate telephony specification (e.g., GSM) and on the device manufacturer's product design criteria, especially when it comes to incorporation of elective features.

Most of the functions that the telephony manager controls will be inbuilt features of the RF modem, and we expect the software to interact with the control chip in the modem chipset. This arrangement is shown in Figure 10.19, which also helps us to understand how it might be implemented in a typical mobile device.

We can see from the figure that the telephony software runs as a process sitting on top of the RTOS that runs on the main processor in the mobile device. The microprocessor is physically connected to the controller chip in the modem chipset. The nature of the connection may either be on a shared bus or via some kind of dedicated communications pathway, like a serial link or a shared memory space. There is no preferred way of implementing the communication pathway: it is dictated by the chipset manufacturer who also decides which interface options are possible in a particular device design.

It is possible that the telephony application could use low-level programming techniques to access the communication pathway directly, but this is unlikely. What is more probable is that a device driver is written which offloads this concern from the telephony software. The device driver will include all the necessary low-level software and start-up routines to initialize the communications pathway to the controller in the modem chipset.

The controller itself will run various software processes according to the tasks it is carrying out at any particular time. It is likely that it can carry out many tasks on behalf of the mobile device software. The controller will run a message queue handler that can receive messages across the pathway from the microprocessor, interpret them and subsequently

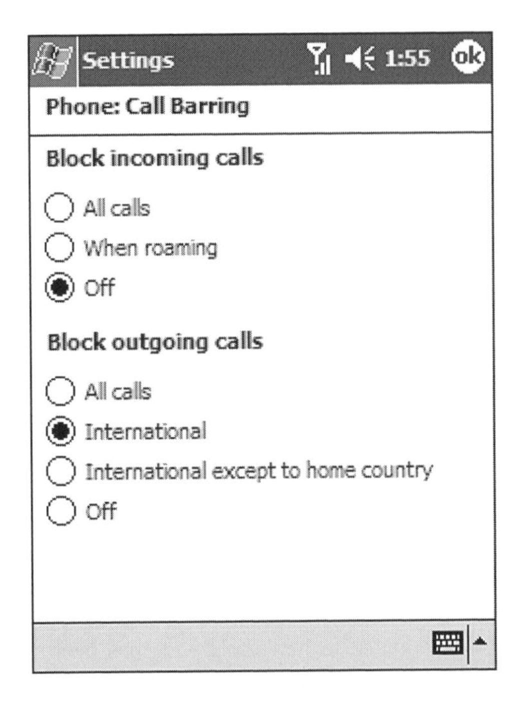

Figure 10.20 Call control via a Pocket PC (Mobile Edition) graphical interface. Reproduced with permission from Microsoft ®.

route them to the appropriate software process, such as the telephony control manager in this instance.

The device driver on the microprocessor side will be familiar with the low-level message formatting that the controller queue expects, including how to mark the message destination to be the telephony control manager.

On the microprocessor the telephony software will access the device driver via an API, a predefined set of software function calls that can be successfully interpreted by the device driver and converted to the appropriate messages destined for the telephony control manager (via the controller message queue). With this arrangement, it is relatively straightforward to implement applications with rich graphical interfaces to control the telephony functions, such as call barring (see Figure 10.20), without the application developer having to know much about the interface details for the telephony controller and its low-level messaging protocol.

This low-level messaging process works both ways. A message from the telephony control manager, such as an incoming call alert, will be sent back to the microprocessor via its appropriate physical communications pathway, such as the bus or other options mentioned earlier.

What happens on the microprocessor is that the peripheral that handles the communications pathway will cause an alert to the RTOS that something has happened on the bus; this alert is called an interrupt. The RTOS is responsible for handling interrupts. It can interpret any interrupt from the processor, identify its source and then cause the appropriate software routine to be scheduled to run on the processor and subsequently handle the interrupt. In our example the device driver for telephony handling is scheduled to respond to the interrupt, and the appropriate event-handling software routine within the driver will then be executed. This in turn will raise a software-generated event within the telephony software, which should have the appropriate means to service the event. For example, in the case of an incoming call the event handler in the telephony software would, say, update the LCD display to indicate the presence of an incoming call and perhaps include the caller ID (number).

Increasingly, it is becoming blurred as to where in the mobile device the telephony management takes place. In the case of some chipsets the ring tone generator is included as an integral part. This can be driven by the core software on the mobile device: from the telephony software running on our microprocessor. Therefore, the device driver will also be expected to provide a low-level interface for sending messages back to the chipset controller message queue, in this case to a different software process on the controller, one that is more than likely running on the integrated DSP, particularly if the tones can include sound samples (e.g., MP3 files or similar). Alternatively, the controller itself can take care of the entire call-handling function, switching on the appropriate ring tone automatically, without any intervention from the core microprocessor on the mobile device.

In terms of how these functions are implemented in software, it is sensible to enquire if the telephony device driver API is accessible to any mobile application running on the microprocessor. There is no reason why this shouldn't be so, but it is not that common in many devices. This is all related to how the device has been implemented and what the manufacturer decides in terms of providing third-party access to embedded device functionality.

We will examine this issue in more depth later; but, for now we might reflect on the fact that most mobile telephones either have been historically developed using proprietary embedded software solutions or have, in any case, not been made open for other applications to be run on – other than 'factory-fitted' ones. On the other hand, PDAs have been designed to run third-party applications from the outset, but they have not grown up with any history of integrated telephony functions. The issue of telephony APIs was not addressed until relatively recently with the advent of the 'smart phone' concept – a PDA and mobile telephone combined (although there is no exact definition for a smart phone). Specialized 'smart phone' or 'mobile-aware' operating systems have since emerged, like Symbian. Efforts to produce standard telephony APIs have also emerged, like the JavaPhone API (albeit not specifically aimed at mobile telephony).

Phone Settings and Control while considering telephony management, we can also look at general phone settings and their control. Phone settings include things like adjusting display settings, interface language (i.e., display character set, such as English, French, German, Arabic, etc.) and similar housekeeping functions. These might include:

- time and date settings;
- language;
- display contrast/brightness;
- power-saving modes;
- PIN code protection;

- hands-free control;
- voice control;
- factory reset.

Such functions would be present in any form factor of mobile device with a user interface, but not necessarily grouped like this nor presented as 'phone' settings (given that the device might not be called a phone nor recognized as one). Some of these features would be common utility services bundled with the operating system, such as time and date settings, language, display control, etc.

As with the telephony software, some of these features may well be accessed through the modem chipset, such as voice control dialling that might utilize an embedded DSP to execute voice recognition algorithms. We would face similar issues in relation to just how far these services are implemented on the main processor and what is the relationship of the interface to them in terms of API provision. There is a similar question about the extent to which these services are deemed operating system services, bundled with the RTOS, or as third-party add-ons. The architecture of a mobile device is no longer a simple issue (if it ever were)!

10.10 PERSONAL INFORMATION MANAGEMENT (PIM) UTILITIES

Mobile devices are personal devices; therefore, they naturally take on some of the characteristics of other personal objects, like diaries. Early on in their history, PDAs were glorified diaries and address books, a more convenient and powerful version of the pocket book or Filofax™. This is probably still the number one reason for buying one. However, email has become a very common personal communications tool, and there has been a growing interest to access email from a PDA, especially now that wireless connectivity is more widespread and affordable.

Mobile phones have always had the concept of an address book, or contacts list. This is an expected feature for a telephony device where number management is essential. Not so common is a diary feature, although many phones do incorporate one. On devices with limited data input abilities (i.e., with only a 10-digit keypad), the addition of extra features like a calendar has been slow to catch on. Perhaps it was felt that those people requiring calendars would most likely own a PDA.

Since the advent of 2.5G data solutions, like GPRS, some phones have started to appear with email clients. Email clients have existed on PDAs for some time, but in the earliest incarnations were limited to carry-and-go clients for desktop email solutions, like Microsoft's Outlook™, which did not have any real time email collection capabilities between the PDA email client and a remote email server. Emails could only be sent and received on the PDA during a synchronization session in the cradle connected to the desktop PC. Such a solution was of limited use, even more so in the mobile context.

In newer 2.5G phones and wireless-enabled PDAs, email has become more important and is a standard offering. We have seen these two converge. Some mobile phones have expanded in functionality to take on PIM features, while some PDAs have taken on wireless connectivity and even telephony. It is interesting to consider the key features that we might expect on either type of device. Typical PIM features are:

- diary with multiple viewing options (day, week, month);
- contacts management with multiple fields, such as name, number (mobile, home, office, fax), postal address (home, office), email address (multiple entries);
- to-do list (tasks);
- notes;
- email client (POP3, IMAP4 support);
- PIM store (storing contacts, appointments and emails).

The existence of an email client usually implies the ability to connect with an IP network directly using TCP/IP, so the IP stack mentioned earlier needs to be implemented on the device. Most devices these days will be using an always-on wireless data connection, and there is no longer the need to dial into ISP (Internet Service Provider) modem banks. Some older devices support this, and even some of the newer ones still offer dial-up IP protocols as a fallback solution – sometimes the easiest route for corporate access (i.e., into an existing remote access server solution). Some devices using proprietary wireless networking, like some two-way paging networks, may not offer IP stack support and rely on gateways to go between a proprietary networking protocol and one of the IP email protocols like POP3.

It is possible to access email services without an email client: this can be done using Web-based solutions, which is a common enough technique on the Internet, even without wireless access. Some operators offer email services via WAP, but often these solutions require a dedicated email account with the operator's email server network, so the solution is limited in terms of accessing other accounts. With some services, it is possible to implement POP3 collection from another account into the user's operator account, but this is still a limited solution because it requires that the user still sends emails from the operator account and not the user's regular account. As with any solution involving remote access to a server, once there is more than one way of accessing the mail (such as from a desktop and a mobile client), the problem of synchronization arises. For example, sending an email via the mobile will probably mean that the message does not get stored in the 'sent items' folder on the desktop. Similarly, reading a message on the mobile probably does not flag the message as read on the desktop; there are many synchronization issues that arise.

We do not wish to discuss particular solutions in too much depth in this section as we are simply trying to highlight the basic PIM functions and the value of their inclusion on a basic mobile device. We would not expect PIM functions on all devices, but convergence is well under way. The further level of convergence is with mobile telephony itself. If a telephony function is present on the device, then we expect it to be fully integrated with the PIM functions: for example, we could dial a number directly from a contact entry, etc.

10.10.1 Text Messaging

In terms of the messaging capabilities of the PIM client suite on a device with integrated telephony, we expect a text-messaging client. Of course, all mobile phones on the common digital cellular networks already have text-messaging support; clients on devices with richer interfaces can be more user-friendly.

In terms of integrating the text-messaging client into the device's user interface, this raises an interesting point as to how we group functions on a device. There is one school of thought that says we should use a messaging metaphor and gather all types of message into one client. This used to be called 'unified messaging' when the concept of handling all message types using one client was first proposed. This was originally conceived of as a means of unifying communications on the desktop, so that faxes, voicemail and emails would be gathered together into one client with a consistent messaging metaphor. For the wired world, there has always been an issue of how disparate messaging sources get integrated into one solution; this problem has been addressed by the production of dedicated infrastructure products to support a unified messaging paradigm. The challenge has been to create bridges between telephony and computing so that messages can be understood and processed by software, whether presented in a single client or not.

With digital cellular networks, this problem is not so acute. Voicemail messages have had visual indication for some time, courtesy of text message alerts. On mobile devices, text messaging has been present from very early on, so in one sense we have always had a unified approach to text messages and voicemail.

Text messaging and email are natural cousins, and it has made sense in some cases to integrate handling of these message types into one interface. It is usually done by taking the email metaphor and extending this to text messaging, which is reflected in the evolution of naming conventions. Text messages used to be available under a 'received' menu option which has now become an 'inbox', whether or not it is also used for emails. Other naming parallels have been adopted.

We have the choice to adopt a unified messaging metaphor if we like. It is really a matter of software implementation on the handset where all the different messages types already converge.

The convergence process has been further provoked, or challenged, by the emergence of MMS on many mobile handsets. The same goes for WAP Push messages, yet another messaging type in the digital cellular world and one rather peculiar to it.

It is really a matter of design choice as to how the mosaic of message types get handled by the mobile device in terms of their presentation and handling via a user interface. With modern interface techniques, such as XUL (Extensible User interface Language), it is perfectly feasible to imagine giving users a choice of interface presentation. Some users are not keen on having all their message types grouped into one client, especially if it entails a single view of the message space through one inbox; this can become cumbersome to manage. With some users potentially receiving hundreds of texts and emails a day, the prospect of handling them all through one view is offputting. We could probably benefit from some serious usability studies in this area, but then why bother if we can offer users the option to configure the interface to operate how they would like it to? For those that want separate menus, the interface keeps them separate. For those that want to combine, they get to choose what they combine with what. This way, everyone's happy.

The type of unification that does make sense is a unified and consistent view of information, like message addressing and contact identification. For example, if a user wants to send an email or a text message, then they should be able to look up their target contacts from the same database. If a user receives a text message, then the sender's number should be checked against the contacts database to find a match. If matched, then the sender's name should be displayed instead of or in addition to their phone number. This is standard stuff on mobile phone text-messaging clients these days, but has taken a while to penetrate text-messaging solutions on PDAs (and even desktop PCs where text-messaging clients are increasingly being used[13]).

To do this implies that all messaging clients have access to the PIM store or, more likely, to an API that provides PIM store management. We therefore would hope that a PIM API is present on our mobile device and not just the PIM client software. This leaves us free to develop new messaging applications based on the inbuilt PIM capabilities of the device. In fact, we would ideally like an API for the PIM client primitives, not just to access the PIM store. So, for example, we could use an API call to send an email message from our software via the PIM client, which means we can maintain an audit of messages in the client. This is potentially more powerful than just sending email messages (e.g., from a field service application) which are not recorded in the email client or have to be assigned to a unique 'sent messages' folder with our custom application rather than the standard 'sent messages' folder.

[13] Text messaging is increasingly finding its way into the enterprise, especially with products that enable databases to be easily accessed and updated via a simple text message. For example, see http://www.xsonic.com

11

Mobile Application Paradigms

11.1 INTRODUCTION

In Chapter 10 we looked at the possible architectures for mobile devices. We identified a basic software/hardware structure that provided us with the essential capabilities that a mobile service would need from a mobile device. Reminding ourselves of the key components, we reproduce a key diagram from the previous chapter, shown here in Figure 11.1.

In this chapter we want to consider the following ideas and concepts, some in far more detail than others:

- application topologies – J2ME, browser, embedded;
- service topologies – client/server, peer to peer, stand-alone;
- device networking architectures – personal mobile gateways (PMGs);
- Device networking technologies – Bluetooth™, WiFi and Infrared (IrDa);
- device types – PDA, smart phones, laptops;
- platform options – Pocket PC, Symbian OS, Linux, Palm OS, among others;
- J2ME approach;
- interfaces and usability.

Next Generation Wireless Applications P. Golding
 ISBN: 0-470-86986-0 (HB)

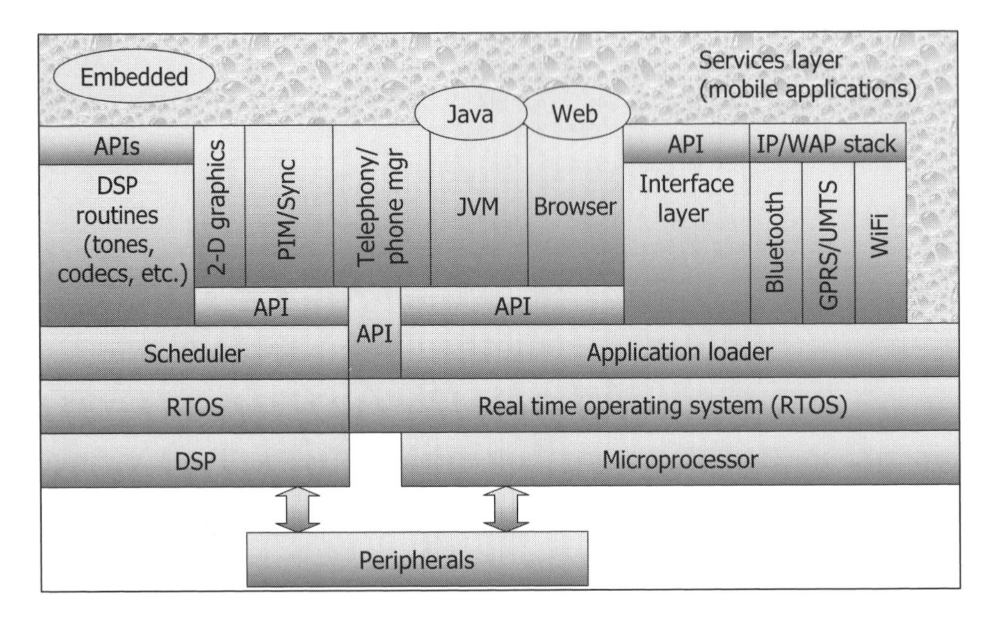

Figure 11.1 Typical mobile device architecture.

11.2 APPLICATION TOPOLOGIES

In Chapter 10 we suggested that there are three basic approaches to developing mobile applications in terms of what runs on the mobile device itself. These are:

1. *Java* – implement a Java program that runs in a Java run-time environment (JRE) and uses Java APIs (application programming interfaces) to access mobile device resources.
2. *Browser* – provide access to a server-based application via a browser, where the browser accesses the mobile device resources for us, such as the network connection to the server.
3. *Embedded* – implement a program that runs natively on the mobile device real time operating system (RTOS) and uses the device APIs to access mobile device resources.

Java and embedded are similar approaches, but the Java approach is significant enough to warrant separate consideration.

Perhaps the first question that comes to mind is why three methods are available or even necessary: can't we just make do with one? A related question is whether the options are mutually exclusive? The short answer is that these three approaches are the three most obvious and prevalent approaches to software development. In fact, the Java and browser approach are really extensions of the embedded approach. Ultimately, any application on a mobile device has to run on the device's operating system and use the low-level resources of the device. This is the most generic solution. A browser is a particular embedded application and a special kind of application that can subsequently enable an alternative general purpose programming model: namely, the Web-based client/server model that we have discussed and analysed so extensively throughout this book. The Java model is also based on an embedded application, called the Java Virtual Machine (JVM), which subsequently enables a different type of general purpose programming model whose appeal is that the model can be implemented in a consistent manner across a whole range of mobile devices. This works by abstracting a generic programming environment and a generic set of APIs built on top of specific, underlying APIs offered by the device's operating system.

An in-depth understanding of the service topologies needs to be developed before answering the question, 'which one should I use?' First, which one should I use for what? The 'what' is important, because these programming paradigms are not the same and cannot be used interchangeably for all application scenarios. For example, if we hope to implement a peer-to-peer (P2P) application so that one device can swap files directly with another, then most likely we will not choose a browser approach because in its unmodified form it requires a Web/WAP (Wireless Application Protocol) server and an Internet Protocol (IP) infrastructure to bring it all together. A network of just two devices cannot easily or economically support such an environment. This means straight away that we are already excluded from using the browser approach and forced to use an embedded approach or a Java one. Now, if we attempt to access files on a device using the Java approach (in particular J2ME, or Java 2 Micro Edition), then we more than likely cannot accomplish this because of the sandbox Java environment that does not support low-level APIs accessing the underlying file system.

Our decision may also be influenced by the APIs available on the device. For file swapping, there may already be API services that are easily adapted to this application. Protocols like Object Exchange (OBEX[1]) protocol may be included on the mobile device and exposed via an API.[2] OBEX is optimized for *ad hoc* wireless links and can be used to exchange all kind of objects like files, pictures, calendar entries (vCal) and business cards (vCard). OBEX is similar to HTTP (HyperText Transfer Protocol), but does not require the resources that an HTTP server requires, making OBEX perfect for devices with limited resources. If an OBEX API is present, then that may well influence our decision to use it, notwithstanding the issue of considering how widespread the protocol is on other devices we want to network with.

We might even be guided in our decision by our own technical competencies and understanding of the options and technologies. This is not unreasonable, though it may be inefficient. However, trade-offs are inevitable, even if we have good knowledge and proficiency in all approaches.

What we should consider first is making sure that we have understood our requirements for our service. Throughout this book we have looked at many aspects of mobile service delivery. The benefit of our wide-ranging and integrated approach is to enable us to appreciate the wide variety of options for building mobile services. There is no one way to implement a service; this is something we should feel confident about. To a degree, the richness of device capabilities and software technologies, combined with the increasingly powerful, underlying processing power, means that we can probably afford to be more flexible in our approach to implementation. To be frank, considering the amount of time we might spend evaluating different approaches and API documentation, it is tempting to advise just jumping in feet first with the initial approach that seems to fit the service needs, but software design aficionados would frown at such a suggestion.

[1] http://www.irda.org/standards/pubs/IrOBEX12.pdf

[2] There is an API for OBEX provided in Symbian OS V7.0, in Pocket PC and in some others.

As we remarked in the opening chapters, always take what already exists in terms of APIs, open standards and even open source and learn to become a 'smart integrator'. Market events are probably going to move too quickly to spend a long time investing in writing code where something suitable already exists, even if it only approximates what is required. As we have stated before and will state again in more depth, usability is a key deliberation that should not be circumvented. It is probably the most singular design consideration; so, it would be better to spend resources in this area rather than programming a low-level software function that already exists and can be integrated with ease. A compelling reason to be smart about integration and development is the need for the agility necessitated by the heterogeneous and possibly fickle nature of the future mobile services market. Accordingly, applications providers will almost certainly have to find many ways of generating revenue, and they will need to think in terms of groups of applications rather than just one application, each highly customizable to end user requirements, recognizing that requirements might change frequently.

As we will see, there is scope for rapid implementation and high degrees of personalization across all the approaches – whether browser, Java or embedded. Let's now look at the fundamental characteristics and attributes of each approach.

11.3 EMBEDDED APPLICATIONS

This is without doubt the most flexible of all the options; we could say it is the 'king' of approaches. Anything that a particular device is capable of being programmed to do is theoretically possible using the embedded approach. There are no restrictions on access to the underlying device resources, as compared with the other approaches, which are altogether much more limited. This is because the embedded approach provides the ability to get right down to the guts of the device and programmatically access its low-level functions and peripherals. The reason that this programming power is only theoretically possible is that in practice the ability of a third party to gain low-level programming access is limited. The provisions that the device manufacturer makes available to third-party programmers, such as public API access, documentation and related programming tools, determine the limitations. These considerations will become clearer as we progress.

Let's consider coming up with a new way of implementing handwriting recognition. In theory, we could program the solution as an embedded application; whereas, using a browser approach, we do not have this kind of access to the device resources. Although, to be fair, the browser paradigm is not intended to allow for this degree of flexibility – the programming and interface model is deliberately simplified for good reasons. Functional extensions to the browser might be possible,[3] but this would involve a hybrid approach that still requires that the extended functions be programmed as an embedded application.

An embedded application works by running on the main device processor courtesy of the RTOS, which manages how the application runs and how it gains access to the device

[3] The External Functionality Interface (EFI) specifications in WAP 2 provide means that enable WAP applications to access 'external functionality' (embedded applications running on the same device) in a uniform way through the EFI application interface (EFI AI). See http://www.openmobilealliance.org/wapdocs/wap-231-efi-20011217-a.pdf

resources. Via bundled software pre-installed on the device, an embedded application has a plethora of APIs that it could use to harness the services of peripherals and other applications that may already be bundled on the device (such as a telephony manager, PIM and so on). For example, as already considered in Chapter 10, an email client is probably a standard feature on a mobile device and its internal functions may well be accessible via an API. An embedded application could utilize this API whereas a browser certainly could not. A Java application may or may not be able to access the API depending on factors we will discuss later – to do with how Java support is implemented on the device.

11.3.1 What Do We Need to Develop an Embedded Application?

Developing embedded mobile applications is not that different from writing applications for desktop platforms. We basically need similar tools and resources:

- a compiler for the chosen platform;
- a debugger for the chosen platform;
- a software developer's kit (SDK) for the chosen platform and possibly for a particular variant within a platform family;
- a workstation (e.g., Windows™ PC) to run the tools on.

In addition to the above requirements we will also need:

- an actual target mobile device on which to try out our solution or possibly several devices to test various, intended mobile products;
- a live network connection to test the application within a mobile network context.

This book is not about how to write software nor about the development process. Therefore, we will not discuss the need for a requirements specification, test plan and so on. However, we will allude to the important aspects of the software process where necessary to emphasize certain mobile considerations.

We have not yet mentioned which language should be used to develop the applications. Usually, most development kits specify a set of APIs and a toolset that support C or C++, this being a language that is particularly suited to embedded solutions for a variety of reasons, not least of which are its features that allow low-level manipulation of data at the hardware level. However, we should not think that C is the only option. We need to dispel some misconceptions about the use of portable languages like Java, to which we assign a category of application development in its own right with respect to a particular Java approach called MIDlets, which we will later explain. However, as we will discuss in a moment, Java can also be used for general purpose embedded programming.

Language support depends mostly on the device's OS and the programming tools that are available to it. Ultimately, the device vendor determines the language support, and it is common for either the mobile OS vendor or the device vendor (or both) to provide the tools required for the embedded development process.

11.3.2 C Is Not the Only Choice

The .NET Way As already mentioned, C enjoys the most widespread language support, but there are some notable exceptions. In the case of Microsoft's Windows CE 3.0 (the underlying OS layer for Pocket PC 2000/2 variants), there is support for Visual C++ (VC) in line with the general support for C++ from Microsoft across all their platforms. Indeed, this is part of the Microsoft ethos. Even though Pocket PC has been built from the ground up as a brand new operating system for mobile devices, its programming model and language support has been made deliberately familiar to Microsoft's Windows programmers: this is why VC is supported in addition to Visual Basic (VB).

The reason that Microsoft developed a new operating system from scratch is that the microprocessor families for mobile devices are very different to the Pentium family which Microsoft has intimately supported for so long. In fact, there is no single processor family for mobile devices, but a collection of families including MIPS, ARM, SH3/4 and the PowerPC. This is what probably led Microsoft to develop a solution for the hand-held PC variant of Windows CE (now marketed as 'Pocket PC'); this utilizes a common executable format (CEF) in which the VC or VB code is compiled and then run on the operating system, via a software mechanism that interprets the CEF code.

The CEF idea appears to have been the inspiration for Microsoft's latest Windows™ platform overhaul – the programming framework called .NET ('dot net') – which uses a common language run-time (CLR). This is a major revamp of the Microsoft Windows™ platform, largely to make it easier to develop applications that are Internet-connected and Internet-responsive; this is why .NET includes technology and rich APIs to support such IT modernity as Web Services and XML handling.

With the .NET framework comes two new languages called C# (pronounced 'C sharp') and VB.NET; they are both supported by the newest mobile OS called *Windows CE .NET 4.2* – which is the basis for OS variant Pocket PC 2003 and the new Smart Phone variant. These both support the Microsoft Windows™ .NET Compact Framework that is integrated in ROM. The Compact Framework is the mobile edition of the .NET framework and has been ported to enable a common programming model and toolset approach across all Microsoft platforms.

It is still possible to develop using C++ (using the embedded VC toolset), but new applications in VB are no longer supported. Programming in the newer languages enables access to a much richer set of APIs, so it is both tempting and sensible (advisable) to migrate to the .NET toolset. Furthermore, programs that are written using the newer languages benefit from advanced OS features like program management. With program management, the OS prevents managed applications from freezing up the device or acting maliciously.[4] It is also possible to load these applications over the air (OTA), a technique that we will discuss later and one that is very important within the mobile context.

... Visual Basic Reprise It is worth noting that VB is supported on other mobile operating systems using the innovative App Forge[5] solution. According to its promotional material:

[4] 'Acting maliciously' means doing something on the device that would be disruptive to its overall behaviour, such as one program corrupting the data of another.

[5] http://www.appforge.com

AppForge MobileVB software integrates directly into Visual Basic 6.0 enabling you to immediately write applications, using the Visual Basic programming language, that will run on over 90% of the world's handheld, mobile, and wireless devices.

MobileVB™ uses Microsoft VB V6.0 (VB6) – the last version of VB before Microsoft switched to VB.NET. VB6 was and is a highly popular language among developers, with a large and loyal following. MobileVB™ has been ported to Palm OS, Pocket PC and Symbian (though not all device types).

The advantages of MobileVB™ are its ongoing support for a familiar language with all its fondly associated rapid development characteristics. However, API support is currently very limited. Native APIs on a given mobile platform are designed to be accessed via the language they have been designed for. Therefore, MobileVB™ is forced to attempt to map native APIs to VB. Because of some fundamental differences between VB and a language like C (which is used to implement native APIs on most mobile platforms), it is not always possible to successfully map the device APIs. When it can be done, performance suffers and we would expect MobileVB™ apps to run slower than native ones. That said, there are plenty of mobile applications that would probably work fine when programmed in MobileVB™; so it does seem to have its place in mobile application development.

For some devices – like the Sony-Ericsson P800 running Symbian – MobileVB™ has APIs to support features such as telephony and SMS (texting), thereby enabling very powerful applications with telephony and message features.

Besides being usable for commercial applications development, where the API support is the limited factor, MobileVB™ makes rapid prototyping a possibility, which is an attractive prospect; VB has often played this role in the desktop world. For example, it is feasible for an application to send field-collected data via a form, and text messaging could be implemented within a matter of hours on a P800 by an experienced VB programmer who otherwise may be a complete novice in mobile device development. For such a person to learn the Symbian OS and APIs and then to struggle with C or Java (more on this later), the learning curve would be relatively steep, even prohibitive in some cases (usually due to time constraints).

11.3.3 'Native' Java Support

RIM's BlackBerry We will shortly see that any mobile device can potentially support programming applications in Java via the porting of JRE to the device as an embedded application itself in the services layer; Chapter 10 has already introduced us to this concept.

However, on some devices, Java is the only way to program applications in the services layer. This is because the whole OS is built from the ground up either to make use of Java or to offer intimate Java support. This is not a surprising development given the increased capacity of newer mobile devices with greater processing power and memory, yet still within an acceptable price/performance curve (including power consumption).

A notable 'native Java' device is RIM's BlackBerry 6210™ (a GSM/GPRS device), as shown in Figure 11.2. Apart from restricting the language choice to Java, the development process for the BlackBerry is similar to the general embedded approach: we still have to use a developer toolset, in this case the Java Developer Environment (JDE), and we still access system resources using APIs. The APIs on the BlackBerry are a combination of standard

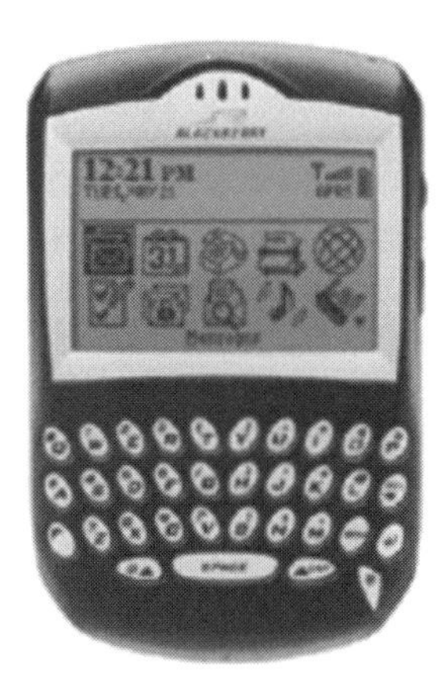

Figure 11.2 RIM's Blackberry 6210. Reproduced with permission from RIM.

JRE APIs (as applicable to the type of JRE running on the BlackBerry) and extension APIs provided by RIM (not a standard part of Java).

It is important to understand that Java is not just a language, it is also a set of APIs and an associated run-time environment, which together form a Java *platform*. When implementing the Java platform on the BlackBerry, RIM implemented their own API set to suit the architecture of their device, thereby exposing the underlying resources. This is what we would expect when looking back at our proposed generic device architecture shown in Figure 11.1. All the APIs would have to be made available in the Java platform so that they are made accessible to the Java programming language when writing applications to run in the services layer of the device. To implement a powerful application that exploits the potential of the device, we have to use the custom Java API set. This is going against the Java portability ethos, which is to use a consistent set of Java APIs to standardize an abstraction of device capabilities, thus confining development to this set in order to achieve portability objectives. We will discuss this issue in more detail below.

If we consider the 2-D graphics capabilities of the device, then we see that the API for this is made available in Java, as we would expect. As we will see later, the various, standard Java platforms have their own concepts of graphical user interface (GUI) widgets and related visual metaphors, and they provide APIs accordingly. This is supported on the BlackBerry; but, additionally, RIM has devised a custom GUI metaphor for the BlackBerry, including the infamous track wheel for tactile response. The API for this is in the extended set, not the standard Java set. Similarly, to access some of the radio resources in the network layer, proprietary RIM Java APIs are provided – such as a mechanism to enquire about radio signal strength.

With custom Java APIs there seems to be a contradiction in using a language and platform in a way that goes against its intended benefit of application portability. This is only partly true, as a large amount of the Java code would probably remain portable; but, the main intended benefit here is to provide a mobile development platform that is attractive to Java developers, simply because there are now so many of them that providing a Java solution makes sense. The other aspect not to be overlooked is that Java remains an open standard, truly portable to any device, unlike more proprietary languages like the .NET family.

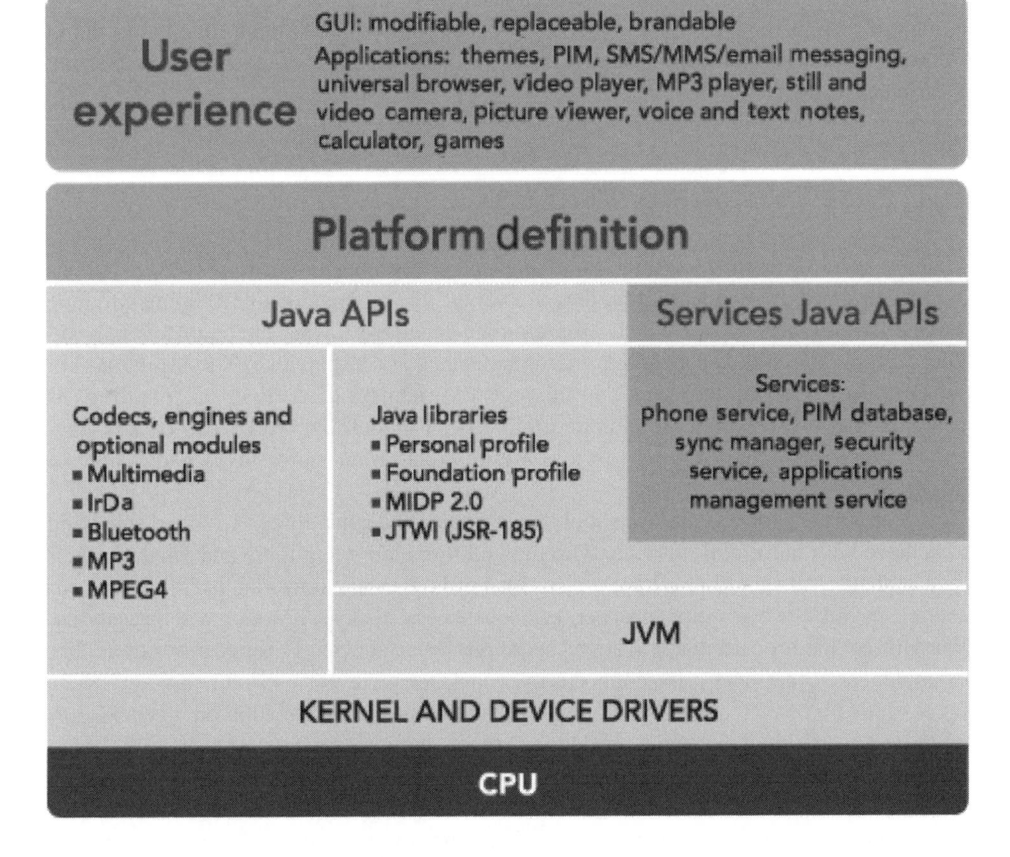

Figure 11.3 SavaJe OS 2.0 architecture.

We have only looked at the programming support for applications on the BlackBerry, not its detailed architecture and features. For more information see the RIM website.[6]

SavaJe OS Another native Java OS worthy of note, not least because Orange is an investor, is SavaJe OS 2.0 (pronounced 'savij'). Like BlackBerry, the entire OS API set is presented in Java. However, support for the Java platform is more comprehensive than with the BlackBerry. A larger subset of the standard Java platform is supported and the extended API set is quite extensive, including the provision of a large number of applications that are ready to use 'out of the box'.

The inner kernel of the OS actually supports C/C++, but the programming model presented through the API set is pure Java, even for library functions that have been developed to run natively on the kernel (i.e., not in Java). This hybrid approach has a performance advantage due to the ability of the kernel to run native applications quicker than Java ones; therefore, the computationally intensive tasks, like video compression, are all written in C and executed natively, as shown in Figure 11.3.

[6] http://www.rim.net

For applications that need the extra performance boost, SavaJe OS supports the development of applications in C/C++ for native execution; this is a facility that is offered in the higher end Java platforms through a Java API called the Java Native Interface (JNI), but SavaJe has its own approach.

11.4 EMBEDDED DEVELOPMENT TOOLS

We have spent some time looking at the language options open to us for embedded applications development and learnt that they are varied. We reiterate that developing a mobile service is not a question of which programming language to use, although this may influence our chosen path depending on competencies and experience. The available API set will also influence the decision, although the richness of APIs on all platforms is continually evolving, as is the generic architectural model that we developed in Chapter 10; hence, new and exciting services are possible on an increasingly wide range of devices.

A point made earlier is that we should have a clear understanding of what we are trying to achieve with our mobile service, who the end users are going to be and what are their requirements. We should also have a clear idea of how to maximize the usefulness of our service, paying close attention to key criteria for success, not least of which will be usability. We will see that for certain applications some platforms are clearly better than others. For example, for secure enterprise connectivity, BlackBerry currently has an edge due to its focus on this type of feature. Developing a secure enterprise application on a BlackBerry will almost certainly dictate an embedded approach, particularly if we want to push data to the device asynchronously of user interaction. There is an API to do this and a programming model on the device that supports the concept of listening for push messages – a service that the OS manages for services layer applications; so, it is natural that we would want to take advantage of this API for this type of requirement. However, if we did this, then the implication is that all the users of the proposed application would be required to own BlackBerry devices.

Whatever the reason to use an embedded approach, the development process is usually the same for all devices; this is summarized in Figure 11.4.

Let's briefly examine the development process and outline its constituent parts:

- design;
- configuring the integrated development environment (IDE);
- editing the program(s);
- compiling and building the project;
- testing and debugging via a simulator (emulator);
- testing on a target mobile device(s);
- conducting formal usability testing;
- pilot trials and deployment.

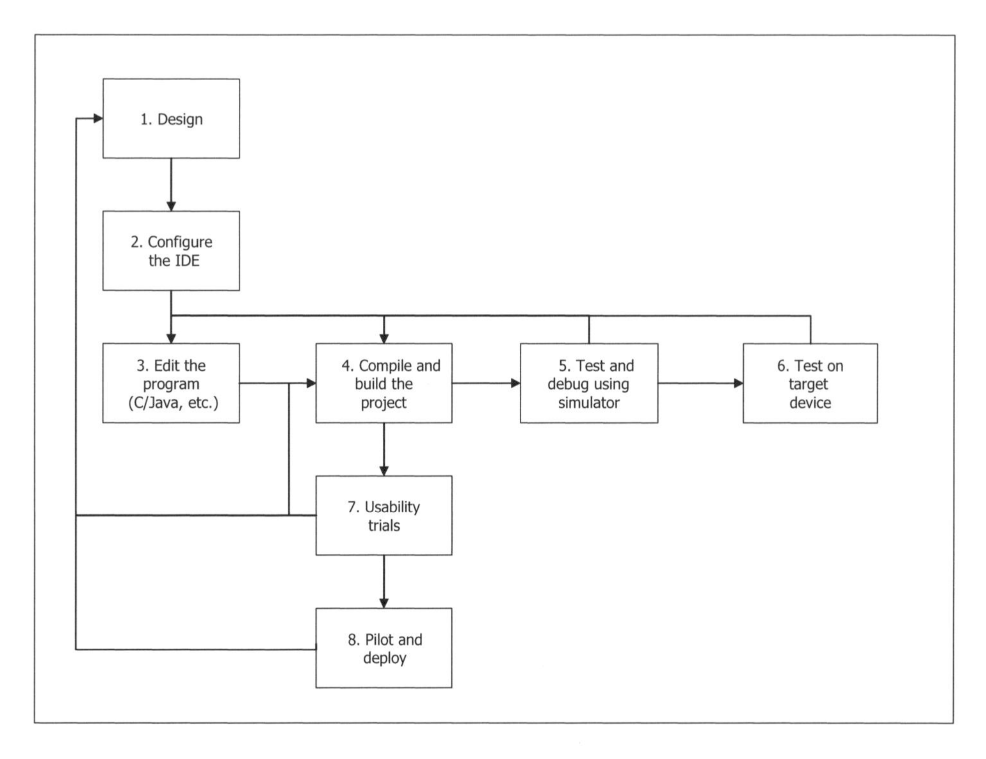

Figure 11.4 Typical process flow for embedded mobile application.

Design Remember, this book is not about how to develop software in terms of the design process and coding practices, so we do little justice to this subtask by devoting only a few lines to it. Software designers will have their own preferred approach. The things to be aware of in the mobile context are the constraints of the mobile environment, such as:

- potentially intermittent existence of a networked connection;
- bandwidth variability of the networked connection;
- the existence of mobile push mechanisms and the different platform-specific techniques for handling push;
- constrained and varied user interface (UI) possibilities;
- limited memory capacity;
- limited processor performance;
- wide variety of devices and likelihood of differences in running an application, even on a supposedly 'standard' device model (e.g., a particular Java platform);
- restricted ability to update software once deployed;

- minimizing unnecessary drain on battery;
- optimizing the communications paradigm at the application level in order to conserve bandwidth (and implied costs).

With all these considerations, we may wonder whether a standard design approach exists; unfortunately there isn't one, as there is so much room for variation. We emphasize the need for careful consideration of the service requirements and, having done that, a thorough review of the device possibilities and the implications for the target user community.

If we come from a desktop software background, we may be totally unprepared for the vagaries of mobile development, especially those spurred by the constraints that the mobile context places on us. We might be tempted to think that, with newer generation processors, batteries and RF networking peripherals, performance constraints are no longer an issue. This would be a nice assumption if it were true, but unfortunately it isn't. The upshot is that we need to relearn many of the software design and coding practices that the programming collective abandoned many years ago, once memory and processing power were no longer major problems in terms of resource constraints. Perhaps we should not have forgotten these techniques, but it is more than likely that many of us may not have ever known them, especially in the post-GHz processor era.[7]

Consider memory allocation as an example. If an application exceeds the available SRAM,[8] then the application might start to use flash memory, which is comparatively slow, and performance will noticeably decline. This implies that we have to design applications carefully to avoid exceeding the available memory. For example, we should not allocate large arrays and we should use techniques to create arrays of an exact size when possible. It is also possible for Java applications on a BlackBerry to leak memory, even though Java has a facility called garbage collection to remove unwanted processes from memory – after they have been used and done with. However, an object[9] cannot be garbage-collected if references to it still exist, so the programmer must take care to release objects when finished with them, especially when objects are shared between different areas of code. Similarly, network connections should be closed when finished with.

These types of performance concerns do not really plague our development process in the desktop world, but they can kill a mobile application. I have used many applications that falter under certain mobile conditions. I recall visiting a GPRS (General Packet Radio Service) applications testing lab run by Ericsson, made available for third parties to test their applications using a GPRS network simulator to evaluate application performance under different conditions. At the time, I was told that over 80% of the applications failed to work when the RF (Radio Frequency) network became particularly intermittent. This was simply because there were no timeouts in the software to bring things back under control if the network failed to respond within a certain time period. On some devices this failure could lock up the device altogether rendering it useless until switched off (and then on again).

[7] I do not want to expose my age here, but there was a time when we thought that processors reaching gigahertz speeds was a futuristic fantasy of unfathomable proportions.
[8] The volatile memory space available for running an application.
[9] A piece of code in Java is called an *object*.

There are so many strategies that can be deployed to reduce code size and to improve execution speed, some of them general and some of them specific to the device and the toolset supporting it (like the compiler). In Java a *long* data type is a 64-bit integer. BlackBerry hand-helds, like many others, use a 32-bit processor, so operations run two to four times faster if we use an *int* instead of a *long*. We will also use less memory by using *ints* instead of *longs*.

Another example of the interaction between design decisions and the target execution environment is related to system design rather than just coding technique. Perhaps our overall system design leads us toward the use of XML (eXtensible Markup Language) for mobile message passing, a not unreasonable approach, particularly if it fits in with an 'XML bus' approach in our IP network, where we generalize all message-passing formats to one XML vocabulary or another.

However, we might not find an XML API on the target device, thus forcing us to implement our own XML processing software. Straight away, this will cost us memory and execution time. It will also cost us bandwidth, which may or may not be a problem depending on how much we have, how much it costs and whether or not we are able to implement a compression scheme, such as gzip.[10] In a very constrained environment the use of XML may not be appropriate, such as where the most reliable transport mechanism is text messaging, where the message payload size is very limited (160 characters or bytes).

Possible conclusion about mobile software design There are so many low-level details to take into account or that can be taken into account, when developing software for a particular target device, that really the only option is to take time to consult the device documentation in the associated SDK, a laborious yet necessary process. We really need to get our hands dirty when it comes to appropriate design optimizations for mobile devices. Some general principles are applicable to all designs, but this whole area of optimization is the subject of an entire book[11] in its own right (not one that I intend to write).

Configuring the IDE/Program Editing/Compilation and Build The IDE is an environment in which we can write code, compile it and test it within one toolset with a common UI. Sometimes we are able to get the entire toolset from the platform vendor, such as the JDE from BlackBerry or Visual Studio .NET from Microsoft. Other times we can use a third-party toolset, like CodeWarrior Wireless Studio (formerly from Metrowerks, now owned by Motorola). Again, this is something that needs careful consideration, although our options may be limited or dictated to us.

The crucial matter is that the toolset we use should directly support the device or device family (or software platform) we wish to code for. Perhaps this is not obvious because if we can code in C or Java with a particular tool, then surely that's good enough. Well, we should understand by now that the language is only one element of the embedded environment: we have to also consider the platform itself and, particularly, its inclusion of APIs.

[10] Gzip is an openly available lossless data compression technique specifically developed to be free of any patent infringement.
[11] See http://www.smallmemory.com for such a book.

An IDE that is aware of the particular APIs we are coding to can provide direct support for programming to these APIs, such as syntax checking and checking that we are passing the right parameters to the API. We would also probably like to have built-in documentation support (help system). Another aspect is UI programming. Many development tools support visual construction of the UI using drag-and-drop interface widgets like menus, option boxes, buttons and so forth. We would therefore prefer that we can build interfaces to suit the particular target platform we are coding for.

Some devices are supported by third-party toolsets. Where this is so, it is probably a good idea to invest in a well-established IDE product because it may pay off in the long run by enabling better productivity. Toolset specialists naturally put more effort into fine-tuning the tools, whereas a toolset in an SDK is often designed with time to market in mind and can be very basic, perhaps not supporting advanced features like visual construction of interfaces.

Testing and Debugging with a Simulator This is a crucial part of the design flow that is not generally found or required during desktop development. Once we have written our software and compiled it, we need to test the program. The obvious thing to do is to load the packaged program onto the mobile device and then try it out for real. However, this turns out not to be very useful initially.

Commonly, we may not have a device; that may sound crazy, but there are perfectly valid reasons for developing mobile software without having the target device; the first is that the devices aren't available yet; this is a common scenario. The device manufacturer has not released any devices, but they are obviously keen to get developer support for them. Therefore, they may release an SDK before the device becomes available. In this case the solution to testing without a device is to use a simulator.

A simulator provides a graphical view of the target device, including all its physical interfaces, within a desktop-executable environment. An example of the BlackBerry simulator is shown in Figure 11.5. We can see from the figure that the entire device is visually displayed, including the interfaces such as the keyboard and buttons. The graphical display of the device is faithfully reproduced and looks, feels and behaves identically to the actual device. Furthermore, the simulator will run our developed mobile application and behave in the same way as it would on the device.

The uses of a simulator are not just as a device replacement: it is a valuable productivity and debug tool. It is far quicker to load an application onto a simulator for testing than to load it onto a device where extra steps are required, including physically uploading the program to the device. In some cases the simulation can take place in lockstep with execution of the code within the debugging environment of the IDE. For example, we could step through the code in the IDE, watching where we are in the code and then seeing the results on the device simulator. This is a powerful feature that has a lot of benefits for debugging.

Some of the time we will only get a generic simulator that makes crude form factor assumptions about the target device. These may be in contradiction to how the target device interface actually behaves; this may predicate earlier testing on the target device than otherwise anticipated or desired.

There is a difference between a simulator and an emulator. The difference can be an important one and so we need to understand it, especially in the context of testing our application. A simulator attempts to mimic the general behaviour of the target device by using features from the software environment of the desktop platform to replace equivalent

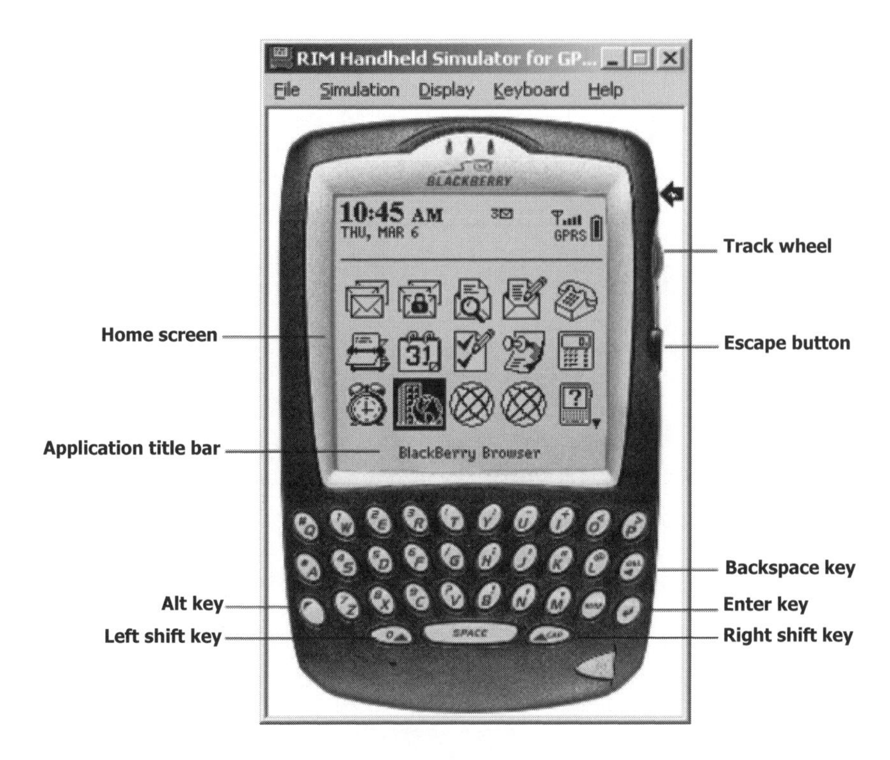

Figure 11.5 BlackBerry desktop simulator. Reproduced with permission from RIM.

features on the device. For example, if the desktop platform supports a windowing graphical tool kit, then this will be used to mimic the one on the device as closely as possible. Similarly, if the desktop platform has in-built HTTP networking functions, then these will be used to substitute the ones on the device. Overall, we can appreciate that simulation is trying to map one feature set onto another on a different platform in order to reach a useful approximation of the target device, as shown in Figure 11.6. In the end, only testing on the device itself will enable complete testing to be carried out.

An alternative approach to simulation is emulation. This is not so common because it is a complex task to achieve. Emulation does not map software functions at the higher layers of articulation and execution; instead, it attempts to model the device hardware and low-level software structures and runs this model on the desktop platform as a complete self-contained process. In other words, the emulator is a virtual device.

If we think about emulation technology, then we will understand why it is a complex process and can be difficult to implement. With emulation, we are proposing to take the actual compiled software image that we would want to load onto a physical device and load it onto a desktop computer instead. Software running on the desktop PC – i.e., the emulator – will attempt to run the application image as if it were the actual target device. We have moved beyond the level of mimicking APIs at a functional level: when the software image runs, it is expecting to run natively on the target processor and to gain access to chunks of code that are images of the APIs themselves, as they would be stored and accessible on the target device. These chunks of code will exist in the emulation environment, and when they

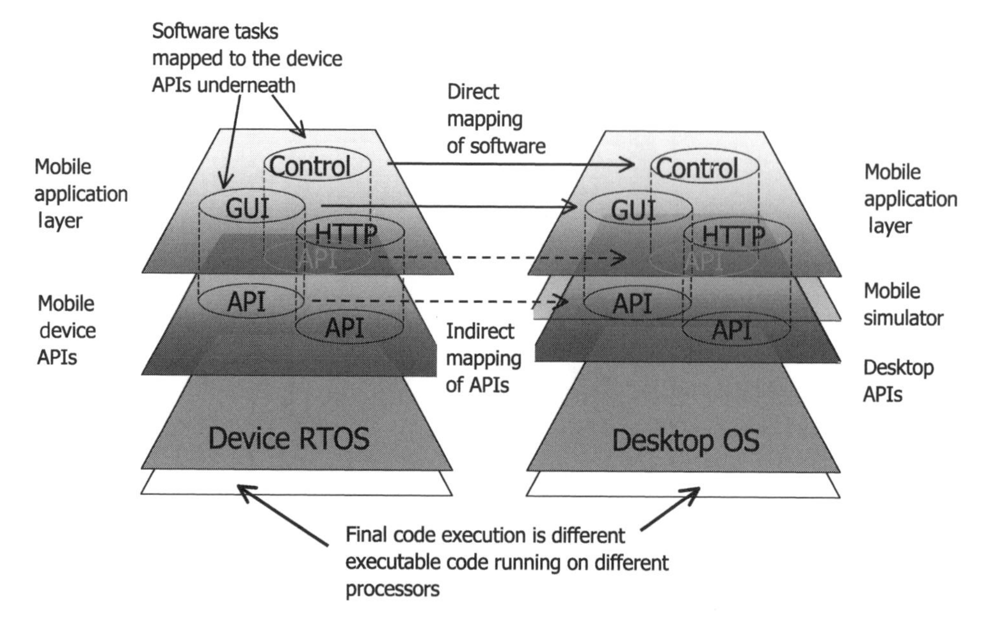

Figure 11.6 Mapping mobile software to a device simulator.

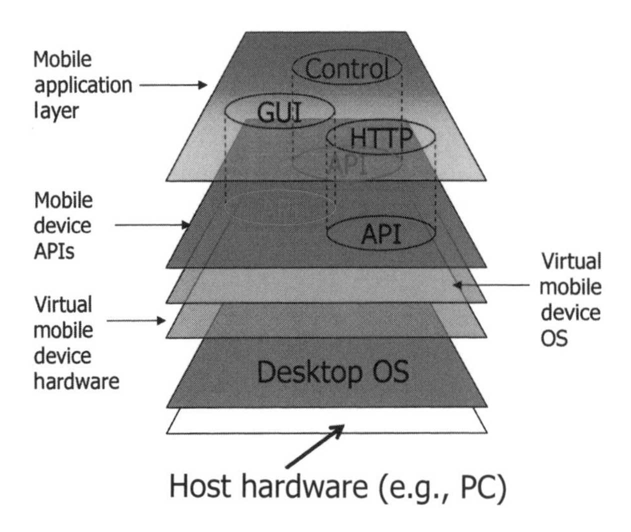

Figure 11.7 Mobile device emulation on a desktop PC.

get called they will want to interact with real hardware, like an LCD controller chip or a serial port. The emulator responds to the required interaction by faithfully reproducing the low-level hardware and device driver environment that will be present on an actual device. In other words, we manage to create a virtual hardware device and virtual mobile OS that runs on the host system, as shown in Figure 11.7.

In the case of the Microsoft Windows™ CE 4.3 .NET emulator, it provides emulated access to the following hardware elements:

Emulated hardware	Description
Parallel port	Provides a parallel port that you can map directly to an LPT port on your development workstation.
Serial port	Provides two serial ports that you can map to communications (COM) ports on your development workstation.
Ethernet support (optional)	Allows the device emulator to support a single Ethernet card using shared IP. When Ethernet support is enabled, the device emulator uses the Ethernet card in your development workstation to emulate a DEC 21040 Ethernet card. This means that your development workstation does not need a specific driver for Windows™ CE .NET to emulate a physical Windows CE-based target device.
Display	Allows you to specify the emulated display size, from 80 × 64 pixels to 1,024 × 768 pixels. You can specify a colour depth of 8, 16 or 32 bits per pixel.
Keyboard and mouse	Supports a standardized keyboard and a PS/2 mouse.
Audio	Provides basic support for audio, including audio input, audio output, microphone input and line output as well as support for full duplex audio, which enables Voice over IP (VoIP).

Emulating the general purpose device features is useful, but emulating the RF networking and telephony features would be even more useful. It is not clear the extent to which this is available, but it ought to be possible to emulate access to other hardware elements. For example, it would be useful to emulate a serial connection directly to an actual Bluetooth™ modem attached to the desktop PC and then to interact with other devices in a device network via a real Bluetooth™ link. Such emulators can be built, but their complexity usually rules them out for general release as development tools. Such test bench systems are probably limited to device manufacturer labs and similar places.

The challenge that remains both with simulators and with emulators concerns the reproduction of RF networking and telephony features. Probably, when we reach this stage, we have to move to the real device to complete our testing scheme.

Testing on the Target Device Eventually, we will have to move to the target device to test our application properly – as it is intended to function on the hardware and in the real world – including connectivity with other devices (device network) or with the RF network.

Even at this stage of testing, it is feasible that we can enhance our test strategy by the inclusion of test software within the application, accessible via a test mode. This might include software to help us gain access to internal messages and operational states of various software elements. This information could be written to a file on the device, made accessible via the screen or any manner that we choose, such as via a serial link.

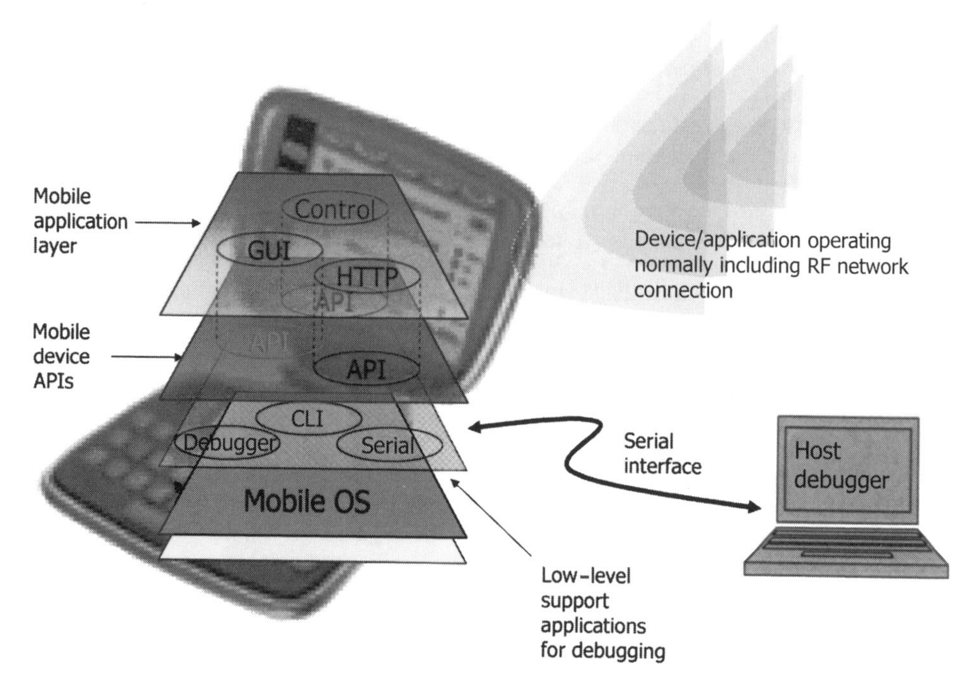

Figure 11.8 On-target application debugging via host PC interface.

The use of on-target debugging is supported by some mobile device operating systems, such as the Symbian OS.

Symbian ships with a debugger (GDB GNU Debugger), which supports on-target debugging of mobile applications. This is a particularly powerful feature as it means we can fully test our application within its intended environment, particularly the RF networking aspect, and gain access to useful debugging information. A special piece of code (called a stub[12]) is installed on the target Symbian OS phone and communicates over a serial link with the debugger running on a host system (i.e., a desktop machine locally networked to the device), as shown in Figure 11.8. On the host the command line interface is supported. Metrowerks includes an application debugger in its CodeWarrior IDE toolset for Symbian OS, called MetroTRK. The standard GDB command line interface is extended to support Symbian OS-specific features, such as:

- downloading files from the host to the target;
- selecting the program to debug and setting its command line interface (as a further debug access point).

This is one of the advantages of the embedded software approach: we can interface at a very low level with the device to enable debugging monitors and host communications to take place. This is because we have access to the device resources to do this. With other approaches, such as the sandboxed Java approach we will look at soon, this is not so easy and may even be impossible in many cases.

[12] The stub here is not the same as the remote method invocation stubs we looked at in J2EE in Chapter 9; these are totally different.

One of the problems with testing software on a mobile device is attempting to produce a range of test scenarios that represent the expected variation in usage and operational context. This is particularly so for RF networking. A useful feature would be the ability to put the device itself into some kind of engineering mode that can facilitate some of the test cases required. For example, the RF modem could deliberately throttle bandwidth or simulate radio signal variation.

Engineering modes are available on mobile devices and have been in existence for a long time, but they are usually undocumented features that are only supposed to be accessible to authorized test personnel. These modes usually enable access to special test features as well as measurement and reporting information from the RF modem subsystem. Sometimes, the engineering features enable the device behaviour to operate outside the bounds of normal device interaction with the RF network; this may be an undesirable feature to place in the hands of the developer community, in case of deleterious side effects within the RF network.

A problem that needs to be tackled is how to test the application on many devices. If we have implemented a solution that is portable or otherwise targeted at many different devices, then it is essential to test on all possible target devices. This usually presents the problem of having all the devices available. This can sometimes be addressed by participating in development forums that provide a testing service, usually at a fee. Some of the operators provide such facilities, although test devices are often in short supply in relation to the size of the developer community being supported. Otherwise, we face the prospect of buying all the devices ourselves, assuming that they are available. There is no easy solution to this problem.

Conducting Usability Tests Something that I learnt very early on in mobile applications design is that the UI design can be very misleading. To the designer, what appears to be a perfectly sensible interface model and metaphor often baffles the unprepared and unsuspecting novice user. The novice probably better represents the average user mindset, despite all our protestations that they are 'stupid' because 'they don't get it'. Perhaps overfamiliarity with the background concepts and even the application itself, especially after much playing with it, is what leads to such gross mismatches of expectations from application behaviour.

Fortunately, there is an easy and enlightening solution to this problem: let the users have a go! It sounds obvious, because it is. Nevertheless, it is remarkable how often designers overlook, dismiss or ignore usability testing.

The topic of usability for mobile devices is slowly being addressed by the usability gurus. An instructive study was conducted by Jakob Nielsen, who is somewhat of a sage in the general field of usability, particularly Web interfaces. To quote from one of Nielsen's Alertbox newsletters on usability called 'Test with 5 users':[13]

> *Some people think that usability is very costly and complex and that user tests should be reserved for the rare web design project with a huge budget and a lavish time schedule. Not true. Elaborate usability tests are a waste of resources. The best results come from testing no more than 5 users and running as many small tests as you can afford.*

In essence, usability testing is very straightforward. The idea is to assign tasks that the users are usually expected to carry out while using the application. For example, if the application is a mobile email service, then we would set tasks like:

[13] http://www.useit.com/alertbox/20000319.html

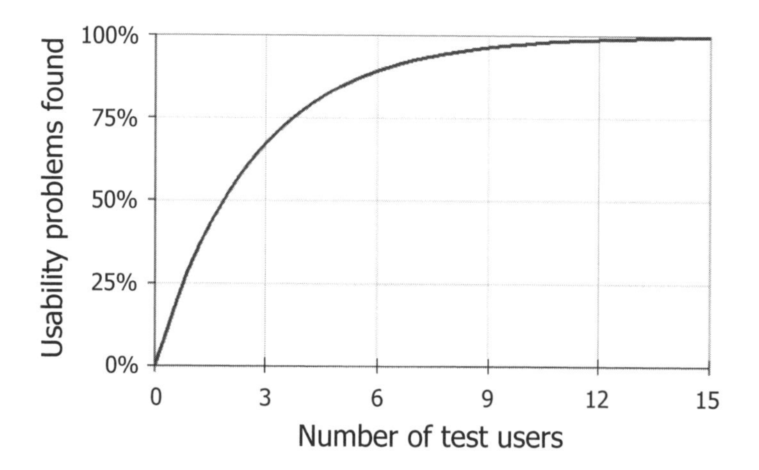

Figure 11.9 Curve to show usability effectiveness dependent on users.

- compose and send an email to 'Joe Bloggs' – his name is in the address book;
- add an appointment to the diary for one day next week;
- find a message from 'Fred Smith' about opportunities in wholesale coffee.

In other words, we set typical tasks that we expect end users to carry out. We then conduct surveys to get the users' assessment of how easy it was to carry out each of the tasks. This is all done without any unnecessary intervention in the process. If a user struggles to carry out a task, that's useful information that we need to know. As tempting as it may be to intervene, if we are there to supervise the testing process, we should not give hints on how to do the tasks. After all, in the 'real world' they won't have us standing over their shoulder giving hints, so let them sink or swim according to their abilities to fathom the application.

The curve shown in Figure 11.9 gives us a nominal metric for how effective usability testing is in relation to the number of testers. As Nielsen points out:

> *The most striking truth of the curve is that **zero users give zero insights**. As soon as you collect data from a **single test user**, your insights shoot up and you have already learned almost a third of all there is to know about the usability of the design. The difference between zero and even a little bit of data is astounding.*

Nielsen's brutal bluntness is refreshing and badly needed. While it is obvious that zero testers will tell us absolutely nothing about usability, this is the shocking reality of many mobile projects, with operators often the worst offenders. I have personally reviewed many commercially available mobile applications and found that usability was dreadful. Until recently, the topic of formal usability testing was noticeably absent from most formal software processes (despite what the official project representatives might say when quizzed on this topic). The tidal wave of Web pages that have surged into view has caused a new focus on this important topic.

Usability in mobile applications is even more important than with the Web, as the margin for 'error' is much narrower. The lack of tolerance of dysfunctional design is acute. These are reasons enough to give usability its due rights in our development process.

Pilot Trials and Deployment At the end of the development process we need to deploy the application. For embedded applications, this can take a variety of forms:

- Web download and manual install via desktop PC;
- OTA installation;
- pre-installation in ROM or other nonvolatile memory;
- packaged distribution with device or separate – as on CD-ROM.

In the future there may be other distribution formats, such as removable memory devices like MMC cards, which is more akin to the games cartridge distribution model.

The method of distribution may be dictated, depending on whether or not there is any intervention in the process from a mobile network operator. Indeed, we could well have chosen to deploy our application via an affiliated scheme with the operators. Such a scheme is likely to have its own preferred distribution requirements. Let's briefly examine each option.

Web download via PC This is the most common method for distributing any software solution today. It is already commonly used for PDA (personal digital assistant) applications, so there is a legacy of experience with distributing mobile applications in this manner. The actual loading mechanism onto the device is via a cable, either USB or serial, sometimes via a cradle. For PCs with Bluetooth™ connectivity (e.g., some laptops or via a USB Bluetooth™ dongle) it is possible to establish the serial connection wirelessly.

The advantage of the direct installation method, where available, is that it is relatively quick and less prone to errors than alternatives, such as OTA. Using the Web for distribution has obvious benefits, not least of which is its prevalence. The existence of various marketplaces to offer software solutions is also an advantage. There is an increase in marketplaces for mobile applications, especially games, with some of the big names like Yahoo offering a mobile games arcade in addition to what many of the mobile operators are already doing.

The disadvantage of downloading from the Web is that the device needs to be connected to the PC. In some cases this is not practical as the user may not even have a PC. In may also be the case that users' expectations are such that once they have a connected mobile device, the need to use a wired connection (i.e., via a desktop PC) may seem at odds with the underlying mobile service concept. Thus, manual installation via a Web download may frustrate users.

At this point we might sensibly enquire as to why we have to bother with a Web download via the PC if we do indeed have a connection available via an RF network. There are two reasons. The first is that on some devices there simply is no mechanism for installing applications other than via a desktop PC, such as the ActiveSync 3.1 companion for Pocket PC and the Palm Conduit for Palm devices. This is because these solutions were devised a long time before compatible wireless modems were available, so these host synchronization solutions were not accessible via a wireless connection. At the same time, the browsers that were made available for mobile solutions did not have a mechanism for downloading files, so the idea of buying an application via the Web and downloading it directly is not supportable on many devices, even today. This type of file download feature was notably absent from earlier WAP implementations, for example, and is still missing in most current implementations.

The other reason for downloading via a PC, which may still be valid, is the potential size of the file. Mobile applications could easily be several hundred kilobytes in length. Over a very slow RF link, this may be too much to expect a user to download.

Even if it were possible, we would probably still want to ensure that we had a purpose-designed mechanism for handling software installations via this method. For example, we would want to ensure that chunking of downloads is possible with error detection for each chunk. Should the link become disrupted, then we would not have to start the whole download process again – something that until recently was required on desktop implementation until some downloads became truly massive. This prompts us to also consider the need for application compression, but this topic is discussed in Section 11.5.

Over the air (OTA) There is nothing new about remote loading of software over a connection. This technique is used all the time in embedded applications. Ironically, it has been used in cellular base stations for some time to avoid the cost of deploying service engineers for upgrades.[14] In other words, the techniques for remote software installation are not rocket-science and are well understood. They are not that complex either. Remote installation is not even new for wireless, the most obvious example being satellites (where it's the only option, of course, unless you can reach them via the shuttle and have a few million to spend on the service plan).

However, it has taken some time for the concept to arrive on consumer mobile devices. It has really been advanced by the recent developments in mobile Java (more on this later), where initially some recommendations were made for OTA provisioning of Java applications on the device; however, recently, the technique has become formally part of the specification.[15] There are a number of technical issues that need addressing relating to OTA provisioning:

- an agreed and available mechanism on the device to cope with an OTA installation;
- an agreed file format, if applicable, for encapsulating downloads;
- a compression method(s);
- a suitable file download protocol;
- a network protocol for establishing the OTA link with the application source.

Let's look at each of these issues in turn to see what the details might entail.

In terms of a mechanism for coping with OTA, this clearly implies that the device itself has a software application that is able to receive application files and then install them. We don't really need anything new because the existing application-loading mechanism could be used; it simply receives the installation file from an RF modem connection rather than a serial port, both being serial streams in any case, so the adaptation is not that challenging.

[14] Note that the software downloaded to a base station is done over the landline connection to it, like a 2-Mbps leased line or something similar. This is worth stating in case we thought that base stations were somehow upgraded wirelessly – due to their obvious wireless nature. But, we shouldn't forget that they usually have a wired connection too, which is the backhaul to the rest of the cellular network.

[15] The so-called MIDP 1.0 version of Java did not have a formal OTA policy, just a recommended practice, whereas OTA has been fully integrated into the MIDP 2.0 specification.

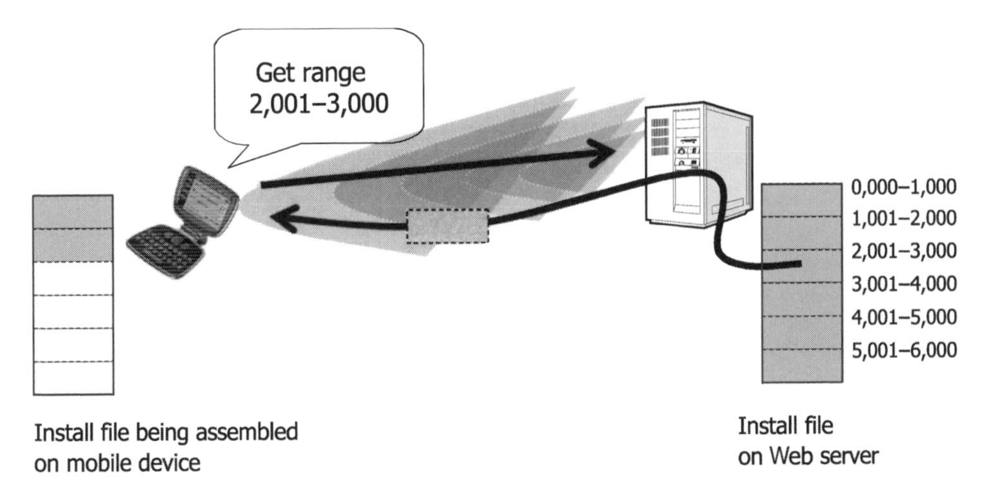

Figure 11.10 Segmentation and reassembly process for downloading applications.

However, there are some subtle differences, which may seem trivial, but can end up being significant if not addressed. It may have escaped our attention that when installing software via a cable and PC, the PC itself is running part of the overall installation application. One of its functions is to provide back-ups, this being particularly important before we install a new application. If something goes wrong during the installation, then we would like a means to get back to a previous known working state. The PC can provide this assistance for us. Potentially, the PC installation could even back up the entire memory image from the device so that we could reliably get back to a known working condition whenever things go wrong.

A backing up procedure over a wireless connection may not be feasible, though in general it is probably something we are going to see emerge with faster RF connections, not just for applications but for user data too. If backing up over the network is not possible, then we probably want to take a more conservative approach toward installing new applications. For example, we may want to provide local back-up via spare memory, on the device. Applications may contain APIs or shared files that are already on the device and need updating for the new application to run properly. It would be a good idea for shared files to be backed up first. This would enable a previous state to be retrieved should the new application not install properly or should any of the currently installed applications stop working.

SAR (or WTP-SAR) stands for segmentation and reassembly. It's an optional feature of the Wireless Transport Protocol (WTP) within WAP.[16] SAR defines a means for a WAP gateway to break a large message (e.g., an application we're downloading) into smaller chunks (the segmentation, or 'chunking') and for the phone to piece it back together (the reassembly), as shown in principle in Figure 11.10; this is discussed in Chapter 7.

[16] We should note that SAR here is not to be confused with how it gets mentioned in the context of IP datagram communication, where packet sizes can be made small for 'small packet' networks. Even using packet-based IP communications, we can still lose the link, no matter how well the IP layer is functioning. If anything goes wrong in the download session, then we want to be able to recover from the problem without restarting the entire download process; hence we need SAR at the transaction level, not the packet level. (See Chapter 7 for more insight into this issue.)

Not all mobile devices use SAR, but many will access Web servers through a WAP gateway. Nokia WAP phones with Java games support use SAR, while Motorola devices apparently use features of HTTP 1.1 to retrieve small chunks of a file one at a time and then reassemble them, as we will discuss shortly.

The sender can exercise flow control by changing the size of the segments depending on the characteristics of the network. Selective re-transmission allows for a receiver to request one or multiple lost packets. Without this, in the event of packet damage or loss, the sender would be forced to re-transmit the entire message, which may include packets that have been successfully received. This could rapidly become very tedious for the user.

One problem that arises with WAP gateways concerns MIME (Multipurpose Internet Mail Extension) types returned by the Web server. If a user requests a JAD or JAR file (a type of Java program file) and the server returns the wrong MIME type, the gateway and subsequently the phone will handle the transfer incorrectly. This reminds us that our mobile device application is only one piece of the puzzle. There are many other participating elements in the mobile network that need to be considered, which is why wide systems knowledge is important (the reason for this book).

An optimization introduced in the HTTP 1.1 specification is something called *byte range* operations. We've all experienced trying to download software from a website, only to have the connection fail with a few kilobytes left – out of 10 Megabytes, of course! At that point, we initiate the download again, hoping for the best, maybe giving up after a few tries. Using HTTP 1.1 the client software application can request just the last few kilobytes of the resource instead of asking for the entire resource again. This significantly enhances usability and general user satisfaction for web Downloads. This technique can also be used for downloading applications, not just data files. When requesting a byte range, a client makes a request as normal, but includes a *range header* field specifying the byte range the resource is to return, as shown in Figure 11.10.

The client may also specify multiple byte ranges within a single request if it so desires, causing a queue of requests to be aggregated into one response. In this case the server returns the resource as a *multipart/byteranges* media type. This seems to go against the ethos of ranged requests, but it might be a useful mode to deploy dynamically during a download, should the client detect that the RF network connection is robust enough to support it.

The use of byte ranging is not limited to recovery of failed transfers or for application OTA provisioning. Certain clients may wish to limit the number of bytes downloaded prior to committing a full request. A client with limited memory, disk space or bandwidth can request the first so many bytes of a resource to let the user decide whether to finish the download. This will depend on whether or not the nature of the proposed download allows for this pre-emptive interaction with the application prior to finishing the download.

It should be noted that Web servers are not required to implement byte range operations, but it is a recommended part of the protocol and fairly widely implemented in any case. Clearly, if it is not implemented, then the fallback is to attempt downloading the entire application. This may suggest that it is a good idea to examine carefully the application file format for OTA, taking into account compression optimizations for OTA and any inherent chunking of the installation files and process.

To deploy an OTA application, we possibly need a new file format for the transfer process – a file format that is optimized for OTA deployment and can be used to encapsulate the installation file. In some cases we should even consider an implementation for installation files which is optimized or built for OTA provisioning from the ground up, rather than as an afterthought.

Table 11.1 Possible compression gains for OTA provisioning of CAB files.

	CAB file size (KB)	AirSetup installer size (KB)	Compression level (%)
Sample CAB (file #1)	757	188	24
Sample CAB (file #2)	2690	1090	40
Sample CAB (file #3)	901	424	47
Sample CAB (file #4)	1470	640	43

Source: SPB Software Warehouse[17].

In the case of an encapsulated file format, it should simultaneously support OTA-optimized compression and software chunking. Compression is usually a feature of software installation files, but this is not always the case. We cannot assume that the default approach is the best method, as it may not have been optimized for OTA provisioning. General purpose compression techniques applied to a binary file will never be as good as ones specifically designed for a particular purpose, such as compressing application files for OTA provisioning. Table 11.1 gives us some food for thought in this regard: it is a table reporting possible improvements on compressed CAB[18] installation files for Microsoft's Pocket PC devices. These improvements are the claims of the SPB Software House when using its AirSetup installer technology, a third-party OTA solution.

Irrespective of the file format required, we need a protocol for managing the download process. We would like to incorporate the ability to establish that the download has been successful. This will enable us to support our end users in a reliable fashion. Furthermore, the ability to determine that an application has been successfully downloaded and installed provides an opportunity to complete a financial transaction with the user, assuming that we want to bill them for the downloaded application.

As we will see, a mechanism for doing this has already been proposed by the Java Community Process while formulating the specification for the Mobile Internet Device Profile (MIDP) version 2.0, a version of Java aimed at mobile devices. Using the MIDP mechanism as a model, the Open Mobile Alliance (OMA) has proposed a similar method for the downloading of any content to mobile devices.[19] We emphasize here that this mechanism is intended for all types of content, such as multimedia clips, ring tones and so forth. However, it is a generally applicable model that can be used for any type of file, so it would be sensible to presume that this method be adopted for downloading embedded applications. Even if such a provision does not exist natively on a device, as part of the underlying operating system bundle, there is no reason why it should not be installed on the device first and then used to fetch the target application(s).

OMA download mechanism (see Figure 11.11) Step 1: **browsing (discovery process)** – the method for perusing downloadable content is presumed to be via a browser-based presentation layer. The browser picks up Web (WAP) pages from a server and the user

[17] http://www.spbsoftwarehouse.com/products/airsetup/index.html

[18] CAB is short for cabinet, which is a metaphoric name to indicate many files placed into one cabinet (one place) for download purposes.

[19] OMA-Download-OTA-v1_0-2002121920030221-C, Version 19-Dec-200221-Feb-2003 available from http://www.openmobilealliance.org

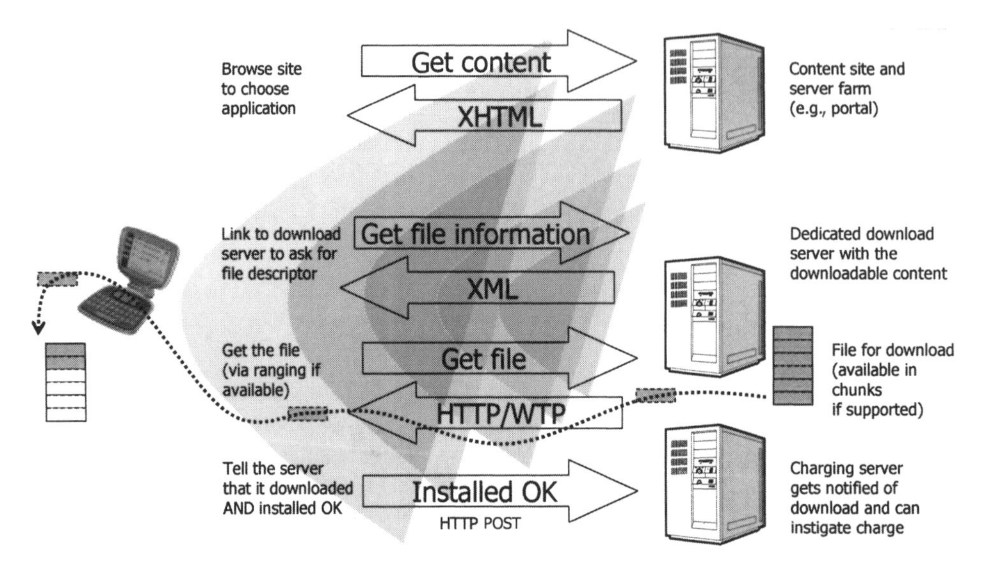

Figure 11.11 Possible content download mechanism as per OMA guidelines.

eventually selects an embedded link to take them to the next step, which is to download (and/or view) a descriptor file about the proposed target download application (or file).

Step 2: **evaluation (content negotiation)** – prior to downloading the required application install file, we would like to be able to evaluate the application in terms of its resource requirements. To do this, we first fetch an object descriptor file. This is done using HTTP; so, any sever capable of delivering the file would suffice, not necessarily the same one used for the browsing in step 1. The reliance on HTTP for this process does not preclude nonbrowser solutions. Any mobile device application that can handle HTTP would be able to fetch the descriptor information from the server; so, we could even fetch the descriptor from within another application, such as an embedded one. This may be useful for software upgrades.

The download descriptor (DD), so called by the OMA, is written in an XML vocabulary and contains information about the application attributes, such as file size and file type. The mobile device can use this information to ensure that the proposed download is compatible with the device and its available resources at that time (e.g., memory space). An example of a DD is shown here:

```
<media xmlns="http://www.openmobilealliance.org/xmlns/dd">
  <type>image/gif</type>
  <objectURI>http:/download.example.com/myapp.app</objectURI>
  <size>100</size>
  <installNotifyURI>http:/download.example.com/
  gotis.asp?id=ab35612</installNotifyURI>
</media>
```

If we examine the anatomy of the DD, we can understand some key elements of the download process:

Table 11.2 Status codes to indicate content download success.

900 Success	Success indicates to the service that the media object was downloaded and installed successfully.
901 Insufficient memory	Indicates to the service that the device could not accept the media object, as it did not have enough storage space on the device. This event may occur before or after the retrieval of the media object.
902 User cancelled	Indicates that the user does not want to go through with the download operation. The event may occur after DD analysis or instead of the sending of the installation notification (i.e., the user cancelled the download while the retrieval/installation of the media object was in process).
903 Loss of service	Indicates that the client device lost network service while retrieving the media object.
905 Attribute mismatch	Indicates that the media object does not match the attributes defined in the DD and that the device therefore rejects the media object.
906 Invalid descriptor	Indicates that the device could not interpret the DD. This typically means a syntactic error.

A. *Type* – this is the MIME type for the resource. This may or may not be useful, but indicates something about the file type to the client, which it may find useful or simply ignore. It could be used to confirm that the file type is as expected.
B. *objectURI* – this is the Web address of the downloadable resource and is where the mobile device can now go to get the file via HTTP or WTP, accordingly.
C. *Size* – this is the size of the resource in bytes. Note that this only indicates the size of the download file; it does not mention how much space is actually required by the resource once it is installed. After decompression and installation, the size of an application may be bigger (this does not appear to have been considered by the OMA).
D. *installNotifyURI* – after the resource has been downloaded a status code can be sent back to the server via this address; this can be any server, not restricted to the same one from where the download just came. The advantage of this process is that the downloading process can be state-monitored, so we should always know what state the download is in and what went wrong with it, should it go wrong. This 'download success' acknowledgement mechanism is a unique addition to the downloading of content. It is not included in the HTTP protocol or associated WAP protocols. The codes used to indicate the download status are to be submitted via HTTP POST to this web address. Some of the codes suggested by the OMA download specification are shown in Table 11.2.

Step 3: **installation/notification** – after downloading the file, the mobile device attempts to make use of the downloaded resource. The actual downloading process has already been discussed in terms of the possibilities for compression, file encapsulation and chunking. We are now assuming that we are at the state where a usable resource file is now available

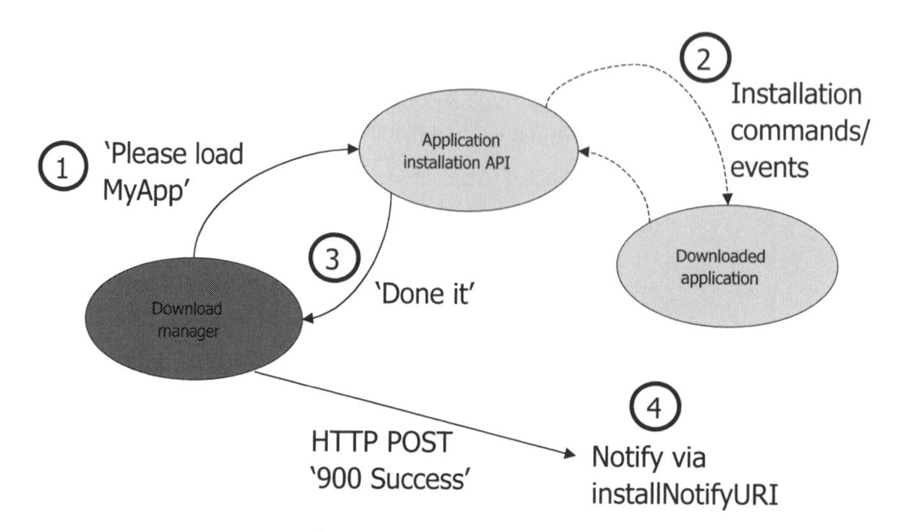

Figure 11.12 Download manager manages both the installation and the reporting back to the download server.

on the mobile device. Clearly, we need the installation process to be initiated. The program that initiated the download, which may have been a dedicated OTA-handling application or another embedded application, would also take care of the installation. In either case the program will call the appropriate API for installing and registering applications with the OS, bearing in mind that we have already shown how there needs to be a part of the device platform that can load applications.

An aspect of the OTA process is that the installation process must have a means to report back to the download server (or the designated install notify server) via the installNotifyURI that we said is included in the DD. There seem to be several ways of managing this process on the device itself, and we mention two of them here.

The first option, as shown in Figure 11.12, makes the assumption that a separate download manager exists to control the installation of the downloaded application and that it does this via an API call. After a series of events to process the installation, there would be an API callback to the manager to notify it that the installation has finished and to indicate how successful it has been. The download manager can then interpret the callback information and use this to initiate the HTTP POST of an appropriate download status code to the installNotifyURI.

The second option is to let the actual downloaded application initiate the communication back to the installNotifyURI, either directly or via an API call to an appropriate application capable of carrying out an HTTP POST (such as a browser on the device that has its inner workings exposed via APIs, like Pocket Internet Explorer on Windows CE). This would simply require that the installed application be passed to the installNotifyURI as a parameter, so that it knows where to respond.

The key difference between the two approaches is that the second approach enables a greater degree of flexibility in reporting back meaningful status information. For example, a user could be asked to fill out a licence acceptance form for the application before a status code is returned. This may seem a bit obtuse, as presumably we could use the application itself to communicate anything we want back to a Web server in order to implement charging or some other process in relation to the download and installation operation. This is entirely true. We should understand that the need for a separate download manager and status-reporting function is largely dictated by the need to support a wide range of downloadable content. For example, a ring tone could be downloaded using this mechanism. A ring tone does not have the ability to communicate back to a server, so we can see the need for a separate reporting mechanism in this case.

The final observation about the installNotifyURI is that is should contain a unique identifier that can be used to identify each download. This is so that installations can be associated uniquely with users and is desirable for the purposes of charging and support.

We might wonder whether there are other ways of managing the OTA process. In fact, there are probably many ways it can be carried out. We have discussed a procedure that is generic in nature and has been supported by a large number of mobile industry players via the OMA initiative. The procedure is deliberately modelled on a similar process for OTA provisioning of MIDP Java applications, which is something we will discuss shortly.

What we could argue is that the OMA process is aimed at single-file downloads or content that is not that complex, such as a single ring tone file or a photo; the process of application installation is obviously more complex than this. Potentially, a very complex application could involve a considerable amount of files being installed or updated on the device, including some system files being modified. We have said that altering critical system resources may require selective back-up procedures to be undertaken first. It is feasible that this can be done over the RF network. Such a process would involve an intelligent OTA manager on the device which could record device state information, file changes and other device alterations, almost like a transaction monitor that in the event of failure could undo all the changes. Clearly, certain changes might preclude backing up over the network should they impact on the networking mechanisms themselves, like critical device drivers and so forth. The OTA manager should have the ability to detect such implications and flag them to the user, giving the option to back up files to a memory device instead of over the network, which may subsequently become unreachable if the installation process fails.

Even the process of backing up would need to be made more suitable for the OTA environment than perhaps we are used to in the wired world, where everything gets backed up prior to starting the installation. We would not want to ask the user to sit through 5 Mbyte of back-up across the network only to discover that the first files we download for installation are corrupted or that one of them is not compatible with the device. There may be opportunities for incremental approaches, gradually backing up and replacing files so that any download failures result in a more graceful degradation or quicker recovery. We may still be able to benefit from some of the updates to our device system files, even though the entire installation process was unable to complete; therefore, we should apply 'undo' operations in an intelligent manner.

Looking at the approaches to remote software installation and maintenance that have been developed over many years in other fields, it is entirely feasible that robust OTA mechanisms could be implemented. This is a major necessity for next generation devices, simply because if we are to achieve mass penetration of services and devices we need to remove as much as possible any technical support activity from the users. Ideally, the users should not even be aware of what an application is, except in terms of its service offering expressed via a user interface. Low-level concerns with installation, code signing and so forth should not be a required part of the user's sphere of concern. This seems entirely possible given the level of technological resource that we can now deploy and apply toward this objective in the mobile device world.

Packaged distribution Returning to our options for installing software onto our mobile device, the final option is to simply distribute the application physically to the user, usually through a CD. This may be useful for over-the-counter sales, such as in a mobile device shop. It also has its place in the usual distribution channels and in some promotional channels, such as magazine covers. In future, we may also find that pluggable nonvolatile memory devices, such as MMC cards, are used to distribute applications.

The distribution of applications in this manner brings us to a related topic: namely, the digital signature of an application by its creators. This will become increasingly important in the pursuit of guarding against surreptitious and malicious practices, such as viruses.

We are often prepared to delegate our choices on a basis of trust. We will do something because we trust the people we are dealing with; software is no exception. If a vendor claims that their software is free from viruses, then we have the option to trust them or not. The problem is verifying that we are in fact dealing with a vendor whom we trust. For example, if we download an application or load it from a CD, then we want to be sure that it is from a particular vendor purporting to be the creator of the application. We are willing to trust their software to run on our device, so long as we can be sure that the software has come from them.

The ability to determine authenticity is not unique to mobile applications and has been tackled already using digital certificates and code signing; a similar method can be applied to mobile applications.

Trials Before we engage in full-scale deployment a trial of our application is a useful undertaking and probably worth the investment in time and effort. This is not to be confused with usability testing, which, as we have already discussed, is actually a part of the development phase, *not* the deployment phase. However, we would of course welcome any feedback from trials that further enhances our understanding of usability.

A trial is often confused for what we sometimes call a beta test, but they are different. A trial is really about rating the acceptance of our application according to its intended usage and market. We want to see how users get on with it. It is confirmation of our concept and its realization, not testing of the software *per se*. Beta-testing is about uncovering the inevitable latent bugs and flaws in the system that we could not expose during testing. However, it is perfectly valid for these two activities to be combined, but the point is that end users should be aware of what's going on.

11.5 BROWSER-BASED APPLICATIONS

The second method of accessing a mobile service via our mobile device is to use a browser-based approach. This is where just the interface layer is executed on the device and the rest of the application's services are run somewhere on the network, most likely within the IP network on a Web server, but not exclusively so, as we will see.

The advantages of the browser approach are well known; we have already discussed them in some depth in our earlier discussions on HTTP and client/server architectures, along with HyperText Markup Language (HTML) and its cousins. Today, we have wireless-optimized variants called WAP. What we are interested in here is a discussion of the characteristics of a browser-based approach, especially in contrast to the other approaches discussed in this chapter: embedded and Java platform applications.

Following on from our previous discussions on the distribution methods for embedded applications, we have no such problems with browser applications. Simply, there is no need for distribution. This is the beauty of the Web: we could have a hundred applications, but we don't need a hundred installations on our device, just the single browser. What we do need is the means to access those applications, which is the browser and a uniform resource locator (URL), or address, where we can find each of them.

Finding myriad applications via their URLs is perhaps the most cumbersome aspect of the Web within the mobile context. We have two problems:

- application discovery;
- address (URL) entry.

Application discovery is about finding the application in the first place or simply knowing that it exists. This sounds strange, but in a mobile device we have limited options for how we discover applications, simply because we tend to browse infrequently and, in general, there are fewer ways to find out about a site we might be interested in surfing – although *WAP PUSH* could change this, as we will see.

Address entry is the second problem: an application is accessed via its URL, but this is lengthy and can be cumbersome to enter on a mobile device, especially small form factor devices with fiddly keypads. This problem in itself could possibly be circumvented. It should be possible to type only the unique portion of the URL and arrive at the site by auto-completion of the address, a feature found on some desktop browsers, but sadly lacking on many mobile devices (though this is improving all the time). We could simply type 'trains', where we mean http://www.trains.com or http://wap.trains.com, there being a trend for some site designers to use 'wap' as the subdomain in place of the conventional 'www'.

There are so many other usability enhancements to this problem – like selective text prediction – so that 'www' and '.com' are auto-completed in any case, but not the actual domain name.

As an aside, a well-implemented website should be able to redirect automatically to the mobile-compatible pages from a single entry point (home page), so there is really no need for the 'wap' subdomain. This redirection is done by detecting the browser accessing the home page and then steering it to an appropriate page to match its content-handling abilities.

Ultimately, the best approach is to have a home page that is used as the initial entry point into the applications space on the mobile Web. With the advent of colour devices and support for colour graphics in newer browsers, it is possible to have a home page with colourful and attractive icons, such as those shown on Three's home page (Figure 11.13).

It is an easy matter to introduce new services to the user by simply adding more options to the home page. In the case of Three's home page the user is able to configure which icons appear on the main initial view. The rest of the applications are accessible via a 'more' option to load another page into the browser. Allowing the user to personalize the start page is an enhancement to usability that makes a lot of sense. It is like having a configurable 'favourites' page. Even though the options here are constrained to applications within the operator's portal, there is no reason why they couldn't also point to affiliated third-party applications if necessary.

Figure 11.13 Home page of Three's mobile portal. Reproduced by permission from 3. (Hutchinson Telecomm).

To access applications not in the operator portal, the user has to manually enter the addresses. This can be circumvented thereafter by adding the address to a 'shortcuts' or 'favourites' menu on the device, usually a feature of the browser software itself, although on some mobile device implementations this feature is construed as being a 'phone settings' attribute and so is accessible via a different menu than the browser's, which causes all kinds of user frustrations.

On many devices, managing bookmarks can be overly cumbersome and an abuse of the word 'shortcut'. On some devices the route to enacting a shortcut can be circuitous and hard to fathom. What makes most sense is to include the ability to display a shortcuts page as a personal home page, so that useful sites are only one click away from starting the browser.

Discovering an application is a particularly difficult problem for mobile browsers. With installed (embedded) applications, the application is easily accessed through its launch icon. There have been some suggestions proposed for making services easier to access via browsers, especially to overcome the entry of cumbersome addresses. Bango.net,[20] for example, has pioneered the use of unique numbers to access services, to circumvent the addressing problem.

With Bango.net the service assumes that the user can at least access the Bango.net Web page in the first place, which itself would be bookmarked. Thereafter, entering the Bango number for a particular service takes the user straight to it. So, instead of entering http://www.usefulservice.com to access a useful service not previously known by the user, they could simply enter the Bango number, like '1224', to navigate straight to the same destination on the mobile Web. It suffers from the problem of ensuring that the Bango service itself is sufficiently widely known and publicized that the Bango numbers become widespread and subsequently meaningful to a large user population.

This shorthand method of accessing browser-based services is a useful principle, but it could be enhanced by its adoption as an Internet standard or by some other industry body,

[20] http://www.bango.net

Figure 11.14 A special access key to make browsing easier.

instead of a commercially owned and controlled addressing space. Of course, for Bango the control of an address space is useful commercially as it compels users to use their service, which is exploitable in various ways and enables premium 'short code' addresses to be rented at a premium.

If there were a standard way of using short codes and of registering them, then it would be possible to add a new key to devices: the 'master shortcut' key. Instead of dialling a number to access a phone line, we could 'dial' a number to access an application. We could use a reserved character – like '@' – to indicate the 'access code' for applications, just like we have country codes or area codes: dialling '@1221' would take us to the website application for code 1221 (see Figure 11.14).

It is relatively easy to implement this as a standard by adopting a convention for translating domain names to numbers or vice versa. For example, if we introduce a new high-level domain name, called .mob, then we can allow the use of numbers as the domain, by convention. For example, if we have an application registered at http://www.usefulapp.com, then we can register for an alias, like http://www.1221.mob, which would be accessible via standard domain mechanisms (DNS). All that we then require is that mobile device manufacturers implement the access shortcut in software on the devices. So, if we now dial '@1221', the device is programmed not to attempt dialling this number, but to open the browser at page http://www.1221.mob, which is a simple procedure to implement in the device software.

The purpose of having the '@' character is that users would become familiar with this moniker, which is already strongly symbolic of Internet metaphors. After some time, users would instinctively know that dialling '@1221' (or any number) means they will be accessing an application rather than making a voice call.

On billboards and in adverts we could simply add the line 'dial @12213 for more information on your mobile phone' or however we want to say it – simple and effective. Over time the prefix will become widely understood.

We could consider extensions to the idea that make subsites accessible from a single domain: for example, '@12213.1' and '@12213.2' would translate to http://www.12213.mob/1 and http://www.12213.mob/2 and so on. More likely, as we would want to use this for targeted advertising, we would use a parameter, like http://www.12213.mob?param=1 and http://www.12213.mob?param=2. This would be useful for location-specific adverts that do not need to rely on location-finding processes in the mobile network. So, an advert for train times could have different postfixes that map to different timetables on the website.

11.5.1 Limited Local Processing

A browser-based application has limited local processing capabilities compared with embedded applications and Java platform applications. The browser has two main functions:

- display Web pages from the Web server;
- accept user input in forms for sending back to the Web server.

The browser metaphor is really one of information exchange with a remote content source, with very little local processing of the fetched information. It would not be possible to implement a typical spreadsheet or a word processor as a browser application, at least not in any conventional sense. The only processing is the interpreting of the presentation information-encoded page, followed by the subsequent visual rendering to the screen.

The extent to which the browser can carry out local processing tasks is very limited on mobile devices, unlike with desktop counterparts where the browser has evolved to include powerful scripting capabilities (e.g., JavaScript) and the ability to host embedded applications within its framework, something that originated with the Netscape plug-ins. A well-known example of a plug-in is Macromedia's Flash™ player.

The browser framework does not support a general purpose programming model, unlike the embedded approach. The browser is an embedded application itself, written in a general purpose programming language like C or C++, and runs natively on the underlying mobile OS. It interprets text files that contain mark-up languages of one type or another – there is no such concept as running compiled code, like assembly code that has been compiled from C.

It is possible for browsers to enable compiled applications to run, using the OS as the execution environment, with the browser as the display and communications framework. The conceptual link is already available via the *object* tag in XHTML-MP (eXtensible HyperText Markup Language Mobile Profile), which is a mechanism to refer to content in a Web page that requires an external program (helper application) to handle it. This is primarily aimed at handling multimedia content types, but it could be used to launch a custom application written to run natively on the mobile OS. In practical terms, this may not be possible on many devices, especially where there is no support to install custom applications in the first instance. Where Java is supported on a device, this mechanism may be used to launch a Java application; though, in some environments, such as the MIDP Java Platform, this mechanism is not directly supported by the MIDP specification.[21]

In the specification for WAP 2.0 there is no support required for browser-based scripting, unlike the original WAP 1.0 where WMLScript was proposed and supported. This is a limited capability scripting language that enables some processing to be done locally, but this is constrained to handling data in the published pages and rudimentary checking of data submitted via user input forms. Like all browser-scripting approaches, the scope of execution of a script is also constrained to running only when the container page is displayed in the browser, so general purpose scripting is not possible; therefore, we should not think

[21] There is no API to handle the communications between the browser and the Java MIDP program.

that browser scripting provides a container for general scripting on a device. In fact, the scripting only has access to APIs that enable access to the page (document) components, not to any other resource on the device.

ECMAScript is a new addition to Openwave Mobile Browser 7.0 which enables content developers to build more powerful mobile applications, the same way that JavaScript™ enabled more powerful applications for the desktop browser.

11.5.2 Necessity of An Available Network Connection (Caching)

An essential requirement for the browser approach is the availability of a network connection to the IP network where the content server is situated (whatever that server happens to be). This is an integral part of the client/server model on which the browser paradigm is built. Therefore, if we do not have a network connection at a particular time, then we have a problem with access to the required application. We are simply unable to fetch the pages into the browser. This is where the vagaries of an RF network might let us down. As much as we would like ubiquitous network coverage, there will always be circumstances where radio coverage is intermittent or even nonexistent, for whatever reason (see Chapter 12 on the RF network).

The absence of a network connection not only affects retrieving information but also in submitting information back to the network from the device. In the latter case the absence of a link would be problematic for any application that required information gathering from the user, such as a field service application.

In both cases – retrieving and submitting – the use of temporary storage may help us, a technique known as caching. The principle of caching is that a copy of everything we fetch from a server is stored in a local memory space on the mobile device, usually nonvolatile memory. If the RF connection is lost, then we can view pages previously fetched by fetching them from the cache, as shown in Figure 11.15.

Caching in Web browsers is a technique originally used for performance enhancement to speed up page access; but, in mobile scenarios it can have an even more useful purpose, which is to enable information to be displayed when the connection is absent. In this situation we are obviously constrained to viewing information that is only as recent as when last placed in the cache. This might be a problem, depending on what the application does and on the user's particular needs (which may vary in time). It would be useful for the mobile device browser to make it clear that cached and, therefore, potentially stale information is being accessed, just to avoid any usability issues that might arise should the user mistake old information for current information.

A recent development in mobile browsers is to implement a cache in the uplink direction, so that forms can still be filled out and submitted, even without an RF connection. In this case an embedded cache management program will watch out for an RF connection to be re-established and then upload the form as soon as a connection with the server is present.

Caching uploaded forms is only workable when no immediate feedback is expected or required from the server. It can be implemented as a transparent feature whereby the application does not need to be specially designed with the caching mechanism in mind. As far as the application and, possibly, the user are concerned the connection appears to be present and the form appears to have been submitted as usual. In reality, there is a delay before the form actually gets processed.

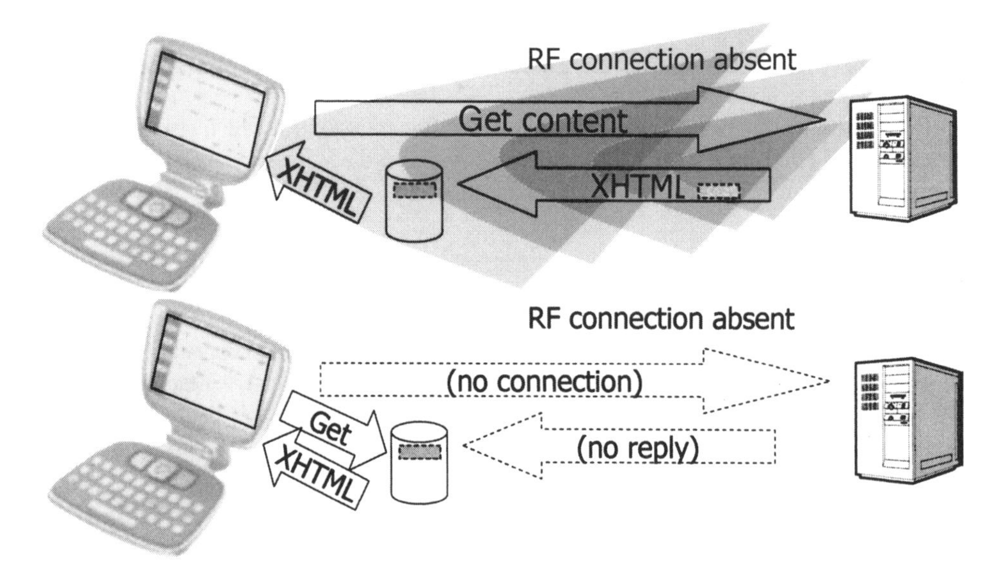

Figure 11.15 Caching enables page to be fetched even if it has already been accessed before.

Under ordinary form-processing arrangements, there will be a response from the server as part of the normal sequence of form submission under the HTTP protocol. It is the job of the cache management program to supplant this response with its own response, both to assure the user that the form has been 'submitted' and that everything is running 'normally' despite the lack of a connection.

Clearly, this method of caching is only appropriate for certain application scenarios: namely, wherever the submission of a form by itself is a meaningful operation without being part of an atomic sequence of page and information exchanges that require server contact. Caching might be suitable for a number of job-reporting applications in the field where an agent is required to submit an onsite report. Whenever this is the case, this caching process has value. It effectively enables a browser-based solution to operate in stand-alone mode for a period, in order to maintain a useful level of service for its user.

11.5.3 User Interface Constraints

Browser-based applications use mark-up languages to allow the annotation of data for the purposes of directing what the data look like in a suitable browser. We have already discussed this process and its variations for wireless devices at length throughout the book (e.g., see Chapter 9).

The mark-up language paradigm uses a set of primitives (e.g., paragraphs, tables, inline images) for presenting data, the construction of which dictates the level of design freedom that the application designer has in determining the richness of the UI. At the same time, we should recognize that the browser paradigm itself places certain constraints on the UI. For example, the HTTP model uses a request and response method of fetching pages of information to the browser. As such, there is no concept of persistence in the UI, other than how much of the browser area we choose to dedicate to an aspect of the interface that we

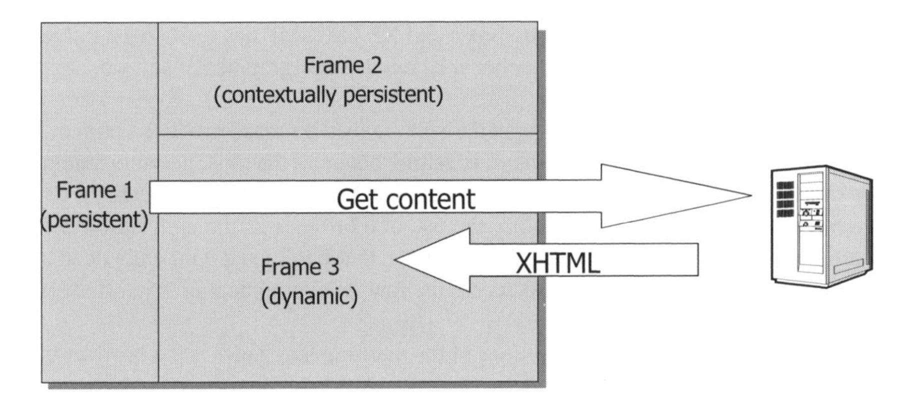

Figure 11.16 Use of framesets so that display areas persist in a browser.

wish to remain constant across browser sessions. We might want to emulate a persistent display in order to provide certain UI elements that would be useful to constantly display throughout the user's engagement with the service, such as menus. This is quite normal for a software application and is the normal design approach for visual interface design in any embedded application that has a UI.

Certain constructs in some mark-up languages can potentially support persistent display areas. In desktop browsers, framesets have long been supported and can be deployed to divide the browser into different display areas. Some of these can be dedicated to persistent UI elements while the others can display dynamic content, such as the scheme shown in Figure 11.16. The figure shows the browser divided arbitrarily into three areas, called frames. Frame 1 is persistent throughout the duration of the occupancy of the browser by our application. It provides a set of mark-up data to provide a menu of application options. Whenever we select an option – like 'show reports', for example – the corresponding response from the server is targeted toward frame 3, which is the area used for dynamic display purposes. In the example shown there is a third frame (frame 2) which provides a subcontext menu that is specific to the current subsection of the application being displayed. For example, in the area concerned with 'viewing reports', frame 2 can provide a persistent set of options that relate to manipulation of the report data. When we move into another context – like 'analyse sales', for example – the page in frame 2 changes so that it can aid in the navigation of options pertinent to the new context.

Persistence in UIs is important for a consistent user experience. However, the method just described assumes that sufficient screen space exists that we can afford to dedicate a portion of it to persistent display areas. This may not be possible, or even sensible, in situations where we have very small displays. In such circumstances we may find that we only have room to maintain the navigation interface or the dynamic information, but not both; so, we are forced to dedicate our valuable real estate to the dynamic information and thus enable the application to function at all.

For this reason the mark-up languages for display-limited devices, like XHTML-Basic, do not support framesets. Moreover, even if they did, the ability to utilize the persistent area is still limited by the relatively crude display widgets. For example, the use of pull-down (or slide-up) menus is not supported by XHTML-Basic or XHTML-MP, because it would be too onerous to insist that all devices should be able to support such a rich interface primitive. The design ethos of XHTML-Basic and XHTML-MP has really been influenced by gravitating toward a sensible *lowest common denominator* graphical display context, not a rich one.

The increasing use of colour, bitmapped displays, even on mass-produced phones, suggests that the lowest common denominator is getting better all the time, notwithstanding the legacy of devices that remain in circulation with older displays. However, even on devices with a much enhanced display capability the use of a browser as the application delivery mechanism will probably still impose UI limitations that don't exist with embedded applications, where it is usually possible to access the low-level graphical user interface (GUI) APIs on the device.

The exception to the limited capabilities of the mark-up languages is if a browser is able to support a plug-in feature that provides a richer display capability to the content provider. A notable example of this is the Macromedia's Flash™ player, which is very prevalent in the desktop browser world and now also available on certain mobile device platforms, such as Pocket PC and some Symbian devices. Flash™ allows vector-based graphics to be displayed within the browser, thus giving pixel-level control to the design, which is a level of design freedom far greater than XHTML-MP can provide. However, Flash™ is not available on most mobile devices, so it cannot be relied on for general (wide audience) usage, and so XHTML-MP is still required.

11.6 JAVA PLATFORM APPLICATIONS

So far, we have looked at embedded applications and browser-based applications. We are now going to explore the remaining option, the Java Platform, which in some sense might be the best of both worlds, especially in its newest variant called Java 2 Micro Edition (J2ME), which is going to be the focus of our attention. We will now examine the use of Java for mobile devices, highlighting the key attributes of this approach while comparing it with the alternatives.

To understand the use of Java for mobile devices, we need to first summarize what is meant by Java. What is this technology all about? We have already looked in some detail at Java 2 Enterprise Edition (J2EE); but, we did so from a purely functional perspective, focusing on its ability to support highly scalable applications based on the client/server architecture. We did not examine the Java-programming environment itself, only its features within the J2EE environment, like remote method invocation (RMI) and Java messaging (JMS).

It seems obvious that J2EE is not a suitable technology for implementing applications on the mobile devices themselves – almost certainly overkill! For example, the need to cluster mobile devices and support distributed processing seems at odds with the mobile device paradigm, which is very much a personal device concept with an implied single user per processor (device), not millions of users vying for back end resources, as we saw in the J2EE world. We don't expect a mobile device to host a large database either. Clearly, circumstances are very different!

At this stage, before exploring the general concepts behind Java, we might sensibly question why bother at all with running Java on mobile devices. Certainly, most of the heavy duty features in J2EE are not required, so why not use a different technology altogether; indeed, what does it mean to support Java on a small device?

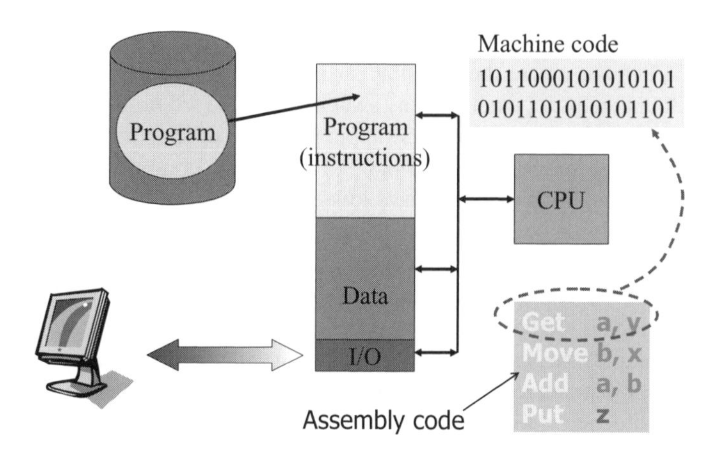

Figure 11.17 Basic computer architecture.

To answer this question, we need first to look at the origins of Java in terms of its unique software-engineering approach that popularized its appeal when first introduced into the computer world back in the glory days of the early Internet boom.

11.7 THE JAVA ETHOS - A TALE OF TWO PARTS

First, we need to remind ourselves of how a computer works! Computers have more or less stuck to an architectural theme that was popularized by the PC (what we used to call the IBM-compatible PC).

Figure 11.17 shows the basic computer architecture that revolves around the central processing unit (CPU), like a Pentium or Power PC, which can crunch numbers loaded from memory and output them back to memory across a bunch of printed circuit wires called a bus.[22] Certain memory locations are special because they also enable the CPU to interface with peripherals, like a graphical displays. Writing certain values to the display memory will affect what appears on each pixel on the screen, and so in this way a software program can write a whole sequence of pixel values in order to control what the screen displays. As discussed earlier in the chapter, the OS is a program that runs on the CPU and can run utility code to manage how this area of memory gets updated and in doing so also provides a more meaningful interface, like a windowing one, to other programs running on the CPU. These applications, instead of writing to the display memory, will write values to a portion of memory monitored by the window utility part of the OS, which in turn writes to the display memory in a way that creates a windowed look and feel.

For the CPU to know what to do with memory values in order to carry out useful work, it has to follow a set of instructions: a program (or an *executable*). A CPU can only run one set of instructions at a time and hence one program at a time. However, the OS is a special program that can swap other programs in and out of memory and give them turns

[22] The Von Neumann architecture introduced the idea of a stored program in the same writable memory that data were stored in.

on the CPU. In this way, if the programs are swapped often enough and given long enough to run on the CPU, we create the 'illusion' of running many programs at once. This all takes place without the user perceiving any swapping or delayed responses, simply because the CPU runs so fast that it can do the swapping and the time-shared processing without anyone noticing. That may seem incredible, but that's how computers actually work. We might think that it would be easier to have lots of CPUs, so we can dedicate one to each program and avoid all the swapping. As elegant as this solution seems, it turns out to be a nontrivial and expensive solution. We still only have one set of peripherals, so we have to manage how the CPUs would compete for such single-instance resources. This is a difficult problem to solve, and it is easier to use one processor with a swapping paradigm, which also saves a lot of money! If we have an efficient swapping technique, enough memory and CPU power, then it is as if we have multiple processors in any case – they're just virtual ones.

The CPU runs (executes) instructions one at a time while stepping through a program. But, how do we, as programmers, tell the CPU what to do? First, a CPU is a machine that is exceptionally good at manipulating binary numbers. However, it turns out that this capability is useful for automating a good deal of processes once we can articulate them as numerical problems; we can do this for most problems either directly or by modelling the real world using numbers. A CPU contains number-crunching devices,[23] like an adder, a subtractor and a multiplier; it also has the means to get the operands from memory and store them back to memory. There are all kinds of mathematical operators that can be used to manipulate numbers usefully, some of them not obvious to us unless we understand binary logic.

It turns out that we can build very tiny switches called transistors, which when combined into various configurations can act as logical operators that underpin most of the mathematical functions we want to perform. There are various useful combinations of switching patterns that get assigned to an instruction which the CPU can repeat when it gets the command to do so from the appropriate bit pattern loaded in from memory. Fortunately, we do not have to remember these bit patterns as programmers (though we used to have to learn them once upon a time); instead, we can use labels, called mnemonics, such as LOAD, ADD, STORE, MOV. Using these mnemonics we can program a CPU (computer); this is called assembly level programming, the mnemonics collectively being known as an assembly language.

It turns out that programming a CPU using assembly language is highly prone to mistakes (human error), as we are not very good at thinking in the same low-level mnemonics that the CPU understands. We naturally want to express a problem and a solution in an unconstrained natural language, like English, or some other higher-level symbolic form, like the arithmetical, algebraic and calculus expressions of mathematics. A CPU simply does not understand these symbols. A fabric of switches, like a CPU, needs the right low-level instructions to operate. To bridge the gap, computer scientists invented what we call high-level programming languages that are sufficiently readable to be understood by humans (who are suitably trained) and are sufficiently constrained by a formal syntax that they can be mapped to the mnemonics by a special program called a compiler. An assembler simply converts the mnemonics into their corresponding binary forms (bit patterns). This hierarchy of symbols, from high-level language representation to binary patterns, is shown in Figure 11.18.

[23] The number-crunching bit is called an arithmetic logic unit (ALU)

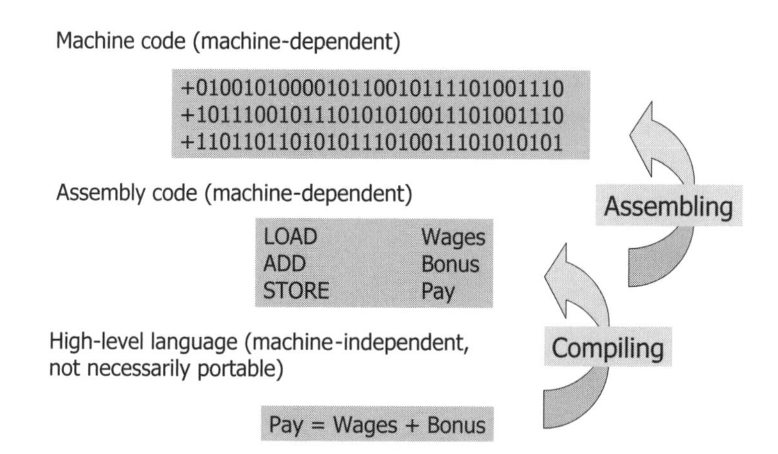
Machine code (machine-dependent)
+01001010000101100101111101001110
+10111001011101010100111101001110
+11011011010101110100111101010101
Assembly code (machine-dependent)
LOAD Wages
ADD Bonus
STORE Pay
High-level language (machine-independent, not necessarily portable)
Pay = Wages + Bonus
Assembling
Compiling

Figure 11.18 The hierarchy of computer languages from high-level to low-level symbols.

Critically, what we can extract from this potted history of computers and software is that most high-level languages – like C, C++ and Pascal – are compiled to an assembly language that is compatible with a particular CPU. Processors can be designed in many different ways, although they have many similarities. The differences in design can account for differences in performance, whether measured by speed of instruction throughput (millions of instructions per second, or MIPS) or some other measure like power consumption, or simply the number of transistors utilized, which will closely determine the cost. So, design is important for product differentiation as far as chip manufacturers are concerned, so we shouldn't expect a common set of mnemonics any time soon, if ever.

The upshot of the differences in mnemonics is that a compiler has to be made for each target processor device. Therefore, if we make a program, like a tax calculator, and want to run it on different devices (presumably in different computer products), then we have to compile it for each processor we want to support. This has important implications. Ideally, we would prefer to write a program only once and then be able to run it on any suitable device, thereby facilitating an efficient, productive and cost-effective software production process.

This is what the Java language concept was devised for. With Java, the process of symbolic conversion is modified so that the high-level language – Java – is compiled first in an intermediate form that we can think of as a set of virtual mnemonics. In this form we call the converted code 'byte code' (see Figure 11.19).

Imagine that, instead of a physical CPU with a hard-coded set of mnemonics, we have a virtual machine that supports byte code. This virtual processor is called the Java Virtual Machine (JVM), and we can imagine unplugging the actual CPU and plugging in the JVM. Of course, this is impractical, but we will soon see how we go about solving this problem. Later, we will also see how developing a processor to run byte code is not as crazy at it seems.

Let's consider for a moment the wider process of writing a software program to run on a particular computer. There are other aspects to the process to take into account besides variation in the CPU and its corresponding mnemonics. We spent a good deal of time in previous chapters explaining that a useful computing environment includes APIs, which

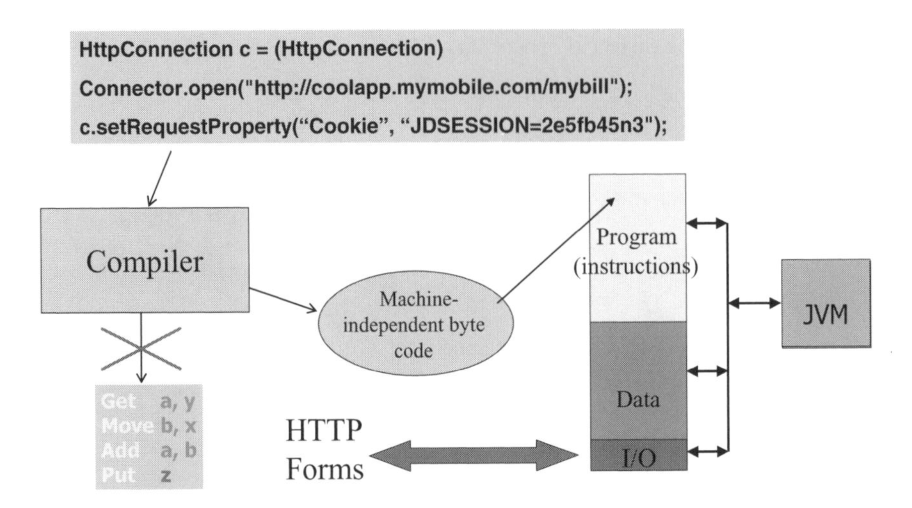

Figure 11.19 The Java language concept – from Java to byte code, not assembly code.

are utility programs that enable our main program to call on powerful pre-written chunks of code that can carry out routine tasks common to many applications, such as supporting a windowed GUI. Java has an identical concept: Java programs benefit from APIs that make general purpose programming a lot easier and more productive.

The APIs in Java are implemented using the same byte code as our main application; therefore, they will run wherever a JVM is available. This is an essential and powerful part of the Java concept, and it means that we are now talking about something greater than just a programming language. We now have the language, some accompanying standard APIs and the JVM to host them, so we can think of this as a language and a platform: this is a vital concept to understand when talking about Java. In the official parlance, we have the Java language and we have the Java Runtime Environment (JRE), which is the JVM and the associated APIs and, of course, the ability for all these elements to interact cohesively.

The Java APIs are blocks of code already available in byte code format and pre-installed[24] on the target machine; they are readily accessible to the JVM to call whenever the main program requires one of their code routines. This concept is shown in Figure 11.20. Here we can see that the program in Java still gets compiled to byte code to run on the JVM. The code snippet in the figure shows an important keyword in the Java language called *import*, which is a link that points to an API we wish to use in our main program. Once we have referenced the API, we can use its software routines, which are called *methods* in Java parlance. The JVM recognizes when we refer to a method that is not in our code, but sitting in an API somewhere in the computer memory. The JVM uses a convention[25] to find the compiled APIs, which are called class files, and loads them into memory for execution by the JVM whenever it needs to run one of the referenced methods.

Of course, we don't swap the real CPU for a virtual one (the JVM) as suggested. The JVM is actually an embedded software program itself, written using mnemonics compatible

[24] Pre-installed usually means that the APIs are already sitting in nonvolatile memory, which would be a disk on a desktop PC and a ROM or Flash memory on a mobile device.

[25] There is usually a system-wide pointer, called *classpath*, that the CPU can access to find where to look for referenced class files (APIs).

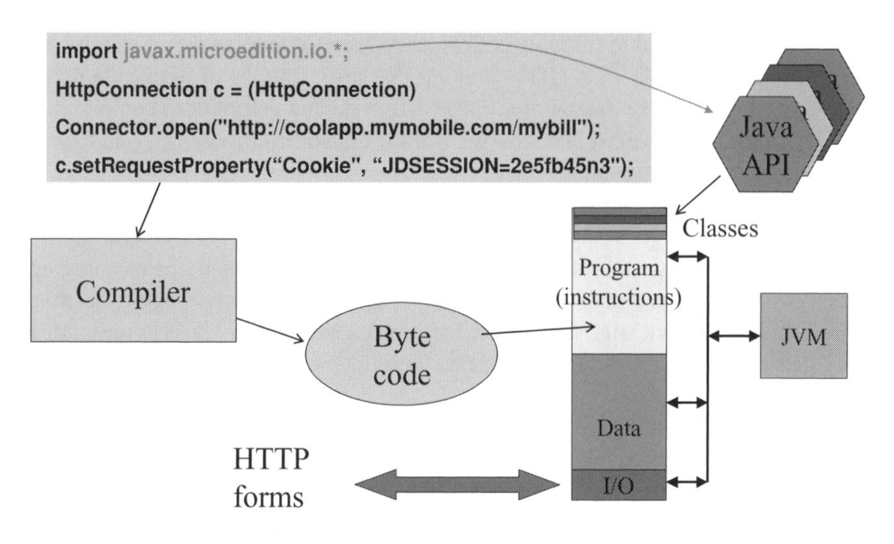

Figure 11.20 The Java API concept.

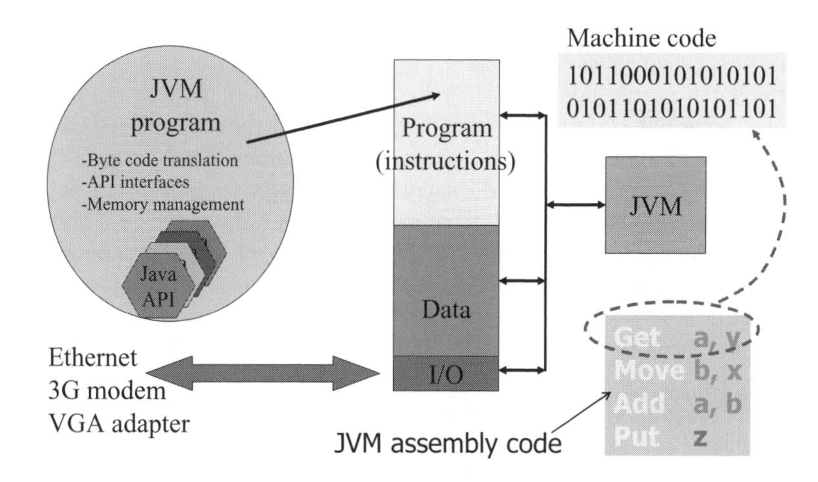

Figure 11.21 The JVM is a program written for the target CPU.

with the target CPU (see Figure 11.21). Once the JVM is running on the CPU, it appears to take over from the CPU to start execution of byte code rather than native mnemonics, like a translator or interpreter. What the JVM does is convert incoming byte code from our program in memory to native mnemonics, mapping the byte code 'mnemonics' to the equivalent native ones. Clearly, this is an added processing overhead not present in a native software program running on the CPU, so we expect Java programs to run slightly slower than natively programmed ones. Moreover, in the translation from byte code to native code, there is not always a convenient one-to-one mapping from byte code commands to mnemonics, so there might be additional overhead in the execution itself after the translation has taken place due to nonoptimal translation.

Techniques can be used to mitigate performance losses due to byte code interpretation, such as the use of a just-in-time (JIT) compiler that takes chunks of byte code on the fly and converts them to native mnemonics, rather like a second compiler to compile the byte code. Other optimizations exist and later we will look at some of those relevant to wireless device operation.

The JVM is a program that carries out byte code translation to native code and manages the loading of reference APIs. Additionally, the JVM performs housekeeping tasks to keep things running smoothly. One of these tasks is memory management, or cleaning up unused memory after our program finishes with it – a process called *garbage collection.* We don't need to know the details here, but in the process of executing a Java program some memory gets assigned to store data associated with the creation of software entities called *objects.*

For example, we might create an object called *VideoPlayer* that plays a video clip which we have downloaded from the RF network. We can appreciate that there is a chunk of byte code in our program or, possibly, in an API which contains all the instructions for the video player. This chunk of code gets loaded into volatile memory when it is needed to play a video clip. We have to ensure that the code is deleted from the volatile memory space once we have finished with it. Otherwise, the memory will become filled with code that is not actively required and, eventually, the JVM will not be able to create any more objects and will grind to a halt. In other words, we will gradually lose our available memory to dead code – a process called *memory leakage.*

Object deletion is best done explicitly by the programmer because the programmer knows best when an object is finished with; this is inherent in the software design. However, programmers often forget to include the necessary instructions to inform the JVM that an object can now be discarded. The JVM can periodically check for objects that appear to be unused and delete them automatically from memory. This turns out to be a highly useful process, especially on a desktop PC or server that is carrying out lots of programming tasks, particularly where mission-critical applications are being run and the disruption of memory leakage would be severe.

In addition to memory management the JVM checks the byte code inside class files to ensure data integrity. This is to detect file corruption and thereby avoid running any code that might subsequently go haywire when being executed.

Let's consider the process of developing a program in Java and then running it; we can divide the life cycle into two distinct phases – design time and run-time – as shown in Figure 11.22. At design time the programmer is actually writing the software in the Java language. A small snippet is shown in Figure 11.23 along with the output screen showing what happens when we run it. The Java language is written using keywords that are expressed in English, using Latin symbols (like { }) and arithmetic operators. A Java file is saved to a text file with a name that corresponds to the name of the software object (user-defined class) being defined in the file. This file is then compiled to a corresponding byte code file called a class file, which is a binary representation of the meaning of the Java code translated into byte code instructions.

The class file is installed onto a mobile device that already contains the JRE, which includes the JVM and the associated APIs. When the program is invoked the JVM will load our class file into volatile memory and run it, interpreting the byte code into the native CPU instructions for the target CPU. Prior to execution, byte code integrity is checked to verify that the class file is not corrupt in any way.

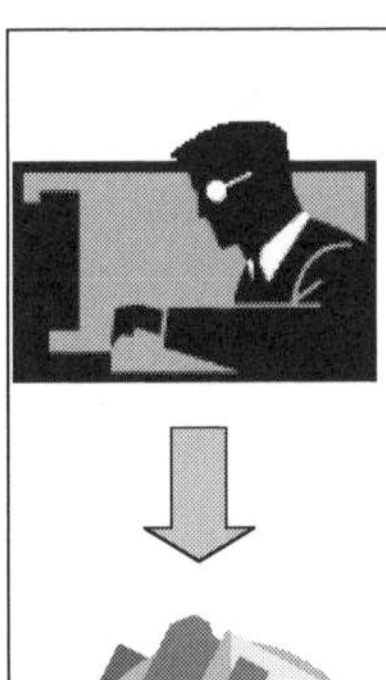

Design time
1. Edit program (.java) on disk using IDE
2. Compile to class (.class) file (byte code)
Run-time
3. Classloader loads class into memory
4. Byte code verification takes place
5. Byte code gets executed – interpreted line by line by the JVM

Figure 11.22 Java development life cycle.

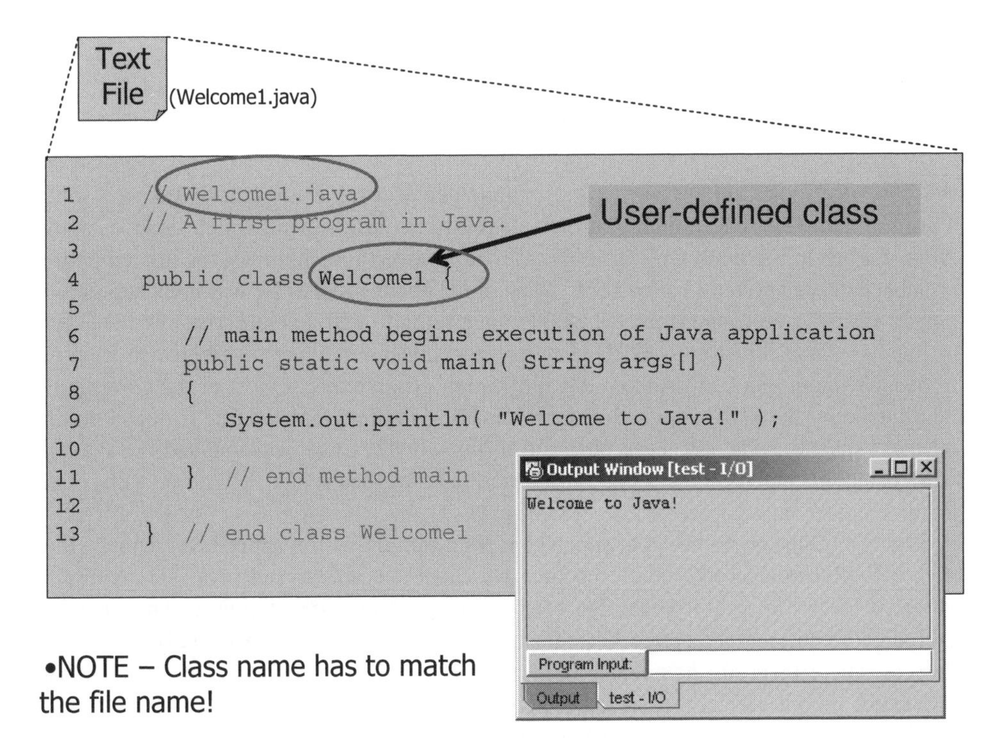

Figure 11.23 An example Java program.

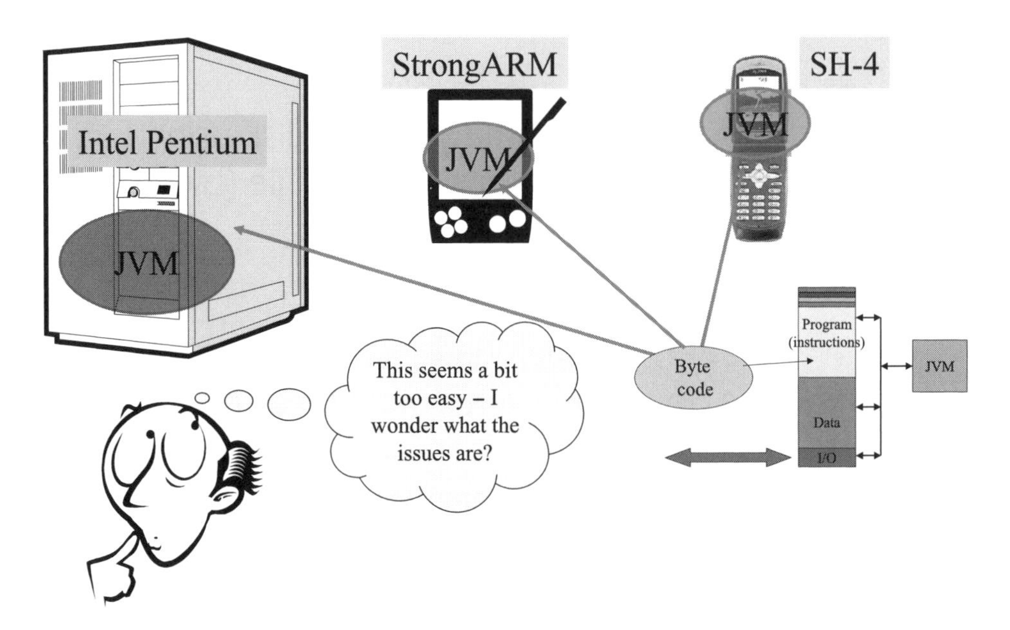

Figure 11.24 Will Java really run the same everywhere?

11.8 THE 'WIRELESS JAVA' RATIONALE (J2ME)

We have just looked at Java at its most fundamental level. We learnt that Java is both a programming language and a platform, or run-time environment, which includes some APIs. Earlier in the book we learnt about a very powerful technology for implementing scalable back end systems called J2EE. Most of the goodness of J2EE was in its huge set of powerful software mechanisms that enable interfacing to myriad enterprise systems, such as databases and directories. J2EE also has mechanisms to allow distributed processing to occur without much effort from the application programmer – most of the clever stuff being done by the J2EE application server, leaving the programmer free to focus on the business-specific code. All these powerful utilities in J2EE are implemented in Java and work using the underlying processes just described – by means of a JVM, Java APIs, class loading and so on.

Figure 11.24 shows a possible conclusion from our discussion about Java: namely, that with a JVM available on a target machine we should be able to run Java. This is highly attractive to developers as it means they can focus their efforts on mastering one language that will allow them to create applications for any type of environment, especially for mobile devices where there are potentially huge numbers ripe and ready for running all those interesting mobile services we have been contemplating throughout the book. However, as the figure shows, our thoughtful developer is rather curious that this all seems a bit too easy. Perhaps, something is not quite right!

If, like our curious developer, we ponder for a moment on the implications of what is shown in Figure 11.24, we might start to realize some of the apparent contradictions with the 'run anywhere' philosophy of Java, especially when taking into account the differences between the devices shown in the figure.

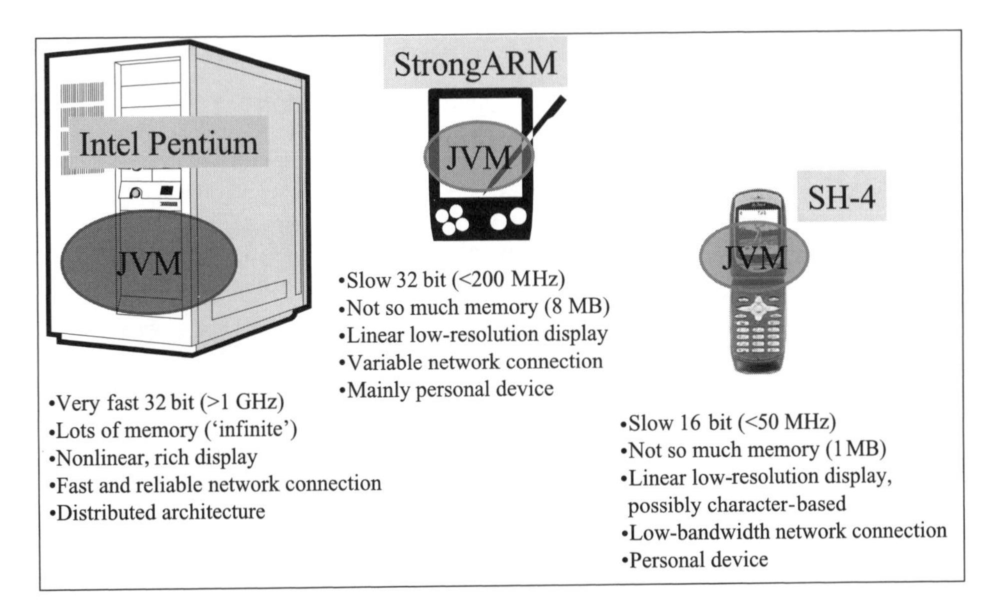

Figure 11.25 Physical differences across device categories.

Two issues spring to mind. First, the devices themselves are very different. They are different categories of device altogether. Servers, like the ones needed for J2EE clusters in a mobile service delivery platform, probably have massive amounts of memory, both volatile and nonvolatile, huge displays capable of rich colours, high resolutions – collectively big enough to display many applications (or windows) at once. In other words, the first set of issues relate to the physical characteristics of the devices in each category.

While device capabilities are changing all the time, Figure 11.25 presents a snapshot of some typical device characteristics for servers, PDAs and mass-manufactured cellphones. The key differences are in processing power (CPU), memory size, and connectivity and display characteristics.

The second set of issues is to do with the types of applications that we might want to run on each type of device. Clearly, on the servers that are supporting our mobile delivery platform we need all those wonderful J2EE interface capabilities to hook up with database servers, directories, legacy mainframes, messaging systems, mail servers and so on. But, on a mass consumption handset such interfaces would be preposterous. We can hardly imagine installing[26] an enterprise class relational database application on our handset, never mind finding something useful to do with it.

And what about display capabilities? On a server or desktop it is normal to offer a rich UI based on the windows metaphor that is now so pervasive in human–computer interfaces. To make life easier for the Java programmer, one of the APIs that was made available by the inventors of Java is a rich set of code to do all the windowing, leaving the programmer free to concentrate on what the windows are to be used for, rather than constructing and maintaining them (a nontrivial programming task as it happens).

[26] A particular instance of a well-known relational database server requires over 1 GB of disk space just to install it!

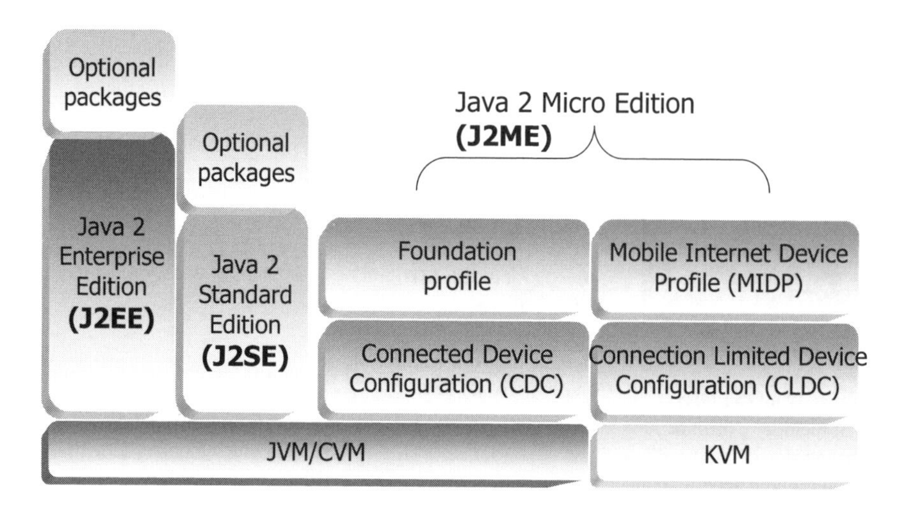

Figure 11.26 The family of Java editions.

Again, on a small mobile device with a tiny display the use of windows is not only inappropriate it is also mostly infeasible. This leads us to an idea that involves chopping out the APIs that we don't need and thereby enable a slimmer version of the JRE for resource-limited devices.

Recognizing the increasing support from developers for the Java language, Sun took the step in 1999 to revise the Java strategy. Acknowledging that 'one size doesn't fit all', Sun announced the splitting of Java into three versions:

- Java 2 Enterprise Edition (J2EE);
- Java 2 Standard Edition (J2SE);
- Java 2 Micro Edition (J2ME).

The '2' was not coincidental with this split, but rather with an earlier decision to release a significant upgrade to the APIs.

Figure 11.26 shows the family of Java editions. On the left side of the diagram we see the already familiar J2EE version of the Java platform. It has all the powerful enterprise APIs that we have already discussed throughout the book, especially in relation to their potential usefulness in implementing a mobile services delivery platform. Java 2 Standard Edition (J2SE) is closest to the original version of Java, and we can consider it a very feature-rich implementation suitable for general purpose programming on desktop PCs.

Both the J2EE and J2SE editions can run on the same design of JVM, which is also available in the C programming language as the CVM, so that those who want to port the JVM to their own environment can do so by a suitable compiler to the target native code (mnemonics).

The Java 2 Micro Edition (J2ME) is of most interest to our current concern – utilizing Java for mobile applications running on mobile devices. J2ME was targeted at a range of consumer and embedded electronic devices with constrained resources, and it is this version that we will concentrate on for the remainder of our discussion about Java in this chapter.

Because the remit of J2ME covers a wide range of devices in various market segments, J2ME itself is subdivided into configurations targeted at particular hardware families, or types.

We can think of a configuration as a core set of Java APIs and an associated virtual machine specification, which are grouped together in a way that is optimized as far as possible for a particular group of target devices and possible application scenarios. More specifically, a Java configuration specifies:

- the Java programming language features supported;
- the JVM features supported;
- the basic Java libraries and APIs supported.

Configurations still use the Java programming language, trying to maintain as much of the basic language support as possible. This is an obvious point perhaps, but it is worth remembering that the primary benefit of J2ME is that Java programmers do not have to learn another language to 'go mobile'. But they do have to make the effort to learn a new set of APIs and new skills related to designing applications for resource-constrained environments, some of which we mentioned in Chapter 10.

The configuration aimed at the lower end of mobile devices (e.g., mass-produced phones) is the Connected Limited Device Configuration CLDC, which aims at having 160–512 KB of memory available for storing the JRE. Other considerations are that the environment will probably be battery-powered most of the time, most likely will have a very slow CPU (less than 20 MHz) and will have an intermittent network connection. This is clearly appropriate for the typical mobile device, remembering that we are talking about mass-produced hand-held devices, including mobile phones – and not just PDA-like devices, which can easily have a much heftier resource allocation.

To make the CLDC available for particularly resource-constrained devices, the JVM has been cut down in terms of its functionality: it has become the KVM,[27] the 'K' standing for 'kilo' to indicate that we are dealing with memory spaces in kilobytes (as little as 160 KB of total memory are available for the Java implementation,[28] whereas desktop memory budgets are measured in megabytes).

Sitting on top of the CLDC is the Mobile Information Device Profile (MIDP) which specifies an API set appropriate for mobile information devices, such as phones. However, we could simply use the CLDC core API set as it is: so, why the further need for a profile to state exactly the designated API definitions intended for mobile devices?

The profile is more akin to an industry agreement that states how supporters of the agreement agree to use CLDC to support a Java implementation on a particular device category within their industry. With a particular set of devices and application scenarios in mind, the profile defines appropriate APIs for the target market. All hardware supporters of the profile agree to implement these APIs on their devices running on top of the KVM. The MIDP profile has the backing of a large group of mobile phone operators, phone manufacturers, and affiliated technology and service companies, such as toolset vendors.

[27] The KVM technology came from a research system called Spotless, developed at Sun Microsystems Laboratories. More information on Spotless is available in the Sun Labs technical report *The Spotless System: Implementing a Java system for the Palm Connected Organizer* (Sun Labs Technical Report SMLI TR-99-73).

[28] The K virtual machine is very small – starting from approximately 70 KB in size.

Key J2ME definitions (taken from Sun):

Configuration – a J2ME™ device configuration defines a minimum platform for a 'horizontal' category or grouping of devices, each with similar requirements on total memory budget and processing power. A configuration defines the Java language and virtual machine features and minimum class libraries that a device manufacturer or a content provider can expect to be available on all devices of the same category.

Profile – a J2ME™ device profile is layered on top of (and thus extends) a configuration. A profile addresses the specific demands of a certain 'vertical' market segment or device family. The main goal of a profile is to guarantee interoperability within a certain vertical device family or domain by defining a standard Java platform for that market. Profiles typically include class libraries that are far more domain-specific than the class libraries provided in a configuration.

The term 'Wireless Java' has no useful meaning *per se*; this marketing phrase has become prevalent, having been initially promoted by Sun Microsystems. In actuality, it is a means of labelling the dedicated activities and technologies of the Java community effort dedicated to the important area of wireless (which we can interpret more accurately as meaning 'mobile', as principally the objectives of the MIDP are mobility, portability and ubiquity, not simply wireless connectivity).

Of course, we already understand that Java is a wide set of technologies for developing applications, having first examined its benefits in this book when we looked at J2EE; so, we should not be under the misapprehension that only J2ME is used for mobile applications. We already know that the mobile network is a network of networks, and Java can be applied at each service point with varying degrees of appropriateness and success.

11.9 USING THE MOBILE INTERNET DEVICE PROFILE (MIDP)

Now that the principle of J2ME has been described, we can return to our main consideration for this chapter: the ways and means of developing mobile applications, not just the technical processes underlying the technology. What we need to do is understand what it means to use MIDP to develop mobile applications, especially compared with the other two methods we have already identified and discussed: namely, the embedded approach and the browser approach.

In essence, the Java approach is very similar to the embedded one: we write 'any' software application we like using a familiar programming language, compile it, install it and then let our users use it. Well, we could already do this without MIDP; so, before we get confused about why we need the Java approach at all, let's remind ourselves of some key points.

In terms of developing mobile services, we are generally interested in ways of maximizing ubiquity. We have discussed this in detail, especially regarding how we go about defining a mobile service and how we might realize it, such as through the provision of service delivery platforms, IP networks and all the myriad mechanisms that seem to have been useful and necessary on our journey across the mobile network thus far.

We will shortly be looking at mobile application topologies to see how these programming techniques can be used to implement different types of application, but let's finish our survey of the programming techniques first by acquainting ourselves with some of the implications and intricacies of using MIDP.

The criterion of ubiquity implies that we need to support as many devices as possible to become potential access points to our mobile services. There would be little point in building the 'splash messaging' (virtual notice boards) services discussed in Chapter 13 on mobile location services if it only worked on one device type and accounted for less than 1% of the device population in our target market area.

The mobile device market is very competitive because it is so large. The principles and ingredients we have described for building a mobile device, such as dedicated processors, special OSs, utility applications and so on, are within the reach of many vendors to implement or integrate. Differences in design detail are bound to surface in order to find a competitive edge that each supplier is hunting for or to find an extra ounce of profit margin in the implementation.

The bottom line is that there is a huge array of devices in the market, especially in the mass-manufactured mobile phone category, which accounts for hundreds of millions of devices. If we want to develop applications to run across all the devices available, then we would probably find that the classical embedded approach is not that open to us in terms of providing end user applications. Embedded solutions will get developed for utility applications, like the WAP browser itself – Openwave's Phone Tools being a good example. This is a set of utility applications including the browser, messaging client and download file manager; but, this is aimed at device manufacturers to license from Openwave to install on a large device population. The numbers (economies of scale) add up in such cases as probably millions of licences are involved at a time.

If we attempted to use the embedded approach for a particular end user application, such as a game, then in order to get a large number of users we would have to develop the game for many different types of device. Each device probably has a unique programming environment, such as a specific OS and associated system APIs. Each device probably also has a unique way of installing the software (see above discussion on installation methods). This would be prohibitively expensive and cumbersome, as there are potentially hundreds of devices on the market, taking into account the legacy of devices already in circulation. At any one time, there are probably at least one hundred different devices on sale in any one market.

The MIDP solution addresses this problem quite elegantly, as shown in Figure 11.27. This is its main selling point: by standardizing on a single language and associated API set, independent of the underlying OS and processor, developers can now write an application once only and then make it available to a heterogeneous and potentially huge device family. We had previously argued that the 'write once, run anywhere' philosophy of Java broke down across the abundance of computer platforms in the entire (horizontal) spectrum of CPU-enabled products. However, we should not forget that the concept is still very valid and very powerful *within* a vertical device family, like mobile devices, which also happens to be a huge potential market. What MIDP offers is:

- a standard API set that is guaranteed to be available on all MIDP-compliant devices;
- the possibility to use a single toolset (IDE) to develop any application;

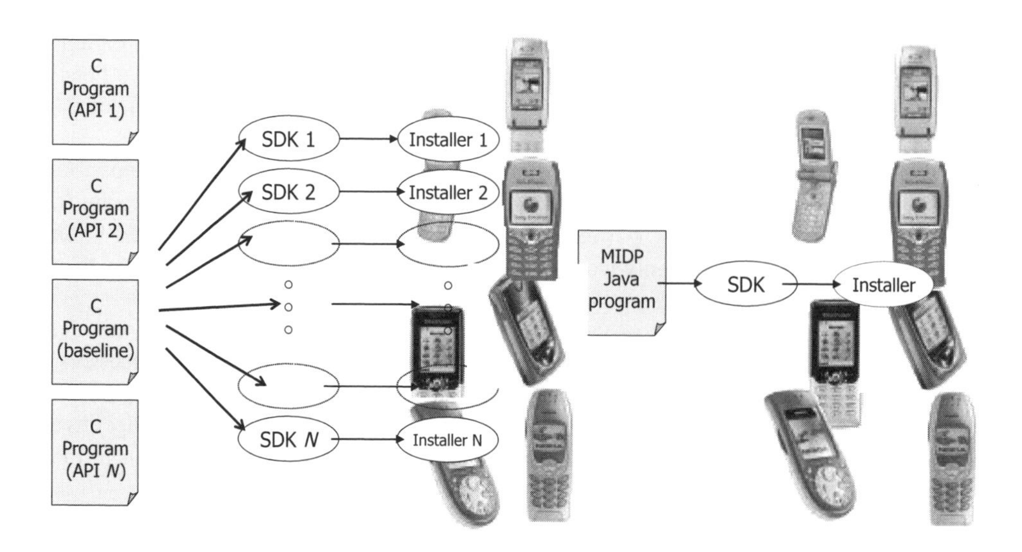

Figure 11.27 Embedded headache versus MIDP 'walk in the park'.

- a single way of packaging and distributing applications (more on this later);
- a standard network communications model (IP-based) that is also guaranteed to be available on all MIDP-compliant devices (and can utilize HTTP, should we want to).

The unified programming environment across all devices in a family is a huge step forward for the mobile data future, especially ubiquitous access. There's little point in building ubiquitous RF network coverage if we can't deploy applications easily across the mesh of connected devices.

It is reasonable to ask if there are any downsides to this approach. We should be instinctively wary of Utopian solutions and be looking out for the catch. First, even if there are downsides, we should reflect on the reality that without MIDP there are probably no credible alternatives in any case, not if we want the 'write once, run anywhere' advantage. Until MIDP devices came along, it wasn't even possible to develop applications for most mobile devices regardless of Java support or not, unless we opted for browser-based applications (primarily WAP 1).

It seemed as if the entire mobile industry was facing a chicken-and-egg problem. The reluctance to develop applications for myriad devices using myriad SDKs and API sets was also the same reason no mobile vendor bothered offering them, except in rare cases. Furthermore, the skill required to offer a developer kit (SDK) is not to be underestimated; so, the idea of generating one of these for each and every mobile device type was (and still is) unattractive. The emergence of the KVM and CDLC caused huge interest among device vendors in offering programmable devices – suddenly this became possible and viable, for all the reasons we have just been discussing. Hence, there was a rush to adopt MIDP, resulting in plenty of MIDP devices on the market within a relatively short space of time.

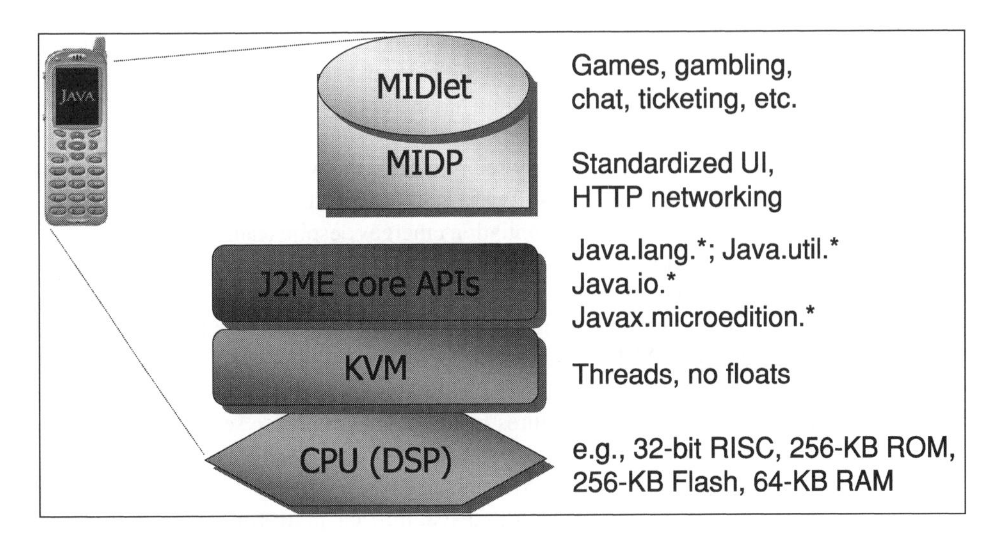

Figure 11.28 The MIDP software 'stack'.

That said, there are some limitations to the MIDP approach; we briefly mention some of them here, although realistically they are not serious problems and don't really present roadblocks to the potential success of MIDP.

To provide a unified programming environment for MIDP, we have to agree on a single set of APIs that constitutes the profile. This means that all devices will behave the same as far as the programming model is concerned, even though one device might be more capable than another in terms of its underlying API set, were we to develop for it in an embedded fashion. For example, a particular device OS might offer powerful system-level APIs for cryptography, whereas MIDP (1.0 in any case) does not offer any cryptographic APIs. It is also not possible to access the underlying APIs, different from the J2SE case where the Java Native Interface (JNI) is supported to allow underlying native code to be called from within the JVM. This is a deliberate limitation of the CLDC core set of Java functionality, remembering that the MIDP is built on the CLDC and has to inherit its limitations, as shown in Figure 11.28.

Figure 11.28 shows the microcosm of an MIDP program (MIDlet). The MIDlet is restricted to the API set available in the MIDP, unless we use an extension set of APIs provided by the handset manufacturer and already pre-installed on the device. This is possible, although it is moving away from a standard profile and making our development process more customized, which is possibly going to limit our application's appeal if it only works well on one device that supports the extension APIs.

As an example, let's consider the Motorola A008 device – probably the first to market with MIDP 1.0 support; the UI of the device supported a limited windowed appearance, albeit only one window at a time, but the concept of title bars and close buttons was apparent. This UI metaphor is not present in the basic UI API for MIDP 1.0, but it is possible to implement using the lower level graphics APIs in the CLDC. Therefore, on the A008, by sticking strictly to the default MIDP implementation, we produce an application that essentially has a look and feel inconsistent with the A008 UI, which we could say is a limitation. However,

Motorola made an extension API available to support the limited windowing capability of the device, thereby making it possible to avoid the inconsistency. The downside is that we end up with a particular instance of our program having to be supported specifically for the A008, should we choose to use the extension.

The idea of offering extension APIs may seem to sully the wonderfully standard world of MIDP. Well, that's what the world of software is like! Everybody has their own idea of what is useful and what is 'cool', so differentiation emerges despite standard approaches or even because of them. The problem with MIDP 1.0 is that, like all early versions of software technology (or any technology), things were missed out in order to get the product quickly into the market. On balance and with hindsight (foresight even), that was probably a good thing to do; the emergence of MIDP came at the right time and quickly created an interest among developers in developing applications for the potentially lucrative mobile market. However, attempts to promote better features in some devices and to see them utilized in MIDP applications has caused various API extensions to be released.

The good news is that many of the API enhancements became redundant with the introduction of MIDP 2.0, an altogether more feature-rich platform. Therefore, rather than summarize the features in MIDP 1.0, we should jump straight to MIDP 2.0 in our discussion.

11.10 WHAT DOES MIDP 2.0 OFFER?

Let's now examine what we can do when developing a mobile application using MIDP 2.0 technology. Where possible, in keeping with the ethos of this chapter, we will compare and contrast the available capabilities with other means of developing mobile applications, so as to keep our discussion relevant to the theme of mobile service delivery, not 'Wireless Java' *per se*, as there are plenty of books in this topic.[29]

The MIDP 2.0 specification is based initially on the MIDP 1.0 specification and provides backward compatibility with MIDP 1.0 to the extent that MIDlets written for MIDP 1.0 can execute in MIDP 2.0 environments. The momentum behind MIDP has been gathering in the marketplace, so backward compatibility was an important consideration. This is a problem particularly pertinent to the mobile devices market because upgrades in the field are very unusual – unlike the desktop PC environment, where they are the norm.

As already discussed, the MIDP is designed to operate on top of the CLDC. While the MIDP 2.0 specification was designed assuming only CLDC 1.0 features, it will also work on top of CLDC 1.1. However, it is probable that most MIDP 2.0 implementations will be anchored in CLDC 1.1.

Because Mobile Information Devices (MIDs) capable of running MIDP Java will be in abundance and the number of expected applications will be vast, rather than trying to provide capabilities for all eventualities, the MIDP 2.0 specification defines a limited set of APIs, attending to those functional areas that were considered essential to achieve broad portability goals for typical mobile application scenarios.[30]

[29] Enrique Ortiz, C. and Giguere, E., *Mobile Information Device Profiles for Java 2 Micro Edition*. John Wiley & Sons, Chichester, UK (2002).

[30] This was done by industry 'consensus' though the Java Community process, so the final specification is not without considerable input from those with a view of the market. Whether it is the 'correct' view or not remains to be seen.

MIDP 2.0 capabilities include:

- application delivery and billing;
- application life cycle (i.e., defining the semantics of a MIDP application and how it is controlled);
- application code-signing model and privileged domains security model to control deployments within a sanitized scope of behaviour;
- end-to-end transactional security (HTTPS, as discussed in Chapter 9);
- MIDlet PUSH registration (a server push model, like WAP PUSH, but for MIDlets);
- IP-based networking;
- persistent storage;
- sound support;
- timers;
- UI (including general display and input, as well as some unique requirements for games).

Let us examine each of these capabilities in turn, summarizing the key attributes and keeping in mind our concern for understanding the array of applications options open to us on the device, not the specifics of Java programming (MIDP or otherwise).

11.10.1 Application Packaging and Delivery

As discussed earlier in this chapter, an effective means to deploy applications is essential, especially when we consider the potentially huge and diverse set of users and devices involved. It is no surprise, then, that the MIDP expert group focused quite heavily on this aspect: it is a testament to the benefits of approaching mobile software deployment holistically.

MIDP supports application loading via any networking route onto the device, but, most importantly of all, it supports an OTA mechanism. This is actually contained within the MIDP 2.0 specification – unlike MIDP 1.0, where the OTA mechanism was a recommended practice, seemingly an afterthought. This led to variations in implementation and, hence, the need to formalize the technique within the specification.

The MIDP OTA mechanism should look familiar to us: earlier in the chapter we described the OMA method, which, as mentioned then, was deliberately based on the MIDP OTA model. Therefore, as with OMA, the MIDP method is based on HTTP and is practically identical to the OMA except that the object descriptor is a specific file type called a Java Application Descriptor (JAD) and the downloadable file is the MIDlet itself, which is encapsulated in a file type called a Java Archive (JAR), which is like a ZIP™ file.

Since we have already described the OMA method, we defer the explanation of the MIDP OTA mechanism to the end of this chapter.

What we want to focus on here is the application packaging, especially the new developments for MIDP 2.0 that warrant extra attention, such as code signing.

MIDP 2.0 introduces the idea of trusted MIDlets. This is a means to authenticate that a MIDlet is from a particular vendor, and thereby enable the user to benefit from its features with confidence and without fearing any malicious misappropriation of device resources. For example, a virus could be constructed that surreptitiously sends messages from the phone to recipients in the device phone book. With MIDP 2.0 it is possible not only to check that a MIDlet is from a trusted source but also to control access to sensitive phone resources, such as the address book.

Presumably, we don't expect deleterious software features from a well-respected and known MIDlet vendor, so we would like to be able to verify an application is from that vendor in order that we can trust it and go ahead and run it. To do this, MIDP introduces the concept of code signing using digital certificates. The Java community, motivated to promote open standards and support existing ones wherever possible, elected to incorporate an existing digital certificate method for code signing. Recalling our discussions earlier in the book about HTTP, public key cryptography is a powerful method for supporting digital certificates, the principle being that with certain unique one-way key associations we have a means of digital identification.[31]

The mechanisms defined in the MIDP 2.0 specification allow signing and authentication of MIDlets based on the X.509 Public Key Infrastructure (PKI) specification.[32] In Chapter 8 we looked at the principles of public key cryptography. However, the means of sharing keys and authenticating whom they belong to and so on is what PKI is all about.

We should note that this mechanism is open to any application developer to adopt for code signing, so we could similarly use such a method for distribution of embedded applications for mobile devices, whether or not we are using CLDC, MIDP or even Java. However, the use of code signing has been standardized for MIDP 2.0, which in itself is a useful achievement, but its use is also intimately tied up with the concept of permissions-based access to APIs and MIDP functions, as we will discuss next when examining API support in MIDP 2.0.

11.10.2 API Summary

We are not too concerned with the details of MIDP 2.0's APIs, as that is the concern of the developer. We are interested in the general capabilities, especially so that we can evaluate how applicable MIDP 2.0 is for a particular mobile service requirement compared with classically embedded and browser-based applications.

The good news is that API support of MIDP 2.0 is constant across all MIDP-compliant devices, at least in terms of the core API features that are deemed mandatory by the MIDP specification. Of course, vendors can offer APIs as optional packages, should they wish to do so. We have already argued that this might be a good thing or it might not. There is no easy answer.

The API set is aimed at general purpose programming, not system programming. The difference is that with system programming we are concerned with implementation of

[31] In Chapter 8 we discussed public key cryptography and mentioned that by being able to open an encrypted message using someone's public key, we would then know that they must have at one time secured it with the corresponding private key, which provides a means for digital identification.

[32] RFC 2459 – Internet X.509 Public Key Infrastructure (http://www.ietf.org/rfc/rfc2459)

applications that are to be incorporated into the device ecosystem itself, not run as an end user application on top of the system. For example, a telephony management program (e.g., dialler and associated phone book) is an essential system program on a mobile device that supports telephony. Another example would be intelligent battery charge management programs. These types of program are characterized by the need to gain low-level access to system resources on the device, such as right down at the device driver level (see earlier discussions to understand this in the context of a mobile device architecture).

There are very few APIs that support low-level access to the mobile device and would make MIDP suitable for system programming. The effort in supporting such an implementation of Java and in porting it to a device is more akin to the J2SE level of functionality; this would be too resource-intensive and detract from the original design ethos for CLDC and MIDP.

As we might expect, the API set is really focused on small-device UI realization and on IP networking. We will now look at these and briefly mention the other APIs.

11.10.3 User Interface APIs

MIDP's UI capability is contained within two APIs: high-level and low-level graphics APIs. The high-level API is designed for applications where a workable and basic level of interactivity is required to be guaranteed on all devices. Interaction at this level is at its most basic and is more akin to the forms capabilities and limited layout options of XHTML-MP in a browser than a full-blown API capable of controlling every pixel on the screen. The interface widgets are typically text boxes, buttons, selection lists and so on.

For applications using a high-level graphics API, portability across devices is important. To achieve this portability, the high-level API employs a high level of abstraction by providing very little developer control over the look and feel. In fact, the look and feel is left to the MIDP implementation on the device itself and so, typically, will differ from device to device, although the widgets will all have a similar morphology. This very high-level abstraction is further evident in the following ways:

- The actual drawing of components on the device display is performed entirely by the underlying MIDP implementation. In other words, the appearance of, say, a button on the screen is dependent on the system code that the device manufacturer has written to support it, notwithstanding the actual capabilities of the display itself. MIDlets do not define the visual appearance (e.g., shape, actual colour, font, etc.) of any of the components in this mode.

- Navigation, scrolling and other interactivity mechanisms are entirely encapsulated by the MIDP implementation. The MIDlet application is not aware of these interactions and has no control over their manifestation on the device.

- Applications cannot access concrete specific devices, like a particular key on the keypad. When using a high-level API, it is assumed that the underlying implementation will do the necessary adaptation from interface widgets to the device's hardware and native UI style.

A low-level API, on the other hand, provides very little abstraction and allows for detailed control of the display via software. This API allows exact positioning and management of graphical elements, as well as access to low-level input events.

Some applications also need to access special, device-specific features; a typical example of such an application would be a game. Using a low-level API a MIDlet can:

- control entirely what is drawn on the display (though confined to the portion of the display given over to the KVM or MIDlet by the device's OS);
- listen for programmable primitive events like key presses and releases;
- access specific keys and other input devices, where available.

Because of the variation in display sizes, colour depths and variations in keypad design, the freedom to program to the limit of the device capability clearly means that applications that utilize a low-level graphics API are not automatically going to be portable: low-level APIs only enable device-specific interface design. For example, a program with interface elements sized to 300 pixels across is not going to work too well on a device with only 200 pixels width. This might not be a problem depending on what the programmer is trying to achieve; there are ways of writing low-level graphics software that can allow for display variations.

If the application does not use low-level API features, it is guaranteed to be portable. It is recommended that applications use only the platform-independent part of a low-level API whenever possible. This means that applications should not directly assume the existence of any keys other than those predefined in the API, and they should not depend on a specific screen size. For something like a game, it is possible to use a generic game key-mapping technique rather than 'hard-code' the keys (see Box 11.1: 'Generic game action keys'). This is clearly advisable to enable portability. Similarly, assumptions about screen size should be avoided where possible, so that the design approach is adaptable to the available display characteristics. Alternatively, the MIDP designer has to take care to code for different devices, but this will add complexity and increase the cost of developing an application, as well as making it more burdensome to move to new devices as they are rolled out in the marketplace.

Box 11.1 Generic game action keys

Earlier we discussed the Java language concept and introduced the unit of software called a *class*. As we now know, some classes come already written for us in the Java implementation, and that is what we have been discussing – they are called APIs. In MIDP there is a class called 'Canvas', which enables us not only to control the screen but also to get events from the keypad.

Rather than figure out on a device-by-device basis which keys map to what, the Canvas class defines a number of constants for commonly used keys. In particular, it defines abstract game actions (UP, DOWN, LEFT, RIGHT, FIRE, GAME_A, GAME_B, GAME_C and GAME_D) whose key code mappings can only be determined at run-time.[33]

[33] This means that until the MIDlet is running on the target device we cannot tell which actual keys on the device are mapped to the abstract key names. Let's say we wanted to provide on-screen help instructions to convey which key is the 'jump' button in our game (to be assigned, say, to GAME_A). We could not hard-code such an instruction as 'press the 1 key to jump' in the software, because we would not know in advance which key is going to be assigned.

What this means is that as the MIDlet developer, we deal in our software in terms of keys like 'UP' and 'RIGHT', not '2' or '6', or even other codes relating to input devices like joysticks.

Making assumptions about specific keys can be problematic and in the worse case render our application unusable, even in ways that we don't expect when writing software! For example, an operator in the UK running a MIDP games arcade from its portal found that its arcade became disrupted by the introduction of the Nokia 3650, which, although a wonderful phone, has a circular keypad. Had we hard-coded the keys, perhaps quite sensibly '1' and '3' for left and right, then on the circular pad we clearly have a problem, as the above illustration testifies!

In addition to high-level and low-level APIs for graphics, MIDP 2.0 introduces a new graphics API for gaming. This API supports graphical metaphors that work well for games where common graphics operations are to be expected, such as complex and movable background images, animation and layering of the scene to aid the visual effect of movement. For example, in a game with complex scenery the effect of motion can be produced by constructing the scene from layers and then animating (panning) them at different speeds to get a feeling of depth. Foreground elements, such as action characters can be animated as single entities (*sprites*) rather than pixel by pixel.

Some of these gaming concepts are shown in Figure 11.29: a deconstruction of an arcade game programmed as a MIDlet. We can clearly see logical layers in the graphics, which, if they were treatable as programmable entities, would make our programming lives a lot easier. This also enables the device manufacture to take advantage of any underlying hardware that is capable of handling graphics in this manner directly, such as graphics accelerator chips.

There are two movable sprites shown in the figure that represent the characters in the fight scene. We could have different fixed sprite shapes to represent their major action poses in the game. We are showing a zoomed image on the right of the 'jumping position' sprite for one of the characters. We can move this sprite as it is (i.e., a predefined and self-contained block of pixels) wherever we like on the canvas simply by calling the relevant Java method in the API. The sprite will be defined by a complex polygonal shape, as we have shown (though this distinct boundary is not visible in the game, it is just for illustration purposes).

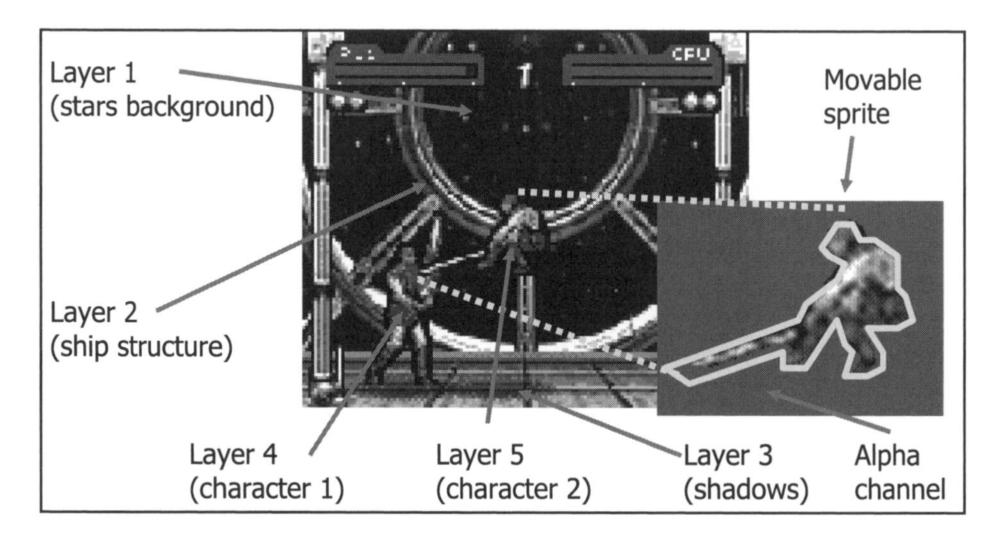

Figure 11.29 Game graphics layering, sprites and alpha channels.

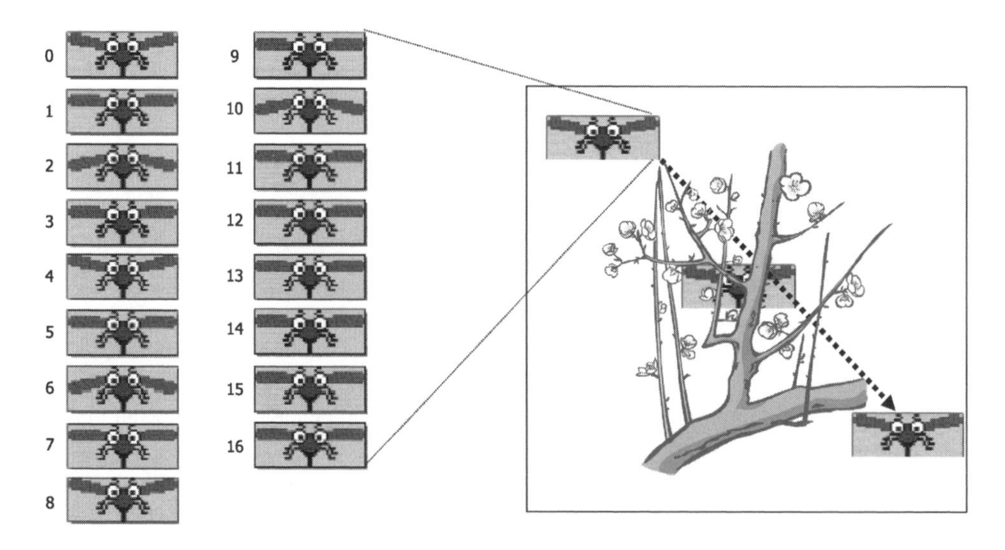

Figure 11.30 Sprite animation is achieved by using subframes in the sprite image object.

If the entire sprite were the large rectangle, then we can nominate the background colour of the sprite to be transparent. This is called alpha channelling and is how we achieve layering with visual continuity between overlapping layers.

We can move a sprite's container frame around in order to create panoramic movement in the game. To achieve animation of a sprite local to its frame, we can move between sprite frames within a composite image object divided into subimages of equal size (and equal to the frame size), as shown in Figure 11.30.

11.10.4 Networking API

As we would hope, the MIDP API set does not fall short of expected networking support. The underlying CLDC platform has a powerful class of connection methods, and, quite sensibly, a particular focus of the MIDP interpretation is in providing an HTTP API. Significantly, and something that was lacking in MIDP 1.0, we now have a secure connection option via an API that enables us to establish HTTPS communications pathways (i.e., HTTP over TLS, or Transport Layer Security). This is in line with everything we have discussed in the book thus far to do with security concerns for various mobile service scenarios. As we have already established, the only viable solution for secure transactions is the provision of an end-to-end encrypted pipe, and this is now possible with MIDlets.

It is also possible to work at a lower layer in the Transmission Control Protocol (TCP)/IP stack, with the ability to form datagram streams from our MIDlet. This would be useful for media-streaming applications, although it remains to be seen what kind of media processing any given device could sustain. Something complex like MPEG-4 decoding might not fly on many slower devices.

A new addition to the MIDP 2.0 specification is the provision of an API to connect with a logical serial port on the device, whether that manifests as an actual physical port, such as an RS232 link,[34] or an on-board port to a peripheral, like an Infrared modem, or remains a logical serial port (emulated by underlying system software, such as a Bluetooth™ driver). Presumably, we could use a serial link to talk with a Bluetooth™ peripheral, although the Java Community is working on a separate CLDC-based API set (JSR-82) specifically to handle Bluetooth™ peripherals.

11.10.5 Securing the APIs

With MIDP 1.0, all APIs were available all the time to the MIDlet. The design philosophy of MIDP 1.0 was also a sandbox approach whereby the MIDlet could not access system resources directly, only indirectly via the basic API set. It would be a useful facility to be able to access other resources on a device via extension API sets, whether vendor-specific or not, or via the expected evolution of the MIDP itself, or via other APIs that the Java Community process might make available for vendors to optionally include in their Java implementations. We can think of MIDP 2.0 as enabling access to the device from the KVM, but in a sanitized and controllable manner. While this initially seems a bit odd, the idea is nothing new. In the desktop world it is unusual that all programs running on a computer are allowed by the OS to access all its resources for any user. The concept of restricted access rights has been a necessary and essential part of the software industry for some time. This concept has now been carried across into the mobile world in order to allow various usage and security policies to be implemented. How they will end up being implemented and by whom is another matter. But, as an example, let's say that a particular device has an API to allow location information to be accessible to MIDlets; this API may be something that the mobile operator introduced to a device with a view to controlling its usage. With MIDP 2.0, this API could be restricted to be only accessible to MIDlets provided by the operator.

[34] An RS232 link could be used to connect a mobile device with a PC, such as via a cradle. It may be useful for a MIDlet to gain access to this link, despite the OTA ethos of MIDlets.

11.10.6 Push Mechanism

An exciting new feature of MIDP 2.0 is the *push registry*. This is a means to allow an otherwise dormant MIDlet (i.e., one not currently running on the device) to start running following an event on the phone: namely, the receipt of a PUSH message or the occurrence of a timer event. Any of the MIDlets installed on a device can register with the push registry to ask for notifications of particular events of interest. An API is used to carry out the registration process. There are many interesting applications for this technology. An obvious example is the implementation of an email or PIM client that can receive alerts of changes to a central mailbox or PIM database, thereby allowing real time synchronization to occur. For example, if an email message arrives at the central mailbox, such as one hosted by Microsoft Exchange™, then some kind of server mechanism could push an alert to the mailbox user's mobile device, perhaps via text messaging. Using the push registry, the alert can activate a particular MIDlet on the user's device, even if the MIDlet was not currently running on the device. This would allow a constantly synchronized PIM application to be implemented using MIDP technology. Not only could email be notified but so could changes to a central calendar or contacts database. The technology is useful for customer relationship management (CRM) applications, such as maintaining real time inventory information for field sales representatives.

The significance of the push registry is that it allows MIDP to take on new types of application tasks not formerly possible. It also means that MIDP has moved yet another step closer to the capabilities of the classical embedded approach, although this particular ability – to bring a dormant application to attention – is not necessarily possible on all mobile device platforms.

11.11 MIDP OTA DOWNLOAD MECHANISM

The mechanism for downloading a MIDlet to a device takes place using HTTP or WSP (Wireless Session Protocol), as shown in Figure 11.31. The objective of the process is to download MIDlets in the form of JAR files from a central source, such as a database or file server, placing them into the JAR memory space in the device. As previously mentioned, a portion of device memory belongs to the JRE and is used for storing MIDlets. The file format is a compressed form of the compiled class file, using the ZIP™ file format.

The Java Application Management System (AMS) is responsible for the downloading and the process, which also includes installation of the MIDlet class file on the device. This involves the JRE becoming cognizant of the fact that it has a new class file in its memory, courtesy of the application manager, and making it available to the user via the part of the UI from where MIDlet applications can be launched. This gives the user the ability to run the newly installed MIDlet. The Java application manager will provide a means for the user to run the MIDlet.

Figure 11.31 shows us the server component responsible for managing MIDlets as a centralized resource, which might well be a J2EE platform. This 'MIDlet manager' or 'vending machine' could be a bespoke application or an off-the-shelf *service delivery platform* (SDP) aimed at content management, including games 'vending' (assuming the MIDlets are games[35]).

[35] The OTA download principles are the same for games as for any other type of MIDlet. An SDP that specifically caters for games will have additional features that enable payment to be made and extra levels of gaming difficulty to be charged for, etc.

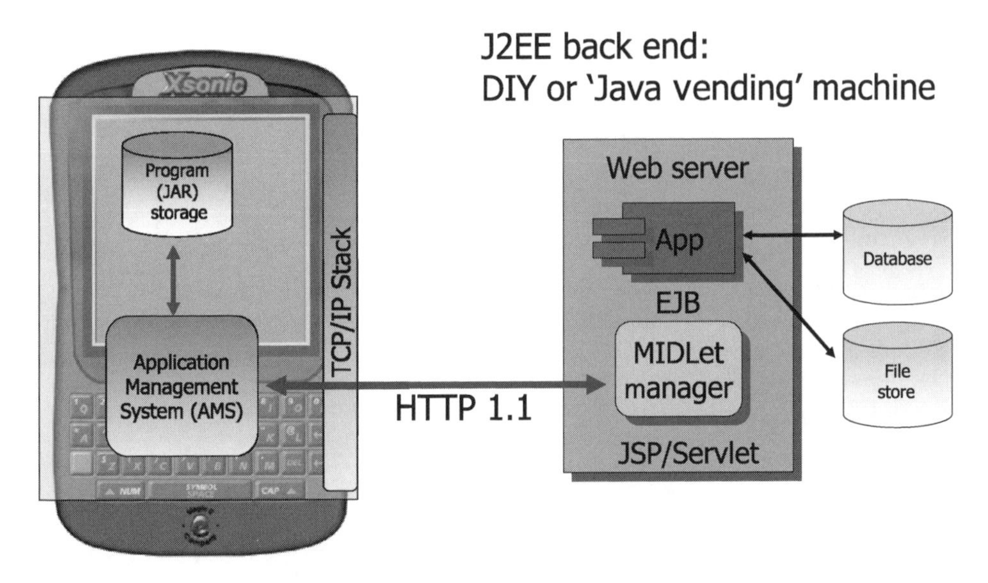

Figure 11.31 Basic MIDlet download architecture.

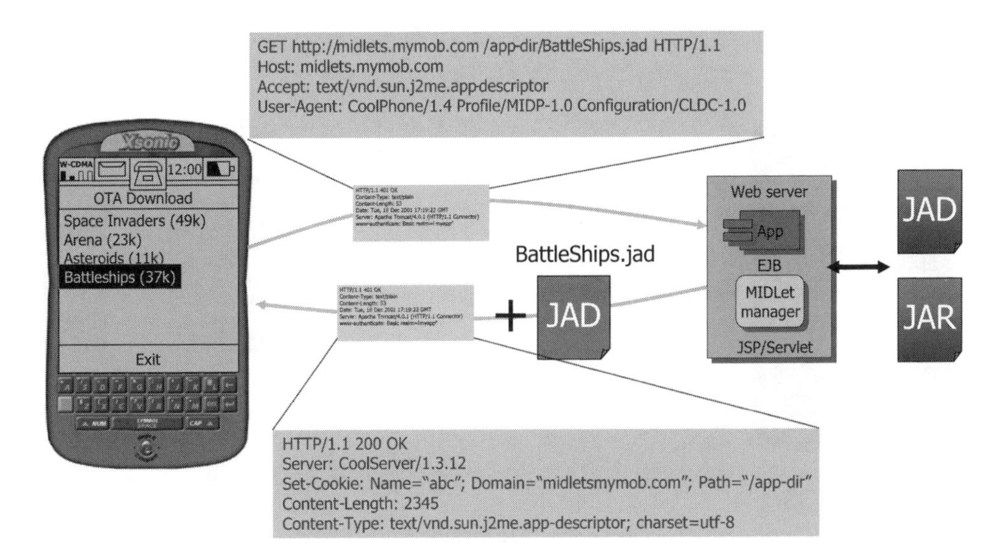

Figure 11.32 Downloading the JAD file first.

As Figure 11.32 shows, the first thing that the application manager does is to download a JAD file. The JAD describes the essential attributes of a uniquely associated JAR file that contains the MIDlet. One of the functions of the JAD is to inform the user and the device about the size of the JAR file. The application manager should then indicate to the user whether there is enough spare memory on the device to download and install the referenced MIDlet. The JAD also contains the URL that points to the location of the JAR file. If the user decides to download the JAR file, the application manager uses HTTP GET to fetch the MIDlet file, as shown in Figure 11.33. In Chapter 7 on IP protocols and earlier on in

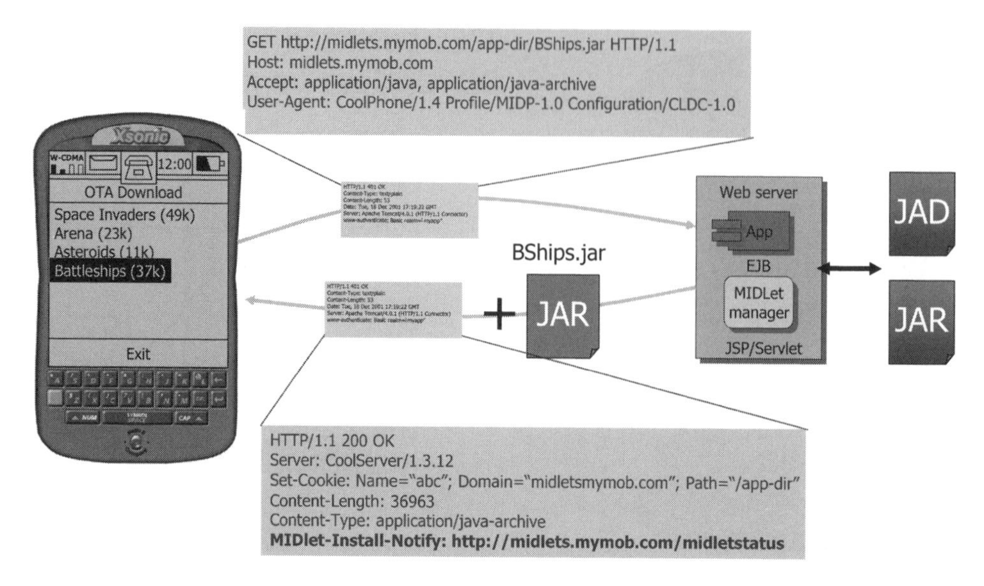

Figure 11.33 Downloading and installing the JAR.

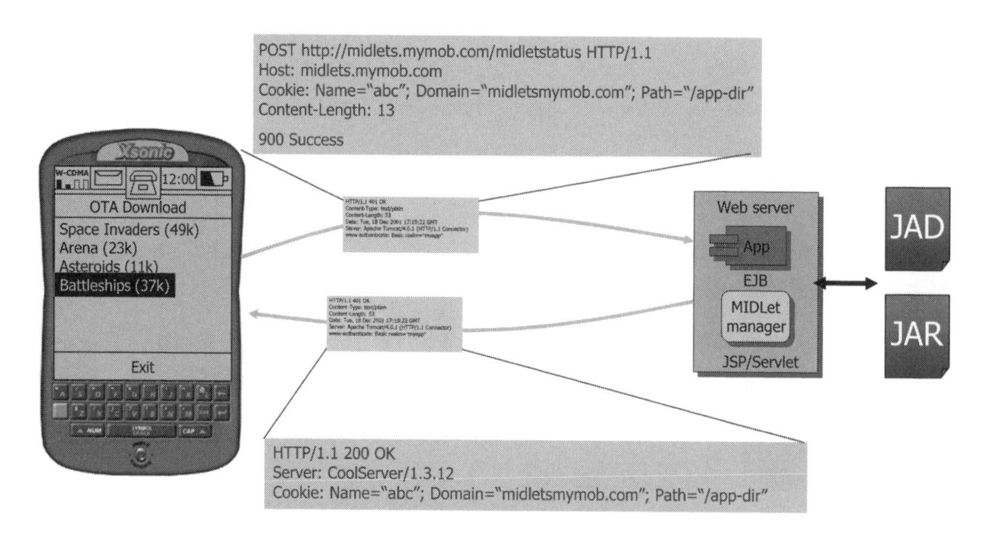

Figure 11.34 Reporting the status of the JAR installation attempt.

the current chapter we saw that the download process can use SAR to enable a successful download in the face of data errors on the link.

Figure 11.33 shows the HTTP header that the 'vending' server returns as part of the JAR download response. Within the header the field *MIDlet-Install-Notify* contains a reference to a URL. This location is where the application manager should send (POST) a status code[36] back to the server using HTTP POST, as shown in Figure 11.34; some of the possible status codes for the download process are shown in Figure 11.35.

[36] Note that these codes are similar in principle to the status codes used by HTTP, although they are entirely different codes with independent meanings.

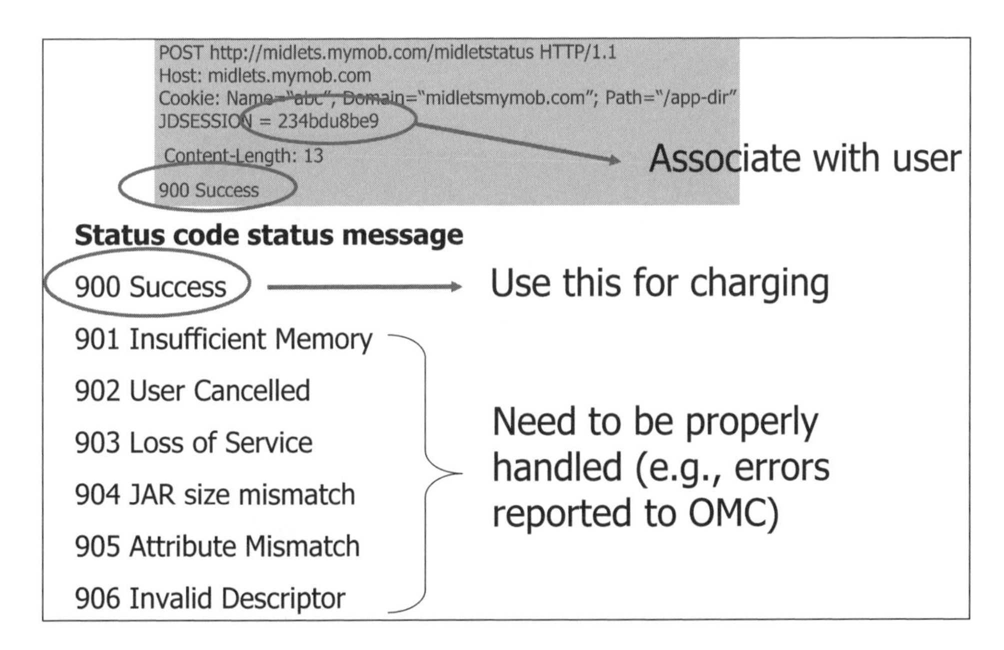

Figure 11.35 Importance of status codes when downloading a MIDlet.

Notice that a cookie value is part of the POST response from the application manager, reflecting a cookie sent earlier by the 'vending' server during the download process. This cookie enables the 'vending' server to identify uniquely a particular download session and its user, allowing the association of download status reports with their users. This was the recommended practice for MIDP 1.0; but, MIDP 2.0 states that a cookie is not mandatory (nor recommended) for this process. Instead, the URL used to post the status code should be rewritten to include unique session information, as per the techniques discussed in Chapter 7 on IP protocols (e.g., http://www.midletsmymob.com/midletstat?session=a23sdn32n9fhu).

It is an important part of the overall mobile service to report the download status codes back to the server. The common assumption is that the value of the status codes lies in being able to detect a successful download in order to charge the user. However, this only makes use of one of the codes. A fault management system should log these codes and make them accessible to application support staff and help desk staff. It is likely that throughout the course of potentially thousands of downloads some errors will occur during the download process, including human errors, such as trying to download a file to a device with insufficient spare memory.[37] In such a situation a user may place a call to a help desk, and, ideally, the help desk operator should have meaningful access to an interpretation of the status codes. This will greatly assist with customer care initiatives by helping to identify and eliminate common errors.

Earlier in the book we looked at an example of a J2EE system for managing a video-streaming application. Figure 11.36 reminds us of this application in its basic form; but, this time we have added an MIDlet download facility to enable MIDP-capable devices to

[37] On modern devices it is expected that the AMS itself will prevent the download of a MIDlet that is too large for the available memory space.

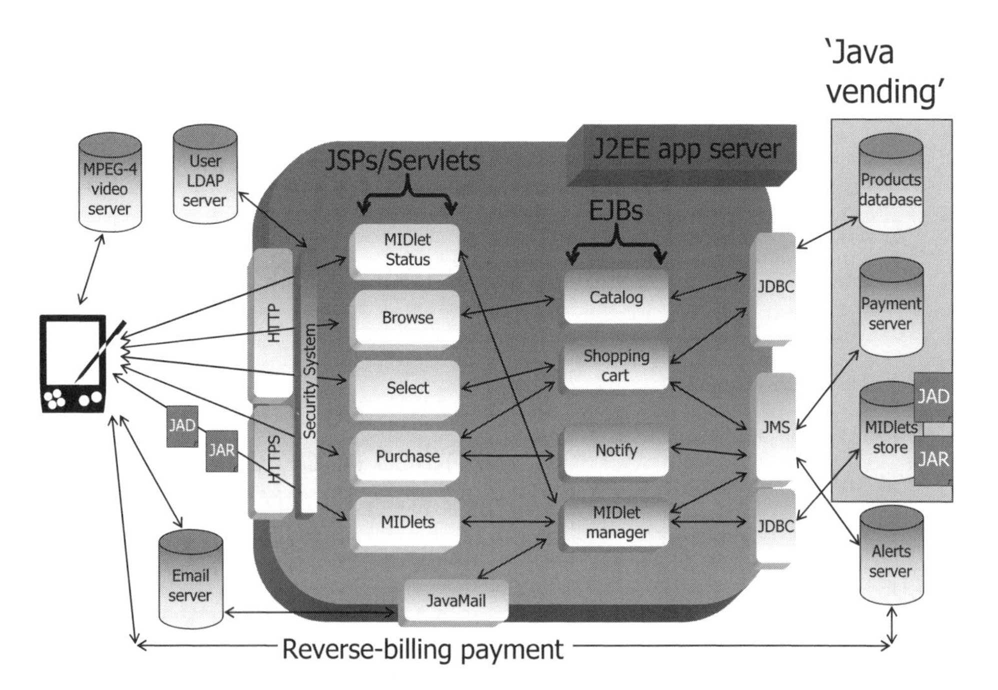

Figure 11.36 Possible architecture of J2EE system with MIDlet download.

download an MPEG4 video player in order to use the video juke box service. This would be useful for users with devices that don't have an MPEG4 player.

Notice that a text-messaging server facilitates payment for the service via reverse-billed text messaging. Otherwise, the application is the same as before. The ability to first offer a MIDlet to the potential user shows how new services can be offered to users without being concerned about whether their device can handle the information or media format, such as MPEG-4 video in this case. With the use of MIDlets, it is entirely feasible to provide the user with the application needed to enjoy the service. Furthermore, this automated process can be carried out remotely OTA. On RF networks with high-speed access this download process would not incur excessive delays. This may add to the success of OTA provisioning in the future, allowing for applications that are more complex to be downloaded. However, the sensitivity (intolerance) of users to download delays should not be underestimated: it is certainly the case that in the wired world the success of application downloads declines with size. This is usually because the user gets frustrated with a large download taking too long and subsequently cancels it or they are loath to retry if it all goes wrong. SAR can ameliorate the latter case, but care is still required to avoid overly large file sizes.

12

The RF network

Computer networks are now able to utilize a variety of wireless connection technologies. In this chapter we are interested in wireless connectivity that facilitates portable or mobile computing,[1] although our main interest is in wide-area (cellular) solutions. An increasingly diverse range of wireless solutions provides users with potentially ubiquitous access to services, including the Internet. As already discussed throughout this book, the evolution of Internet-oriented software technology, like J2EE, XML and Web Services are helping to make the Internet become an incredibly powerful delivery mechanism for services. The browser paradigm (or client/server, or even HTTP-centric paradigm), clearly has much strength as a means to deliver software services, an assertion amply discussed and elaborated in this book. It is time to look at how to sustain such paradigms while mobile, thus fulfilling a vital part of our mobile services definition from Chapter 3, wherein we wrote:

> The ability to interact successfully, confidently and easily with interesting and readily available content, people or devices *while freely moving anywhere* we are likely to go in conducting our usual day-to-day business and social lives.

Continuous wireless access to the Internet is now possible on a wide scale thanks to cellular networks. Providing a cellular connection that has a semblance of IP (Internet Protocol) compatibility has been possible since the advent of second-generation (2G) networks, like GSM. Better still, in their adapted versions the so-called 2.5G networks (e.g., GPRS), IP communications has been possible in a much more efficient manner. However, these attempts to scale up to a true IP-like environment were only ever going to be short term and insubstantial. The underlying infrastructure and its technical basis came about in response to the need to talk, not the need to exchange data. For large amounts of IP-based traffic, 2.5G

[1] Wireless has plenty of other applications, such as connecting one building to another.

Next Generation Wireless Applications P. Golding
 ISBN: 0-470-86986-0 (HB)

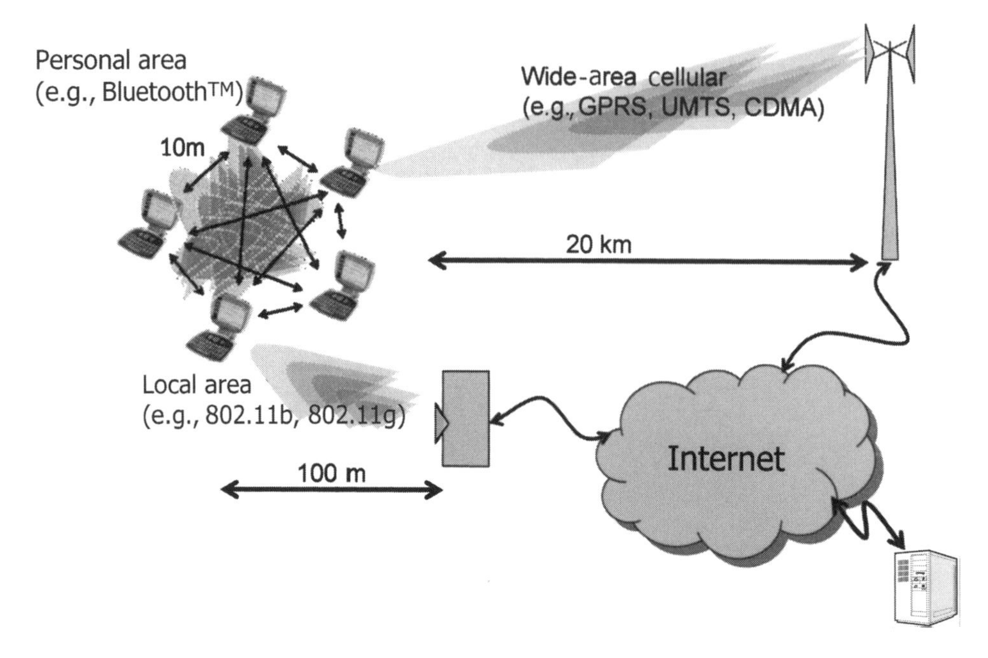

Figure 12.1 Heterogeneous and ubiquitous wireless access.

networks are very limited. However, cellular technology has entered into its third generation of standards and technology: the so-called *3G networks*. These are wide-area solutions built for voice and data. Devices can communicate over distances of many kilometres while maintaining useful battery life and portability. In many countries, roaming devices will always be within range of a 3G *base station* – the access points to the 3G network. The 3G network itself can carry data back and forth between the devices and the Internet or any other packet data network.

Imagine an area being 'illuminated' with RF coverage from 3G base stations. Underlying that umbrella coverage will be pockets of local area wireless-networking coverage, like 802.11b (WiFi) or 802.16 (WiMax). The lower range of these localized islands means a much faster connection speed than with current 2G standards, shifting greater quantities of data with similar power and portability constraints as 3G. These technologies complement each other and can run side by side, as shown in Figure 12.1, even on the same mobile device. At an even greater level of intimacy, devices can directly talk to each other within the vicinity of a few metres, as offered by personal networking technologies, like Bluetooth™ and WiFi.

Throughout this book we have looked at a wide range of technologies. In doing so we have examined various topologies, paradigms and possible modes of operation of those technologies. It should have become clear and we should feel 'comfortable' with the fact that there is no 'right way' to solve a mobile services problem. Indeed, since for the most part we don't really yet know the nature of many future mobile services, it is not possible to say which approach is more 'right' than another. There are many possible definitions of mobile services and there are many possible configurations, including as just alluded the Radio Frequency (RF) network. In the remaining two chapters of this book I have chosen to focus on the wide-area coverage offered by cellular networks. This is justified because cellular networks offer much more than just wireless connectivity and it is important to understand the wider potential of the cellular network within the context of other topics discussed throughout this book. For example, accessing the cellular network's core network assets, such as voicemail, text messaging, call processing, charging mechanisms, will become increasingly important for next generation services. These assets will form part of the service delivery platform that we have attempted to define at various points – albeit loosely because that's probably an honest reflection of the fluidity of this business right now. Therefore, we need to understand how the availability of these assets might happen. We have already alluded to it on several occasions, such as our discussion of Web Services and Parlay, a topic that we will explore more thoroughly toward the end of this chapter.

One of the assets of the cellular network is the ability to locate (geographically) the mobile device. This potential seems so powerful that it warrants special treatment and, therefore, is the main theme of Chapter 13, which closes with a discussion of how the various ideas developed throughout this book might come together to enable exciting new service themes in the near future. This seemed an apt way to end the book for two reasons. First, it is a gathering together of ideas discussed throughout the book in order that we might gain some feel for how to 'connect the dots' on our stellar map of next generation technologies. Second, it also underlines a previous and important point about becoming 'smart integrators'. Perhaps we might be fooling ourselves if we look too hard for the killer app or if we think that such an app (or even the 'killer cocktail') might revolve around a single or dominant technology. I cannot say whether this will end up being the case or not, but it does seem to me that we should be preparing ourselves to integrate many technologies in order to realize interesting services; thus the book ends by returning to the theme of becoming 'smart integrators'.

We start the conclusion of our journey through the constellation of next generation technologies by examining what a cellular network does and what it can do for our mobile service aspirations.

12.1 THE ESSENCE OF CELLULAR NETWORKS

At its simplest the RF network layer of our mobile applications network is simply a big switch. It can route information, voice or data from fixed information appliances to moving ones, as shown in Figure 12.2, or from one moving appliance to another. Wired data communications connections are on the fixed network side, while on the mobile side the connections are made using RF.

The degree of movement we are interested in supporting should ideally support the scope of the mobile applications and services we have been considering thus far. This is the reason we restated at the start of this chapter the aspect of our mobile services definition in terms of its implications for what we might require from the RF network.

The freedom to move anywhere (ubiquity) is a motivating concern. We have vividly observed that allowing voice calls to be placed anywhere at anytime has amply proven itself as a liberating service for a large number of people. Those 'crazy' inventors of cellular telephony have vindicated themselves and their audacity. The extent to which ubiquity is important is not really a matter of debate, but how to achieve ubiquity often is. Some pundits argue that WiFi hot spot coverage is sufficient for many applications, whereas others argue that the contiguous coverage of wide-area cellular systems trumps WiFi in most cases.

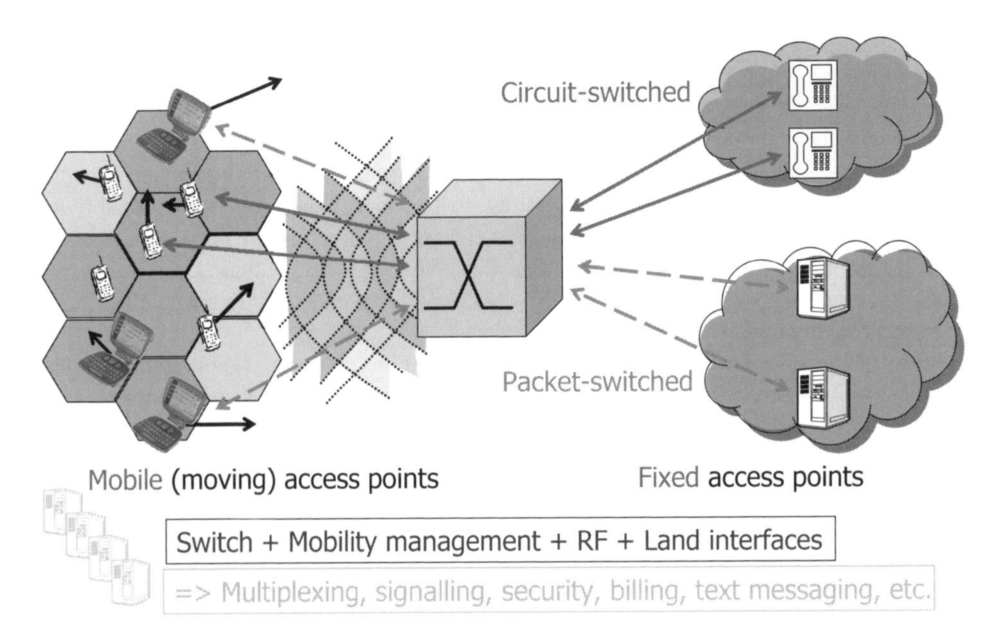

Figure 12.2 Essence of a mobile network.

New voices with new ideas, like WiMax, have yet more opinions. In this book we are not interested in the debate, but we are interested in the implications, which are to do with being able to cope both with any of the access methods and with any possible interaction between them, especially regarding how this might affect our use of mobile applications technologies.

Being faithful to our mobile services definition, our philosophy and approach toward building applications should encompass any access method if this is what it takes to enable successful end user engagement with mobile services. It also seems entirely possible and likely that mobile services will develop from both ends of the spectrum (no pun intended) and then often meet in the middle; WiFi applications and 3G applications will converge, overlap and augment each other. Undoubtedly, successful WiFi applications will emerge – especially as it is relatively straightforward to deploy WiFi infrastructure. It doesn't require a government licence for transmission and it is cheap as well as extremely low-tech compared with a cellular solution (to WiFi's credit). On the other hand, there is a lot of appeal in developing applications to deploy on 3G networks, irrespective of the ubiquity consideration, simply because the potential user base is so vast and there are clear, well-entrenched precedents for charging users for services, not forgetting the existence of mechanisms to actually make the charges!

12.1.1 RF Network Convergence

These two access methods – WiFi and 3G – are bound to converge, more than collide. The attributes of WiFi services that might make them attractive will emerge and become familiar to users, such as their high speed and apparent cost-effectiveness for large data transfers. Similarly, users will also come to understand the meritorious attributes of 3G

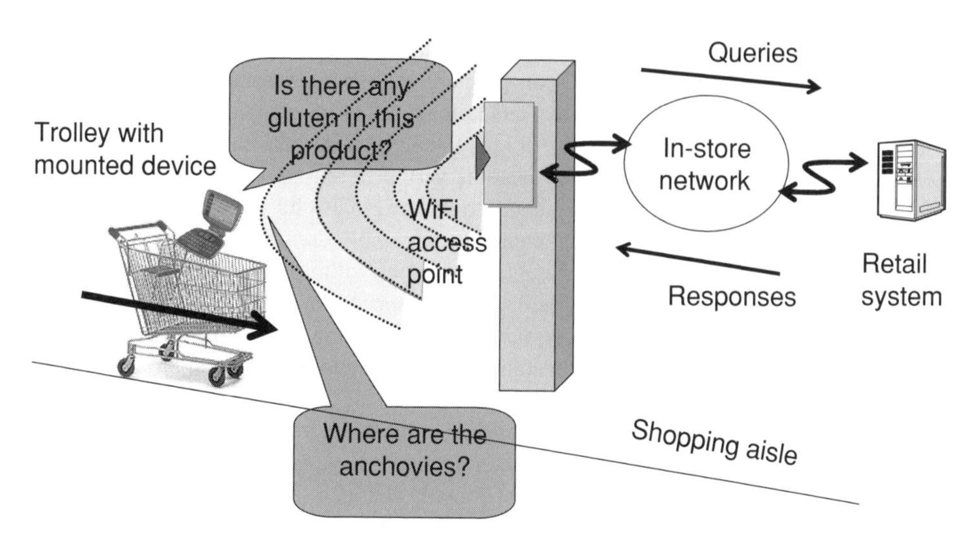

Figure 12.3 A wireless shopping assistant system using WiFi.

that will emerge, such as its ability to determine user location. It often goes unnoticed by the pundits that it is entirely possible to have the best of both these worlds. Dual-mode devices are not only technically possible they are both economical and entirely desirable as well.

Let's ponder for a moment on the idea that each technology has its place. As an example, imagine that we are interested in providing wireless shopping assistant services to mobile terminals while inside a supermarket.[2] As the shopper moves around the shop, we know their exact location (to within a particular aisle) and we can offer all kinds of services accordingly. We could build such a system using the browser paradigm as well as many of the solutions and technologies discussed in this book, even though the freedom to move anywhere in this case starts and ends at the supermarket door. A cost-effective solution for this system would be to use WiFi to connect the terminals to the retail back end system, as shown in Figure 12.3. With WiFi access points at regular intervals inside the shop, it is possible to locate shoppers quite accurately.[3] The best that a wide-area solution could do is place the shopper inside the shop, but nothing finer in terms of spatial resolution.

Many solutions like this will probably emerge; WiFi services are limited in coverage to a particular place or collection of similar places (e.g., all stores in a chain), although not necessarily for the reason just mentioned (i.e., fine resolution location finding). However, we live in a changing world – a world of convergence, to use that often-ambiguous expression. Despite its overuse (abuse), it does have some value and relevancy here. For example, no sooner have we defined and built a service that works in our supermarket when along comes a bright spark who suggests extending it into the host shopping mall itself. Perhaps someone has identified that the commercial interests of the supermarket converge with serving the

[2] There are various ideas like this under review by major supermarket chains, such as providing services via terminals attached to shopping trolleys.

[3] Indoor positioning is an area of intense research and is being carefully scrutinised by retailers. For a useful summary see 'Retrieving position from indoor WLANs through GWLC', by G. Papazafeiropoulos *et al.* which can be found at http://www.polos.org/reports.htm

general interests of shoppers in the mall. We can think of this as a convergence of opportunity, in this case influenced heavily by common geography (co-location) and common consumer interest (shopping). However, the original technological solution scoped for the supermarket suddenly might need a rethink in terms of how to extend it into the shopping mall. At this point, perhaps the economics suddenly become very different and the coverage area and footfall[4] no longer lend themselves so readily to the original solution.

Our degree of movement (mobility) changes, potentially dramatically, between the relatively constrained yet partially predictable movement within a shop and the wider less knowable movement in the shopping mall. After the success of the shopping mall service, someone might then make the connection with trying to lure potential shoppers into the retailer's den, an altogether different technical challenge and now more than likely in 3G territories.

On the other hand, a critical examination of the dynamics of the in-shop service might suggest that such a service can only work in conjunction with supplementary services, like an associated and already established Internet shopping channel. Perhaps the implementation of electronic coupon redemption on phones has become more widespread than anticipated, forcing our in-shop service to support such a scheme; this is convergence at the service level. One service, like mobile electronic coupons, ought to work with another, like wireless shopping trolley assistants. The availability of coupons may be a key driver that attracts shoppers; so, the discovery of available coupons cannot be limited to shoppers already inside the shop: coupons need to be discoverable and available outside the shop – which means another degree of mobile freedom is required in our application.

There are yet other areas of possible convergence. Cost-effective delivery of the in-shop service requires enough customers to reach critical mass; so, it is probably only viable to offer such a service on customers' *existing* personal devices or at least in conjunction with them. However, newer devices – with better displays, the ability to watch video-streaming adverts and to receive images – will make them more suitable for the shopping service. In other words, we require technological convergence. Perhaps it is also possible to reward shoppers with credits on their mobile phones in proportion to the amount spent in the shop, so we have further convergence; this time between the service fulfilment ends of the two systems, perhaps executed using Web Services.

Our shopping assistant service now seems to require accessibility from many different places and devices and to a variety of back ends. From these considerations, possible system architectures begin to emerge, at least at its highest level. We end up with a system (see Figure 12.4) that looks remarkably like the RF network model we proposed earlier (see Figure 12.2). However, the access methods are heterogeneous. If a single operator is able to offer both access methods (a likely development, already seen in places like the USA), then a common core network (routing function, authentication) will already be in place. However, if there is no common service provider (an equally likely scenario in many cases), then we need a different means of service convergence, such as Web Services across the Internet.

In our shopping example, mobility has two dimensions or, more accurately, two levels of resolution: fine-scale movement within locations and the more coarse-scale movement between those locations. We are going to need to support this type of mobility in the future. However, this challenge is not without other considerations. We have to take into account

[4] Footfall is a retail term for the number of feet (i.e., people) who walk into the shop (or shopping area).

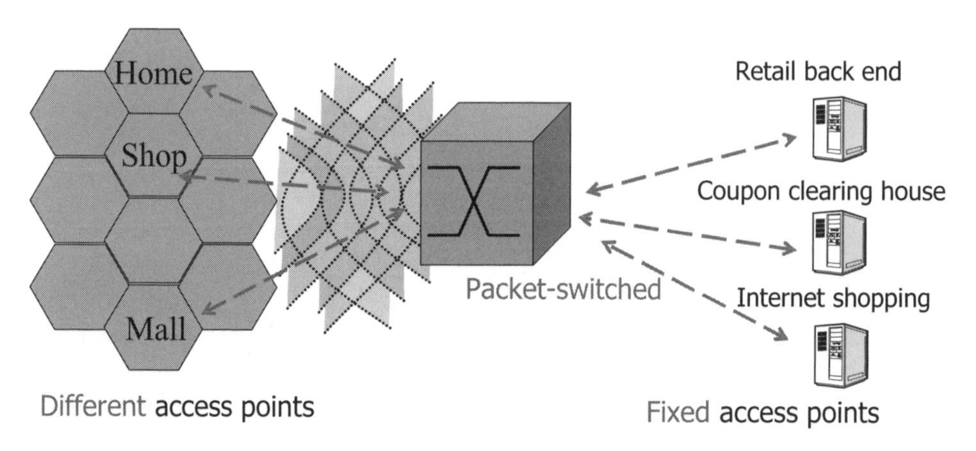

Figure 12.4 Useful services require many access points.

not just the freedom of movement but also the qualitative criteria that the opening part of our mobile services definition suggests: 'The ability to interact *successfully, confidently and easily . . .*'

Would requiring the user to own several devices, one for each location (in-store, in-mall and out-of-mall), facilitate an ability to interact easily? It seems unlikely. Multimode devices do seem inevitable in future mobile service architectures.

The foregoing discussion has highlighted the likely emergence of multimode RF services and devices. However, for reasons mentioned earlier, such as the importance of other capabilities of the cellular network, we will focus on cellular networks for the rest of this chapter. We need to expand on the notion of an RF network as a giant switch, as shown in Figure 12.2, in order that we can more fully appreciate its capabilities. Before doing so, we ought not to forget that we may need to support mobile application topologies that do not require a wide area network, such as P2P topologies and the close-range inter-device communications within the device network itself (personal networking). The RF networks we can use to support this are mainly Bluetooth™ and WiFi, but used in *ad hoc* mode they largely function as cable replacements. Given such an obvious function, all that remains is to understand the technical details of such connections. However, such details are not within the scope of our enquiry.

Looking again at Figure 12.2, the core of our network appears to be little more than the switching (routing) function and a means to support the RF connections themselves. However, this is an oversimplified view. Not only is it simplified in terms of how these components of the network actually work but, more critically, it in addition overlooks a good deal of mobile network assets and a host of other support functions that are key to many types of mobile service. These network assets either have been included since day 1 of 2G networks or slowly accrued over long periods of network evolution. Digital cellular networks are now very mature and most of them in their second decade of operation. They have therefore a wealth of assets that feature very strongly in the anatomy of the network, beyond an RF connection and a switch. As Figure 12.5 shows, cellular networks are no longer just wireline replacements with an ability to support roaming. Text messaging and multimedia messaging are prime examples of important mobile network assets that have become important service platforms and services in their own right.

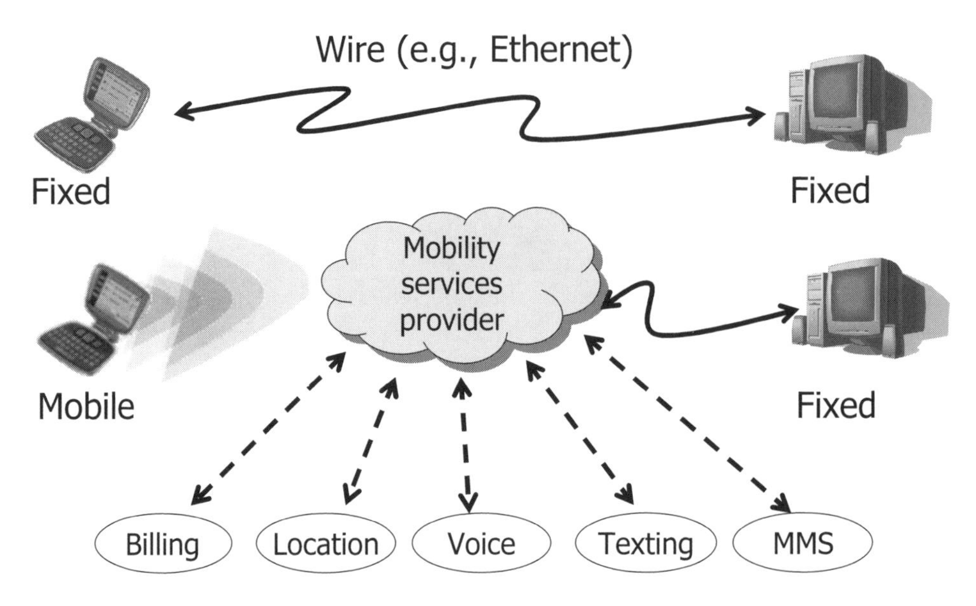

Figure 12.5 A mobile network is not just a wireline replacement.

Before we discuss these network assets, let's first look at how the RF and switching function is enabled. What is a cellular network?

12.2 THE RADIO PART

Some books tend to place a lot of emphasis and spend a lot of time on discussing the radio part of a cellular network. This is perhaps because the RF is what naturally feels like the essence of a mobile application or a mobile network. However, from an applications perspective, understanding the actual RF part is perhaps the lowest priority in understanding the entire network of networks that we have been traversing throughout this book. As we pointed out in Chapter 4, the overriding challenge for the RF component is for it to be as transparent as can be, meaning that it is actually as 'wirelike' as possible in terms of its information-handling attributes. To an extent, one could treat the RF component as such and ignore its characteristics and vagaries. As we will see, that would probably be a foolhardy approach to mobile service design. Perhaps the most crucial parts of the RF network to understand are its boundary with the IP network. This often seems to be a source of confusion: how do we get from the IP world to the RF one?

We do not wish to belittle the RF network. As we will come to appreciate, the processes and mathematics that underpin RF transmission technology are by no means trivial. The information theorists and communications experts have invented something quite incredible. However, given a functioning RF network, the greater challenges in application system design are seldom to do with the RF: they are probably more likely to do with the configuration of the network assets (like the Short Message Service Centre) and with how to connect to them.

That said, an appreciation of the RF is useful, especially the ways in which its behaviour can affect application performance and, ultimately, our approach to design. Understanding

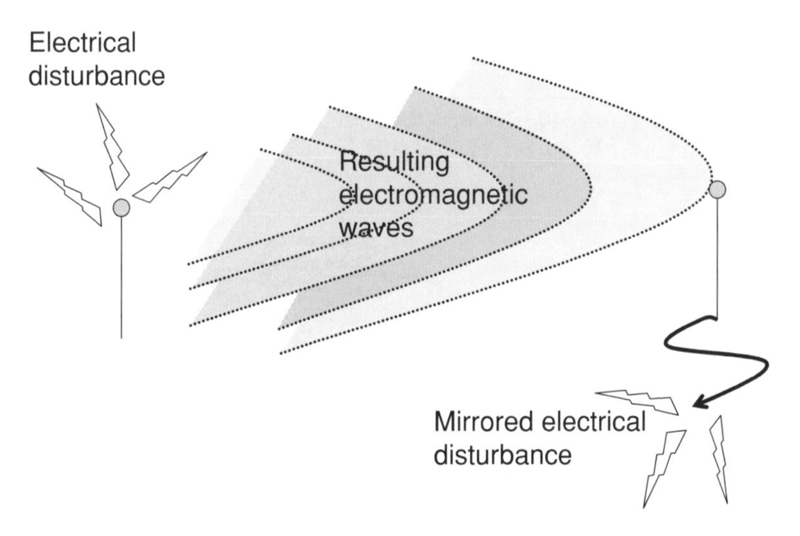

Figure 12.6 Electronic disturbance (excitation) felt at a distance.

the essential characteristics of a cellular network is the emphasis we will place on understanding the RF network, rather than a rote description of the different cellular standards, such as GSM, GPRS, CDMA, UMTS, etc. There are plenty of good books on these networks and their details.

12.2.1 Basic RF

To cut a long story short and to avoid physics history lessons, certain electrical disturbances in a metallic conducting rod (antenna) can produce an effect at a considerable distance from the point at which they occurred, as shown in Figure 12.6. The conveyance of the effect is by electromagnetic energy that travels outward from the disturbance source as invisible waves moving at the speed of light. We all know this effect well by tuning into a radio station on our transistor radio; we benefit from the effect in the receiver by converting it into sound energy that comes out of the loudspeakers. I make special mention of the word 'transistor', because this is possibly a much more profound scientific invention than that which makes modern RF communications possible. This is particularly so when millions of them are combined to make a silicon chip that can process RF signals using digital signal processing (DSP) techniques. We made mention of the importance of DSP in Chapter 10 on devices.

Thinking of the radio is probably a good place to start. We probably already appreciate that the radio effect can take place within a band of possible frequencies (the frequency being the rate at which the disturbance takes place, such as 900 million times per second, or 900 Mega[5] Hertz[6], or 900 MHz). We divide RF frequencies into bands[7] that we use for transmitting information from one source to its destination or destinations.

[5] 'Mega' is from the Greek word *megas*, which means large, but in scientific measurement refers to 'times one million'.

[6] Hertz was someone's name (Heinrich Hertz) and is a unit of measurement that indicates one cycle in a periodic event, like a repeated RF disturbance (excitement).

[7] Although we tune our radio into a precise frequency to get a particular radio station, the information transmission process actually requires a small band of frequencies either side of the main frequency we tuned in to.

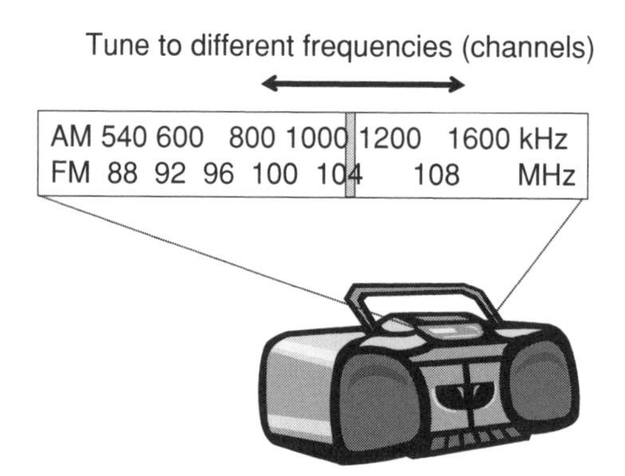

Figure 12.7 Radio spectrum divided into channels.

From basic experience and perhaps common sense, we know that by occupying a band with the output of a particular station prevents its use by another station. We can easily imagine what happens if two radio stations attempt to transmit on the same band ('frequency[8]'). Chaos would ensue and we would hear garbled noises, if we were indeed able to hear anything meaningful at all. Interference is the name given to this basic limitation of RF transmission. Most of what radio engineers do when designing radio transmission systems, especially cellular radio ones, is to invent schemes for maximizing capacity (i.e., the number of stations in our case) while mitigating interference; in effect, reducing interference and maximizing capacity are the same thing. We could say that this is the number one challenge – the RF *information capacity* challenge.

Those familiar with Citizens' Band (CB) radio (or any two-way radio system) already know one obvious solution to interference. We simply avoid more than one user transmitting on the same frequency at the same time. We divide the spectrum up and allow each user (or radio station, say) to use their own band, or *channel*, exclusively (see Figure 12.7). This is how we achieve more than one user accessing the RF spectrum or multiple concurrent access – hence why we call it frequency division multiple access (FDMA).

Exclusive use of a given band is either permanent, as with broadcast radio stations, or temporary, as with two-way radio. With two-way radio, we generally use a system on a first come, first served basis. If I'm already on channel 8 (say), then you can't use it until I finish. This procedure is what we call a *protocol*, and we need *channel reservation* protocols like this in order for a communications system to function in its entirety. Clearly, we don't remember checking to see who's on our channel when we place a cellular voice call. In this instance the reservation protocol is automatic, carried out by software running in the device and in the RF network.

The use of most bands in any given country requires a licence from the government in order to transmit. This requirement came about in order to stop chaos from ensuing in a 'free for all' situation, particularly because the chaos would affect emergency service transmissions with possibly onerous consequences. However, some bands are unlicensed. The chaotic situation can be managed using appropriate RF transmission techniques, such

[8] Although RF communications does take place across bands of frequencies, we say 'frequency' because the band is usually referred to by its centre frequency, which is why we hear radio stations as being on a particular frequency, like '99.8 FM'.

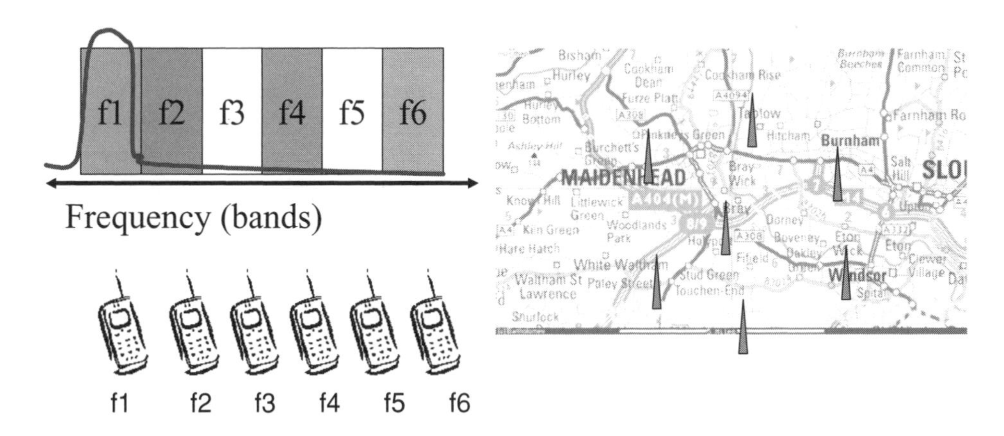

Figure 12.8 Assigning bands to each device.

as the spread spectrum method we will discuss shortly. It is because of these technologies or the ability to implement them cost-effectively with today's electronics, that some people in the USA, in particular, are contesting the need to have a licence. Their motivation in the main seems to be about removing restrictions that might stifle innovation and competition or about removing restrictions in principle on the unfettered use of a public common, such as the RF spectrum.

12.2.2 Building an RF Network

Given what we know so far, we can see that if we want to use radio signals to support a mobile phone network, then our starting point might be to assign a unique frequency to each mobile, as shown on the left of Figure 12.8. Here we can also see that each mobile actually transmits on a small range (band) of adjacent frequencies. In practice, due to the physical limits of RF circuit components, it is very difficult to confine the transmitter to a distinct, vertically edged band. As the figure shows for our device on band f1, there is some residual 'spill' into adjacent bands; so, there is a tiny bit of interference experienced by f2 in this case. This interference is called *adjacent channel interference* (for obvious reasons).

To enable our mobiles to communicate with the fixed network, we can install base stations in our service area, as shown on the right-hand side of Figure 12.8. Thinking of each base station as being like a broadcast radio station, we can assign a frequency for each base station to handle. What we end up with is a grid of base stations, each transmitting on their own frequency – the situation shown on the right-hand side of Figure 12.9. Radio waves propagate like ripples on a pond after a stone causes a disturbance. In the figure we can see these 'ripples' propagating out from the base on frequency f6. Clearly, each base station is also causing a similar radiation of waves from the centre; the hexagonal cells are an idealized way of representing each region served by a particular frequency. We can see that f6 radiates way beyond its immediate vicinity: the exact distance related to the power or the transmission is measured in Watts.

In the geographical region immediately next to f6, it would be folly to use the same frequency (f6 again) as clearly the signals would interfere with each other. This is why adjacent cells transmit on a different band. However, as we can see from the figure, the waves

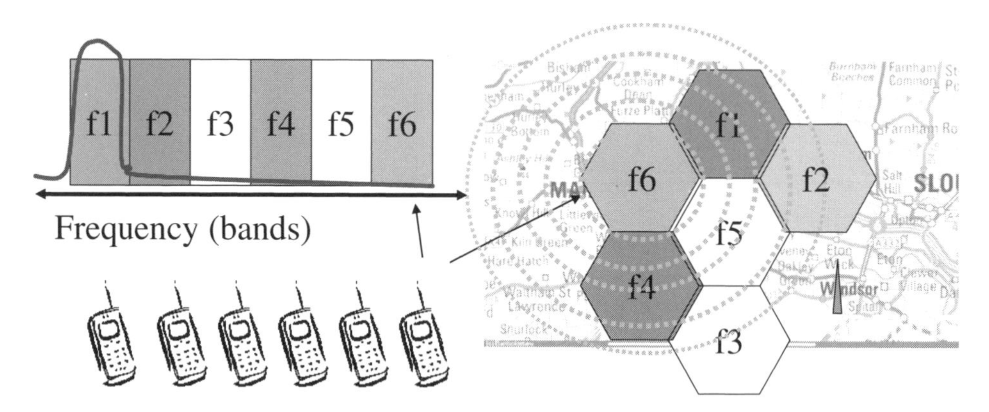

Figure 12.9 Allocating frequencies to regions.

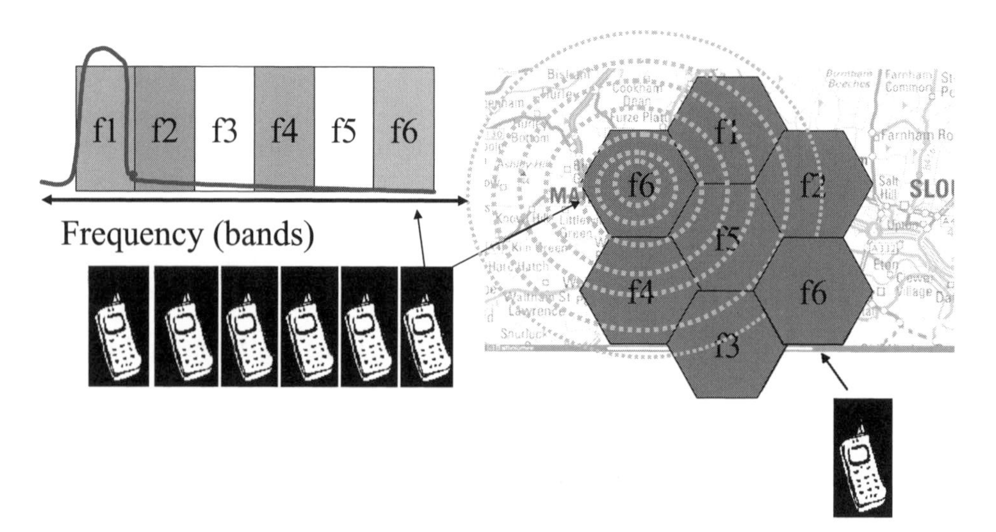

Figure 12.10 Frequency reuse at a safe distance (notice f6 is used again).

spread out a long way from the base station. Just as light from a torch weakens over distance, the reach of the RF signal causes them to weaken until the signal eventually becomes quite weak. If we go far enough away from the base station transmitting on f6, we would eventually be able to use the same frequency again, without any fear of detrimental levels of interference on the same band (*co-channel interference*). There would still be a residual level of co-channel interference, but we can design our network in a way that seeks to minimize it.[9]

Moving far enough away from the first station using f6, we can install a second station on f6 and use that frequency all over again; this is called frequency reuse and is how we arrive at a patchwork of cells across a wide area, like a country – it is also the inspiration behind the name 'cellular' network. This frequency reuse concept is shown in Figure 12.10.

[9] Co-channel interference in a cellular network is one of the limiting factors that affects the capacity and performance of the network.

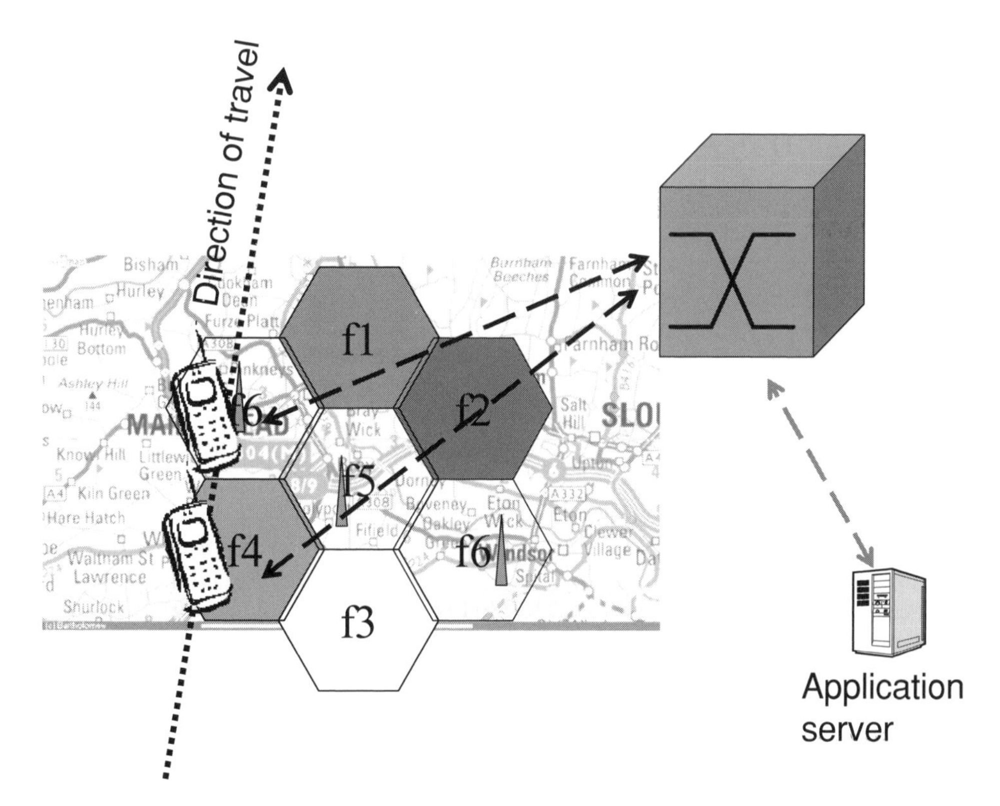

Figure 12.11 Mobile network switches (hands over) from one cell to another as the mobile moves through the network.

Our system would be extremely limited if we didn't have frequency reuse because after using our six bands our system would run out of bands and therefore capacity.

The basic operational scenario of our cellular system is that as a device moves through the service area, it swaps from one frequency to another, called *handover*, to keep a continuous transmission active. Clearly, this has implications:

- devices have to support operating on more that one frequency and have the ability to swap frequencies as required.
- the network has to support a mobile staying in touch from one cell to another, which means it has to hand over (i.e., from one cell to the next), as shown in Figure 12.11.
- if the mobile moves into an area where there is already a mobile transmitting on that band, then it cannot transmit on that frequency at the same time, otherwise it will cause potentially catastrophic co-channel interference.

With sophisticated software we can track mobiles and facilitate handover; this is not a problem. It is clearly important to track where mobiles are anyway, because if we want to initiate a communication session with one, then the cellular network needs to know where it is in order to route information to the right cell. Tracking mobiles for routing purposes is part of something called *mobility management*.

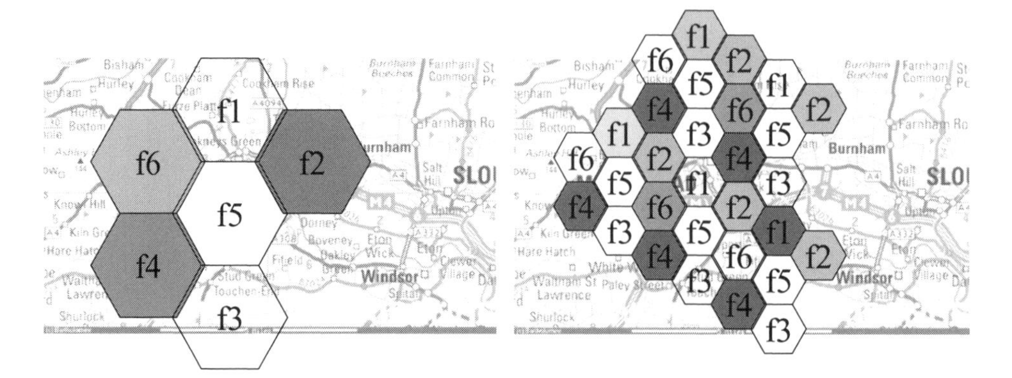

Figure 12.12 Smaller cells means more mobiles means higher capacity network.

Reflecting on what we have achieved thus far, we now have a basic design for a wide-area mobile device network, using cells to allocate frequencies and by preventing devices from transmitting on the same frequency within the same area (FDMA). However, if we look carefully at our cellular layout as shown in Figure 12.11, we see the system has some obvious and severe limitations: namely, a lack of capacity. If we can only handle one device per frequency per cell, then the people of Maidenhead (the town on our map) are not going to be very excited about cellular technology, as our service can apparently support only a handful of users. And the population of Maidenhead is more than a handful of people! There are two ways to solve this problem:

1. Make the cell sizes smaller.
2. Find a way for several devices to share the same frequency in a cell.

Shrinking the cell size seems a sensible proposition. In actual cellular systems, this turns out to be the main way of increasing capacity, everything else being constant (which, as we will see, it probably is once we have built the network).

We can cope with a higher density of mobiles (i.e., more customers) if we have a denser cell pattern, as shown in Figure 12.12. Clearly, the higher capacity does not come without a cost: namely, the costs associated with installing more base stations and connecting them back (the *back haul*) into the core of the mobile network (*core network*), such as the switch.

Smaller cells bring some added benefits to the end users. The amount of power needed to send an RF signal back to the base station is much less, simply because it is much closer. This means lower power consumption for the device, which in turn means longer battery life. In fact, cell size reduction is one of the major reasons that cellular handsets have improved so much in terms of improved battery life. The other reason for better battery life is the dramatic improvements in the electronics inside the handsets, mainly the silicon chips running off lower electricity levels (voltage), which means less energy consumption.[10] This also significantly benefits the standby time of the device as lower operating voltage causes a reduction in draining of the battery while in standby mode (from reduced *current drain*).

[10] Integrated circuits (silicon chips) are made of lots of transistors; these are tiny switches. Turning a transistor on requires energy in proportion to the voltage level it has to rise to. For lower voltages less energy is therefore required; hence, the battery drains that little bit less.

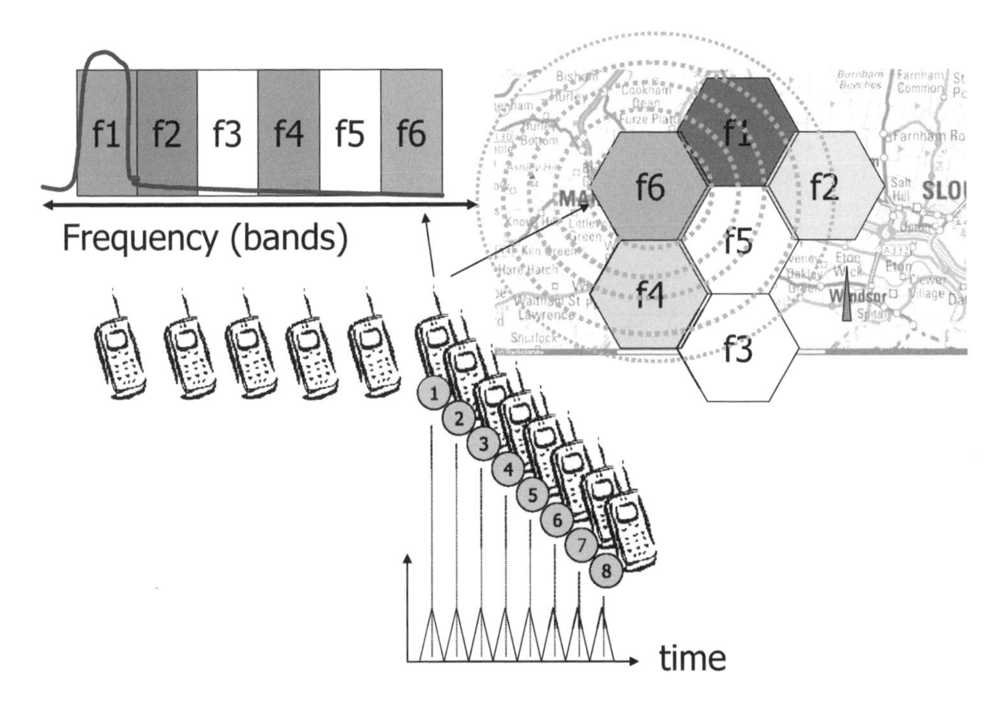

Figure 12.13 Mobiles take turns to transmit on the same frequency.

As mentioned, the second way of increasing capacity is to find a means to share the same frequency between several mobiles. This seems to contradict our earlier concern for interference, but it turns out there are some neat tricks to get round that problem – one of them almost mechanical in nature and the other very mathematical, almost magical.

12.2.3 Increasing Capacity using TDMA

TDMA stands for Time Division Multiple Access: the principle is simple and is shown in Figure 12.13. The figure shows that eight mobiles are all transmitting on f6 in the cell over Maidenhead, but not at exactly the same time. They take turns to transmit in quick succession in a continuous cycle: 1, 2, 3, 4, 5, 6, 7, 8, 1, 2, 3, ...

The cyclic nature of sharing the same frequency seems strange at first. Looking again at the radio station analogy, it would be like listening to our favourite station as a stream of quick bursts: an audio stream punctuated mostly by silence. If that were the case, then clearly such a proposition is unworkable. However, with digital cellular, this is not what happens. Because everything takes place digitally, the digitized speech can be compressed using digital processing techniques. We can imagine the mobile device as having a digital recording function (which, in effect, it does) and then passing the recorded digital samples through a compression algorithm, as shown in Figure 12.14. We can think of this process as chopping the recorded sound into lots of tiny sound files. A compression program then compresses each file down to a smaller size. In our example we use a compression technique that manages to squash any given sound file into one eight times smaller. For example, if we are recording in 1-second chunks, each input file might be 24 KB in size. Our compression

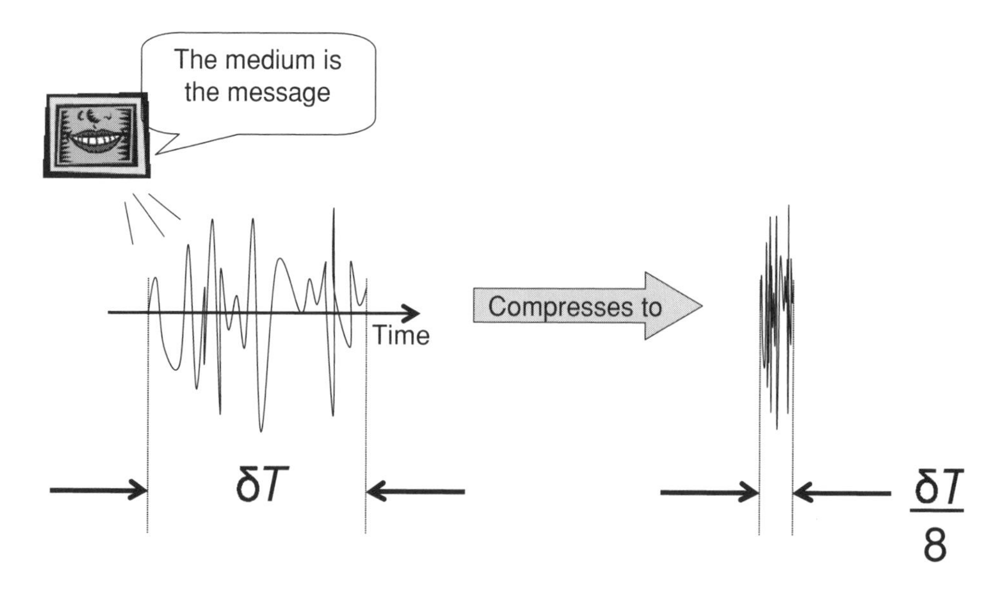

Figure 12.14 Speech compression means we take less time to transmit a given sample.

process takes these files and makes them only 3 KB each (which is 24 divided by 8). Our device continually takes input files, squashes them and then puts them into a queue in the mobile device's modem, ready for transmission.

What have we achieved with this compression process? Well, each compressed file now only needs one-eighth of the transmission time across the RF connection. If the mobile transmits each compressed sound file and waits its turn while seven other mobiles transmit (in succession), then by the time its transmission slot comes round again it will be ready with the next file. As long as the recording and compression process keeps up with the arrival of each transmission slot, the whole process will work smoothly: well, almost. What we need is a reverse process at the receiving end that can take the compressed files, uncompress them and then *seamlessly* stitch them back together into a continuous audio output stream. Using digital processing, this process is easy. After all, if dinosaurs can be made to appear in films,[11] then voice files can be stitched back together again.

The process of taking it in turns to transmit on the same frequency band is called TDMA; this is used for GSM (Global System for Mobile Communications) and GPRS (General Packet Radio Service). The principles of TDMA have been established for a long while and are not the preserve of radio systems; the technique applies equally to wired (or even optical) transmission systems. TDMA allows digital information streams to share a common channel; the streams do not have to contain speech. There is also nothing magical about the eight slots in our example either; I chose eight, as this is the number of subchannels per frequency originally supported by GSM,[12] which is probably the world's most prevalent and successful digital cellular system.

With GSM, a better voice compression technique emerged after the design, construction and delivery of live systems. Consequently, GSM evolved to support 16 mobile devices

[11] An altogether different proposition, I know; but, once we can get information into the digital domain and apply enough processing power, we can do some incredible things.
[12] Why GSM uses eight bursts per channel is a deeper technical consideration to do with a variety of parameters.

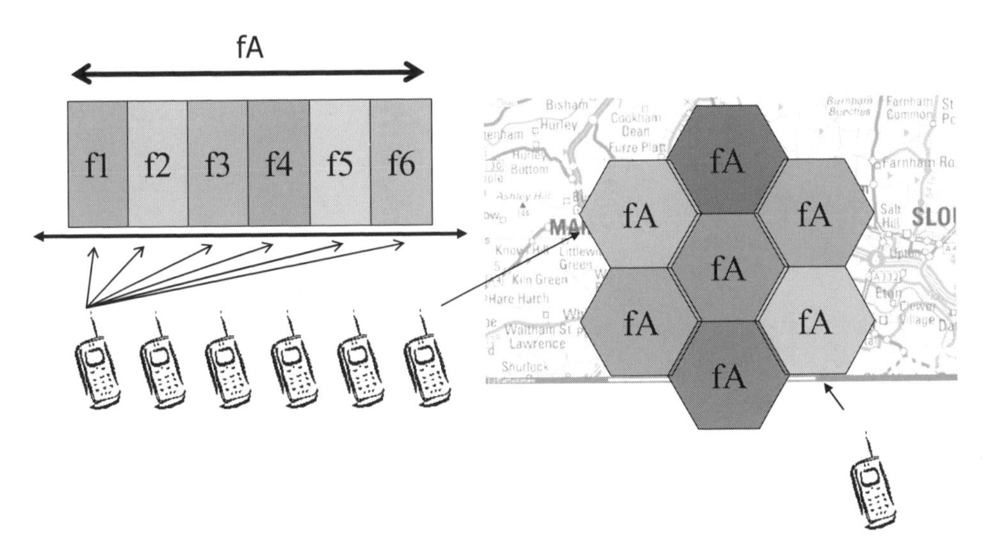

Figure 12.15 With CDMA, mobiles blast out on all frequencies.

per channel (frequency), but had to stick with the original 8-slots arrangement to maintain backward compatibility. However, this was not a major problem. Simply by transmitting every other turn, two mobiles could share the same transmission slot (sometime called a *burst*) by taking turns. Effectively, that means a pair of devices alternately use the same slot; but, that's not a problem, so long as they remain synchronized and don't try transmitting at the same time.

12.2.4 Increasing Capacity Using CDMA

With CDMA (Code Division Multiple Access), things get a bit confusing very quickly, so we have to resort to a degree of poetic licence in explaining the essence of the technique, just so that you are aware that the following description is not a literal explanation of the technique. Having earlier said that we can't transmit on the same frequency, we've already discovered that TDMA does provide a way out. The benefit is that more mobiles can use a single frequency within a given cell, which is good for capacity and good for the users.

TDMA doesn't really allow mobiles to use the same frequency at the same time. Taking turns means they never transmit at the same time,[13] thus avoiding the dreaded interference that we said was a fundamental limitation on RF communications. However, CDMA seems to defy these principles and violate common sense: mobiles do indeed transmit at the same time on the same 'frequency' (or band).

The inverted commas around the word 'frequency' give the clue about the reality of CDMA, so let's proceed with an explanation of how it works.

As Figure 12.15 shows, the CDMA set-up is a strange one at first glance: each mobile within a cell transmits across the entire range of available frequencies and there is no

[13] TDMA mobiles have to be careful to ensure they don't end up transmitting at the same time, so synchronization is important; however, this isn't always easy. An added problem is that the signal from a distant mobile will arrive late at the base station, with the chance that it can still be arriving when a nearby mobile transmits on the next slot. To avoid these problems, there is a guard time at the end of the slot during which mobiles should not transmit.

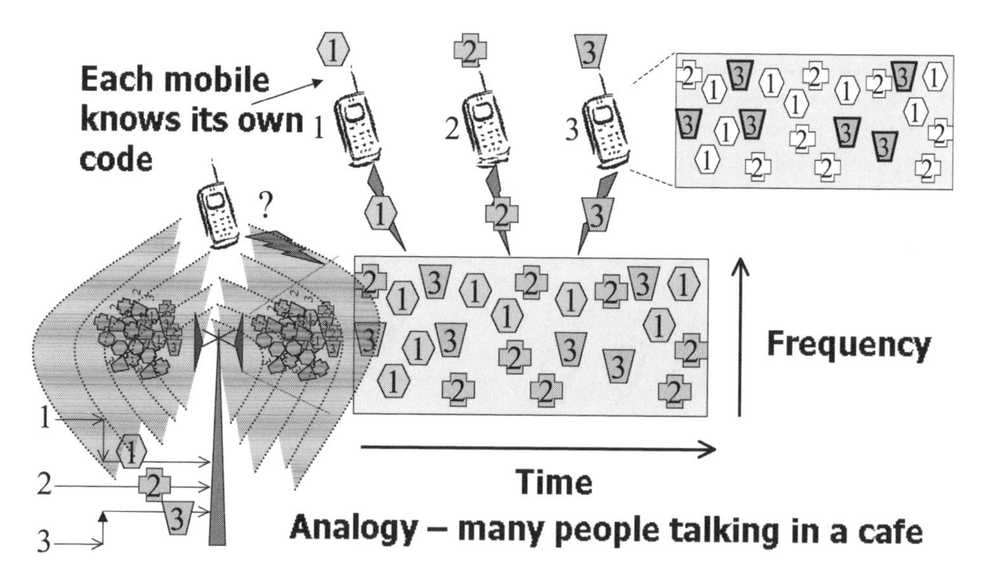

Figure 12.16 How CDMA works.

division of the spectrum into bands, like our radio station analogy. It is perhaps an odd thing to consider in the first place: instead of tuning in to a particular frequency on the dial, our radio station is now transmitting everywhere on the dial 'at the same time' (almost). This is a by-product of the transmission technique called *spreading* or *spread spectrum*, so-called because quite literally we spread the signal across the entire spectrum. This is not a book about RF communications theory, so we spare ourselves the mathematical details of spreading,[14] but we can take a closer look to figure out what's going on without having to consider the maths. To reiterate, our explanation does have a hint of poetic licence, but it serves us well enough for our purposes in this book.

Figure 12.16 gives an indication of what is happening in a CDMA system: here we have three information streams going into the base station. The transmitter takes each stream and transmits it in an apparently scrambled fashion so that at any particular point in time the signal can be found transmitting somewhere[15] in the frequency spectrum. Returning to our radio set again, we can think of this as our radio station jumping around, apparently randomly, all over the dial on our radio. This analogy is to help us avoid the mathematical explanation, but allows an appreciation of the inner workings of CDMA. Each input stream is scrambled, so on the output of the base station we appear to have a complete mess, with everything mixed up. To an extent, this is true; but, it's a bit like mixing lots of coloured beads in a bowl. The big picture is messy, but using the colours, we can extract each set again and recover from the chaos. Knowing the colour in this case is like knowing where to

[14] Ultimately, all communications theory is mathematical; but, with CDMA, in particular, a mathematical explanation is the only satisfactory way of explaining how it works because in a mathematical sense (or in the language of mathematics) there is nothing odd or counter-intuitive about the explanation. When we start thinking of what's going on physically, it becomes a little less clear.

[15] The amount of information in speech (or data) that we wish to transmit does not warrant a frequency band as wide as gets used by CDMA, which is why we can think of the narrower and required amount of frequency as jumping around somewhere within the entire spectrum that gets used as a result of the spreading process.

go on the radio dial in order to pick up the signal for that moment in time. To know where to go on the dial, we use the spreading code that was used to scramble the signal in the first place.

With the scrambling of any particular input stream, the output becomes a pseudorandom mess; however, it is not truly random. In fact, if it were, then our signal would no longer contain useful information because complete randomness is meaningless: it conveys no information.[16] Our information stream is actually deeply embedded in the apparent randomness. It can be extracted using a special code to reverse the pseudorandom spreading – the same code that was used to spread the input stream. As long as our receiver has the right code, as shown in Figure 12.16, it is as if it can see its signal clearly among the mess made up from the other 'randomly' spread signals in the mix. A receiver without the right code just sees a mess, which is actually an advantage as it offers a degree of immunity to spying, similar to the effect of ciphering.[17] This is hardly surprising given that the historical context of CDMA was clandestine military use.

A popular analogy for CDMA is to imagine lots of people in a room who are all talking different languages. Overall, the cacophony of the crowd is a noisy and unintelligible mess; however, when you stand close to the person speaking your language, suddenly everything becomes clear and makes sense.

The language analogy serves us well as a means of elaborating on some of the challenges of running a CDMA network. For example, if there are too many people in the room, then the overall noise level goes up and we probably have to either move closer to the person we are listening to or ask them to talk louder. If people are unable to move, then talking louder is the only option. The downside to talking louder is that others nearby get drowned out and they too have to move closer. For those unable to move, the net effect is that the range of the speaker reduces and, eventually, we will no longer be able to hear them clearly enough to understand what they are saying. In CDMA cells this is equivalent to the cell actually shrinking and those on the fringes being 'pushed' out. When the problem goes away the cell size can grow again, a process called *cell breathing*, as shown in Figure 12.17. One solution is for an adjacent cell to take over the dialogue with the muted mobile.

The codes used to scramble the signals in CDMA have special properties. The result is that the signals spread in such a way (*orthogonally*) that they are least likely to interfere with each other in practice, so that we can still extract them. As long as each mobile knows the code of the signal it wants to receive, then it can extract the wanted signal with only a tiny residual of interference from the other signals, due to their orthogonal properties. Using the cafe analogy, the more unintelligible the other languages the less they can cause interference. However, some languages – modern Punjabi, for instance – utilize many English words. Therefore, in the case of an English conversation, a nearby Punjabi speaker would occasionally break through into the conversation because their English words would be detectable by the unintentional listener. However, a nearby Chinese speaker would not cause any such interference because the sounds of Chinese words would be very different (orthogonal) from anything the English speaker is able to detect.

[16] Ironically, this is what we are trying to approach with CDMA. The more random we make our signal the less it can be 'understood' by a receiver that doesn't know what to look for. No matter how hard we try, a receiver is always contaminated by unwanted signals entering its input. The less these contaminants are understood the better, as they can't interfere with the informational coherence of the wanted signal: something that has no meaning can hardly interfere with something that does have meaning.

[17] Although, we should be clear that spreading is not the same as ciphering.

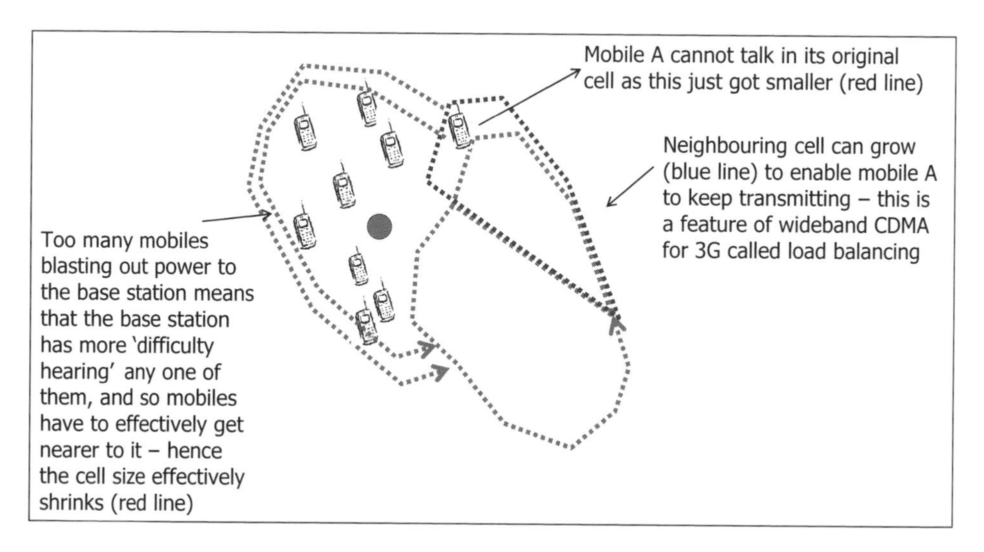

Figure 12.17 Cell breathing in CDMA.

CDMA is the technique used for 3G systems like the Universal Mobile Telecommunications System (UMTS). Theoretically, it offers the highest possible capacity.[18] Its mathematical complexity requires a lot of computer-processing power to extract the signal, but advances in silicon technology have made such techniques accessible to consumer products. Costs are kept low enough and power consumption is manageable within the capabilities of current portable battery technology.[19]

CDMA is also the technique used for 802.11b, the very popular wireless replacement for Ethernet (local area networking).

12.3 THE HARSHER REALITY OF CELLULAR SYSTEMS

Our brief tour of cellular has shown us that, despite the problem of RF transmission being susceptible to interference, we can find different ways around the problem in order to gain enough capacity for lots of users to benefit within a realistic population density. However, unlike the entirely controllable and predictable world of using cables and wires, the wireless environment still offers a degree of hostility to our attempts to beam information to our roaming mobiles, especially to the need to keep up with mobiles' every move.

The reality of modern digital cellular communications is that we are constantly operating at the limit of available implementation technology. Therefore, under certain conditions, wireless systems start to degrade. There are three main difficulties:

[18] Compared with TDMA; but, there are other theoretical multiple access techniques that promise even better capacity than CDMA, such as Spatial Division Multiple Access (SDMA).

[19] There are always signs of breakthroughs in battery technology, like fuel cells; but, battery technology has advanced the least of all the electronics technologies used in mobile devices, so we are still very much reliant on improvements in silicon technology (Moore's Law).

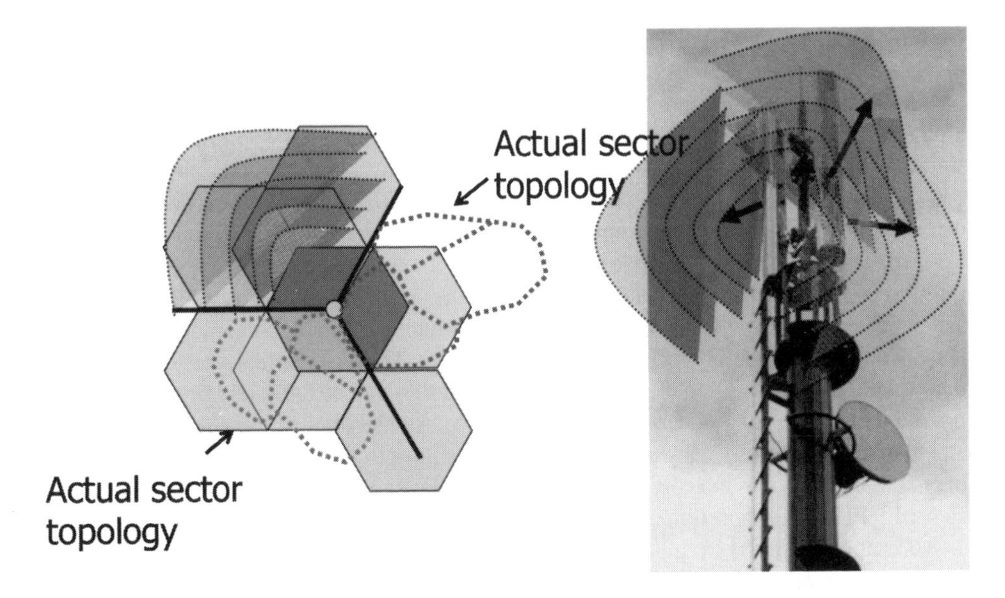

Figure 12.18 Cells are divided into sectors and the edges are fuzzy.

- unpredictable terrain;
- unpredictable mobile location;
- unpredictable traffic loads.

A brief examination of these challenges will highlight the difficulties in meeting our original challenge for the RF network: to achieve transparency or to make it as 'wirelike' as possible.

The problem with cellular systems lies simply with the RF waves not reaching the target directly, either from the mobile to the base station (uplink) or vice versa (downlink). RF signals do have their limitations, bound by certain physical laws. Despite seeming counter-intuitive, RF signals can go round corners (diffraction) and can go through materials (refraction), like buildings. They can also bounce off certain materials (reflection). However, there are limits to this, and the signals eventually end up dispersed and possibly unable to reach certain points. Powerful signal-processing techniques are used to mitigate these effects in the margins, but nothing can be done about the laws of physics once the limits are reached; so, RF systems do fail (as do wired systems if you push them to the limit[20]).

In an actual implementation of a cellular system, rather than the base station sitting in the middle of the cell, as we originally intimated, it often sits on the edge and beams out in several directions, as shown in Figure 12.18. In GSM a typical cell site would utilize three directional antennae equally spaced at 120 degrees; these areas of RF coverage are called *sectors*. It is unlikely that the cell sites sit on a nice rectilinear grid suggested by the idealized hexagonal pattern. This is because cell coverage is related to where the users need service, following the patterns of population distribution, as well as static and migratory patterns.

[20] After all, Ethernet was stuck at 10 Mbps for years before engineers figured out fancier transmission techniques (made possible by better electronics). However, there is still a point at which a signal going along a wire will no longer properly travel down it.

In terms of static populations we would expect to find tightly packed clusters of cells in densely populated areas, particularly in busy commercial zones, like the business districts of cities and so forth. In terms of migratory patterns we also expect to find many cells along major commuter routes, both road and rail, and associated terminals, like railway stations and car service stations. Within the areas of coverage the actual position of the cell sites is determined by a range of factors. Their idealized positions are first calculated using sophisticated cellular planning tools that attempt to arrive at optimal locations to achieve the best frequency (or code) reuse patterns, maximizing capacity while minimizing interference. The planning tools take into account the terrain of the land as much as possible. This is done using 3-D models, the accuracy of which is not always certain, particularly if it involves the modelling of buildings and so forth. However, actual site locations will then be chosen according to restrictions as to where they can be built, initially governed by finding suitable physical locations in the first place. Site location is influenced by the cost and feasibility of getting backhaul communications lines to the sites in order to connect the base stations back into the core network of the cellular system (i.e., the cellular switch and so forth). Planning restrictions also affect site location.

The net result of the irregular distribution of sites, combined with the irregular terrain, is an irregular cell pattern, as also shown in Figure 12.18. In fact, the true shape of a cell is difficult to know without a detailed survey using sophisticated RF measuring equipment and painstaking coverage detection, typically by driving around with equipment, hoping to find the extent of coverage. Despite all the sophistication of RF planning tools, a degree of drive testing is usually required in a network, especially to fine-tune the coverage. In terms of understanding usage patterns that determine cell locations and density, this is well understood thanks to the experience of running 2G systems, although these patterns are mostly related to voice service, not data.

As far as design challenges in a cellular network are concerned, the constant movement of mobiles is not a particularly nice problem for a communications system to handle. This imposes various limitations. First, given the cellular pattern that we need to handle capacity,[21] roaming mobiles need to switch from one cell to another, possibly during an active session (voice call or data connection). This is not an easy situation to cope with, as it opens the door to possible performance disruption and even erroneous operation, such as connections not making it from one cell to the next (being *dropped*). Second, we can intuitively sense that there must be a lost opportunity in having to beam out an RF signal over a wide sector when the target is actually sitting in a very narrow part of the signal's beam. The signal that is spilling out everywhere else is nothing more than 'wasted' signal, a source of interference for whatever else lies in its beam. This problem could be considered as the limiting factor to capacity within a given cell radius. The net effect is that we still will experience capacity problems when too many mobiles vie for the RF resource, even with the more gracefully degrading CDMA. Overcrowding of the cell is not always easy to plan for as it is related to unpredictable numbers of mobiles trying to make a call.

12.3.1 Data Rate Variation

In understanding the actual limitations to the capacity of a cell, let's start with the theoretical best performance that our RF network is designed to cope with at its peak and then work

[21] CDMA still has cells, but the mosaic is staggered around different code sets, not frequencies.

our way down from there. So, using GSM slots to send packets of data rather than voice (i.e., *GPRS*), we start with the theoretical best performance of using all eight slots[22] for one mobile on one frequency in a cell with ideal radio performance and we get a maximum data rate of 160 kbps. However, a user will seldom get, if ever, this kind of performance. First, the statistical likeliness of getting all eight slots is remote,[23] never mind the fact that most operators are highly unlikely to configure any system to use eight slots because they will always reserve a minimum number of slots for voice calls. Moreover, most mobiles cannot handle eight slots: typically, they can handle four to receive data (downlink) and two to transmit (uplink). The reason for this is that we begin to defeat the processing advantage we gained with the slot mechanism. If a mobile only transmits, or receives, on one slot per frame, then it has ample time to do all the processing before the next slot. If we now ask it to process twice, or four times as much, it begins to run out of steam even using cost-effective processors. More challenging is the greater rate of energy consumption, worsened to such an extent that an eight-slot system would most likely not have a very useful battery life.

If the RF conditions begin to get more hostile due to interference, poor signal strength or other artefacts, then we have to use some (or more) of the available traffic capacity to carry redundant information which helps combat errors in the transmission. As we noted before, most data transmissions are extremely sensitive to errors – a single corrupted character in a bank statement can make a huge and unacceptable difference. To protect against errors, *error control coding*[24] is used; this means that our theoretical 20-kbps capacity on one time slot can be reduced to 8 kbps.

The vagaries of cellular RF networks inevitably lead to variable levels of data rate performance, possibly suboptimal for some applications – at least some of time. Therefore, our mobile service should be able to cope with the various levels of performance while staying within the bounds of acceptable end user experience (not forgetting our mobile services definition is qualified with words like 'successfully, confidently and easily').

Many of the performance optimizations we discussed when looking at WAP 1 and WAP 2 begin to make sense when we take on board these very real limitations in the RF network. Some scenarios are particularly challenging, like 3G operators that allow roaming to GPRS (or other 2.5G systems) wherever 3G coverage is poor or non-existent. Dropping from a 200-kbps link speed to something like 16 kbps is a dramatic discontinuity. How is the application going to cope? We need strategies to overcome these problems, or else we probably shouldn't be launching certain services at all. WAP 1 suffered this on some networks (though not all), where the RF performance overall was not able to carry the service with an acceptable level of performance.

We should expect variable data rate performance and then design our application to cope with it. Figure 12.19 gives us a feel for what may occur in practice. We can think of the coverage area as being a constantly changing patchwork of data rates.

The core services of cellular networks, such as voice, are already designed to cope with this variance in quality of service. A voice compression algorithm like Adaptive Multirate

[22] With GPRS, mobiles can try to use as many of the eight time slots as possible when transmitting a burst of data, such as sending an email. Mobiles are not limited to one time slot and can use up to the maximum assigned to GPRS, provided they are not being used by another device.

[23] This can be properly estimated using statistical estimation techniques and queuing theory, but we can safely use words like 'remote' in this context.

[24] There are four levels of stringency in the error coding scheme for GPRS, called class 1 to 4 (CS-1,CS-2, CS-3, CS-4), which give us 8, 12, 14.4 or 20 kbps throughput per slot.

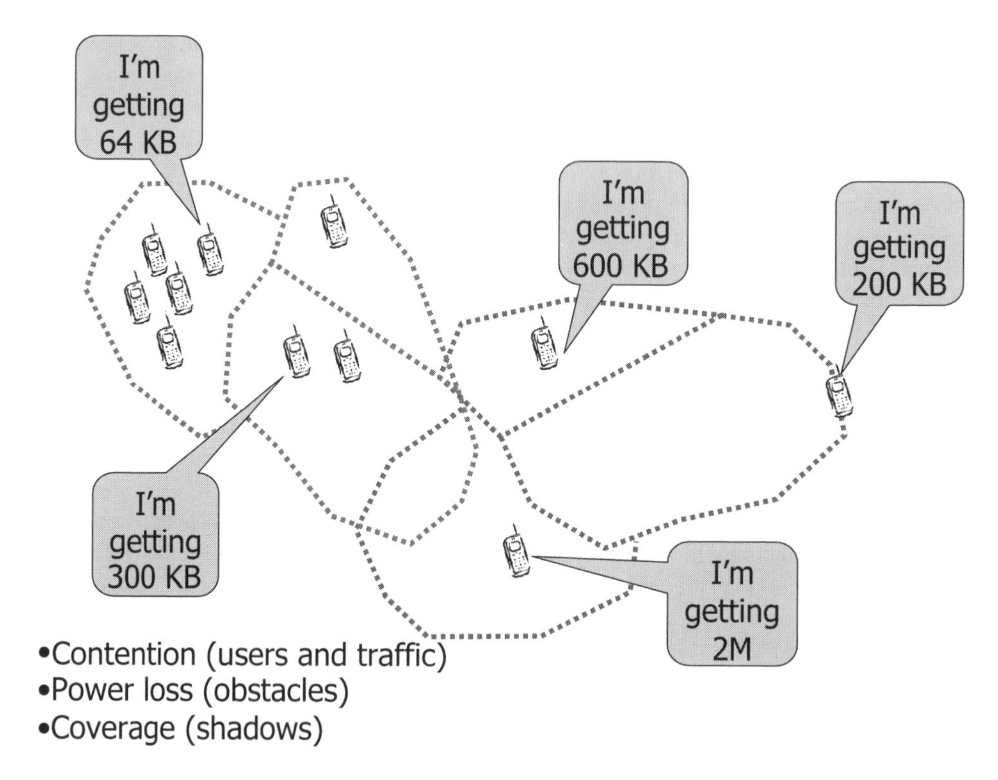

Figure 12.19 Don't think that we get the 'advertised' speed everywhere.

(AMR) is able to adjust dynamically the level of compression according to the prevailing network conditions.

Compression techniques for audio and video signals are what we call *lossy*: this means that in order to achieve compression one of the underlying principles is that some of the source information is not going to make much difference to the user's perception of quality if omitted, so it gets removed to achieve the compression. In an image this is easy to conceptualize. The fine detail in a collection of trees is difficult for the eye to perceive. Similarly, detail in an image with *motion blur*[25] is also almost imperceptible. Converted into a digital data stream, this fine detail would require a lot of network capacity to transmit. Getting rid of some of it (the finer details) saves on capacity and makes little difference to the result in the eyes or ears of the recipient.

The advantage of a lossy compression technique is that more and more detail can be thrown out in order to achieve greater levels of compression, provided we are prepared to accept a reduction in quality, consequently. Obviously, we can keep removing more detail and, subsequently, produce a coarser output, but, eventually, quality will suffer and can even reach a point of becoming intolerable. In the case of the AMR voice codec,[26] voice reproduction quality is traded against capacity. If we have many mobiles vying for a

[25] Motion blur occurs while panning the camera across a scene, such as following a football in a soccer match: as the grass whizzes by on the screen it becomes blurred and detail gets lost.
[26] Codec stands for coder–decoder, which normally is taken to mean compressor/decompressor.

Table 12.1 Different service levels according to data rate.

Data rate available	Level of service for multimedia-based location-based service
2 Mbps	Full-quality video clips for local cinema can be streamed to the user. Product advertisements can be viewed (e.g., for a nearby car under offer)
300 kbps	The buyer can hold an online video conference with a sales assistant specialist and view product videos before buying the product
56 kbps	High-quality, colour, animated advertisements sent to user, possibly with some audio streaming available. Images could be accessed when looking at restaurants, attractions, etc.
40–600 kbps (variable)	For user drill-downs into a tourist application, the system will attempt to return highest quality data (images, audio, etc.) and switch back to low grade if not possible
28 kbps	Ordinary, baseline, text-driven and map-driven service with basic advertisements
SMS alerts	Push adverts; buddy notifications still get sent. User can make limited requests that are better fulfilled when full service becomes available again

limited resource in an overcrowded cell, we can back off the demand for capacity by simply lowering the voice quality on some or all of the mobiles.

This adaptation principle should also be present in our application design. If we are not used to such a notion, then at first it might be difficult to think about how to design for it. Perhaps for some applications it is not necessary, but let's examine some possible scenarios.

Table 12.1 gives us an idea of the range of service levels for a particular application, depending on the available data rates. The transition from one data rate to another is not just a matter of service variance within a particular RF technology – with a multi-standard device the access method itself may change, such as a transition between GPRS/UMTS and WiFi, and the potential contrast in speeds could not be starker (see Figure 12.20). It is feasible for a GPRS user to access services on a train at a measly 40 kbps, only to jump to 1.5 Mbps (or higher) on a WiFi access point in the station itself.

12.4 TECHNIQUES FOR ADAPTATION

We may decide to design our application to allow different levels of service depending on the data rates available to our devices. There are two approaches to the problem:

- manual approach – ask the user to make choices;
- automatic approach – use network information to guide choices about content type and formats.

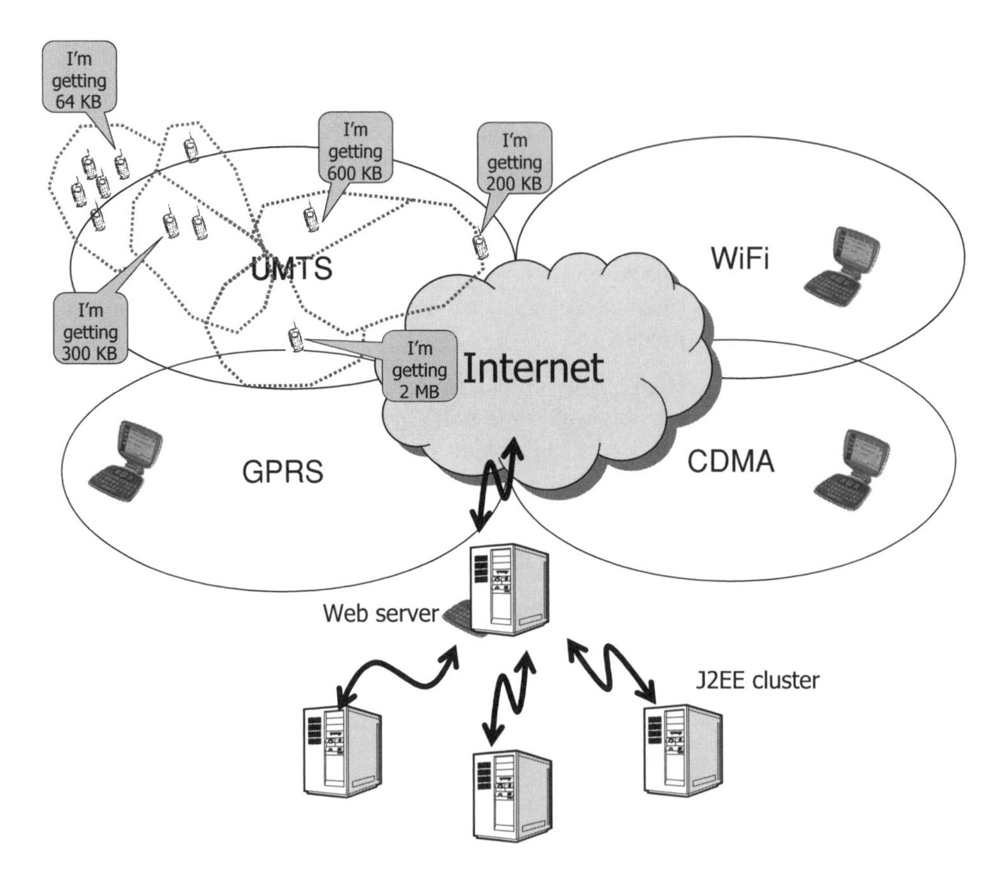

Figure 12.20 Data rate is also affected by service type.

The manual approach can include:

- Giving the user a choice to view different versions of the site ('low bandwith' and 'high bandwidth' versions);
- designing the site in a way that data-intensive information is separate from the main flow and can be optionally pulled in by the user (e.g., not showing images and asking the user to 'reveal' those images).

The manual approach entails the user making choices about what they want to see and then having to specify their preferences. The problem with this is that the user may be intolerant of such approaches, especially because offering more than one choice may often seem counter-intuitive. A user may get confused as to why they are presented with such options.

The manual approach can also end up masking the fact that data rates change, keeping in mind that the rate of change compared with the typical length of a session might be noticeable. Therefore, a user may start a session with a particularly slow link and elect to view the low-bandwidth view. The link speed may subsequently improve, but the user is still using the current view, needlessly foregoing the benefits of the richer content. An automatic approach would be better (which is not to say that a manual override should not be present).

Designing the site to be tolerant of variable bandwidth access is clearly a good idea, although it possibly involves extra work to formulate a suitable design. However, the mechanism for handling more than one version of the content has already been discussed (Chapter 8) when we looked at the J2EE (Java 2 Enterprise Edition) presentation layer and the general issue of device-based adaptation; this mechanism would accommodate nicely a solution to the rate adaptation problem.

As an example of such an approach we could replace pictures in a page with links or low-bandwidth spacers.[27] On some pages, pictures can easily constitute a large proportion of the overall data size, so treating them in this way makes sense. Some browsers give the user the option not to display images. This means that the user agent does not go back to the Web server to fetch any images. These two approaches differ and can produce different results. The main limitation of allowing a browser to exclude image fetches is that we cannot rely on it as a solution: not all browsers support such a mode. Therefore, we need to implement the solution at the origin server, not the browser. Furthermore, browser-based omission of pictures can produce undesirable results, such as adversely affecting the page layout. For example, if the browser does not know the size of an image,[28] then we will not know how to lay the page out to allow the rest of the content to flow around the image space that remains. However, if we purposely built a page without the image, then the layout could be fashioned accordingly.

Whatever the design approach used to adjust the visual appearance of pages according to available data rates, the use of an automatic means to control the content stream is preferable. However, identification of a suitable mechanism for this remains to be done. Fortunately, there is an ideal opportunity available using the UAProf (User Agent Profile) mechanism discussed in Chapter 9 when we looked at how to vary content according to fixed device capabilities, such as screen size. In the UAProf file included in the browser's HTTP request header it is possible to insert semantic information about the connection speed. This is an ideal solution because it enables a potentially high degree of control, especially as each response from the server can adjust its output according to each request, rather than a blanket control across the entire session (i.e., opting to view the 'low bandwidth' version for the current session).

The challenge is how to insert the required semantic information and what to insert. What we need is some kind of estimation of the bandwidth available. It is relatively easy for a mobile to estimate the downlink bandwidth to the device. This is available from the device itself by consulting the control software on the modem chipset (see Chapter 10). The browser is also able to estimate bandwidth itself by timing how long HTTP or WSP (Wireless Session Protocol) response streams take to arrive at the device. Dividing the stream size (page size, image size, etc.) by the time, we get the estimated bandwidth. If the browser estimates bandwidth, then the measurement will include all network delays, which is a more accurate reflection of the user experience. If it takes 5 seconds before an HTTP request generates a response, then this delay should factor in the measurement in order to calculate the *effective bandwidth*. For example, a 100-KB page that takes

[27] A spacer is some replacement image that is low resolution or, possibly, just blank which can be clicked on to reveal (fetch) the actual image.

[28] There are techniques to ascertain the image size. Ideally, it is included as a parameter within the mark-up itself, so the browser has the information. Failing that, the browser can attempt to fetch the image while using HTTP (HyperText Transfer Protocol) *byte ranges* to fetch only a limited amount of the image. In some file formats, like PNG and GIF, image size information is embedded at the start of the file.

8 seconds to stream, plus 5 seconds delay, gives 100 divided by 13 KB per second, which is 61.5 kbps.[29]

As discussed on numerous occasions throughout the book, usability is affected by delays. Generally, users do not have enough patience to wait around for a Web document with long delay. One may expect to receive a Web document in a reasonable waiting time t: say, 10 seconds. In our adaptive system the browser reports a bandwidth b to the server: say, $b = 10$ KB per second. Then the Web server will use content adaptation to send the Web document with size not greater than $b \times t$: i.e., 100 KB in this case.

The browser can also take into account its current downloading activities when making a new request. For example, the user could be in the middle of downloading a large MP3 file to listen to a new music track. A browser request made during the download interval should take into account that bandwidth is already being used by the device, which possibly means that extra bandwidth allocated to the new request will be less than expected. For example, if the device is operating in a cell where it could sustain 120 Kbps, then the MP3 download might use most of this bandwidth. Therefore, if 120 kbps is the bandwidth estimate for a concurrent browser request, this might result in the server producing an adapted file that is much too large for the bandwidth that the browser assigns to downloading the page (assuming that the MP3 download is not to be affected).

The strategy to allow the browser to estimate available link speed seems a useful one. However, it does have some potential flaws. First, if there is a significant delay between the last HTTP response stream and the current request, then the network conditions may have changed compared with the last response. Perhaps there is more congestion and, consequently, the network is much slower, in which case the browser might be over-optimistic in its measurement report and, consequently, the content adaptation will be nonoptimal. The other potential flaw is that the browser is not the only application on the device: other applications are potentially vying for bandwidth. For example, video-streaming or audio-streaming applications may work independently of the browser. If the activities of such applications do not figure in the link speed estimate, then the browser report may not be realistic.

A solution to the problem of link speed estimation is to use transport layer estimation techniques that take into account all the available resources end to end in real time. To achieve this, equipment in the RF network should be accessible to provide the required estimates. There are probably a number of strategies to achieve this, but the general principle is to ask the RF network itself what resources are available at a particular moment for streaming information to a particular mobile device in a particular cell. As Figure 12.21 shows, there are various network elements used to build a cellular network. We will soon explain the elements in more depth, but for now observe that each cell is served by its own base station and a group of base stations are concentrated into a controlling element. The controlling element, such as the Radio Network Controller (RNC) in a UMTS network or the Base Station Controller in a GSM/GPRS network, is aware of how much RF network resource is available in each cell under its management. It is its job to keep track of the devices and allocate resources. To put it crudely, if a particular cell has a total capacity of X Mbps and our RNC is aware that there are N devices using Y Mbps between them at a particular moment in time, then to stream a response to our $(N + 1)$th mobile, the RNC could potentially report that we have $X - Y$ Mbps of capacity available. This is a somewhat simplified suggestion, but, ultimately, the potential to do something like it does exist within the network.

[29] Remembering that bandwidth is measured in bits, not bytes (i.e., kbps means kilobits per second).

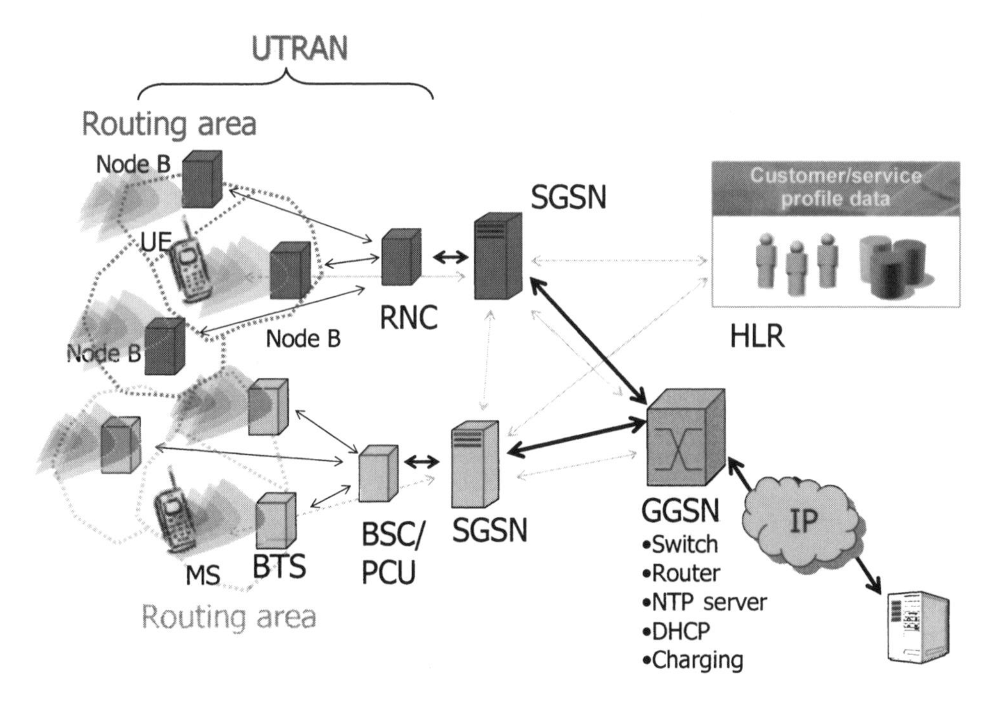

Figure 12.21 Simplified network architecture (GPRS/UMTS).

At the device end, instead of asking the browser to report measurements (which it still might do, as this gives us extra information on which to base decisions), we can involve the control software in the RF modem, as it has the best view of the RF link characteristics. It will know how much data are passing through the modem at any one time and it will know about the link quality measurements (e.g., power, signal quality, etc.) that it is required to make as part of its operation in the network.

The best thing about using UAProf as our means to convey the link estimate is that any entity in the network can add profile headers (*profile-diff*), not just the user agent itself – although with UAProf we have the potential to combine both. In terms of network-generated UAProf information a gateway or proxy sitting somewhere between the device and our origin server could add the UAProf to the requests, as shown in Figure 12.22. This is very convenient, as it means we have a way of inserting link capability information without requiring any special protocols or modified origin server solutions.

Having attained our link estimate, we can calculate how much data we should attempt to stream in the response. In deciding on the optimal page size to suit available bandwidth it is debatable whether the size we are interested in should be the size of the document in question or the aggregate size, which means the document plus all the resources it refers to also need to be fetched from the server. Recalling our discussion of HTTP 1.1 (including WSP and W-HTTP), the initial document can be streamed in and displayed before or while the remaining content is being streamed, not forgetting that such a process is pipelined (overlapping). Of course, this is a somewhat finer point that doesn't need much deliberation. Using an aggregate will suffice and any fine-tuning can be applied in light of operational experience.

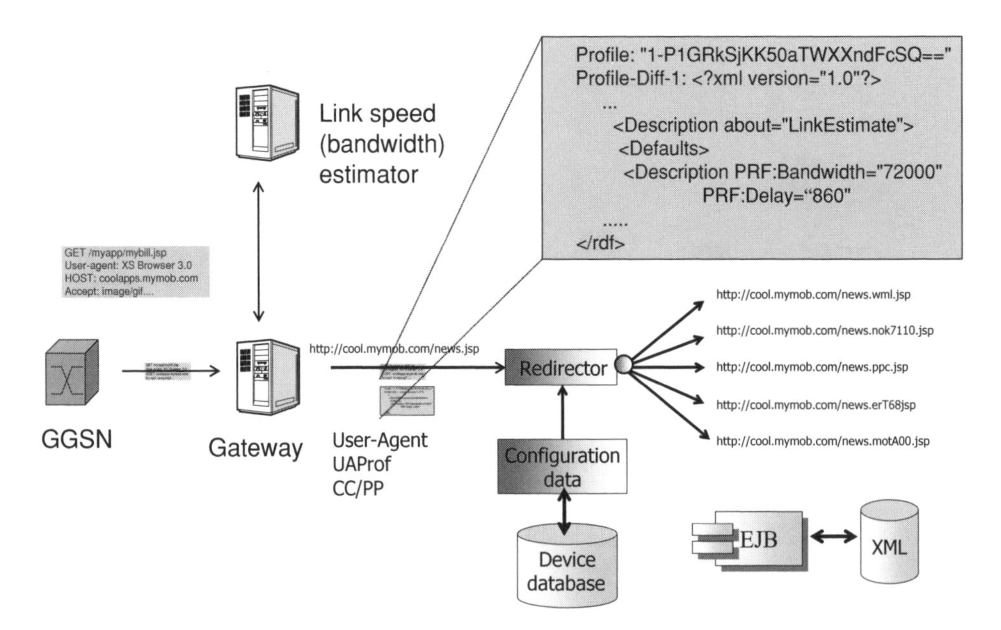

Figure 12.22 Inserting link quality information via a gateway using UAProf.

What are the content-sizing strategies that we can apply to take advantage of dynamic adjustment? We have already considered the somewhat dramatic options of text-only sites, which could be the lowest common denominator. But how might the content be adapted otherwise? In addition to image removal we can consider image resizing and using different levels of compression, especially for JPEG images, where the compression technique is lossy, which as we discussed means that we can sacrifice image quality for data size. In future, more images will be constructed using vector drawing techniques, using standards like Scalable Vector Graphics (SVG). Complex drawings can be quite verbose in their description, so we could consider dynamically adjusting the amount of detail. For example, in a mapping application where there is lots of bandwidth we can send a very rich schematic with detailed map features, such as boundaries, rivers, roads, etc. If bandwidth is restricted, then we can pare down the schematic to reveal only the major tributaries and so on. An example of this is shown in Figure 12.23, where the area of interest (UK) is shown in the map with and without the neighbouring countries. By not showing the neighbouring countries and removing the cities, we can show how the amount of data to be streamed is reduced.

12.5 CELLULAR NETWORK OPERATION

In the previous discussion we introduced the simplified network architecture of a UMTS or GPRS cellular network, as shown in Figure 12.21. We are concerned in this section with the basic network architecture, particularly how we interface the IP network with the RF network, so that we can begin to see how we will connect our J2EE content world with the device network.

Let's elaborate on the architecture (see Figure 12.21). We have already discussed how the cells are supported by base stations in cellular systems; base stations are the sources of

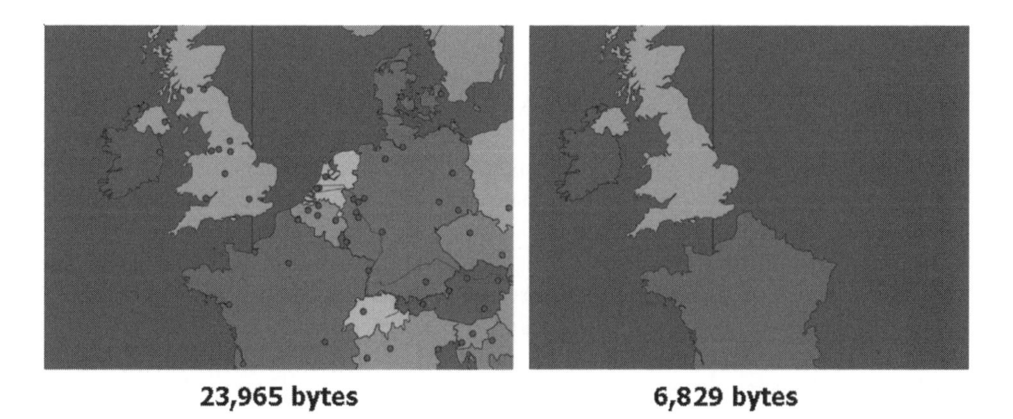

Figure 12.23 SVG files with different levels of details.

Figure 12.24 Cellular antenna disguised as tree.

RF connectivity in each cell (or sector). In a GSM/GPRS network the name given to the base station is *Base Transceiver Station* (BTS). In UMTS the equivalent unit is a *Node-B*. These units actually contain the RF transmission systems and the required RF antennae to enable the RF waves to propagate in the required sector pattern. Some antennae disguise themselves as trees or other objects, to avoid unpleasant eyesores on the landscape, such as the 'Scots pine' shown in Figure 12.24.

Base stations are housed in equipment cabinets that can accommodate several radios, one per cell (sector). The number of radios depend on the nature of the site and come in a variety of sizes, but are loosely categorized as macro-cells (up to 20 km radius for GSM), micro-cells (a little as a few hundred metres) and pico-cells (usually limited to indoor coverage, such as the floor of a large building). Installing smaller and smaller cells is the main way of achieving higher traffic capacity; this is the origin of the microcellular concept – small cells to fill 'hot spots' in dense urban areas, such as central business districts, main road intersects and so forth. Pico-cells extend the same concept indoors[30] (where other services

[30] It should be noted that RF from cellular systems does propagate into buildings in any case, but here we are talking about the very specific idea of increasing capacity inside a building by installing a dedicated base station.

are also feasible, such as a cellular private branch exchange[31]). In a cellular network all the cells are uniquely identifiable by their cell ID, which is an address that the network uses to keep track of where cells are in the network.

Base stations transmit and receive the RF signals on one side (antenna) and connect back to the network via wireline signals on the other. A base station needs to connect the information stream to the core of the RF network (e.g., switch), so a data connection is required – usually a leased line. Occasionally, it is difficult to route a suitable wireline to the station, so microwave wireless connections (or short-range optical equivalents) are sometimes used. The signals from the base stations are concentrated into a controlling element higher up the network chain toward the eventual boundary with the outside (IP) world. In GPRS systems the BTS connects back into a Base Station Controller, or BSC. A BSC can handle as many base stations as it is designed to handle.[32] Similarly, in a UMTS network the RNC manages a multitude of Node-Bs, the exact number being manufacturer-dependent.

We are mostly concerned with packet data support in the network, not voice; so, we will not dwell on the voice communications features. In both 2.5G and 3G networks (e.g., GPRS and UMTS) the rest of the network, behind the RF front end elements, is very similar and is called the *core network*. The RF component in GPRS is collectively called the *Base Station Sub-system* (BSS), while the RF component in the UMTS network is called the *Universal Terrestrial Radio Access Network* (UTRAN).

In the voice network (*circuit-switched domain*) the signals work their way up the pyramid of network elements toward the public telephone network via a switch (*mobile switching centre*, or MSC); in the data network (*packet-switched domain*) the data packets work their way from the Node-Bs toward the IP interface of the network support via the *Gateway GPRS Service Node* (GGSN).

Beneath the GGSN sit the *Serving GPRS Service Nodes* (SGSN). These take control of packet data support for a group of cells, each group known as a *routing area*. In terms of handling data coming into the network from the IP world – perhaps a WAP-PUSH message – the underlying IP protocols only know IP addressing, they do not know about cell IDs or mobile IDs, or any other type of cellular routing information. The SGSN and the GGSN provide the necessary routing of IP packets to the correct cells where the target devices reside and carry packets back again to the GGSN, shuttling between the internally routable packets to the externally routable IP packets with their IP addresses. Part of the shuttling involves dynamic routing due to the constant movement of the mobiles, thus requiring *mobility management*. The SGSN and GGSN also take care of various security features in the network, including encrypting data exchanged between these network entities.

12.5.1 Getting Data in and Out

The GGSN is the key network element as far as our IP layer is concerned. As its name suggests, it is the gateway to the RF network from the IP network. It is where IP data enter and leave the RF network, going to and/or coming from the mobile devices. It is where data

[31] Cellular PABX is where mobile phones can also act as internal phone extensions, allowing such features as short-code dialling and extension-to-extension calls free of charge.

[32] Meaning that the number is somewhat arbitrary and product-dependent – there is no specified number.

physically pass from our RF network into the IP network, usually to a local area network (LAN) run by the operator on which sits the GGSN as an IP-addressable entity. Internally, the cellular infrastructure itself is not IP-based. It has its own set of internal protocols optimized for the support of mobile devices in an RF environment, taking into account such requirements as security and mobility-aware data routing. We can think of the GGSN as providing an IP overlay onto the cellular network, presenting the outside world with an 'IP view' of the cellular network, such that devices have IP addresses and can receive and transmit IP traffic.

Such are the independent habits of users that they are not all using the infrastructure at the same time. Usage has a statistical flow that largely reflects the fact that humans and their patterns of behaviour (business and social habits) are erratic or unpredictable. A user might access their email, check a few messages and then do nothing for a few hours: during that time, perhaps a few WAP-PUSH messages arrive. Meanwhile, another user is busy streaming MP3 files all day, listening to whatever takes their fancy. The consequence of diverse user habits is that different users require different network resources at different times. Therefore, our core network works in a way that assigns resources on an as-needed basis in order to achieve an efficient sharing of what is ultimately a limited resource (the RF spectrum itself). Setting up data connections for devices in each cell is part of the *connection management* role of the core network. We can think of this as allocation of physical resource – actual packets on the RF interface that assign codes in the case of CDMA or slots in the case of TDMA.

The assignment of services at the RF level depends on the type of service requested by the user. For streaming applications that require a repetitive flow of packets with a fixed delay, the network takes a certain approach to assigning RF resources – one that is suited to real time services, such as videoconferencing. For non-real time applications, like the flow of packets in an HTTP request, the core network takes another approach to assignment of RF resources. These activities are all part of connection management.

Session management is the assignment of complete packet-switched circuits once the connection management has taken place. A critical aspect of session management lies in giving a mobile device an *active PDP context*, where PDP stands for Packet Data Protocol. We can think of this as the cellular equivalent of activating an IP device by giving it an IP address. Packets through the core network flow as *packet data units* (PDUs), for which we need an active PDP context in order for our mobile to exchange PDUs. The user requests a PDP context, although their device software makes the request implicitly whenever an application on the device requires an IP connection, such as the WAP browser. We can think of the mobile requesting a PDP context as bringing it 'to life' on the cellular data network, ready for IP connectivity. As soon as the device has a PDP address, it is ready to send and receive IP data. Its routing and location information are known to the GGSN and relevant SGSN, which both collaborate to get data to and from the appropriate devices according to IP address allocation, as shown in Figure 12.25.

This has been a very simplified discussion of the cellular network, but the detail is beyond the scope of this book. We have omitted the other features of the cellular network that lay in its *service domain*, and we will come back to some of these. However, we need to focus more on how we connect our IP network to the RF network. Throughout the book, we have examined IP-based protocols in some depth, such as HTTP and WSP. We have assumed that devices have IP addresses and are contactable via IP. We have seen how the cellular network can support IP-addressable devices that roam while maintaining a communications

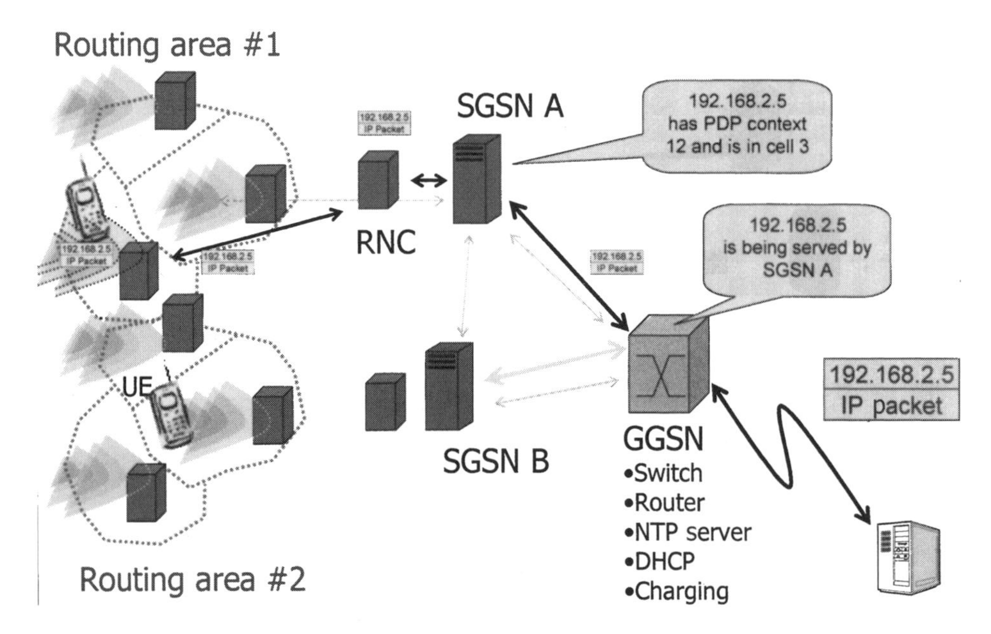

Figure 12.25 Routing packets via session and connection management.

pathway to and from the IP network. We now look in more detail at the interface between the IP network and the RF network.

12.5.2 Gateway GPRS Service Node

The GGSN is clearly an important node in our cellular network because it controls the IP view of the network. This book has dealt mainly with IP-related service architectures; therefore, of all the RF network nodes, it seems most appropriate to understand the GGSN in more depth, especially in terms of how it might affect our considerations for application development.

Figure 12.26 shows a possible architecture for a GGSN. To an extent, we can think of a GGSN as a kind of 'all-in-one' IP system, providing many of the crucial IP networking services that we need for an IP network to function, such as:

- routing;
- IP address allocation;
- authentication;
- firewall;
- caching;
- IP switching;
- virtual private networking (VPN) support.

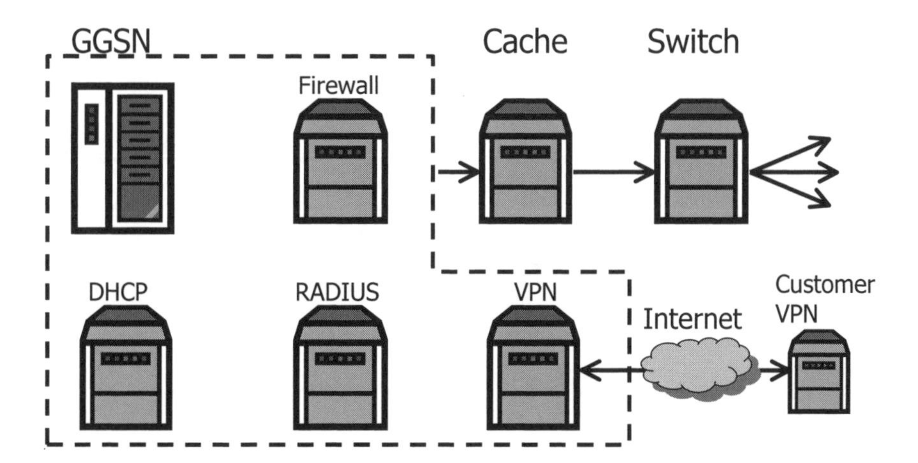

Figure 12.26 GGSN architecture.

The GGSN is essentially like a router in an ordinary IP network. Routing is concerned with taking IP packets and forwarding them from one router to another using the most efficient (or available) path to get the packets to their final destination. All kinds of criteria exist for routing, such as minimizing congestion, minimizing delay and so on. However, in the case of the GGSN the routing tables will not have any of these criteria when constructing input (IP address) versus output (mobile address) mappings or vice versa. The routing algorithm is altogether more straightforward: it is simply to route the packet to whichever SGSN is serving the device with the destination IP address on the packet – in other words, it is 'location-based' routing.[33] Clearly, the SGSN and GGSN need an internal protocol to manage the routing information. The GGSN needs to know where to send packets to, bearing in mind that devices move and the distinct possibility that they may move from one SGSN to another.

In order to route an IP packet to a mobile we first need to have an IP address assigned to a device. There are several strategies for doing this, all under the control of the GGSN:

- assign an address from a configurable pool of addresses in the GGSN database;
- use the Internet's Dynamic Host Control Protocol (DHCP);
- use an address taken from the Home Location Register (HLR).

Within these methods there are degrees of flexibility. For example, the IP address pool in the GGSN configuration database can be set to any range of addresses. Different pools may exist that contain IP addresses reserved for different users. This allows segmentation of the user base into different IP address pools, something that might be useful for a variety of reasons. Allocation from the pool can be dynamic, based on a cyclical first-in, first-out allocation of addresses, or static, if a fixed IP address is required on a per-mobile basis.

[33] One could argue that ordinary IP routing is also location-based. After all, that is the whole point of routing: to send packets to a device *located* somewhere on the Net. However, differently from the Internet, the core network does not possess a multitude of potential routes to forward packets, so the performance criteria that routers are designed for simply do not apply.

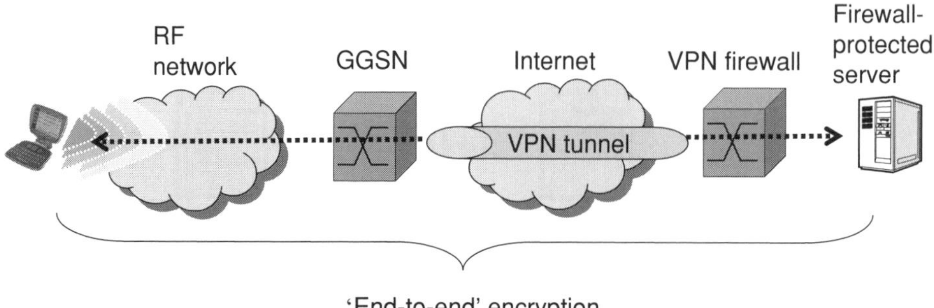

Figure 12.27 VPN enables end-to-end encryption across the Internet.

Static IP addresses can also be stored in the user's customer record found in the network's master directory, known as the HLR, which already holds other user information.

When using the Internet's DHCP to assign IP addresses, the DHCP server can be internal to the GGSN or externally hosted as part of the operator's local IP network infrastructure. Alternatively, any DHCP server can supply IP addresses on a user-by-user basis. Perhaps an Internet Service Provider (ISP) has its own IP requirements for users accessing its servers using the RF network, in which case the ISP's DHCP server can allocate its own IP addresses.

Regardless of the IP address allocation method, the management of IP addresses by the operator can take advantage of *network address translation* (NAT), if required – another feature of the GGSN. To access the operator's services on internally hosted server farms, the operator will assign internally routable IP addresses from one of the private address ranges specified in RFC 1918.[34] However, these addresses do not route across the Internet,[35] so when a device requires access to an externally hosted service the GGSN translates the packets to another IP address taken from a pool of public addresses assigned to the operator.

Using the VPN feature of the GGSN, it is possible to route traffic securely across the Internet to another network that is also using a VPN, a process called *secure tunnelling*. Most GGSNs support a variety of VPN configurations; this allows the support of secure services, such as mobile access to enterprise networks, including Intranet access, CRM access and email. The core network also supports encryption internally – node to node and across the RF interfaces – thereby securing the entire data path from the device to the enterprise, as shown in Figure 12.27. However, it is not true to say this is end-to-end encryption unless the application itself supports encryption on the device and the application server, such as HTTP or W-HTTP using SSL (the Secure Sockets Layer), as discussed when we looked at IP protocols in Chapter 5.

The GGSN can also support RADIUS[36] authentication. This has nothing to do with authenticating users onto the RF network or onto any downstream application, such as the 'basic' or 'digest' authentication schemes we discussed earlier in Chapter 9 when looking at how to secure J2EE applications. The RADIUS protocol and the support for it at the

[34] http://www.faqs.org/rfcs/rfc1918.html

[35] Because they are allocated for internal usage only, routers are supposed to be programmed not to route these addresses onto the public Internet.

[36] RADIUS stands for Remote Authentication Dial In User Service – a protocol specified by the Internet Engineering Task Force (IETF) working group.

GGSN allows authentication onto another network, like an IP network used for hosting applications, or a separate IP network, like one run by an ISP that might want to make sure only its customers are accessing its resources. RADIUS is currently the de facto standard for remote authentication. It is very prevalent in both new and legacy systems, so its support at the GGSN is a useful provision.

In addition to these essential IP networking functions some GGSN products provide other integrated functions, like HTTP caching, which can be useful for accessing Web pages. We discussed earlier in the book how caching in the browser can provide improved performance when accessing Web pages that have not changed since last viewed; this function may also be available by proxy in the network. Network-side caching in itself does not really improve performance that much for the end user in terms of faster page loading across the RF connection (which was the motivation for browser-based caching), but is more useful for limiting the amount of Web traffic coming into the core network, warding off congestion and so forth.

Other functions, such as IP switching and firewall protection, are housekeeping functions – albeit important ones that may still have an impact on our applications, as we will now discuss.

The core network is a tightly controlled network. Effectively, the operator of the network owns the connections to the mobile devices and is fully at liberty to manage the connections how they see fit. Sometimes, this might include apparently punitive network policies: for example, some operators have been known to limit the flow of IP traffic only to internally hosted servers, thus 'garden-walling' their network, preventing others from utilizing it for their own external applications. Other operators may provide access to external networks, but in a restrictive manner: for example, it may only be possible to access external WAP sites via the operator's WAP gateway. This has the effect of preventing certain applications that wish to use HTTP for non-Web applications from operating properly, which is not such a strange idea. The firewall may block inbound IP connections on certain port numbers, which again restricts flexibility in application design. Rightly or wrongly, many IP restrictions might apply in a cellular network. The implication for developing mobile services is to ensure that the operator can support the proposed design strategy, ensuring that no IP restrictions will adversely affect the service offering. The operator should provide details about their network configuration: many operators have developer forums that developers can join and, subsequently, gain access to network configuration information.

12.6 ACCESSING NETWORK ASSETS

In addition to the RF network's support for packet communications, there is a range of other services that a cellular network can offer. Depending on the type of application we want to develop, some of these services could be essential; it would therefore be a good idea to gain programmatic access in order to allow integration of these services into the application being developed. The types of asset that an operator has are:

- voicemail services;
- call control (e.g., call forwarding, call conferencing, etc.);
- charging and billing mechanisms;

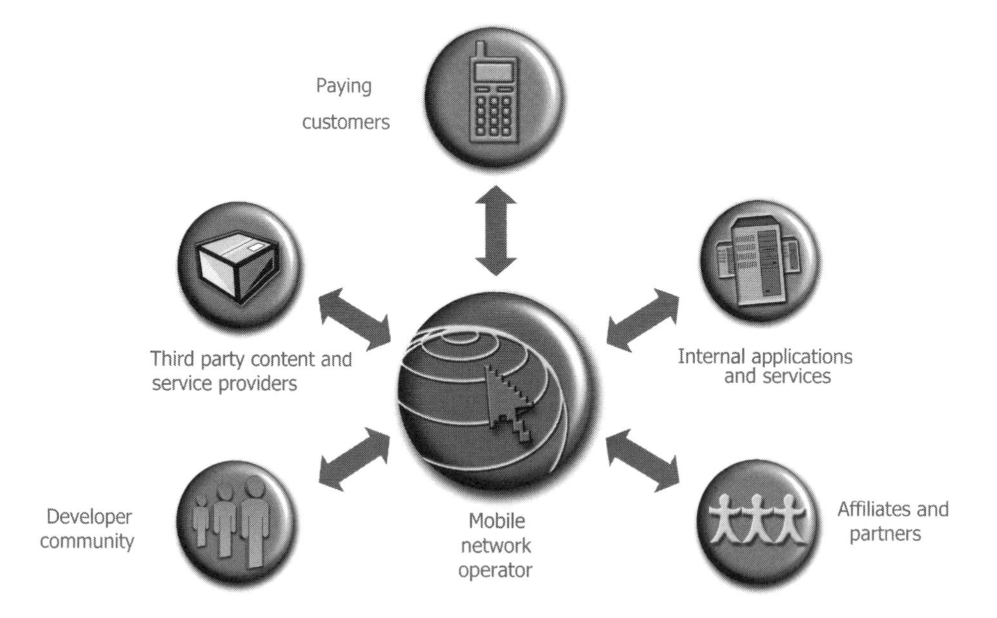

Figure 12.28 Operator context of supporting services and applications.

- bidirectional text messaging, including sophisticated mechanisms for micro-charging, like reverse-billing (i.e., paying for receiving messages);
- multimedia-messaging systems;
- WAP gateways;
- WAP-PUSH gateways;
- fine-grain IP management of mobile devices (e.g., GGSN capabilities);
- VPN and other security mechanisms (courtesy of GGSN again);
- interactive voice response systems;
- location-finding infrastructure to locate mobiles;
- numbering management schemes (e.g., short codes, alias code, multiple device numbering, etc.)

First, we should examine the problem of exploiting these assets from the perspective of the operator, since it is the operator who owns the network and has to find a way of making its assets available to third-party developers and other interested parties. Figure 12.28 shows us a possible context of the mobile network operator's business.

As the figure shows, the mobile operator view is one that places their network at the centre of the mobile cosmos: all other networks are tributaries to their network and business. The operator probably views other entities in terms of commercial relationships, not technical ones, though clearly some technical interfacing has to occur; this is the problem we are going to look at.

The primary external network is the network of customers: these are the paying customers. This is an obvious point, but its nakedness is worth examining for a moment. Historically, the development of mobile voice networks relied on customers paying for access to services. No aspect of operators' offerings has ever been free of charge, thus setting the precedent that mobile services cost money. This is a key consideration for wanting to gain access to an operator's customer base: access to paying customers. Furthermore, the mechanisms for charging, billing and collecting dues are already in place in an operator's network. A partnership with operators based on offering them a means to increase the *average revenue per user* (ARPU) might generate cash if they are willing to yield a share of the revenue to the partner. This is why the new world view of the operators includes all the relationships shown in Figure 12.28.

It is with this world view in mind that we should proceed to discuss how we exploit the technical possibilities and opportunities covered in this book within a revenue share model. Let's recap the challenges facing the operator, so that we have a context for proposing technical solutions. Here is a list of some of the challenges facing the operator.

- *Give users something interesting to use and pay for:*
  - Not necessarily what they 'want', as they don't know what it is they want yet!!! Creativity is required!!
  - Compelling enough to generate revenue.
- *Balance service delivery and subscriber demand with capital expenditure on the service delivery platform.*
- *Cope with uncertainty in next generation:*
  - Content and service types.
  - Pricing models.
  - Consumption patterns.
- *Providing a rich, 'open' end-to-end service package:*
  - Building a third-party service platform that enables any application to be launched from the network, meeting both the operator needs and the third party's.
  - Embrace a technological approach that the developer community will welcome.
  - Reach the 'event horizon' of a leading IT company, not a utility company.
- *Entering into 'networked business' models with:*
  - Affiliates.
  - Third-party service providers.
  - And even competitors.

This is a list of challenges that we first proposed and discussed in Chapter 4, a discussion that we will now elaborate on.

The first challenge in this list is one we have discussed several times already. We even went to great lengths to ponder definitions for mobile services and ways to validate a particular idea, such as using Ahonen's 5M characterization of service attributes. There is nothing prescriptive about these definitions or approaches, they are just tools to generate thoughts and reflections during the development of mobile service ideas.

Uncertainty seems an appropriate theme for the current discussion. As the first point elucidates, users probably *do not know* what it is they want until they see it, feel it, touch it, use it, abuse it or modify it. However, the 'build-it-and-see' approach is almost regarded as taboo in the post dot.com crash; cast-iron revenue has to be demonstrable or else! A conservative financier's grip may strangulate many good ideas because they lack a proven business model or are not a finished product. However, we should be honest that there are no proven business models for mobile services and the fantastic amount of 'expert' analyst reports has gotten us nowhere. However, no one likes to throw money at something with no apparent scope for success, so a means to mitigate risk would definitely be worth obtaining.

Lessening of uncertainty would come as a great relief to the operators, who already face too many variables, most of them with unfamiliar roots. Looking at the second point in our list, the impact of not knowing what services to launch causes a delicate balancing act to be entertained. On the one hand, investment in new infrastructure is required, such as building the 3G network. Current revenues from 2G have bottomed out, so we need to move somewhere. However, new infrastructure is expensive and operators would like to see a demand for new services gathering pace, so that investment is balanced by new revenue, if not now, then at least at some foreseeable and hopefully predictable point in the future. The absence of knowing what types of services are going to prove popular compounds the problem of lack of demand and perpetuates inexperience and lack of precedents in the pricing of new services.

Pricing uncertainty actually frightens operators and can even cause extreme adversity to risk, which is not what we want. They talk of not wanting to launch services that 'kill the network': presumably, where something becomes overly popular at the 'wrong' price point (i.e., too cheap). The apparent need for novel and complex pricing schemes heightens this fear. There is a concern that this might lead to pricing anomalies that could pave the way for the operator's precious resources to be exploited at the said wrong price point. For example, if operators attempt to create a so-called *event-based pricing* structure, then it has the potential to be exploited by users interested in minimizing how much they pay per chunk of data (e.g., per kilobyte). Let's say we will get charged 10p per email, the operator having figured out that this is a pricing strategy that users can relate to; thereby, positively assessing their willingness to spend money in this fashion. However, this might prove to be so popular that the operator ends up setting a precedent for service price expectation that in the end is too demanding on network resources at the service price point. With a multitude of services, there is also the danger that pricing anomalies become vulnerabilities. For example, for simply transferring data, there may be a volume-based 'transfer tariff' offered to users. However, it is perfectly feasible that some users might design a system that uses email as its transport mechanism and, subsequently, exploit the better pricing model. More surreptitiously, an enterprising user could design a modified email client that concatenates email messages, so that they end up getting five for the price of one[37] or something like that.

These examples may not be very realistic or even likely, but the point is that the operator has a finite resource (RF capacity); so, they would not want to act in a way that undercuts

[37] It is unlikely that they could get all their emails in one message, as presumably the operator will impose an upper limit in any case – such as 100 KB per message.

their ability to exploit it. This is obvious, but the way to avoid it while still launching many new services with different pricing schemes is not obvious. Paradoxically, to achieve success the operator needs a vast and diverse set of services in order to attract users and their money in the first place. This will entail opening up their network assets to third-party access, which will present all kinds of pricing headaches due to the high number of service permutations possible in offering these assets.

This last consideration relates to the final points on the list of challenges facing the operators; these are to do with opening the network up to third parties. Given the anxiety that comes from uncertainty, it would seem natural for operators to baulk at letting third parties in the door to offer their wares while the operator has such a clouded understanding of mobile services. To an extent, such conservatism is justified, except for the fact that required service diversity is impossible without calling on the creativity of as many third parties as possible to build applications. This is the most likely path to success and one that has a good deal of consensus. It seems the best approach toward finding the 'killer app' or 'killer cocktail' that will usher in the new era of mobile services.

Given the need to involve other parties in building the next generation mobile services success story, strategies are needed for building a suitable partnership framework. Furthermore, it is not enough just to build relationships on paper, creating 'partnership schemes' that are not able to empower real change. Real mechanisms have to emerge that allow other players to trade their creativity and willingness to take risk in return for a share of the revenue; invariably, these players are software experts. This is a new world where the giant money-raking billing machine of the operator has to court quick-minded software gurus who have at their disposal an insight into many of the technologies discussed throughout this book and, hopefully, an instinct for what users might want. They are people who instinctively know or can find the unfettered time in figuring out how to get the best from technologies like J2EE, J2ME and the entire gamut of techniques that enable powerful services to arise from our network of networks, even including some of the newly emerging paradigms, like peer to peer.

There should be as few restrictions as possible placed on the creativity of software engineers and their adroitness in exploiting new technologies and the assets of the operators. Ideally, they need 'full access' to the RF network. If we think of the RF network as simply a set of capabilities and assets, then surely the paramount approach is to offer these assets to the best minds who know how to create the right mix that leads to something spectacular in the hands of the end users. To do this, we need to find a way to let these minds into the front and back doors of the network. The exploitation of software techniques in the mobile world is leading to the following trends:

- *componentization of the service delivery platform (SDP)* – with chunks of next generation mobile services being wholesale developed as platform components by third parties, such as content management systems for downloadable content (i.e., ring tones, wallpaper, screen savers, video clips, etc.);
- *rationalizing the platform* – transforming the entire RF network into a kind of 'cellular operating system' that is intimately accessible to external software processes;
- *componentization of RF network assets* – realizing that hitherto embedded network assets, like voicemail, call forwarding and so on, should become components in this 'cellular operating system'.

On various occasions throughout the book, we have discussed the idea of an SDP. There is no official definition of an SDP, but the idea is to build a software platform to host common mobile services, the platform itself providing the means to offer a range of ancillary software functions common to mobile services. The first point above describes a trend toward third parties providing pre-built parts of this platform as 'service in a box' components.

A common target for platform components is the development of a *download server*. This is a system to enable the management of downloadable content to devices, such as ring tones, games and screen savers. The point is that these platform products do not solely provide a narrow part of the solution, like the content database and download mechanism itself, but the entire gamut of functions that an operator would need to run a 'downloads business' as well. What's more, the functions are extensible or programmatically accessible; this is so that other services can be built that need a 'downloads component', while not being pure content services in their own right. In the case of a download server we might anticipate 'out of the box' functions like:

- content management;
- enabling content providers to submit their own content and suggest prices;
- enabling the operator to review the content and override pricing;
- enabling the content to be published to the discovery server;
- content control, including digital rights management;
- adding digital rights management to the content to prevent unlicensed usage;
- determining who is able to access content;
- classifying content into genres, irrespective of the owner's categorization;
- content discovery;
- enabling the content to be discoverable via a Wireless Application Protocol (WAP) or Web portal;
- enabling searching and ranking of content;
- enabling download rates to influence promotion of content on the portal;
- content billing;
- providing mechanisms for billing the users;
- charging mechanisms for per-play or one-off charges;
- token-based billing;
- time limit-based operation;
- content analysis;
- tracking the success of different content types;
- tracking the popularity of each specific item;
- monitoring which devices request which content.

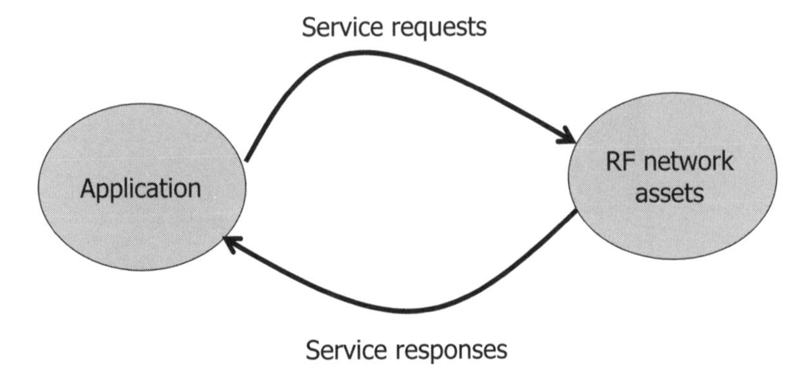

Figure 12.29 The RF network is just another 'software' asset.

In addition to these functions some vendors might also deliver pre-packaged content with the platform, mostly through syndication deals already struck with content providers, such as games programmers, ring tone providers and so on.

There are some key challenges with the SDP approach: one deals with ways of enabling flexible and rapid deployment of new services at any point in the future. Let's stick with our download example and focus on games for a moment, and then examine the issues.

A service to provide first-generation Java games is relatively easy to build and support because the games are mostly stand-alone with simple pricing models. The main challenge for the operator is in provisioning the user with the games. In most cases this involves dedicating part of the operator's portal to the games, probably calling it an arcade, with some basic ranking of titles according to popularity or maybe just a simple catalogue based on genre (e.g., puzzle games, platform games, retro games, etc.). Then, a download mechanism is required, which is nothing more than the ability to point the phone's browser to downloadable files that are subsequently downloaded using WSP or HTTP (as described at the end of Chapter 11). Then, there's the billing aspect, which for most first-generation games systems is based on a one-off charge implemented with a reverse-billed text message.

This is all relatively straightforward. However, what about providing games that offer greater levels of sophistication, requiring access to operator assets? For example, say we want to allow content providers to supply games that take advantage of the location-finding capability of the RF network. Perhaps we want to allow multi-player games where messages can be sent to each other in real time, possibly using the operator's instant messaging (IM) infrastructure. Perhaps we want games to be able to send multimedia messages to other users or even voice messages. In fact, the variety of ideas and options is quite vast if we ponder on what is possible by making network assets available to developers. There is no doubt that this would significantly augment their ability to provide interesting services to end users.

As far as accessing the network's resources is concerned, we can think of the network as just another software application that can respond to requests, as shown in Figure 12.29.

At the lowest physical level, this is made possible thanks to the GGSN in the core network, which facilitates an IP-based view of the RF network; this means we can apply any of our IP networking principles and software services to accessing resources in the RF network – thus the RF network 'plugs into' our IP network. From a software services perspective, we

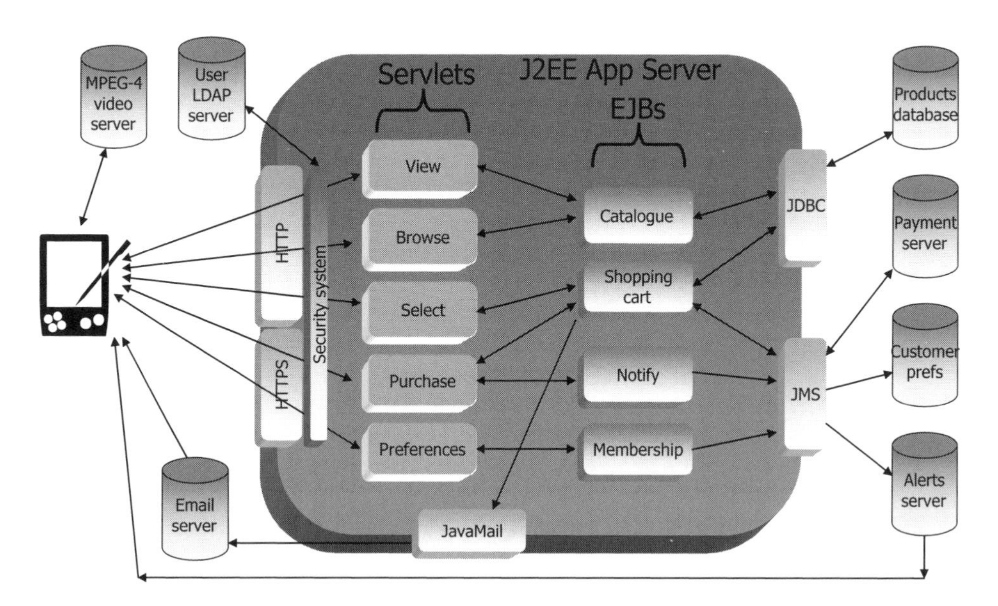

Figure 12.30 A possible J2EE mobile application architecture.

could even think of mobile devices as just another part of the entire RF network portfolio of software services and assets, enabling us potentially to map the entire RF network domain into the J2EE world – almost considering the RF network as just another EIS tier. The RF network is now a connectable service, accessible via APIs, Java Messaging or even Enterprise Java Beans (EJBs). What makes this possible is the SDP, the software layer that converts raw network assets into forms that can integrate into our J2EE world.

It would require a whole book to discuss the assets of an RF network in any depth. Instead, we will focus on the means of accessing and integrating with these, especially in relation to the J2EE and Web Services worlds. We will look at trends in this area and see how 3G mobile networks can accommodate this idea as an integral part of their design, rather than an embarrassing afterthought. As far as the assets themselves are concerned, we will look at just one of them in depth, in order to open our minds to the possibilities for next generation applications. Our focus will be the location-finding capabilities of a network, the basis for developing *location-based services* (LBSs), which will be described in Chapter 13. In Chapter 13 we will also attempt to put forward application ideas that bring together all of these concepts, if only as a thought experiment to encourage ambitious ideas for future mobile services.

12.6.1 J2EE Revisited

Figure 12.30 shows a possible J2EE system for a mobile video-on-demand application. Perhaps it would be useful for streaming old B-movies[38] to otherwise bored commuters on the train journey home. The diagram is a very high-level view, but it is conceptually plausible. On the left, we have our servlets and Java Server Pages (JSPs) which take care of

[38] B-movie-streaming services are becoming increasingly popular on the Internet.

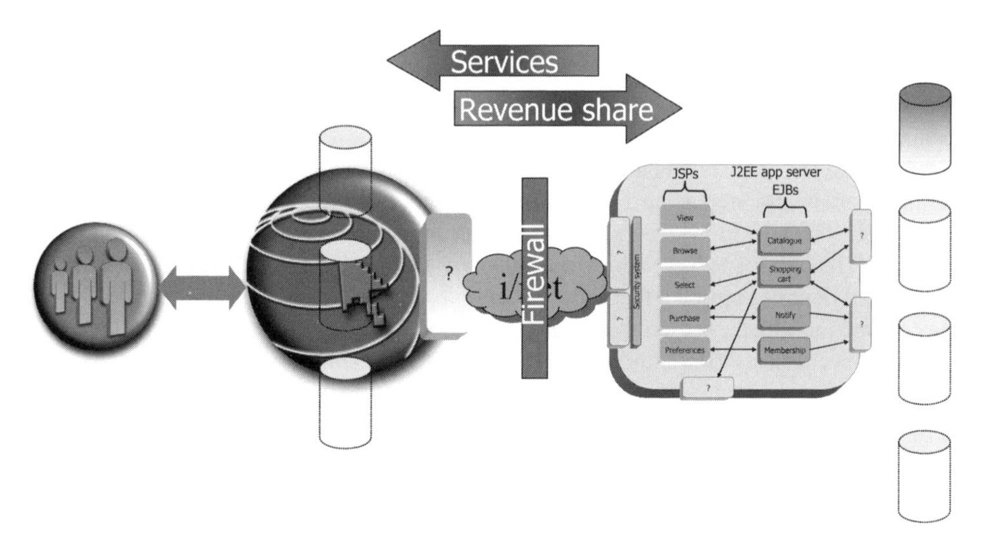

Figure 12.31 Problems with running a J2EE system outside the operator's LAN.

the presentation layer within a model–view–controller pattern, as discussed in Chapter 8. On the right, we have the guts of our system – not just the EJBs but also the external network entities with which the EJBs interact using the powerful J2EE APIs (application programming interfaces). These entities include a database server (accessed via the Java Data Base Connector, or JDBC) which contains the catalogue of movies, a payment server (accessed via the Java Messaging Service, or JMS) to collect payment, a preferences database (JDBC) which stores user-personalized information, such as a 'wish list' of movies to see in the future. There's also an alerts server (JMS) to send an alert when a selected movie becomes available, or when payment clears, or to remind the user to view the film within the specified time limit, should they not view it straight away. Finally, using the JavaMail API, we can send the customer a receipt for payment, if they request one, or we can send them a newsletter about the service.

This possible architecture for an application works well within the operator's environment, as presumably all of the different resources are available within the reach of the LAN. However, what happens when we want to develop a similar system and, subsequently, host it outside of the operator's LAN?

Running the application platform outside the operator's LAN presents us with the problem shown in Figure 12.31. On the right-hand side of the diagram, we have the J2EE system; but, some of the required external entities (to the right) are not present on the LAN hosting the application server – these are the ghosted parts in the diagram. For example, the operator has a payment gateway that implements a micro-payment system. The operator also has a server for submitting text messages and another for sending WAP PUSH messages. However, all these assets are sitting in the operator's network behind a firewall, as shown in the figure.

What we would like to do is find a way for the J2EE system in the figure to work by somehow accessing the missing resources that are present within the operator's network. In return for providing services to the end users, we would also like our application owner to receive a share of the revenue. As far as accessing the operator's hidden assets is concerned, we do not want to have to implement private leased line circuits to connect with the operator's network. The most obvious solution is to access them securely over the Internet, given its prevalence and all that we have said about the power of Internet-based protocols and software solutions.

The diagram indicates some of the problems with this approach. First, no one connects to the Internet without placing a firewall at the boundary between the LAN and the Internet itself. The job of the firewall is to prevent any unauthorized Internet traffic from penetrating the LAN. Techniques to button down the boundary are numerous; they include, for example, closing all IP ports except for those few that are subsequently closely guarded by the firewall in conjunction with the servers that sit on these ports on the inside (i.e., which have their own security mechanisms to add further levels of control to monitoring and defeating unwanted traffic).

A port that is typically open, at least for outbound traffic, is port 80 – the default port for HTTP traffic. For organizations that host their own Web servers, port 80 is also open for inbound connections. Using HTTP and port 80 safely and securely is now a very well-understood process and many software and hardware tools have become available to make it as secure and efficient as possible, mostly motivated by the obvious need to access the World Wide Web. Very often, companies will host their own Web servers on their LAN; so, they already understand the process of opening up port 80. Most Web-related and firewall infrastructures have powerful tools to manage this process safely.

The popularity of HTTP traffic on the Internet has led to the emergence of Web Services, which, as discussed in Chapter 7, proposes the use of HTTP for inter-software traffic rather than human–machine traffic, thereby enabling software to talk across firewalls courtesy of port 80 being open. We can think of Web Services as the emergence of a parallel Web space that is a network of machines, not people. Clearly, Web Services is not concerned with visualization of information, such as Web pages, and there is no marriage between HTTP and visual mark-up languages like XHTML (eXtensible HyperText Markup Language): the objective of Web Services is information interchange. Therefore, as we might expect, XML (eXtensible Markup Language) is the primary means of marking up the data flows; so, where we have the HTTP/XHTML partnership for browsers, we have HTTP/XML for software services. Technically, there is little difference between the two uses, especially as XHTML is in fact an XML vocabulary.

Figure 12.32 shows the principle of Web Services. The systems at either end exchange XML messages. We can see a sample message indicating a request from our application to the location-finding platform in the operator's network. We can imagine an XML tag pair called <getlocation>, which indicates to the operator network that we want to get the location (i.e., coordinates) of a particular mobile, the identity of which we indicate in a nested tag pair <num>, wherein we would insert the phone number of the mobile device. In response to this request we would expect the operator network to return the location results in the response HTTP stream.

The HTTP protocol allows our applications to enjoy the same fetch–response paradigm for software communication as is used for Web pages; however, we are not concerned with fetching Web pages. We may not even be concerned with fetching data *per se.* More specifically, we are concerned with asking software services on a remote machine to do work for us and return an outcome. In fact, with J2EE and programs like servlets, this is what fetching Web pages is really about – running a remote service, like a servlet on the J2EE Web engine, but with a particular target result: namely, generating a return content

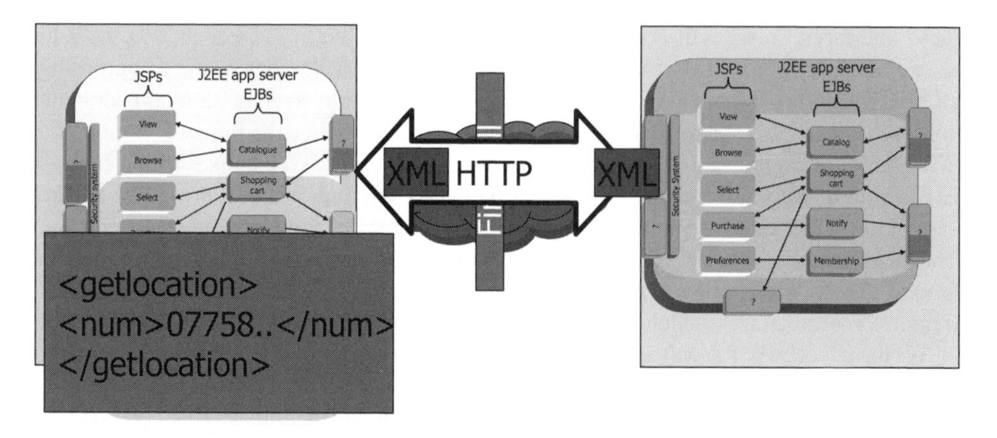

Figure 12.32 Principle of using Web services to access operator assets.

stream that articulates a Web page interface in a browser. With Web Services, we perhaps should think of the process conceptually as more like enabling remote method invocation (RMI) over the Internet, although this is not what we actually do.

This latter point is an interesting one; we could have stuck to RMI or perhaps any of the J2EE APIs, as they are all IP-compatible at the lowest level. However, they do not run very well over the Internet, especially as most of the ports they use are not typically open in default firewall configurations. Furthermore, Figure 12.32 also shows that the two ends in the Web Services dialogue could be black boxes. We do not need to know how the operator implements their system, as long as it can 'talk' Web Services over HTTP. The same applies to the other end, although we have focused on using J2EE throughout this book, for reasons justified elsewhere. Therefore, Web Services is software technology-agnostic – it cares not how the services and requests are fulfilled, just that the dialogue abides by the agreed protocols embedded within an XML stream running over HTTP.

Apart from the XML vocabulary, we need to implement Web Services. The remaining part of the puzzle from Figure 12.31 is the big question mark representing the entity that handles the inbound Web Services requests. Perhaps it is not obvious, but we cannot rely on operator network assets having Web Service interfaces, mostly because these assets are legacy resources that never had Web Services as a design consideration. Moreover, it is likely that the design of most internal resources did not take into account the possibility of external (third party) access, hence the mechanisms required to support third-party access are probably missing. In addition to APIs to enable programmatic access to an asset, plenty of other functions are required, such as charging mechanisms (e.g., per text message submitted or per location request) and auditing – mechanisms that are essential to facilitate third-party access within a workable commercial framework.

Before we address the problem from the operator's point of view, we should consider the implications for our J2EE system architecture. In our discussion of mobile services using J2EE we have emphasized the usefulness of the powerful J2EE APIs in accessing the support tier. The use of JDBC, JMS or other powerful J2EE APIs has been essential to enabling dialogue with external entities. However, these APIs do not work over firewalls, so we must now turn to HTTP and XML and utilize Web Services. Fortunately, as we have already examined in some depth, the J2EE platform is more than capable of handling these interfaces, so we will not dispense with its powers; but, it is clear that we have to re-engineer our approach slightly. How we do this depends on whether we are re-purposing an existing system to use Web Services or designing one from scratch, although common approaches may also apply. It is not that difficult to conceptualize the implementation of a 'Web Services engine': it is a software layer on top of the J2EE Web engine which enables EJBs to make Web Services requests via an API. The function of the layer will be to take care of the HTTP dialogue and the translation of messages to and from the XML domain. The exact nature of the API will depend on which of the proposed Web Services protocols is used. The two favourites are SOAP (Simple Object Access Protocol) and XML-RPC (remote procedure call), which we will introduce briefly.

We have looked at the way that our J2EE system can interact with the operator network to request services over the Internet. We will return to some of the features of the Web Services protocols (SOAP and XML-RPC), but first we should examine higher level system considerations, especially from the network operator perspective, in order to see if there is a unified or systematic approach that can be taken to 'Web-enable' their assets. Furthermore, although we may have a mechanism for making remote service requests, be it SOAP or XML-RPC, we have not said anything about what those messages might sensibly contain and whether there are any emerging standards in this regard. It would seem an important area to standardize across the industry; otherwise, the blossoming of applications and services being sought is unlikely to happen. In most markets there is little incentive to develop mobile applications unless the entire market (i.e., all networks) is accessible in a consistent manner.

12.6.2 Service Delivery Platforms Based on Web Services

Providing a Web Services interface to the operator's network is a good idea; it provides a unified and standardized interface for third parties to gain access to the 'cellular operating system'. The unification comes from enabling a common means to access any of the network assets. Whether it be requesting the location of a mobile or submitting a text message, we use HTTP/XML. This means that developers can implement powerful applications that potentially can access a disparate range of powerful network assets, but within a single software paradigm. This simplifies implementation and allows the application of systemwide software services where necessary, such as a common authentication apparatus. For example, we could insist that all requests come from the same registered source address or we could apply a common ID token to all requests, masking it from others using the SSL to encrypt the message transfers. The standardization comes from the Web Services initiative itself, which is subject to open standards agreement and is built on the open standards of XML and HTTP. This means that J2EE systems will increasingly be Web Services-cognizant by default, either via existing applicable APIs or newly emerging Web Services APIs.

Web Services dialogues are very loosely coupled: the two ends need make very few, if any, assumptions about each other and operating characteristics and implementation details can remain hidden. This is due mainly to the fact that Web Services access is via a URI (uniform resource identifier), like http://www.mymobileoperator.com/locationfinder. Any Web-connected software application can therefore access what lies behind this URI. It is the ultimate in global and unfettered access to a programmatic resource. There is no inkling of what lies behind this URI. As long as something meaningful (and expected) returns from the URI call, the Web Services client can go about its business and should be happy.

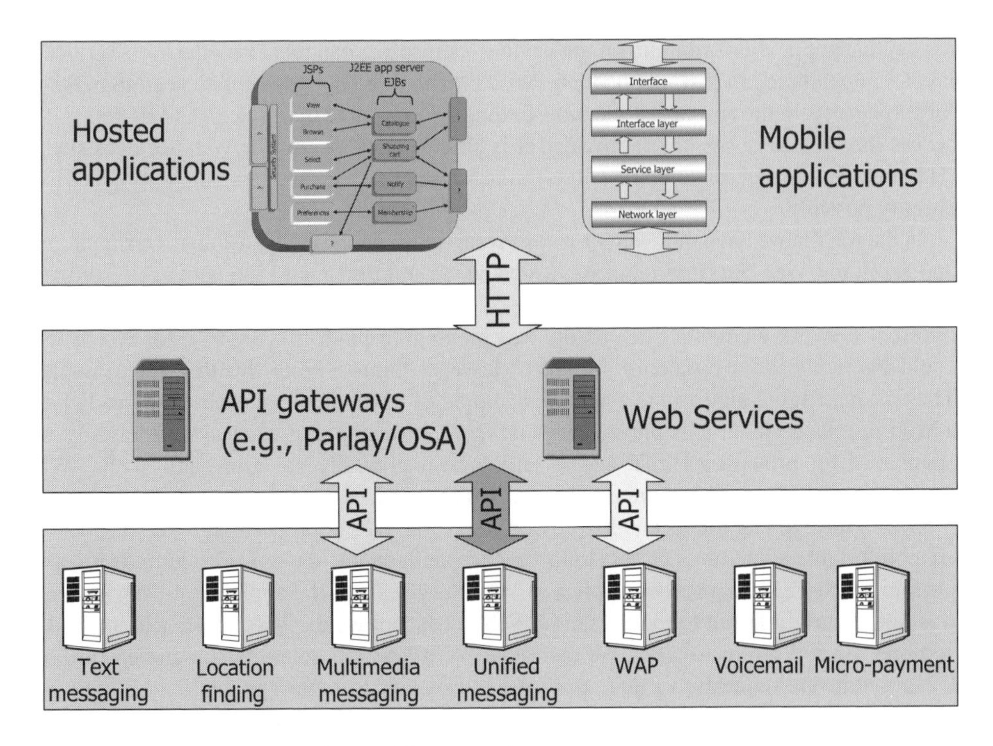

Figure 12.33 Service layer for mobile networks.

This loose coupling is a fantastic characteristic of Web Services. It means that the operator can choose to implement whatever they want that 'sits behind' the URI, and they can also change it whenever they feel like it, without any fear of upsetting clients, so long as the same results are returned (i.e., in their correct XML format). The network operator can take advantage of this fact because it enables the network assets to be available in a controllable and predictable manner without dangerously exposing lower level details and mechanics to potential third-party users. For example, a location-finding platform may have an expansive API capability, but only a subset of which makes sense to be publicly accessible. Moreover, perhaps the fine-grain control of its features needs re-purposing for public access. For example, perhaps the default interface primitive for a location query returns a lot of information in its response, some of which should not be accessible by third parties. This is not a problem: the Web Services layer can tailor responses as the need arises, such as removing or adding information to the responses as required. The response to a query could contain a restricted set of fields in the XML stream, omitting much of the data that are inappropriate for inclusion in a third-party query response.

Access to the operator's network might not use Web Services standards all of the time; some of the network assets may already have historical HTTP interfaces implemented by gateways and these might remain in place for backward compatibility, probably alongside a Web Services equivalent that will gradually take over from the legacy interface. Whether using HTTP or Web Services, Figure 12.33 shows us a possible architecture for our application-centric RF network.

At the top we have our application layer where the third-party or operator-hosted applications sit. This is not limited to applications connected directly to the Internet: there is no reason that applications running on the devices themselves cannot access the Web Services layer. For example, a J2ME application (MIDlet) running on a device may need to make a location query. In the absence of location-finding APIs local to the device, the MIDlet could access the Web Services layer, provided it is able to converse over a Web Services stack (HTTP/XML), which in some resource-limited cases may be an overreach, but otherwise entirely possible.

In the middle we have the Web Services layer itself, which is responsible for terminating and servicing Web Services requests. It acts as an interlocutor to the network resources that sit in the services layer below. A variety of APIs are in abundance for interfacing with different network elements, such as the text messaging platform (SMSC), for which we could internally use a protocol like Short Message Point-to-Point (SMPP), for example. The text-messaging platform is a notable example of a resource that probably already has a Web interface via an existing gateway server. Text-messaging gateways are already in abundance for providing HTTP access, although historically not using any of the Web Services protocols (although the trend toward Web Services has accelerated).

One of the exciting things about Web Services is the URI approach: with Web pages, we are comfortable about the idea that links (anchor tags) enable the reader to jump from one page to another. The same idea applies to Web Services and the application-centric Web. It is particularly relevant to the Semantic Web, a concept we discussed earlier in the book and one we will return to soon. We can make Web Services requests and embed further URIs within the requests, so that applications can talk to other applications as they go about their business. It is perfectly feasible for a Web Services response to tell the calling service where to go next in order to get further assistance with a particular request. For example, a location request could redirect the caller to a URI where mapping information is available for the requested location, tourist information or some other geo-coded data. These reference resources could sit anywhere on the Web, not restricted to the operator's network; so, powerful, additional Web Services can be easily incorporated into a mesh of services that enable a powerful mobile service to come to fruition.

12.6.3 Standards for the Service Layer APIs – Parlay/OSA

We have been looking at ways of providing external applications to gain access to mobile operator network resources via the Internet. Although we have discussed how this could be achieved using Web Services (HTTP/XML), we have not addressed the issue of what messages we can send using any of the Web Services protocols (e.g., SOAP, XML-RPC, etc.)

Intuitively, the messages supported by the Web Services layer must surely reflect what the underlying APIs can support. Long before Web Services emerged in the rapidly developing world of Internet solutions, the Parlay Group[39] was already addressing the issue of standardizing on a set of universal APIs to access commonly found features in telephony networks (fixed or mobile). This would allow applications developers to develop applications that could run in any Parlay-compliant environment. This would allow operators to host third-party applications, which is after all our objective: the primary challenge of allowing a 'killer cocktail' of applications to flourish on the operator's network – a necessary step toward reaching the tipping point with mobile services. Hence, we can think of Parlay API calls as the API interface into our 'cellular operating system'.

[39] http://www.parlay.org

OSA (Open Services Architecture) refers to the agreed architecture for mobile services developed by the Third Generation Partnership Program, or 3GPP[40] (also adopted by 3GPP2). These are the industry forums responsible for developing the specifications for 3G cellular systems, like UMTS. These bodies selected Parlay as the basis for providing the API for OSA.

Parlay/OSA is a set of APIs. The functions covered by these APIs include the following:

- mobility;
- location;
- presence and availability management (PAM);
- call control;
- user interaction;
- messaging;
- content-based charging;
- policy management.

These APIs are divided into common functional areas: PAM, for example, is to do with functions relating to detecting the state of the user. For example, we could have states such as:

- user is busy – in a call;
- user is busy – out to lunch;
- user is available via IM;
- user is available to receive phone calls.

We discussed these ideas in action in the prelude to the book (Chapter 1), showing how a user can take advantage of such abilities to allow more effective management of their time and tasks using these various modes of state management. PAM as an API is designed to assist the development of all kinds of interesting applications and services using potentially any of the common communication systems (IM, email, text messaging, voicemail, etc.) – mobile or fixed.[41] It is probably true that, currently, most of the telecommunications networks do not provide such features for third-party access, so Parlay APIs represent an exciting development for mobile services.

If we combine PAM with something like call control, then we get some very exciting possibilities. For example, we can use our presence information to control call handling, such as automatically diverting all voice calls to voicemail while in a meeting, allowing access to the recipient only via IM provided the user is logged in with an IM client. If not logged in with an IM client, then invitations could be sent to join an IM session, the invitation being sent via WAP PUSH. The possibilities and permutations are vast, and we are only talking about using two of the APIs in the Parlay/OSA set! By incorporating features across all the APIs, which is easily done thanks to the loose coupling of Web Services, some potentially amazing services are surely possible!

[40] http://www.3gpp.org

[41] The 3GPP OSA is obviously concerned only with mobile services, but the Parlay Group is concerned with all modes of communication, including wireline (fixed).

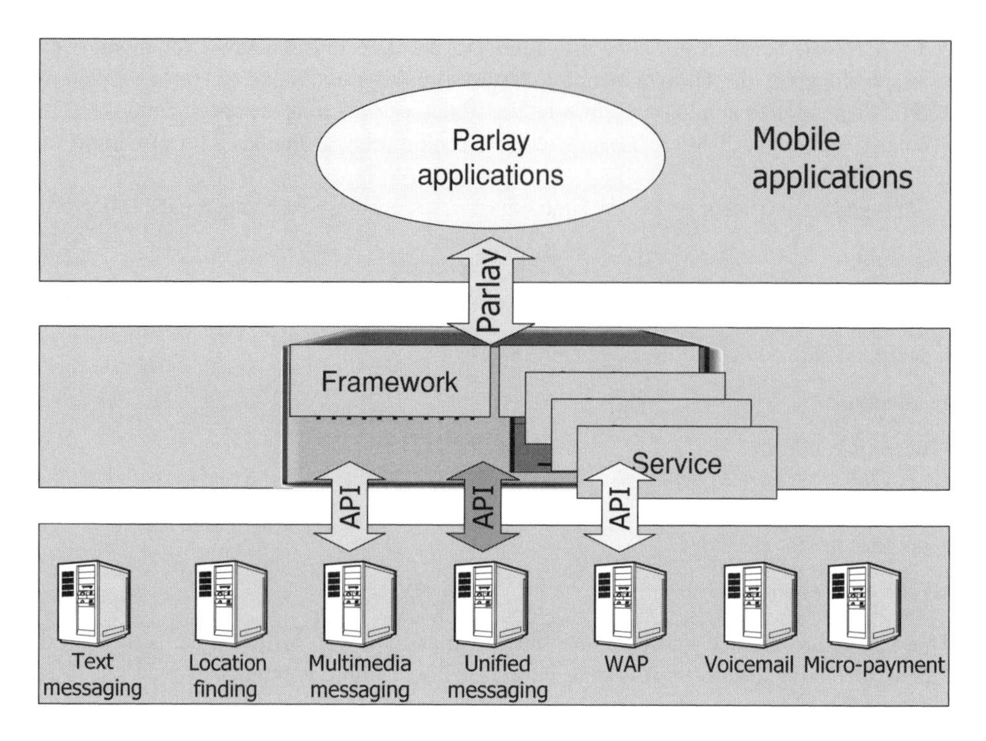

Figure 12.34 Parlay gateway.

In addition to the APIs mentioned above there is the Framework API (shown in Figure 12.34), which provides some of the housekeeping functions we mentioned earlier which any unified approach would need to have. For example, it enables applications to identify themselves, to authenticate and to discover what capabilities (services or APIs) are available via the Parlay/OSA gateway – the name of the network element that implements Parlay APIs. This is a layer of abstraction that acts similarly to the way the Web Services layer functioned in Figure 12.33. Telecoms products and platforms do not necessarily come with Parlay APIs built in. Therefore, we need to add them retrospectively, which means adding a new network element to translate Parlay/OSA API calls into APIs that are natively understood by the underlying infrastructure.

The existence of the Parlay/OSA gateway means that applications are isolated from the specific protocols used within the network by the network's assets. This allows networks to change without affecting existing applications and services. Networks can be upgraded without any impact on applications accessing Parlay/OSA APIs. The Parlay/OSA gateway is the element that implements the Parlay Framework. Parlay/OSA gateways are generally not built by the operators, but by specialist software vendors, like AePONA,[42] Telcordia[43] and Incomit.[44]

Parlay/OSA APIs are designed to enable the creation of advanced telephony applications. Formerly, we would have referred to some of this capability as Intelligent Networking (IN).

[42] http://www.aepona.com/
[43] http://www.telcordia.com/
[44] http://www.incomit.com/

However, perhaps more significantly for the future of mobile services, the way to think of this powerful capability is that Parlay/OSA allows for the 'telecom-enabling' of any IT application or service – and not just 'telecoms-enabling' by 'mobile enabling', too! All we need is to be able to access Parlay/OSA using Web Services from our third-party application; the potential to develop powerful Net-based mobile services is truly awesome. This is the objective of the Parlay X initiative – a Web Services presentation of the Parlay/OSA API.

12.7 PARLAY X (PARLAY WEB SERVICES)

Parlay X is the name of the interface for accessing Parlay/OSA APIs using Web Services. The essence of the project can be seen by quoting from the 'Parlay 4.0 Parlay X Web Services Specification' from the Parlay Group:

> *The Parlay X Web Services are intended to stimulate the development of next generation network applications by developers in the IT community who are not necessarily experts in telephony or telecommunications. The selection of Web Services should be driven by commercial utility and not necessarily by technical elegance. The goal is to define a set of powerful, yet simple, highly abstracted, imaginative telecommunications capabilities that developers in the IT community can both quickly comprehend and use to generate new, innovative applications.*

In addition to being a Web Services interface, the Parlay X interface is a much-simplified presentation of the full-blown Parlay/OSA APIs. This is to help meet the objective of attracting widespread usage and facilitating quick development cycles. It should be relatively easy to utilize the powerful features in the operator's network from our J2EE application (or handset application).

It is a truly exciting prospect that, with just one request from a URI, it is possible to initiate a phone call between two parties (*third-party calling*) or to forward an incoming call to another number if the user is unavailable (*network-initiated third-party call control*). An example of how we could utilize Parlay X in a Web-based context is to use the call control feature to indicate to our J2EE application (via a Web Service request) whenever a call is made to our device during office hours. The types of features that will be available initially via Parlay X are as follows:

- *third-party call* – this is the ability to initiate calls from one caller to another, such as putting a trader in touch with his stock broker automatically when a price threshold is reached on a stock being watched;
- *network-initiated third-party call* – this is the ability to control calls that are initiated by a caller (mobile phone): for example, to detect out-of-hours product support calls to a particular number and forward to the appropriate support person by first looking up from a database who is on call that night;
- *Short Message Service (SMS)* – to submit and receive text messages, check for delivery, etc.;
- *multimedia message* – to submit and receive MMS messages etc.;

- *payment* – to initiate payment sessions according to different charging methods, such as one-off payments, regular payments, etc.;
- *account management* – checking account status, such as account balance, account credit expiration, etc.;
- *user status* – this is similar to the presence management features of Parlay discussed above, but a very limited subset;
- *terminal location* – this subset of the API enables the location of a mobile to be ascertained.

Let's just look at one of these Web Services components in slightly more detail, in order to appreciate the power of simplicity of the Parlay X solution and to examine briefly the anatomy of a Web Services message using SOAP.

The Parlay API is constructed from service calls. The way they are defined is shown in the following example, which is taken from a network-initiated call control API:

```
handleCalledNumber(EndUserIdentifier callingParty, EndUserIdentifier called-
Party, out Action action)
```

What this function specifies is a Web Services message that will be initiated by the Parlay/OSA gateway when a call takes place to the *calledParty*.

The message *handleCalledNumber* requests the application (our application) to notify the gateway how to handle the call between two numbers, the *callingParty* and the *calledParty*. The method is invoked when the *callingParty* tries to call the *calledParty*, but before the network puts the call through to the *calledParty* (i.e., the called party will not know they are being called at this stage). It is interesting to reflect that the *calledParty* does not have to refer to a human receiver; it could be an auto-answering service (Interactive Voice Response, or IVR). The application in the return HTTP stream (reply SOAP message) is expected to return the *action*, which tells the gateway to perform one of the following actions:

- *continue* – resulting in normal handling in the network (i.e., the call will be routed to the *calledParty* number, as originally dialed);
- *endCall* – resulting in the call being terminated (the exact tone or announcement that will be played to the *callingParty* is operator-specific);
- *route* – resulting in the call being re-routed to a *calledParty* specified by the application.

This is an extremely powerful facility, easily accessible to any competent Web programmer. These days it is possible to implement Web Services with simple scripting languages, thus not even requiring any knowledge of a complex J2EE environment. Contrast this with the complex IN programming that would have been required otherwise, out of reach of most programmers.

We could use this feature as a simple filtering process on calls to a particular number. For example, we could check the recipient's activities via their online calendaring application (e.g., Lotus Notes or Microsoft Exchange). If the calendar API is flagging a meeting for that particular moment, then we could route calls to a voicemail number. Alternatively, we could look in the recipient's online employee profile to extract the number of their personal assistant or department secretary and route the call to their number.

We cannot overestimate the power of this type of service being made available via a simple API. Ordinarily, with or without Parlay/OSA (but not Parlay X), the complexity of determining and handling a call in a mobile network is well beyond the ability of an average software programmer unfamiliar with a telecommunications environment. The standard IN approach requires a high degree of network expertise, not just programming knowledge. Using the Parlay X Web Services approach, even someone with rudimentary programming skills and practically no telecoms knowledge can rapidly create a powerful mobile service. Someone perhaps skilled in programming for groupware applications, like Lotus Notes, is suddenly able to 'telecoms-enable' their wares with flare. This is perhaps one of the most exciting aspects of the emerging trends in mobile applications construction and is an oft-overlooked aspect of the '3G platform'.

12.7.1 What Does a Parlay X Message Look Like?

Here, we briefly touch on the anatomy of SOAP, which stands for Simple Object Access Protocol, or at least it used to. As SOAP actually has nothing to do with *objects* in the object-oriented programming (OOP) sense, the current owners[45] of the specification abandoned the acronym or at least its meaning. It now, apparently, stands for nothing and is just a name for a particular Web Services protocol that we will now examine in the context of its usage in Parlay X.

By now, we are familiar with such HTTP methods as GET, the one most often used to request a resource from the origin server. However, with Web Services, we are interested in passing information back and forth; therefore, we use the POST method in the request so that we can include the SOAP message in the request stream.

It is easier to demonstrate a SOAP message by repeating the example from the SOAP specification:

```
POST /StockQuote HTTP/1.1
Host: www.stockquoteserver.com
Content-Type: text/xml; charset="utf-8"
Content-Length: nnnn
SOAPAction: "Some-URI"

<SOAP-ENV:Envelope
  xmlns:SOAP-ENV="http://schemas.xmlsoap.org/soap/envelope/"
  SOAP-ENV:encodingStyle="http://schemas.xmlsoap.org/soap/encoding/">
  <SOAP-ENV:Body>
    <m:GetLastTradePrice xmlns:m="Some-URI">
      <symbol>DIS</symbol>
    </m:GetLastTradePrice>
  </SOAP-ENV:Body>
</SOAP-ENV:Envelope>
```

[45] SOAP is under the auspices of the World Wide Web Consortium (http://www.w3.org/TR/SOAP/)

In the SOAP request we can see the familiar HTTP headers, including the POST method and the path (/StockQuote); we can also see the host header: www.stockquoteserver.com. It is the responsibility of the resource at the implied URI (http://www.stockquoteserver.com/StockQuote) to extract the SOAP body (envelope) and process it. This could easily be a Java servlet using its built-in HTTP API to extract the body from the request. It could then use one of the Java APIs for XML-processing (JAXPs) to parse the XML and extract the fields.

There is an additional HTTP header in the request – one we have not seen before – called 'SOAPAction'. This header is a server-specific URI to indicate the intended nature of the SOAP request. This allows the server to determine what the request is all about without having to open the XML message to get the details. This might be useful for filtering SOAP requests: for example, to ensure that certain requests are unable to get through or to enable routing of requests in a distributed implementation (more than likely[46]).

The extra header here is peculiar to sending SOAP messages using HTTP, which, as we have discussed, is an attractive solution. However, just like many other XML payload-messaging systems, SOAP is not tied to a single transport protocol (i.e., HTTP): it could just as easily work over Simple Mail Transport Protocol (SMTP), Java Messaging Service (JMS) or one of the proposed modern alternatives to HTTP, like Blocks Extensible Exchange Protocol[47] (BEEP). Recognizing that HTTP was indeed designed for Web pages and not things like Web Services and given the prevalence of the concept, based on globally addressable resources (i.e., via URIs), alternatives have been proposed that are not so closely tied to Web pages. We can imagine asking, 'How would we design something like HTTP, if we had to do it all over again, knowing what we know now?' Of course, that's an obvious question to ask about anything, but in the case of HTTP something like BEEP is what the answer might be. According to the IETF's task force on BEEP,[48] a more precise description of BEEP is:

> *... a standards-track application protocol framework for connection-oriented, asynchronous request/response interactions.*

Let's return to the SOAP message itself: it comes in two main parts, the header (i.e., not the HTTP header, but the SOAP header) and the body, as shown in Figure 12.35; the body may optionally include information about reporting fault conditions.

The critical component is of course the body, which tells the Parlay X gateway what we require from it. In the example above we can see the crux of the request, ignoring the XML paraphernalia: a service-request called 'GetLastTradePrice'. Where the gateway can go to find the XML vocabulary definition for this request is the 'Some-URI' reference, which is a parameter in the request tag itself. Parlay X defines Web Services calls using a language called Web Services Description Language (WSDL), which itself is formulated using XML. This is similar to the Document Type Definitions we met earlier in the book when looking at XHTML and its cousins.

[46] We have not discussed the architecture of a Parlay gateway, as it is beyond the scope of this book; however, it is a networked software system just like any other. It has to be able to scale and access all kinds of network resources, etc. Therefore, not surprisingly, it is common for Parlay gateways to use the J2EE platform, as it is the ideal candidate for supporting such a product. Big J2EE vendors, like BEA Inc., are not surprisingly very heavily involved in supporting Parlay initiatives.
[47] http://www.ietf.org/rfc/rfc3080.txt
[48] http://www.ietf.org/html.charters/beep-charter.html

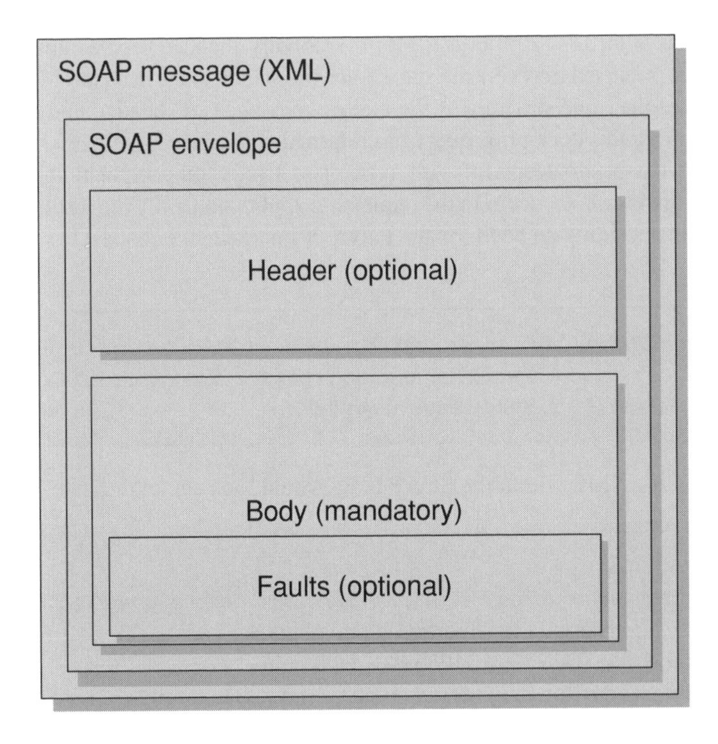

Figure 12.35 Anatomy of a SOAP message.

For the 'GetLastTradePrice' request, the WDSL would tell us that there is a field called 'Symbol'. So, the Parlay gateway will search for this field (<Symbol> tags) in the body of the SOAP message.

Having extracted the fields, the Parlay gateway routes the request to the appropriate handler in the gateway platform, which could be any part of a scalable software system – there are no restrictions on how Web Services requests are processed.

The results from the request are returned in the HTTP response, which itself contains a SOAP message, just as an HTTP stream from a Web server contains an XHTML payload:

```
HTTP/1.1 200 OK
Content-Type: text/xml; charset="utf-8"
Content-Length: nnnn
<SOAP-ENV:Envelope
  xmlns:SOAP-ENV="http://schemas.xmlsoap.org/soap/envelope/"
  SOAP-ENV:encodingStyle="http://schemas.xmlsoap.org/soap/encoding/"/>
    <SOAP-ENV:Body>
      <m:GetLastTradePriceResponse xmlns:m="Some-URI">
        <Price>34.5</Price>
      </m:GetLastTradePriceResponse>
    </SOAP-ENV:Body>
</SOAP-ENV:Envelope>
```

If the mechanics of the HTTP process itself run smoothly, then we get the familiar '200 OK' response status header. However, the bit we are most interested in is the SOAP message itself, which has the same structure as the request message (i.e., header, body, etc.). As we can see, a result for the stock price query has returned with a price – <Price>34.5</Price> – so the process has been successful. Any type of software could be behind servicing this request – it really is of no concern to the requesting application, only the result is of interest.

If we consider sending an SMS via the Parlay X interface, the request has the following structure:

```
sendSms   (EndUserIdentifier[] destinationAddressSet, String
          senderName, String charging, String message, out
          String requestIdentifier)
```

The corresponding XML within the SOAP body would look something like:

```
<sendSMS>
  <destinationAddressSet>4487718776351</destinationAddressSet>
  <senderName> EmailAlerts</senderName>
  <charging>Bulk5000 01283940</charging>
  <message>You have a new email from Joe Somebody</message>
</sendSMS>
```

The SOAP message *sendSms* requests the Parlay X gateway to send an SMS, specified by the *String* message to the specified address (or address set, which is an array of addresses), specified by *destinationAddressSet*. The application can also indicate the sender name (*senderName*), which the user's terminal displays as the originator of the message.

Charging arrangements (*charging*) can also be specified; this is the charging scheme defining how the SMS is going to be charged. The application can receive notification of status of SMS delivery: to do this, it must make a separate SOAP request using the *getSmsDeliveryStatus* message. In this case the *requestIdentifier* (see above definition), which was returned by SOAP response to the above message, can be used to identify the SMS delivery request.

We have only touched briefly on SOAP and its mechanics, but the principles are straightforward. The detail is really in understanding the message formats. However, once we have a software function that can take care of Parlay X messaging on our J2EE platform, we can then set about figuring out what Parlay X APIs consist of and how best to use them in creating interesting mobile services: our real goal.

13

Mobile Location Services

13.1 'I'VE JUST RUN SOMEONE OVER'

In the USA there were an estimated 6,356,000 car accidents in 2000.[1] There were about 3.2 million injuries and 41,821 people were killed, based on data collected by the Federal Highway Administration.

Using a cellphone to call for emergency assistance can be an awkward affair. The biggest problem is pinpointing the location of the accident. 'It's near a lamp post, near a postbox, near a turning' is not quite good enough for the emergency services to find where the accident has taken place. Because of the number of deaths and serious injuries that are aggravated by lack of a timely response from the emergency services, the Federal Government (Federal Communications Commission, or FCC) in the USA introduced a law, called the wireless *Enhanced 911* edict (E911[2]). This stipulated that a mobile phone operator must be able to locate physically (geographically) a mobile making an emergency call. According to the E911 website:[3]

> *The wireless E911 program is divided into two parts – Phase I and Phase II. Phase I requires carriers, upon appropriate request by a local Public Safety Answering Point (PSAP), to report the telephone number of a wireless 911 caller and the location of the antenna that received the call. Phase II requires wireless carriers to provide far more precise location information, within 50 to 100 meters in most cases.*

What made the FCC ask a mobile operator to carry out this location task and not some other agency? Figure 13.1 reveals the answer: by virtue of its cellular nature, the mobile network

[1] http://www.car-accidents.com/pages/stats.html
[2] http://www.fcc.gov/911/enhanced/
[3] http://www.fcc.gov/911/enhanced/

Next Generation Wireless Applications P. Golding
 ISBN: 0-470-86986-0 (HB)

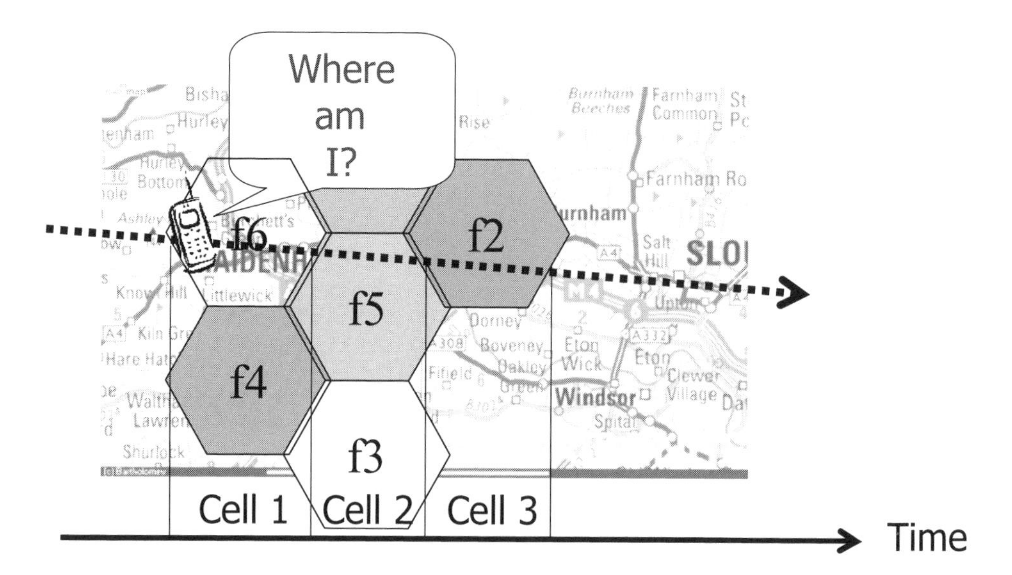

Figure 13.1 By default, a cellular network knows our whereabouts.

already knows the 'whereabouts' of each subscriber; this is the *mobility management* function inherent in a cellular network.

As Figure 13.1 also shows, this information may well be useful to the subscriber, not just the emergency services. The user might ask, 'Where am I?' This question turns out to be not only an interesting question that but one that most of us ask frequently for one reason or another.

13.2 WHERE AM I?

At the edge or centre of each cell or sector in a cellular network, there is a base station. Operators know where these base stations are, because they were installed at fixed, known locations. The mobility management function of a cellular network constantly keeps track of mobiles that are *switched on* (they don't need to be in a call). In fact, mobiles are in regular contact with the nearest base station just to keep tabs on such parameters as signal strength to ensure that the best station is being accessed and to periodically tell the network which of the reachable base stations seems best to 'camp onto' in readiness for making or receiving calls, or exchanging data. It is imperative that the network knows where a mobile is so that any inbound calls can be routed to the appropriate cell. The same applies for inbound data, such as a WAP PUSH or text message. Hence, if we know which base station a mobile is currently camped on, then we know its rough location with respect to its proximity to the serving cell site (base station). At the very least, we can have the network report, 'mobile *X* is currently attached to base station *Y* which is at our Maidenhead site'. Hence, mobile *X* is in Maidenhead. This is the basis of the *cell ID* approach to *location-based services* (LBSs).

With the ability to determine the location of a mobile, all kinds of applications and services become possible. The UMTS (Universal Mobile Telecommunications System) Forum suggested a classification of LBS into four service areas:

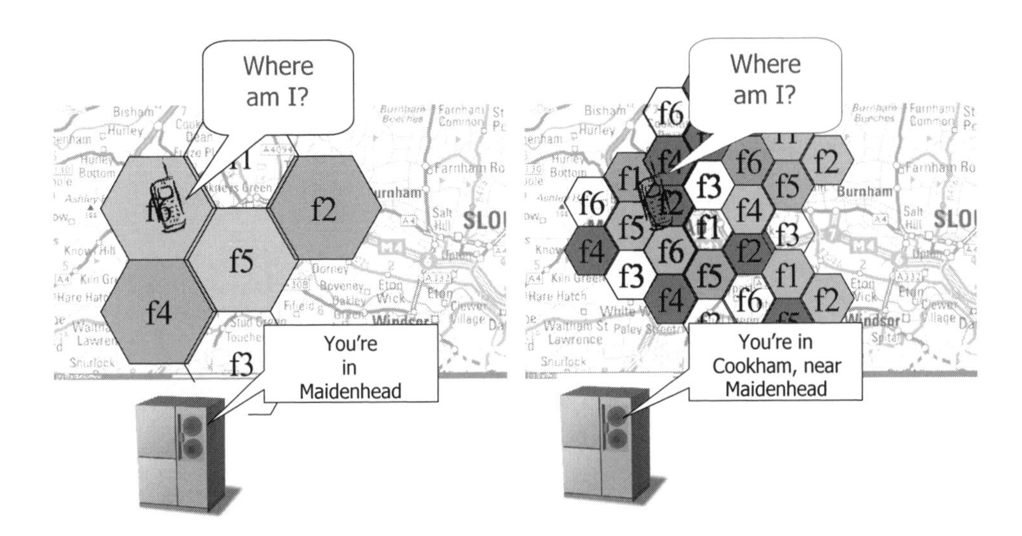

Figure 13.2 Cell ID is only as good as the cell size.

- *the commercial LCS (or value-added services)* – value-added services that typically the user will pay money for, such as 'Where is the nearest . . . ?', 'Where's my buddy?', maps, directions, location-enhanced games and entertainment, etc.;
- *the internal LCS* – the use of the location information for internal operations, such as location-assisted billing, usage patterns, network measurement, operations and maintenance (O&M) tasks, supplementary services, etc.;
- *the emergency LCS* – this assists subscribers who place emergency calls and may be mandatory in some jurisdictions, such as E911 in the USA;
- *the lawful intercept LCS* – the use the location information to support various legally required or sanctioned services, such as covert security force operations.

After the initial grievances[4] of having to deliver a tracking system to suit FCC ideals, the belated excitement about the potential of location services in the commercial domain took over. It seems there is no end of ideas for location-enabling services. Some services are possible because of location finding, while it is a means of augmenting others, leading to a rich potential for location enabling of many applications. We will discuss accuracy and its implications later, but a range of applications are possible, some better than others given a certain resolution. As Figure 13.2 shows, the accuracy of cell ID is susceptible to cell size. Smaller cells are clearly better, as it means the mobile can be more accurately located.

Once we have a location-finding system in the mobile network, subscribers' positions can be charted on a map; an exciting possibility is 'buddy finding', as shown in Figure 13.3.

[4] The operators protested that the costs of fulfilling E911 should be met by the FCC, not by the operators themselves.

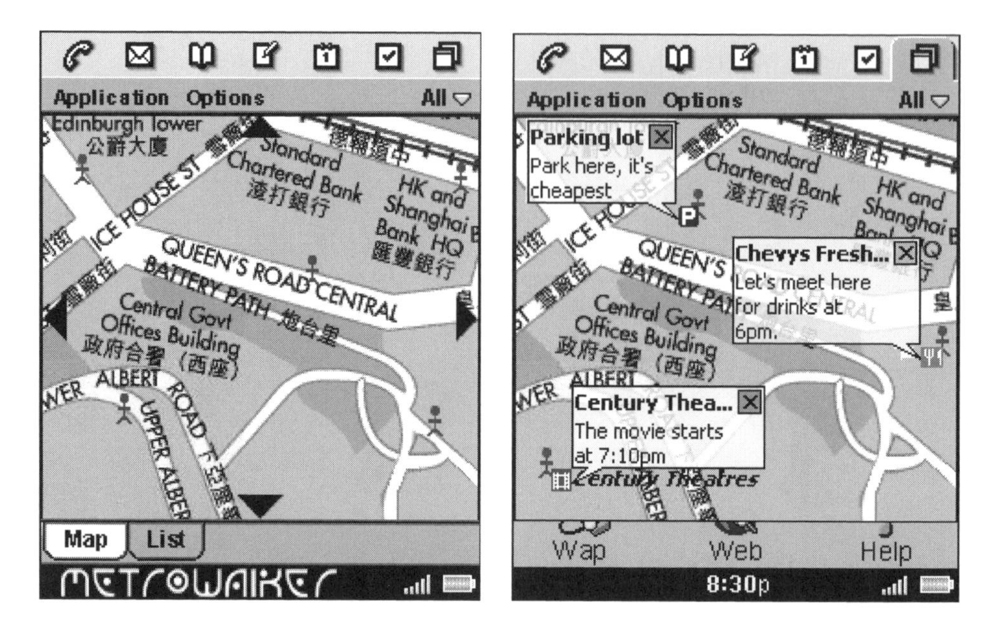

Figure 13.3 Buddy finding and instant messaging. Reproduced by permission of Magic E.

This mapping application (showing Hong Kong) appears to be locating users with high accuracy. In a dense, urban area, cell sizes can be quite small, such as 100 metres; so, even with cell ID, this type of accuracy might be possible. However, the requirement for accuracy depends on the application. It may be sufficient in many cases just to know the general proximity of someone with respect to either a fixed point or another subscriber. Two applications that use this approach are 'child minding' and 'mobile tag'.

The concept behind a child-minding application is the monitoring of the whereabouts of a child in relation to a home zone or 'free to roam' zone. If the child leaves the zone, then someone is notified – perhaps the parents who have gone out for the evening. The parents could specify the 'curfew' zone and the action to be taken if the curfew is broken, such as sending both parties a text message indicating the breach of trust, as shown in Figure 13.4.

Here we can see that Johnny is supposed to remain within the 'Happy Valley 3' region. The map here shows each of the districts on Hong Kong Island, but we can imagine that each one has its own cell and so this resolution is achievable using cell ID (we can discuss other methods of location finding later). If the LBS system in operation has a resolution limitation, then it is better to indicate this visually to the users so that they properly specify a reasonable curfew zone. There would be little point in letting the parents draw their own zone boundary, which may end up being smaller than the resolution of the system. This would be misleading and would contravene good usability practice.

There are a variety of ways that the application might operate, all configurable by the parents, who probably have password access to the application management console, which might be a desktop application or a mobile one. If the child strays from the zone, then the child could be warned, giving him or her a chance (time limit) to return to the zone before the parents get notified. The exciting thing is that with Parlay X Web Services (as discussed in Chapter 12) the application could first detect the location violation using the location API

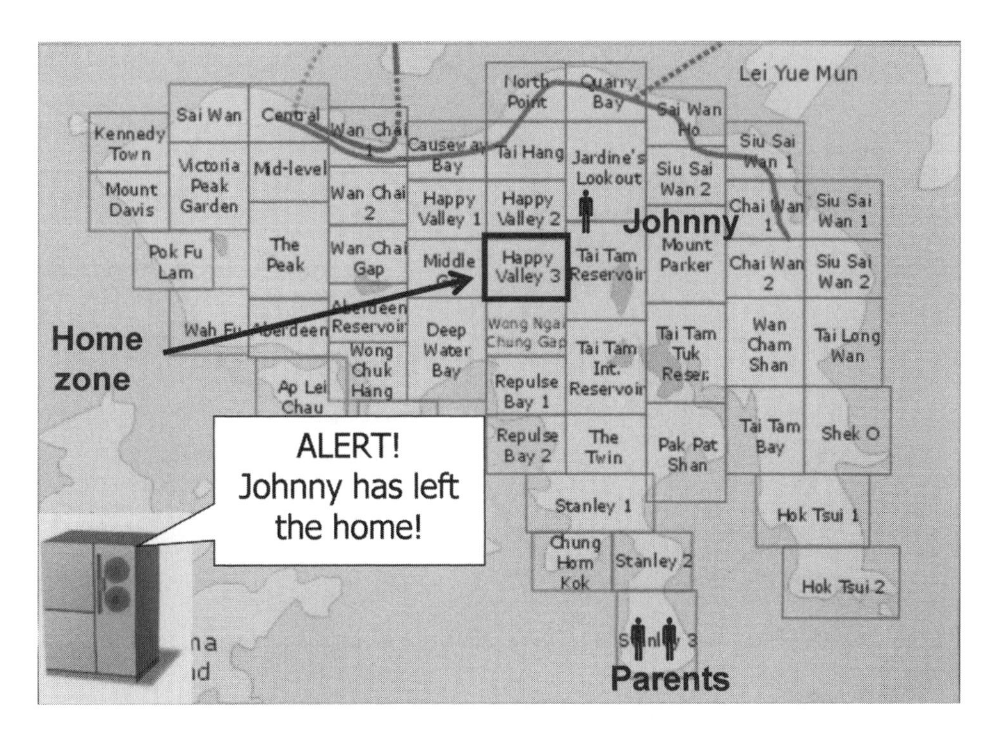

Figure 13.4 Child minding using LBS.

(application programming interface) and then use the text-messaging API to send the warning message. In other words, it is easy to build a system like this using the power of Parlay X.

If the child persists in roaming outside the zone, then a while later, after another warning has been fired off, the parents are notified. This could be followed by a third-party call being initiated by the Parlay gateway using the third-party call API. An alert could also be sent to a friend or a neighbour.

This all sounds wonderful stuff (except to the would-be errant child, of course), but such a service is contingent on several things:

1. The child has to be carrying the mobile.
2. The mobile has to be switched on (an obvious point, but an obvious limitation too!).
3. The child has to consent to being tracked (a possible and likely legal requirement in many jurisdictions, but we will get to this issue later).
4. The resolution has to be within a usable limit.
5. The monitoring of location has to be sufficiently frequent to achieve the desired result (no use if the child can zip off out of the zone for 15 minutes and come back before the next location query).

In terms of location querying, there are a number of approaches, but Figure 13.5 illustrates the general problem. We can see from Figure 13.5 that, in terms of detecting location events, we actually have four primary variables to track:

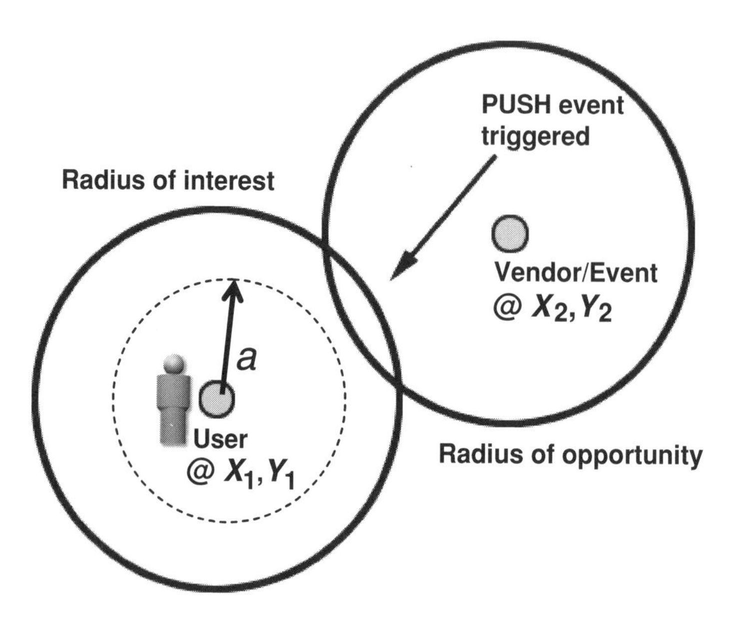

Figure 13.5 Detecting location events. Reproduced by permission of Magic E.

1. The location of the primary user (X_1, Y_1, a).
2. The location of the event or place (X_2, Y_2) that we wish to compare with the location of the user.
3. The radius of user interest (R_i)
4. The radius of opportunity (R_o)

These variables need some elaboration: they underline concepts to help model the location-finding problem, but should not be taken as a definitive articulation of the location problem.

The location of the user is the most obvious measurement or variable. The primary coordinates are the x, y position (X_1, Y_1), but it is necessary to take into account that location-based systems, like all measurement systems, have a degree of error or uncertainty, which here is articulated as a radial quantity a. So we can think of the user as being represented by a virtual circle that roams across the map's surface, bearing in mind that the size of the circle may well change as the user roams from place to place, depending on the available LBS resolution in their locality.

The location of the place of interest is also an understandable variable and, presumably, most of the time we know its whereabouts with a decent level of accuracy, so we don't really expect a margin of uncertainty in real terms, even though there might be one in actuality (due to human reasons).

The radius of interest specifies a zone around the user that we may want to use in order to understand the geographical extent of the user's interests. Say we have a tourist application and the user is on foot, then perhaps a good assumption is that they are willing to walk no more than two kilometres to get to any attraction of possible interest. This radius can be set to any value, of course. Say a user is driving along a major road and we need to notify them

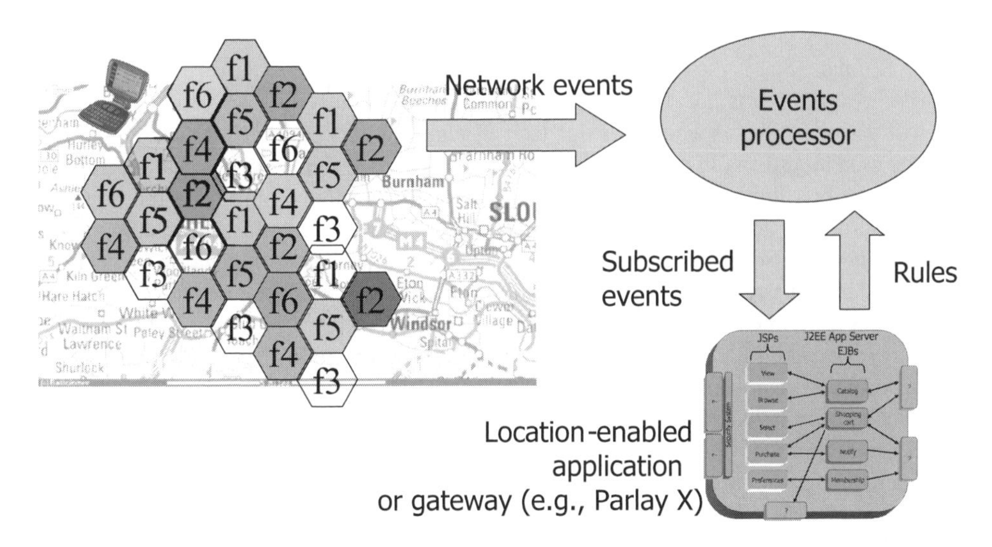

Figure 13.6 Processing of location events.

of service stations, then the radius of interest might extend to tens of kilometres, perhaps even as much as a hundred, all depending on the circumstances.

The radius of opportunity really represents the extent to which a particular place should be of interest to a user. For example, if a particular eatery were running a buy-one-get-one-free deal on a soft drink, or some other low-value good, then probably only people within the immediate vicinity would respond to that information, were it 'beamed out' from the eatery. However, for a science fiction buff a science fiction book fair might be worth beaming out over a few kilometres or even over a whole town.

In general, it would seem that most location-finding problems can be articulated with these four parameters: X_1, Y_1, R_i and R_o. A key challenge for large LBS systems is to constantly monitor all four of these parameters for each user and for each potential other user, place or event of interest – all in real time. This presents a significant real time data collection and sifting problem.

The objective of tracking these radii is to monitor when they intersect. The coalescence of radii is the cause of location events, and being able to detect and respond to these events is at the heart of many location-based systems. The more fine grain the location information becomes (or higher resolution the LBS gets) the higher the frequency of events, compounded by the increasing numbers of users of LBS applications. The processing requirement can become a formidable task.

To keep track of events in real time is a major computing and software challenge that falls outside what we have looked at in this book thus far, such as the link–fetch–response paradigm of the Web (HTTP/XHTML-MP). Figure 13.6 illustrates the essence of a system for tracking location events.

Figure 13.6 shows a generic architecture of an events-driven software platform to facilitate LBS applications. We have our now-familiar J2EE (Java 2 Enterprise Edition) platform at the bottom, which could be hosting the actual location-enabled application itself, or it could be a gateway, like Parlay X; it doesn't really matter which at this stage of the discussion.

Two primary problems need addressing. First, what is the nature of an events processor itself? Second, what is the means of receiving events into the J2EE platform?

The first question is not something we will address deeply, but it is worth dwelling on for just a short while, especially as this type of real time processing is not something we may be used to, especially when approaching the software design from a Web background. Perhaps the first thing to point out is the nature of the events coming from the network or, more crucially, the volume of events. If we are planning to monitor continually all the mobiles in our network, then the frequency and volume of location updates will be vast, especially in a fine-resolution system. A user walking along a street in a major urban area could walk from one cell to another every few minutes. However, if we are using satellite tracking (assisted GPS, or A-GPS, which we will discuss in Section 13.4), then we may well be tracking updates much more frequently than this. Let's consider a user with a walking speed of something like 4 km/h, which means 4,000 m every hour. If we assume something like 25 m accuracy (even less is possible), then we potentially have 160 location updates per hour for a constantly walking user. This has to be multiplied by the number of users, modulated of course by their own speed and resolution parameters. We can see that we are talking about large volumes of events, especially if they are concentrated into a single application (platform) for monitoring.

The problem of monitoring large volumes of events is not just the raw data flow but also the number of comparisons needed in order to interpret the data. At its crudest, this depends on how many opportunities are in the database. If we imagine a database with 100 opportunities, then we have to compare the location of these opportunities against all the incoming location update events to see if we get a match, depending on what the condition is – at its simplest, it could be the overlap of the two circles we looked at: the circle of interest and the circle of opportunity. Of course, we can envisage all kinds of optimization of the events-monitoring solution. We wouldn't necessarily have to compare all opportunities against all events if we pre-arrange the data into geographical zones.

No matter the means of comparing location, the crux of the matter is that a hefty processing task is at hand and the whole process needs to take place in real time. It is a difficult design problem to achieve the minimum level of real time responses across a large volume of transactions. For example, a surge in the volume of events (e.g., as more and more users start moving faster) should not slow the performance of the system to the detriment of the timeliness of the output. In other words, if our sci-fi buff is approaching the sci-fi fair in their car, then they do not want to be notified several kilometres after they passed the entrance to it because a whole batch of other events crowded out the processing queue. If we have a location-enabled game, like Mobile Tag (see Figure 13.7), then real time performance is critical to the game. We have to be able to constantly detect the whereabouts of players with respect to each other, so that players can tag an opponent – the object of the game. There are many variations on this simple theme, all of them sensitive to timely processing of the location variables.

Solutions to handling vast amounts of input data and comparisons tend to use in-memory database technology. In-memory applications run entirely in the dynamic memory (RAM) of the computer, thus avoiding hard disks, which are orders of magnitude slower than RAM. Companies like Times Ten™ produce solutions based on this approach that are specifically aimed at events processing, originally for financial transaction applications and more recently targeted at mobile applications.

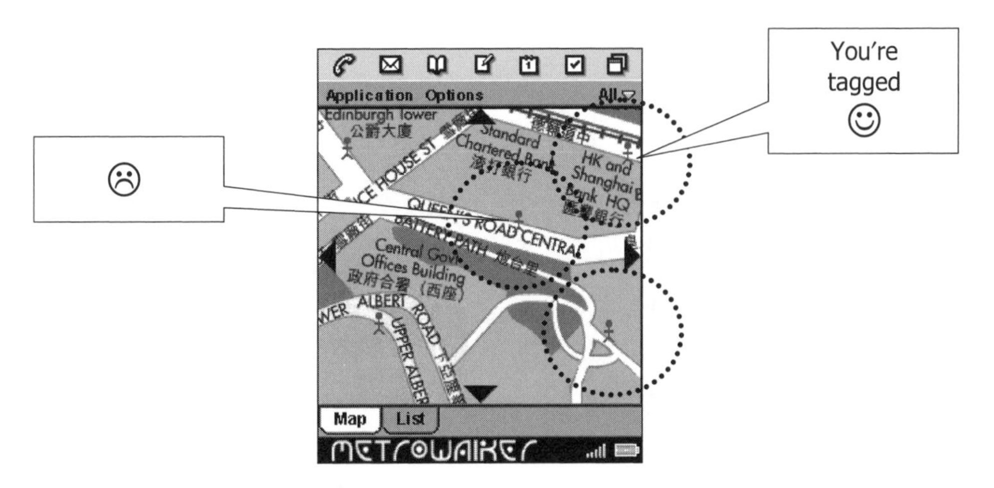

Figure 13.7 Mobile tag. Reproduced by permission of Magic E.

Real time events handling has been a challenge in the finance industry for a long time, just like many other problems from the finance industry that prompted solutions within the J2EE specification. Therefore, not surprisingly, the designers of J2EE included features that make J2EE suitable for asynchronous events processing.

When examining how J2EE mobile applications can interface with the events processor, throughout the book we have focused on a design pattern for J2EE mobile applications that divides the logic into two parts: presentation logic and business logic. It is possible to feed events into a J2EE platform via HTTP (HyperText Transfer Protocol), such as attaching event messages to HTTP requests using POST methods: to an extent, this is a viable solution. Indeed, it has to be viable if Web Services is going to work, especially in a Parlay X context where we can expect the Parlay gateway to send event-based SOAP (Simple Object Access Protocol) messages to our application in just such a manner. However, we will return to this consideration after we examine a feature of J2EE that is more ideally suited to message handling: namely, the Java Messaging Service (JMS) and a particular type of Enterprise Java Bean (EJB) called a *message-driven bean* (MDB).

13.3 MESSAGE HANDLING USING J2EE

The concept of software collaboration via message passing is sometimes confused with collaborating via remote procedure calls (RPCs). The latter is what we discussed in depth when we looked at remote method invocation (RMI). Unlike RMI, messaging is a very loosely coupled process: the sender and receiver remain totally disassociated from each other. In fact, the sender of a message may not even know who consumed it. It could be a general message, like 'this just happened in cell 12', and any software process (subscribed) listening to events about cell 12 can pick it up. The fire-and-forget nature of message sending is a useful communications paradigm that we call *asynchronous messaging*. An example of the usefulness of asynchronous processing in general may serve us well while thinking about location services.

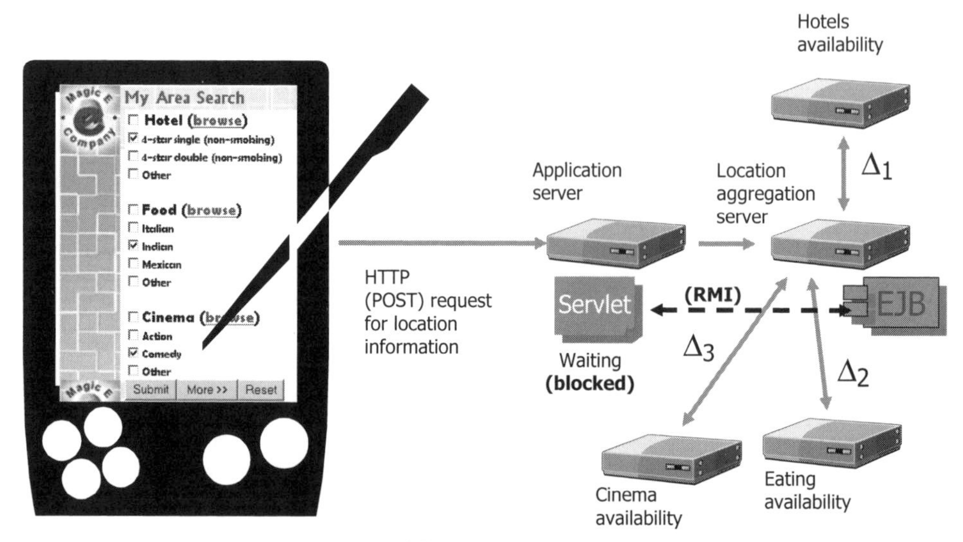

Figure 13.8 Traditional communications can get bogged down waiting for responses.

As Figure 13.8 shows, a user has unexpectedly arrived in a foreign place (perhaps for a last minute business meeting) and wishes to find and arrange accommodation, food and entertainment. Using the 'My Area Search' application, the user selects the required search fields and then specifies some particulars: 4-star nonsmoking hotel room, somewhere to eat Indian food and somewhere to watch a comedy film that evening. This type of request is a common requirement for travellers of all kinds. LBSs have a lot to offer the traveller, particularly if the information service is available in the user's language and uses familiar terminology.

The emphasis in this example is on the query being a real time concern for actual venue information, not just a static directory of venues (like *Yellow Pages*). For example, the user wants to know where there are *available* rooms, not just hotels in general. The user would also ideally like to book a room, book the restaurant (unlikely[5]) and book the cinema ticket. The back end required to fulfil a request would probably involve interfacing with a variety of systems. For example, to understand hotel room availability, we need to find access to a hotel information system, perhaps several of them, covering a variety of hotelier chains. To access cinema-seating availability, we would certainly need to interact with a cinema back end, perhaps for several cinema chains too. The problem is that each of these queries could take a long time to process, especially if we are reliant on gathering of results from several sources. It is conceivable that parts of the process could even take minutes to complete, depending on the back end processes, many of which could be on legacy machines.[6]

[5] It is unlikely that restaurants would manage information to this degree. It is conceivable and at some point likely that major eateries would advertise some real time data, like special offers and menus, possibly daily, but enabling seating availability to be placed online is perhaps too much to expect.

[6] We often forget that a large proportion of information systems continue to run reliably on legacy platforms that emerged long before the heady world of J2EE and Web Services, etc.

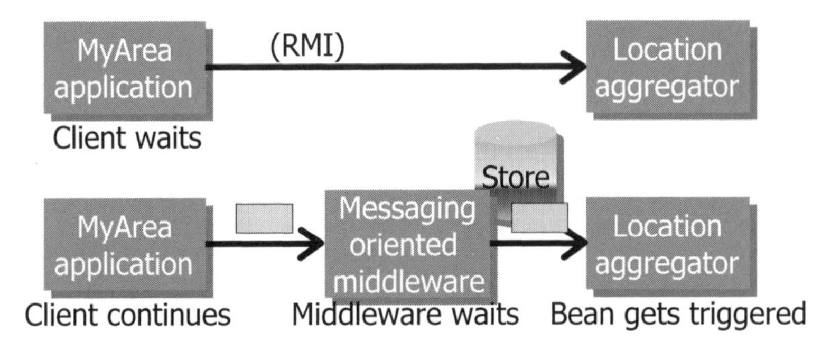

Figure 13.9 Messaging acts like an 'interlocutor' for the software call.

Without examining the details of our system design, we can well imagine that at some point in the process a software call is needed to gather one element of the required information set. Subsequently, the calling software may have to sit idle waiting for a response (while a process somewhere on the end of the command chain gathers its data, such as on the hotel information system). Figure 13.8 shows a servlet on our application server making an RMI call to a location aggregation server and being blocked, which means it is waiting for a response and cannot continue until it gets it. In other words, the servlet is blocked from completing its task, which is to hand the result back to a JSP for rendering the user interface (XHTML, or eXtensible HyperText Markup Language) via a reply HTTP stream. The consequence of this blocking is that the user, having submitted their Web-based query (i.e., a form via HTTP POST), has to endure the agony of an irksome egg timer or other animated 'progress' icon, while the whole process waits to complete. The user has no clue what is going on, would probably assume that the entire process has died and be tempted to abort (press 'stop') and start again or get frustrated, possibly considering the entire service as dysfunctional.

Figure 13.9 presents an alternative to the blocked RMI call (top of diagram). Instead of making an RMI call to an EJB that makes the query and waits, the EJB fires off a message to all the information sources asking for the queries to be answered. The EJB then returns control back to the servlet, which can hand off to a JSP (Java Server Page) to generate a courtesy Web page for the user. This page might state that the query is under way and that the user may have to wait for a certain time period to receive a result, which will be notified to them. This means that the user at least knows what is happening and can use their browser for other tasks while waiting, perhaps to check email, local tourist guides or just finding a taxi from the airport to the town (where the hotel is hopefully going to be available). Back at the application server, reply messages will eventually arrive from the information brokers. Once enough information gathers or a timeout interval elapses, the software can push out a WAP PUSH message to the user, steering them to a URL (uniform resource locator) where the results are available (having been stored somewhere in a local database after the message-gathering process).

As Figure 13.10 shows, J2EE has an API called JMS for sending software messages. But, where do these messages go? Just as email messaging has an infrastructure of email servers, messaging has an infrastructure of *message-oriented middleware* (MOM). Among other functions, MOM provides the critical *store-and-forward* function of any messaging system that requires resilience to message loss. Messages are labelled with topics and then

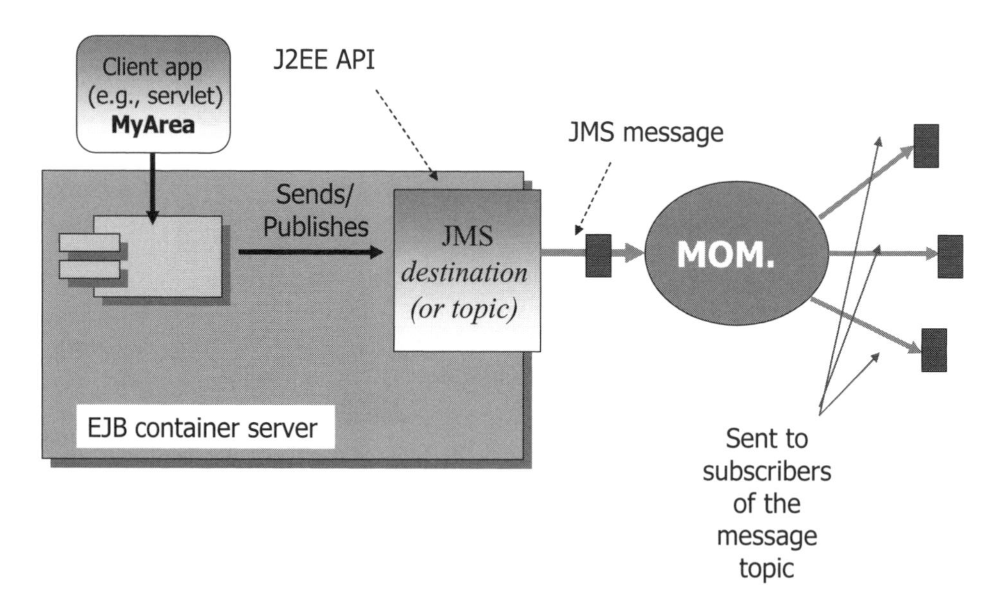

Figure 13.10 Sending messages using JMS.

broadcast to potential message recipients, and it is up to the recipient to take its message according to the message topics it is interest in (*subscribed* to).

We will not discuss the detailed mechanics of MOM and messaging in general, only mention the basic principles. MOM itself is provided by a number of vendors – popular products on the market include:

- Rendevouz by Tibco;
- MSMQ by Microsoft (now part of the .NET infrastructure);
- MQSeries by IBM;
- Tuxedo by BEA Inc.

The method for receiving a message into a J2EE application is via a special type of EJB, called an MDB, as shown in Figure 13.11. An MDB does not interact with any software entity (e.g., another EJB or a servlet) to receive and process requests for its services. It can only 'come alive' by receiving a message with a subject (topic) that correlates with the topics it has subscribed to on the J2EE server. For example, if it subscribes to all messages with the moniker 'Message about hotels in Munich', then in response to such a message passing through the MOM from some source the J2EE server will pass the message to the bean and then cause the bean to execute. The bean has only one method (or function), and this is run when the bean acts in response to an inbound message of interest. The MDB is a J2EE program, so it can access any of the J2EE APIs it likes, including RMI. Therefore, it is able to call on the help of other beans to achieve any programmable task possible on the J2EE server(s). In our earlier example an MDB would subscribe to several topics in relation to an initial enquiry from the user, such as hotel availability, cinema tickets, etc. Each time the bean 'wakes up' to a messaging event, it could amass the information (e.g., in

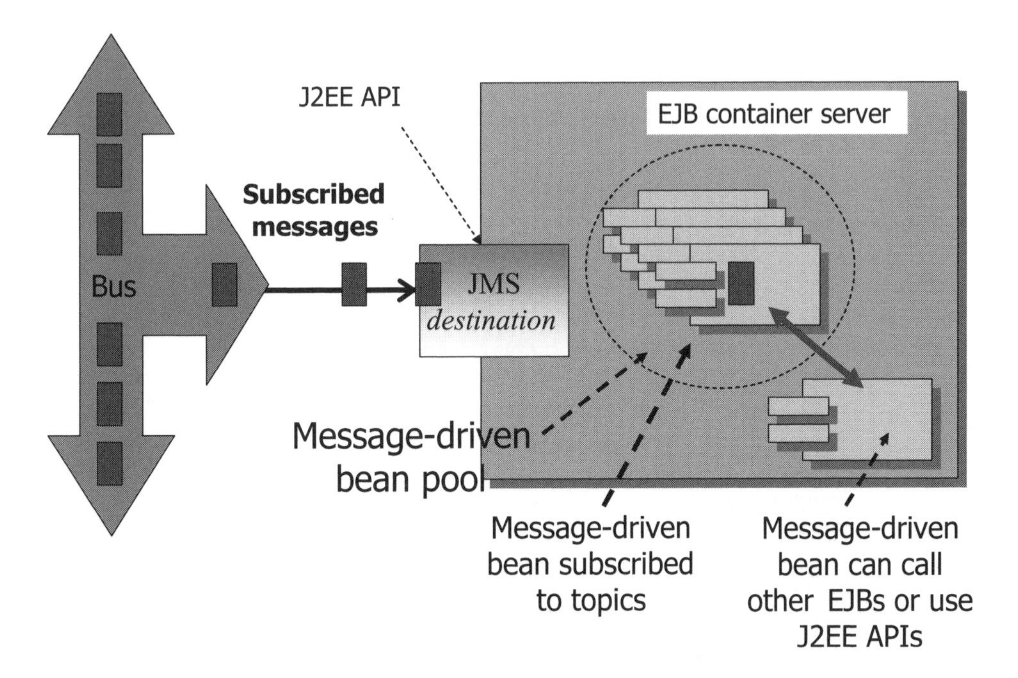

Figure 13.11 Intercepting messages using MDBs.

a database) until all expected messages have been received to enable the user's initial query to complete. It could then use HTTP or JDBC (Java Data Base Connector) to submit a WAP PUSH message to the user, informing the user that the query is finished. The message would contain a URL to reach an application (e.g., servlet) that can present the MDB's aggregated messages (from the database) in a meaningful manner.

It has become increasingly popular for mobile service delivery platforms to incorporate support for messaging systems, largely because their asynchronous 'fire-and-forget' nature enables systems to be constructed that are scalable in the context of high-volume real time event processing.

When thinking about location events, it is easy to see how a message-driven software infrastructure maps neatly onto the event-driven world of LBSs. We can envisage a generalized mechanism for location-based applications that uses a messaging bus to pass round location events. These events could be a mixture of general events, of wide interest to many subscribers, and specific events, of interest to a particular user or group of users. The message topic can be used to discern intended recipients (if there are any). In addition, MDBs have a message selection logic that enables conditional execution, managed by the J2EE application server. For example, messages could be in constant circulation regarding shopping offers from a chain of supermarkets. The participating J2EE server could bring alive a particular MDB using a condition like, 'If the offer involves audio goods and is in the Maidenhead area'. If this condition becomes true, then the relevant message arrives at the MDB, which then runs, consumes the message and does something useful in response.

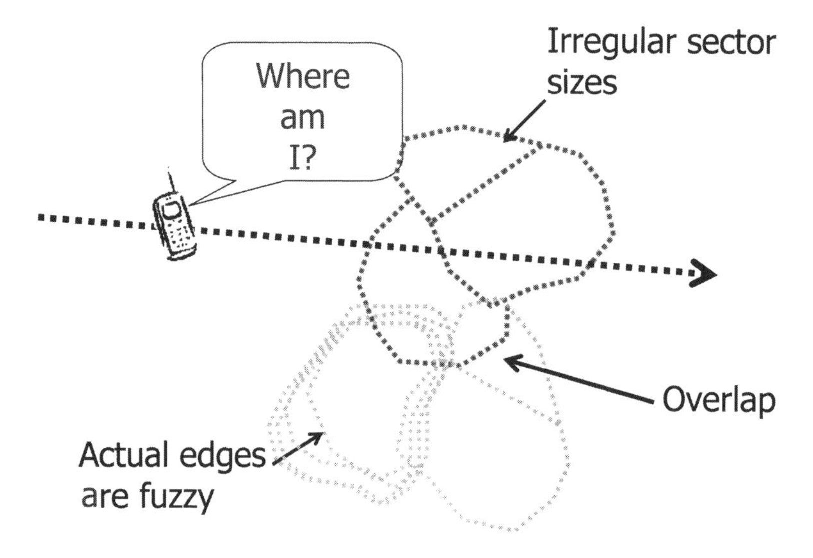

Figure 13.12 Uncertainty of location based on cell boundaries.

13.4 ACCURACY OF LBSs

Let's return to the methods of location finding. A major concern for any service or application is the accuracy of the location-finding process. Perhaps it is not just the accuracy but also the consistency quality of location finding to a certain minimum level of guarantee that is more important: otherwise it is easy to envisage a service where the user has to suffer hit-and-miss performance.

Figure 13.12 shows some of the problems associated with cell-based ID. Cell boundaries do not follow the neat honeycomb structure suggested by an idealized explanation of cellular systems. The edges suffer distortion due to the irregular physical topology of the cell and its physical influence on the strength of the RF (Radio Frequency) signal throughout the area. Cells also overlap, as they are required to do – to a certain degree – in order for mobiles to smoothly hand over from one cell to another as they move around the network area.

Figure 13.13 gives us an indication of the effect that cell uncertainty could have. Let's say our user urgently needs to use a temporary office facility (such as the Regus Chain in the UK), probably to host a last-minute meeting. The user consults their location-finding application to ask, 'Where is the nearest Regus?' The figure indicates that the system determines the nearest location using a simple cell overlap technique. In other words, the Regus in the same cell serving the mobile device is the one that is determined as being nearest. However, as the figure shows, the actual nearest Regus office is in a neighbouring cell and doesn't get picked up by the system.

This example of inaccuracy seems contrived and we might feel able to posit various ideas for attempting to lessen the problem. While this is true, however, the use of cell-based finding techniques remains problematic and the above example is still a representative illustration of the issues. Cellular systems are notoriously unpredictable and difficult to tune – the RF goes wherever it pleases and it is not possible to tightly control cell site locations, never mind the obviously impossible-to-control variations in the terrain. The vagaries of

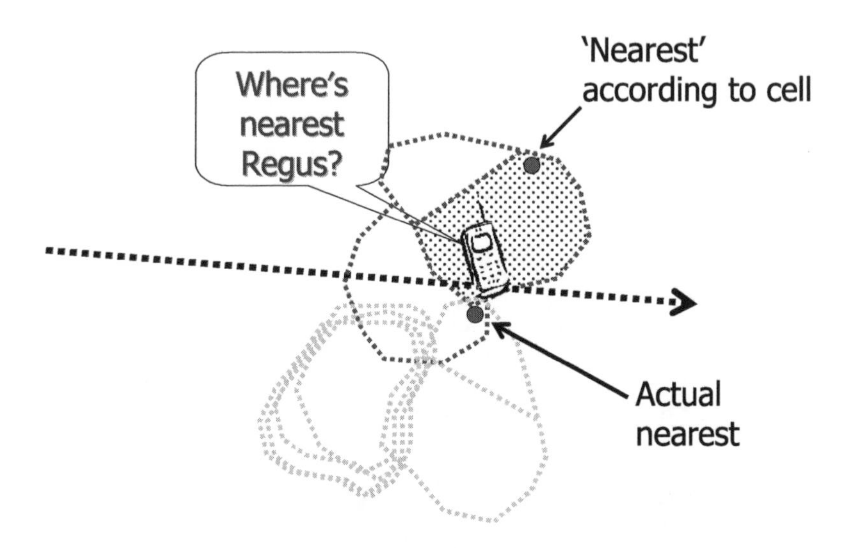

Figure 13.13 Uncertainties in cell size can distort location finding.

cellular communications remain apparent to all those who continue to experience less-than-satisfactory service from time to time, such as dropped calls or garbled speech.

However, perhaps the greatest problem with the cell ID approach is variation in cell size. We acknowledge that small cell sizes in dense urban areas are probably sufficient for many types of application, notwithstanding the above problems. However, the existence of small cells is incidental to the needs of location finding. They are small because there are many people in the areas they need to cover. This coincidentally allows accurate location services in those areas. However, the accuracy needs of an application will often have nothing to do with the density of people. For example, if we want to get the detailed whereabouts of another user (with their permission), then the small cells may be sufficient to support this if the user is in a small cell. However, if a user wants to use the same application, but in a more sparsely populated area with larger cells, then it no longer works well or, possibly, not at all.

Figure 13.14 shows some actual cell site locations (green triangles) near some of the water parks in the Cotswold Hills in England. Let's assume we have an E911 type of service requirement. If a tourist needed emergency assistance at one of the lakes, then trying to locate them using cell ID is going to be problematic. The same would be true of any service that needed to locate a particular spot within the mesh of lakes.

Clearly, if we need high levels of accuracy, then we need to look at alternative solutions to compliment cell ID.

Figure 13.15 shows us a possible improvement. The concentric rings radiating from the base stations indicate intervals of propagation distance at regular intervals of time, reminding us that RF waves travel at a finite speed, albeit very fast. If we can measure the time that signals take to get from the mobile device to the nearest base stations, then we have an idea of how far the mobile is from the cell site. A simple geometric calculation (*triangulation* or similar) based on the observed time difference (OTD) of transmissions from each base station will enable the calculation of the whereabouts of the mobile device, as shown by the shaded area on the map close to where the darker rings overlap.

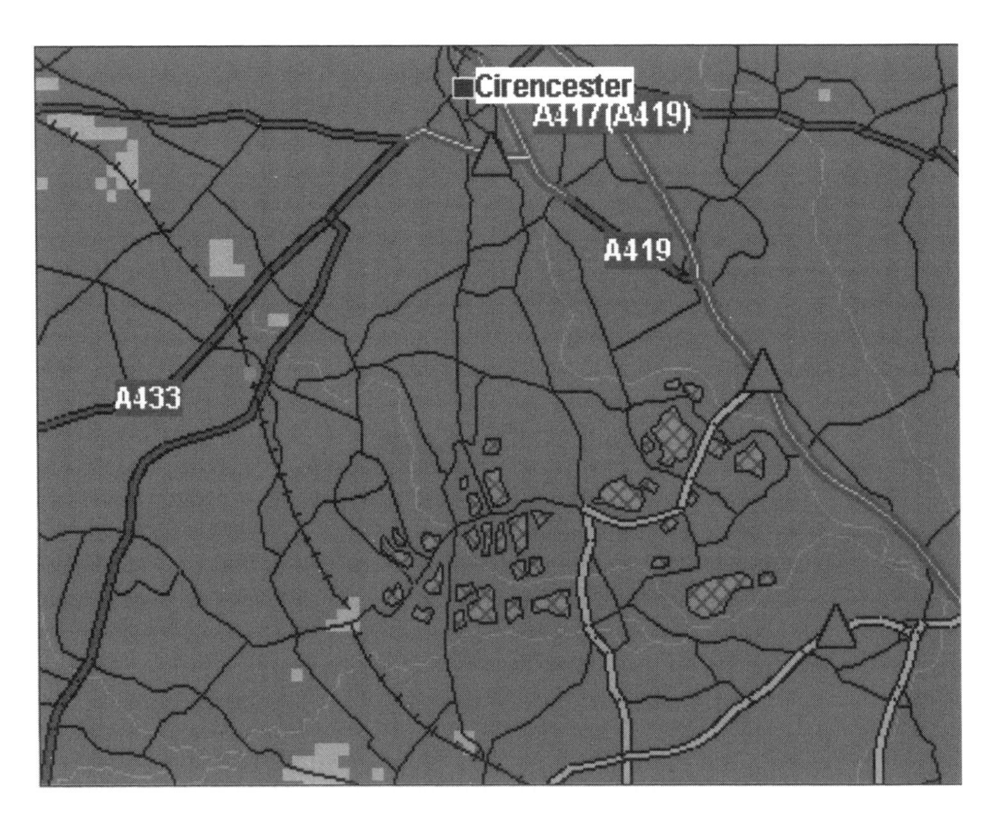

Figure 13.14 Cell locations near Cotswold water parks.

For this system to work, we need the mobile device to report the propagation measurements to the triangulation system for it to perform its calculations. However, by taking other measurements in the network, triangulation accuracy can improve by making better use of these reports. In GSM or UMTS a *Location Measurement Unit* (LMU) makes radio measurements to support the positioning of mobiles (see Figure 13.16). In the case of OTD the LMU measures the exact timing of transmissions in order to provide an accurate reference for making comparative measurements. This enhancement to the measuring process, as well as enhancements to the handsets to better measure the timing differences, is an improvement over standard OTD, called Enhanced-OTD, or E-OTD.

Measurements from the mobile and the LMU are made available to a higher entity in the network that can perform location interpolation and decide where the mobile is, based on whatever coordinate system is used (e.g., Cartesian). This is called the *Serving Mobile Location Centre* (SMLC).

There are many variations on E-OTD, most of them using a variety of possible measurements and other metrics in the network to achieve the same basic triangulation approach. The accuracy of most of these systems improves as the number of fixed references available increases, to give us an indication of any distortion in the measurements made by the mobile. We can think of the LMU as being a fixed timing reference that enables reduction of uncertainty in mobile measurements. Corrective techniques mainly attempt to overcome the fact that the RF seldom reaches the mobile directly (*line of sight*, or LOS) and that any

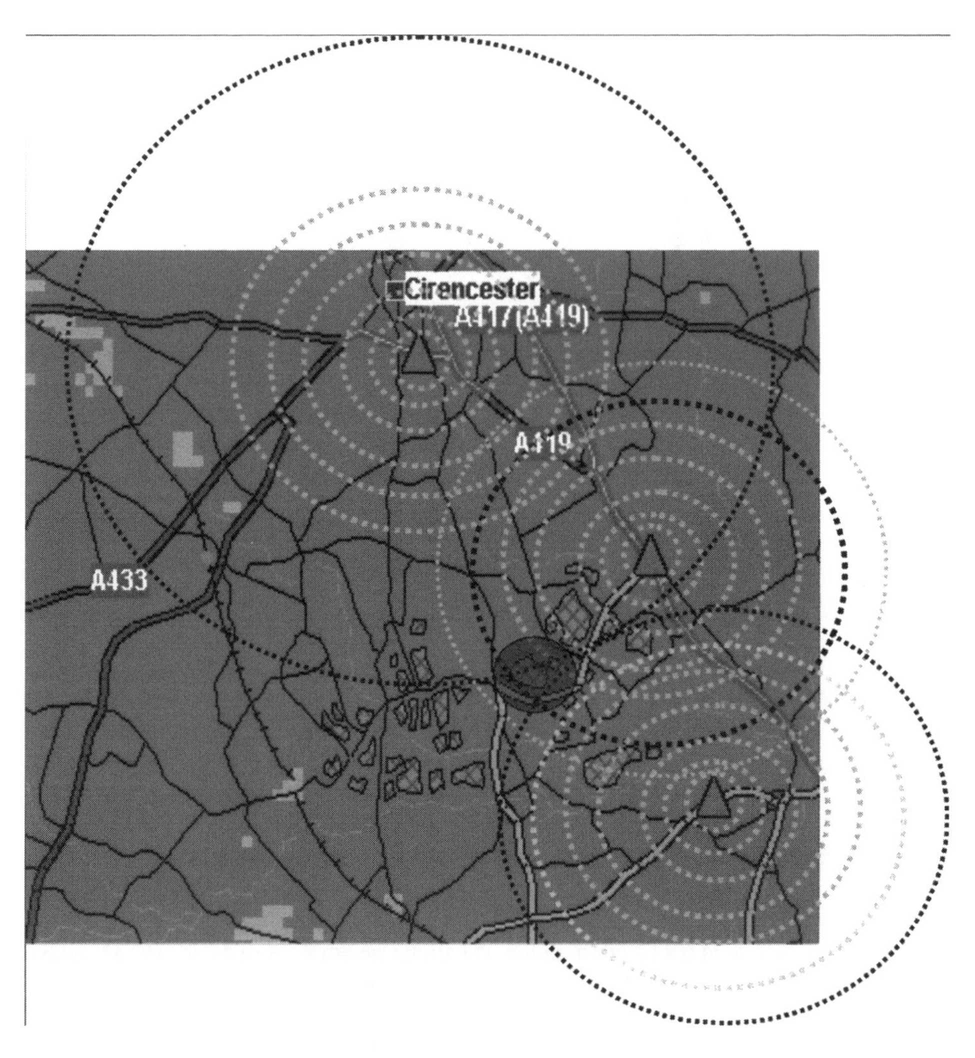

Figure 13.15 Using timing delays to improve accuracy.

measurement by the mobile cannot be relied on to be a true indication of its distance from the cell site. The reported distance and the actual distance are seldom the same, and some gross distortions are possible under certain landscape topologies, both in urban or rural environments.

The most accurate method of location finding is the Global Positioning System (GPS) network of satellites (see Figure 13.17). This is essentially still a triangulation technique, but with a direct LOS from the mobile to the satellites and a superior ability to measure timing offsets. The offsets allow measurement to a fine degree, translating into an on-the-earth resolution of a few metres. Due to its military origins and uses, publicly accessible signals are deliberately conditioned to introduce measurement errors, a process called *selective availability*. However, after 11 September 2001, this was switched off and now the full accuracy of GPS is available for civilian use, including mobile applications.

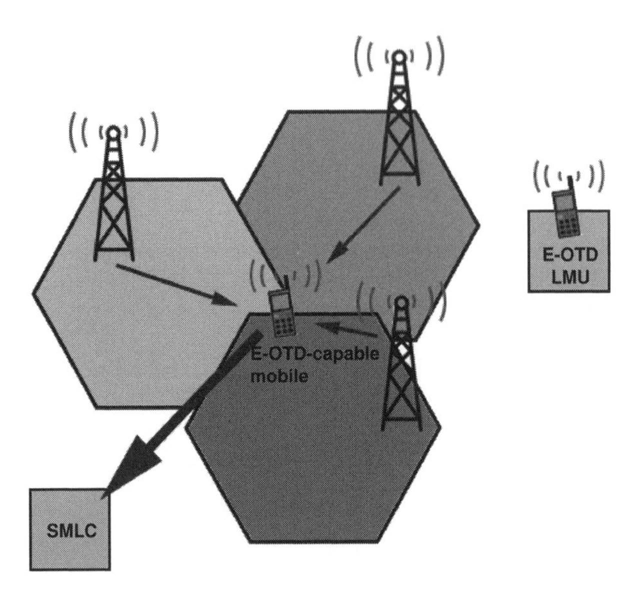

Figure 13.16 Observed time difference used to improve location-finding accuracy. Reproduced by permission of Magic E.

With a basic GPS receiver, we may expect better than 20-m horizontal accuracy on the ground. Performance will vary depending on the particular receiver and the level of solar disturbance of the ionosphere. However, there are several techniques to improve accuracy. One is called differential GPS, or D-GPS; this uses receivers in known locations to measure their location and apply various models of GPS network parameters that would work to correct the reported measurement, bringing it back to the actual location. A well-tuned GPS system could reach accuracies of 3–5 metres, whereas a well-tuned D-GPS could reach accuracies of 1 metre.

The system proposed for location finding in cellular networks is *assisted GPS* (A-GPS), which sends GPS measurements made at the base stations to the mobiles so that they can better perform GPS calculations. Demodulation of signals from the GPS satellites occurs at a relatively slow rate and requires satellite signals to be relatively strong; this can be perturbed by difficult terrain, such as the user being in a dense urban area with lots of buildings. To address this limitation, an A-GPS receiver receives context data from an A-GPS unit in the network (e.g., at a cell site). Context data provide the receiver with information it would normally have to demodulate as well as other information, which decreases start-up time (i.e., to make the first satellite location fix) from possibly 2 minutes down to somewhere in the region of 5–10 seconds.

Although A-GPS sounds fantastic, it still has limitations, especially in its coverage, because it is subject to the required satellites being in direct LOS from the mobile. This is not always possible, especially in built-up areas and when indoors.

A possible solution is to adopt a hybrid approach using a combination of location solutions in effective unison, to gain the best of both worlds – the coverage of cell ID and the accuracy of A-GPS. Working with DaimlerChrysler, Dr Cannon of Calgary University (Canada) is exploring ways[7] of combining GPS with existing, commercially available

[7] http://www.eng.ucalgary.ca/Press_Release/2002/SteacieFellowship2002.htm

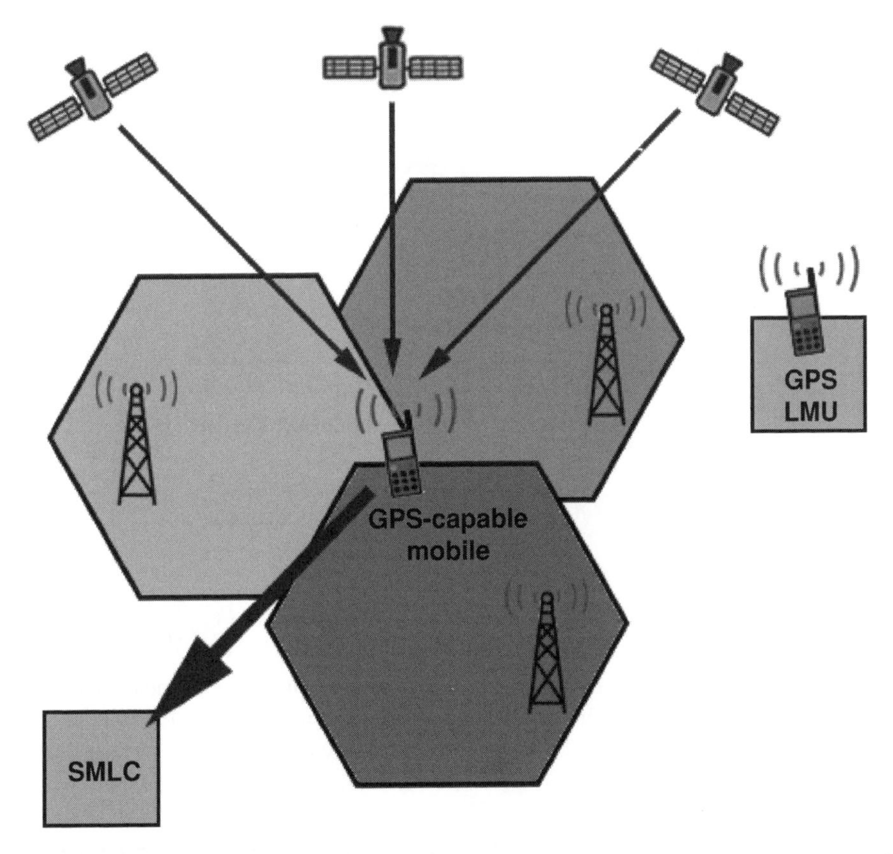

Figure 13.17 The global positioning system (GPS) uses satellites. Reproduced by permission of Magic E.

inertial navigation systems – sensors inside a vehicle that record its rate of acceleration and direction. Her work focuses on developing algorithms and error modeling that will provide the best mathematical ways for merging the two different types of information provided by GPS and inertial systems. She says:

> *We want to develop a car-based system that would allow it to continuously position the vehicle to centimetre-level accuracy in real time.*

While this extremely precise positioning is perhaps excessive for buddy finding or E911, it's the kind of ability that opens whole new realms of possibilities, such as autonomous driving and ultra-accurate location services. 'Accuracy is addictive,' Dr Cannon notes. 'People start to think, I could do this or that.' Perhaps in the near future, jet-powered camera drones will be able to fly straight to the scene of an accident to assess the situation ahead of emergency personnel arriving.

This is a nice segue into the issue of what accuracy do we actually need from a location-finding solution? There is no single answer, of course. It depends on the application. However, this comment of Dr Cannon's seems an important observation and perhaps a motivation for pursuing the most accurate systems possible from the outset. Were we to apply Ahonen's 5Ms to a proposed LBS, we may well find that accuracy is a key driver to all the parameters

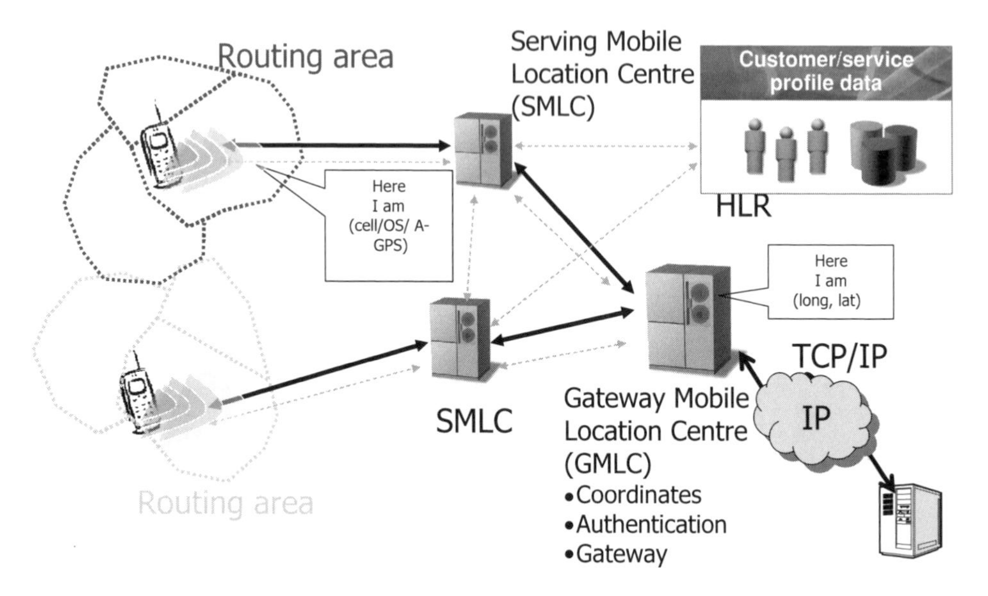

Figure 13.18 LBS architecture in the cellular network.

of success. At the end of this section we will examine possible service ideas that examine many of the technologies we have investigated thus far in the book. We will see in the example that very accurate location finding is a necessity for certain types of service.

13.5 INTERFACING LBS APPLICATIONS WITH THE CELLULAR NETWORK

The remaining issue is how to get location information into our location-enabled application. We have previously discussed accessing location information via Web Services APIs, like the 3G Open Services Architecture (OSA), which is an implementation of Parlay – Parlay X being the Web Services presentation.

Figure 13.18 shows the basic architecture of the location infrastructure in a cellular network. Just as we saw that clusters of adjacent cells formed routing areas for SGSN data concentration, there are similar routing areas of the network, each served by a SMLC. This is responsible for gathering the actual location measurements from devices and for the presentation of this information in a useful and agreeable form to the higher entities in the network, including, eventually, our location-enabled applications – J2EE or not.

The information from the SMLCs is concentrated into a *Gateway Mobile Location Centre* (GMLC), which is a secure boundary point for external entities to attain mobile location information about a particular subscriber. Apart from providing the main interface with the outside world, the GMLC makes location information available in terms of subscriber ID, not device ID; hence, why it has a relationship with the Home Location Register (HLR), which stores user information that enables device ID to be associated with user ID. This is necessary because within a cellular network a phone number does not identify a device, but an International Mobile Subscriber Identity (IMSI) does. Therefore, when asking for a

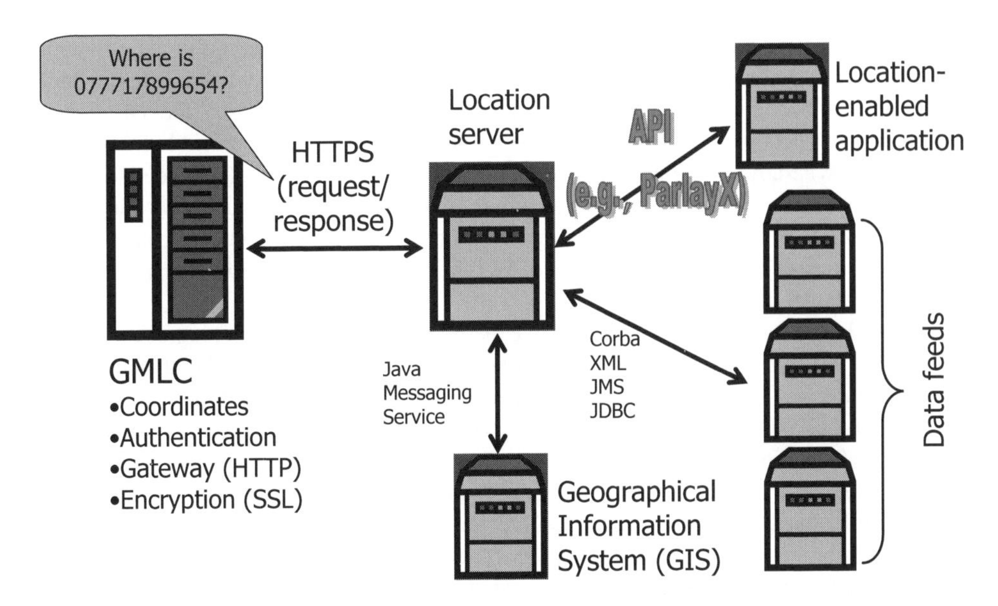

Figure 13.19 Possible LBS application support architecture.

mobile's location, an application probably expects to enquire using a phone number, and so translation needs to take place to the IMSI.

The GMLC can present location information in certain formats. It is up to the operator how they want to make location services available to third parties, but clearly a good deal of housekeeping functions are also required, such as authentication, account management, charging management and so on. These could all well be taken care of via a Parlay or Parlay X gateway that sits in front of the GMLC and becomes the main interface point for location applications, as shown in Figure 13.19. It is unlikely that a location-enabled application would ever interface directly with the GMLC.

Figure 13.19 shows several features of a possible LBS application environment or service delivery platform. The GMLC itself is the main point of interface to the cellular network and provides:

- HTTP access to coordinate information;
- authentication to protect unauthorized use;
- encryption (SSL, or Secure Socket Layer) to protect location queries.

As the figure shows, it is unlikely that the GMLC is ever directly available to the applications that need to use its facilities. Intermediate 'middleware' servers will provide additional services and present them via suitable APIs to the applications that require location information, be they internal to the mobile operator or external (i.e., third-party applications).

Typically, we would expect a Web Services API for LBSs, although whether this is Parlay X is another matter. Perhaps, eventually, most Web Services service points in mobile networks will become Parlay-compatible and the Parlay X API will grow in its sophistication. However, it has to be said that a large number of more specialized 'nontelecoms' services will emerge from operator gateways: these will have a Web Services presentation, but probably not within the Parlay X framework initially. For example, multimedia content management systems, photo blogs[8] and other such services will probably become standard to the point of becoming utility functions that other services could utilize – the reason they will eventually become available for third-party access. Later on, an extensible framework may emerge whereby the Parlay framework may allow for the addition of extended features. This is the beauty of the loosely coupled nature of Web Services, in general: it hinges on accessing services via URIs, and this makes it possible for service providers and application developers to easily aggregate a wide set of Web Services resources into a single service or application portfolio.

ATGIS is an example of a location service that an operator might want to offer as a utility service to third parties. This provides many of the basic services associated with geography, such as maps and place locations – a function beyond the Parlay X API set. A critical element of a GIS is the ability to perform *geo-coding* or the reverse process. Geo-coding is the transformation of geographical data into polar or Cartesian coordinate format. For example, let's take a directory of restaurants. It is more than likely their postal addresses will be in the directory, but not their geographical coordinates, as this would probably be an unusual parameter in a standard directory of restaurants. Therefore, the directory needs to be geo-coded: identifying postal addresses in terms of coordinates. This will then enable a location-enabled application to process the whereabouts of restaurants.

Products already exist on the market, such as ESRI's ArcWeb, which is a hosted GIS solution that offers a Web Services (SOAP) interface, with features like:

- access dynamic street, demographic and topographic maps, shaded relief imagery and more;
- determine the locations of street addresses;
- generate routes and multilingual driving directions between multiple locations;
- provide place finder capabilities;
- perform census demographics mapping and reporting;
- upload user-defined points of interest for geo-coding – ESRI hosts these data for the client, so that the client does not have to host them;
- perform reverse geo-coding (going back from coordinates to postal addresses);
- provide thematic mapping;
- display maps using a choice of projections.

ArcWeb for Developers works with Internet standards, including HTTP and XML (eXtensible Markup Language). It uses the XML-based SOAP to communicate, making it compatible with the majority of Web Services frameworks available today, including Parlay X. This means that our developer community using J2EE design patterns centred on SOAP Web Services could easily integrate GIS features into their Parlay X application; this should greatly enhance what can be done using Parlay X, as we will examine shortly.

The Web Services from ArcWeb is constructed from components divided into the following functional areas:

[8] Photo blogs are weblogs that enable a photo diary to be kept by sending pictures from a mobile camera phone.

- *place finder* – ranks a candidate list of place names and associated latitude/longitude coordinates for a given input place name;
- *address finder* – determines the latitude and longitude coordinates for street addresses;
- *route finder* – returns textual multilingual driving directions for a multipoint user-defined route;
- *map image* – provides access to a wide variety of dynamic maps;
- *proximity* – returns all points of interest (POI) locations within a user-defined distance of a specified location (i.e., find all POIs within 5 miles of x, y) or determines the nearest specified number of POI locations to a specified location (i.e., find nearest three POIs to x, y);
- *POI manager* – allows you to upload a custom set of points to ESRI, where they are geo-coded and stored for later use;
- *query* – determines the physical, environmental or cultural characteristics of a specific location;
- *utility service* – allows you to change the projection of your map image;
- *account info* – lets you use a SOAP interface to access account information (such as usage statistics).

ArcWeb is part of a wider suite of products, called ArcLocation, which can be used to construct a service delivery platform (SDP) for LBSs. One part of the suite involves the solutions connector, which is an API suite that handles integration between ArcWeb Services and industry standard mobile applications and Web servers. It also handles integration between ArcWeb Services and the GMLC and currently supports the Location Interoperability Forum (LIF) Mobile Location Protocol (MLP).

As we might expect, MLP uses an XML-based vocabulary. The service level elements are XML messages and there is no specified transport, but it could be HTTP, SOAP or any available method for exchanging XML messages. The baseline MLP specification suggests how to use HTTP, using the POST method to submit queries. While this sounds similar to SOAP, recall that the XML messages in SOAP are themselves packaged in SOAP envelopes (also XML) before appending to the HTTP stream, even though SOAP is also transport-agnostic. SOAP is commonly seen in conjunction with HTTP because of the latter's prevalence.

What an operator may provide in their SDP is a connection into various information services, such as place directories, events directories and so on. This information could be used by third parties or by operator applications, most likely on a per-enquiry charging structure.

A plethora of exciting mobile services can be envisaged thanks to the availability of a powerful SDP for LBS, especially with a Web Services presentation – even more powerful if it is available alongside a Parlay X interface. The application developer can pull in resources very easily, seamlessly integrating features of many Web Services components into one application or service portfolio. Clearly, the Web Services paradigm works well with our application based on J2EE, but it is also likely that much of the platform itself is implemented using J2EE, such is its appeal to platform developers.

Some interesting possibilities arise when J2EE is used in conjunction with J2ME (Java 2 Micro Edition) clients on mobile devices. For example, ESRI have a J2ME API available that seamlessly integrates into their Web Services platform. The ArcLocation J2ME tool kit is used to build mobile client applications that effortlessly consume ArcWeb Services. It is based on Connection Limited Device Configuration (CLDC) and the Mobile Information Device Profile (MIDP). Applications built using the tool kit will run on all devices that support MIDP, so this is an exciting possibility for MIDP developers.

13.6 INTEGRATING LBS APPLICATIONS

In the introduction to the book (Chapter 2), we talked about becoming 'smart integrators', being able to take the new technologies and rapidly integrate them to provide exciting mobile services. It is highly likely that this is the future of mobile services: a constant stream of services that users selectively subscribe to according to their needs. It is even possible that users could construct their own services to meet their precise needs. Although the means to do this requires some further thought, there should be no doubt that such an eventuality is possible given the highly conducive nature of the underlying technologies we have been discussing throughout the book in terms of facilitating flexible and rapidly changeable solutions.

Earlier in this chapter we talked about the nature of location services being very much event-driven. There are so many possibilities for location services, and many lead to the generation of events. For example, it has become increasingly popular to buy items via online auctions like eBay™. For some items, the postage and packing costs are sometimes prohibitive. Also, there is a tendency to have a passing interest in certain goods, but not a compulsion to buy them. Under these circumstances, the use of LBSs may assist the buyer by introducing them to goods available within their locality, wherever that happens to be. Perhaps the user has a casual interest in acquiring a certain type of old car, just for fun, perhaps as a restoration project.

We can imagine a service whereby the user indicates their interest in a particular car, as shown in Figure 13.20. Here we have the buyer submitting their details about the car to an 'agent' – a fancy word for the concept of a software process doing work for the user, on their behalf, as an ongoing background task while the user does other things. The user might not be actively using any service on their device; it could be dormant. Nevertheless, the agent in the network carries on its task, which in this case is looking for the car, but within the context of the geographical movements of the user, searching for suitable cars within reasonable proximity of the user at any one moment in time as they travel around. Of course, the use of the word 'agent' tends to indicate a process that is permanently engaged in hunting on the user's behalf; this is not quite the case. What is more likely is that a particular condition is set up in an events processor platform, an idea that we discussed earlier when looking at detecting the overlap of the two radii: the radius of opportunity and radius of interest. We will shortly look at possible manifestations of the application at the J2EE level.

The agent's task is to facilitate a service that alerts the user whenever they come within reasonable proximity of a car for sale, as shown in Figure 13.21.

As the user proceeds down a major route, like the one shown on the map, the service might send alerts at various intersection points. We are able to combine many powerful features from the operator's SDP to make this application work.

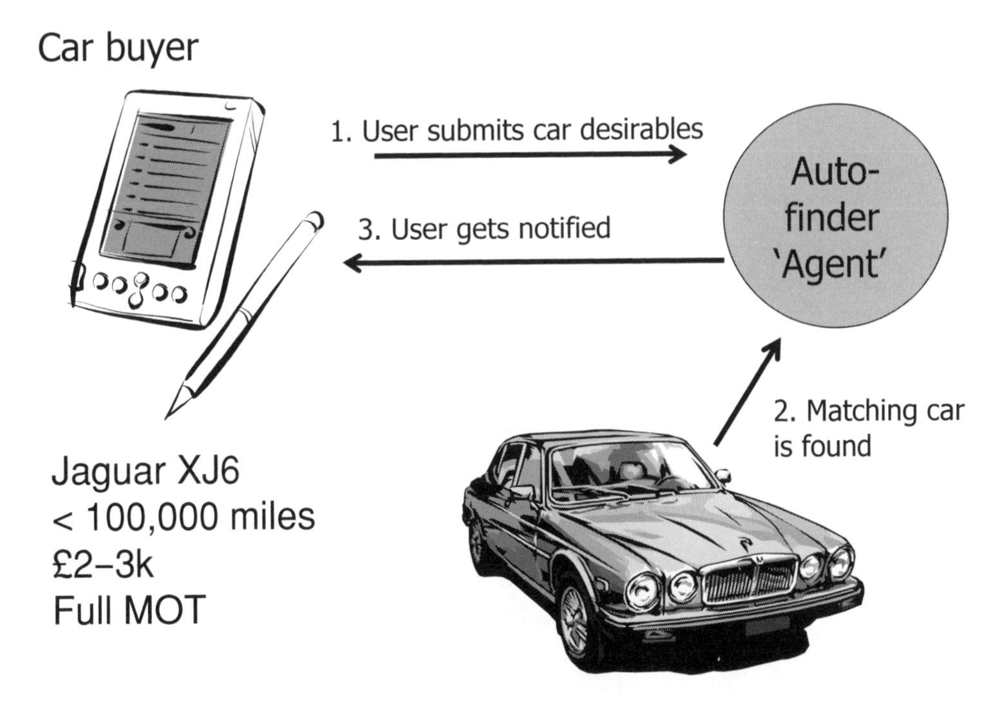

Figure 13.20 Using an 'agent' to find a product.

We need initially to detect that the user is within range of a car for sale; this happens by setting an event condition. This part of the service relies heavily on asynchronous messaging across several loosely coupled platforms. We are interested in two types of event: first, an indication that a car has become available for sale which fits the user's criteria, irrespective of where it is located; and, second, the movements of the user from one place to another, as well as a stream of location updates. We can imagine the first type of event as causing messages to be sent from the car database to the application, causing an MDB to insert the item into the location event processor. Now, if the processor works by comparing the proximity of coordinate pairs, whatever their format,[9] we need to first geo-code the car's position, because it will enter the system with a postal address and we need to calculate proximity using coordinate information. This conversion could be done via a Web Services call to the location SDP, utilizing a 'geo-coding API'. The coordinates can then be submitted to the location event processor, which is a real time database, like TimesTen™. The location events of the user are subsequently fed into the location events processor, and whenever a proximity match occurs the processor fires out a software message via MOM.

Part of the agent is implemented as an MDB that subscribes to message topics related to our user's requests. The exact construction of topics is a matter of system design, but it could be that we have instances of MDBs that listen for messages labelled with the user's phone number as the message topic. Various addressing techniques and design patterns need to be considered for implementing such a system efficiently, but even a rudimentary paper design will reveal a variety of possible approaches very quickly. The kind of trade-offs that

[9] Coordinates in the MLP have different forms, such as ellipsoidal and Universal Transverse Mercator (UTM).

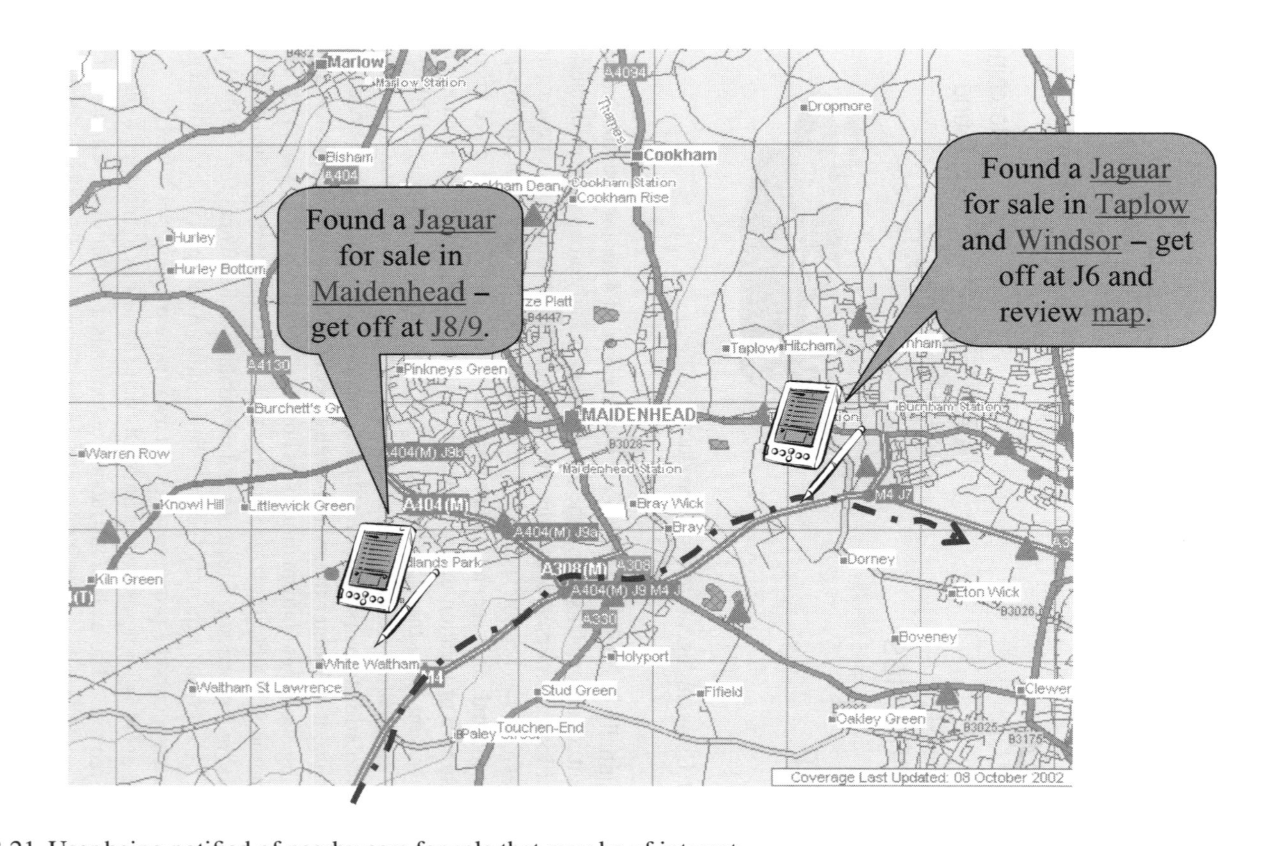

Figure 13.21 User being notified of nearby cars for sale that may be of interest.

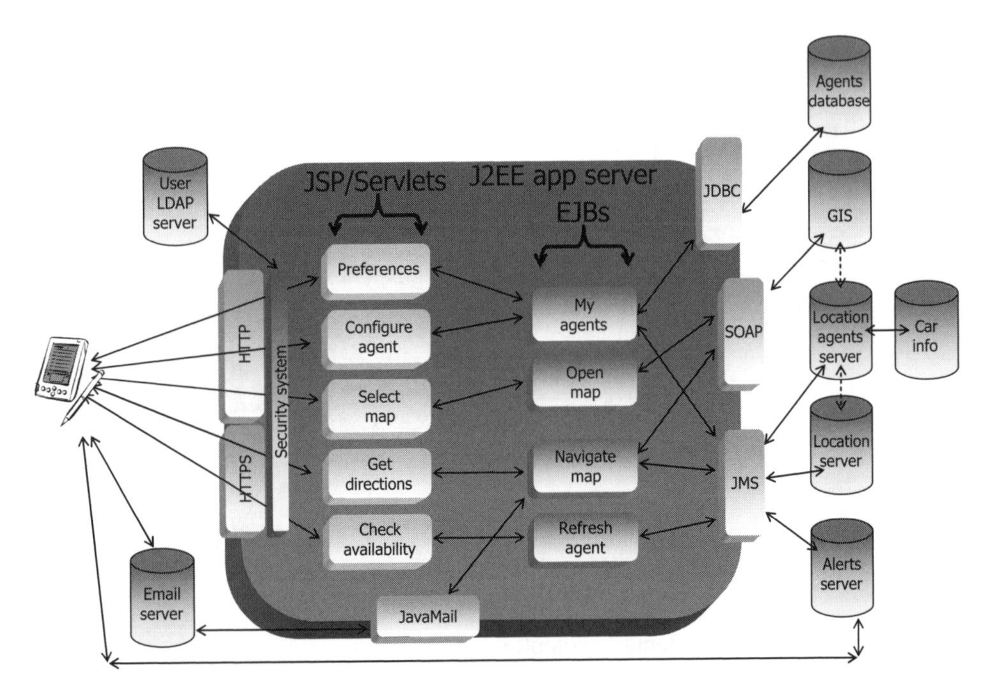

Figure 13.22 Possible J2EE architecture for the car agent application.

the designs will have to make will be system resources versus real time performance (as ever).

By designing our system in this way, the MDB will come alive when the user enters the proximity of an available car. What we could envisage is that the message itself contains the car's details and whereabouts. The bean's task is to encode the location and car details into a WAP PUSH message, using anchor tags (hyperlinks) to enable the user to access details about the car and its location.

Using the GIS platform, the WAP PUSH message can link to mapping information and directions to reach the car. However, it is likely that the user will first want to understand more about the car. This could be done in several ways. The user could link to a URI (uniform resource identifier) that has the car's details. The user could also be offered the chance to place a call with the car owner. This could be done via the browser itself, via the inbuilt dialler application on the device which picks up the number for a link or via a Parlay X third-party call API.

The J2EE components shown in Figure 13.22 are by now familiar to us. The presentation logic, constructed from servlets and JSPs (as per the model–view–controller pattern), takes are of the user interface. We have elements to configure the application, such as setting preferences, and to configure the agent. We have elements to view the results, like selecting maps, getting direction and checking the availability of the car by viewing its online catalogue entry. On the back end the EJBs are responsible mainly for integrating the services from the SDPs into our application. We have a messaging interface (via JMS) into the location server to set up the agent events and a messaging bus to receive alerts from the location agent's server (events processor). There is also a JDBC into the agents database itself, as we will need somewhere to store all the various agents' configurations, not just for each

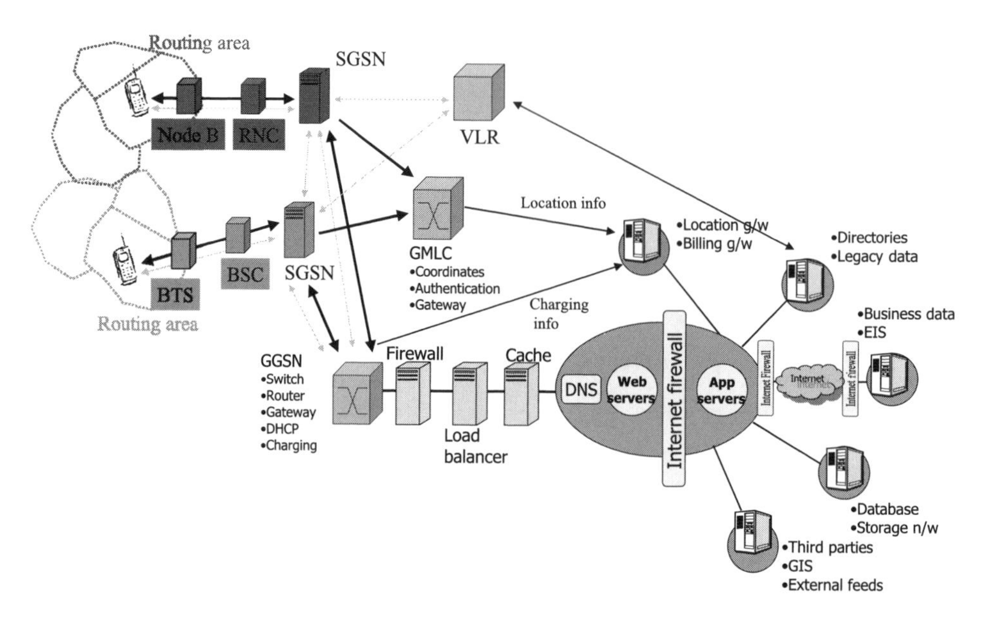

Figure 13.23 The entire system is a complex one.

user but also for each product or service of interest – there could potentially be hundreds per user, especially for a general purpose location-enabled agent that is not confined to searching for cars.

It is possible that the user will not be able or want to respond to a location alert. Perhaps our user didn't have sufficient time to go and investigate or they were unable to contact the seller, for whatever reason. A variety of follow-up responses is possible in that scenario. In the current example we use the JavaMail API to send an email to the user with all the event details included, so that a record of the event is then stored in the user's inbox for later reference. This can then be accessed by the user later on as a non-real time reminder about the event, so that they can return to it when more convenient, either from the mobile device or via desktop access.

The alerts server shown in the figure is an aggregated platform that can support WAP PUSH, text messaging and even multimedia messaging. The latter is a possible option for including a photo of the car.

Although greatly simplified and contrived, we can appreciate from this brief application example how the various technologies and software services can be brought together and smartly integrated, especially taking advantage of the loose coupling of both Web access and that of event-driven software design. The ability to access some major, very powerful APIs via Web Services and Java Messaging make for a much easier implementation path and a greatly enhanced end user experience. If the Web Services and other SDP aspects are not available, then it is difficult to see how an application like this could be successfully delivered, unless there is a significant investment of resources. Were this investment and level of low-level design actually required in the absence of 'smart integration' technologies, the likelihood of lots of these sorts of applications arising would be very low.

Figure 13.23 reminds us that our location-based service is actually taking place across a very complex network of powerful entities. The application of J2EE technology and the SDP concept enables our application development to be free from knowledge of or involvement with the inner complexities of the network of networks that make up the mobile network we have been touring throughout this book.

New possibilities are emerging all the time. For example, it is now possible, using products like iBus™ from Softwired,[10] to send software messages across the MOM 'bus' directly to MIDlets running on the mobile devices, which is an exciting possibility. It is an interesting idea that probably offers major benefits toward enabling some powerful design patterns to emerge for event-driven applications, especially those that seem to arise in LBS portfolios, due to their common nature. For location services in general, there is no single best fit solution, so it is not possible to say that wireless Java messaging is in any way an ideal approach. Nevertheless, knowing that this is possible is a major guiding star to add to our constellation of possibilities within the mobile service technology cosmos.

We will shortly look at another example of an LBS, one that ties together even more technologies than the example we have just considered. However, before doing that, we should briefly investigate what the multimedia messaging service (MMS) is all about.

13.7 MULTIMEDIA MESSAGING

Here we will examine a brief overview of the MMS. This is so we can consider alternative delivery mechanisms to the Web-centric approach we have looked at so often in this book, even though MMS and Web technologies are closely but unsurprisingly related.

The push paradigm is clearly a powerful part of mobile service considerations and is almost unique to mobility, although we could pretend that email is effectively a push technology in one sense. However, the power of push technology in the mobile context is its immediacy, given that we carry our mobile devices with us most of the time. Thanks to the intimate relationship with a mobile device, it is possible to interrupt the user at any time with information that may be of use to them. In this chapter we have been building both the 'business case' and technological framework for being interrupted by location-sensitive events, even making the claim that the ability to monitor and notify events in this context is probably a key enabling resource in a mobile network. Its importance cannot be overstated, which is why we have dedicated this entire chapter to examining this aspect of the RF network in isolation and its close ties to the other networks in our mobile network model.

The ability to send a text message is commonplace and barely needs explaining, other than our earlier glance at the ability for our application to interface with the text-messaging infrastructure via Web Services APIs, like Parlay X. However, the ability to send multimedia messages needs some explaining, largely because the content is more complex and the delivery mechanisms are different from those for plain text messaging, so the possibilities need exploring. Our earlier example of pushing out location information, such as cars for sale, seems to lend itself to a multimedia message. The thumbnail photos of cars is a common feature of car-selling activities, so to extend this to the mobile world seems a natural fit. Pushing the picture of the car, as opposed to asking the user to browse to it, seems to have an immediacy and an appeal that enhance the user experience. This is probably true of many similar event-driven scenarios. We now need to see what type of content can be embedded

[10] http://www.softwired.com

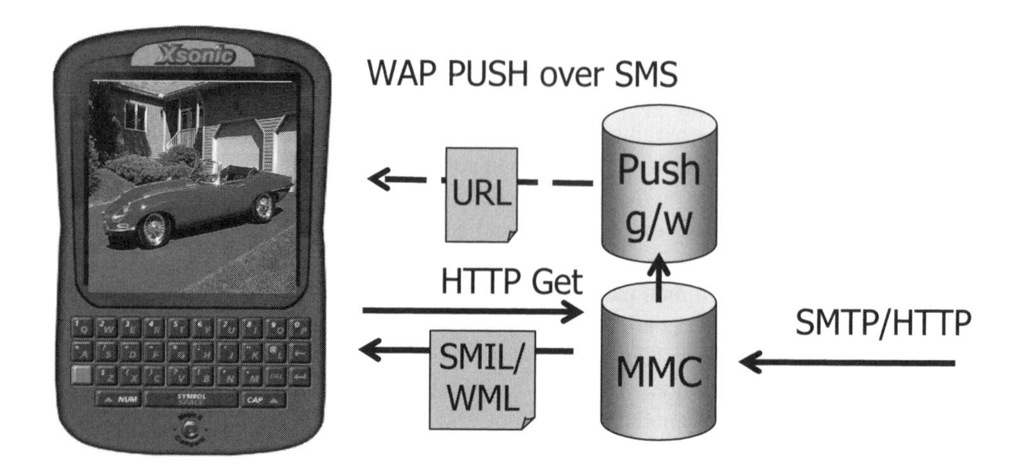

Figure 13.24 Mechanism for sending an MMS message.

in an MMS message and how to do it. We also need to look at how this facility relates to the Web design patterns we have been considering thus far throughout the book.

The easiest way to conceptualize of MMS messages is to think of pushing simplified Web pages to the user. In fact, this is rather an apt description. The way that MMS delivery works is by sending an initial *service indication* (SI) message to the device, which contains a URL embedded in a text message, as shown in Figure 13.24. The MMS client on the device then pulls down the contents from the URL automatically via HTTP or WSP (or W-HTTP). The SI is written as a WBXML-tokenized message so that its contents are compressed enough to fit within the 160-character text-messaging limit.

The MMS client fetches the content from a server that is storing the message that was sent to it via HTTP or SMTP. As the figure shows, the encoding of the contents of the message is in one of two formats: WML or SMIL. We have met WML (Wireless Markup Language) before in our discussion on mark-up languages and WAP (Wireless Application Protocol). We have not yet met SMIL (pronounced 'smile'), but will examine this shortly.

According to the UMTS (Universal Mobile Telecommunications System) specification [3G TS 23.140 V3.0.1] for MMS, there is no definitive presentation format for a multimedia message (MM). The visual rendering on the device of an MM is known as *presentation*. Various types of data may be used to drive the presentation. For example, the MM presentation may be based on a WML deck or Synchronized Multimedia Integration Language (SMIL).[11] Other presentation models may include a simple text body with image attachments. WAP has not specified any specific requirements for MMS presentations. User Agent Profile (UAProf) content negotiation methods should be used for presentation method selection; earlier in the book we described how this can be done when pushing data to a device.[12] However, for MMS to be successful, clearly a degree of conformance is required between devices; we will discuss this in the next section.

[11] http://www.w3.org/TR/smil20/

[12] The point being that UAProf information is normally only available when a user agent requests Web resources, not when pushing to a hitherto unknown user agent with hitherto unknown display capabilities.

Since its adoption for MMS, the use of SMIL has gone through several interpretations, just as we saw with XHTML, XHTML-Basic and then XHTML-MP. However, before we examine them, let's briefly look at SMIL and what it can do.

13.7.1 Composing MMS Messages

SMIL is an XML vocabulary. It enables multimedia elements to be presented within a compatible client (SMIL browser or SMIL player). We are already used to multimedia elements being possible within a Web environment. We can view images, read text, play videos (via players) and listen to sounds. So, why the need for yet another markup language? Multimedia authors, especially from the world of animated and sequenced multimedia presentations, have long been used to the concept of a time base. This is the idea that within the presentation of information there is a need to stipulate what happens when. This is the most critical aspect of SMIL. In short, SMIL allows authors to write interactive multimedia presentations. Using SMIL an author can control the timing of events within a multimedia presentation, associate hyperlinks with multimedia elements and control the visual layout of the presentation on the viewer's screen (both 2-D and across time).

In moving from version 1.0 to 2.0, SMIL can now be used to instigate a time-based control to XHTML elements and another language called Scalable Vector Graphics (SVG). This combination enables nearly any kind of visual presentation to be constructed and delivered across the Internet using open standards.

With SVG, instead of using graphics files to represent images, XML commands can be used to indicate drawing primitives, such as rectangles and circles. With SVG, a rectangle would be rendered by the XML description as follows:

```
<rect fill="#69ABFF" y="200px" x="16px" width="200px" height="100px"
  style="stroke: #69ABFF" />
```

This will render a rectangle that is sky blue in colour (as indicated by the RGB values #69ABFF) and would be 200 pixels wide, 100 high, placed on the screen 16 pixels across and 200 pixels down (from the top left-hand side, which is the origin).

Using the animation module of the SMIL language, it is possible to add animation to the rectangle by controlling its attributes over time (see Figure 13.25):

```
<rect fill="#69ABFF" y="200px" x="16px" width="200px" height="100px"
  style="stroke: #69ABFF">
  <animate attributeName="width" from="100px" to="200px" begin="0s"
  dur="3s" />
  <animate attributeName="y" from="15px" to="200px" begin="0s"
  dur="3s" />
</rect>
```

The <animate> tags stipulate that the attribute named within the tag should be ranged from one value to another over time, the duration being specified by the 'dur' attribute.

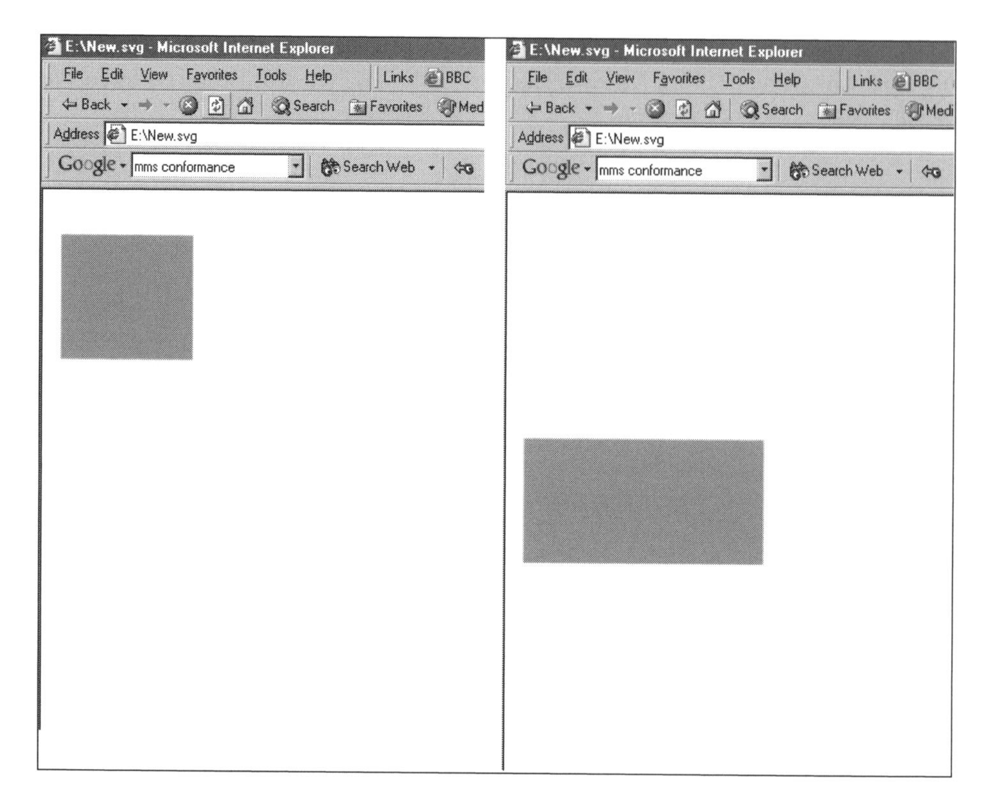

Figure 13.25 Animation of a rectangle using SVG and SMIL.

The ability to describe any number of primitives with SVG and to animate them gives us a very powerful graphical presentation technique. To create this animated sequence using XHTML and standard images is difficult and consumes a lot of memory. Clearly, with SVG it takes just a few lines of XML in a compatible viewer.

There are many drawing primitives available in SVG, as shown in Figure 13.26. It is also possible to have layers in the rendering, similar to SMIL, so primitives can appear on top of each other as required.

SVG provides some powerful possibilities for mobile services. SVG 1.1 is modular in the same way that XHTML is modular. This makes it possible to pick and choose modules to form new profiles. Naturally, lightweight profiles of SVG have emerged for wireless devices (and devices other than the desktop computer). The central effort in this regard is via the World Wide Web Consortium (W3C) project *SVG Mobile*. From SVG Mobile we have two new SVG profiles:[13] *SVG Basic*, which is a subset of SVG-targeting PDA devices, and *SVG Tiny*, a subset of SVG Basic-targeting mobile phones.

The 3G Partnership Project (3GPP), in their definition[14] of requirements for codecs and standards for mobile phone presentation formats, state that SVG Tiny should be supported by next generation phones and should have full SVG Basic compliance.

[13] W3C Recommendation: 'Mobile SVG profiles: SVG Tiny and SVG Basic', available at http://www.w3.org/TR/2003/REC-SVGMobile-20030114/, January 2003.
[14] 3GPP TS 26.234 V5.5.0 (2003-06).

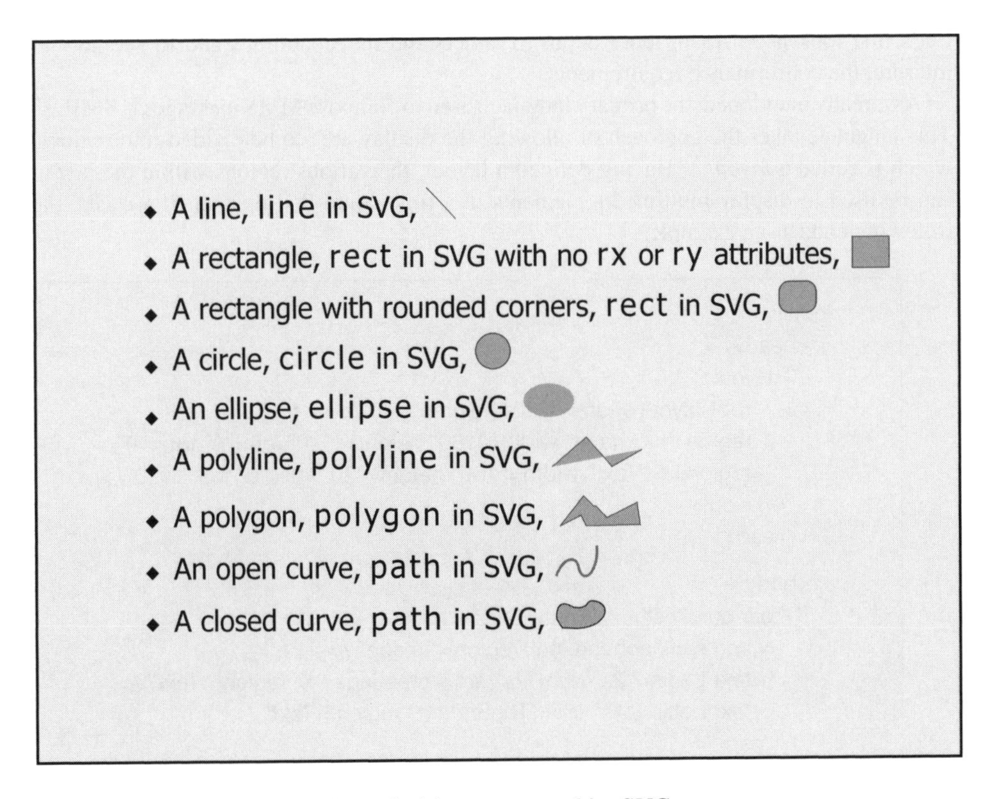

A line, line in SVG,
A rectangle, rect in SVG with no rx or ry attributes,
A rectangle with rounded corners, rect in SVG,
A circle, circle in SVG,
An ellipse, ellipse in SVG,
A polyline, polyline in SVG,
A polygon, polygon in SVG,
An open curve, path in SVG,
A closed curve, path in SVG,

Figure 13.26 Different drawing primitives supported by SVG.

SVG Basic is very close to full SVG in terms of its features, but leaves off some of the more complex elements that require a lot of CPU resource. SVG Tiny goes further in stripping scripting, filters, gradients, patterns, opacity and cascading style sheet (CSS) support. However, perhaps crucially for mobile applications, the *Animation Module* is preserved in SVG Tiny: this is in keeping with the trend toward animated messaging content, as supported by SMIL in MMS.

As we previously learnt when looking at XHTML and XML (its foundation), the modularized versions of XML-based languages enable any one of them to act as a host language. This means, for example, that an XHTML document could include SVG elements. Similarly, SMIL, which is also XHTML-based, could contain SVG elements. SMIL itself could also contain XHTML elements, although that possibility is not yet supported (or suggested) by 3GPP.

While SVG Tiny is a mandatory media type stipulated by 3GPP, this does not mean that all phones will support its inclusion in MMS messages. The basic minimum standard for MMS is defined by the MMS Conformance Document[15] maintained by the WAP Forum (now the Open Mobile Alliance, or OMA), which states:

> *The MMS conformance document defines the minimum set of requirements and guidelines for end-to-end interoperability of MMS handsets and servers.*

[15] OMA-IOP-MMSCONF-2_0_0-20020206C – MMS Conformance Document available from http://www.openmobilealliance.org

Let's first look at MMS in more depth to understand its capabilities and to see how to interpret the conformance requirements.

As already mentioned, the primary language used to compose MMS messages is SMIL.[16] This language takes the approach of allowing the display area to be divided into regions, which is called a *layout*.[17] Having defined a layout, the various regions within the layout can be used to display multimedia elements in a time-controlled manner. If we take the following code as an example:

```
<smil>
  <head>
    <layout>
      <root-layout width="160" height="140"/>
      <region id="Image" width="160" height="120" left="0" top="0"/>
      <region id="Text" width="160" height="20" left="0" top="120"/>
    </layout>
  </head>
  <body>
    <par dur="6s">
      <img src="pghead.jpg" region="Image" />
      <text begin="2s" dur="2s" src="greetings.txt" region="Text" />
      <text begin="4s" src="TheEnd.txt" region="Text" />
    </par>
    <par dur="10s">
      <img src="message.jpg" region="Image" />
    </par>
  </body>
</smil>
```

The code clearly shows two blocks encased by the <head> and <body> tags – which is very similar to XHTML – with the <layout> tags nested in the <head> tags and defining the screen regions we referred to a moment ago. In the XML we can find two <region> tags that define areas we label 'Image' and 'Text', where we are eventually going to place image and text *media objects*, as shown in Figure 13.27. On the top we see the 'Image' region and on the bottom we see the 'Text' region. These are arbitrary regions as far as SMIL is concerned; so, we could define many regions all over the viewing space, to be used by whatever media objects we like. However, the MMS Conformance Document indicates a limited structure, like the one shown here. This is in order that messages get correctly rendered on as many devices as possible in a way that is consistent with the original message composition, thereby facilitating interoperability between devices within what is otherwise a wide range of capabilities.

If the layout is divided into four quadrants, like the flag of England, then this may prove problematic for some devices; it may be that many devices can support the layout in

[16] SMIL 2.0: W3C Recommendation 7 August 2001 available at http://www.w3.org/TR/smil20/
[17] A *layout* is similar in concept to a *frameset* in HTML.

Figure 13.27 MMS message is compartmentalized into regions using SMIL layout tags.

principle, but may have to display it in a modified form if the media objects are larger than the screen size apportioned to each quadrant, as shown in Figure 13.28.

Examining the SMIL, we can see that there are two sub-blocks in the body with <par> tags. We can think of these as being like slides in a slide show, except each slide is a multimedia container rather than a static image.

For the first 'slide' the duration (dur) is 6 seconds. On loading this slide the first media object is inserted; it is an image (<img> tag) linked to an external image file addressed by its URI (relative to the document). The <img> tag specifies that the image is going to be loaded into the 'Image' region (aptly named, but the name is in fact arbitrary in SMIL as long as it conforms to the XML rules of being well formed). Hence, we can see that the first image (pghead.jpg) gets loaded into the region labelled 'Image' (as seen in Figure 13.27).

We would like to add some text to the first image, and this is done using a <text> tag to refer to an external file, which in this case is a plain text file containing the string 'Totally distracted by . . . ' In the 3GPP specification the <text> tag can refer to an external XHTML file that contains text with XHTML formatting, but we have not shown this here. Referencing XHTML is not part of the MMS Conformance Document recommendation. The critical parameters in most of the tags are the 'begin' and 'dur' tags, which have their fields populated with time values in seconds, such as '3s' to indicate 3 seconds. Using these parameters, we can step through a series of media objects and slides. The above SMIL snippet produces four primary events in our sequence, as shown in Figure 13.29.

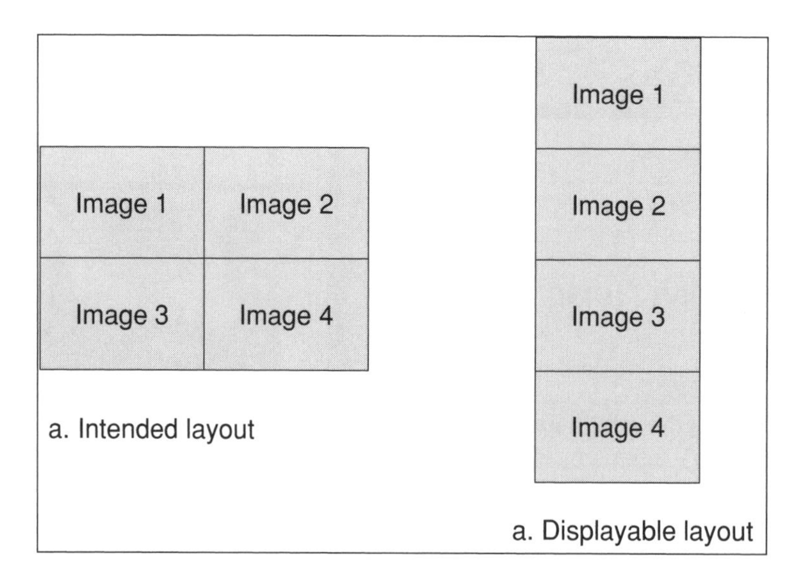

Figure 13.28 Intended layout modified by a device with a narrow display.

Figure 13.29 Using SMIL, MMS messages can cycle media objects in a 'slide show'.

Displaying images and text is not the only capability of SMIL, we can also insert sounds and movie clips. The 3GPP-supported media types are:

- *speech/audio*
  - AMR speech encoder (narrow and wideband),
  - MPEG-4[18] AAC low complexity (AAC-LC),
  - MPEG-4 AAC long-term prediction (AAC-LTP);
- *synthetic sound*
  - scalable polyphony MIDI (SP-MIDI) content format[19];
- *video*
  - H.263 profile 3, level 10 decoder,[20]
  - MPEG-4 visual, simple profile, level 0 decoder[21];
- *bitmap images*
  - GIF87a, GIF89a,[22]
  - PNG[23];
- *still images*
  - JPEG, baseline DCT, nondifferential, Huffman coding, as defined in table B.1, symbol 'SOF0',[24]
  - JPEG, progressive DCT, nondifferential, Huffman coding, as defined in table B.1, symbol 'SOF2'[25];
- *scalable graphics*
  - SVG Tiny profile,
  - SVG Basic profile;
- *text*
  - XHTML Mobile Profile;
- *character-coding formats*
  - UTF-8,[26]
  - UCS-2.[27]

[18] ISO/IEC 14496-3:2001: 'Information technology – Coding of audio-visual objects – Part 3: Audio'.
[19] *Scalable polyphony MIDI Specification Version 1.0, RP-34*, MIDI Manufacturers Association, Los Angeles (February 2002).
[20] ITU-T Recommendation H.263 – Annex X (2001): 'Annex X: Profiles and levels definition'.
[21] ISO/IEC 14496-2:2001: 'Information technology – Coding of audio-visual objects – Part 2: Visual'. ISO/IEC 14496-2:2001/Amd 2:2002: 'Streaming video profile'.
[22] CompuServe Incorporated, *GIF Graphics Interchange Format: A Standard Defining a Mechanism for the Storage and Transmission of Raster-based Graphics Information*, Columbus, OH (1987). CompuServe Incorporated, *Graphics Interchange Format: Version 89a*, Columbus, OH (1990).
[23] IETF RFC 2083: 'PNG (Portable Networks Graphics) Specification Version 1.0'. Edited by Boutell, T. *et al.* (March 1997).
[24] ITU-T Recommendation T.81 (1992) | ISO/IEC 10918-1:1993: 'Information technology – Digital compression and coding of continuous-tone still images – Requirements and guidelines'.
[25] ITU-T Recommendation T.81 (1992) | ISO/IEC 10918-1:1993: 'Information technology – Digital compression and coding of continuous-tone still images – Requirements and guidelines'.
[26] The Unicode Consortium, *The Unicode Standard, Version 3.0*. Addison-Wesley Developers Press, Reading, MA (2000, ISBN 0-201-61633-5).
[27] ISO/IEC 10646-1:2000: 'Information technology – Universal multiple-octet coded character set (UCS) – Part 1: Architecture and basic multilingual plane'.

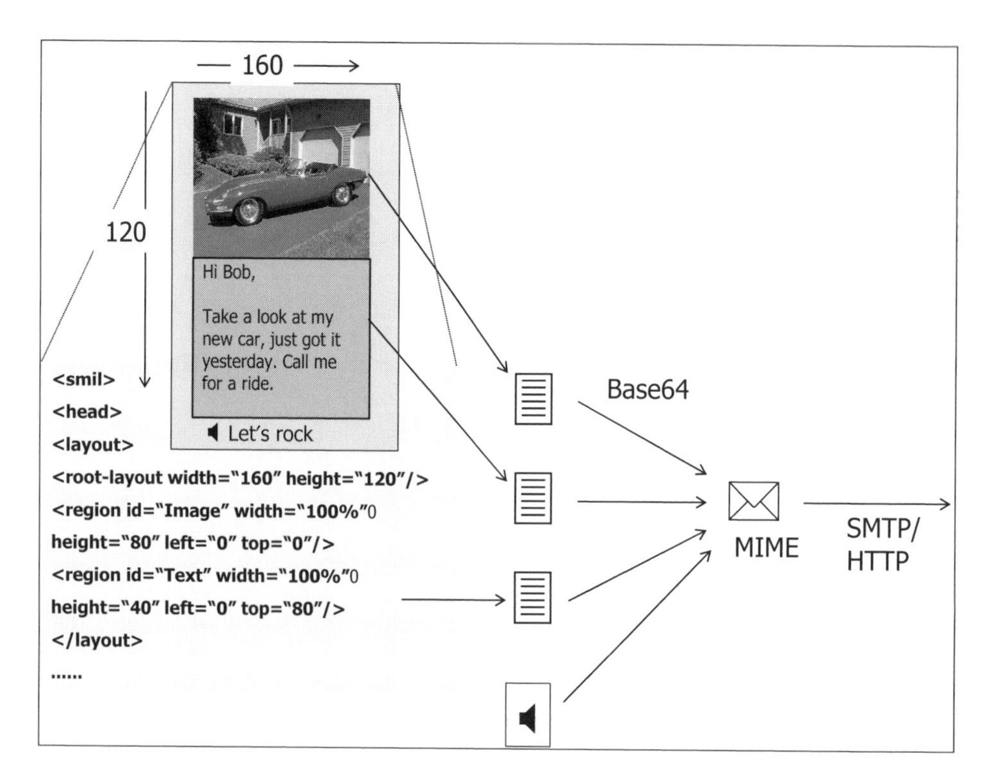

Figure 13.30 Encapsulating MMS components into a MIME message.

As far as sending the MMS is concerned, the complete message has to be encapsulated into a *protocol data unit* (PDU) to be sent to the Multimedia Messaging Services Centre (MMSC). From the mobile a composed message is usually submitted using HTTP POST or its semantic equivalent in WSP (Wireless Session Protocol). From a nonmobile client a message is usually submitted via HTTP or SMTP (Simple Mail Transfer Protocol), but could also be submitted using FTP (File Transfer Protocol).

When being submitted and transferred through the messaging infrastructure, the MMS components are encapsulated[28] into one entity, as shown in Figure 13.30. This aggregated structure is based on the well-known message structure of Internet email which is defined in RFC 822,[29] RFC 2045[30] and RFC 2387.[31] For transport over WSP the WSP specification provides mechanisms for binary encoding of such messages and serves as a basis for the binary encoding of MMS PDUs.

The PDU is one complete data structure and suitable for transmitting over WSP or HTTP, as shown in Figure 13.31. It has a header section that contains header fields; these fields give the message an identity and enable it to be transported meaningfully through the messaging

[28] OMA-MMS-ENC-v1_1-20021030-C: 'Multimedia messaging service encapsulation protocol', Version 1.1, 30 October 2002.
[29] 'Standard for the format of ARPA Internet text messages'. Edited by Crocker, D. (August 1982). URL: http://www.ietf.org/rfc/rfc822.txt
[30] 'Multipurpose Internet Mail Extensions (MIME) Part 1: Format of Internet message bodies'. Edited by Freed, N. (November 1996). URL: http://www.ietf.org/rfc/rfc2045.txt
[31] 'The MIME multipart/related content type'. Edited by Levinson, E. (August 1998). URL: http://www.ietf.org/rfc/rfc2387.txt

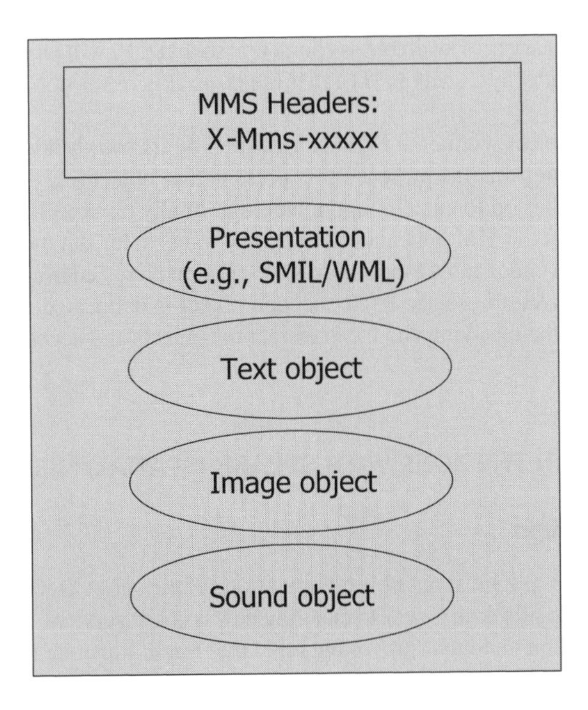

Figure 13.31 MMS PDU structure.

network. Otherwise, the MMS message itself is just a collection of media objects. We need to know who is sending the message, to whom, what it is about and other such parameters that we would ordinarily expect in an email message (on which the PDU transport method is based). Examples of the headers are:

- *X-Mms-Message-Type* – this specifies the PDU type;
- *X-Mms-Transaction-ID* – a unique identifier for the PDU;
- *X-Mms-MMS-Version* – the MMS version number;
- *date* – date and time of submission of the PDU (if the field was not provided by the sending MMS client, the MMS Proxy-Relay inserts the time of arrival of the PDU at the MMS Proxy-Relay);
- *from* – address of the originator MMS client. The originator MMS client sends either its address or an *Insert-address-token*. When a token is used, this indicates to the MMS Proxy-Relay that it is required to insert the correct address of the originator MMS client.

Within the message itself each of the referenced objects is first converted to a format that makes it suitable for sending as character-encoded text. This is done using Base64 encoding – something we came across before (see Chapter 9) when we looked at obfuscating the user name and password tokens used in basic authentication in HTTP. Differently from basic authentication, Base64 in the case of MMS is being used for its originally intended purpose: to make binary files available as characters so that they will pass character-oriented communications pathways, such as HTTP. It is not used here for obfuscation, although it has that benefit.

Each encoded object within the PDU is identified by its own header. The MMS specification actually states that the presence of a presentation object (e.g., a SMIL file to bind everything together) is optional, although it would normally be expected. In the absence of a presentation object the MMS client decides how it wants to lay out the remaining objects; so, clearly, this is an undesirable aspect of MMS if we are interested in controlling the visual appearance of received messages. Each encoded object is delineated with its own header (separator), so that the receiving client can extract the objects and decode them in readiness for rendering.

13.8 GETTING IN THE ZONE WITH SPLASH (SPATIAL) MESSAGING

13.8.1 Introduction

Finally, we turn now to look at possible combinations of the various technologies discussed in this chapter, especially with a view to enabling new types of services. Part of the intention is to close the book on a thought-provoking topic that might illustrate new possibilities for next generation mobile services. Having started the book with a tour of possibilities, the book ends with an examination of the possibilities, not so much a tour as an elucidation of a particular theme.[32] The ideas explored include some of the technologies explored throughout the book. I hope that it provokes thought and leaves you with higher aspirations for next generation mobile services than when you started reading this book.

13.8.2 Connectedness of Things

Location finding, peer-to-peer (P2P) technology and social networking – what do they have in common? I'm not sure that I have the answer(s), but I instinctively feel there are some connections and possible emergent themes for mobile services. I am going to attempt to identify some here.

Leaving messages in space containing pictures and video clips that are searchable according to who's in them might be a powerful concept worth enacting. To do this, we need a few current technologies and ideas to converge, as is probably going to be the case for most of the exciting apps that will emerge in the coming 3G era.

13.8.3 Making a Splash

Imagine walking down the road and, as you turn a street corner, up pops a message on your mobile phone. Using fine-resolution location sensing, a real time application has detected you entering an area where messages have been pinned to a virtual notice board on that very street corner. Using software on your mobile device you can read the messages, including

[32] The bulk of the text for this portion of the book is taken from *Enmesh* (Volume 1 Part 1) by Paul Golding, which is an infrequent publication of ideas for the mobile industry to ponder and available at Paul's weblog: http://radio.weblogs.com/0114561/2003/09/24.html

viewing pictures and watching video clips, all deposited by previous occupants of the same physical space who left messages there at some bygone time.

This is the concept of *spatial messaging* (SM): the ability to post a message on a virtual notice board anywhere on the Earth for someone else to pick up. The idea has many interesting variants and possibilities, and these will be examined here as a vehicle for exploring some of the myriad opportunities for next generation mobile technologies.

Sometimes called air graffiti,[33] it may also be useful to think of and refer to this phenomenon as 'splash messaging', as in 'making a splash', wherever we go. (If we were trying to be clever, we could also intimate Pollock as a source of inspiration, because of the order that comes from the chaos of splashing paint; though, it isn't obvious that any order would necessarily emerge from SM.)

We should not limit our view of this messaging paradigm just to text, obsessed as we currently seem with texting, judging by the ongoing large volumes of messages and the increasingly attractive deals to tempt us further into the messaging habit. Leaving text messages pinned in space was an early idea, one that was originally discussed when location-finding technologies began to emerge and well before MMS had arrived on the scene. Strictly speaking, 3GPP[34] includes video clips as a supported media type in MMs, so video is a possible format for SM, although text, photo and sound elements seem more likely for now.

Interestingly, the concept of associating synthetic information with the real world is, in its general form, much older than the text message, dating back to those early days when some of us got excited about virtual reality (VR seems a very departed era). The idea of VR now oddly feels dated, like something from a very old run of *Tomorrow's World* on the BBC, even though it remains a futuristic idea yet to happen (in any mainstream sense, never mind mobile[35]).

However, VR may be about to make a come back in another guise called *augmented reality*: the overlay of real world images (what we see) with virtual ones (what we project). This is how the idea behind SM surfaced: in order to augment the real world view with synthetic information a spatial coalescence between the two worlds is desirable (though not necessary). The likelihood of cheap, wearable displays[36] is now just around the corner, and with location technology and ubiquitous high-speed mobile networks, who knows what's possible?

Being as comfortable as we are now becoming with picture messaging, the suggestion of leaving a picture message in space, be it a logotype, manufactured image or an actual photo, is probably not so hard to imagine taking hold. Even simpler, leaving a voice message for the next passer-by is even more plausible, perhaps an addendum to push to talk services. We can easily imagine leaving voice messages when we exit the front door of our homes, perhaps to let our families know where we've gone. On their return the generation of an alert when nearing the home would prompt them to listen to the message.

With the location-finding technology that exists now in mobile networks and with picture messaging infrastructure in place, this concept is entirely possible to implement today.

[33] This was a commonly used term a few years ago for the concept of pinning messages in space, but I have been unable to find its origins.
[34] 3GPP stands for 3G Partnership Project, a collection of interests to develop standards for the next generation of mobile device technology, such as UMTS.
[35] I attempted to research mobile VR concepts for a PhD back in the early 1990s. The sponsor discouraged the work, considering it too esoteric.
[36] For example, see Xybernaut's POMA, URL:
http://www.xybernaut.com/Solutions/product/poma_product.htm

However, there are several challenges. One of them is the current accuracy of the location-finding system. With the most rudimentary forms of this system, such as associating location with the cell the user is in, the best accuracy is somewhere around 150 metres in the high density of cells found in crowded urban areas. Elsewhere, this can deteriorate significantly, as cells begin to span kilometres rather than metres.

However, with assisted-GPS (A-GPS) and other techniques, including hybrid solutions, the accuracy tightens to less than 10 metres. Some futuristic ideas are also under consideration, such as the use of inertial sensors to track movement from known fixed points;[37] the idea of pinpoint accuracy is no longer a pipe dream. Other inertial technologies, or the use of 3-D visualization technologies, also enable a directional element to the augmented reality experience. For example, standing on a street corner, we could be alerted differently according to which direction we are facing (or holding the mobile device).

Leaving a video message pinned 'in the air' is another challenge, primarily because we would need some heavy-duty wireless networking to sustain the video streams. However, with 3G networks or supplemental hot spot technologies (e.g., WiFi or 3G TDD[38]), this is not that incredulous: using streaming, both up and downlink, this seems possible.

We will shortly look at things we could do to make multimedia splash messaging more interesting and more personalized. However, let's look first at the potential social connectivity we can build around a network of splash messages, their contributors and consumers.

If we are going to leave messages pinned in space, then we might start thinking about ways to connect these messages with people, software agents and possibly machines. How do we engage usefully with an emerging fabric of messages in the virtual world which also have a spatial existence, splashed across the physical world, on street corners, in shopping malls[39] and just about anywhere we may care to roam, locally, nationally or globally[40]?

We can think of splash messages themselves as being active objects that we can send software messages to and expect replies from, as illustrated in Figure 13.32. The idea that we can ask about messages, about their contents, their owners and recipients brings an important dimension to the SM concept which puts it into the cyberspace realm, rather than a huge collection of inert media objects, like a draw stuffed with old photo albums in need of explanation.

Proxies, acting on behalf of real people, could generate the software messages (queries). Alternatively, these messages could be from software agents or even machines: a navigation application in a car could follow a trail of messages to find something interesting at their source (at the discretion of the trail setter).

To encourage and facilitate an open SM network, we need to use open standards as much as possible, as well as commonly established methods of data interchange (e.g., HTTP, SOAP, etc).

[37] http://www.eng.ucalgary.ca/Press_Release/2002/SteacieFellowship2002.htm

[38] TDD stands for time division duplexing, a technique for alternating the transmission and reception of signals by assigning time slots. Used in conjunction with Wideband CDMA, it has some interesting performance attributes that make it suitable for high-capacity 'hot zones' of RF coverage, in contradistinction to WiFi which has a much smaller coverage potential (hence 'hot spots'). For example, see IP Wireless: http://www.ipwireless.com

[39] Indoor positioning is an area of intense research, carefully scrutinized by retailers. For a useful summary, see 'Retrieving position from indoor WLANs through GWLC'. Edited by Papazafeiropoulos, G. *et al.* Can be found at http://www.polos.org/reports.htm

[40] It is conceivable for splash messaging to be accessible anywhere on the planet, whether the local network operator supports it or not. This should provoke some interesting thoughts regarding its socio-technical, legal and political implications.

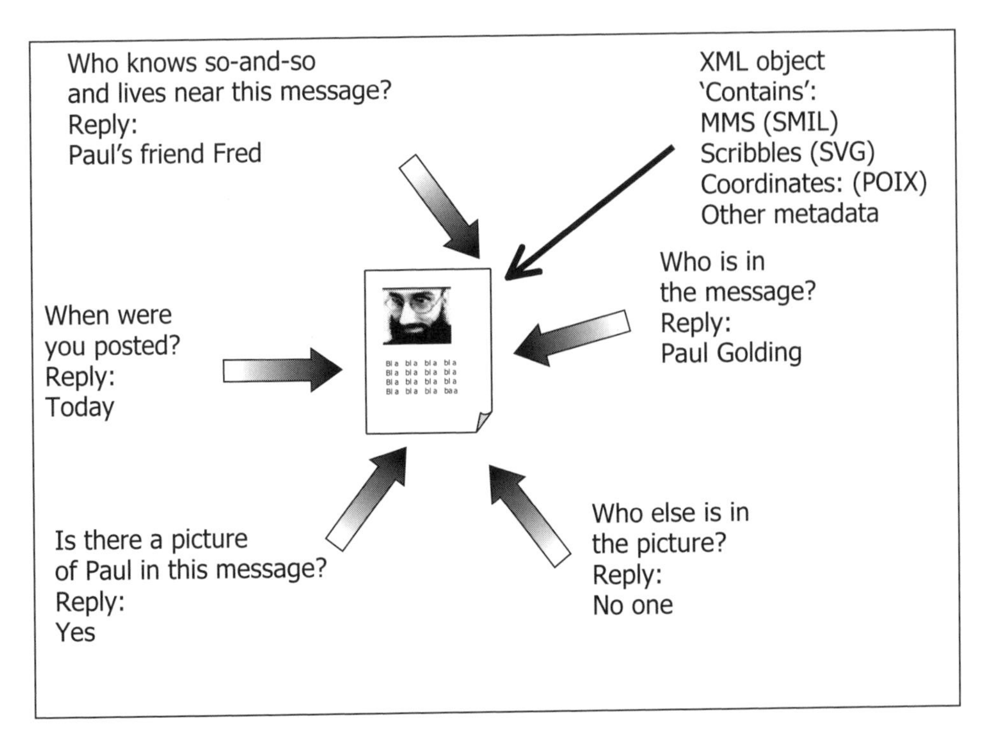

Figure 13.32 Concept of active multimedia messages.

XML is a useful way to describe a splash message. We could think in terms of spatial envelopes formed by XML documents that could describe the nature of spatial posting (e.g., recipients, topics, etc.) and could host the presentation part of the MM, which itself is XML (such as SMIL[41]). If the Mobile Location Protocol[42] XML vocabulary were modular, then we could use this to describe location. We also need other XML vocabularies to describe information such as the message owner. The proposed XML format for vCards might do for this.

The SM world is different from the text-messaging world, which is altogether more ethereal. Splash messages could potentially have a very long lifetime, a lifetime presumably controllable by the messenger, perhaps even retrospectively so; this might make for interesting real time dynamics in various messaging scenarios. Leaving information pinned at a location for a limited time may be a useful means to attract people to a place within a certain period. This has obvious uses to bring people to venues, such as shops. The idea can be used to tempt people to convene at a certain place and time, for whatever reason. Taking this a stage further, using a string of linked messages with limited lifespans, we can encourage people to move from one place to another, like an old-fashioned paperchase.

How might we start exploiting social connectivity in the SM world? One attractive feature of mobile phones is that they each contain an address book that is constantly available, probably updated frequently and a potentially valuable source of information, not just to its 'owner' but also to others, were the contents made accessible for sharing.

[41] 'Synchronized Multimedia Integration Language (SMIL 2.0)', W3C Recommendation 07 (August 2001). URL: http://www.w3.org/TR/smil20/

[42] See Open Mobile Alliance Location Forum: http://www.locationforum.org

What if address book *X* talks to address book *Y*, which talks to so-and-so's *Z* address book and so on at every node in the mobile network? We could find out whom *X* and *Y* both know, whom *X* knows that *Z* might want to know and so on. P2P technologies (e.g., JXTA[43]) could facilitate the unhindered sharing of address books, allowing software agents to search through intersecting peer groups to find an interesting person to call or intercept in cyberspace. Permissions-based protocols would help to avoid fears of surreptitious Trojans accessing anyone's precious 'black book' of numbers.

Clearly, phone numbers offer us a unique identifier in the mobile phone network, but to establish connections with the Internet world, email addresses are probably a more useful identifier. Increasingly, we will find email addresses appearing in mobile phone books anyhow, thus readily enabling connections between the two worlds. This option is also eligible in a P2P application, thus avoiding interrogation of centralized directories to match numbers with email addresses. One of the great things about this approach is that we can use text messaging for P2P communication, even with devices that aren't running a client. We could simply get humans to respond to P2P text messages!

However, how could we make social connections more meaningful, other than by simple association of address book entries? In the Internet world an emerging standard for identifying relationship semantics between net users is the Friend Of A Friend (FOAF[44]) file, an XML file describing relationships of one person to their declared contacts. We would probably want our mobile peers to formulate an FOAF file from our phone book. We can then use this as our semantic bridge to link people whenever we need to.

Currently, in MIDP[45] 1.0 of the shrunk-to-fit Java version for phones the Java run-time environment is sandboxed, preventing any access to the phone book from a MIDlet Java program.[46] With MIDP 2.0 this is no longer the case, and since MIDP 2.0 readily supports HTTP (as did its predecessor) we have the potential foundation for a wide-area relationship-mining network running on a grid of mobile phones. FOAF could provide the method of describing our connections that we find in address books.

In itself FOAF and its mobile embodiment, whether facilitated by P2P or not, is a powerful networking tool. It is worth noting that MIDP 2.0 has policy-based control over how MIDlets access phone resources, like the address book, thereby providing us with a privacy control mechanism to protect unwitting users from those Trojan activities we feared earlier. In addition to its own networking potential (for social and business networking activities in a trusted network), this wide-area relationship mining can be used as an enabling technology in its own right, and we are entertaining such ideas here.

One idea then is to connect relationships with the people in photos left hanging in our SMs. Another is simply to connect relationships with SM senders and recipients. Yet another is to connect all these monikers in a buzzing network of social interaction, where we get people, place, time, opinion and serendipity all in one package.

[43] JXTA is an open initiative to develop P2P protocols and technologies, URL: http://www.jxta.org

[44] 'FOAF vocabulary specification: RDFWeb namespace document' (16 August 2003), URL: http://xmlns.com/foaf/0.1/

[45] The Mobile Information Device Profile (MIDP), combined with the Connection Limited Device Configuration (CLDC), is the Java run-time environment for today's mobile information devices (MIDs), such as phones and entry level PDAs, URL: http://java.sun.com/products/midp/

[46] Address book access is possible if the phone vendor provides an API to enable access, but this would not be a standard MIDP 1.0 feature.

Figure 13.33 An annotated picture message.

The ability of *X* and *Y* to conjure up interesting uses from this available cocktail is surely going to be limited, but, in the hands of millions of users, exciting things will unquestionably emerge. Let's familiarize ourselves with what splash messaging can consist of. We can leave 'splash' messages hanging at a particular place: anywhere where we get network coverage and anywhere we visit. We can:

- leave video clips;
- leave text messages;
- leave pictures;
- leave audio (recorded there and then or pre-recorded);
- leave links to any other resource on the Internet, including clickable phone numbers;
- leave any combination of the above;
- leave annotated videos and pictures, as shown in Figure 13.33 (annotations can be via hand sketching, using a stylus);
- use post labels on images showing whom they depict, linking the depiction to an address book entry if necessary;
- store annotations and labels SVGs[47] to maintain our desire to use open standards in order to ease widespread processing of our messages (SVG is a media format supported by 3GPP).

[47] W3C Recommendation: 'Mobile SVG profiles: SVG Tiny and SVG Basic', (January 2003), URL: http://www.w3.org/TR/2003/REC-SVGMobile-20030114/

Lest we forget to think 'out of the box', let's ponder for a second on the thought that we don't need to go anywhere to splash a message. There is no reason to assume we have to visit a place to send a message there. We could do it remotely. We could send a video clip to a particular location and wait for some passer-by to pick it up. It could be anything, like a goodwill message for a friend as they arrive in a certain place.

This does raise some interesting legal issues, many of which frighten me, to be honest. Perhaps we have to think about spatial authentication as a means to verify that we actually went to a place to leave a message, although the possible separation of the device from the person is an obvious obstacle. That needs some thought, but it is worth pursuing, as it is probably important to some applications. For example, mobile ticketing and coupons using pictures is now a distinct possibility, but open to various types of fraud. The possibility of verifying message receipt according to location has some interesting applications.

Another legal issue that comes to mind is the possibility that spatial 'publishing' in public places actually constitutes broadcasting in a legal sense and would be subject to broadcasting laws or publishing ones, including due care against libel. Is this the case? The answer is not clear, nor is it clear that existing legal frameworks can cover what may turn out to be some very strange social and commercial activities, hitherto never contemplated by law. Even if existing laws do apply, such as broadcasting laws, one immediately suspects that legal loopholes will exist and their existence will emerge in unexpected ways.

The legal issue is clearly a serious one, an obvious concern being offensive content, such as adult material posted where kids can pick it up. There may be nothing new about the nature of the threat, but the Internet is notoriously difficult to police and its invisible connection with space may bring undesirable consequences. The daubing of public places with potentially anonymous messages has obvious negative connotations: yet more ramifications to ponder in this brave new connected world. It would seem that digital identification is required, coupled with a reliable method of establishing trust. This is so much easier to state than to implement.

To avoid information overload (or lessen it), we can leave news of our splashes using a suitable form of message summaries, such as Really Simple Syndication (RSS). It's certainly possible. Thus, we are able to add the ability to subscribe to spatial content, which may be useful for some applications. Perhaps we would like to receive messages from a particular shop that we regularly pass, but only about a particular topic, if there's something new to say about it.

With all the semantic information in place, there are so many ways of connecting. To name just two obvious ones, we have:

- co-depiction[48] (friends or FOAFs shown in the same picture);
- coalescence (friends or FOAFs in the same place).

With co-depiction, a nomadic user would be alerted to MMs with pictures of people they know or people their friends know. With coalescence, just the coincidence of any metadata to do with friends (or FOAFs) could cause an alert. The nomad can access the MMS messages associated with the alerts, add replies, make contact, etc. They would also be free to make their own annotations of the messages.

[48] 'Photo metadata: The co-depiction experiment'. Edited by Brickley, D., URL: http://rdfweb.org/2002/01/photo/

13.8.4 Splash-Messaging Summary

The location-finding capabilities of a mobile network enable messages to be location-stamped, not just time-stamped. If the messages are MMs, then we have the added dimensions of a shared visual and audio experience. Allowing the sender to add annotations in a suitable format (e.g., XML), powerful semantic references become possible and we move from the flat messaging realm into a multidimensional geo-social realm. With a suitable infrastructure, messages are no longer inert, but can become active participants in cyber-conversations, able to respond to queries about their contents, authors and recipients.

Within the global collection of mobile address books there is a powerful network of social connections. Using P2P technologies, users can tap deep veins of social capital. Conjoined with the geo-social aspects of SM and additional FOAF semantics, powerful possibilities seem likely to emerge.

In many parts of the world, text messaging has proven to be a cultural phenomenon[49] with many unexpected characteristics. Among other things, social scientists have noted its effect on how people meet and gather.[50] It seems obvious that the ability to add spatial awareness and involve the extra senses of sight and sound will surely lead to hitherto untold uses with increasing profundity.

While the technology to achieve this extension to our communications reach is not all that sophisticated, the possible challenges to our sense of privacy and judicious use of new technologies may well be accompanied by a disproportionate plethora of legal and social implications.

[49] Rheingold, H., *Smart Mobs – The Next Social Revolution*. Perseus Books, Cambridge, MA (2002).
[50] Katz, J.E. and Aakhus, M. (eds), *Perpetual Contact. Mobile Communication, Private Talk, Public Performance*. Cambridge University Press, Cambridge, UK (2002).

Index